Crystals, Defects and Microstructures

Materials science has emerged as one of the central pillars of the modern physical sciences and engineering, and is now even beginning to claim a role in the biological sciences. A central tenet in the analysis of materials is the structure–property paradigm, which proposes a direct connection between the geometric structures within a material and its properties.

The increasing power of high-speed computation has had a major impact on theoretical materials science and has permitted the systematic examination of this connection between structure and properties. In this textbook, Rob Phillips examines the various methods that have been used in the study of crystals, defects and microstructures and that have made such computations possible. The author presents many of the key general principles used in the modeling of materials, and punctuates the text with real case studies drawn from recent research. A second key theme is the presentation of recent efforts that have been developed to treat problems involving either multiple spatial or temporal scales simultaneously.

This text is intended for graduate students and researchers in science and engineering with an interest in the theoretical constructs that have been devised to undertake the study of materials.

Crystals, Defects and Microstructures
Modeling Across Scales

ROB PHILLIPS
Brown University, Providence RI

CAMBRIDGE
UNIVERSITY PRESS

PUBLISHED BY THE PRESS SYNDICATE OF THE UNIVERSITY OF CAMBRIDGE
The Pitt Building, Trumpington Street, Cambridge, United Kingdom

CAMBRIDGE UNIVERSITY PRESS
The Edinburgh Building, Cambridge CB2 2RU, UK
40 West 20th Street, New York, NY 10011-4211, USA
10 Stamford Road, Oakleigh, VIC 3166, Australia
Ruiz de Alarcón 13, 28014, Madrid, Spain
Dock House, The Waterfront, Cape Town 8001, South Africa

http://www.cambridge.org

First published 2001

Printed in the United Kingdom at the University Press, Cambridge

Typeface Times 11/14pt.   *System* LATEX 2_ε [DBD]

A catalogue record of this book is available from the British Library

Library of Congress Cataloging in Publication data

Phillips, Rob (Robert Brooks) 1960–
Crystals, defects and microstructures: modeling across scales / Rob Phillips.
p.   cm.
"April 9, 2000."
Includes bibliographical references and index.
ISBN 0 521 79005 0 (hc) – ISBN 0 521 79357 2 (pb)
1. Crystals. 2. Crystals–Defects. 3. Crystal lattices. I. Title.

QD921.P44 2001
548′.81–dc21 00-058591 CIP

ISBN 0 521 79005 0 (hardback)
ISBN 0 521 79357 2 (paperback)

Dedicated to Sonic's accomplices

Contents

Contents

Preface

Materials science as a formal discipline has quietly emerged as one of the central pillars of the physical sciences. Whether one interests oneself in the creation of lighter, more durable surfboards, or in the invention of new technologies to free up the traffic jam on the world wide web, in the end, these questions will always imply demands which must be met by new classes of materials.

Though the study of materials is largely rooted in the enlightened empiricism of traditional metallurgy, the advent of high-speed computers and the emergence of robust quantum simulations suggests new engineering strategies in which mechanism-based understanding might be hoped to lead to new materials. As a result of the maturation of theoretical materials science, it has become increasingly possible to identify a corpus of central results which serve as the basis for the analysis of materials. In addition, increasingly complex materials with structures at many different scales have led to the emergence of methods built around the explicit consideration of multiple scales simultaneously. As a result of these advances, this book attempts to take stock of the present capacity for modeling materials, with the word modeling used in the broadest sense.

The book is divided into four basic parts. It opens with an overview of some of the key issues concerning materials that one might hope to succeed in modeling. Special reference is made to the notion of material parameters as characterized by the broad variety of data about materials that may be found in any databook. Though my comments on material response will be rather generic, my efforts to model such response will be restricted almost exclusively to *crystalline* solids. Though I am ashamed to have ignored so many classes of important and interesting materials, in the end, even restricting myself to crystalline materials left me with a project that exceeds my competence. After the introduction to material response, I present something like a modelers toolkit with a review of the key ideas from continuum mechanics, quantum mechanics and statistical mechanics that one might want to bring to the evaluation of a particular problem in the mechanics of materials. My argument here is that these three broad fields of study serve as the cornerstone for the rest of what we will do. The second main section of the book is entitled 'Energetics of Crystalline Solids' and aims to describe the various tools that may be brought to bear on the problem of understanding the properties of perfect crystals. The argument made here is that the understanding we will require later concerning the role of defects in materials will have more impact when measured against the backdrop of the perfect solid. Particular emphasis in chaps. 4–6 is placed on computing the total energies of a given solid and using these energies to deduce the thermal and elastic properties of solids and the structural stability of different materials as revealed by phase diagrams. It should be noted that though I have restricted my attention to crystalline solids, much of what we will say in these chapters can be borrowed without alteration in the context of noncrystalline solids (and even liquids).

Part 3, 'Geometric Structures in Solids: Defects and Microstructures' is orga- nized according to the dimensionality of the various defects that populate materials and considers point defects (chap. 7), line defects (chap. 8) and interfacial defects (chap. 9). The organization of the material along dimensional lines is in the end somewhat artificial, but provides an organizational thread which has been useful to me if not the reader. Once an understanding of these various defects is in hand, I make an attempt at describing the assembly of these defects into a material's microstructure (chap. 10). The final part of the book, 'Facing the Multiscale Challenge of Real Material Behavior', should be seen as the culmination of the efforts set forth in the preceding chapters with the aim being to see just how far the modeler can go in the attempt to concretely consider material behavior. In addition, I have had a go at trying to seek the generic features of those models in which a deliberate attempt has been made to eliminate degrees of freedom. Indeed, my contention is that the enormous current interest in 'multiscale modeling' is born in large part of a self-conscious attempt to construct theories in a way that will lend them computational efficiency. Though there has been an incontrovertible increase in computational power, it has carried with it an attendant perception that almost always our appetite will outstrip our abilities. As a result, modelers throughout the physical sciences have had to seek to reformulate existing theories and to create new ones in a way that is ever mindful of the need to reduce computational burden.

My reasons for writing this book are primarily selfish. I have found that the only way I can really learn something is by self-study, and in particular, by seeking understanding at the level of the details – factors of 2 and π included. I consider myself one of the lucky few, namely, those who are able to earn a living as a college professor. Indeed, my privileges are greater yet. Almost daily I count my good fortune for having landed in the Solid Mechanics group at Brown University. One of the challenges in entering such a group has been to grope my way to some modicum of understanding in the area of the mechanics of materials. This book represents one part of that groping and reflects my own sense of some of the key strategies that may be brought to bear on modeling material behavior. As part of the process of bringing myself up to speed for attempting to write this book, I have also undertaken the writing of a number of review articles with which the perceptive reader will undoubtedly see some overlap. This process was intentional in the hope that by working with various coauthors a useful vision of much of the material included here could be built up. Similarly, many of the ideas set forth here have been tested out in courses given at Brown, Caltech, the National University of Singapore and the Institut National Polytechnique de Grenoble.

My intention in writing this book has not been to produce an encyclopedic compendium of all that is known concerning the modeling of materials and their properties. Rather, I have attempted to consider a sufficiently diverse set of problems to reveal the *habit of mind* that can be brought to the study of materials. Indeed, the book is largely anecdotal, with what I perceive to be fundamental ideas punctuated by more speculative 'case studies' that may not stand the test of time. As a result, the book has mixed character, alternating back and forth between text and monograph mode. In his outstanding book *The Gift of Fire*, Richard Mitchell notes the distinction between *knowing* and *knowing about*. The aims of the present work move back and forth between these objectives, with my hope being that after reading those sections having to do with general principles (and working the corresponding problems), the reader will 'know' these ideas. By way of contrast, the case studies are put forth more in the spirit of 'knowing about' with the hope that such studies will familiarize the reader with the implementation of the general ideas as well as the attendant literature, and will embolden him or her to set out to carry through a detailed case study themselves. I should also note that although many a preface emboldens the reader with the claim that various chapters can be read independently, I have written the present work very much as narrative, with the same idea presented in a different light several (or even many) different times, each contrasted with the other. For example, the discussion on Bridging Scales in Microstructural Evolution in chap. 12 presupposes a knowledge of what has gone before in chap. 10. Though it is true that one can read one without the other, it is certain that the message I wish to convey about microstructure and its evolution can only be gleaned by reading them both. For me, the book was written as a single entity and its logic is intertwined accordingly.

In keeping with this general philosophy, the book is populated by a minimum of numerical tables in which specific data are presented. Rather, I have opted for pictorial and graphical representation of numerical results with the aim being the presentation of trends rather than detailed quantitative insights. On the other hand, there are a number of instances in which I present tables of data, but in the more sympathetic form of pictures. The calculations of the group of Skriver come to mind in this regard since they have done the service of providing the energies of a number of defects (vacancies – fig. 7.15, surfaces – fig. 9.5 and step energies – fig. 9.29) for a whole host of different materials, using the same theoretical analysis in each instance raising the possibility of a meaningful assessment of the trends. In addition, at times I have resorted to presenting collages of experimental data since I often feel it is beyond my competence to bring expert judgement to one experiment vs another and thus leave it to the reader to explore several sources and thereby decide for his or herself. My selection of material is also based in part upon a sinking feeling I sometimes get both while attending meetings and while cranking out the next calculation at my desk: it seems that so much of 'research' (mine included) is slated for obsolescence from the moment of its inception. It has always seemed to me that pedagogy is an ideal litmus test of significance. Is a particular piece of work something that I would want to tell my students? The difficulty in answering this question is further exacerbated by the pedagogical uncertainties that attend the increasing reliance on numerical solutions and computer simulation. How exactly are we to communicate what is being learned from computational approaches? I have tried to follow a basic formula in many of the later chapters in the book. An attempt is first made to illustrate the fundamental principles involved in contemplating a given class of material properties. These generic ideas, which are assumed to be largely uncontroversial, are then illustrated in an iterative fashion with a given case study examined from as many different angles as possible. For example, in discussing structural stability of Si, I make a point of showing how these questions can be addressed both from the perspective of empirical potentials and using first-principles quantum mechanical treatments. Similarly, in my treatment of grain growth, I show how the Potts model, phase field models and sharp interface models may all be used to examine the same basic process. My logic in these instances is to encourage a breadth of perspective filtered by a critical eye that sees the shortcomings of all of these approaches. In the end, this book reflects my quest to unearth some of the exciting developments in the modeling of thermomechanical properties which have the power to either instruct or predict or entertain.

In addition to commenting on what is in this book, it is with a trace of sadness that I also note what is not. With the narrowing of time between a given day of writing and the date of my self-imposed deadline, as well as with the increasingly swollen zip disks and hard copies of my book, it became increasingly clear that I must abandon many of my original intentions and resort to wholesale slashing. It is with particular regret that I eliminated my discussions of thermal conductivity, electromigration, the use of path integral methods to study diffusion and the electronic structure of quantum dots. In the end, I came to the view that I must stop and return to more active daily participation in family, sleep and my own research. I should also note that despite its length, in the end experts will recognize that my book has provided only a caricature of their fields. My aim was to highlight what seemed to me to be either the most general ideas or the most intriguing from each broad area.

A word of explanation concerning references and further reading is in order. Ultimately, I decided to reference the works of others in only two ways. First, the *Further Reading* sections at the end of each chapter make reference only to those books or articles that form a part of my working and thinking vocabulary. Each time reference is made to one of these works, I attempt to provide some editorial comment as to why it has appeared on my list. My taste is idiosyncratic: if given the choice, I will always seek out that author who tells me something other than the party line, who has forged his or her own vision. I hold to this same mean in graduate teaching with the conviction that students are not interested in a watered down regurgitation of what they can read in the textbook – a unique perspective is what interests me.

In addition to the suggestions for further reading, I have also compiled a single bibliography at the end of the book. This list serves an entirely different purpose and there is only a single way to get on it. These books and articles are those from which I have borrowed figures, followed derivations or turned the analysis into a 'case study'. One of my abiding aims has been to try to figure out which part of the work on modeling has earned some level of permanence. Certain models (for example the quantum corral discussed in chap. 2) are pedagogically satisfying and at the same time permit immediate comparison with experiment. This is the ideal situation and, wherever possible, I have attempted to include them. In addition, I have also sought those steps forward that illustrate a certain approach that, in my opinion, should be imitated.

As is evident from the preceding paragraphs, my bibliographic acumen is severely limited. The reference list is not encyclopedic and reflects my taste for browsing the index in a few favorite journals such as *Physical Review B* and *Physical Review Letters, Journal of Mechanics and Physics of Solids, Metallurgical Transactions A, Philosophical Magazine* and *Acta Metallurgica*. I am a particular fan of the 'Overview' sections in *Acta Metallurgica* and the articles written by award recipients in *Metallurgical Transactions*. Further, my choice of references is biased in favor of friends and acquaintances, due largely to the fact that in these instances I have had the benefit of private discussions in order to learn what *really* happened. This is especially true in those cases where I sought a picture to illustrate a concept such as, for example, a polycrystalline microstructure. In these cases, it was most convenient to tap acquaintances, resulting in an over representation of work carried out at Brown. Hence, while my choice of references should be seen as something of an endorsement for those that are here, it should not be seen as a rejection of those that are not.

Another feature of the book deserving of comment is the problem sets that are to be found at the end of each chapter. Many of these problems have already had a tour of duty in the various courses that I have taught at Brown and Caltech. On the other hand, there is a reasonable fraction of these problems that have not seen the light of day before now and thus risk having bugs of various types. The problems come in all types ranging from the fleshing out of material covered in the chapters themselves to the construction of full-fledged computational solutions to problems that are otherwise not tractable.

A few stylistic comments are also in order. For the most part, I have written the book in the first person plural, making constant reference to 'we' this and 'we' that. My reason for selecting this language is an honest reflection of my presumed relation with the reader – I imagine us both, you the reader and me, side by side and examining each question as it arises, almost always in a questioning tone: can we do better? I have reserved the more personal and subjective 'I' for those places where it is really my opinion that is being espoused as, for example, in the sections on *Further Reading*. In addition, the book is written in an informal (and at times light) style with various asides and observations. I fear that I will probably be charged with both flippancy and arrogance, but wish to note at the outset that my hope was to produce a book that was to some extent fun to read. But even more importantly, I have tried to have fun in its writing. Three hours a morning for four years is a lot of hours to sit by oneself and I confess that on many occasions I could not resist certain turns of phrase or observations that amused *me*.

Rob Phillips
Barrington RI
2000

Acknowledgements

As a voracious reader myself, I am a veteran of many a preface and its attendant list of acknowledgements. What I realize now that it is my turn to make such a list is that for one such as me, who owes so much to so many, acknowledgements are deeply felt and yet are probably irrelevant to the anonymous reader. As said above, I wrote this book largely for myself in the hope that I would learn something in the act of so doing. This learning can be divided along two main lines, one emanating from the isolated groping I did at my desk, and the other from the pleasant meaningful interactions I have had with the people it is the business of this section to acknowledge. These paragraphs are for me and my family, my friends and my teachers. To all of you who have taught, tolerated, scolded, encouraged, discouraged, challenged and befriended me, let me state it simply: thank you for enriching my life.

I now see the dangers inherent in making a list of names, but have decided to risk it on the grounds that I wish to express my indebtedness to some people publicly. I am grateful to William Martin and Andrew Galambos who literally opened up a new world of scientific thinking to my teenage mind, and life has never been the same. I am also grateful to Solomon Deressa, Chuck Campbell and Anders Carlsson, all of whom gave me a chance when I probably didn't deserve one, and from whom, I have learned much. It is a pleasure to acknowledge my various friends at Brown from whom I have taken the vast majority of the courses offered in our Solid Mechanics curriculum (B. Freund – Stress Waves, Thin Films; R. Clifton – Advanced Continuum Mechanics; K.-S. Kim – Foundations of Continuum Mechanics; A. Bower – Plasticity; F. Shih – Finite Element Method; C. Briant – Mechanical Properties of Materials; and Mig Ortiz who taught Dislocations in Solids and much else through private consultation). Indeed, the inimitable Professor Ortiz has been the kind of friend and collaborator that one is lucky to find even once. My good fortune at Brown has also been built around the chance to learn from a series of outstanding graduate students and postdocs. Ellad Tadmor, Ron Miller, Vijay Shenoy, David Rodney and Harley Johnson all were a key part of my first five years at Brown and I can only hope they learned a fraction as much from me as I did from them. More recently, I have benefited enormously from my interactions with Nitin Bhate, Dan Pawaskar, Kedar Hardikar and David Olmsted.

As I said above, much of the fun in writing this book came from the interactions I had with others. I must single out Kaushik Bhattacharya, Rick James, Art Voter, Mig Ortiz, Chris Wolverton, Vivek Shenoy and Mark Asta, all of whom need special acknowledgement for running a veritable intellectual babysitting operation. I have also benefited from expert advice on a number of different chapters from Janet Blume, Jane Kondev, Craig Black, Saryn Goldberg, Deepa Bhate, Anders Carlsson, K.-J. Cho, John Jameson, Tim Kaxiras, Stephen Foiles, Fabrizio Cleri, Mark Asta, Chris Wolverton, Art Voter, Luciano Colombo, Ladislas Kubin, Lyle Roelofs, Lloyd Whitman, Clyde Briant, Ben Freund, Dan Pawaskar, Karsten Jacobsen, Perry Leo, Kaushik Bhattacharya, Rick James, Simon Gill, Kevin Hane, T. Abinandanan, Peter Gumbsch, Long-Qing Chen, Georges Saada, Peter Voorhees, Alan Ardell, Emily Carter, Alan Needleman, Sam Andrews, Didier de Fontaine, Jakob Schiotz, Craig Carter, Jim Warren, Humphrey Maris, Zhigang Suo, Alan Cocks, Gilles Canova, Fred Kocks, Jim Sethna, Walt Drugan, Mike Marder, Bob Kohn, and Bill Nix and Mike Ashby indirectly as a result of a series of bootlegged course notes from excellent courses they have given. My book has also been read cover to cover by a number of hearty souls (Ron Miller, David Rodney, David Olmsted, Rob Rudd, Nitin Bhate, Walt Drugan and Bill Curtin) who made immensely useful suggestions that took my manuscript from an infantile performance to something that might be considered presentable. Of course, any errors that remain are strictly my own responsibility and indeed, at several junctures throughout the text, my arguments are suspect. It is also my great pleasure to thank Maryam Saleh who literally sketched, traced, drew and redrew over 300 figures, and Lorayn Palumbo who took care of everything from making sure I didn't forget to teach my class to taking care of all manner of miscellaneous tasks. I am happy to extend my warmest thanks also to Amy Phillips who submitted literally hundreds of requests for permission to reprint figures. I would also like to thank Florence Padgett from Cambridge University Press who has supported me through the various stages of frustration that attended the writing of this book and, similarly, Simon Capelin. I would also like to acknowledge the outstanding editorial efforts of Maureen Storey who suggested all manner of improvements. The opportunity to write this book was supported generously through my wonderful NSF Career Award.

As will be evident to any reader of my book, I have tried to present as many of my conclusions as possible in pictorial form. The task of amassing the various figures that form the backbone of this work was lightened enormously by the generous help of many friends, namely, Sharvan Kumar, Mike Mills, Peter Voorhees, Dan Fridline, Allan Bower, Alan Schwartzman, Harley Johnson,

Karsten Jacobsen, Jakob Schiøtz, Rolf Heid, Rajiv Kalia, Sanjay Kodiyalam, Rick James, Kedar Hardikar, Dan Pawaskar, Charles McMahon, Vasily Bulatov, David Vanderbilt, Ralf Mueller, Lindsey Munro, Luciano Colombo, Tomas de la Rubia, George Gilmer, Christophe Coupeau, Marc Legros, Kevin Hemker, Sean Ling, Ian Robertson, Torben Rasmussen, Nasr Ghoniem, Klaus Schwarz, Marc Fivel, Bill Gerberich, David Rodney, Vivek Shenoy, Hans Skriver, Lloyd Whitman, Furio Ercolessi, Martin Bazant, Dieter Wolf, David Seidman, Craig Carter, Eric Chason, Stuart Wright, Brent Adams, Bob Hyland, Subra Suresh, Clyde Briant, Alan Ardell, Mitch Luskin, Liz Holm, Simon Gill, T. Abinandanan, Long Chen, Michael Ortiz (da Mig), Eberhard Bodenschatz, Art Voter, Chris Roland, Chris Wolverton, Charles-Andre Gandin, Michel Rappaz, Nik Provatas, John Hamilton, Farid Abraham, Harry Bhadesia and David Clarke. Thank you all again.

My parents and sisters provided me a rich and interesting family life for which I count my blessings more each day and for which I thank them. Though the writing of this book took much time and effort spent in solitude to the sounds of Santana, Yehudi Menuhin, Bobby Caldwell, Eric Clapton, Sarah Brightman and others, in the end, it is but a small thing when measured against the things that really count such as a family. Many authors close their acknowledgements with accolades for their families who tolerated their absence and fatigue, etc. My temptations are different and can be stated succinctly by noting that it is my greatest pleasure to round out this list of acknowledgements by thanking Amy, Casey and Molly, who make coming home even more fun than going to work.

Notes on Units, Scales and Conventions

In the chapters to follow, one of our aims is to try to recast many of our results in numerical form, giving an impression of the many interesting scales (length, time, energy, stress, etc.) that arise in discussing solids. A prerequisite to such discussions is to make sure that a correspondence has been constructed between the units favored here and those encountered in the everyday experience of the reader.

Geometric and Structural Notions. One of the most significant single ideas that will appear in what follows is the important role played by geometric structures at a number of different scales within materials. Indeed, the coupling between structure and properties is an element of central orthodoxy in materials science. In characterizing the structures within materials, it is clearly overly pedantic to hold to any single set of units for characterizing such structures, and we will comfortably interchange between a few different units of measure. The smallest scales that we will touch upon are those tied to the existence of atoms and their arrangements to form crystalline solids. Two fundamental units of length are most popular in this setting, namely, the ångstrom (1 Å $= 10^{-10}$ m) and the Bohr radius ($a_0 \approx 0.529\,177 \times 10^{-10}$ m). The ångstrom should be seen as a member in the series of scales which are all simple factors of 10 away from the meter itself. On the other hand, the Bohr radius arises as a natural unit of length in assessing atomic-scale processes since it is built up as a combination of the fundamental constants characterizing these processes, namely, $a_0 = \hbar^2/m_e e^2$, where $\hbar$ is Planck's constant, m_e is the mass of the electron and e is the charge on an electron.

We will also occasionally resort to the use of nanometers (1 nm $= 10^{-9}$ m) as an alternative unit of length when characterizing atomic-scale processes. When characterizing structures at the microstructural level, we will try to hold to a single unit of length, the micron (1 μm $= 10^{-6}$ m). Note that a human hair has dimensions on the order of 50 μm. The use of both nanometers and microns will be seen as characteristic of the dimensions of many structural features such as precipitates and inclusions as well as the grains making up polycrystals. In our ambition of providing meaningful geometric classifications of solids, it will also be necessary to make reference to units of area and volume, both of which will be reported either using ångstroms or meters as the base unit, resulting in either $Å^2$ or m^2 for areas and $Å^3$ or m^3 for volumes.

Time Scales and Temporal Processes. One of our recurring themes in the pages to follow will be the role of disparate scales in characterizing processes and structures within materials. In addition to the diversity of spatial scales, there are a host of processes in solids which are themselves characterized by widely different time scales. As a result, we will be forced to adopt a series of different measures of the passage of time. One of the elemental processes in solids that will occupy much of our attention is that of the vibrations of solids. As will become evident in later chapters, atomic vibrations are characterized by frequencies on the order of 10^{13}–10^{14} Hz, which corresponds to a time scale on the order of one-tenth of a picosecond. As a result, we will make use of both 1.0 fs $= 10^{-15}$ s and 1.0 ps $= 10^{-12}$ s in describing atomic motions.

Force and Energy. Just as with the characterization of the various geometric structures that populate solids we will stick to a few standard units of measure: we will adopt a strategy of characterizing the various measures of force, stress and energy in terms of a few basic units. The two units of force that will occupy centerstage in the pages that follow are the standard MKS unit, the newton, and the less familiar choice of eV/Å (eV is a unit of energy) which is much more representative of the types of forces that arise in characterizing atomic-scale motions. When translated into units of stress, we will find it advantageous again to adopt two distinct sets of units, namely, the MKS unit of stress, the pascal (1 Pa $=$ 1 N/m^2) and the atomic-scale analog, eV/$Å^3$. In addition, we will find it convenient to interchange between units of energy reported in both joules and electron volts (eV). In particular, we recall that the conversion between these two sets of units is given by 1.0 eV $\approx 1.602 \times 10^{-19}$ J. On the other hand, from time to time we will also invoke the rydberg as a unit of energy. This unit of energy arises naturally in characterizing the energy levels of atoms and can be expressed in terms of fundamental constants as Ry $= me^4/2\hbar^2$. Note that the conversion between Ry and eV is, 1 Ry $=$ 13.6058 eV. In our analysis of activated processes, we will repeatedly find it advantageous to reckon our energies in dimensionless form, where the energy scale is determined by $k_B T$, here $k_B \approx 1.38 \times 10^{-23}$ J/K is Boltzmann's constant. We note that a useful rule of thumb is that for room temperature, $k_B T \approx 1/40$ eV. For the remainder of the book, we will suppress the subscript on k_B.

Material Parameters. The key means whereby material specificity enters continuum theories is via phenomenological material parameters. For example, in describing the elastic properties of solids, linear elastic models of material response posit a linear relation between stress and strain. The coefficient of proportionality is the elastic modulus tensor. Similarly, in the context of dissipative processes such as mass and thermal transport, there are coefficients that relate fluxes to their associated driving forces. From the standpoint of the sets of units to be used to describe the various material parameters that characterize solids, our aim is to make use of one of two sets of units, either the traditional MKS units or those in which the eV is the unit of energy and the ångstrom is the unit of length.

Conventions. During the course of this work we will resort to several key conventions repeatedly. One important example is that of the summation convention which instructs us to sum on all repeated indices. For example, the dot product of two vectors may be written

$$\mathbf{a} \cdot \mathbf{b} = a_1 b_1 + a_2 b_2 + a_3 b_3 = a_i b_i. \tag{1}$$

A second key notational convention that will be invoked repeatedly is the definition of the dyadic product given by $(\mathbf{a} \otimes \mathbf{b})\mathbf{v} = \mathbf{a}(\mathbf{b} \cdot \mathbf{v})$. Another key convention that will be used repeatedly is the notation that $\{a_i\}$ refers to the set $a_1, a_2, \ldots, a_N$. When carrying out integrations, volumes will typically be denoted by Ω while the boundary of such a volume will be denoted as $\partial\Omega$.

Part one

Thinking About the Material World

ONE

Idealizing Material Response

1.1 A Material World

Steel glows while being processed, aluminum does not. Red lasers are common-place, while at the time of this writing, the drive to attain bright blue light is being hotly contested with the advent of a new generation of nitride materials. Whether we consider the metal and concrete structures that fill our cities or the optical fibers that link them, materials form the very backdrop against which our technological world unfolds. What is more, ingenious materials have been a central part of our increasing technological and scientific sophistication from the moment man took up tools in hand, playing a role in historic periods spanning from the Bronze Age to the Information Age.

From the heterostructures that make possible the use of exotic electronic states in optoelectronic devices to the application of shape memory alloys as filters for blood clots, the inception of novel materials is a central part of modern invention. While in the nineteenth century, invention was acknowledged through the celebrity of inventors like Nikola Tesla, it has become such a constant part of everyday life that inventors have been thrust into anonymity and we are faced daily with the temptation to forget to what incredible levels of advancement man's use of materials has been taken. Part of the challenge that attends these novel and sophisticated uses of materials is that of constructing reliable insights into the origins of the properties that make them attractive. The aim of the present chapter is to examine the intellectual constructs that have been put forth to characterize material response, and to take a first look at the types of models that have been advanced to explain this response.

1.1.1 Materials: A Databook Perspective

What is a material? The answer to this seemingly nonsensical question strikes right to the heart of some of the key issues it is the aim of this book to examine.

From the most naive of perspectives, questions surrounding the defining qualities of a particular material are easily answered in terms of our everyday perceptions: weight, luster, color, hardness, susceptibility to heating. However, these simple notions are a reflection of a deeper underlying identity, an identity that is revealed quantitatively the moment one poses the question of precisely how a given material replies when affected by some external probe. If we subject a material to a force, it changes shape. If we apply a potential difference, electrical current might flow. And if the temperatures of the two ends of a sample are different, a flow of heat results. In each of these cases, these experiments reveal something further about the identity of the material in question.

One of the overarching conceptual themes that has emerged from such simple experiments and that rests behind the quantitative description of materials is the idea of a material parameter. For example, Hooke's original efforts, which were aimed at uncovering the relation between a body's extension and the applied force that engendered it, led to the recognition that there exist a series of numbers, namely the elastic moduli, that characterize the elastic response of that material under different loading conditions. Similarly, there is a well established tradition of subjecting materials to applied fields which result in the emergence of various fluxes such as the electrical and thermal currents mentioned above. Allied with these fluxes are material parameters that link the response (i.e. the flux) to the applied field. In these cases and many more, the central idea is that a particular material can be identified in terms of the specific values adopted by its material parameters. In fact, for some purposes, a particular material may be idealized completely in terms of a set of such parameters. For the elastician, single crystal Al is characterized by a density and three elastic constants, namely, $C_{11} = 106.78$ GPa, $C_{12} = 60.74$ GPa and $C_{44} = 28.21$ GPa (data for Al at 300 K taken from Simmons and Wang (1971)). By way of contrast, to the engineer concerned with the application of Al in thermal environments, Al is specified in terms of a density and a thermal conductivity $\kappa = 2.37$ W/(cm K) (data for Al at 300 K taken from Shackelford *et al.* (1995)). This type of idealization of a material in which its entire identity is represented by but a few numbers is one of far-reaching subtlety. In the context of the elastic constants, all of the relevant atomic bond stretching and bending has been subsumed into the three material parameters introduced above. Similarly, the full complexity of the scattering of phonons giving rise to the thermal properties of a material has also been subsumed into just one or a few numbers. One of our primary missions in the coming chapters will be to explore how such effective theories of material behavior may be built strictly on the basis of such material parameters and to examine what gives rise to the difference in these parameters from one material to the next.

From the standpoint of the idea given above, a particular material is characterized by a set of numbers that can be unearthed in a databook. For example,
important parameters include the density ρ, the yield stress σ_y, the fracture
toughness K_{IC} and the diffusion constant D. We note that in each case there is
a number that can be looked up that characterizes the weight of a material in some
normalized terms (i.e. the density), the resistance of the material to permanent
deformation and fracture (yield strength and toughness), the ease with which
mass can be transported within the material at elevated temperatures (diffusion
constant) and any of a number of other possibilities. Our main point is to illustrate
the way in which a given number (or set of numbers) can be used to introduce
material specificity into continuum treatments of material response. For example,
in considering the continuum treatment of mass transport, it is held that the flux
of mass is proportional to the gradient in concentration, with the constant of
proportionality being the diffusion constant. This same basic strategy is exploited
repeatedly and always hinges on the fact that the complexity of the atomic-level
processes characterizing a given phenomenon can be replaced with a surrogate in
the form of material parameters.

The significance of the notion of a material parameter is further clarified by
putting the properties of different materials into juxtaposition with one another.
In figs. 1.1 and 1.2, we follow Ashby (1989) with some representative examples
of the range of values taken on by a few prominent material properties, namely
the Young's modulus, the yield strength, the fracture toughness and the thermal
conductivity. The basic idea adopted in Ashby's approach is to allow the contrasts
between the properties of different materials to speak for themselves. One of our
aims in the chapters that follow will be to develop plausible explanations for the
range of data indicated schematically in fig. 1.1, with special attention reserved for
thermomechanical properties.

Despite the power of the idea of a material parameter, it must be greeted with
caution. For many features of materials, certain 'properties' are not *intrinsic*. For
example, both the yield strength and fracture toughness of a material depend upon
its internal constitution. That is, the measured material response can depend upon
microstructural features such as the grain size, the porosity, etc. Depending upon
the extent to which the material has been subjected to prior working and annealing,
these properties can vary considerably. Even a seemingly elementary property
such as the density can depend significantly upon that material's life history. For
a material such as tungsten which is often processed using the techniques of
powder metallurgy, the density depends strongly upon the processing history. The
significance of the types of observations given above is the realization that many
material properties depend upon more than just the identity of the particular atomic
constituents that make up that material. Indeed, one of our central themes will

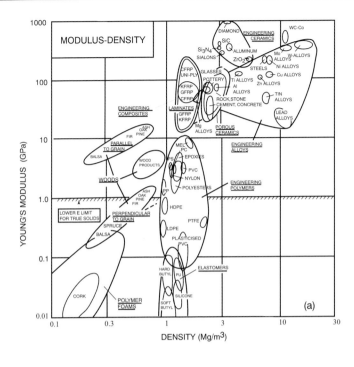

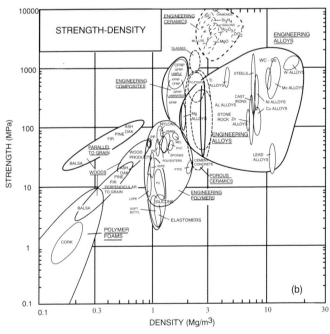

Fig. 1.1. Elastic and plastic properties of a wide class of materials (adapted from Ashby (1989)).

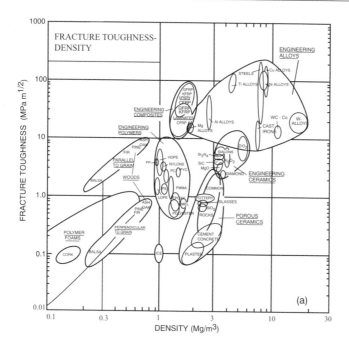

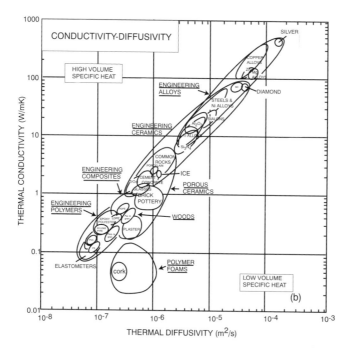

Fig. 1.2. Fracture and thermal properties of a wide class of materials (adapted from Ashby (1989)).

be the argument that microstructural features such as point defects, dislocations and grain boundaries can each alter the measured macroscopic 'properties' of a material.

One of our primary concerns in the pages that follow is to understand the emergence of material properties on the basis of the geometric structures that populate materials. This critical link between structure and properties has been canonized through the structure–properties paradigm which elevates the analysis of structure as a prerequisite to understanding properties. We have tried, in this section, to present something of the backdrop against which we will develop models of material behavior, especially with reference to thermomechanical properties. As an introduction to such models we first examine the role played by geometric structure in dictating material properties, followed by an overview of the ways in which materials may be tailored to yield particular values of these material parameters.

1.1.2 The Structure–Properties Paradigm

In the previous section we noted that, in the abstract, Al (or any other material) may be characterized by a series of numbers, its material parameters, to be found in a databook. However, as we already hinted at, because of the history dependence of material properties, the description of such properties is entirely more subtle. There is no one aluminum, nor one steel, nor one zirconia. Depending upon the thermomechanical history of a material, properties ranging from the yield strength to the thermal and electrical conductivity can be completely altered. The simplest explanation for this variability is the fact that different thermomechanical histories result in different internal structures.

A fundamental tenet of materials science hinted at in the discussion above is the structure–properties paradigm. The claim is that by virtue of a material's structure many of its associated properties are determined. Structure is an intrinsically geometric notion and one of the abiding themes in this book will be the constant reminder that it is structure on a variety of different length scales that gives rise to many of the well-known properties of materials. From the atomic-scale perspective we will constantly return to the implications of the fact that a material has a given crystal structure, whether it be for its role in dictating the properties of interfacial defects such as antiphase boundaries or the elastic anisotropy that must enter elasticity theory if the crystal symmetries are to be properly accounted for. Next, we will devote repeated attention to the lattice defects that disturb the uninterrupted monotony of the perfect crystal. Vacancies, interstitials, dislocations, stacking faults, grain boundaries and cracks will each claim centerstage in turn. At yet lower resolution, it is geometry at the microstructural scale that comes into relief and will occupy much of our attention. At each of these scales we will return to

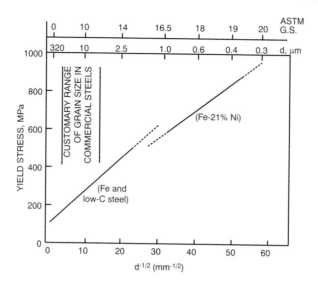

Fig. 1.3. Yield strength data for steel as a function of grain size *d* (adapted from Leslie (1981)). Plot shows dependence of yield stress on inverse power of grain size. The legends at the top of the figure show the actual grain size as measured in both μm and using the ASTM units for grain size.

the question of the structure–properties paradigm, always with a critical eye, to see just how far it may take us in our desire to unearth the behavior of real materials.

A celebrated example of the coupling of structure and properties is exhibited in fig. 1.3 in which the relation between the yield strength and the grain size is depicted. In particular, the Hall–Petch relation posits a relation between the yield stress and the grain size of the form $\sigma_y \propto 1/\sqrt{d}$, where d is the grain size of the material. The Hall–Petch relation leads us to two of the most important notions to be found in contemplating materials: the existence of microstructure and its implications for material properties, and the development of scaling laws for characterizing material response. An immediate consequence of the results depicted here is the insight that not only are structures at the atomic scale important, but so too are the geometric structures found at the microstructural scale.

We have noted that the attempt to understand materials demands that we confront a hierarchy of geometric structures, starting with the atomic-level geometries presented by the crystal lattice and increasing in scale to the level of the isolated defects that exist within materials to their assembly into the material's microstructure itself. The quest to understand the structure of a given material inevitably commences with the phase diagram. Phase diagrams are one of the primary road maps of the materials scientist. Such diagrams represent a summary of a series of tedious analyses aimed at determining the *equilibrium* atomic-scale structure of a given element or mixture of elements for a series of temperatures (and possibly

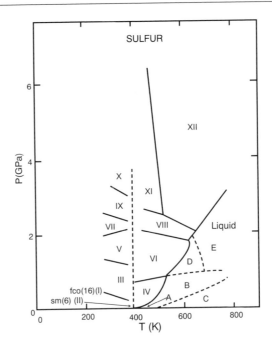

Fig. 1.4. Phase diagram of elemental sulfur (adapted from Young (1991)).

pressures). In figs. 1.4 and 1.5, we show the equilibrium phase diagrams of elemental sulfur and the iron–carbon system. The particular choices of materials shown here are meant to give a feel for the existence of the rich atomic-level complexity that is found in both elemental systems and their alloy counterparts. Note that in the case of elemental sulfur, there are not less than ten different equilibrium structures corresponding to different values of the temperature and pressure. These structures are built around molecular S_8 and range from orthorhombic to monoclinic lattices. The iron–carbon phase diagram illustrates a similar structural diversity, with each phase boundary separating distinct structural outcomes.

Phase diagrams like those discussed above have as their primary mission a succinct description of the *atomic-level* geometries that are adopted by a given system. However, as we have already mentioned, there is structure to be found on many different length scales, and one of the surprises of deeper reflection is the realization that despite the fact that phase diagrams reflect the equilibrium state of a given material, they can even instruct us concerning the *metastable* microstructures that occur at larger scales. The simplest example of such thinking is that associated with precipitation reactions in which an overabundance of substitutional impurities is frozen into a system by quenching from high temperatures. If the material is subsequently annealed, the phase diagram leads us to expect a two-phase

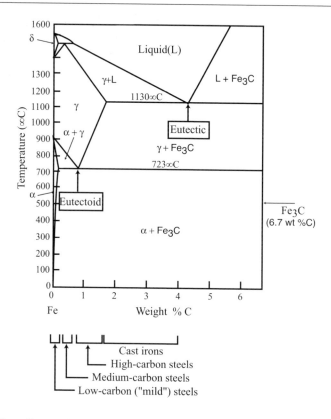

Fig. 1.5. Phase diagram for the iron–carbon system (adapted from Ashby and Jones (1986)).

microstructure in which the overabundance of substitutional impurities is now taken up in the relevant equilibrium phase. What this suggests is that strategies can be concocted for preparing particular metastable states with desirable properties. It is this insight that has led to many of the heat and beat strategies that have attracted pejorative attention to the materials engineer. A broad description of the significance of phase diagrams to materials science may be found in Massalski (1989).

As noted above, the phase diagram instructs our intuitions concerning the atomic-level geometries of materials. At the next level of geometric complexity in the hierarchy of structures that exist within a material, we must confront the defects that populate materials. Indeed, one of the key realizations that we will revisit from a number of different perspectives is that of the role of defects in the determination of material response. What this means is that the structure–properties paradigm makes an ambiguous use of the word 'structure' since in different situations, the structures being referenced can range all the way from atomic-scale structure to that of the nature of the grains making up a polycrystal. At the level of the defect geometries found within materials, our present argument is that if we trace yield in crystals to the mechanisms which engender it for example, it is the motion of dislocations which will be implicated. Similarly, if we pursue the stress–strain curve to its limits, the material will fail either through crack propagation or void coalescence or some other defect mechanism. The implication of this insight is that much of the work of modeling the thermomechanical behavior of materials can be reduced to that of carrying out the structure–properties linkage at the level of defect structures within materials.

In later chapters, we will adopt a hierarchical approach to constructing models of material response. First, we will examine the way in which atomistic models can be used to uncover the structural rearrangements in defect 'cores' which are a signature of the nonlinear interatomic interactions that are difficult to capture within continuum theory. Once these tools are in hand, we will turn to an analysis of the dominant defect types themselves, with a classification scheme centered on the dimensionality of these defects. Having successfully captured the structure and energetics of single defects, our next mission will be to understand the role of defects as conspiratorial partners in the emergence of observed macroscopic behavior. The uncovering of this conspiracy will in the end always lead us to questions of averaging. The challenge here is in forging the connection between the behavior of single defects, on one hand, and the macroscopic material response, on the other, which often emerges as a synthetic response of many defects in concert.

1.1.3 Controlling Structure: The World of Heat and Beat

In the previous section we considered the critical role played by structures at a number of different scales in determining material properties. From the standpoint of the materials engineer this insight may be recast as a challenge: how may the various structures within a material be tailored so as to yield desired properties? Metallurgy has a longstanding reputation as an empirical subject founded upon the twin pillars of heating and beating a material to some desired form. Whether one resorts to cold working a material or subjecting it to a high-temperature anneal, the outcome of these *processes* is a change in the internal constitution of that material at one or more scales. Because of these structural changes, there is a concomitant change in the properties of the material. Indeed, these strategies for altering the structure of materials should be seen as the answer to the challenge posed above.

By carefully selecting the thermal and mechanical history of a material, it is possible to tailor a number of different features of that material. Low-temperature anneals can induce precipitate reactions that increase the yield stress. Cold working changes the dislocation density, and this too alters the yield stress. If the material

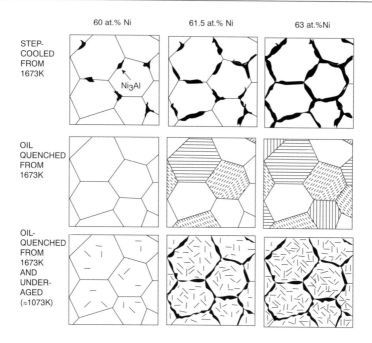

Fig. 1.6. Schematic of the different microstructures for the Ni–Al system that are obtained using different thermomechanical processing schedules (adapted from Kumar *et al.* (1992)). The figure illustrates precipitation at grain boundaries and within grains.

is exposed to a particular chemical atmosphere it can result in a thin surface layer that can increase the material's hardness, a strategy that is adopted in the production of carburized steels. In each of these examples, through annealing (heat) or deformation (beat) or both, the internal structures within the material are changed and as a result, so too are the macroscopic properties.

As an explicit example of how a material changes when subjected to the type of processing program introduced above, fig. 1.6 shows a schematic of the microstructures that arise from slightly different chemical compositions and for different thermal histories for Ni-rich Ni–Al alloys. The figure illustrates the significant variety in microstructure as a function both of slight ($<3\%$) changes in chemical composition and in the aging process. As a result of the aging process, second-phase particles of either Ni_3Al or Ni_5Al_3 can develop. In addition, if the sample is initially slow cooled, Ni_3Al precipitates at grain boundaries, while quenching and subsequent aging induces precipitation in the matrix. As is also indicated in the figure in the striped regions in the middle row, another microstructural outcome is the development of Ni–Al martensite. Out of the series of specimens indicated schematically in the figure, each of which had a different microstructure, the fracture toughnesses varied by as much as nearly a factor of

2. The role of this example in the current discussion is twofold: first, we wish to illustrate the significant structural differences that attend different processing schedules, and second, we aim to show the inextricable link between measured properties and structural features at the microstructural level.

The richness of the microstructures that result from different thermomechanical processing histories is perhaps best illustrated through the example of steels. As was indicated in fig. 1.5, at temperatures below roughly 720 °C, the equilibrium constitution of an Fe–C alloy in the low carbon concentration limit is a two-phase alloy consisting of both α-Fe (ferrite, in the bcc structure) and Fe_3C. However, this description merely tells us which crystal structures are present. The microstructure that attends the presence of these structures is known as pearlite and consists of alternating plates of α-Fe and Fe_3C. For the purposes of the present discussion we wish to note that there are a wide variety of metastable microstructures that can be attained which feature the different phases revealed in the phase diagram. For further details the reader is urged to consult Ashby and Jones (1986) or Honeycombe and Bhadeshia (1995).

In this first section of the book, our intention has been to illustrate the abstract representation of materials by different material parameters and to show how via the structure–properties linkage, material properties can be tuned by controlling the structures themselves. From a quantitative viewpoint, material response is characterized in terms of a series of material parameters that we have argued depend heavily on the internal constitution of materials at a number of different scales. These internal structures can be controlled by a number of different processing strategies. This series of insights now leaves us with the main substance of the type of questions that must be tackled in constructing viable models of material response.

1.2 Modeling of Materials

1.2.1 The Case for Modeling

Modeling has quietly become a ubiquitous part of daily life. Each evening's weather forecast leans heavily on the existence of well-defined theoretical models that are built around the vast quantities of data that are recorded at worldwide weather stations each day. Wall Street thrives on the existence of financial derivatives based largely on stochastic models of pricing and value. In the manufacturing arena, the release of the Boeing 777 exemplifies a reliance on modeling heretofore unknown: structural mechanics and aerodynamics, flight dynamics, operations research, process modeling, modeling of manufacturing, all were part of this huge project. Modeling has assumed an increasingly important role in the materials context as well.

The role of modeling in the materials setting is quite diverse. On the one hand, in the design process, the uses to which materials will be put must be evaluated critically with an eye to performance, reliability and safety. In addition, modeling can play a role in the quest to control materials and the processes used to produce them. At the most fundamental level (and the primary focus of the present work), modeling serves as the basis for an understanding of materials and their response to external stimuli. Broadly speaking, the demands placed on a model strongly depend upon its intended use. Bronze Age practitioners were successful in their material choices without detailed understanding. The challenge of modern alloy design, by way of contrast, is to see if the quantitative understanding from the modern theory of defects may be used to suggest new material strategies. On the one hand, phenomenological models may be entirely satisfactory if the goal is to test the response of a given material to casting in different shapes, for example. On the other hand, if the goal is to produce mechanistic understanding with allied predictive power a phenomenological model may not suffice and might be replaced by detailed insights concerning the underlying mechanisms. From the perspective of the engineer, the pinnacle of the modeling approach is the ability to alter engineering strategies through either the design of new materials or the institution of new processes for exploiting existing materials.

From the perspective of understanding why materials are the way they are, two of the most compelling examples that can be given concern the deformation and ultimate failure of solids. Though we will take up both of these issues again later from a quantitative perspective, our present aim is to illustrate the conceptual leap in understanding that attended the solution to the puzzles of plastic deformation and fracture. The simplest models of both of these processes consider homogeneous solids. In the case of plasticity, deformation was first posited to result from the uniform sliding of adjacent crystal planes, while fracture was envisaged to arise from a similar homogeneous process in which adjacent planes are uniformly separated. The problem with these simple models is that they lead to critical stresses for these processes that are well in excess of those observed experimentally. In both cases, the resolution of the paradox came from modeling insights concerning the role of defects in reducing the critical stresses for these processes. In the case of plastic deformation, the introduction of the dislocation resolved the discrepancy, while in the case of fracture, it was the presumed presence of preexisting cracks.

1.2.2 Modeling Defined: Contrasting Perspectives

As is perhaps already evident, the idea conveyed by the word 'modeling' is ambiguous. This ambiguity is ultimately linked to fundamental questions concerning the aims of science itself. At the most fundamental level, there is an age old debate concerning the true object of science: are we describing some underlying reality in nature, or rather is the aim to elucidate rules that yield successively better approximations to what we observe in the world around us? From the latter perspective, even the elaboration of Maxwell's equations or the edifice of classical thermodynamics might be seen as profound exercises in modeling. These types of advances must be contrasted with those in which it is clear from the outset that what is being done is picking off some fundamental feature of what has been observed and casting it in mathematically tractable terms with the aim of making falsifiable predictions.

These arguments are perhaps best illustrated by recourse to examples which serve to demonstrate the ways in which fundamental laws must be supplemented by models of material response. Whether we discuss the electromagnetically busy vacuum of interplanetary space, the absorption of radiation by an insulator or the ceramic innards of a capacitor, our expectation is that Maxwell's equations are the appropriate theoretical description of the underlying electromagnetic response. This part of the 'modeling' process was finished in the last century. On the other hand, the description in terms of these equations alone is incomplete and must be supplemented by constitutive insights which provide a description of the electromagnetic *properties* of the medium itself. In this context we refer to quantities such as the dielectric constant and the magnetic susceptibility. For example, with reference to optical absorption, the classical model of Lorentz assumed that the charges within a material could be thought of as tiny harmonic oscillators with a particular natural frequency. The electromagnetic field impinging on the material has the effect of forcing these oscillators, with the result that one can determine the absorption as a function of incident frequency. Note the character of this model. It is noncommittal with respect to any *fundamental* description of the material. Rather, it aims to reflect some element of the reality of the material in a way that can be calculated and compared with experiment, and, if successful, used as the basis of design.

As another example, this time drawn from the realm of classical thermodynamics, we may consider the thermal state of a neutron star, a high-temperature superconductor or a dense gas. In each case, there is little doubt as to the validity of thermodynamics itself. On the other hand, if we wish to make progress in the description of the dense gas, for example, the laws of thermodynamics by themselves do not suffice. This is where modeling in the sense that it will be used primarily in this book comes in. In addition to the fundamental laws that apply to *all* thermodynamic systems, we must characterize those features of the problem that are nonuniversal. That is, we require an equation of state which has nowhere near the same level of generality as the laws of thermodynamics themselves. Again, one turns to models whose aim is a characterization of the properties of matter. Different gases in different density regimes are appropriately described by different equations of state.

These examples, and others like them, allow us to discern three distinct levels of model building, though admittedly the boundary between them is blurred. In particular, the level of such modeling might be divided into (i) *fundamental laws*, (ii) *effective theories* and (iii) *constitutive models*. Our use of the term 'fundamental laws' is meant to include foundational notions such as Maxwell's equations and the laws of thermodynamics, laws thought to have validity independent of which system they are applied to. As will be seen in coming paragraphs, the notion of an 'effective theory' is more subtle, but is exemplified by ideas like elasticity theory and hydrodynamics. We have reserved 'constitutive model' as a term to refer to material-dependent models which capture some important features of observed material response.

To make the distinction between effective theories and constitutive models more clear, we consider both elastic and hydrodynamic theories in more detail. The existence of elasticity theories is an example of what we mean by an effective theory. The central thrust of such theories is that some subset (or linear combination or average) of the full microscopic set of degrees of freedom is identified as sufficing to characterize that system, or alternatively, the system is described in terms of some new phenomenological degrees of freedom (i.e. an order Indeed, one of the threads of recent scientific endeavor is the contention that reductionistic dogma is unfit to describe *emergent* properties in which it is the synthetic properties of the many-particle problem itself that yield many of the fascinating phenomena of current interest. Generally, we do not undertake a structural analysis of the Golden Gate Bridge or the beautiful convective patterns seen in the clouds from an airplane window on an atom by atom basis. These examples lead more naturally to effective descriptions in which one imagines the identification of an order parameter that characterizes the emergent property, and for which there is some appropriate continuum description.

Once the degrees of freedom have been identified, a dynamics of these degrees of freedom is constructed. To continue with our elaboration of the sense in which both elasticity and hydrodynamic theories serve as paradigmatic examples of such thinking, we note that in the case of elasticity (we have yet to say precisely which elastic *constitutive* model we have in mind) the characterization of the system is in terms of kinematic quantities such as displacements and strains which are themselves surrogates for the full atomic-level description of the system. Similarly, in the hydrodynamic context, velocities and strain rates replace an atom by atom description of the system. What all of these examples have in common is their reliance on a truncated description of material response in which the underlying discrete nature of the material is abandoned and is replaced by an effective theory in which microscopic processes have been subsumed into material parameters.

A key feature of the types of effective theories introduced above is that they depend upon the existence of a set of parameters that the theory itself is unable to determine. The elastic modulus tensor arises in describing linear elastic materials and the viscosity serves to characterize the hydrodynamic response of fluids. Similarly, the thermal conductivity arises in contemplating continuum models of heat conduction, while the magnetic susceptibility and the dielectric constant (to name a few) reflect a material's response to electromagnetic fields. What we learn from this observation is that normally effective theories like those described above must be tied to constitutive models which serve to distinguish one material from the next. Hence, we see that in addition to the level of modeling that is done in constructing the effective theory in the first place, there is a second key step in the modeling process in which *material specificity* is introduced. The elaboration of these various levels of modeling is one of the primary missions of the remainder of the book.

1.2.3 Case Studies in Modeling

One subbranch of materials science that has especially benefited from cross fertilization from different fields is that of the mechanical behavior of materials, itself one of the primary thrusts of the modeling efforts to be described in this book. The traditional disciplines of mechanical and civil engineering draw from the repository of information concerning the behavior of structural materials under both thermal and mechanical loading, with resulting structures from the Eiffel Tower to spacecraft that carry out 'fly-by' missions to distant planets. At more human size scales such as in the use of materials in applications ranging from lubricants on magnetic recording disks to the tungsten filaments that light millions of homes each evening, the interface between chemistry, materials science and condensed matter physics is obvious. Some of the issues that arise in the treatment of problems like these are the role of chemical impurities in either hardening materials or weakening grain boundaries, the ways in which microstructural size and shape influence both yield strength and ultimate resistance to fracture and the dazzling number of processing steps that attend the development of key technologies such as the Pentium chip which presides over the computer on which I am writing these words. Our discussion thus far has emphasized the mechanics of constructing models of material response without illustrating the outcome of using them. In the present section, our aim is to show in qualitative terms the different sorts of models that might be set forth for modeling materials and what is learned from them.

Modeling Phase Diagrams. Earlier we noted that phase diagrams form one of the main backbones of materials science and are a fertile setting within which to pose many of the most interesting questions about materials. Dirac argued that once the Schrödinger equation was in hand, the rest of chemistry was just a matter of implementing sufficiently powerful numerical schemes for solving the relevant governing equations. To be exact, he noted in 1929:

> The underlying physical laws necessary for the mathematical theory of a large part of physics and the whole of chemistry are thus completely known, and the difficulty is only that the exact application of these laws leads to equations much too complicated to be soluble. It therefore becomes desirable that approximate practical methods of applying quantum mechanics should be developed, which can lead to an explanation of the main features of complex atomic systems without too much computation.

Indeed, Dirac could have extended the scope of his claim to many problems in the study of materials, including phase diagrams, as problems in which the underlying governing equations are known, but which are at the same time characterized by oppressive complexity. On the other hand, an alternative argument can be made in that much of the most interesting physics present in treating problems with a high level of complexity such as phase diagrams is exactly that which we learn from trying to *avoid* the brute force calculation hinted at in Dirac's assertion. In particular, the physics of effective theory construction is precisely the business of replacing the brute force solution of the governing equations with some simpler description. Indeed, to enliven the discussion with dueling quotes we note that P. W. Anderson (1972) has remarked on this hierarchy of theories as follows:

> But this hierarchy does not imply that science *X* is 'just applied *Y*'. At each stage entirely new laws, concepts, and generalizations are necessary, requiring inspiration and creativity to just as great a degree as in the previous one. Psychology is not applied biology, nor is biology applied chemistry.

To which we might add that modeling complex many degree of freedom systems is not just an application of what is already known about single degree of freedom systems. To my mind, one of the most exciting current activities in the physical sciences is precisely this self-conscious attempt to systematically eliminate degrees of freedom so as to construct theories demanding minimal information.

In chap. 6, we will take up the subject of phase diagrams, with special attention being given to the computation of such diagrams on the basis of atomic-level insights. One of the insights that will become evident when we delve into these questions more deeply is the large extent to which computing phase diagrams represents the confluence of ideas and methods from many distinct sources. On the one hand, these calculations demand the total energies of all of the imagined

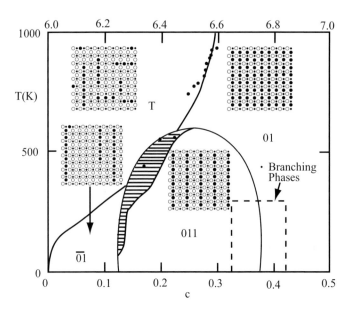

Fig. 1.7. Phase diagram for oxygen ordering in the YBCO class of high-temperature superconductors (after Ceder (1994)).

structural competitors which is the playing out of solving the Schrödinger equation referred to above. The calculation of such energies will be one of the centerpieces in the chapters to follow. In addition, once these energies are obtained, they must be supplemented by statistical arguments for assessing the entropy associated with both configurational disorder and the presence of thermal vibrations. It will be seen that in some instances it is possible to use an effective Ising representation of the various structural competitors to replace the laborious case by case search over different structural competitors.

An example of the computed and measured phase diagrams for oxygen ordering in the Cu–O high-temperature superconductors is given in fig. 1.7. The basic idea is the construction of an effective description of the energetics of the various structural competitors and to rank order these competitors as a function of some control parameters such as the composition and the temperature. For each point in parameter space, the victor in this competition is the equilibrium structure. We will come back later to some of the explicit details involved in computing the phase diagram of fig. 1.7 and note for the time being only that the calculation of such phase diagrams is a key part of computational materials science.

Modeling Material Parameters. In section 1.1.1, we argued that in many instances an extremely powerful idea is that of a material parameter. Whether discussing elasticity, heat conduction, mass transport or magnetism, we can capture the particulars of a given material in terms of such parameters. As already alluded to, the continuum models of heat conduction or mass transport cannot instruct us as to why the values of transport coefficients in some materials are large while in others they are small. Consequently, an important mission in the attempt to figure out what makes materials tick is the ambition of deducing material parameters on the basis of microscopic models.

As a first reflection on calculations of this type, fig. 1.8 shows examples of the correspondence between macroscopic material parameters and a subset of the associated microscopic calculations that might be used to determine them. One example of the use of microscopic calculations to inform higher-level models concerning material parameters is that of the diffusion constant. One of the challenges posed by data on diffusion is its strong variability in the presence of 'short-circuit' diffusion pathways such as surfaces, grain boundaries and dislocations. From a microscopic perspective, the essential idea is to examine the energetics of the diffusing particle as it passes the saddle point connecting two different wells in the energy landscape. In this case, a conjecture is made concerning the dominant reaction pathway in terms of a reaction coordinate which provides a measure of the extent to which the system has passed from one state to the next. The energy at the saddle point is used to determine the activation energy for diffusion. This activation energy, in conjunction with a model of the frequency with which the diffusing species attempts to cross the barrier leads to the diffusion constant itself. The key point made in fig. 1.8 is the idea that the atomic-level calculations can be used to inform our understanding of a higher-level material parameter.

A second example revealed in fig. 1.8(c) and (d) is that of the thermal conductivity. The left hand frame shows the phonon dispersion relation for Ge (a subject we will return to again in chap. 5) as computed using microscopic analysis. The right hand frame shows the measured thermal conductivity itself. Without entering into details, we note that on the basis of a knowledge of both the phonon dispersion relation and the anharmonic coupling between these phonons, it is possible to build up an analysis of the thermal conductivity, again revealing the strategy of using microscopic calculations to inform our understanding of higher-level continuum quantities. These case studies are meant to exemplify the logical connection between microscopic calculations and material parameters. Much of the business of coming chapters will surround the details of such calculations.

Modeling Dislocation Cores. In an earlier section, we belabored the critical coupling of structure and properties, and this is another arena within which modeling can produce valuable insights. One of our insights concerning the importance of structure was that of the role of lattice defects in governing many

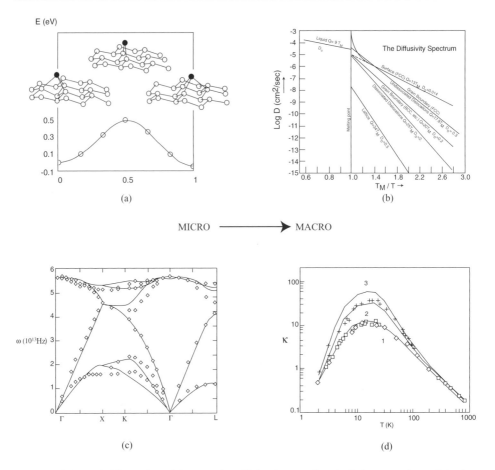

Fig. 1.8. Schematic illustration of the ways in which microscopic calculations may be exploited to model macroscopic material properties. The first example (frames (a) and (b)) illustrates the use of microscopic calculations to examine surface diffusion, while the second example (frames (c) and (d)) illustrates the analysis of phonons in Ge as the basis of an analysis of thermal conductivity. Figures adapted from (a) Kaxiras and Erlebacher (1994), (b) Gjostein (1972), (c) and (d) Omini and Sparavigna (1997).

of the properties of materials. There exists a mature elastic theory of such defects. However, this theory is plagued by the existence of singularities which preclude its use in the immediate vicinity of the dislocation itself. For understanding of the dislocation core region one must turn instead to insights from atomic-scale calculations. As an example of the types of successes that have been achieved in this vein, we consider the core structure for the Lomer dislocation in Al in fig. 1.9. In this instance, an embedded-atom model for the interactions between Al atoms was invoked in order to compute the forces between atoms which were then used in turn to find the energy minimizing dislocation core structure. As will be discussed

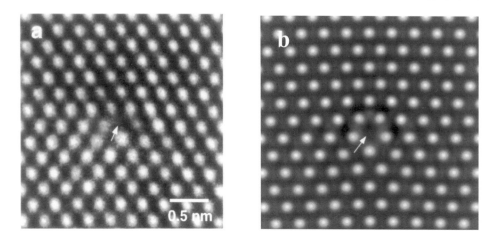

Fig. 1.9. Dislocation core in Al as seen experimentally using high-resolution transmission electron microscopy (a) and image simulation of results of atomistic simulation using embedded-atom potentials (b) (adapted from Mills *et al.* (1994)).

in due time, the significance of simulations like this is that they may even afford the opportunity to draw conclusions about the macroscopic plastic behavior of a material. In the present context, we merely note that atomic-level calculations of the atomic positions within the dislocation core are susceptible to comparison with positions deduced from high-resolution microscopy. Once confidence in the atomic-level structure has been attained, then one can begin the harder work of trying to uncover the implications of this structure for observed properties.

Modeling Microstructure and its Evolution. Until this point, we have emphasized the virtues of modeling efforts built from the smallest scales up. However, as we have already noted, in many instances it is preferable to construct our theoretical vision on the basis of continuum models. We have already made reference to the critical role of microstructure in determining material properties and it is in this arena that we will consider a case study in continuum modeling. In particular, we consider the problem of a two-phase microstructure in which second-phase particles are dispersed through a matrix and the question under consideration concerns the temporal evolution of this microstructure. In particular, the modeler can ask questions both about the temporal history of particle size and shape as well as about its privileged terminal shape.

The continuum basis of models of this type of microstructural evolution really amounts to a consideration of the competition between surface and elastic energies. Both the elastic and interfacial energies depend upon the shape and size of the second-phase particles, and may even depend upon the relative positions of

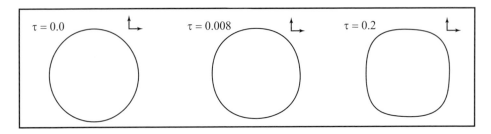

Fig. 1.10. Temporal evolution of a circular particle as a result of diffusive processes driven by the competing influences of interfacial and elastic energy. The parameter τ is a dimensionless measure of the elapsed time (adapted from Voorhees *et al.* (1992)).

different particles which are linked to particle–particle interactions. It is argued that a given particle can change both its size and shape in time as a result of diffusive processes that transport mass from one particle to another, or from one part of a given particle to another. An example of the temporal evolution of an initially circular particle as a function of time is shown in fig. 1.10. The calculations leading to this figure are based upon a treatment of the coupled problem of solving the equations of mechanical equilibrium in conjunction with a treatment of mass transport. One of the aims of coming chapters will be to set up enough machinery concerning elastic and interfacial energies as well as diffusive processes so that we can delve more deeply into the theoretical structure that makes these calculations possible.

As was shown above, the modeling of observed phenomenology concerning the properties of materials need not necessarily adopt an atomistic perspective. In addition to the microstructural evolution associated with both shape changes and coarsening of two-phase microstructures, one can also imagine the evolution of microstructures under conditions in which the driving force for structural change is more palpable. In particular, electromigration is a problem of abiding interest in the context of materials exploited in the microelectronics setting. In this case, mass transport is biased by the presence of a symmetry breaking charge flow which couples to the underlying microstructure. In fig. 1.11, a series of snapshots revealing the evolution of a void in the presence of an electric field are shown. The numerical engine used in these calculations is the finite element method.

In this section, we have attempted to give various snapshots of the modeling process, primarily with the aim of giving a sense both of what can be accomplished by recourse to modeling, and the different mindsets that attend different modeling paradigms. The work of coming chapters will largely concern the question of how to undertake problems like those sketched above from a systematic quantitative perspective.

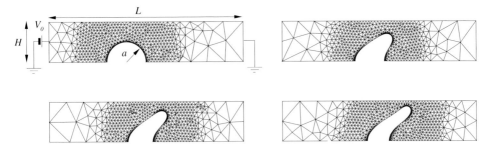

Fig. 1.11. Void evolution under conditions of electromigration (adapted from Fridline and Bower (1999)).

1.2.4 Modeling and the Computer: Numerical Analysis vs Simulation

Our spin thus far on the subject of modeling has been primarily complimentary. On the other hand, the advent of powerful computers as a tool for the evaluation of physical problems has not been universally beneficial. As we will describe presently, the establishment of computer-based models is attended by both conceptual and computational difficulties. It is true that the act of creating a numerical incarnation of a given theory requires an unambiguous statement of that theory itself. It is not possible to 'code up' a theory unless the mathematical statement of that theory can be clearly articulated. But these virtues are accompanied by vices as well. Because of the ease with which computational models may be advanced, the act of validation, both conceptual and logical, becomes increasingly uncertain.

In the context of materials science, there is an increasing desire to 'simulate' one's way to understanding. But the notion of simulation is far reaching. On the one hand, one might imagine the possibility of taking a well-defined partial differential equation, built in turn on the basis of well-known physical laws, which remains out of reach because it is not tractable analytically. In this instance, the appeal of numerical approaches is immediate and their power is indisputable. By way of contrast, there are a number of problems for which the proper physical laws have not yet been elucidated and often researchers are reduced to the act of formulating models based upon *ad hoc* assumptions of questionable validity. For example, in the context of the formation of dislocation patterns, it remains unclear whether techniques based upon minimization of some energy functional provide a proper theoretical context or whether a completely different strategy based upon reaction–diffusion equations, for example, is more suitable.

Another difficulty that attends the increasing use of computational models is that of pedagogy. The increasing reliance on simulation suggests that the problems of current interest are so complex that they defy description in proper analytic terms. If it emerges that this scenario is correct (which I expect it is not), then this raises serious questions concerning conventional notions of pedagogy. Here I refer to the difficulty of passing useful information from one generation of researchers to the next. As a teacher, I find it entirely unsatisfactory to tell a student 'it is so because computer simulation says it is so'. Such utterances do not normally constitute understanding. As a result, even in those cases where numerical analysis has been resorted to in order to solve well-posed problems based upon physically reasonable models, there is much yet to be done to learn how to incorporate these results into the canon of ideas that are passed from one generation of researchers to the next.

An additional intriguing challenge posed by computer models is that of handling large space and time scales. Historically, statistical mechanics has served as the workhorse for handling the many-body problem. However, the advent of powerful computers has opened the door to actually simulating huge numbers (10^{10}) of particles without entering into a debate as to how the many-particle aspects of the problem are assembled to yield macroscopic observables. It is this question of *emergence*, how macroscopic behavior is built up from microscopic motions, that remains as one of the prominent challenges for modelers, and will serve as one of the philosophical cornerstones in the chapters to follow.

1.3 Further Reading

Made to Measure by Philip Ball, Princeton University Press, Princeton: New Jersey, 1997. This semipopular book examines some of the key materials that are making headlines at the time of this writing. I found the book both enjoyable and very enlightening.

Designing the Molecular World by Philip Ball, Princeton University Press, Princeton: New Jersey, 1994. This book places a more chemical emphasis on some of the materials of interest here, with special attention given to the molecular origins of various processes.

Molecules by P. W. Atkins, W. H. Freeman and Company, New York: New York, 1987. Though the emphasis here is on molecules rather than materials *per se*, Atkins does a fantastic job of connecting structure and function and illustrates the type of thinking that materials science should emulate in the quest to explicate this connection.

Metals in the Service of Man by A. Street and W. Alexander, Penguin Books, London: England, 1998. I find this book fun and interesting and full of useful insights for courses. Part of their humor is revealed in their *Dramatis Personae* where such 'Principals' as IRON – the most important metal, ALUMINIUM – the light metal, second in importance to iron, MAGNESIUM – the lightweight champion, TITANIUM – the strong middleweight, etc. are introduced.

Engineering Materials Vol. 1 (second edition: Butterworth-Heineman, Oxford: England, 1996) and *Engineering Materials* (Pergamon Press, Oxford: England, 1986) Vol. 2 by M. F. Ashby and D. R. H. Jones. These two books provide, at a very readable level, a broad description of the properties and uses of materials. I particularly enjoy these books because they dare to give an all-encompassing view of the role of materials in a variety of settings.

Introduction to Engineering Materials: The Bicycle and the Walkman by C. J. McMahon, Jr and C. D. Graham, Jr, 1994 (private publication, available from author: http://www.seas.upenn.edu/mse/fac/mcmahon.html). This book is a general introduction to many of the properties of materials illustrated through the medium of a few key examples, namely, the bicycle and the walkman. I always enjoy looking up new topics in this book.

Strong Solids by A. Kelly and N. H. McMillan, Clarendon Press, Oxford: England, 1986. This book provides a number of intriguing insights into the origins of many of the mechanical properties of solids.

Atlas of Stress–Strain Curves edited by H. W. Boyer, ASM International, Metals Park: Ohio, 1987. Given that our mission is dominated by the quest to understand mechanical properties, this book gives a sense of the variety of data to be encountered on the most immediate of mechanical tests. Indeed, to understand the subtlety of real stress–strain curves should be seen as one of our primary goals.

Metals Handbook edited by Taylor Lyman, American Society for Metals, Metals Park: Ohio, 1972. Volume 7 of this series is *Atlas of Microstructures of Industrial Alloys* and provides a series of fascinating pictures of microstructures in metals. Our reason for including it as the basis for further reading associated with this first chapter is that it gives a sense of the diversity the modeler must face, even within the limited domain of metallurgy.

Understanding Materials Science by Rolf E. Hummel, Springer-Verlag, New York: New York, 1998. I find this book particularly appealing in that Hummel has attempted to interweave technical insights into materials with a discussion of their historical uses.

Materials Science and Engineering – An Introduction by William D. Callister, Jr, John Wiley and Sons, New York: New York, 1994. I have included this selection as a well-known example of an undergraduate materials science textbook in which many of the phenomena of interest in the present setting are introduced.

'On the Engineering Properties of Materials' by M. F. Ashby, *Acta Metall.*, **37**, 1273 (1989). This article gives an overview of many of the most important thermal and mechanical properties of materials. The presentation of data is provocative and the allied discussions are enlightening. This article is a personal favorite that I hand out in most of my classes.

Physical Metallurgy edited by R. W. Cahn, North Holland Publishing Company, Amsterdam: The Netherlands, 1965. Cahn's books are made up of a series of articles by leading experts in the field. Topics ranging from phase diagrams to fatigue are represented from a modern perspective and, though the emphasis is on metals, the ways of thinking about materials are more general. There is a more recent version of this excellent series.

Statistical Mechanics of Elasticity by J. H. Weiner, John Wiley & Sons, New York: New York, 1983. Weiner's book has a number of interesting and useful insights into the meeting point between continuum mechanics and microscopic theories.

Computational Materials Science by Dierk Raabe, Wiley-VCH, Weinheim: Germany, 1998. Raabe's book is included here as an example of the high level to which a discussion of modeling materials may be taken. Raabe covers many topics similar to those covered in this book, though the emphasis is often different.

Thermophysical Properties of Materials by G. Grimvall, North-Holland Publishing Company, Amsterdam: The Netherlands, 1986. Grimvall has written an outstanding book that is largely neglected. The idea is to make a concrete and quantitative attempt to use the knowledge of solid-state physics to understand the properties of real materials. This is one of my favorite books.

Solid State Physics by N. W. Ashcroft and N. D. Mermin, Saunders College, Philadelphia: Pennsylvania, 1976. My relationship with this fine book reminds me of Mark Twain's quip that as he passed from the age of 15 to 20 he couldn't believe how much his father had learned. Indeed, with each passing year I am amazed at how their book improves. Should the reader wish to see a clear statement of the body of solid-state physics as it stood fifteen years ago and as it is used in the present work, Ashcroft and Mermin is where I recommend he or she look.

TWO

Continuum Mechanics Revisited

2.1 Continuum Mechanics as an Effective Theory

Materials exhibit structural features at a number of different scales, all of which can alter their macroscopic response to external stimuli such as mechanical loads or the application of electromagnetic fields. One of the fundamental difficulties faced in the modeling of materials is how to extract those features of the problem that are really necessary, while at the same time attaining some tolerable level of simplification. Traditionally, one of the favored routes for effecting the reduction of problems to manageable proportions has been the use of continuum theories. Such theories smear out features at 'small' scales such as the discrete phenomena that are tied to the existence of atoms and replace the kinematic description of materials in terms of atomic positions with field variables. Adoption of these strategies leads to theories that can describe material behavior ranging from deformation of polycrystals to mass transport to the behavior of domain walls in magnetic materials.

The continuum mechanics of solids and fluids serves as the prototypical example of the strategy of turning a blind eye to some subset of the full set of microscopic degrees of freedom. From a continuum perspective, the deformation of the material is captured kinematically through the existence of displacement or velocity fields, while the forces exerted on one part of the continuum by the rest are described via a stress tensor field. For many problems of interest to the mechanical behavior of materials, it suffices to build a description purely in terms of deformation fields and their attendant forces. A review of the key elements of such theories is the subject of this chapter. However, we should also note that the purview of continuum models is wider than that described here, and includes generalizations to liquid crystals, magnetic materials, superconductors and a variety of other contexts.

As noted above, one of the central advantages that comes with the use of continuum models is that they reduce the complexity of description of the material

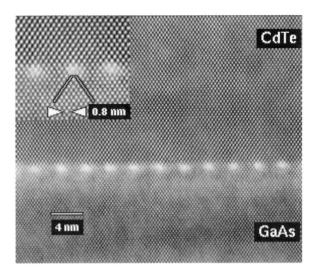

Fig. 2.1. High-resolution electron micrograph showing a periodic array of Lomer misfit dislocations for an annealed CdTe/GaAs interface. Inset magnifies the two half-planes associated with the perfect pure edge Lomer dislocation (courtesy of Alan Schwartzman).

system. To see this clearly, consider the treatment of a dislocated region of a crystal with dimensions on the order of a micron on a side. This situation is shown from the perspective of high-resolution transmission electron microscopy in fig. 2.1. The figure illustrates the presence of a series of lattice defects at an interface in a semiconductor alloy. What we are to note is that in the immediate neighborhoods of the dislocations there is a significant departure of the atomic positions from their preferred crystalline positions. For the moment, we consider only the geometric description of these deformations without reference to additional complexities that would arise if we chose to consider the chemical disposition of the system as well. Explicit reckoning of the atomistic degrees of freedom in the area of interest would involve on the order of at least 10^{10} atoms. In principle, this implies, in turn, the necessity to solve 10^{10} coupled differential equations of motion. By way of contrast, when viewed from the perspective of continuum mechanics, the deformation problem necessitates the use of only three distinct displacement fields which satisfy three coupled partial differential equations. In most instances (at least in the far field regions), the post-processing that would be involved in discovering the relevant linear combinations of the atomistic degrees of freedom is implicit in the existence of the continuum fields themselves. That is, the field variables reflect the collective behavior of the atomic degrees of freedom that is germane to the observed material response. We begin with an analysis of how such kinematic reduction is accomplished.

Broadly speaking, our description of continuum mechanics will be divided along a few traditional lines. First, we will consider the kinematic description of deformation without searching for the attributes of the forces that lead to a particular state of deformation. Here it will be shown that the displacement fields themselves do not cast a fine enough net to sufficiently distinguish between rigid body motions, which are often of little interest, and the more important *relative* motions that result in internal stresses. These observations call for the introduction of other kinematic measures of deformation such as the various strain tensors. Once we have settled these kinematic preliminaries, we turn to the analysis of the forces in continua that lead to such deformation, and culminate in the Cauchy stress principle.

With the tools described above in hand, we can then turn to an analysis of the various balance laws (i.e. linear and angular momentum, energy) that preside over our continuum and to exploring the implications of these balance laws for the solution of the boundary value problems that arise in the modeling of materials. To fully specify a boundary value problem, a constitutive model which relates forces and deformations must be supplied. One of the significant meeting points of the types of continuum and microscopic models we are attempting to link in this book is that of constitutive models. The equations of continuum dynamics are sufficiently generic to describe continuum problems ranging from the stretching of rubber to the breaking of waves. As a result, there is a freedom of description in the generic continuum setting that must be filled in by models of material response. Here, the point is to determine models of material behavior that are sufficiently rich to reflect the roles of the various experimental and microstructural parameters that govern material behavior, without at the same time becoming mired in computational intractability. Historically, the paradigmatic example of such a model is provided by the linear theory of elasticity which is a natural outgrowth of Hooke's law, and often suffices to describe solids that have suffered small deformations. Indeed, later in this chapter we will illustrate the ways in which linear elastic analyses can even be applied to the treatment of highly *nonlinear* phenomena such as plasticity and fracture. There, we will see that linear elastic analyses provide a wealth of insights into the behavior of the key defects governing plasticity and fracture, namely, dislocations and cracks.

2.2 Kinematics: The Geometry of Deformation

Continuum theories have a number of generic features, one of which is the use of some set of field variables for characterizing the disposition of the system of interest. In the context of the thermodynamics of gases, pressure, volume and temperature may suffice while for electromagnetic media, the electric and magnetic fields play a similar role. For the purposes of constructing continuum models of the mechanical response of deformable bodies, a crucial subset of the required field variables are those describing the geometry of deformation. The business of kinematics is to provide the tools for characterizing the geometric state of a deformed body.

2.2.1 Deformation Mappings and Strain

To formalize our treatment of deformation in materials, we introduce the notion of a deformation mapping which is based upon the selection of some reference state which serves as the basis for analysis of the deformed state. The definition of local regions within the continuum is predicated upon the slightly ambiguous notion of a 'material particle' which is imagined to be a sufficiently large region on the scale of the subscale geometries being integrated out (e.g. the crystal lattice or the various grains making up a polycrystalline microstructure), but still small on the scale of the spatial variations in the relevant field variables. As hinted at above, we find it convenient to label material particles in the reference configuration by their position vectors $\mathbf{X}$. The idea of seeking to label each material particle is the continuum analog of the labeling of particles already familiar from discrete particle Newtonian mechanics. In discrete particle mechanics, each and every particle is labeled by an integer, while by contrast, since a continuum by definition has an infinite set of degrees of freedom, we label the particles here by their positions in some reference state. The position of such a material particle at time t after deformation is given by a vector valued function, $\mathbf{x} = \mathbf{x}(\mathbf{X}, t)$, known as the deformation mapping. The relation between the reference and deformed configurations is shown in fig. 2.2. It is convenient to define displacement fields $\mathbf{u}(\mathbf{X}, t)$ such that $\mathbf{x}(t) = \mathbf{X} + \mathbf{u}(\mathbf{X}, t)$. We will see below that often, it is more transparent to center attention on the displacement fields themselves instead of the deformation mapping, though the descriptions are equivalent.

For the purposes of examining the geometry of deformation locally, one strategy that is particularly enlightening is to focus our attention on some small neighborhood of the material particle of interest and to examine the changes in the infinitesimal vectors separating it from nearby material points. This strategy may be quantified by centering our attention on the linear part of the deformation mapping as follows. Prior to deformation, the infinitesimal vector that separates neighboring material points is given by $d\mathbf{X}$. After deformation, the point at $\mathbf{X}$ is carried into the point $\mathbf{x}(\mathbf{X})$, while the point initially at $\mathbf{X} + d\mathbf{X}$ is carried into the point $\mathbf{x}(\mathbf{X} + d\mathbf{X})$, which may be linearized to yield the mapping

$$\mathbf{x}(\mathbf{X} + d\mathbf{X}) \simeq \mathbf{x}(\mathbf{X}) + \mathbf{F}d\mathbf{X}. \tag{2.1}$$

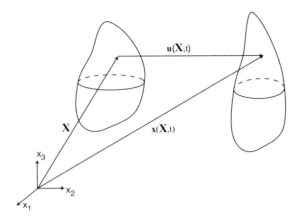

Fig. 2.2. Schematic indicating the relation between reference and deformed states of a material. The vector **X** characterizes the position of a material point before deformation while **x** characterizes the position of the material particle after deformation. The displacement field **u**(**X**, *t*) characterizes the displacement of the material particle at **X**.

Here we have introduced a tensor **F** known as the deformation gradient tensor whose components reflect the various gradients in the deformation mapping and are given by

$$F_{ij} = \frac{\partial x_i}{\partial X_j}. \tag{2.2}$$

Use of the fact that the deformation mapping may be written as $\mathbf{x}(t) = \mathbf{X} + \mathbf{u}(\mathbf{X}, t)$ reveals that the deformation gradient may be written alternatively as

$$\mathbf{F} = \mathbf{I} + \nabla \mathbf{u}, \tag{2.3}$$

or in indicial notation as $F_{ij} = \delta_{ij} + u_{i,j}$.

Once we have determined **F**, we may query a particular state of deformation as to the disposition of the vector $d\mathbf{X}$ as a result of the deformation. For example, the simplest question one might ask about the vector between two neighboring material particles is how its length changes under deformation. To compute the length change, we compute the difference

$$dl^2 - dL^2 = d\mathbf{x} \cdot d\mathbf{x} - d\mathbf{X} \cdot d\mathbf{X}, \tag{2.4}$$

where dl is the length of the material element after deformation and dL is the length of the same material element prior to deformation. This expression may be rewritten in terms of the deformation gradient tensor by making the observation that $d\mathbf{x} = \mathbf{F}d\mathbf{X}$ and hence,

$$dl^2 - dL^2 = d\mathbf{X} \cdot (\mathbf{F}^T \mathbf{F} - \mathbf{I}) \cdot d\mathbf{X}, \tag{2.5}$$

where $\mathbf{F}^T$ denotes the transpose of the deformation gradient tensor. One of our aims is the determination of a measure of deformation that fully accounts for the relative motions associated with internal stresses while at the same time remaining blind to any superposed rigid body motions. This inspires the definition of an auxiliary deformation measure known as the Lagrangian strain tensor and given by

$$\mathbf{E} = \frac{1}{2}(\mathbf{F}^T\mathbf{F} - \mathbf{I}).$$ (2.6)

The strain tensor provides a more suitable geometric measure of relative displacements, and in the present context illustrates that the length change between neighboring material points is given by

$$dl^2 - dL^2 = 2d\mathbf{X} \cdot \mathbf{E} \cdot d\mathbf{X}.$$ (2.7)

In addition to the use of the strain to deduce the length change as a result of deformation, it is possible to determine the ways in which angles change as well. The interested reader is invited either to deduce such results for him or herself or to consult any of the pertinent references at the end of the chapter.

For both mathematical and physical reasons, there are many instances in which the spatial variations in the field variables are sufficiently gentle to allow for an approximate treatment of the geometry of deformation in terms of linear strain measures as opposed to the description including geometric nonlinearities introduced above. In these cases, it suffices to build a kinematic description around a linearized version of the deformation measures discussed above. Note that in component form, the Lagrangian strain may be written as

$$E_{ij} = \frac{1}{2}(u_{i,j} + u_{j,i} + u_{k,i}u_{k,j}).$$ (2.8)

Here we have used the fact that $F_{ij} = \delta_{ij} + u_{i,j}$. In addition, we have invoked the summation convention in which all repeated indices (in this case the index k) are summed over. For the case in which all the displacement gradient components satisfy $u_{i,j} \ll 1$, the final term in the expression above may be neglected, resulting in the identification of the 'small strain' (or infinitesimal strain) tensor,

$$\epsilon_{ij} = \frac{1}{2}(u_{i,j} + u_{j,i}).$$ (2.9)

As we will see in subsequent chapters, for many purposes (e.g. the linear theory of elasticity) the small-strain tensor suffices to characterize the deformation of the medium.

The ideas introduced in this section provide a skeletal description of some of the ways in which the geometric deformations in solids are considered. To provide a more concrete realization of these ideas, we now introduce several case studies

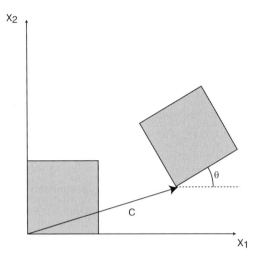

Fig. 2.3. Rigid body deformation involving a rotation through an angle θ and a translation by the vector **c**.

in the geometry of deformation that will be of particular interest in subsequent chapters.

2.2.2 Geometry of Rigid Deformation

Many of the most important deformation mappings in solids can be reduced to one of a few different key types of deformation. One example of a deformation mapping that is enlightening is that of a rigid body motion in which a body is rotated through an angle θ about a particular axis and suffers a translation characterized by the vector **c**. An example of this type of deformation is shown in fig. 2.3. The deformation mapping in the case of a rigid body deformation may be written as

$$\mathbf{x}(t) = \mathbf{Q}(t)\mathbf{X} + \mathbf{c}(t), \tag{2.10}$$

where $\mathbf{Q}$ is an orthogonal tensor (i.e. $\mathbf{Q}\mathbf{Q}^T = \mathbf{I}$) which results in a rigid body rotation, while **c** is a constant vector and corresponds to a rigid body translation. For the particular case shown in fig. 2.3, the matrix $\mathbf{Q}$ is given by

$$\mathbf{Q} = \begin{pmatrix} \cos\theta & -\sin\theta & 0 \\ \sin\theta & \cos\theta & 0 \\ 0 & 0 & 1 \end{pmatrix}, \tag{2.11}$$

a familiar rotation matrix for the case in which the z-axis is the axis of rotation.

Application of the definition of the deformation gradient introduced earlier to the case of rigid body deformation yields $\mathbf{F} = \mathbf{Q}$. Recalling that $\mathbf{Q}$ is orthogonal and hence that $\mathbf{Q}^T\mathbf{Q} = \mathbf{I}$ reveals that the Lagrangian strain vanishes for such rigid body motions. On the other hand, the infinitesimal strain tensor is nonzero, an observation that is taken up in detail in the problems at the end of the chapter. This example is of interest in part since it shows that despite the fact that the displacement fields are unable to distinguish between uninteresting rigid body motions and deformations involving *relative* motions of material particles, the Lagrangian strain weeds out rigid body deformations. Our reason for making this distinction is the claim that it is the relative motions of material particles that induce internal forces that correspond to our intuitive sense of the stiffness of a material. Despite the existence of nonzero displacements (potentially large), the deformation mapping given above implies no relative motions of material particles and is thus irrelevant to the production of internal stresses.

2.2.3 Geometry of Slip and Twinning

In addition to the importance that attaches to rigid body motions, shearing deformations occupy a central position in the mechanics of solids. In particular, permanent deformation by either dislocation motion or twinning can be thought of as a shearing motion that can be captured kinematically in terms of a shear in a direction **s** on a plane with normal **n**.

For simplicity, we begin by considering the case in which the shear is in the x_1-direction on a plane with normal $\mathbf{n} = (0, 0, 1)$, an example of which is shown in fig. 2.4. In this case, the deformation mapping is

$$x_1 = X_1 + \gamma X_3, \tag{2.12}$$

$$x_2 = X_2, \tag{2.13}$$

$$x_3 = X_3. \tag{2.14}$$

Using the definition of the deformation gradient (i.e. $F_{ij} = \partial x_i / \partial X_j$) we see that in this case we have

$$\mathbf{F} = \begin{pmatrix} 1 & 0 & \gamma \\ 0 & 1 & 0 \\ 0 & 0 & 1 \end{pmatrix}, \tag{2.15}$$

which can be written as

$$\mathbf{F} = \mathbf{I} + \gamma \mathbf{e}_1 \otimes \mathbf{e}_3. \tag{2.16}$$

Here we make the key observation that it is only the perpendicular distance from the shearing plane that determines the magnitude of the shearing displacement. In the more general setting characterized by shear direction **s** and shearing plane with

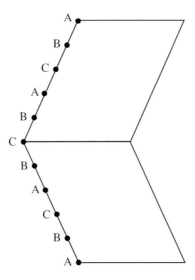

Fig. 2.4. Shear deformation.

normal **n**, we see that the deformation gradient may be written as

$$\mathbf{F} = \mathbf{I} + \gamma \mathbf{s} \otimes \mathbf{n}, \tag{2.17}$$

where γ represents the magnitude of the shear. We note that the outer product used in the equation above is defined through its action on a vector **v** through the expression $(\mathbf{s} \otimes \mathbf{n})\mathbf{v} = \mathbf{s}(\mathbf{n} \cdot \mathbf{v})$. From a crystallographic perspective, the type of shearing deformations introduced above are most easily contemplated in the context of twinning in which the material on one side of a twinning plane is subject to a uniform shear of the type considered above. The reader is encouraged to work out the details of such twinning deformations in fcc crystals in the problems section at the end of the chapter.

2.2.4 Geometry of Structural Transformations

Yet another example of the geometry of deformation of interest to the present enterprise is that of structural transformation. As was evidenced in chap. 1 in our discussion of phase diagrams, material systems admit of a host of different structural competitors as various control parameters such as the temperature, the pressure and the composition are varied. Many of these transformations can be viewed from a kinematic perspective with the different structural states connected by a deformation pathway in the space of deformation gradients. In some instances, it is appropriate to consider the undeformed and transformed crystals as being linked by an affine transformation. A crystal is built up through the repetition

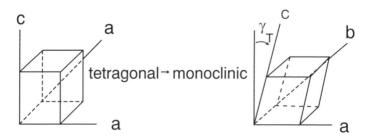

Fig. 2.5. Representation of the tetragonal to monoclinic transformation in ZrO_2 as an affine transformation (adapted from Budiansky and Truskinovsky (1993)).

of basic units with a given lattice point given as an integral linear combination of the form $\mathbf{R}_{m_1 m_2 m_3} = m_i \mathbf{a}_i$, where the set $\{\mathbf{a}_i\}$ are the so-called Bravais lattice vectors which serve as the basic template for building the entire crystal lattice. The deformation of such lattices is built around the idea that the Bravais lattice vectors before, $\{\mathbf{a}_i'\}$, and after, $\{\mathbf{a}_i\}$, transformation are linked via $\mathbf{a}_i' = \mathbf{F}\mathbf{a}_i$.

There are a number of different examples within which it is possible to describe the kinematics of structural transformation. Perhaps the simplest such example is that of the transformation between a cubic parent phase and a transformed phase of lower symmetry such as a tetragonal structure. We note that we will return to precisely such structural transformations in the context of martensitic microstructures in chap. 10. If we make the simplifying assumption that the transformed axes correspond with those of the parent phase, then the deformation mapping is of the form

$$x_1 = (1 + \alpha)X_1, \tag{2.18}$$

$$x_2 = (1 + \alpha)X_2, \tag{2.19}$$

$$x_3 = (1 + \beta)X_3. \tag{2.20}$$

This corresponds in turn to a deformation gradient tensor of the form

$$\mathbf{F} = \begin{pmatrix} 1+\alpha & 0 & 0 \\ 0 & 1+\alpha & 0 \\ 0 & 0 & 1+\beta \end{pmatrix}. \tag{2.21}$$

Another concrete example of a structural transformation is that of ZrO_2 in which the cubic, tetragonal and monoclinic phases are obtained with decreasing temperature. The immediate continuum treatment of this problem is to represent this sequence of structures as being linked by a series of affine deformations. A schematic of the tetragonal to monoclinic transformation is shown in fig. 2.5. On the other hand, in the case of transformations such as that of ZrO_2 between the cubic and tetragonal structures, there are additional internal rearrangements

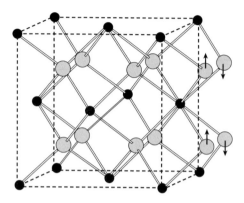

Fig. 2.6. Illustration of the internal rearrangements that attend the structural transformation in ZrO_2 (adapted from Finnis *et al.* (1998)). Although the overall shape of the unit cell can be described in terms of an affine deformation characterized by a constant deformation gradient **F**, the individual internal atoms do not transform accordingly.

that cannot be captured by the type of description advocated above, as shown in fig. 2.6. In this case, the key idea is that in the process of the cubic to tetragonal transformation, not only is there an overall change in the Bravais lattice vectors of the type described above, but in addition, there are internal shifts of the oxygen atoms which are above and beyond the simple affine deformation considered already. A full treatment of the kinematics of structural transformation demands a more complete set of degrees of freedom than that envisioned above.

2.3 Forces and Balance Laws

2.3.1 Forces Within Continua: Stress Tensors

As yet, we have given little attention to the forces that lead to the deformations we have worked to characterize geometrically. Remarkably, forces within continua can be treated in a generic way through the introduction of a single second-rank tensor field. The surprising feature of the continuum stress idea is that despite the severely complicated atomic configurations that are responsible for a given state of stress, the interatomic forces conspire to produce a net response that can be summarized in terms of the six independent components of a tensor quantity known as the Cauchy stress tensor.

We now examine the question of how in continuum mechanics the forces due to material external to the region Ω are communicated to it. Note that we will adopt the notation $\partial\Omega$ to characterize the boundary of the region Ω. In simplest terms, forces are transmitted to a continuum either by the presence of 'body forces' or via 'surface tractions'. Body forces are those such as that due to gravity which

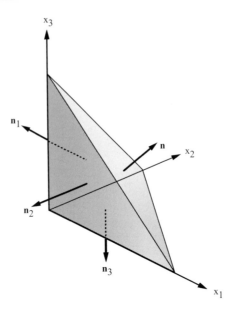

Fig. 2.7. Tetrahedral volume element used to illustrate the Cauchy stress principle.

communicate with the interior of the body via some field, and can be written as $\int_\Omega \mathbf{f(r)}dV$, where $\mathbf{f}$ is the force per unit volume at the point of interest. More typically, forces are communicated to bodies by grabbing them on the surface. Such forces are known as surface tractions and yield a net force $\int_{\partial\Omega} \mathbf{t}dA$, where $\mathbf{t}$ is a vector known as the traction vector and has units of force per unit area. In both the cases of body forces and surface tractions, we note that the total force acting on the region of interest (call it Ω) is obtained by summing up the contributions due to body forces and surface tractions and may be written as

$$\text{Net force on } \Omega = \int_{\partial\Omega} \mathbf{t}dA + \int_\Omega \mathbf{f}dV. \tag{2.22}$$

The Cauchy stress principle arises through consideration of the equilibrium of body forces and surface tractions in the special case of the infinitesimal tetrahedral volume shown in fig. 2.7. Three faces of the tetrahedron are perpendicular to the Cartesian axes while the fourth face is characterized by a normal $\mathbf{n}$. The idea is to insist on the equilibrium of this elementary volume, which results in the observation that the traction vector on an arbitrary plane with normal $\mathbf{n}$ (such as is shown in the figure) can be determined once the traction vectors on the Cartesian planes are known. In particular, it is found that $\mathbf{t}^{(n)} = \boldsymbol{\sigma}\mathbf{n}$, where $\boldsymbol{\sigma}$ is known as the stress tensor, and carries the information about the traction vectors associated with the Cartesian planes. The simple outcome of this argument is the claim that once one has determined the nature of the tractions on three complementary planes, the tractions on all other planes follow routinely via $\mathbf{t}^{(n)} = \boldsymbol{\sigma}\mathbf{n}$. The details of this argument may be worked out by the reader in the problems section at the end of this chapter.

Now that we have the notion of the stress tensor in hand, we seek one additional insight into the nature of forces within solids that will be of particular interest to our discussion of plastic flow in solids. As was mentioned in section 2.2.3, plastic deformation is the result of shearing deformations on special planes. Certain models of such deformation posit the existence of a critical stress on these planes such that once this stress is attained, shearing deformations will commence. To compute the resolved shear stress on a plane with normal $\mathbf{n}$ and in a direction $\mathbf{s}$ we begin by noting that the traction vector on this plane is given by

$$\mathbf{t}^{(n)} = \boldsymbol{\sigma}\mathbf{n}. \tag{2.23}$$

To compute the component of this traction vector in the direction of interest (i.e. the resolved shear stress), it is necessary to project the force along the direction of interest via

$$\sigma_{rss} = \mathbf{s} \cdot \boldsymbol{\sigma}\mathbf{n}. \tag{2.24}$$

As will become evident shortly, calculation of the resolved shear stress will serve as a cornerstone of our analysis of plasticity in single crystals.

2.3.2 Equations of Continuum Dynamics

Now that we are in possession of a reckoning of the forces that act within a continuum, it is of immediate interest to seek the continuum analog of the second law of motion which presides over the mechanics of discrete particles. It is not obvious on first glance whether the laws of discrete particle mechanics must be somehow supplemented to account for the smearing out of degrees of freedom which takes place in defining a continuum. However, it has been found that by strictly generalizing the notions of discrete particle mechanics to the continuum setting, this forms a sufficient conceptual backdrop for most purposes. To make this generalization it is first necessary to define the linear momentum of the volume element of interest. We restrict our consideration to a subregion of the body which is labeled Ω. The linear momentum $\mathbf{P}$ associated with this region is defined as

$$\mathbf{P} = \int_{\Omega} \rho \mathbf{v} dV, \tag{2.25}$$

where ρ is the mass density field and $\mathbf{v}$ is the velocity field. The principle of balance of linear momentum is the statement that the time rate of change of the linear momentum is equal to the net force acting on the region Ω. This is nothing more than the continuous representation of $\mathbf{F} = m\mathbf{a}$. In mathematical terms the continuous representation of this idea is

$$\frac{D}{Dt} \int_{\Omega} \rho \mathbf{v} dV = \int_{\partial \Omega} \mathbf{t} dA + \int_{\Omega} \mathbf{f} dV. \tag{2.26}$$

It is important to emphasize that Ω itself is time-dependent since the material volume element of interest is undergoing deformation. Note that in our statement of the balance of linear momentum we have introduced notation for our description of the time derivative which differs from the conventional time derivative, indicating that we are evaluating the material time derivative. The material time derivative evaluates the time rate of change of a quantity for a given material particle. Explicitly, we write

$$\frac{D}{Dt} = \frac{\partial}{\partial t} + \mathbf{v} \cdot \nabla, \tag{2.27}$$

where the first term accounts for the explicit time dependence and the second arises in response to the convective terms.

   This notation is not an accident and represents a novel feature that arises in the continuum setting as a result of the fact that there are two possible origins for the time dependence of field variables. First, there is the conventional explicit time dependence which merely reflects the fact that the field variable may change with time. The second source of time dependence is a convective term and is tied to the fact that as a result of the motion of the continuum, the material particle can be dragged into a region of space where the field is different. As a concrete realization of this idea, we might imagine a tiny ship sailing in a stream, armed with a thermometer which measures the temperature of the water at the material point being convected with the boat itself. Clearly, one way in which the temperature T can vary with time, even if the fluid is at rest, is in the presence of heating which will change the temperature of the fluid and result in a nonzero DT/Dt because of the $\partial_t T$ term. By way of contrast, we can similarly imagine a time-independent temperature distribution but with the added feature that the temperature of the water is varying along the downstream direction. In this instance, the temperature associated with a given material particle can change by virtue of the fact that the material particle enters a series of spatial regions with different temperatures. In particular, if we imagine a steady flow in the x-direction with velocity v_x, then DT/Dt will be nonzero because of the term $v_x \partial_x T$.

   Though the integral form of linear momentum balance written above as eqn (2.26) serves as the starting point of many theoretical developments, it is also convenient to cast the equation in local form as a set of partial differential equations. The idea is to exploit the Reynolds transport theorem in conjunction with the divergence theorem. For further details on various transport theorems, see chap. 5 of Malvern (1969). The Reynolds transport theorem allows us to rewrite the left hand side of the equation for linear momentum balance as

$$\frac{D}{Dt} \int_\Omega \rho \mathbf{v} dV = \int_\Omega \rho \frac{D\mathbf{v}}{Dt} dV. \tag{2.28}$$

The logic behind this development is related to the equation of continuity which asserts the conservation of mass and may be written as

$$\frac{\partial \rho}{\partial t} + \nabla \cdot (\rho \mathbf{v}) = 0. \tag{2.29}$$

This result insures that when we pass the material time derivative operator through the integral as in eqn (2.28), the result is of the form given above. Working in indicial form, we note that using the divergence theorem we may rewrite the term involving surface tractions on the right hand side of eqn (2.26) as

$$\int_{\partial\Omega} \sigma_{ij} n_j dA = \int_\Omega \sigma_{ij,j} dV. \tag{2.30}$$

Thus, eqn (2.26) may be rewritten as

$$\int_\Omega \left(\sigma_{ij,j} + f_i - \rho \frac{Dv_i}{Dt} \right) dV = 0. \tag{2.31}$$

If we now recognize that the choice of our subregion Ω was arbitrary we are left with

$$\sigma_{ij,j} + f_i = \rho \frac{Dv_i}{Dt}. \tag{2.32}$$

This equation is the main governing equation of continuum mechanics and presides over problems ranging from the patterns formed by clouds to the evolution of dislocation structures within metals. Note that as yet no constitutive assumptions have been advanced rendering this equation of general validity. We will find later that for many problems of interest in the modeling of materials it will suffice to consider the static equilibrium version of this equation in which it is assumed that the body remains at rest. In that case, eqn (2.32) reduces to

$$\sigma_{ij,j} + f_i = 0. \tag{2.33}$$

Just as it is possible to generalize the notion of linear momentum to our continuum volume, we can also consider the angular momentum associated with such a material element which is defined by

$$\mathbf{H} = \int_\Omega \rho \mathbf{r} \times \mathbf{v} dV. \tag{2.34}$$

Following the prescription adopted above, we assert that angular momentum balance implies

$$\frac{D}{Dt} \int_\Omega \rho \mathbf{r} \times \mathbf{v} dV = \int_{\partial\Omega} \mathbf{r} \times \mathbf{t} dA + \int_\Omega \mathbf{r} \times \mathbf{f} dV. \tag{2.35}$$

In indicial form, this may be simplified once we make the identification that $t_k = \sigma_{km} n_m$ and that the Reynolds transport theorem is still in effect. Bringing all terms to one side and carrying out the appropriate derivatives yields

$$\int_\Omega \epsilon_{ijk} \left(\rho v_j v_k + \rho x_j \frac{Dv_k}{Dt} - x_j \sigma_{km,m} - x_{j,m} \sigma_{km} - x_j f_k \right) dV = 0. \tag{2.36}$$

We have adopted the convention that the position vector be written in component form as $\mathbf{r} = x_i \mathbf{e}_i$. Now we recognize that upon factoring out x_j, three of these terms are a restatement of linear momentum balance, and are thus zero. In addition, the term involving $v_j v_k$ also clearly yields zero since it is itself symmetric and is contracted with the antisymmetric Levi-Cevita symbol, ϵ_{ijk}. The net result is that we are left with $\epsilon_{ijk} \sigma_{kj} = 0$ which immediately implies that the stress tensor itself is symmetric. Hence, in the present setting we have seen that the balance of angular momentum implies the symmetry of the Cauchy stress tensor. We also note that in addition to the balance laws for mass, linear momentum and angular momentum presented above, the conservation of energy can be stated in continuum form and the interested reader is referred to texts such as that of Gurtin (1981) referred to at the end of the chapter.

2.3.3 Configurational Forces and the Dynamics of Defects

Earlier in this section we introduced the stress tensor as the tool for describing forces within materials that have been represented as continua. The aim of the present section is to describe a second class of forces within solids that are of particular interest to the mechanics of materials. Materials science is replete with examples of problems in which the notion of a 'driving force' is invoked to explain the temporal evolution of a given configuration. For example, in the context of nucleation, one speaks of a driving force underlying the tendency of a given nucleus to continue its growth. This driving force derives from the reduction in the bulk free energy due to the particle being nucleated. On the other hand, there is an opposing driving force tending to reduce the particle size since the interface will move in a way that minimizes the overall interfacial area, thereby lowering the interfacial contribution to the free energy. Examples of this type can all be brought within the same theoretical fold by recourse to the theory of configurational forces.

The notion of a configurational force is entirely in keeping with our aim of developing effective theories for characterizing the behavior of materials. In particular, the use of configurational forces allows us to remove our attention from the point by point reckoning of forces demanded both in atomic-level descriptions and continuum notions of stress and, instead, to focus on the effective degrees of freedom introduced to describe defects. As a result, the dynamics of the continuum which demands that we account for the relevant field variables everywhere within the body is replaced with an explicit dynamics of defects in which the kinematics of the defected body is tied only to a set of degrees of freedom that characterize the configuration of the defects themselves. For example, we will see later that despite the fact that dislocations carry with them long-range stress fields, an effective theory of dislocation dynamics can be built in which rather than keeping track of stresses and strains at all points in the continuum, the kinematic description is reduced to a set of nodal coordinates describing the positions of the dislocations. Similarly, in treating the dynamics of interfacial motion, our entire kinematic description can be built around the consideration of coordinates describing the position of the interface.

The central observation associated with the definition of configurational forces is that the total energy of the body of interest and associated loading devices depends explicitly on the positions of the various defects within that body. A small excursion of a given defect from position x_i to $x_i + \delta \xi_i$ will result in an attendant change of the total energy. The configurational force on that defect associated with that motion is defined via

$$\delta E_{tot} = -F_i \delta \xi_i. \tag{2.37}$$

This definition may be elaborated alternatively as

$$F_i = -\frac{\partial E_{tot}}{\partial \xi_i}. \tag{2.38}$$

As noted above, the notion of a configurational force may be advanced as a basis for considering the dynamics of defects themselves since, once such forces are in hand, the temporal evolution of these defects can be built up in turn by the application of an appropriate kinetic law which postulates a relation of the form $\mathbf{v} = \mathbf{v}(\text{driving force})$.

To further elaborate the underlying idea of a configurational force, we appeal to the examples indicated schematically in fig. 2.8. Fig. 2.8(a) shows an interface within a solid and illustrates that by virtue of interfacial motion the area of the interface can be reduced. If we adopt a model of the interfacial energy in which it is assumed that this energy is isotropic (i.e. γ does not depend upon the local interface normal $\mathbf{n}$), the driving force is related simply to the local curvature of that interface. Within the theory of dislocations, we will encounter the notion of image dislocations as a way of guaranteeing that the elastic fields for dislocations in finite

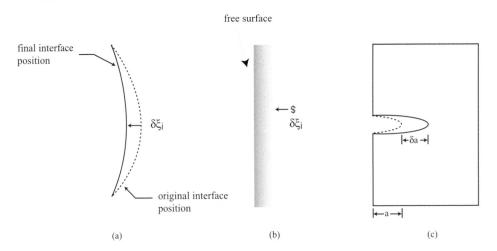

Fig. 2.8. Representative examples of the origins of the concept of configurational forces: (a) curved interface, (b) dislocation near a free surface and (c) solid with a crack.

bodies satisfy the free surface boundary conditions. What is noted in fig. 2.8(b) is that because of the free surface, a small motion of the subsurface dislocation will result in an attendant energy change and hence in the existence of an image force. Within the present setting, we will see that such image forces are a manifestation of the more general notion of a configurational force. Yet another example of the energy changes that attend the motion of defects is shown in fig. 2.8(c) in which a crack is shown to have suffered an excursion by the amount δa. In this instance, there is a competition between the energy cost to create new free surface, which leads to an energy penalty for incremental crack extension, and the reduction in the elastic strain energy associated with such motions.

Though we have discussed the idea of a configurational force primarily through appeal to specific examples and the generic definition of eqn (2.37), Eshelby pioneered (for example, Eshebly (1975b)) a way of recasting these ideas in an abstract form that is indifferent to the particulars described above. His argument is made on the basis of the observation that because of the broken translational invariance in the types of systems described above, there is an *explicit* dependence of the total energy on the position of the defect. Continuing with the dislocation example given in fig. 2.8(b), we first note that if this single dislocation were embedded in an infinite medium, the elastic energy would not depend explicitly on the position of the dislocation. In particular, we note that

$$\int_{\Omega_\infty} \frac{1}{2}\sigma_{ij}^{dis}(\mathbf{x}-\mathbf{x}_0)\epsilon_{ij}^{dis}(\mathbf{x}-\mathbf{x}_0)dV = \int_{\Omega_\infty} \frac{1}{2}\sigma_{ij}^{dis}(\mathbf{x}-\mathbf{x}_0-\delta\boldsymbol{\xi})\epsilon_{ij}^{dis}(\mathbf{x}-\mathbf{x}_0-\delta\boldsymbol{\xi})dV,$$

$$(2.39)$$

which is the assertion that whether the dislocation is positioned at position $\mathbf{x}_0$ or $\mathbf{x}_0 + \delta\boldsymbol{\xi}$, the stored energy is the same. To be explicit about notation, Ω_∞ refers to the fact that the dislocation is in an infinite body and $\sigma_{ij}^{dis}(\mathbf{x} - \mathbf{x}_0)$ refers to the ij^{th} component of stress due to a dislocation located at position $\mathbf{x}_0$, while $\epsilon_{ij}^{dis}(\mathbf{x} - \mathbf{x}_0)$ are the corresponding linear elastic strains. By way of contrast with the single dislocation embedded in an infinite medium described above, in the presence of the free surface, the total elastic energy of the system will depend explicitly on the distance of the dislocation beneath the free surface. As a result of this explicit dependence on configuration, there will be a corresponding configurational force.

In light of the definition of the configurational force given above, we may exploit this explicit dependence of the elastic energy on the position of the defect to rewrite the force as

$$F_l = \int_\Omega \left(\frac{\partial W}{\partial x_l}\right)_{explicit} dV. \tag{2.40}$$

W is the elastic strain energy density, and the subscript *explicit* reiterates our strategy of picking off the explicit dependence of the total elastic energy on the defect coordinates. As our expression stands, it is nothing more than a schematic representation of the physical origins of the configurational force. To give the equation substance, we must compute $(\partial W/\partial x_l)_{explicit}$ itself. To do so, we note that the change in the strain energy with position has implicit contributions in addition to those that arise from the symmetry breaking. In particular, the derivative of the strain energy with respect to the position of the defect involves three terms (at least for a material in which the strain energy density depends only upon the displacements and their first derivatives)

$$\frac{\partial W}{\partial x_l} = \frac{\partial W}{\partial u_i} u_{i,l} + \frac{\partial W}{\partial u_{i,j}} u_{i,jl} + \left(\frac{\partial W}{\partial x_l}\right)_{explicit}. \tag{2.41}$$

Note that the first two terms on the right hand side tell us that there is a change in the strain energy density that is incurred because the defect has moved and it has dragged its elastic fields with it. However, these terms do not reflect the presence of broken translational invariance. By exploiting equality of mixed partials, and by rearranging terms via a simple application of the product rule for differentiation, this expression may be rewritten as

$$\frac{\partial W}{\partial x_l} = \left[\frac{\partial W}{\partial u_i} - \frac{\partial}{\partial x_j}\left(\frac{\partial W}{\partial u_{i,j}}\right)\right] u_{i,l} + \frac{\partial}{\partial x_j}\left[\frac{\partial W}{\partial u_{i,j}} u_{i,l}\right] + \left(\frac{\partial W}{\partial x_l}\right)_{explicit}. \tag{2.42}$$

We now note that the term multiplying $u_{i,l}$ is actually the Euler–Lagrange equation associated with the variational statement of the field equations for the continuum, and thereby vanishes. The net result is that the explicit dependence of the strain energy density upon position may be computed in terms of the so-called Eshelby tensor as follows:

$$\left(\frac{\partial W}{\partial x_l}\right)_{explicit} = \frac{\partial P_{lj}}{\partial x_j}, \tag{2.43}$$

where the Eshelby tensor (P_{lj}) is defined as

$$P_{lj} = W\delta_{lj} - \frac{\partial W}{\partial u_{i,j}}u_{i,l}. \tag{2.44}$$

The configurational force may now be rewritten as

$$F_i = \int_{\partial\Omega} P_{ij}n_j dA, \tag{2.45}$$

where we have used the divergence theorem in conjunction with eqn (2.40). This result tells us that by integrating the Eshelby tensor over any surface surrounding the defect of interest, we can find the driving force on that defect. This result can be used in a variety of settings ranging from the derivation of the famed J-integral of fracture mechanics to the Peach–Koehler force which describes the force on a dislocation due to a stress.

The theory of configurational forces set forth above is incomplete without some attendant ideas that allow for a determination of the dynamics of defects in the presence of these driving forces. One powerful and convenient set of ideas has been built around variational approaches in which the energy change resulting from the excursion of defects is all dissipated by the mechanisms attending defect motion. Two particularly useful references in that regard are Suo (1997) and Cocks *et al.* (1999). In many ways, the development is entirely analogous to that of conventional analytic mechanics in which for those problems featuring dissipation, the Lagrangian is supplemented with a term known as the Rayleigh dissipation potential. In the present setting, we consider a system whose dissipative excitations are pararameterized with the coordinates ($\{\mathbf{r}_i\}$). For example, this may be a discrete set of points that characterize the position of a dislocation line or an interfacial front. A variational function (or functional for systems whose dissipative excitations are characterized by a continuous set of degrees of freedom) is constructed of the form

$$\Pi(\{\dot{\mathbf{r}}_i\}; \{\mathbf{r}_i\}) = \Psi(\{\dot{\mathbf{r}}_i\}; \{\mathbf{r}_i\}) + \dot{G}(\{\dot{\mathbf{r}}_i\}; \{\mathbf{r}_i\}), \tag{2.46}$$

where the first term on the right hand side is the dissipative potential and the second term is the rate of change of the Gibbs free energy. The set ($\{\mathbf{r}_i\}$) is the complete set of parameters used to specify the configuration of whatever defect(s) are being considered, and the set $\{\dot{\mathbf{r}}_i\}$ is the associated set of velocities which serve as our variational parameters. What this means is that at a fixed configuration (specified by the parameters ($\{\mathbf{r}_i\}$)), we seek those generalized velocities which are optimal with respect to Π. Note that we have replaced the description of the system either in terms of the atomic coordinates ($\{\mathbf{R}_i\}$) or the field variables $\mathbf{u}(\mathbf{x})$ with a new set of degrees of freedom that make reference only to the 'coordinates' of the defects themselves. A dynamics of the degrees of freedom ($\{\mathbf{r}_i\}$) is induced through an appeal to the variational statement $\delta \Pi = 0$. What this statement means is that the dynamics of the ($\{\mathbf{r}_i\}$) is chosen such that the rate of working on the system, as represented by $\dot{G}$, is entirely spent in operating the various dissipative mechanisms that are accounted for in Ψ. Since our aim is to find velocity increments to step forward in time from a given configuration, for a discrete set of dissipative excitations, $\delta \Pi = 0$ amounts to the condition that $\partial \Pi / \partial \dot{\mathbf{r}}_i = 0$ for all i.

This idea can be illustrated in schematic form through the example of the motion of a single curved interface in two dimensions under the action of curvature-induced forces. For convenience, we will represent the interface as a function (rather than via a discrete set of parameters as in the discussion of the variational principle above) of a parameter s as $\mathbf{r}(s)$. In this case, the rate of change of Gibbs free energy is given by

$$\dot{G} = \int_{\Gamma} \gamma_{gb} \kappa(s) v_n(s) ds, \tag{2.47}$$

where Γ is the boundary curve, γ_{gb} is the grain boundary energy and specifies the energy cost per unit area of interface (which for the moment we assume is isotropic, i.e. no dependence on the grain boundary normal $\mathbf{n}$), $\kappa(s)$ is the local curvature of the interface at position $\mathbf{r}(s)$ and $v_n(s)$ is the normal velocity of the interface at the same point. The energy dissipation associated with this process is in turn given by

$$\Psi = \int_{\Gamma} \frac{v_n^2(s)}{2M} ds, \tag{2.48}$$

where we have introduced the parameter M which is a mobility and relates the local velocity and the local driving force through the relation $v_n = Mf$. Indeed, f is precisely the configurational force described above, and this part of the argument amounts to making kinetic assumptions. What we note now is that by evaluating the variation in Π we end up with equations of motion for the configurational degrees of freedom that enter in our expressions for the configurational forces.

To drive the point home, simplify the discussion above to the case in which the interface is a circle, thus described by a single degree of freedom R. For simplicity, we ignore all contributions to the driving force except for the curvature-induced terms. The Gibbs free energy is given by $G = 2\pi R\gamma$. In this case, all of the expressions given above collapse to the form

$$\Pi = 2\pi \dot{R}\gamma + 2\pi R \frac{\dot{R}^2}{2M}.$$ (2.49)

We are interested now in determining the dynamical equation for the temporal evolution of this circular interface. To obtain that, we consider variations in $\dot{R}$ at a fixed instantaneous configuration, which demands that we consider $\partial \Pi / \partial \dot{R} = 0$, resulting in the equation of motion

$$\frac{dR}{dt} = -\frac{M\gamma}{R}.$$ (2.50)

Not surprisingly, this equation of motion implies the curvature-induced shrinkage of the particle.

This section has provided two related tools that we will return to repeatedly throughout the book. The first of those tools was the idea of a configurational force, which tells us that if a defect is characterized by some set of coordinates ($\{\mathbf{r}_i\}$), then there is a set of forces conjugate to these coordinates such that the change in energy when the defect suffers an excursion ($\{\delta\mathbf{r}_i\}$) is given by $\delta E_{tot} = -\sum_i \mathbf{F}_i \cdot \delta\mathbf{r}_i$. By virtue of the elastic energy–momentum tensor, we have also argued that there is a generic formalism that allows for the determination of these configurational forces, although we wish to emphasize that all that is really needed in the consideration of a given problem is the definition of configurational forces as the work-conjugate partners of the various coordinates used to characterize the position of the defect. We have also seen that there is a variational formalism for inducing a dynamics on the coordinates used to describe the configuration of a given defect. This variational principle selects those trajectories such that the rate of working on the system is balanced by the rate of dissipation that attends the dissipative excitations of the defects. We note again that the ideas presented here serve as the foundation of the construction of effective theories of defects since they allow us to choose the degrees of freedom in a given problem as the defect coordinates themselves, allowing us to abandon a description in terms of either atomic coordinates or continuum fields. The framework described here will be used to consider the motion of cracks and serves as the basis of the development of the J-integral, the motion of dislocations serving as the basis of the deduction of the Peach–Koehler force, the motion of grain boundaries resulting in expressions for both curvature- and diffusion-induced motion of boundaries and will serve as the basis of our discussions of the dynamical simulation of both dislocation dynamics and microstructural evolution.

2.4 Continuum Descriptions of Deformation and Failure

2.4.1 Constitutive Modeling

Both the power and limitation of continuum mechanics derive from the same origins. On the one hand, the beauty of the continuum approach is its ability to renounce microscopic details. On the other hand, because of this, it is necessary to introduce material specificity into continuum models through phenomenological constitutive models. Recall that in writing the equation for the balance of linear momentum, we have as yet made no reference to whether or not our equations describe the deformation of a polycrystal with a highly anisotropic grain structure or the moments before a wave ends its journey across the ocean in a tumultuous explosion of whitewater. Whether our interest is in the description of injecting polymers into molds, the evolution of Jupiter's red spot, the development of texture in a crystal, or the formation of vortices in wakes, we must supplement the governing equations of continuum mechanics with some constitutive description.

The role of constitutive equations is to instruct us in the relation between the forces within our continuum and the deformations that attend them. More prosaically, if we examine the governing equations derived from the balance of linear momentum, it is found that we have more unknowns than we do equations to determine them. Spanning this information gap is the role played by constitutive models. From the standpoint of building effective theories of material behavior, the construction of realistic and tractable constitutive models is one of our greatest challenges. In the sections that follow we will use the example of linear elasticity as a paradigm for the description of constitutive response. Having made our initial foray into this theory, we will examine in turn some of the ideas that attend the treatment of permanent deformation where the development of microscopically motivated constitutive models is much less mature.

2.4.2 Linear Elastic Response of Materials

The linear theory of elasticity is the logical extension and natural generalization of the insights into material behavior that are embodied in Hooke's law. The basic idea is suggested by the types of experiments that were originally considered by Hooke and are depicted in fig. 2.9. The fundamental idea of experiments of this type is to determine a relation between the geometry of deformation and the forces that produce it. The central result for linearly elastic materials is that the strain and stress are linearly related. For the types of geometries shown in the figure, a single scalar relation suffices. For example, in tension experiments, Hooke's law may be stated as

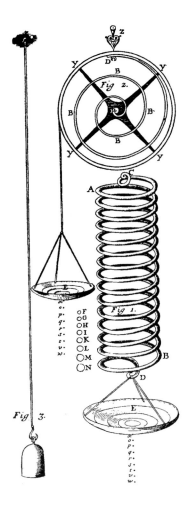

Fig. 2.9. Apparatus used by Robert Hooke in his investigation of the relation between the displacement of a body and the load which produces it (adapted from Bell (1973)).

$$\frac{F}{A} = E\frac{\Delta l}{l}, \tag{2.51}$$

where the left hand side of the equation is the force per unit area, and the ratio on the right hand side is the fractional change in length of the specimen. The constant E is the so-called Young's modulus and for this simple test geometry is the point of entry of material specificity into the problem.

Under more complex states of loading, Hooke's law required elaboration. The generalization of Hooke's law is the assertion that an arbitrary component of the stress tensor may be written in terms of the infinitesimal strains, ϵ_{ij}, as follows:

$$\sigma_{ij} = C_{ijkl}\epsilon_{kl}. \tag{2.52}$$

What this equation tells us is that a particular state of stress is nothing more than a linear combination (albeit perhaps a tedious one) of the entirety of components of the strain tensor. The tensor C_{ijkl} is known as the elastic modulus tensor or stiffness and for a linear elastic material provides nearly a complete description of the material properties related to deformation under mechanical loads. Eqn (2.52) is our first example of a constitutive equation and, as claimed earlier, provides an explicit statement of material response that allows for the emergence of material specificity in the equations of continuum dynamics as embodied in eqn (2.32). In particular, if we substitute the constitutive statement of eqn (2.52) into eqn (2.32) for the equilibrium case in which there are no accelerations, the resulting equilibrium equations for a linear elastic medium are given by

$$C_{ijkl}u_{k,lj} + f_i = 0. \tag{2.53}$$

As usual, we have invoked the summation convention and in addition have assumed that the material properties are homogeneous. For an isotropic linear elastic solid, the constitutive equation relating the stresses and strains is given by

$$C_{ijkl} = \lambda\delta_{ij}\delta_{kl} + \mu(\delta_{ik}\delta_{jl} + \delta_{il}\delta_{jk}), \tag{2.54}$$

where λ and μ are the Lamé constants. Simple substitution and a few lines of algebra yield the so-called Navier equations that serve as the equilibrium equations for an isotropic linear elastic solid,

$$(\lambda + \mu)\nabla(\nabla \cdot \mathbf{u}) + \mu\nabla^2\mathbf{u} + \mathbf{f} = 0. \tag{2.55}$$

Note that these equations are a special case of the equilibrium equations revealed in eqn (2.53) in the constitutive context of an isotropic linear elastic solid.

In addition to the characterization of elastic solids in terms of the stress–strain response given in eqn (2.52), linearly elastic materials may also be characterized in terms of a local strain energy density that reveals the energy per unit volume stored in the elastic fields. To the extent that we are interested in describing the energetics of deformations that are sufficiently 'small', the total energy stored in a body by virtue of its deformation is contained in the parameters that make up the elastic modulus tensor. In particular, the total strain energy tied up in the elastic fields is given by

$$E_{strain} = \frac{1}{2}\int_\Omega C_{ijkl}\epsilon_{ij}\epsilon_{kl}dV. \tag{2.56}$$

As we will see in coming chapters, the logic that rests behind this model is a quintessential example of the construction of an effective theory within which the microscopic degrees of freedom have been subsumed into a series of parameters whose job it is to mimic the atomic-level response of the material, but in situations where $\epsilon_{ij}(\mathbf{x})$ has spatial variations on a scale much larger than the mean interatomic spacing. From the standpoint of modeling a material from the linear elastic perspective, a given material is nothing more than a particular set of elastic moduli. These numbers, in conjunction with the density, are all that are needed to solve boundary value problems concerning small-scale deformation in a particular material.

One of our primary objectives is the development of constitutive models of the permanent deformation of materials. At first glance, both deformation and fracture involve deformations that exceed the limits of validity of the theory of elasticity. Interestingly, we will find that many ideas for describing plasticity and fracture can nevertheless be constructed on the basis of the linear theory of elasticity. In particular, with regard to plasticity, it will be seen that the linear theory of elasticity serves as the basis for the elastic theory of dislocations which in turn provides insights into the many strengthening mechanisms central to plasticity. Similarly, the analysis of failure via fracture will be largely built around the ideas of linear elastic fracture mechanics. It is an amusing twist that some of the best ideas for capturing the highly nonlinear processes of plasticity and fracture emerge from a linear constitutive description.

2.4.3 Plastic Response of Crystals and Polycrystals

From the standpoint of a stress–strain curve, the constitutive behavior discussed in the previous section corresponds to a limited range of loads and strains. In particular, as seen in fig. 2.10, for stresses beyond a certain level, the solid suffers permanent deformation and if the load is increased too far, the solid eventually fails via fracture. From a constitutive viewpoint, more challenges are posed by the existence of permanent deformation in the form of plasticity and fracture. Phenomenologically, the onset of plastic deformation is often treated in terms of a yield surface. The yield surface is the locus of all points in stress space at which the material begins to undergo plastic deformation. The fundamental idea is that until a certain critical load is reached, the deformation is entirely elastic. Upon reaching a critical state of stress (the yield stress), the material then undergoes plastic deformation. Because the state of stress at a point can be parameterized in terms of six numbers, the tensorial character of the stress state must be reflected in the determination of the yield surface.

As will be revealed in more detail in chap. 8, the initiation of plastic deformation can, in most instances, be traced to the onset of dislocation motion. From the

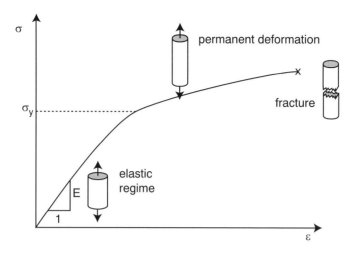

Fig. 2.10. Schematic of the stress–strain response of a cylindrical specimen under tensile loading. Elastic response is obtained only in the initial linear region of the loading curve.

perspective of the forces that induce plastic flow, it will be seen that it is the resolved shear stress on a particular slip plane in the direction of the Burgers vector that induces dislocation glide. The Burgers vector is the increment of relative displacement carried by each such dislocation. Our present aim is to show that the yield surface must have buried within it some reflection of the notion of a critical shear stress. In single crystals, shearing deformations occur only along privileged slip directions on privileged slip planes. By way of contrast, the simplest model of a polycrystal holds that because of the existence of slip planes of all different orientations, a more isotropic picture of yield is appropriate.

One plausible yield surface for polycrystals is built along the following lines. We recall that because the stress state at a point is described by a symmetric Cauchy stress tensor, we may find three principal stresses at a given point within the medium of interest which are associated with three orthogonal principal directions. We now suppose that associated with this point are an infinity of available slip planes along which glide can occur. What this means is that unlike the situation in a real crystal with slip planes present along certain special directions determined by the underlying crystal lattice, we imagine that every direction can support such shear deformations. Given this assumption, our aim is to discover within which planes the shear stress will be maximum, and to formulate a criterion for the initiation of plastic flow in which yield will commence once the maximum shear stress reaches some critical value.

The calculation goes as follows. If we define the principal axes as $\mathbf{e}_1$, $\mathbf{e}_2$ and $\mathbf{e}_3$ and the corresponding principal stresses σ_1, σ_2 and σ_3, the state of stress may be conveniently written as

$$\boldsymbol{\sigma} = \sum_{i=1}^{3} \sigma_i \mathbf{e}_i \otimes \mathbf{e}_i. \tag{2.57}$$

With respect to this coordinate system, we aim to find the plane (characterized by normal $\mathbf{n} = n_i \mathbf{e}_i$) within which the shear stress is maximum. The traction vector on the plane with normal $\mathbf{n}$ is given by $\mathbf{t}^{(\mathbf{n})} = \boldsymbol{\sigma} \cdot \mathbf{n}$. To find the associated shear stress on this plane, we note that the traction vector may be decomposed into a piece that is normal to the plane of interest and an in-plane part. It is this in-plane contribution that we wish to maximize. Denote the in-plane part of the traction vector by $\boldsymbol{\tau}$, which implies $\mathbf{t}^{(\mathbf{n})} = (\mathbf{t}^{(\mathbf{n})} \cdot \mathbf{n})\mathbf{n} + \boldsymbol{\tau}$. It is most convenient to maximize the magnitude of this quantity, which may be written as

$$\tau^2 = \sigma_1^2 n_1^2 + \sigma_2^2 n_2^2 + \sigma_3^2 n_3^2 - (\sigma_1 n_1^2 + \sigma_2 n_2^2 + \sigma_3 n_3^2)^2. \tag{2.58}$$

If we eliminate the unknown n_3 via the constraint $\mathbf{n} \cdot \mathbf{n} = 1$, then minimizing τ^2 of eqn (2.58) with respect to n_1 and n_2 results in two algebraic equations in these unknowns. What emerges upon solving these equations is the insight that the planes of maximum shear stress correspond to the set of {110}-type planes as defined with respect to the principal axes, and that the associated shear stresses on these planes take the values $\tau = \frac{1}{2}(\sigma_i - \sigma_j)$, which are the differences between the ith and j^{th} principal stresses.

Our geometric model of the crystal is most appropriate for polycrystals since we have hypothesized that any and all planes and slip directions are available for slip (i.e. the discrete crystalline slip systems are smeared out) and hence that slip will commence once the maximum shear stresses have reached a critical value on *any* such plane. This provides a scheme for explicitly describing the yield surface that is known as the Tresca yield condition. In particular, we conclude that yield occurs when

$$\frac{1}{2}(\sigma_{max} - \sigma_{min}) = k, \tag{2.59}$$

where k is a critical parameter characterizing the stress at which plastic flow commences.

An alternative scheme is the von Mises yield condition. In this case, one adopts an approach with a mean-field flavor in which plastic flow is presumed to commence once an averaged version of the shear stresses reaches a critical value. To proceed, we first define the deviatoric stress tensor which is given by,

$$S_{ij} = \sigma_{ij} - \frac{1}{3}\delta_{ij}\, \mathrm{tr}(\boldsymbol{\sigma}). \tag{2.60}$$

The von Mises condition holds that plastic flow begins when the second invariant of the deviatoric stress attains a critical value. That is,

$$J_2 = \frac{1}{2} S_{ij} S_{ij} = k^2. \tag{2.61}$$

This condition may be rewritten once it is recognized that, in terms of the principal stresses, the deviatoric stress tensor is diagonal and its components are

$$S_1 = \frac{2}{3}(2\sigma_1 - \sigma_2 - \sigma_3), \tag{2.62}$$

$$S_2 = \frac{2}{3}(2\sigma_2 - \sigma_1 - \sigma_3), \tag{2.63}$$

$$S_3 = \frac{2}{3}(2\sigma_3 - \sigma_1 - \sigma_2) \tag{2.64}$$

and hence that the yield condition is equivalent to

$$\frac{2}{3}\left[(\sigma_1 - \sigma_2)^2 + (\sigma_3 - \sigma_1)^2 + (\sigma_2 - \sigma_3)^2\right] = k^2. \tag{2.65}$$

Note that when the state of stress is purely hydrostatic, the deviatoric stress will vanish (as will the shear stresses that arise from the differences in the principal stresses) and hence there will be no plastic flow.

Our treatment until now has served as the basis for models of perfectly plasticity. What we note is that once yield is attained, we have made no provision to account for subsequent hardening. Hardening refers to the fact that once plastic deformation has begun, there is an increase in the stress needed to maintain subsequent plastic deformation. Inspection of a generic stress–strain curve indicates that hardening is a crucial part of the story. For example, in fig. 2.10, note that once the stress exceeds the 'yield stress', the stress to continue plastic deformation increases. We will undertake an analysis of *kinematic* and *isotropic* models of the hardening of a material. When we speak of hardening, in the language developed above what this implies is some evolution of the yield surface as the deformation history proceeds. Hence, we seek some simple representation of the coupling between the 'state' of the material and the criterion for commencement of plasticity.

For the type of polycrystal yield surfaces described above, the elementary treatments of hardening are highly intuitive. The fundamental physical idea behind such laws is the recognition that as plastic deformation proceeds, it becomes increasingly difficult to inspire subsequent yield. In chap. 11, we will attempt to describe the origins of this phenomenon in the evolution of the distribution of dislocations, while here our aim will be to advance phenomenological approaches to this effect.

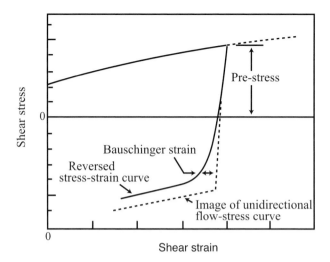

Fig. 2.11. Schematic of a stress–strain curve illustrating the anisotropy of work hardening (adapted from McClintock and Argon (1966)).

The simplest model of isotropic hardening contends that hardening behavior can be accounted for by assuming that with increasing deformation, the critical value of k (see eqn (2.59)) increases. In particular, if we idealize the yield surface as a sphere in stress space, the isotropic hardening model would assert that the radius of this sphere increases with increasing deformation. One significant insight afforded by this idea is the recognition that certain properties of a material are a function of that material's history. In particular, such observations cloud the idea of material parameter itself since not only the material type must be specified, but we must also say something about how the material got to its current configuration.

The isotropic hardening model presented above suffers certain irreparable flaws, one of which is that it cannot properly account for tension/compression asymmetries that are collectively known as the Bauschinger effect. For example, under cyclic deformation of the type indicated schematically in fig. 2.11, the yield point in compression differs from that in tension. In this case, the simplest idea is that of kinematic hardening in which it is assumed that the yield surface center in stress space is shifted as a result of the hardening of the material. Continuing with the simple spherical yield surface considered above, kinematic hardening refers to a shift in the origin of the yield surface away from the point of zero stress. What this implies is a breaking of the tension/compression asymmetry with the relevant symmetry breaking parameter being the magnitude of the shift itself.

As we have already belabored, the ideas presented above are predicated on the fundamental isotropy of the possible slip systems in the material of interest. This isotropy, in turn, derives from the polycrystalline microstructure in which it is claimed that every possible plane is a possible slip plane. On the other hand, for single crystals, these ideas are insufficiently refined. In the following paragraphs, we take up a quick overview of the ideas of single crystal plasticity. We will return to these ideas in more detail in chap. 8. The starting point for models of single crystal plasticity is the kinematic acknowledgement that in a single crystal, slip can occur only on certain well-defined crystal planes. We imagine a series of slip systems which we will label with the Greek index α. A slip system refers to a particular slip plane characterized by a normal $\mathbf{n}^{(\alpha)}$ and an allied slip direction characterized by the unit vector $\mathbf{s}^{(\alpha)}$. Activation of a given slip system means that plastic deformation (i.e. a relative displacement across particular planes) has begun on that slip system. The motivation for introducing such strong kinematic assumptions is revealed already at the level of the optical microscope in which a plastically deformed material is seen to have surface steps that run along certain preferred directions corresponding to the terminus of the slip planes at the crystal surface. It is now imagined that each such slip system has a critical resolved shear stress, τ_c^α, at which plastic flow will commence. Hence, in the case of single crystal deformation, the yield condition may be written as

$$\mathbf{s}^{(\alpha)} \cdot \boldsymbol{\sigma} \cdot \mathbf{n}^{(\alpha)} = \tau_c^\alpha. \tag{2.66}$$

This equation is nothing more than the statement that plastic flow will begin on the αth slip system when the resolved shear stress (i.e. $\mathbf{s}^{(\alpha)} \cdot \boldsymbol{\sigma} \cdot \mathbf{n}^{(\alpha)}$) on that slip system exceeds the critical stress τ_c^α.

   The notion of yield advanced above may be elaborated by considering the possible hardening of various slip systems. Here the physical picture is that because of interactions between dislocations, increasing dislocation density on one slip system can have the effect of hardening another. In mathematical terms, this postulate is formulated as

$$\dot{\tau}_c^{(\alpha)} = \sum_\beta h_{\alpha\beta} \dot{\gamma}_\beta. \tag{2.67}$$

This equation says that the critical stress on the αth slip system can change as a result of the accumulation of plastic strain on the βth slip system. The accumulation of plastic strain on the βth slip system is measured in terms of the shear strain γ_β. The matrix $h_{\alpha\beta}$ is the so-called hardening matrix and is the backbone of a hardening law for single crystal plasticity. One of the dominant tasks facing those who aim to build viable models of plasticity on the basis of insights at the dislocation level is to probe the veracity of expressions like that given above.

2.4.4 Continuum Picture of Fracture

The terminal process in the permanent deformation of solids is fracture. For any number of reasons a structure can end its useful life in catastrophic failure of the material composing the structure by fracture. The most naive model of such failure is that in which atomic bonds in a perfect crystal are homogeneously stretched until a critical stress known as the ideal tensile strength. On the other hand, observed strengths are usually much lower than the ideal tensile strength. For the analyses in this book, our conceptual starting point for the examination of fracture is the linear elastic theory of fracture mechanics. Historically, this field has its origins in the recognition that cracks play the dominant role in magnifying stresses beyond their remote values, thereby permitting the failure of a material at macroscopic stress levels that are lower than the ideal strength. Our aim in this section is to give an overview of the key ideas that stand behind the linear elastic treatment of fracture.

An instructive two-dimensional calculation that reveals the stress magnifying effects of flaws is that of an elliptical hole in an elastic solid as depicted in fig. 2.12. The crucial idea is that, despite the fact that the specimen is remotely loaded with a stress σ_0 which may be lower than the ideal strength needed to break bonds in a homogeneous fashion, locally (i.e. in the vicinity of the crack-like defect) the stresses at the termination of the major axis of the hole can be enhanced to values well in excess of the remote load. The exact solution to this problem can be found in any of the standard references on fracture and we will content ourselves with examining its key qualitative features.

Consider an elliptical hole characterized by a semimajor axis of length $2a$ and semiminor axis of length $2b$. The result of a detailed elastic analysis of a plane strain geometry with remote loading parallel to the semiminor axis (call this direction y) is an enhancement of the local stresses well above the remote value of σ_0. The key point that emerges from an analysis of the stresses associated with an elliptical hole are summarized in schematic form in fig. 2.12. In particular, the maximum value of the stress σ_{yy} is given by

$$\sigma_{yy}^{max} = \sigma_0 \left(1 + \frac{2a}{b} \right). \tag{2.68}$$

As a result, we see that the enhancement depends upon the aspect ratio of the hole, and that as the ratio a/b increases, so too does the local value of the stress, though we note that the factor modulating σ_0 is always greater than 1.

The elliptical hole prepares our intuition for the more extreme case that is presented by the atomically sharp crack. For the sharp crack tip, the dominant stresses near the crack tip may be shown to have a singular character with the particular form $r^{-1/2}$ (see Rice (1968) for example). The realization that the crack tip fields within the context of linear elasticity have such a simple singular form has

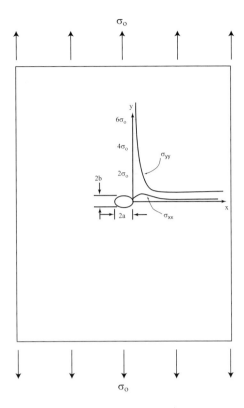

Fig. 2.12. Elliptical hole in a material and the resulting stresses in the vicinity of the hole.

far reaching implications. One remarkable feature of the linear elastic approach to fracture mechanics is the emergence of a new material parameter known as the stress intensity factor which characterizes a material's propensity for fracture. In particular, as noted above, the crack tip stress fields may be written as

$$\sigma_{ij} = \frac{K}{\sqrt{r}} f_{ij}(\theta), \qquad (2.69)$$

where the stress intensity factor K typically scales as $\sigma_0 \sqrt{a}$ with a characterizing the crack length and σ_0 is the remotely applied load. Indeed, as noted above, the spatial dependence of the local crack tip fields is universal and all information on the geometry of the specimen and loading is reflected only through the scalar multiplier K. The term $f_{ij}(\theta)$ captures the angular dependence of the crack tip fields. A commonly used fracture criterion is that once the stress intensity factor reaches a material-dependent critical value, fracture will commence. The subtlety of the idea is that once K_C has been determined for a given material for one test geometry, one is in a position to make predictions about the fracture stress in other geometries.

For the case of brittle solids, the quest to quantify the onset of fracture was undertaken by Griffith. Using the elastic arguments of Inglis concerning the fields surrounding an elliptical hole, Griffith was able to advance an energetic argument for the onset of fracture. Many of the important historic papers can be found in Barsom (1987). The basic energetic idea is an example of the type of argument given earlier in the context of configurational forces. In the language of eqn (2.46), we consider a single parameter description of a crack of length $2a$ and write

$$\Pi(\dot{a}, a) = \dot{E}_{elastic}(\dot{a}, a) + \dot{E}_{surf}(\dot{a}, a) + \Psi(\dot{a}, a), \tag{2.70}$$

where $\dot{E}_{elastic}$ is the elastic energy release rate and is negative for $\dot{a} > 0$, $\dot{E}_{surf}$ is the increase in surface energy and may be written down directly as $\dot{E}_{surf} = 4\gamma\dot{a}$, where γ is the surface energy and $\Psi(\dot{a}, a)$ is the dissipative potential and is of the form $\Psi(\dot{a}, a) = \dot{a}^2/2M$. The factor of 4 in the expression for $\dot{E}_{surf}$ results from the fact that for a crack of length $2a$ such as that under consideration here, there is both a top and bottom face contributing to the overall surface energy. To proceed with the calculation, we must still determine the rate of change of $E_{elastic}$. For a crack of length $2a$ under plane strain conditions and subject to a remote load σ_0, the elastic energy is given by (a result which we apologize for quoting without proof, see Lawn (1993))

$$E_{elastic}(a) = -\frac{\pi a^2 \sigma_0^2 (1 - v^2)}{E}, \tag{2.71}$$

implying that $\dot{E}_{elastic}(\dot{a}, a)$ is given by

$$\dot{E}_{elastic}(\dot{a}, a) = -\frac{2\pi a \sigma_0^2 (1 - v^2)}{E}\dot{a}. \tag{2.72}$$

Note that E is the Young modulus and v is the Poisson ratio for the material of interest. We are now ready to turn to the concrete implementation of the variational statement of eqn (2.70). Evaluation of $\partial\Pi/\partial\dot{a}$ results in the equation of motion

$$\frac{\dot{a}}{M} = \frac{2\pi a \sigma_0^2 (1 - v^2)}{E} - 4\gamma. \tag{2.73}$$

Griffith sought the criticality condition at which $\dot{a} = 0$. That is, his argument (which was carried out completely differently than we have done here) was to determine the precise condition under which the elastic energy release rate would just compensate for the energy that must be donated to create new free surface. Regardless of the route taken to get the result, this condition results in a critical stress for crack initiation of the form

$$\sigma_0^{crit} = \sqrt{\frac{2E\gamma}{\pi a(1 - v^2)}}. \tag{2.74}$$

Much of the discussion given above can be recast in the language of section 2.3.3. In particular, we see that the question posed above was how does the total potential energy of the system change during incremental crack extension? This may be recast in mathematical form as

$$G = -\frac{\partial E_{elastic}}{\partial a}, \tag{2.75}$$

where G is the mechanical energy release rate. The key point to be made presently is the observation that the quantity $-\partial E_{elastic}/\partial a$ is exactly the type of configurational force introduced earlier and, as a result, we may invoke the ideas presented there. In particular, in the fracture context, eqn (2.45) is of such importance it is dignified with a special name, namely, the J-integral and given by

$$J_i = \int_\Gamma \left(W n_i - \sigma_{lm} n_m \frac{\partial u_l}{\partial x_i} \right) ds. \tag{2.76}$$

The integral is framed in the context of a two-dimensional geometry with the contour Γ parameterized by s. The key point is that by evaluating this integral it is possible to reproduce exactly the type of result given in eqn (2.71).

For the purposes of the present discussion, there were three main ideas that we wished to put forth concerning fracture. First, we noted that if fracture is imagined to be a homogeneous process in which bonds in the perfect crystal are stretched until they break, this leads to a critical stress for fracture (the ideal tensile strength) known to be completely out of line with experimental values. As a remedy to this paradox, it was realized that flaws such as cracks could have the effect of locally magnifying the remotely applied stress to levels consonant with that needed to break atomic-level bonds. The linear elastic theory of fracture mechanics teaches us how to assess the elastic consequences of these defects, a subject we will return to in chap. 11. Finally, we wished to illustrate how using energetic arguments like those of Griffith, which we have reformulated in the language of configurational forces and the associated variational principle presented in section 2.3.3, the critical stress for crack initiation may be determined on the basis of microscopic parameters such as the surface energy. It bears emphasizing again that it is particularly interesting how far a linear elastic analysis can take us in evaluating what is an intrinsically nonlinear process.

2.5 Boundary Value Problems and Modeling

From the standpoint of the continuum simulation of processes in the mechanics
of materials, modeling ultimately boils down to the solution of boundary value
problems. What this means in particular is the search for solutions of the equations
of continuum dynamics in conjunction with some constitutive model and boundary
conditions of relevance to the problem at hand. In this section after setting down
some of the key theoretical tools used in continuum modeling, we set ourselves
the task of striking a balance between the analytic and numerical tools that have
been set forth for solving boundary value problems. In particular, we will examine
Green function techniques in the setting of linear elasticity as well as the use of the
finite element method as the basis for numerical solutions.

2.5.1 Principle of Minimum Potential Energy and Reciprocal Theorem

In many of the cases in which we resort to the theoretical analysis of a particular
problem from a continuum perspective, it will be from the point of view of the
linear theory of elasticity. Though we have already introduced the fundamental
governing equations of continuum mechanics as well as their linear elastic incarnation, in many instances it will be useful to assign these ideas a variational
character. In the context of linear elastic solids, the global principle presiding
over equilibrium states (which we assert without proof, for details see Gurtin
(1981), for example) is the fact that they are privileged states characterized by the
property that they minimize the total potential energy. In particular, in the context
of linear elastic boundary value problems, the equilibrium displacement fields are
minimizers of the potential energy functional

$$\Pi[\mathbf{u}(\mathbf{x})] = \int_{\Omega} W(\{u_{i,j}\})dV - \int_{\partial\Omega} \mathbf{t} \cdot \mathbf{u} dA - \int_{\Omega} \mathbf{f} \cdot \mathbf{u} dV, \qquad (2.77)$$

where the first term on the right hand side yields the elastic energy stored in the
body, the second term is associated with the potential energy of external loads
that are communicated to the body at its surface and the third term provides a
measure of the work associated with body forces. The formal statement of the
principle is that of all the kinematically admissible fields (i.e. those fields that
are sufficiently smooth and satisfy the displacement boundary conditions), the
equilibrium fields will minimize the functional of eqn (2.77). The importance of
this theorem will be especially felt in two separate contexts. First, in a number of
instances, it will be most natural to construct models of a given problem which are
couched in the language of energetics. In many such instances, the total energy
will be parameterized in terms of one or several configurational coordinates, and
the resulting minimization will seek their optimal configuration. The principle

of minimum potential energy will also see action as the basis of many of the approximation strategies that are adopted to solve linear elastic boundary value problems, and in particular, the finite element method.

To illustrate the equivalence of the energy minimization principle advanced above and the linear elastic equilibrium equations of section 2.4.2, we resort to evaluating the functional derivative of eqn (2.77) and setting it equal to zero, or in mathematical terms $\delta\Pi/\delta u_i = 0$. Equivalently, working in indicial notation, we find

$$\delta\Pi = \frac{1}{2}\int_\Omega C_{ijkl}\delta\epsilon_{ij}\epsilon_{kl}dV + \frac{1}{2}\int_\Omega C_{ijkl}\epsilon_{ij}\delta\epsilon_{kl}dV - \int_\Omega f_i\delta u_i dV - \int_{\partial\Omega} t_i\delta u_i dA.$$
(2.78)

Exploiting the symmetries of the elastic modulus tensor and the strain tensor, the first two terms may be joined to form $\int_\Omega \sigma_{ij}\delta\epsilon_{ij}dV$. Rewriting the integrand $\sigma_{ij}\delta\epsilon_{ij}$ as

$$\sigma_{ij}\frac{\partial}{\partial x_j}\delta u_i = \frac{\partial}{\partial x_j}(\sigma_{ij}\delta u_i) - \sigma_{ij,j}\delta u_i,$$
(2.79)

and plugging this result back into the expression given above followed by an application of the divergence theorem leaves us with

$$\delta\Pi = \int_\Omega (\sigma_{ij,j} + f_i)\delta u_i dV + \int_{\partial\Omega} (\sigma_{ij}n_j - t_i)\delta u_i dA = 0.$$
(2.80)

The significance of this result emerges once it is recalled that δu_i itself is arbitrary except for the fact that $\delta u_i = 0$ on that part of the boundary for which **u** is prescribed, and hence that the integrands themselves must vanish. But in the case of the integral over Ω, this results in $C_{ijkl}u_{k,lj} + f_i = 0$ which is nothing more than the original linear elastic equilibrium equations for a homogeneous material. Further, applying the same reasoning to the surface integrals compels us to accept that $\sigma_{ij}n_j = t_i$ on that part of the boundary characterized by traction boundary conditions. Again, what all of this means is that there is an equivalence between the original equilibrium equations for linear elasticity and insisting that the potential energy $\Pi[\mathbf{u}(\mathbf{x})]$ be minimized. As remarked above, in most instances in fact, we will favor the energetic interpretation of equilibrium.

In addition to the theoretical advantages offered by the principle of minimum potential energy, our elastic analyses will also be aided by the reciprocal theorem. This theorem is a special example of a more general class of reciprocal theorems and considers two elastic states $(\mathbf{u}^{(1)}, \boldsymbol{\epsilon}^{(1)}, \boldsymbol{\sigma}^{(1)})$ and $(\mathbf{u}^{(2)}, \boldsymbol{\epsilon}^{(2)}, \boldsymbol{\sigma}^{(2)})$, where each state satisfies the equilibrium equations and boundary conditions associated with body force fields $\mathbf{f}_1$ and $\mathbf{f}_2$, respectively. The theorem asserts that the work done by the external forces of system 1 on the displacements of system 2 is equal to the work done by the external forces of system 2 on the displacements of system 1.

Mathematically, this is written as

$$\int_{\partial\Omega} \mathbf{t}^{(1)} \cdot \mathbf{u}^{(2)} dA + \int_{\Omega} \mathbf{f}^{(1)} \cdot \mathbf{u}^{(2)} dV = \int_{\partial\Omega} \mathbf{t}^{(2)} \cdot \mathbf{u}^{(1)} dA + \int_{\Omega} \mathbf{f}^{(2)} \cdot \mathbf{u}^{(1)} dV. \quad (2.81)$$

The significance of this result will be appreciated later (for example in section 8.4.2) when we see that it can be used for the construction of fundamental solutions for the elastic displacement fields of defects such as dislocations.

The proof of the reciprocal theorem is based upon two fundamental realizations. First, we note the relation

$$\int_{\partial\Omega} \mathbf{t}^{(1)} \cdot \mathbf{u}^{(2)} dA + \int_{\Omega} \mathbf{f}^{(1)} \cdot \mathbf{u}^{(2)} dV = \int_{\Omega} \boldsymbol{\epsilon}^{(1)} : \mathbf{C}\boldsymbol{\epsilon}^{(2)} dV. \quad (2.82)$$

The veracity of this expression can be shown by running the divergence theorem in reverse, basically turning the argument given above to prove the principle of minimum potential energy around. This result in conjunction with the application of symmetries of the tensor of elastic moduli allows us to note

$$\int_{\Omega} \boldsymbol{\epsilon}^{(1)} : \mathbf{C}\boldsymbol{\epsilon}^{(2)} dV = \int_{\Omega} \boldsymbol{\epsilon}^{(2)} : \mathbf{C}\boldsymbol{\epsilon}^{(1)} dV. \quad (2.83)$$

Based on this latter result, the reciprocal theorem is immediate.

2.5.2 Elastic Green Function

One of the most powerful means for solving boundary value problems in linear elasticity is that of Green functions. The Green function approach offers the possibility of building up solutions to complex problems on the basis of insights garnered from the solution to the problem of a unit point load. The pivotal assumption exploited in this context is that of linearity which allows us to use the fact that a sum of solutions to a given problem is itself a solution, a result known as the superposition principle. As a result of this principle and a knowledge of the Green function, solutions to generic problems are written as integrals over a kernel provided by the fundamental solution itself.

As noted above, the elastic Green function yields the elastic fields induced by a point load. The ideas here resemble those from the setting of electrodynamics where point sources of charge are the basis of the Green function approach. Recall from earlier that the equilibrium equations that govern the continuum are given by

$$\sigma_{ij,j} + f_i = 0, \quad (2.84)$$

where σ_{ij} is the ij^{th} component of the Cauchy stress and f_i are the components of the body force. In the special case of an isotropic linear elastic solid, we have seen that these equations reduce to the Navier equations shown in eqn (2.55).

To construct the elastic Green function for an isotropic linear elastic solid, we make a special choice of the body force field, namely, $\mathbf{f}(\mathbf{r}) = \mathbf{f}_0 \delta(\mathbf{r})$, where $\mathbf{f}_0$ is a constant vector. In this case, we will denote the displacement field as $G_{ik}(\mathbf{r})$ with the interpretation that this is the ith component of displacement in the case in which there is a unit force in the kth direction, $\mathbf{f}_0 = \mathbf{e}_k$. To be more precise about the problem of interest, we now seek the solution to this problem in an infinite body in which it is presumed that the elastic constants are uniform. In light of these definitions and comments, the equilibrium equation for the Green function may be written as

$$(\lambda + \mu)\frac{\partial^2}{\partial x_i \partial x_m}G_{mj}(\mathbf{r}) + \mu\frac{\partial}{\partial x_l}\frac{\partial}{\partial x_l}G_{ij}(\mathbf{r}) = -\delta_{ij}\delta(\mathbf{r}). \tag{2.85}$$

Interesting strategies have been adopted for explicitly determining $\mathbf{G}(\mathbf{r})$ both in real space as well as Fourier space. Here we follow de Wit (1960) in adopting the Fourier space scheme, leaving the real space solution for the ambitious reader. To find the Green function, we exploit our Fourier transform convention:

$$\tilde{G}_{ij}(\mathbf{k}) = \int G_{ij}(\mathbf{r})e^{-i\mathbf{k}\cdot\mathbf{r}}d^3\mathbf{r}, \tag{2.86}$$

and

$$G_{ij}(\mathbf{r}) = \frac{1}{(2\pi)^3}\int \tilde{G}_{ij}(\mathbf{k})e^{i\mathbf{k}\cdot\mathbf{r}}d^3\mathbf{k}. \tag{2.87}$$

Upon Fourier transforming eqn (2.85) as dictated by the definitions given above, we find

$$(\lambda + \mu)k_i k_m \tilde{G}_{mj}(\mathbf{k}) + \mu k^2 \tilde{G}_{ij}(\mathbf{k}) = \delta_{ij}, \tag{2.88}$$

where k_i is the ith component of the wavevector $\mathbf{k}$. The k-space Green function can be found by multiplying the previous equation by k_i and summing on the repeated index i, which yields the result,

$$k_m \tilde{G}_{mj} = \frac{k_j}{(\lambda + 2\mu)k^2}, \tag{2.89}$$

which may be substituted into eqn (2.88). After carrying out the resulting manipulation, the explicit expression for the Green function in Fourier space is

$$\tilde{G}_{ij}(\mathbf{k}) = \frac{1}{\mu k^2}\left[\delta_{ij} - \frac{(\lambda + \mu)}{(\lambda + 2\mu)}\frac{k_i k_j}{k^2}\right]. \tag{2.90}$$

This result will serve as the cornerstone for many of our developments in the elastic theory of dislocations. We are similarly now in a position to obtain the Green function in real space which, when coupled with the reciprocal theorem, will allow for the determination of the displacement fields associated with various defects.

Following the procedure used in electrostatics where it is found that the Fourier transform of the Coulomb potential yields a k^{-2} dependence, we can invert the expression found above rather easily as

$$G_{ij}(\mathbf{r}) = \frac{1}{8\pi\mu(\lambda+2\mu)}\left[(\lambda+3\mu)\frac{\delta_{ij}}{r} + (\lambda+\mu)\frac{x_i x_j}{r^3}\right]. \tag{2.91}$$

As a result of these various algebraic machinations, we are now prepared to construct solutions to problems in isotropic elasticity through superposition of the fundamental solution constructed above. The Green function enables us to find the displacements at point $\mathbf{x}$ given the presence of a point force at point $\mathbf{x}'$. However, because of the linearity of the underlying governing equations, the solution for an arbitrary force distribution can be built up as a superposition over a collection of such point forces. That is, we claim that the solution to $C_{ijkl}u_{k,jl} = -f_i$ is

$$u_k(\mathbf{x}) = \int_\Omega G_{km}(\mathbf{x}-\mathbf{x}')f_m(\mathbf{x}')d^3\mathbf{x}'. \tag{2.92}$$

The proof of this claim can be sketched in abstract terms as follows. If we are to write the original governing equation for the displacements as $\mathbf{L}(\mathbf{x})\mathbf{u}(\mathbf{x}) = -\mathbf{f}(\mathbf{x})$, where $\mathbf{L}$ is the linear operator associated with the various differentiations in the equilibrium equations, then the equation satisfied by the Green function may be written as $\mathbf{L}(\mathbf{x})\mathbf{G}(\mathbf{x}-\mathbf{x}') = -\mathbf{I}\delta(\mathbf{x}-\mathbf{x}')$. Our assertion is that the displacements are given by $\mathbf{u}(\mathbf{x}) = \int_\Omega \mathbf{G}(\mathbf{x}-\mathbf{x}')\mathbf{f}(\mathbf{x}')d^3\mathbf{x}'$. Application of the operator $\mathbf{L}$ to this equation for the displacements results in

$$\mathbf{L}(\mathbf{x})\mathbf{u}(\mathbf{x}) = \int_\Omega \mathbf{L}(\mathbf{x})\mathbf{G}(\mathbf{x}-\mathbf{x}')\mathbf{f}(\mathbf{x}')d^3\mathbf{x}' = -\int_\Omega \mathbf{f}(\mathbf{x}')\delta(\mathbf{x}-\mathbf{x}')d^3\mathbf{x}' = -\mathbf{f}(\mathbf{x}), \tag{2.93}$$

demonstrating that when we write the displacements as a convolution integral, namely, $\mathbf{u}(\mathbf{x}) = \int_\Omega \mathbf{G}(\mathbf{x}-\mathbf{x}')\mathbf{f}(\mathbf{x}')d^3\mathbf{x}'$, the original governing equations are satisfied.

Eqn (2.92) is the culmination of our efforts to compute the displacements due to an arbitrary distribution of body forces. Although this result will be of paramount importance in coming chapters, it is also important to acknowledge its limitations. First, we have assumed that the medium of interest is isotropic. Further refinements are necessary to recast this result in a form that is appropriate for anisotropic elastic solids. A detailed accounting of the anisotropic results is spelled out in Bacon *et al.* (1979). The second key limitation of our result is the fact that it was founded upon the assumption that the body of interest is infinite in extent. On the other hand, there are a variety of problems in which we will be interested in the presence of defects near surfaces and for which the half-space Green function will be needed. Yet another problem with our analysis is the assumption that the elastic constants

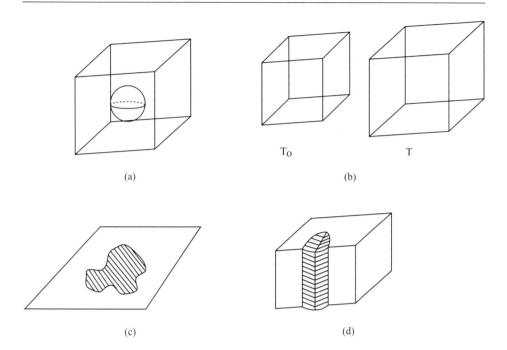

Fig. 2.13. Four schematic examples of eigenstrains: (a) spherical inclusion, (b) eigenstrain due to thermal expansion, (c) dislocation loop and (d) eigenstrain due to wedge of martensite.

are the same throughout the material. On the other hand, when it comes time (see chap. 10) to investigate the elastic consequences of second-phase particles, the moduli of the inclusion can differ from those of the matrix, and again, the particulars (though not the concept) of our Green function analysis will have to be amended.

2.5.3 *Method of Eigenstrains*

A very powerful technique in conjunction with that of elastic Green functions is the method of eigenstrains. The eigenstrain concept allows us to solve elasticity problems for a continuum with internal heterogeneities such as dislocations or inclusions, by replacing these heterogeneities by effective body force fields. An eigenstrain is defined as a strain without any attendant internal stresses. The definition is abstract without the clarification afforded by particular examples. In fig. 2.13, we show four distinct examples of elastic media in the presence of eigenstrains. For example, as shown in fig. 2.13(a), an inclusion may be characterized by a strain ϵ_{ij}^* which carries the information about how the inclusion differs structurally from the matrix material. In addition, we should note that in the absence of the constraining effect provided by the matrix, the eigenstrain ϵ_{ij}^* has

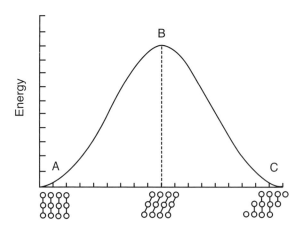

Fig. 2.14. Energy as a function of shear for a generic crystalline solid. The periodicity of the energy profile reflects the existence of certain lattice invariant shears which correspond to the 'stress-free strains' considered in the method of eigenstrains (adapted from Tadmor *et al.* (1996)).

no associated stress. Similarly, as shown in fig. 2.13(b), if a crystal is to undergo thermal expansion, the associated strains can be thought of as eigenstrains since they imply no internal stresses. Dislocations can also be characterized in terms of eigenstrains since the region of the crystal that has slipped (see fig. 2.13(c)) has suffered a lattice invariant shear. In fig. 2.13(d) we imagine the case of a solid, part of which has undergone a phase transformation. Like in the inclusion example introduced above, the phase transformed region is characterized by an eigenstrain.

Further insights into the eigenstrain concept can be obtained on the basis of energetic arguments. We imagine an energy landscape which characterizes the energy of the continuum as a function of the relevant kinematic measure such as the strain. In many cases of interest, the energy landscape is nonconvex, with different wells corresponding to different structural alternatives. The eigenstrain concept allows for the emergence of structures in which one part of the continuum corresponds to one well in the energy landscape while some other part of the continuum corresponds to a different well. In computing stresses, we appeal not to the total strain, but rather the departure of the strain from the well of interest. In many problems, different parts of the body are characterized by different eigenstrains. Clearly, in making a transition from one eigenstrain to another, specific regions will not be associated with a well in the energy landscape, and it is largely these incompatibilities that give rise to internal stresses. A concrete example of these energetic arguments can be made through consideration of simple shear. In generic terms, if a particular crystal is subject to a monotonically increasing shear, the energy as a function of this shear will have qualitative features such as those shown in fig. 2.14. The existence of the periodicity in the energy profile reflects the fact that there exist certain lattice invariant shears that map the Bravais lattice onto itself. The fundamental idea of the method of eigenstrains is the argument that if the system is sheared such that it occupies one of the minimum energy wells, there will be no associated stress.

In mathematical terms, the implementation of the eigenstrain concept is carried out by representing the geometrical state of interest by some distribution of 'stress-free strains' which can be mapped onto an equivalent set of body forces, the solution for which can be obtained using the relevant Green function. To illustrate the problem with a concrete example, we follow Eshelby in thinking of an elastic inclusion. The key point is that this inclusion is subject to some internal strain ϵ_{ij}^*, such as arises from a structural transformation, and for which there is no associated stress.

The outcome of the hypothesis advanced above is that the stress is computed as

$$\sigma_{ij} = C_{ijkl}(u_{k,l}^{tot} - \epsilon_{kl}^*), \tag{2.94}$$

where $u_{k,l}^{tot}$ is the displacement gradient of the total displacement field. Note that the physical meaning of this equation corresponds to our statement above that it is only the departure of the strain from the nearest energy well that leads to the generation of stresses, and not the total strain. The imposition of the condition of equilibrium, namely, $\sigma_{ij,j} = 0$ in this context is then equivalent to

$$C_{ijkl}u_{k,lj}^{tot} = C_{ijkl}\epsilon_{kl,j}^*. \tag{2.95}$$

Since the eigenstrain itself is prescribed, eqn (2.95) can be interpreted as corresponding to a homogeneous elastic body, now in the presence of an effective body force $f_i^{eff} = -C_{ijkl}\epsilon_{kl,j}^*$. Note that in the context of the effective theory concept which we will return to repeatedly throughout this book, the eigenstrain idea allows us to replace some heterogeneity within a continuum by a series of effective body forces resulting in a tractable (sometimes) elasticity problem. As a result of the definition of the effective body force field, the Green function machinery elucidated above may be exploited. In particular, recall eqn (2.92) which expresses the displacement fields as an integral over the force distribution. In the present setting, the unknown displacements may be written as a convolution integral of the elastic Green function and the effective body force distribution as

$$u_k(\mathbf{x}) = -\int_\Omega G_{km}(\mathbf{x} - \mathbf{x}')C_{mrst}\epsilon_{st,r}^*(\mathbf{x}')d^3\mathbf{x}'. \tag{2.96}$$

Of course, there are a number of manipulations (integration by parts, for example) that can be carried out in order to restate this result in a more convenient fashion. On the other hand, our intention at this point is to introduce the conceptual framework, and to reserve the explicit use of the method of eigenstrains until later.

2.5.4 Numerical Solutions: Finite Element Method

Despite the existence of powerful analytical tools that allow for explicit solution of certain problems of interest, in general, the modeler cannot count on the existence of analytic solutions to most questions. To remedy this problem, one must resort to numerical approaches, or further simplify the problem so as to refine it to the point that analytic progress is possible. In this section, we discuss one of the key numerical engines used in the continuum analysis of boundary value problems, namely, the finite element method. The finite element method replaces the search for unknown fields (i.e. the solutions to the governing equations) with the search for a discrete representation of those fields at a set of points known as nodes, with the values of the field quantities between the nodes determined via interpolation. From the standpoint of the principle of minimum potential energy introduced earlier, the finite element method effects the replacement

$$\Pi[\mathbf{u}(\mathbf{x})] \rightarrow \Pi(\{\mathbf{u}_i\}), \qquad\qquad (2.97)$$

where $\Pi[\mathbf{u}(\mathbf{x})]$ is a functional of the unknown displacement fields and $\Pi(\{\mathbf{u}_i\})$ is a function of a discrete set of nodal displacements indexed by the integer i which runs from 1 to N, where N is the number of nodes. We note that the finite element method is in no way restricted to elasticity problems, but that such problems will serve as a convenient vehicle for our discussion.

The central ideas associated with the finite element approach may be summarized as follows. First, the original boundary value problem is stated in weak form (i.e. the relevant partial differential equations are restated in integral form). Next, the continuum is parceled up into discrete regions known as elements, with the unknown field variables considered at the points of intersection of these elements known as nodes. The net result of this discretization is that we have now replaced a problem in which we are searching for an unknown function $\mathbf{u}(\mathbf{x})$ with a search for discrete set of $3N$ variables $\{\mathbf{u}_i\}$ (i.e. the three components of displacement for each of the N nodes). One crucial observation at this juncture is that the elements need not be uniform, thereby allowing for mesh design in which near rapidly varying parts of the solution, one uses high resolution, while in regions of slow variation larger elements can be supported.

From the perspective of variational principles, the idea embodied in the finite element approach is similar to that in other schemes: identify that particular linear combination of basis functions that is 'best' in a sense to be defined below. The approximate solution of interest is built up as linear combinations of basis functions, and the crucial question becomes how to formulate the original boundary value problem in terms of the relevant expansion coefficients.

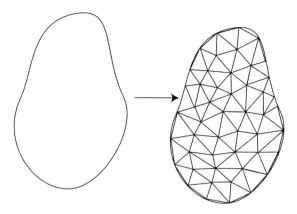

Fig. 2.15. Replacement of a continuous body with a set of finite elements. In this case, the unknown displacement field has been replaced by a discrete set of unknown nodal displacements, $\mathbf{u}_n$.

Recall that within the continuum mechanics setting set forth in section 2.3.2, the equations of equilibrium may be written as

$$\sigma_{ij,j} + f_i = 0. \tag{2.98}$$

Our present task is to rewrite these equations in integral (or weak) form which will simplify the formulation in terms of finite elements. We introduce a series of 'test functions' $w_i(x)$ that satisfy the relevant displacement boundary conditions which are hence denoted as admissible fields and now contract these test functions with the equation given above and integrate over the spatial region of interest. This strategy results in

$$\int_\Omega w_i \sigma_{ij,j} dV + \int_\Omega w_i f_i dV = 0. \tag{2.99}$$

If we now use integration by parts on the first term in this equation, we are left with

$$\int_\Omega w_{i,j} \sigma_{ij} dV = \int_{\partial\Omega} t_i w_i dA + \int_\Omega f_i w_i dV, \tag{2.100}$$

where we have used $t_i = \sigma_{ij} n_j$. As yet, our exercise corresponds to nothing more than a statement of the principle of virtual work. On the other hand, if we now proceed with the concrete consideration of the test functions $w_i(x)$, it is possible to construct a numerical framework.

As yet, our statements have been general and do not reflect our wish to replace the continuous problem with a discrete analog. We now imagine that the body is discretized into a set of elements which meet at vertices denoted as nodes and indicated in fig. 2.15. Within each element, the value of the displacement is

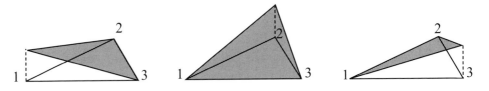

Fig. 2.16. Schematic illustration of the three triangular shape functions tied to the different nodes of a triangular element.

obtained via interpolation according to

$$\mathbf{u}(\mathbf{x}) = \sum_i \mathbf{u}_i N_i(\mathbf{x}), \qquad (2.101)$$

where the subscript i refers to nodes and $N_i(\mathbf{x})$ is the 'shape function' attached to the ith node and defined through $N_i(\mathbf{x}_j) = \delta_{ij}$. That is, the ith shape function is 1 at the ith node and vanishes at neighboring nodes. These shape functions are illustrated in fig. 2.16 in the case of linear triangular elements. In essence what we are saying is that the solution is determined at a discrete set of nodal values, and if we interest ourselves in the solution elsewhere, it is to be determined by interpolation in terms of those nodal values via a series of interpolation functions which are commonly referred to as shape functions. If we substitute the interpolated displacements as denoted above into the integral statement of the equilibrium equations, and furthermore, we take as our test functions the N_is themselves (i.e. $w_i(x) = N_i(x)$), we find a series of equations in the unknown coefficients of the form

$$\mathbf{Ku} = \mathbf{f}, \qquad (2.102)$$

where $\mathbf{K}$ is known as the stiffness matrix (and for nonlinear problems can depend upon $\mathbf{u}$ itself) and $\mathbf{f}$ is the force matrix.

   To make the treatment given above more than just an abstraction, we now consider the case of one-dimensional elasticity. We imagine our one-dimensional elastic medium to have length a and to be characterized by an elastic stiffness k. The traction vector, $t(x)$, is directly related to the gradients in displacements by $t = k\,du/dx$. Our weak statement of the equilibrium equations may be written as

$$\int_0^a k \frac{dw}{dx} \frac{du}{dx} dx = \int_0^a f w \, dx + [t(a)w(a) - t(0)w(0)]. \qquad (2.103)$$

At this point, we are primarily confronted with a series of algebraic manipulations founded upon the finite element edicts described above. First, we approximate the solution $u(x)$ in terms of the finite element shape functions, $u(x) = \sum_i u_i N_i(x)$. Secondly, we make the correspondence between the test functions $w(x)$ and the shape functions themselves. For example, the equation resulting from the assertion $w(x) = N_j(x)$ is

$$\sum_i u_i \int_0^a k \frac{dN_j(x)}{dx} \frac{dN_i(x)}{dx} dx = \int_0^a f(x) N_j(x) dx + [t(a) N_j(a) - t(0) N_j(0)].$$

(2.104)

This equation may now be rewritten as

$$\sum_i K_{ji} u_i = f_j$$

(2.105)

if we make the correspondence

$$K_{ji} = \int_0^a k \frac{dN_j(x)}{dx} \frac{dN_i(x)}{dx} dx,$$

(2.106)

$$f_j = \int_0^a f(x) N_j(x) dx + [t(a) N_j(a) - t(0) N_j(0)].$$

(2.107)

The result of this analysis is a series of coupled algebraic equations in the unknown nodal displacements. For the linear problem we have considered here, these equations are themselves linear (though they are not in general) and our problem reduces to little more than diagonalization of the relevant stiffness matrix which, as seen above, can be computed as products of material parameters and geometric factors.

In this section, we have seen how one may formulate numerical strategies for confronting the types of boundary value problems that arise in the continuum description of materials. The key point is the replacement of the problem involving unknown continuum fields with a discrete reckoning of the problem in which only a discrete set of unknowns at particular points (i.e. the nodes) are to be determined. In the next chapter we will undertake the consideration of the methods of quantum mechanics, and in this setting will find once again that the finite element method offers numerical flexibility in solving differential equations.

2.6 Difficulties with the Continuum Approach

One of the key benefits provided by the types of approaches advocated in this chapter is that the mathematics associated with solving for three unknown fields (for example) is more sympathetic than that of solving coupled ordinary differential equations in a number of variables of Avogradrian dimensions. On the other hand, continuum approaches have their disadvantages as well. We have noted earlier that one of the primary difficulties tied to the use of continuum models is their reliance on constitutive assumptions. In some instances, we will see that the constitutive models adopted are often selected as much for their theoretical convenience as for their phenomenological realism. It is on these grounds that we will appeal to microscopic theories as the proper basis for informing higher-level constitutive strategies.

A second difficulty that attends the use of continuum models is the fact that such models are indifferent to the presence of a minimum length scale as dictated by the typical separation between atoms. This absence of small scale resolution within the continuum framework can have a souring influence on the success of such theories in predicting features in the immediate vicinity of defects. One of our main objectives will be to show how, via the device of incorporating nonlinear information into our models, many of these difficulties may be sidestepped. As a result of the evident appeal of microscopic insights into the ultimate origins of continuum behavior, in the next chapter we undertake a review of the theoretical foundations of microscopic modeling, the partnership of quantum and statistical mechanics.

2.7 Further Reading

Continuum Mechanics

Continuum Mechanics by P. Chadwick, Halsted Press, New York: New York, 1976. My personal favorite, Chadwick's book is direct and to the point, and full of clever insights and interesting examples.

An Introduction to Continuum Mechanics by M. Gurtin, Academic Press Inc., San Diego: California, 1981. Gurtin's book is a classic among solid mechanicians and gives a thorough treatment of much of the material found in the present chapter.

'Mechanics of Solids' by J. R. Rice in *Encyclopedia Britannica* Vol. 23, pg. 734 (1993) is an outstanding overview of the entire realm of solid mechanics, giving coverage not only of the principles of continuum mechanics, but also their applications to deformation and fracture with interesting historical and philosophic commentary.

Nonlinear Continuum Mechanics for Finite Element Analysis by J. Bonet and R. D. Wood, Cambridge University Press, Cambridge: England, 1997. This book provides an interesting discussion of many of the concepts discussed in the present chapter and includes commentary on the numerical implementation of these same concepts.

Mathematical Theory of Elasticity by I. S. Sokolnikoff, Krieger Publishing Company, Malabar: Florida, 1983. One of the classic texts on elasticity theory.

Theory of Elasticity by S. P. Timoshenko and J. N. Goodier, McGraw Hill Book

Company, New York: New York, 1970. My personal taste for this book is founded on the fact that it is full of *solutions* to elasticity problems of interest.

Theory of Elasticity by L. D. Landau and E. M. Lifshitz, Pergamon Press, Oxford: England, 1970. Written in the terse style that the Landau and Lifshitz series is famous for, the book is full of solutions and insights.

Micromechanics of Defects in Solids by T. Mura, Kluwer Academic Publishers, Dordrecht: The Netherlands, 1993. Mura's book is a neglected classic which features an enormous number of examples on the use of Green functions in conjunction with the method of eigenstrains.

Plasticity

The Mathematical Theory of Plasticity by R. Hill, Clarendon Press, Oxford: England, 1967. The definitive work on the mathematical theory of plasticity. This book goes well beyond our treatment regarding continuum descriptions of plasticity without entering into the question of how plasticity emerges as the result of defect motion.

Continuum Theory of Plasticity by A. S. Khan and S. Huang, John Wiley and Sons, New York: New York, 1995. The subject of plasticity is lacking in texts that are directed at interpreting plastic deformation on the basis of the underlying mechanisms. This book provides an overview of much of the machinery that is invoked in current considerations of plasticity.

Fracture

'Mathematical Analysis in the Mechanics of Fracture' by James R. Rice, from Vol. II of *Fracture* edited by G. Sih, Academic Press, New York: New York, 1968. Rice's account of fracture remains, in my opinion, one of the definitive statements of many of the key ideas in the subject. Section E on *Energy Variations and Associated Methods* is especially pertinent to our discussion on configurational forces and gives an in depth application of these ideas in the fracture context.

Dynamic Fracture Mechanics by L. B. Freund, Cambridge University Press, Cambridge: England, 1990. A careful account of many of the fundamental ideas of interest in the analysis of fracture with special reference to dynamics.

Fracture of Brittle Solids by Brian Lawn, Cambridge University Press, Cambridge: England, 1993. I have found Lawn's book quite readable and a useful place to become acquainted not only with much of the phenomenology of fracture, but also with some of the key ideas in that field.

Elementary Engineering Fracture Mechanics by D. Broek, Martinus Nijhoff

Publishers, The Hague: Holland, 1982. A readable account of many of the central ideas in the analysis of fracture.

The Finite Element Method

The Finite Element Method by T. J. R. Hughes, Prentice Hall Inc., Englewood Cliffs: New Jersey, 1987. An exhaustive coverage of the mathematical formalism of finite elements.

Computational Differential Equations by K. Eriksson, D. Estep, P. Hansbo and C. Johnson, Cambridge University Press, Cambridge: England, 1996. A wonderful book with many interesting historical sidelights that describes the numerical tools that arise in the solution of boundary value problems using continuum mechanics.

A First Course in the Numerical Analysis of Differential Equations by A. Iserles, Cambridge University Press, Cambridge: England, 1996. This book, especially in the later chapters, treats many of the numerical issues of interest to those practicing finite elements.

An Introduction to the Finite Element Method by J. N. Reddy, McGraw-Hill, New York: New York, 1993. A readable introduction to the conceptual foundations of the finite element method.

2.8 Problems

1 Deformation Gradient and Volume Change

An additional insight into the significance of the deformation gradient tensor can be gleaned by evaluating the scalar triple product of three noncoplanar vectors $\mathbf{a}$, $\mathbf{b}$ and $\mathbf{c}$ both before and after a homogeneous deformation. In the undeformed configuration, the vectors $\mathbf{a}$, $\mathbf{b}$ and $\mathbf{c}$ span a region whose volume is given by $\mathbf{a} \cdot (\mathbf{b} \times \mathbf{c})$. In the deformed configuration, the transformed region has a volume given by $\mathbf{Fa} \cdot (\mathbf{Fb} \times \mathbf{Fc})$. Show that the reference and deformed volumes are related by

$$\mathbf{Fa} \cdot (\mathbf{Fb} \times \mathbf{Fc}) = (\det \mathbf{F})\mathbf{a} \cdot (\mathbf{b} \times \mathbf{c}). \tag{2.108}$$

Interpret this result and demonstrate that for a pure shear there is no volume change.

2 Finite and Infinitesimal Strains

Consider a deformation mapping that is nothing more than a rigid rotation about the z-axis. First, explicitly write down the deformation mapping and then compute both the Green strain and the infinitesimal strain tensors associated with this deformation mapping. In general, the infinitesimal strains are nonzero. Why?

3 Cauchy Tetrahedron and Equilibrium

In this problem, use the balance of linear momentum to deduce the Cauchy stress by writing down the equations of continuum dynamics for the tetrahedral volume element shown in fig. 2.7. Assign an area ΔS to the face of the tetrahedron with normal $\mathbf{n}$ and further denote the distance from the origin to this plane along the direction $\mathbf{n}$ by h. Show that the dynamical equations may be written as

$$\mathbf{t}^{(\mathbf{n})}\Delta S + \mathbf{f}\Delta V - \mathbf{t}^{(\mathbf{e}_1)}\Delta S_1 - \mathbf{t}^{(\mathbf{e}_2)}\Delta S_2 - \mathbf{t}^{(\mathbf{e}_3)}\Delta S_3 - = \rho\frac{D\mathbf{v}}{Dt}\Delta V. \quad (2.109)$$

Rewrite ΔS_1, ΔS_2, ΔS_3 and ΔV in terms of ΔS, h and the components of the vector $\mathbf{n}$ normal to ΔS and then evaluate the limiting value of the expression given above as $h \to 0$. Now, by adopting the convention that the traction vector on the ith Cartesian plane can be written as

$$\mathbf{t}^{(\mathbf{e}_i)} = \sigma_{ji}\mathbf{e}_j, \quad (2.110)$$

show that $\mathbf{t}^{(\mathbf{n})} = \sigma\mathbf{n}$.

4 Isotropic Elasticity and Navier Equations

Use the constitutive equation for an isotropic linear elastic solid given in eqn (2.54) in conjunction with the equilibrium equation of eqn (2.84), derive the Navier equations in both direct and indicial notation. Fourier transform these equations and verify eqn (2.88).

5 Elastic Green Function for an Isotropic Solid

The elastic Green function is a useful tool in the analysis of a variety of problems in the elasticity of defects in solids. In this problem, we consider the various steps in the derivation of the elastic Green function for an infinite, isotropic solid. In particular, flesh out the details involved in the derivation of eqn (2.90) and carry out the Fourier inversion to obtain eqn (2.91).

---

6 *Displacements for an Elastic Inclusion using the Method of Eigenstrains*

Consider a spherical inclusion characterized by an eigenstrain $\epsilon_{ij}^* = \epsilon\delta_{ij}$. (a)
For an isotropic linear elastic solid, write the effective body forces associated
with this eigenstrain. (b) Use the effective body force derived in part (a)
in conjunction with eqn (2.96) to obtain an integral representation for the
displacements due to the spherical inclusion.

---

7 *Airy Stress Function and the Biharmonic Equation*

The biharmonic equation in many instances has an analogous role in continuum mechanics to that of Laplace's equation in electrostatics. In the context of
two-dimensional continuum mechanics, the biharmonic equation arises after
introduction of a scalar potential known as the Airy stress function ϕ such that

$$\sigma_{11} = \frac{\partial^2\phi}{\partial x_2^2}, \tag{2.111}$$

$$\sigma_{22} = \frac{\partial^2\phi}{\partial x_1^2}, \tag{2.112}$$

$$\sigma_{12} = -\frac{\partial^2\phi}{\partial x_1 \partial x_2}. \tag{2.113}$$

Given this definition of the Airy stress function, show that the equilibrium
equations are satisfied.
The compatibility condition for infinitesimal strains for planar problems may
be written as

$$\frac{\partial^2\epsilon_{xx}}{\partial y^2} + \frac{\partial^2\epsilon_{yy}}{\partial x^2} = 2\frac{\partial^2\epsilon_{xy}}{\partial x \partial y}. \tag{2.114}$$

Compatibility conditions address the fact that the various components of the
strain tensor may not be stated independently since they are all related through
the fact that they are derived as gradients of the displacement fields. Using
the statement of compatibility given above, the constitutive equations for an
isotropic linear elastic solid and the definition of the Airy stress function show
that the compatibility condition given above, when written in terms of the Airy
stress function, results in the biharmonic equation $\nabla^4\phi = 0$.

THREE

Quantum and Statistical Mechanics Revisited

3.1 Background

In the previous chapter we examined the central tenets of continuum mechanics with an eye to how these ideas can be tailored to the modeling of the mechanical properties of materials. During our consideration of continuum mechanics, we found that one of the key features of any continuum theory is its reliance on some phenomenological constitutive model which is the vehicle whereby mechanistic and material specificity enter that theory. As was said before, the equations of continuum dynamics are by themselves a continuous representation of the balance of linear momentum and make no reference to the particulars of the material in question. It is the role of the constitutive model to inform the equations of continuum dynamics whether we are talking about the plastic deformation of metals or the washing of waves across a beach. A key realization rooted in the microscopic perspective is the idea that the constitutive response, used in the continuum settings described in the previous chapter, reflects a collective response on the part of the microscopic degrees of freedom. It is the business of quantum and statistical mechanics to calculate the average behavior that leads to this collective response. In addition to our interest in effective macroscopic behavior, one of our primary aims is to produce plausible microscopic insights into the mechanisms responsible for observed macroscopic behavior. For example, is there a microscopic explanation for the difference in the Young's modulus of lead and silicon? Or, under what conditions might one expect to see creep mediated by grain boundary diffusion as opposed to bulk diffusion? Or, how does the yield strength depend upon the concentration of some alloying element?

As noted above, microscopic modeling is founded upon the two complementary edifices of quantum and statistical mechanics. From the present perspective, quantum mechanics allows us to capture the electronic origins of many of the internal energy differences that accompany processes central to the mechanics of

81

materials. Similarly, statistical mechanics provides a venue within which it is possible to effect a quantitative treatment of the many degrees of freedom that conspire to yield observed macroscopic response. Our purpose in this chapter is to revisit some of the key ideas associated with these two disciplines, with emphasis on features of these theories that will find their way into our later thinking.

3.2 Quantum Mechanics

Quantum mechanics is the appropriate fundamental theory for describing phenomena ranging from bonding in fullerenes to superconductivity to the behavior of neutron stars. For our purposes, quantum mechanics should be seen in its role as the theory which describes the electronic origins of bonding in solids. In many cases, it is found that the very existence of solids is the result of a collective response on the part of the electronic degrees of freedom. In particular, the free energy is lower when the atoms conspire to form a solid than when they remain in isolation, and a key contribution to the total free energy arises from the electronic degrees of freedom.

In the previous chapter, we noted that the principle of linear momentum balance leads to the central 'governing equation' of continuum mechanics. Similarly, within the context of nonrelativistic quantum mechanics, there is another such governing equation, namely that of Schrödinger. Though the conceptual origins of the Schrödinger equation are tied to a long and fascinating history, here we opt for the axiomatic approach in which the governing equation is asserted without derivation and is instead queried for its implications.

3.2.1 Background and Formalism

Quantum mechanics represents one of the cornerstones of modern physics. Though there were a variety of different clues (such as the ultraviolet catastrophe associated with blackbody radiation, the low-temperature specific heats of solids, the photoelectric effect and the existence of discrete spectral lines) which each pointed towards quantum mechanics in its own way, we will focus on one of these threads, the so-called wave-particle duality, since this duality can at least point us in the direction of the Schrödinger equation.

Though classical electromagnetic theory as embodied in Maxwell's equations led to an interpretation of light in terms of waves, the photoelectric effect inspired Einstein to the view that in its interactions with matter, light might be better thought of from a corpuscular perspective. A complementary view of matter which took quantitative form in the hands of de Broglie was the contention that in some cases matter should be thought of in wave-like terms. In most naive terms, the idea is that one should attribute a wavelength $\lambda = h/p$ to a particle with momentum p, where h is Planck's constant. Indeed, the discovery that electrons form interference patterns provided further impetus towards the discovery of a mathematical description of these 'matter waves'. Schrödinger was inspired by an analogy between classical mechanics and the ray theory of light, and hoped that a mechanics could be found that would stand in the same relation to wave optics that classical mechanics does to ray optics. The result of this inspiration is the famed Schrödinger equation. This equation determines the features of the electronic wave function, $\psi(\mathbf{r})$. In turn, the wave function itself has a kinematic character and serves to characterize the 'state' of the system. A convenient analogy is that the wave function characterizes the state of a system within the quantum framework in a fashion that is analogous to the way that the coordinates and momenta characterize the state of a system within classical mechanics. Just as in classical descriptions, once the coordinates $\{q_i, p_i\}$ are determined, all observables that depend upon those coordinates can be evaluated, so too in quantum mechanics, once the wave function is determined, there are well-defined prescriptions for determining the expected values of quantities of interest. The interpretation of the wave function (here given in the one-dimensional setting, though the generalization to higher dimensions is clear) is that $|\psi(x)|^2 dx = \psi^*(x)\psi(x)dx$ is the probability that a particle will be found between x and $x + dx$. Note that we have introduced the notation $\psi^*(x)$ which instructs us to use the complex conjugate of the wave function $\psi(x)$.

When written in time-independent form for a single particle interacting with a potential $V(\mathbf{r})$, the Schrödinger equation takes the form

$$-\frac{\hbar^2}{2m}\nabla^2\psi(\mathbf{r}) + V(\mathbf{r})\psi(\mathbf{r}) = E\psi(\mathbf{r}). \tag{3.1}$$

The quantity E is the energy eigenvalue associated with a given wave function, while m is the mass of the particle of interest. In its role as the governing equation of quantum systems, the Schrödinger equation for a given problem amounts to the solution of a boundary value problem. Depending upon the nature of the potential $V(\mathbf{r})$ that the particle experiences and the associated boundary conditions, one finds different solutions $\psi_n(\mathbf{r})$ with associated stationary energy states, E_n. Unlike the governing equations of continuum mechanics, these solutions involve a fundamental physical parameter with dimensions of action, namely, Planck's constant, which is denoted here by $\hbar = h/2\pi$. Later we will see that Planck's constant sets the energy scale of interest for many problems relating to electrons in solids. As noted above, the Schrödinger equation yields not only the electronic wave functions, but also the spectrum of energy eigenvalues E_n which are available to the system. These eigenvalues, in turn, are the basis for the determination of the total energy of the system, whether it be an atom, a molecule or a solid. It is upon reckoning the relative distributions of eigenvalues for different configurations that we can construct a picture of why some configurations are more favorable than others. For example, if our aim is to determine the ground state geometry of a particular molecule, this question can be unambiguously answered by comparing the total energies of the geometric alternatives.

The relation between a given classical Hamiltonian and its quantum counterpart depends upon the transcription between classical dynamical variables and certain operators that act upon the space of wave functions. The basic idea is to take the classical Hamiltonian, replace the relevant dynamical variables by their operator analogs, and then to solve the resulting differential equation,

$$\hat{H}\psi = E\psi. \tag{3.2}$$

This equation is a restatement of eqn (3.1) and we have introduced the notation that a ˆ above a quantity signals that quantity as an operator. Note that we are considering the time-independent setting and for now have made reference only to a single particle. To make the classical–quantum transcription more complete, the recipe is to replace classical dynamical variables by their operator analogs according to

$$\mathbf{p} \rightarrow -i\hbar\nabla \tag{3.3}$$

and

$$\mathbf{r} \rightarrow \mathbf{r}. \tag{3.4}$$

The Schrödinger equation corresponding to a given classical Hamiltonian is then obtained by replacing all of the dynamical variables in the original Hamiltonian with their operator analogs.

With our realization that dynamical variables are represented as operators, we can also raise the question of how to compute observed quantities within the framework of quantum mechanics. The essential rule is that if we are interested in obtaining the expectation value of the operator $\hat{O}(\mathbf{r}, \mathbf{p})$ when the system is in the state ψ, then this is evaluated as

$$\langle \hat{O}(\mathbf{r}, \mathbf{p}) \rangle = \langle \psi | \hat{O} | \psi \rangle = \int d^3\mathbf{r}\psi^*(\mathbf{r})\hat{O}\psi(\mathbf{r}). \tag{3.5}$$

Here we have introduced the bra ($\langle\psi|$) and ket ($|\psi\rangle$) notation of Dirac which for us will serve as a shorthand way of representing certain integrals. This assertion is of special importance in our enterprise in the context of the energy for which the relevant expectation value may be written

$$\langle E \rangle = \langle \psi | \hat{H} | \psi \rangle = \int d^3\mathbf{r}\psi^*(\mathbf{r})\hat{H}\psi(\mathbf{r}). \tag{3.6}$$

Indeed, one sense in which this expression will be seen again is in its role as the basis of variational approximations for the wave function. Rather than solving for ψ itself, our idea will be to represent it as a linear combination of some set of basis functions, and to choose the coefficients in the linear combination that minimize $\langle E \rangle$.

To explicitly solve a problem using the machinery of quantum mechanics, one specifies the potential experienced by the particle within the Schrödinger equation which results in a corresponding boundary value problem. Upon solving this problem one finds the spectrum of allowed energy eigenvalues. The paradigmatic example of this program is the hydrogen atom, wherein one specifies the Coulombic potential that characterizes the interaction between the nucleus and the electron and finds that the energy eigenvalues are consonant with the well-known Balmer formula of nineteenth century spectroscopy. In addition to the hydrogen problem, there are a variety of model problems that can also be solved analytically. For less tractable problems one can resort to perturbative treatments wherein more complicated potentials can be treated systematically. In the sections that follow, we undertake the solution of a few of these problems that are of particular interest in our attempt to build plausible models of material behavior. Not only will these models illustrate the fundamental concepts of quantum mechanics, but they will indeed prove useful in our quest to build models of materials.

As a quick example to set notation and illustrate a few more points concerning formalism, we consider the problem of a single particle in a one-dimensional box. In this problem, the potential that acts on the particle is of the form

$$V(x) = \begin{cases} 0 & \text{if } 0 < x < a \\ \infty & \text{otherwise.} \end{cases} \tag{3.7}$$

If we like, we can think of this potential as the simplest model of the confining effect of a free surface. The meaning of the potential is that the particle is 'bound' to the region $0 < x < a$. As a result, all that we really need do is solve the equation

$$-\frac{\hbar^2}{2m} \frac{d^2\psi}{dx^2} = E\psi(x), \tag{3.8}$$

subject to the boundary condition $\psi(0) = \psi(a) = 0$. These boundary conditions guarantee that the probability of finding the particle outside of the box is zero. By inspection, the solutions are of the form

$$\psi_n(x) = \sqrt{\frac{2}{a}} \sin\left(\frac{\pi n x}{a}\right), \tag{3.9}$$

where the constant $\sqrt{2/a}$ is chosen to guarantee that $\int_0^a |\psi(x)|^2 dx$ is normalized to unity for reasons that we will see shortly and n is an integer that insures that

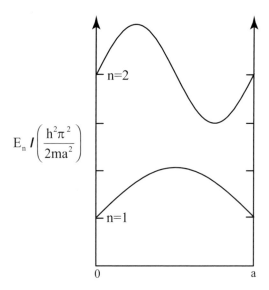

Fig. 3.1. Ground state and first excited state wave functions associated with particle in a box problem. The wave functions are drawn such that their zeroes correspond to the energy of the corresponding state.

the boundary conditions are satisfied. In addition to the wave functions, the energy eigenvalue corresponding to the nth wave function is

$$E_n = \frac{\hbar^2 \pi^2 n^2}{2ma^2}.$$ (3.10)

The ground state and first excited state wave functions for this problem are shown in fig. 3.1. Each wave function has been plotted such that the zeroes of the wave function occur at the value on the vertical energy axis corresponding to that particular state. Thus, in the dimensionless units favored in the figure, $\psi_1(x)$ has an energy $E_1 = 1$, while $\psi_2(x)$ is associated with the energy $E_2 = 4$.

As yet, our quick tour of quantum mechanics has featured the key ideas needed to examine the properties of systems involving only a single particle. However, if we are to generalize to the case in which we are asked to examine the quantum mechanics of more than one particle at a time, there is an additional idea that must supplement those introduced above, namely, the Pauli exclusion principle. This principle is at the heart of the regularities present in the periodic table. Though there are a number of different ways of stating the exclusion principle, we state it in words as the edict that no two particles may occupy the same quantum state. This principle applies to the subclass of particles known as fermions and characterized by half-integer spin. In the context of our one-dimensional particle in a box problem presented above, what the Pauli principle tells us is that if we wish to put two electrons in the box, then in reckoning the total energy we say that one of the particles occupies the ground state and the second occupies the first excited state with a resulting total energy $E_{tot} = E_1 + E_2 = 5$, where the energy is written in the dimensionless units already introduced in fig. 3.1. We confess to ignoring the extra subtleties associated with an explicit inclusion of spin.

3.2.2 Catalog of Important Solutions

The solutions to a few different problems within the quantum mechanical setting will serve as the backdrop for many of the models of materials that will be set forth in later chapters. The quantum mechanical oscillator will serve as our jumping off point for the analysis of the thermodynamics of solids, while the solution to the problem of the hydrogen atom will give us a basis within which to discuss bonding in solids. Consequently, we take up the solution of these problems presently.

Harmonic Oscillator. One exercise that proves central to modeling the thermal properties of materials is that of a particle bounded by a quadratic potential well. Our interest in the harmonic oscillator arises from the following observation. The particles that make up an N-atom solid may be thought of as spanning a $3N$-dimensional configuration space with the potential energy of interaction given by $U(x_1, x_2, \ldots, x_{3N})$. This function gives the potential energy of a given atomic configuration, and regardless of the exact details of the potential energy surface, one may imagine representing this surface in the vicinity of points of equilibrium in terms of a collection of quadratic wells. It is on this basis that the harmonic oscillator serves as a jumping off point for an analysis of the vibrations of solids and reveals itself as of special interest to the modeling of materials. In anticipation of our use of this model in chaps. 5 and 6, we recount the quantum mechanical treatment of a particle in a quadratic potential.

For the case of the one-dimensional harmonic oscillator the potential energy is given by $V(x) = \frac{1}{2}kx^2$, which is parameterized in terms of a stiffness constant k. After some manipulation, the Schrödinger equation associated with this potential may be written as

$$\frac{d^2\psi}{dx^2} + (\alpha - \beta x^2)\psi = 0, \tag{3.11}$$

where $\alpha = 2mE/\hbar^2$ and $\beta = m^2\omega^2/\hbar^2$, and with the constant $\omega^2 = k/m$ corresponding to the classical frequency of vibration. Equations of this generic form suggest a rescaling of the variables motivated by the attempt to isolate the term quadratic in x. In particular, this equation can be further simplified if we make the change of variables $y = \beta^{\frac{1}{4}}x$ which results in a restatement of the original

Schrödinger equation in terms of natural coordinates as

$$\frac{d^2\phi}{dy^2} + \left(\frac{2E}{\hbar\omega} - y^2\right)\phi = 0,$$ (3.12)

where we have defined a new function $\phi(y)$ which reflects the change of variables. Asymptotic analysis suggests a trial solution of the form $\phi(y) = f(y)e^{-y^2/2}$. This form is suggested by the fact that $e^{-y^2/2}$ satisfactorily describes the large-y behavior of the solution, namely, that the particle is bound within the quadratic well. Substitution of this trial solution into the original differential equation results in an auxiliary equation for the unknown function $f(y)$. In particular, we find

$$f''(y) - 2yf'(y) + \left(\frac{2E}{\hbar\omega} - 1\right)f(y) = 0.$$ (3.13)

The details of the solution to this equation are of less immediate interest to us than the nature of the solutions themselves. Suffice it to note that this equation admits of a solution in power series form, and that these solutions guarantee the vanishing of the wave function as $y \to \infty$ only if the series is truncated after a finite number of terms. This large y boundary condition meets with our intuition that the solutions must ultimately vanish at large distances. That is, we expect the particle to be bound by the oscillator potential. The polynomials that result upon truncating the series are the so-called Hermite polynomials. Our requirement that the series solution be truncated after a finite number of terms is imposed by insisting that

$$\frac{2E}{\hbar\omega} = 2n + 1,$$ (3.14)

where n is an integer. Imposition of the boundary condition results in quantization of the allowed energy levels which can be labeled as $E_n = (n + \frac{1}{2})\hbar\omega$. Note that even for the ground state ($n = 0$), the energy is nonzero (the so-called zero point energy). The wave functions that are associated with these various energies are themselves labeled by the integer n (n is known as a quantum number) and can be written as

$$\psi_n(x) = \frac{1}{2^{\frac{n}{2}}(n!)^{\frac{1}{2}}}\left(\frac{m\omega}{\pi\hbar}\right)^{\frac{1}{4}} e^{-\frac{m\omega x^2}{2\hbar}} H_n\left(\sqrt{\frac{m\omega}{\hbar}}x\right),$$ (3.15)

where $H_n(x)$ is a Hermite polynomial of order n. The first few eigenfunctions and their associated energy levels are indicated schematically in fig. 3.2.

One of the key observations concerning our solution is that the probability of finding a particle in the classically forbidden region is nonzero. For example, if we consider the state labeled by $n = 0$, the maximum classical amplitude of vibration is given by equating the fixed total energy, namely $\frac{1}{2}\hbar\omega$, and the

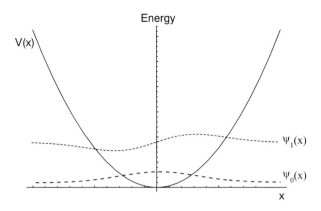

Fig. 3.2. Schematic illustrating the harmonic oscillator potential and the associated wave functions. The wave functions are plotted such that the zeroes of $\psi_i(x)$ intersect the energy axis at that energy corresponding to the energy of the ith state.

classical expression for the energy, which is given by $\frac{1}{2}m\omega^2 x_{max}^2$, where x_{max} is the maximum displacement. This operation implies a maximum classical amplitude of vibration of $x_{max} = \sqrt{\hbar/m\omega}$. As is evident from the figure, the wave function bleeds out of the classical region with nonzero amplitude even at values of x in excess of the classical turning point.

Many of the features of nonrelativistic quantum mechanics have revealed themselves in this problem. First, we have seen that as a result of imposition of the boundary conditions the allowed energy states are discrete. Secondly, we note a progression of associated wave functions, those tied to higher energies having more nodes, and all of which allow the particle to penetrate into the classically forbidden region. For the harmonic oscillator in particular, we have found a series of equally spaced energy levels that will loom important in our quest to understand the entropy of vibration of crystalline solids. In fact, when we undertake our analysis of the thermal motions of crystals, we will need exactly the spectrum of eigenvalues we have uncovered in examining what may at first blush seem an unduly artificial example.

The Hydrogen Atom. The hydrogen atom is the problem against which many other models are measured. In confronting a new subject, often one seeks the 'hydrogen atom' of that field, that is, the illustrative example that at once contains sufficient complexity to capture the key ingredients of the problem without at the same time exacting too high an analytic toll. In this section our aim is to highlight the key features of the quantum mechanical investigation of the hydrogen atom. Our reason for doing so is derived in part from the fact that in attempting to build up models of the bonding in solids, we will find ourselves appealing repeatedly to the hydrogenic wave functions to build intuition concerning the interactions between neighboring atoms.

As with the harmonic oscillator, our starting point is the Schrödinger equation itself. In this case, the equation may be written in the form

$$-\frac{\hbar^2}{2m}\nabla^2\psi(\mathbf{r}) + V(|\mathbf{r}|)\psi(\mathbf{r}) = E\psi(\mathbf{r}),$$ (3.16)

where for the moment we have specified the potential only to the extent that we note it is of the central force form. There is virtue in continuing our analysis in generic terms without yet specifying which central potential will be of interest, though we will do well to anticipate our eventual appeal to the special case of the Coulomb potential pertinent to the hydrogen atom. There are a number of ways to proceed at this point, and we will pursue the classic description in terms of the solution to a differential equation.

By virtue of the presumed spherical symmetry of this problem, we carry out a separation of variables, with the result that the radial and angular terms are decoupled. This may be seen by starting with the Laplacian operator in spherical coordinates which leads to a Schrödinger equation of the form,

$$-\frac{\hbar^2}{2m}\left[\frac{1}{r^2}\frac{\partial}{\partial r}\left(r^2\frac{\partial}{\partial r}\right) + \frac{1}{r^2\sin\theta}\frac{\partial}{\partial\theta}\left(\sin\theta\frac{\partial}{\partial\theta}\right) + \frac{1}{r^2\sin^2\theta}\frac{\partial^2}{\partial\phi^2}\right]\psi(\mathbf{r})$$
$$+ [V(|\mathbf{r}|) - E]\psi(\mathbf{r}) = 0.$$ (3.17)

We now seek a solution of the form $\psi(r,\theta,\phi) = R(r)Y(\theta,\phi)$ which may be inserted into the equation above. Upon so doing and after the requisite reorganization of terms, we find that our problem is reduced to

$$\frac{-\hbar^2}{2m}\frac{1}{R}\frac{\partial}{\partial r}\left(r^2\frac{\partial R}{\partial r}\right) + r^2[V(r) - E] = \frac{\hbar^2}{2m}\left[\frac{1}{Y}\frac{1}{\sin\theta}\frac{\partial}{\partial\theta}\left(\sin\theta\frac{\partial Y}{\partial\theta}\right) + \frac{1}{Y\sin^2\theta}\frac{\partial^2 Y}{\partial\phi^2}\right].$$ (3.18)

The result is an equation in which one side depends upon one set of variables while the other side depends upon a different set of variables and leads us to conclude that both sides are equal to a constant which, with feigned insight, we call $l(l+1)$.

We now further assume that the angular solution may be factored into the form, $Y(\theta,\phi) = P(\theta)u(\phi)$. Substituting this solution into the angular equation results once again in a decoupling of the θ and ϕ dependence and a second separation constant, namely,

$$-\frac{1}{u}\frac{d^2u}{d\phi^2} = \frac{\sin\theta}{P}\frac{d}{d\theta}\left(\sin\theta\frac{dP}{d\theta}\right) + l(l+1)\sin^2\theta = m^2.$$ (3.19)

As a result of these manipulations, the problem of solving the Schrödinger equation for a particle in the presence of a central force potential has been reduced to the solution of three ordinary differential equations which may be written as

$$\frac{d^2u}{d\phi^2} = -m^2u, \tag{3.20}$$

and

$$\frac{1}{\sin\theta}\frac{d}{d\theta}\left(\sin\theta\frac{dP}{d\theta}\right) + l(l+1)P - \frac{m^2}{\sin^2\theta}P = 0, \tag{3.21}$$

and

$$r^2\frac{d^2R}{dr^2} + 2r\frac{dR}{dr} + \frac{2mr^2}{\hbar^2}[E - V(r)] - l(l+1) = 0. \tag{3.22}$$

The ϕ equation is routine and has solutions of the form $u(\phi) = e^{im\phi}$. The requirement of periodicity of the wave function, namely that $u(\phi) = u(\phi + 2\pi)$, restricts m to integer values. The θ equation is simplified by making the change of variables $z = \cos\theta$, resulting in the associated Legendre equation

$$(1 - z^2)\frac{d^2F}{dz^2} - 2z\frac{dF}{dz} + \left[l(l+1) - \frac{m^2}{1-z^2}\right]F = 0, \tag{3.23}$$

which can be tackled via series solution. Rather than examining the strategy for solving the equation, we merely quote the solutions and attempt to learn something of their nature and significance. The general solution to the associated Legendre equation may be written in the form $F(z) = aP_{lm}(z) + bQ_{lm}(z)$, where the functions $P_{lm}(z)$ and $Q_{lm}(z)$ are known as associated Legendre polynomials. The solution $Q_{lm}(z)$ is rejected on the grounds that it is singular when $z = \pm 1$, and hence we keep only the P_{lm}s.

The compositions of the solutions to the θ and ϕ equations, by virtue of their relevance to all problems with spherical symmetry, are so important as to merit a special name, the spherical harmonics. These solutions bear the same relation to spherical problems that the Bessel functions do to those with cylindrical symmetry. The spherical harmonics, $Y_{lm}(\theta, \phi)$, may be written as

$$Y_{lm}(\theta, \phi) = P_{lm}(\cos\theta)e^{im\phi}. \tag{3.24}$$

In the present context, their importance attaches to their role in describing the angular dependence of the hydrogenic wave functions. When we take up bonding in solids, we will see that the angular character implied by the spherical harmonics makes itself known in macroscopic observables as fundamental as those of the elastic moduli. We now take up the specific nature of the radial wave function.

From eqn (3.18), we see that to make further progress requires the specification of a particular form for $V(r)$. For the case of the hydrogen atom, $V(r)$ is of the Coulomb form. A choice must be made with regard to units in treating the electrostatic interaction, and we adopt the esu units within which the Coulomb potential is of the form $V(r) = e^2/r$ and results in a radial equation of the form

$$r^2 \frac{d^2 R}{dr^2} + 2r \frac{dR}{dr} + \frac{2mr^2}{\hbar^2} \left[E + \frac{e^2}{r} - \frac{\hbar^2 l(l+1)}{2mr^2} \right] R = 0. \tag{3.25}$$

The radial equation can be further simplified into a recognizable form if one makes the substitution $f(r) = r R(r)$, leading to

$$\frac{d^2 f}{dr^2} - \frac{l(l+1)}{r^2} f + \frac{2m}{\hbar^2} \left[E + \frac{e^2}{r} \right] f = 0. \tag{3.26}$$

What we have succeeded in doing is to rework the radial equation into exactly the form of a one-dimensional Schrödinger equation in the presence of an effective potential

$$V_{eff}(r) = -\frac{e^2}{r} + \frac{\hbar^2}{2m} \frac{l(l+1)}{r^2}. \tag{3.27}$$

As in our analysis of the θ equation, rather than bothering with the niceties of solving this radial equation, we will quote its solutions and consider their significance. The strategy that is adopted in solving this equation is in many respects similar to that used in solving the problem of the harmonic oscillator. One assumes on the basis of asymptotics that the solution may be written as a product of a term of the form $e^{-\beta r}$ and some as yet undetermined function. What results is a new differential equation, analogous to the Hermite differential equation we confronted in solving the harmonic oscillator, and this equation may be solved in terms of special functions known as the associated Laguerre polynomials, which depend upon the quantum numbers n and l and result in a radial wave function of the form

$$R_{nl}(r) = e^{-r/na_0} \left(\frac{r}{na_0} \right)^l L_{n-l-1}^{2l+1} \left(\frac{2r}{na_0} \right). \tag{3.28}$$

Here we have denoted the associated Laguerre polynomials as L_{n-l-1}^{2l+1}. A few of these radial functions are shown in fig. 3.3. In anticipation of the fact that we will later invoke hydrogen-like orbitals in the context of our discussion of electronic structure in chap. 4, we note that it is conventional to label orbitals using both their principal quantum number, n, and their orbital angular momentum number l. In particular, a hierarchy of states at a given n is built running from $l = 0$ to $l = n - 1$. For example, for $n = 4$ we have 4s, 4p, 4d and 4f states, where $l = 0$ is labeled s, $l = 1$ is labeled p, $l = 2$ is labeled d and $l = 3$ is labeled f.

Like in the case of the harmonic oscillator, our insistence that the wave functions be bounded for large-r forces a truncation of the series in the form of

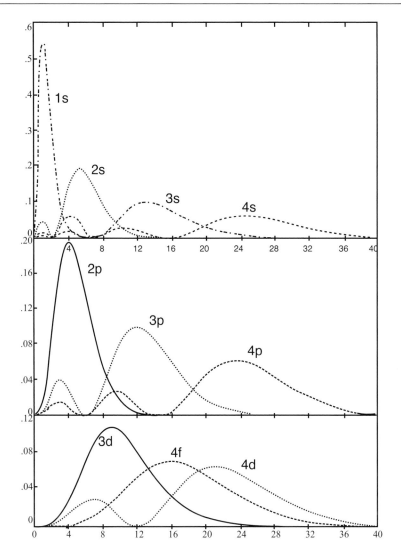

Fig. 3.3. Radial probability distributions for various low-lying states of hydrogen (adapted from Condon and Shortley (1935)).

polynomials. Further, this condition that the wave functions vanish at infinity imposes restrictions on the allowed values of the energy which are

$$E_n = -\frac{me^4}{2\hbar^2 n^2},$$
(3.29)

where n assumes only integral values. We note that the energies depend only upon the principal quantum number n. The implication of this result is that the various states of different angular momenta (labeled by l and m) are degenerate. In numerical form, the series of energy states introduced above is of the form

$$E_n \approx -\frac{13.6}{n^2} \text{ eV}, \tag{3.30}$$

providing a clear idea of the energy scale associated with problems of this type. We have found that the generic hydrogenic wave function takes the form

$$\psi_{nlm}(\mathbf{r}) = R_{nl}(r)Y_{lm}(\theta, \phi), \tag{3.31}$$

with associated energy eigenvalues of the form given in eqn (3.29). This solution will prove of great utility in our later development of the linear combination of atomic orbitals (or tight-binding) method.

3.2.3 Finite Elements and Schrödinger

Recall from chap. 2 that often in the solution of differential equations, useful strategies are constructed on the basis of the *weak* form of the governing equation of interest in which a differential equation is replaced by an integral statement of the same governing principle. In the previous chapter, we described the finite element method, with special reference to the theory of linear elasticity, and we showed how a weak statement of the equilibrium equations could be constructed. In the present section, we wish to exploit such thinking within the context of the Schrödinger equation, with special reference to the problem of the particle in a box considered above and its two-dimensional generalization to the problem of a quantum corral.

Our preliminary exercise will state the problem in one dimension, with the generalization to follow shortly. The one-dimensional Schrödinger equation may be written

$$-\frac{\hbar^2}{2m}\frac{d^2\psi}{dx^2} + V(x)\psi = E\psi. \tag{3.32}$$

In the present case, the weak form is obtained by multiplying eqn (3.32) by some test function and integrating the first term by parts. For the purposes of concreteness in the present setting, we imagine a problem for which the domain of interest runs from the origin to point $x = a$, and for which the wave functions vanish at the boundaries. We note that the Schrödinger equation could have been stated alternatively as the Euler–Lagrange equation for the functional

$$\Pi[\psi(x)] = \frac{\hbar^2}{2m}\int_0^a \left|\frac{d\psi}{dx}\right|^2 dx + \int_0^a V(x)|\psi(x)|^2 dx - E\int_0^a |\psi(x)|^2 dx. \tag{3.33}$$

The finite element strategy is to replace this functional with a function of the N unknown values ψ_n (not to be confused with wave functions related to the hydrogen atom from the previous section) of the wave function at the nodes of the finite element mesh. That is, we imagine a discretization of the region $[0, a]$ by N distinct points which are taken to be the nodes of the finite element mesh. The problem of solving the Schrödinger equation then becomes one of determining not the entire function $\psi(x)$, but rather for the values of ψ (i.e. ψ_n) at the nodal positions, with the understanding that the value of the wave function at points between the nodes is obtained by finite element interpolation.

Specifically, as with our earlier treatment of the finite element method, we imagine shape functions $N_i(x)$ centered at each node, with the subscript i labeling the node of interest. It is then asserted that the wave function is given as

$$\psi(x) = \sum_{n=1}^{N} \psi_n N_n(x), \qquad (3.34)$$

and similarly the potential (which is known) is expanded in terms of the shape functions via

$$V(x) = \sum_{n=1}^{N} V_n N_n(x). \qquad (3.35)$$

What this means is that to write down the interpolated potential $V(x)$, we find the values of the potential at the nodes, $V_n = V(x_n)$ and interpolate elsewhere using eqn (3.35). By virtue of the finite element ansatz, we may now effect a mapping between the functional $\Pi[\psi(x)]$ and the function $\Pi(\{\psi_n\})$. In particular, upon substituting $\psi(x)$ and $V(x)$ from eqs (3.34) and (3.35) into eqn (3.33) and simplifying to the case in which we drop all reference to the fact that the wave function can be complex rather than real, we have

$$\Pi[\psi(x)] \rightarrow \Pi(\{\psi_n\}) = \frac{\hbar^2}{2m} \sum_i \sum_j \int_0^a \psi_i \psi_j \frac{dN_i(x)}{dx} \frac{dN_j(x)}{dx} dx$$

$$+ \sum_i \sum_j \sum_k \int_0^a V_k \psi_i \psi_j N_i(x) N_j(x) N_k(x) dx$$

$$- E \sum_i \sum_j \int_0^a \psi_i \psi_j N_i(x) N_j(x) dx. \qquad (3.36)$$

This result may be simplified considerably if we invoke the definitions

$$K_{ij} = \frac{\hbar^2}{2m} \int_0^a \frac{dN_i(x)}{dx} \frac{dN_j(x)}{dx} dx,$$

$$V_{ij} = \sum_k \int_0^a V_k N_i(x) N_j(x) N_k(x) dx,$$

$$M_{ij} = \int_0^a N_i(x)N_j(x)dx.$$

Note that these matrices are functions only of known quantities, namely, the potential $V(x)$ and the shape functions $N(x)$. It is convenient to further simplify by introducing $H_{ij} = K_{ij} + V_{ij}$, with the result that the function of interest is

$$\Pi(\{\psi_n\}) = \sum_{ij}[H_{ij} - EM_{ij}]\psi_i\psi_j. \tag{3.37}$$

At this point, we need to implement our original variational statement by seeking those ψ_ns that minimize the function $\Pi(\{\psi_n\})$. Imposition of this condition via

$$\frac{\partial \Pi}{\partial \psi_n} = 0, \tag{3.38}$$

results in the reformulation of the original problem as

$$\sum_j [H_{nj} - EM_{nj}]\psi_j = 0. \tag{3.39}$$

As a specific realization of these ideas, consider the celebrated problem of the one-dimensional particle in a box already introduced earlier in the chapter. Our ambition is to examine this problem from the perspective of the finite element machinery introduced above. The problem is posed as follows. A particle is confined to the region between 0 and a with the proviso that within the well the potential $V(x)$ vanishes. In addition, we assert the boundary condition that the wave function vanishes at the boundaries (i.e. $\psi(0) = \psi(a) = 0$). In this case, the Schrödinger equation is

$$\frac{d^2\psi(x)}{dx^2} + \frac{2mE}{\hbar^2}\psi(x) = 0. \tag{3.40}$$

We are searching for the wave functions and associated energy eigenvalues. From an analytic perspective, this problem is entirely routine with the result that the nth eigenfunction is of the form given in eqn (3.9) and the corresponding eigenvalues as given in eqn (3.10). Our objective is to see to what extent we can recover these results using the finite element method. Further, the relevant finite element analysis can be carried out almost entirely by hand.

We imagine discretizing the region between 0 and a with $N + 1$ nodes as illustrated in fig. 3.4. The representation of the wave function is now restricted entirely to the values of the wave function at the nodes, ψ_i, with the value between nodes determined by interpolation via $\psi(x) = \sum_i \psi_i N_i(x)$. The particular choices of shape functions used in the present analysis are shown in fig. 3.4 as well. Because of the fact that the shape functions are highly localized, the matrices K_{ij} and M_{ij} introduced above will be nonzero only for the cases in which $j = i$ or

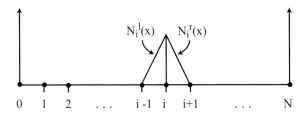

Fig. 3.4. Schematic illustration of the nodes used to discretize the particle in a box problem and the corresponding finite element shape functions.

$j = i \pm 1$. For a given level of approximation (as represented by the number of elements used in the analysis), the solution of the Schrödinger equation has been reduced to the construction of the matrix $H_{ij} - E M_{ij}$ and the determination of the associated energy eigenvalues.

In abstract terms, the relevant matrix elements may be evaluated as follows. Consider the case in which there are $N + 1$ nodes as shown in fig. 3.4. First, we note that since within the region of interest there is no potential $V(x)$, $H_{ij} = K_{ij}$. The matrix element K_{ii} is found as

$$K_{ii} = \frac{\hbar^2}{2m} \left\{ \int_{x_i - \frac{a}{N}}^{x_i} \left[\frac{d N_i^l(x)}{dx} \right]^2 dx + \int_{x_i}^{x_i + \frac{a}{N}} \left[\frac{d N_i^r(x)}{dx} \right]^2 dx \right\}, \tag{3.41}$$

where we have introduced the notation $N_i^r(x)$ to represent the shape function facing to the right of the ith node and $N_i^l(x)$ to represent the shape function facing to the left of the ith node. These shape functions are written as

$$N_i^r(x) = \frac{N}{a}(x_i - x) + 1, \tag{3.42}$$

$$N_i^l(x) = \frac{N}{a}(x - x_i) + 1, \tag{3.43}$$

and are characterized by the property that $N_i(x_j) = \delta_{ij}$. In light of these definitions, the integrals in eqn (3.41) may be evaluated with the result

$$K_{ii} = \frac{\hbar^2 N}{ma}. \tag{3.44}$$

In a similar way, the matrix element $K_{i,i+1}$ (which by symmetry is equal to $K_{i,i-1}$) may be written as

$$K_{i,i+1} = \frac{\hbar^2}{2m} \int_{x_i}^{x_i + \frac{a}{N}} \frac{d N_i^r(x)}{dx} \frac{d N_{i+1}^l(x)}{dx} dx. \tag{3.45}$$

The region of integration is determined by the fact that the only nonzero overlap between the ith and $(i + 1)^{th}$ shape functions is in the element sharing the ith and

$(i + 1)^{th}$ nodes which is the region between x_i and $x_i + a/N$. Again, in light of the definition of the shape functions, it is easily shown that

$$K_{i,i\pm1} = -\frac{\hbar^2 N}{2ma}. \tag{3.46}$$

We also require the various matrix elements of the form M_{ij} which reflect the overlap of adjacent shape functions. The diagonal matrix elements may be written as

$$M_{ii} = \int_{x_i - \frac{a}{N}}^{x_i} [N_i^l(x)]^2 dx + \int_{x_i}^{x_i + \frac{a}{N}} [N_i^r(x)]^2 dx. \tag{3.47}$$

Performing the requisite integrals results in

$$M_{ii} = \frac{2a}{3N}. \tag{3.48}$$

Similarly, the off-diagonal matrix elements may be written

$$M_{i,i+1} = \int_{x_i}^{x_i + \frac{a}{N}} N_i^r(x) N_{i+1}^l(x) dx, \tag{3.49}$$

with the result

$$M_{i,i\pm1} = \frac{a}{6N}. \tag{3.50}$$

Once the various matrix elements are in hand, we can now search for the eigenvalues associated with a given finite element approximation. As a starting point, consider the case $N = 2$ in which there is only a single unconstrained nodal degree of freedom. In this case, the matrix of eqn (3.39) reduces to $(K_{11} - EM_{11})\psi_1 = 0$. As a result, we see that our first approximation to the ground state energy for the particle in the box is given by $E_1^{(N=2)} = K_{11}/M_{11} = 6\hbar^2/ma^2$. At the next level of approximation, we consider the case in which $N = 3$. If we adopt the notation that $K_{ii} = \alpha$ and $M_{i,i\pm1} = \beta$, then $K_{i,i\pm1} = -\alpha/2$ and $M_{ii} = 4\beta$. For the case $N = 3$, therefore, the resulting condition for the eigenvalues is

$$\det \begin{pmatrix} \alpha - 4E\beta & -\dfrac{\alpha}{2} - E\beta \\ -\dfrac{\alpha}{2} - E\beta & \alpha - 4E\beta \end{pmatrix} = 0. \tag{3.51}$$

The resulting energy eigenvalues are $E_1^{(N=3)} = \alpha/10\beta$ and $E_2^{(N=3)} = \alpha/2\beta$ and are represented in terms of $\hbar^2/2ma^2$ in table 3.1, as are increasing levels of approximation. The act of taking our analysis to the next level of approximation (i.e. the use of more elements) requires neither further physical nor mathematical

Table 3.1. *Eigenvalues associated with finite element approximation to solutions for Schrödinger equation for a particle in a one-dimensional box. Energies are reported in units of $\hbar^2/2ma^2$.*

Approximation	E_1	E_2	E_3	E_4
$N = 2$	6.000			
$N = 3$	5.400	27.000		
$N = 4$	5.098	22.44	58.057	113.91
$N = 5$	5.048	21.6	54.0	108.0
Exact	4.935	19.739	44.415	78.896

elaboration. The generic form of the secular equation, the roots of which are the energy eigenvalues, is tridiagonal and takes the form

$$\det \begin{pmatrix} \alpha - 4E\beta & -\dfrac{\alpha}{2} - E\beta & 0 & 0 & \cdots & 0 \\ -\dfrac{\alpha}{2} - E\beta & \alpha - 4E\beta & -\dfrac{\alpha}{2} - E\beta & 0 & \cdots & 0 \\ 0 & -\dfrac{\alpha}{2} - E\beta & \alpha - 4E\beta & -\dfrac{\alpha}{2} - E\beta & \cdots & 0 \\ \vdots & \vdots & \vdots & \vdots & \vdots & \vdots \\ \vdots & \vdots & \vdots & \vdots & \cdots & -\dfrac{\alpha}{2} - E\beta \\ 0 & 0 & 0 & \cdots & -\dfrac{\alpha}{2} - E\beta & \alpha - 4E\beta \end{pmatrix}$$

$$= 0. \tag{3.52}$$

The fact that the matrix is nonzero only in a strip near the main diagonal is an immediate consequence of the locality of the shape functions which implies that the overlap between different shape functions vanishes except in the case of neighboring elements. The result of carrying out calculations with increasing numbers of elements are summarized in table 3.1, and give a feel for the rate of convergence of the results.

In addition to questions concerning the distribution of energy eigenvalues for this problem, we can also demand the implications of our finite element analysis for the wave functions. As a result of our knowledge of the energy eigenvalues, eqn (3.39) can be exploited to determine the wave functions themselves. By substituting the eigenvalues into these coupled equations, we can then solve for the corresponding ψ_is. The simplest realization of this program is for the case $N = 2$ introduced above in which there is only a single free nodal degree of freedom. In this case, the value of ψ_1 is determined entirely by the constraint of normalization. That is, we

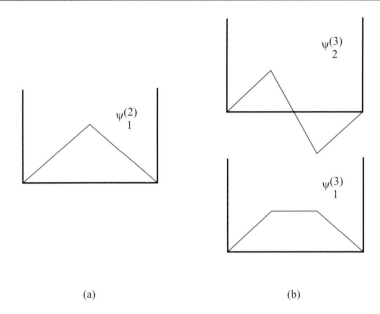

(a)   (b)

Fig. 3.5. Illustration of the finite element interpolated approximate wave functions: (a) ground state wave function for the $N = 2$ approximation; (b) ground state and first excited state wave functions for the $N = 3$ approximation.

require that $\int_0^a \psi^2(x)dx = 1$. In this case, because of finite element interpolation, we have

$$\psi_1^{(N=2)}(x) = \begin{cases} \psi_1 \dfrac{2}{a}x & \text{if } x \leq \dfrac{a}{2} \\[2mm] \psi_1\left(2 - \dfrac{2}{a}x\right) & \text{if } x > \dfrac{a}{2} \end{cases}. \tag{3.53}$$

Our notation $\psi_m^{(N=P)}(x)$ indicates the mth eigenstate associated with the finite element approximation in which $N = P$. In the case above, the constraint of normalization requires that $\psi_1 = \sqrt{3/a}$. The ground state wave function as obtained from our $N = 2$ approximation is illustrated in fig. 3.5.

A more instructive example of the finite element approximation to the wave function is provided by the case $N = 3$. In this case, by substituting the values of the energy eigenvalues back into eqn (3.39), we obtain equations for determining the ratio ψ_1/ψ_2. In particular, we find that

$$\frac{\psi_1}{\psi_2} = -\frac{K_{12} - EM_{12}}{K_{11} - EM_{11}}. \tag{3.54}$$

For the ground state energy, this results in $\psi_1/\psi_2 = 1$ while for the first excited state we have $\psi_1/\psi_2 = -1$. Again, in conjunction with the requirement that the

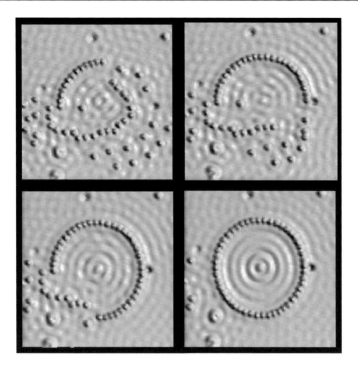

Fig. 3.6. Successive steps in the construction of a quantum corral using the tip of a scanning tunneling microscope (courtesy of D. Eigler).

wave functions be normalized, ψ_1 and ψ_2 can be unambiguously determined, with the result being shown in fig. 3.5.

3.2.4 Quantum Corrals: A Finite Element Analysis

Quantum mechanics is a subject that is well endowed with a series of enlightening schoolboy exercises. However, one of the fascinating features that has accompanied the attainment of atomic-level experimental probes is the ability to carry out experiments that attempt to mimic these very same problems. In this section we will examine the issues surrounding the solution of the problem of the 'quantum corral', in which a series of atoms are brought together on a crystal surface into a closed circular configuration. The presence of these surface atoms has the effect in turn of confining electrons, the states of which can be probed experimentally using scanning tunneling microscopy. In fig. 3.6 a sequence of images is shown that represents the stages in constructing the quantum corral itself.

From the standpoint of the Schrödinger equation, this is another example of a 'particle in a box' problem which in this case is effectively two-dimensional. In particular, the potential within the corral region is assumed to be zero, while the ring of atoms at the boundary is treated in the model as the source of an infinite potential barrier. The infinite barrier is translated into a boundary condition for the wave function at the boundary of the corral ($r = a$), namely, $\psi(a, \theta) = 0$. We begin by writing the Schrödinger equation in polar coordinates,

$$-\frac{\hbar^2}{2m}\left(\frac{\partial^2}{\partial r^2} + \frac{1}{r}\frac{\partial}{\partial r} + \frac{1}{r^2}\frac{\partial^2}{\partial \theta^2}\right)\psi(r, \theta) = E\psi(r, \theta), \tag{3.55}$$

and immediately attempt a solution in the separated form, $\psi(r, \theta) = R(r)u(\theta)$. The differential equation can then be rearranged into two terms, one of which depends only upon r while the other depends only upon θ. As we noted in the case of the hydrogen atom, the equality of two independent functions to one another implies that both must, in turn, be equal to a constant which we denote m^2 in anticipation of the resulting differential equation.

As a result of our separation of variables strategy, the problem of solving the Schrödinger equation for the quantum corral is reduced to that of solving two ordinary differential equations. In particular, we have

$$\frac{d^2 R}{dr^2} + \frac{1}{r}\frac{dR}{dr} + \left(\frac{2mE}{\hbar^2} - \frac{n^2}{r^2}\right)R = 0, \tag{3.56}$$

and

$$\frac{d^2 u}{d\theta^2} + m^2 u = 0. \tag{3.57}$$

The r equation can be massaged into the form of a Bessel equation while the θ equation is elementary. As a result of these insights, the solution may be written as $J_n(\sqrt{2mE/\hbar^2}\,r)e^{im\theta}$. It now remains to impose our boundary conditions. We require that the solution be periodic in the angular coordinate, which implies that m is restricted to integer values, and that the wave function vanish when $r = a$. This second condition means that

$$\sqrt{\frac{2mE_{nl}}{\hbar^2}}\,a = z_{nl}, \tag{3.58}$$

where z_{nl} is the lth zero of the nth Bessel function. We see that each energy eigenvalue E_{nl} is labeled by two quantum numbers, n and l.

A more interesting perspective on the quantum corral problem is garnered from a finite element analysis. Since we have already carried through a detailed implementation of the finite element strategy with regard to the one-dimensional particle in a box, here we only note the salient features of the problem. To illustrate the character of two-dimensional finite elements, the circular region over which the solution is sought is discretized into a series of finite elements. An example of the type of mesh that is generated is shown in fig. 3.7. We should also note

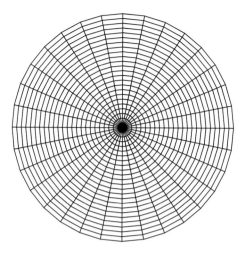

Fig. 3.7. Finite element mesh used to compute the eigenstates of the quantum corral (courtesy of Harley Johnson).

that a full two-dimensional treatment of this problem is an overkill since, as was already shown, the original problem can be reduced to the search for solutions of a one-dimensional radial equation. On the other hand, the two-dimensional solution is enlightening from the finite element perspective.

As with the one-dimensional case described earlier, the unknowns are the nodal values of the wave function, ψ_n. To determine the wave functions between nodes, finite element interpolation is used, in this case with the relevant shape functions taking the form of triangular sheets such that the value of the shape function is 1 at one of the three nodes making up a given triangle, and 0 at the other two nodes and varying linearly in between. Once again, the Hamiltonian matrix is constructed and the resulting secular equation is solved with the result being a series of eigenvalues. In addition, the nodal values of the wave functions associated with the various states are determined as a result of this analysis, and two of the finite element wave functions are shown in fig. 3.8.

3.2.5 Metals and the Electron Gas

So far, our quantum mechanics examples have been rather far afield from questions of immediate interest to our study of solids. In particular, we aim to see what light quantum mechanics sheds on the origin of bonding and cohesion in solids. To do so, in this introductory chapter we consider two opposite extremes. Presently, we consider the limit in which each atom is imagined to donate one or more electrons to the solid which are then distributed throughout the solid. The resulting model is the so-called electron gas model and will be returned to repeatedly in coming

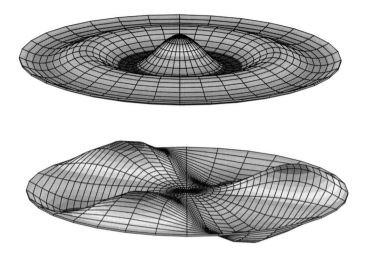

Fig. 3.8. Two different eigenfunctions for the quantum corral as obtained by using a two-dimensional finite element calculation (courtesy of Harley Johnson).

chapters. The other limit that will be considered in the next section is that in which it is imagined that the formation of the solid results in a small overlap between the atomic orbitals associated with different atoms. This picture will lead to the so-called tight-binding method (or linear combination of atomic orbitals).

With our interest in the bonding of solids, we undertake an analysis of the electron gas, which in simplest terms is nothing more than a three-dimensional version of the particle in a box problem which we have already seen in both its one- and two-dimensional guises. Here the physical idea is that some subset of electrons is donated from some host atoms such as sodium or aluminum, and can be thought of as being distributed within a box, with the only confining potential they experience arising from the presence of the boundaries of the solid themselves. In addition, for the time being we advance the assumption that the positive charge associated with the ions is distributed uniformly so as to guarantee charge neutrality. This approximation is the basis of the free electron theory of metals. However, in considering this idealized model, we are immediately confronted with one of the intriguing paradoxes of the electronic theory of solids, namely, the surprising efficacy of independent particle models. Note that we have said nothing of the interactions between electrons. We will revisit this conundrum in later chapters, while for now maintaining our assumption that the electrons behave as though they are independent.

Within the model of a metal proposed above, the Schrödinger equation becomes

$$-\frac{\hbar^2}{2m}\nabla^2\psi(r) = E\psi(r), \tag{3.59}$$

with the additional proviso that the wave functions satisfy boundary conditions to be elaborated below. The solution in this case is immediate and is seen to take the familiar plane wave form, $\psi(\mathbf{r}) = (1/\sqrt{\Omega})e^{i\mathbf{k}\cdot\mathbf{r}}$. Here we have adopted the notation $\Omega = L^3$, where L is the edge length of the confining box. This factor appears in the wave function as a result of our demand that $\int_\Omega |\psi(\mathbf{r})|^2 d^3\mathbf{r} = 1$, a normalization condition. Note that in light of our probabilistic interpretation of the wave function, we see that the electron can be thought of as uniformly spread out throughout the confining box. Associated with a plane wave characterized by a wavevector $\mathbf{k}$, is an energy eigenvalue given by $E_k = \hbar^2 k^2/2m$. Note that the dependence of the energy on the wavevector $\mathbf{k}$ is isotropic. The choice of a finite-sized box imposes restrictions on the allowed values of the wavevector. In particular, if we impose periodic boundary conditions of the form $\psi(x, y, z) = \psi(x + L, y, z)$, we find that the wavevectors can only take the values $k_i = 2\pi m_i/L$, where m_i is an integer. Note that we could just as well have chosen boundary conditions in which the wave functions vanish at the boundary, with no change in our conclusions concerning the energetics of the electron gas. If we wish to characterize these eigenvalues geometrically it is convenient to think of a discrete lattice of allowable points in 'k-space'. Each set of integers (m_1, m_2, m_3) labels a point in this space and is tied to a particular energy eigenvalue

$$E_{m_1 m_2 m_3} = \frac{2\pi^2 \hbar^2}{mL^2}(m_1^2 + m_2^2 + m_3^2). \tag{3.60}$$

The set of allowed points in k-space is depicted schematically in fig. 3.9. The isotropy of the energy spectrum implies a wide range of degeneracies which are of interest as we fill these states.

To probe the implications of our solution, we now need to implement the Pauli principle described earlier. The goal is to assign to the various energy states their relevant occupancies, that is, how many electrons are to be assigned to each state? The Pauli principle demands that only one electron be assigned to each state (or two electrons for each $\mathbf{k}$-vector if we acknowledge the two possible states of spin). This requirement is tied at a deeper level to the existence of certain symmetry requirements for the electronic wave function. For the case of electrons (fermions), one insists that the electron wave function be antisymmetric. In the context of our problem, this demand is equivalent to the assertion that only one electron be allowed to occupy each energy state. The prescription that emerges for computing the total electronic energy of our system, then, is to sum up the energies associated with the lowest eigenvalues, until all of the electrons have been suitably assigned to an associated energy.

For the sake of concreteness, we assume that the confining box has cube edges of dimension L. Further, we assume that N atoms occupy this box and that each has

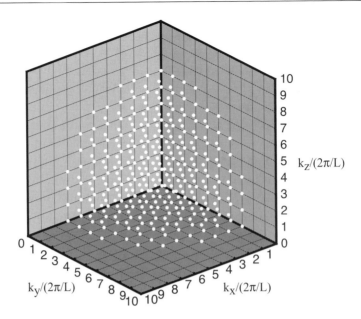

Fig. 3.9. Depiction of allowed k-points associated with energy eigenvalues for particle in a three-dimensional box. The series of k-points shown in the figure are those in the positive octant of k-space falling within a spherical shell of such k-points.

donated a single valence electron which results in an electron density $\rho = N/L^3$. The total electronic energy in our noninteracting electron model can then be found by evaluating the discrete sum over the energy eigenvalues (i.e. $2\sum_{i=1}^{N} E_i$, where the energies are rank ordered $E_1 < E_2 < \cdots < E_N$ and the factor of 2 accounts for spin) and terminating this sum once N states have been occupied. However, in the limit that our box is sufficiently large, the energy levels are so closely spaced we can just as well evaluate this energy by replacing the discrete sum with an integral over a continuous distribution of eigenvalues. To see this, note that adjacent energy levels have an energy difference of order $2\pi^2\hbar^2/mL^2$. The integration that arises in replacing the discrete summation is most conveniently realized in k-space. To effect this integration, it is first necessary to identify the number of states per unit volume of k-space. Recall that the points in k-space are of the form $\mathbf{k}_{m_1,m_2,m_3} = (2\pi/L)(m_1, m_2, m_3)$, and hence occupy the vertices of a simple cubic lattice in k-space. As a result, each such point may be assigned a volume $(2\pi/L)^3$. This observation comes from the recognition that the lattice parameter in k-space is $2\pi/L$. Hence we make the identification $\sum_N = (L/2\pi)^3 \int d^3k$ and it is seen that the total energy is given by

$$E_{tot} = 2\left(\frac{L}{2\pi}\right)^3 \int_{\Omega_k} \frac{\hbar^2 k^2}{2m} d^3\mathbf{k}, \tag{3.61}$$

where the factor of 2 accounts for the fact that each k-point can accommodate two electrons, one for each component of spin. Here we have done nothing more than replace the sum over occupied states with the integral given above, where the prefactor serves as a density of states and insures that our volume element d^3k is properly weighted. Because of the isotropy of the energy spectrum when viewed from the k-space perspective, this integration can be carried out in spherical coordinates, where the radius of the resulting sphere is denoted as the Fermi wavevector, $\mathbf{k}_F$. It is clear that the magnitude of the Fermi wavevector is intimately tied to the electronic density, but how? Here, we see that the total number of states contained within this k-space sphere is the same as half the total number of electrons (again the factor of 2 arising because of spin). The volume of the Fermi sphere is $\Omega_k = 4\pi k_F^3/3$, while the volume per state is $(2\pi/L)^3$. Hence, the ratio of these numbers must yield the total number of electrons and leads to the conclusion that $k_F = (3\pi^2\rho)^{\frac{1}{3}}$. Given the radius of the Fermi sphere, we are now in a position to evaluate the total energy of our electron gas.

Upon explicitly writing the integral given above in spherical coordinates, we find

$$E_{tot} = 2\left(\frac{L}{2\pi}\right)^3 \int_0^{k_F} \int_0^{2\pi} \int_0^\pi \frac{\hbar^2 k^4}{2m} \sin\theta_k \, dk d\phi_k d\theta_k, \tag{3.62}$$

where we include the subscript k as a reminder that our integration is carried out over a sphere in k-space. Performing the relevant integrations leads to the conclusion that

$$\frac{E_{tot}}{V} = \frac{\hbar^2 k_F^5}{10\pi^2 m}, \tag{3.63}$$

which can in turn be converted into an average energy per electron of

$$\frac{E_{tot}}{N} = \frac{3}{5}\frac{\hbar^2 k_F^2}{2m} = \frac{3}{5}\epsilon_F. \tag{3.64}$$

Here we have introduced the notion of the Fermi energy which is denoted by ϵ_F. This energy, like the Fermi wavevector, can be obtained explicitly in terms of the electron density. In fact, since shortly we will use the energy of the electron gas to obtain the optimal density, it is convenient to rewrite eqn (3.64) in terms of the density, recalling that $k_F = (3\pi^2\rho)^{\frac{1}{3}}$ as

$$E_{tot} = \frac{3}{5}\frac{\hbar^2}{2m}(3\pi^2\rho)^{\frac{2}{3}}, \tag{3.65}$$

illustrating that this part of the electron gas energy scales as $\rho^{\frac{2}{3}}$. If we take a typical density such as 0.1 Å^{-3}, the associated Fermi energy is roughly 7 eV, some 300 times larger than the thermal energy scale at room temperature, for example.

The results for the electron gas obtained above can be pushed much further. In particular, the rudimentary calculation developed thus far can serve as a jumping off point for an evaluation of the lattice parameter and bulk moduli of simple metals. As it stands, the energy we have determined for the electron gas is a monotonically increasing function of the density and thus there is as yet no provision for an optimal density. However, we can now supplement the analysis given above by treating the effect of electron–electron interactions perturbatively. In this case, the details of the calculation will take us too far afield, so we will have to be satisfied with sketching what is involved in making the calculation and in quoting the results. The basic idea is that we wish to evaluate the expectation value of the perturbed Hamiltonian

$$H' = \frac{1}{2} \sum_{i \neq j} \frac{e^2}{|\mathbf{r}_i - \mathbf{r}_j|}, \qquad (3.66)$$

which characterizes the electron–electron interaction. Physically what this means is that we are trying to determine the correction to the energy of eqn (3.65) as a result of putting the electron interaction back into the problem. This correction is determined by using the ground state wave function of the *unperturbed* electron gas (this is the normal rule of first-order perturbation theory), which we already know from the previous section is a product of plane waves that have been appropriately antisymmetrized to respect the Pauli principle. In particular, we must evaluate

$$E_{ee} = \langle \psi_{tot} | H' | \psi_{tot} \rangle, \qquad (3.67)$$

where $\psi_{tot}(\mathbf{r}_1, \ldots, \mathbf{r}_N)$ is of the form

$$\psi_{tot}(\mathbf{r}_1, \ldots, \mathbf{r}_N) = \frac{1}{\sqrt{N!}} \begin{vmatrix} \frac{1}{\sqrt{\Omega}} e^{i\mathbf{k}_1 \cdot \mathbf{r}_1} & \frac{1}{\sqrt{\Omega}} e^{i\mathbf{k}_1 \cdot \mathbf{r}_2} & \cdots & \frac{1}{\sqrt{\Omega}} e^{i\mathbf{k}_1 \cdot \mathbf{r}_N} \\ \frac{1}{\sqrt{\Omega}} e^{i\mathbf{k}_2 \cdot \mathbf{r}_1} & \frac{1}{\sqrt{\Omega}} e^{i\mathbf{k}_2 \cdot \mathbf{r}_2} & \cdots & \frac{1}{\sqrt{\Omega}} e^{i\mathbf{k}_2 \cdot \mathbf{r}_N} \\ \vdots & \vdots & \vdots & \vdots \\ \frac{1}{\sqrt{\Omega}} e^{i\mathbf{k}_N \cdot \mathbf{r}_1} & \frac{1}{\sqrt{\Omega}} e^{i\mathbf{k}_N \cdot \mathbf{r}_2} & \cdots & \frac{1}{\sqrt{\Omega}} e^{i\mathbf{k}_N \cdot \mathbf{r}_N} \end{vmatrix}, \qquad (3.68)$$

a construction known as the Slater determinant that instructs us how to build up a properly antisymmetrized wave function as a product of one-electron wave functions. Note that we have been slightly sloppy by omitting the added dependence on the spin coordinates in the wave function. For full details, see chap. 3 of Pines (1963) or chap. 1 of Fetter and Walecka (1971). As is evident from the form of the integrals that will result from evaluating the expectation value demanded in eqn (3.67), the outcome of this analysis depends upon the Fourier transform of the Coulomb potential itself. Careful evaluation of the integrals demanded by eqn (3.67) results in a contribution to the total energy of the electron gas of the form

$$E_{interact} = -\frac{3}{2\pi}\frac{\hbar^2}{2ma_0}(3\pi^2\rho)^{\frac{1}{3}},$$ (3.69)

where a_0 is the Bohr radius and is given by $a_0 = \hbar^2/me^2$. Note that the sign of this contribution to the energy is *negative*, resulting in a competition between the two terms we have found and a consequent emergence of an optimal density.

The result of adding together the two contributions to the energy of the electron gas is

$$E_{egas} = \frac{3}{5}\frac{\hbar^2}{2m}(3\pi^2\rho)^{\frac{2}{3}} - \frac{3}{2\pi}\frac{\hbar^2}{2ma_0}(3\pi^2\rho)^{\frac{1}{3}}.$$ (3.70)

This energy has a minimum at a certain optimal density which may be found by evaluating $\partial E_{egas}/\partial\rho = 0$, resulting in

$$\rho_{opt} = \frac{125}{192\pi^5 a_0^3}.$$ (3.71)

This density may be converted in turn to an estimate for the lattice parameter by assuming that the atom at each lattice site of the bcc lattice has donated a single electron. In particular, the lattice parameter implied by these arguments is

$$a_{eg} = \left(\frac{384\pi^5}{125}\right)^{\frac{1}{3}}a_0 \approx 5.2\ \text{Å}.$$ (3.72)

This number should be contrasted with the experimental lattice constant of a typical free electron metal such as sodium, for which the lattice parameter is 4.23 Å. In addition to the lattice parameter, this model may also be used to estimate the bulk modulus of free electron metals by resorting to the definition $B = V(\partial^2 E/\partial V^2)$. This is left as an exercise for the reader.

In the present context, this example was intended to serve as a reminder of how one formulates a simple model for the quantum mechanics of electrons in metals and, also, how the Pauli principle leads to an explicit algorithm for the filling up of these energy levels in the case of multielectron systems. In addition, we have seen how this model allows for the explicit determination (in a model sense) of the cohesive energy and bulk modulus of metals.

3.2.6 Quantum Mechanics of Bonding

As we hinted at in the section on the hydrogen atom, when we turn our attention to the origins of bonding in molecules and solids, a useful alternative to the

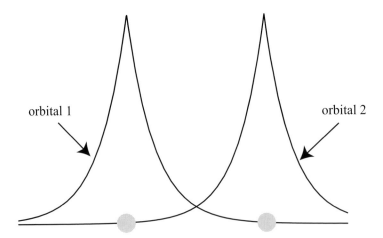

Fig. 3.10. Disposition of wave functions of neighboring atoms during the process of bond formation.

electron gas physics described above is to build our picture around the simple ideas suggested by the H_2 molecule. As shown in fig. 3.10, the hydrogenic wave functions on the atoms of interest begin to overlap once these atoms are sufficiently close. Conceptually, we see that once this happens, the nuclear potential on one of the atoms begins to amend the wave function tied to the other. In this section, our task is to compute this interaction and to demonstrate how it makes possible the reduction in energy more commonly known as bonding. We begin with a discussion of bonding in the abstract, and then turn to an explicit treatment of the H_2 molecule itself, and show how what is learned there can be generalized to systems of increasing size. The ideas developed here provide the conceptual foundations for our development of the tight-binding method in chap. 4.

The intuition leading to this approach is best coached by an analogy. If we are to consider a system of masses and springs such as is shown in fig. 3.11, our various oscillators will vibrate independently, each at its own natural frequency. If we now imagine connecting the oscillators, as also shown in the figure, it is immediate that the vibrational frequencies are altered by the interaction. In particular, if we consider the simple case where the masses and spring constants are all equal, we find that the frequencies are split in the manner indicated in the figure.

The quantum mechanical analog of the spring problem, also illustrated schematically in fig. 3.11, is the idea that when two atoms are sufficiently far apart, their respective atomic energy levels at energy E_a are undisturbed, and the atoms should be thought of as independent. On the other hand, when the atoms are in sufficiently close proximity, the potential due to one nucleus begins to influence the electron centered on the other. This interaction has the effect of splitting the original energy levels as indicated in the figure. For a two-atom molecule in which each atom

noninteracting case

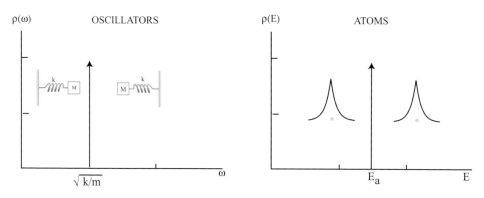

interacting case

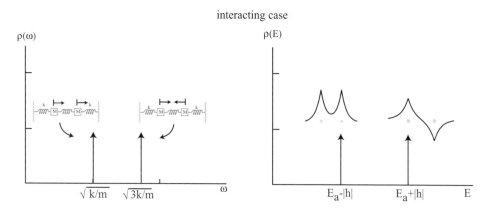

Fig. 3.11. Analogy between problem of coupled oscillators and that of bonding in molecules.

donates one electron, the difference in energy between the original isolated atoms and the molecule is $E_{bond} = -2|h|$, the bonding energy.

From a physical perspective, our strategy in thinking about bonding has certain key similarities to our treatment of the free electron gas. In particular, we assume that the properties of the molecule (or solid) can be obtained by solving a one-electron Schrödinger equation and then filling up energy levels in accordance with the Pauli principle. This means that we completely ignore the interaction between the electrons themselves. Hence, for the two-atom molecule of interest here, the Schrödinger equation may be written

$$-\frac{\hbar^2}{2m}\nabla^2\psi + V_1(\mathbf{r})\psi + V_2(\mathbf{r})\psi = E\psi. \tag{3.73}$$

In this equation, the *total* potential experienced by the electron is obtained by summing up the separate *nuclear* potentials emanating from the two nuclei, V_1 and V_2.

Rather than trying to find those functions which satisfy the Schrödinger equation written above, we adopt a simpler variational approach in which the search for unknown functions is replaced by the search for unknown coefficients. The basis for this strategy is a variational argument. We represent the wave function for the molecule as a linear combination of the atomic orbitals centered on the two nuclei, that is,

$$\psi = a_1\phi_1 + a_2\phi_2, \tag{3.74}$$

where ϕ_1 and ϕ_2 are the atomic orbitals on sites 1 and 2, and the a_is are the coefficients we aim to determine. We have adopted a minimalist approach in which rather than using a basis consisting of all of the atomic orbitals associated with each site, we just borrow a single 1s orbital (i.e. $|1, 1s\rangle$ and $|2, 1s\rangle$, where the first index labels the site and the second labels the orbital type) from each atom and assume that the molecular wave function can, to a first approximation, be represented as a linear combination of these functions. A more sophisticated treatment could include more basis functions such as the 2s and 2p states, and so on. The basis of the determination of the unknown coefficients is the recognition that the expectation value of the energy can be written as

$$\langle E \rangle = \langle \psi | \hat{H} | \psi \rangle, \tag{3.75}$$

and our task is to find those a_is that minimize this expression. In fact, this is exactly the same argument we adopted in our description of the finite element approach to the Schrödinger equation, with the difference being the choice of basis functions.

As a result of the arguments given above, we write the function to be minimized as

$$\Pi(a_1, a_2) = \langle a_1\phi_1 + a_2\phi_2 | \hat{H} | a_1\phi_1 + a_2\phi_2 \rangle - E \langle a_1\phi_1 + a_2\phi_2 | a_1\phi_1 + a_2\phi_2 \rangle, \tag{3.76}$$

in perfect analogy with eqn (3.36). If we write this out explicitly term by term, the result is

$$\begin{aligned} \Pi(a_1, a_2) = {}& a_1^* a_1 \langle \phi_1 | \hat{H} | \phi_1 \rangle + a_1^* a_2 \langle \phi_1 | \hat{H} | \phi_2 \rangle + a_2^* a_1 \langle \phi_2 | \hat{H} | \phi_1 \rangle \\ & + a_2^* a_2 \langle \phi_2 | \hat{H} | \phi_2 \rangle - E(a_1^* a_1 \langle \phi_1 | \phi_1 \rangle + a_1^* a_2 \langle \phi_1 | \phi_2 \rangle \\ & + a_2^* a_1 \langle \phi_2 | \phi_1 \rangle + a_2^* a_2 \langle \phi_2 | \phi_2 \rangle), \end{aligned} \tag{3.77}$$

an equation which can be further simplified if we adopt the notation $E_1 = \langle \phi_1 | \hat{H} | \phi_1 \rangle$, $E_2 = \langle \phi_2 | \hat{H} | \phi_2 \rangle$, and $h = \langle \phi_1 | \hat{H} | \phi_2 \rangle = \langle \phi_2 | \hat{H} | \phi_1 \rangle$. Applying these definitions to the result given above, and exploiting the assumption of orthonormality (i.e. $\langle \phi_i | \phi_j \rangle = \delta_{ij}$), which is, strictly speaking, not true, we now

Energy

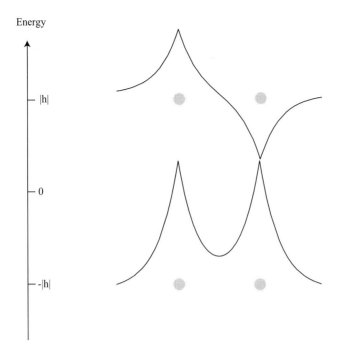

Fig. 3.12. Solutions for H$_2$ molecule using the two-orbital basis.

find

$$\Pi(a_1, a_2) = a_1^* a_1 E_1 + a_1^* a_2 h + a_2^* a_1 h + a_2^* a_2 E_2 - E(a_1^* a_1 + a_2^* a_2). \quad (3.78)$$

For simplicity, we now assume that the two atoms are identical (i.e. $E_1 = E_2$) and further, choose our zero of energy such that $E_1 = 0$.

If we now minimize this function with respect to the a_is, in particular by enforcing $\partial \Pi / \partial a_i^* = 0$ for $i = 1, 2$, we can find the optimal choice for these coefficients. Effecting the minimization suggested above results in two coupled algebraic equation that are more conveniently written in matrix form as

$$\begin{pmatrix} 0 & -|h| \\ -|h| & 0 \end{pmatrix} \begin{pmatrix} a_1 \\ a_2 \end{pmatrix} = E \begin{pmatrix} a_1 \\ a_2 \end{pmatrix}. \quad (3.79)$$

Note that we have used our knowledge of the sign of h to rewrite it as $h = -|h|$, a result of the fact that the potential energy of interaction represented by this parameter is negative. The result of solving these equations is two solutions, one with energy $E = -|h|$ and wave function $\psi_1 = (1/\sqrt{2})(\phi_1 + \phi_2)$ and the other with energy $E = |h|$ and wave function $\psi_1 = (1/\sqrt{2})(\phi_1 - \phi_2)$. These two solutions are shown in fig. 3.12.

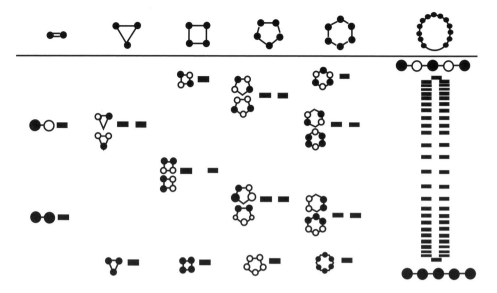

Fig. 3.13. Series of molecules of increasing size and associated molecular eigenstates. The wave functions are indicated schematically with shaded atoms corresponding to one sign of relevant atomic orbital and unshaded atoms corresponding to opposite sign for relevant orbital (adapted from Hoffmann (1988)). Note that the spacing of energy levels is schematic and there is no relation indicated between the energy scales for different size molecules.

Recall that the question we originally posed was that of the origin of bonding. As a result of the solution worked out above, we are now in a position to address this question. With the zero of energy chosen above, the energy of the two isolated atoms is $E_{atoms} = 0$. For the molecule, we fill energy levels according to Pauli's principle, which for the two-electron problem of interest here means one spin-up electron and one spin-down electron are both assigned energy $-|h|$ with the result that the energy difference between the molecule and the free atoms is $E_{bond} = -2|h|$. The energy is lowered by virtue of the partnership struck up between the two free atoms: a bond has formed. This idea is illustrated graphically in fig. 3.12 where it is seen that there is an enhancement in the probability distribution in the region between the two atoms, a signature of the formation of bond charge.

The generalization of this problem to the case of increasingly larger collections of atoms is straightforward and is founded upon the construction of larger matrices of the variety described above. For example, if we consider the problem of three atoms which occupy the vertices of an equilateral triangle, a 3×3 matrix emerges with three resulting eigenvalues. The distribution of energy eigenvalues for increasing system sizes is indicated schematically in fig. 3.13. This figure demonstrates the way in which the energy can be lowered as a result of atoms conspiring to form bonds. In chap. 4, we will undertake a systematic analysis of bonding in solids, using what we have learned of quantum mechanics in the preceding pages.

In this part of the chapter, we have seen that quantum mechanics provides the microscopic foundations for evaluating the disposition of the electronic degrees of freedom. Just as continuum mechanics provides the setting within which one solves the boundary value problems related to the deformation of solids, quantum mechanics provides a scheme for attacking the boundary value problems associated with electrons in solids. Of course, quantum mechanics has a much wider scope than we have hinted at here, but our aim has been to couch it in terms of the subject at hand, namely, the modeling of materials. In this case, it is the role of quantum mechanics in instructing us as to the nature of bonding in solids that is of most importance, and as will be seen in coming chapters, these ideas will be an indispensable part of our modeling efforts.

3.3 Statistical Mechanics

3.3.1 Background

Thermodynamics provides the conceptual cornerstone upon which one builds the analysis of systems in equilibrium. The elemental theoretical statement that can be made about such systems is embodied in Gibbs' variational principle. This asserts that a certain function of state, namely the entropy, is a maximum in equilibrium so long as the system is subject to the additional constraint of constant total energy. The usual intuitive notions one attaches to the analysis of systems in contact with each other such as equality of temperature, pressure and chemical potential emerge from this principle quite naturally. For example, if we consider the box shown in fig. 3.14 with an internal partition that is rigid and impermeable, the variational principle noted above can be used to determine the privileged terminal state (i.e. the state of equilibrium) that results from removing the constraints offered by the wall. The argument is that for the overall closed system, the total energy is given by $E_{tot} = E_1 + E_2 = $ const., implying that $dE_1 = -dE_2$. On the assumption that the entropy S is additive, we note that

$$dS_{tot} = \left(\frac{\partial S_1}{\partial E_1}\right)_{V_1, N_1} dE_1 + \left(\frac{\partial S_2}{\partial E_2}\right)_{V_2, N_2} dE_2. \tag{3.80}$$

These derivatives are evaluated at fixed volumes V_1 and V_2 as well as fixed particle numbers N_1 and N_2. If we now replace dE_2 with $-dE_1$, we see that the condition of maximum entropy is equivalent to the statement that

$$\left(\frac{\partial S_1}{\partial E_1}\right) = \left(\frac{\partial S_2}{\partial E_2}\right), \tag{3.81}$$

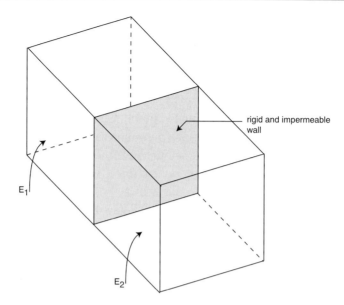

Fig. 3.14. Partitioning of the system under consideration into two boxes which may freely exchange energy.

which because of the definition of temperature as $1/T = (\partial S/\partial E)|_{V,N}$, we recognize as the condition of equality of temperature between the two subsystems. If we consider in turn the case in which the wall is permeable ($dN_1 = -dN_2$) or may move ($dV_1 = -dV_2$), the conditions of equal chemical potential and pressure result.

As it stands, the type of thermodynamic assertions fleshed out above are indifferent to microscopic considerations, though it is certainly of great interest to unearth their microscopic origins. In this part of the chapter our aim is to investigate statistical mechanics. A key insight that emerges from these ideas is the probabilistic nature of equilibrium states. We take the view that statistical mechanics is the microscopic partner of equilibrium thermodynamics. While equilibrium thermodynamics does not concern itself with the mechanistic underpinnings of observed behavior, statistical mechanics teaches how one may deduce macroscopic features of a system as the result of a postulated microscopic model. Such models are widespread, doing service in problems such as the calculation of electric and magnetic susceptibilities, the analysis of the equilibrium concentrations of point defects in solids and the determination of the vibrational entropy of solids and its connection to phase transformations.

Our treatment begins on a philosophical note with an attempt to elucidate the business of statistical mechanics. Next, we will turn to an analysis of the formal apparatus of statistical mechanics with a view of the partition function and how it may be used in the evaluation of quantities of thermodynamic interest. The harmonic oscillator will loom large in our discussion as a result of the contention that this model may serve as a basis for the analysis of material properties, such as the specific heat, the vibrational entropy and the microscopic origins of the diffusion constant. This discussion will then be couched in the alternative language of information theory. These arguments will identify the role of inference in statistical mechanics, and will make more evident how we treat systems in the face of incomplete knowledge of that system's microscopic state. This will be followed by a computational view of statistical mechanics. Here the task will be to see how the use of computing lends itself to a numerical statement of the program of statistical mechanics, with a brief overview of molecular dynamics and the Monte Carlo method.

Our starting point is an assessment of the meaning of a macroscopic observable state. Thermodynamic states are parameterized by a small number of relevant macroscopic variables such as the energy, E, the volume, V, the pressure, p, and the various mole numbers, n_α. As Callen (1985) has eloquently noted in the introductory sections of his book, it is well nigh miraculous that despite the ostensible need of 10^{23} degrees of freedom to fully characterize the state of a system, all but a few of them 'are macroscopically irrelevant.' Indeed, a key outcome of the statistical mechanical approach is an insight into how macroscopic parameters arise as averages over the microscopic degrees of freedom. Our fundamental contention is that a given macroscopic state is consistent with an enormous number of associated microscopic states. This observation lends itself to a nomenclature which distinguishes between macrostates and microstates. For the moment, we do not distinguish between the classical and quantum versions of these microstates, though we will see later that in some cases, such as that of the specific heat of solids, there are dramatic differences in the outcome of classical and quantum analyses. Classical physics envisions a continuum of microscopic states parameterized by the conjugate variables $\mathbf{q}_i$ and $\mathbf{p}_i$, where these are the coordinates and momenta of the ith particle, respectively. For a system composed of N particles, each microstate corresponds to a particular point in a $6N$-dimensional space referred to as 'phase space', with the dynamical variables varying continuously as the system winds about on its trajectory through this space. A given point within the classical phase space corresponds to a particular realization of the coordinates $(\mathbf{q}_1, \mathbf{p}_1, \mathbf{q}_2, \mathbf{p}_2, \ldots, \mathbf{q}_N, \mathbf{p}_N) = (\{\mathbf{q}_i, \mathbf{p}_i\})$. On the other hand, within quantum mechanics, these states are discretely reckoned and each microstate is parameterized by distinct quantum numbers.

The notion of a microstate is perhaps most easily understood by virtue of an example. We imagine a closed box containing an ideal gas. Our use of the term 'closed' is meant to imply that in this case the volume, the number of particles and

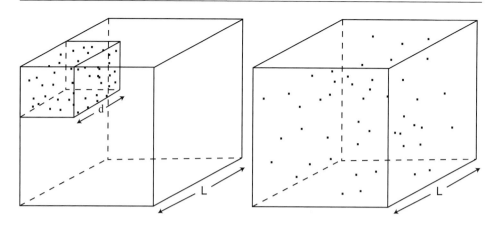

Fig. 3.15. Comparison of two possible microstates for particles in a box.

the energy of the system are all held constant and assume the values V, N and E, respectively. Within either the classical or quantum settings we may immediately identify the nature of the microstates available to the gas. The specification of a particular point in phase space implies knowledge of a geometrical character. In particular, the specification of the $3N$ coordinates $\mathbf{q}_i$ identifies the locations of all of the particles while the $3N$ momenta $\mathbf{p}_i$ characterize their motions. The momenta of the molecules of the gas are imagined to be distributed in some way that is consistent with the assumed total energy, E. In addition, there is a configurational part of the microstate. One possible geometric configuration which corresponds to an allowable microstate of the gas would have all of the molecules bunched in one corner of the box. Intuitively, such a state seems unlikely. On the other hand, there are many more microstates in which the molecules are distributed evenly throughout the box. These two alternatives are contrasted in fig. 3.15.

We can estimate the probability of the state in which the particles are bunched in one corner as $(V_c/V_{tot})^N$, where V_c is the volume of the region where the particles are assumed to be bunched. In keeping with the assumption of noninteracting particles that is the essence of the ideal gas, we have assumed that the probability that a given particle will be in any subregion is given by the ratio of the volume of that subregion to the whole volume. The key feature of interest for the calculation given above is that all such microstates (including the unlikely state in which all the molecules are localized in some small region) are consistent with the overall macrostate characterized by the observables V, N and E.

The postulate that rests behind the explicit calculation of macroscopic observables is that such observables are averages that emerge as a result of the system ranging over the allowed microstates. For our purposes, the central assumption upon which the rest of our development is fixed is that all such microstates consistent with the overall macrostate are equally likely. This statement is sometimes referred to as the principle of equal *a priori* probabilities or the *principle of insufficient reason*. At first sight, it may seem odd that this statement has calculable consequences. But as will be shown below, this postulate provides a logical basis for statistical mechanics and is used to derive the analytic machinery which makes the theory so powerful.

The quantitative basis for the linkage between macro- and microstates is provided by the prescription for computing the entropy $S(E, V, N)$ in terms of the number of microstates, namely,

$$S(E, V, N) = k_B \ln \Omega (E, V, N), \qquad (3.82)$$

where $\Omega (E, V, N)$ is the number of microstates available to the system when it has energy, E, occupies a volume, V, and consists of N particles. This equation is a key example of the type of link between the microscopic and macroscopic perspectives that is our aim. In this case, it is the entropy, S, a property of the macrostate that is related to $\Omega (E, V, N)$, the associated microscopic quantity which characterizes the number of ways of realizing the macroscopic state of interest. The constant that links these two quantities is Boltzmann's constant and will hereafter be referred to simply as k. The immense significance of this equation is tied to the fact that it provides a mechanistic interpretation of the notion of entropy, while at the same time rooting the notion of equilibrium in probabilistic ideas. In particular, the assertion that emerges from this equation in conjunction with Gibbs' variational statement, is that the equilibrium state (the state of maximum entropy) is really the macrostate corresponding to the largest number of microscopic realizations. As a result, we see that the primary business of statistical mechanics has been reduced to that of counting the number of microstates accessible to the system once the energy, volume and particle number have been fixed.

3.3.2 Entropy of Mixing

In this section, we take up the view of the statistical origins of entropy and equilibrium that was introduced in the previous section. We advance the view that equilibrium can no longer be viewed as a matter of certainty, but rather only as a matter of enormous likelihood. An example which is at once simple, enlightening and useful is related to the entropy of mixing, where it is imagined that two constituents are randomly distributed among N sites subject to the constraint that $N_A + N_B = N$, as depicted in fig. 3.16. The analysis of the entropy of mixing serves as a viable starting point for the consideration of problems such as the configurational entropy of a two-component alloy, the distribution of vacancies within a single crystal at equilibrium and the arrangements of magnetic spins.

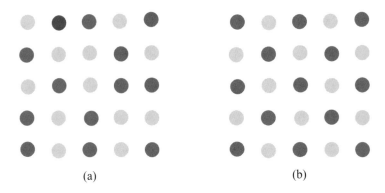

(a)   (b)

Fig. 3.16. Two-dimensional schematic representation of two distinct alloy configurations where different atom types are represented by different colors: (a) disordered state, (b) ordered state.

We note that if we have a total of N sites to occupy, there are $N!$ different ways of occupying these sites on the assumption that each configuration is distinct. However, in our case this logic must be amended on the basis of the realization that there are N_A equivalent A objects and N_B equivalent B objects. This implies that in reality there are

$$\Omega = \frac{N!}{N_A! N_B!} \tag{3.83}$$

distinct configurations. The denominator in this expression eliminates the redundancy which results from the fact that the A objects are indistinguishable among themselves as are the B objects. Following Boltzmann's equation linking the entropy and the number of microstates, the entropy that follows from this enumeration is

$$S = k \ln \frac{N!}{N_A! N_B!}. \tag{3.84}$$

As we can see in this instance, the calculation of the entropy of mixing has been reduced to the combinatorial problem of identifying the number of distinct microstates consistent with our overall macrostate which is characterized by a fixed number of A and B objects.

Further insights into this configurational entropy can be gleaned from an attempt to reckon the logarithms in an approximate fashion. We note that our expression for the entropy may be rewritten as

$$S = k(\ln N! - \ln N_A! - \ln N_B!). \tag{3.85}$$

Following Reif (1965), these sums may be written explicitly as

$$\ln N! = \ln 1 + \ln 2 + \cdots + \ln N. \tag{3.86}$$

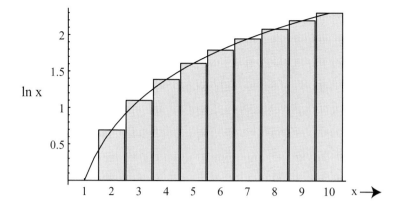

Fig. 3.17. Approximations associated with Stirling approximation.

This sum may be represented graphically as shown in fig. 3.17, which suggests the approximate expression

$$\ln N! \approx \int_1^N \ln x \, dx = N \ln N - N. \tag{3.87}$$

Here it must be noted that we have made an imprecise description of what under more formal terms is known as the Stirling approximation. Within the context of our characterization of the configurational entropy, the Stirling approximation of $\ln N!$ allows us to rewrite the entropy as

$$S = k[(N_A + N_B)\ln(N_A + N_B) - N_A \ln N_A - N_B \ln N_B]. \tag{3.88}$$

If we now define the variable $c = N_A/N$ and rearrange terms it is found that

$$S = -kN[(1 - c)\ln(1 - c) + c \ln c]. \tag{3.89}$$

This entropy is shown in fig. 3.18 as a function of the concentration variable. In chap. 7, it will be shown that this result is one piece of the free energy that reflects the competition between the energy cost to increase the number of vacancies in a solid and the free energy lowering increase in entropy that attends an increase in the number of such vacancies.

The arguments given above were built around the idea of an isolated system. However, there are both mathematical and physical reasons to favor a treatment in which we imagine our system as being in contact with a reservoir at fixed temperature. For example, our later discussion of annealing is a realization of this claim. A particular sample of interest (a chunk of metal, for example), is subjected to external heating with the aim of examining the change in its microstructure. In this case, the model of an isolated system is inappropriate. Rather, we should model

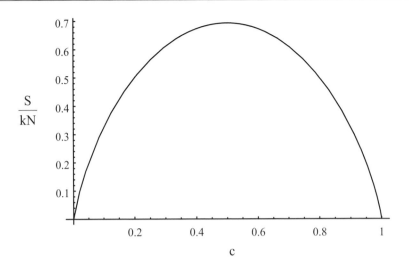

Fig. 3.18. Entropy of mixing associated with binary configurations.

the system as being in contact with a reservoir with which it exchanges energy. This scenario leads to the canonical distribution which will form the substance of the next section.

3.3.3 The Canonical Distribution

For many problems, consideration of the isolated system envisioned above is inadequate. In particular, it is often of interest to consider a subsystem in contact with a reservoir. Such reservoirs can either be a source of energy or particles, and one imagines free communication between the subsystem of interest and its associated reservoir. The conceptual trick introduced to allow for rigorous treatment of this problem, without advancing assumptions beyond those made already, is to treat the entire universe of subsystem plus reservoir as a closed system unto itself. Within this setting, we once again carry out the same sort of counting arguments introduced in the previous section for an isolated system. However, unlike the previous case, we will derive expressions for the underlying thermodynamics which can be couched purely in terms of the degrees of freedom of only the subsystem.

As a first example, we consider the problem in which the subsystem and reservoir are free to exchange only energy, as is indicated schematically in fig. 3.19. In this case, the number of microstates available to the subsystem is $\Omega_s(E_s)$, where the subscript s refers to the subsystem. Similarly, the number of microstates available to the reservoir is given by $\Omega_r(E - E_s)$. From the perspective of the composite universe made up of the system and its reservoir, the total energy is

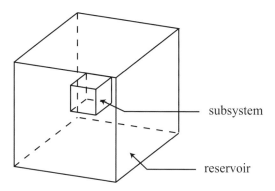

Fig. 3.19. Composite system composed of a subsystem and a reservoir with which it is free to exchange energy.

constant, and the number of microstates available to that composite system when the subsystem has energy E_s is given by $\Omega_{composite}(E; E_s) = \Omega_s(E_s)\Omega_r(E - E_s)$. Here we have noted that the total number of states available to the composite system at a given E_s is the product of the number of states available to the reservoir and those available to the subsystem. Note that the total number of microstates available to the composite system is gotten as $\Omega_{tot}(E) = \sum_s \Omega_{composite}(E; E_s)$. Equilibrium between the reservoir and the subsystem is attained when the entropy is maximized, which corresponds to finding that distribution of energy between the subsystem and reservoir which corresponds to the largest number of total microstates.

Within the microcanonical formalism presented in the previous section we asserted that all microstates are equally likely. In the present situation, we center our interest just on the subsystem. The price we pay for restricting our interest to the nonisolated subsystem is that we no longer find all states are of equal probability. Though the states of the composite universe are all equally likely, we will find that there is a calculable energy dependence to the probability for a given energy state for the subsystem.

To make our analysis concrete, we begin by noting that if we consider the circumstance in which the subsystem is in a nondegenerate state of energy, E_a (i.e. $\Omega_s(E_a) = 1$), then the total number of states available to the composite universe of system plus reservoir is $\Omega_{composite}(E) = \Omega_r(E - E_a)$. What we have done here is to identify the subset of states of the composite system that is consonant with the subsystem having definite energy E_a. We continue with the observation that the probability $P(E_a)$ of finding the system in this state is proportional to the number of microstates available to the reservoir. That is, $P(E_a) \propto \Omega_r(E - E_a)$. The specific dependence of this probability on the energy E_a can be garnered from the following argument. First, we note that as a consequence of the assertion made above we have

$$\frac{P(E_a)}{P(E_b)} = \frac{\Omega_r(E - E_a)}{\Omega_r(E - E_b)} = \frac{e^{\ln[\Omega_r(E-E_a)]}}{e^{\ln[\Omega_r(E-E_b)]}}. \tag{3.90}$$

This equation notes that the ratio of the probabilities of finding the subsystem with two particular energies is equal to the ratios of the corresponding number of states available to the reservoir at the two energies of interest. The remainder of our analysis can be built around the presumed smallness of E_a with respect to E itself, which affords an opportunity for Taylor expansion of the slowly varying logarithms in the form

$$\ln \Omega_r(E - E_a) \simeq \ln \Omega_r(E) - \left(\frac{\partial \ln \Omega_r}{\partial E}\right) E_a. \tag{3.91}$$

Coupled with the thermodynamic identity that relates the temperature to the energy dependence of the entropy, namely,

$$\left(\frac{\partial \ln \Omega}{\partial E}\right) = \frac{1}{k}\left(\frac{\partial S}{\partial E}\right) = \frac{1}{kT}, \tag{3.92}$$

we are left to conclude that

$$P(E_a) = \frac{1}{Z} e^{-(E_a/kT)}, \tag{3.93}$$

where Z is an as-yet undetermined constant. In fact, for the practical purposes of everyday calculation, it is the 'constant' Z that depends on parameters such as T and V which is the basis of many calculations. This quantity, known as the partition function, is found by insisting that our probability distribution be suitably normalized, namely,

$$\sum_n P(E_n) = \frac{1}{Z}\sum_n e^{-(E_n/kT)} = 1. \tag{3.94}$$

This expression states that the partition function itself can be determined from knowledge of the energy levels via

$$Z = \sum_n e^{-(E_n/kT)}. \tag{3.95}$$

Our notation is meant to argue that the subsystem can have any of a number of different energies, labeled by the set of integers n. We would do well at this point to reflect on the preceding argument. Throughout, our derivation has rested on the central assertion upon which the microcanonical distribution is built, namely, that of the equal likelihood of all microstates consistent with the given macrostate. In this case, however, we have seen fit to examine a subsystem in contact with a thermally rigid reservoir which can freely donate and receive energy without altering its temperature. By examining, within a microcanonical setting, the properties of the composite universe of subsystem plus reservoir, we have been able to identify the unequal probabilities attached to the subsystem's energy levels. What has emerged is a scheme in which only the parameters characterizing the subsystem need be considered.

Having established the probability distribution that characterizes our subsystem when in contact with a reservoir, we now seek insights into the thermodynamic implications of this solution. As a first example, suppose we are interested in determining the average energy of the subsystem. On the basis of the familiar notion of determining averages associated with a given probability distribution, it is immediate that

$$\langle E \rangle = \sum_n E_n P(E_n) = \frac{1}{Z} \sum_n E_n e^{-(E_n/kT)}. \tag{3.96}$$

This expression may be considerably simplified on the recognition that we may define a variable $\beta = 1/kT$, and that

$$\sum_n E_n e^{-\beta E_n} = -\frac{1}{Z} \frac{\partial Z}{\partial \beta}. \tag{3.97}$$

Hence, we have that

$$\langle E \rangle = -\frac{\partial \ln Z}{\partial \beta}. \tag{3.98}$$

This result supports our contention that it is the partition function that serves as the analytic engine for calculating quantities of thermodynamic interest.

The specific heat, which is a measure of the stiffness of the system to thermal fluctuations, will emerge later as an additional material parameter which looks to microscopic models for its origins. Within the statistical mechanical formalism we have set up here, the specific heat may be easily determined on the recognition that $C_V = \partial \langle E \rangle / \partial T |_V$. Upon exploiting the definition of β, this may be rewritten as

$$C_V = k\beta^2 \frac{\partial^2 \ln Z}{\partial \beta^2}. \tag{3.99}$$

An alternative and more streamlined approach to the types of results described above emerges once we have made contact between the partition function and the Helmholtz free energy. Recall that thermodynamic quantities such as the pressure and entropy may be evaluated simply as derivatives of the Helmholtz free energy as

$$p = -\left(\frac{\partial F}{\partial V}\right)_T, \quad S = -\left(\frac{\partial F}{\partial T}\right)_V. \tag{3.100}$$

The connection of the partition function to the Helmholtz free energy is effected via the expression

$$F = -kT \ln Z. \tag{3.101}$$

This result is yet another example of an equation (like the Boltzmann expression for the entropy) which connects macroscopic and microscopic quantities. The origins of this equation can be obtained on the basis of thermodynamic reasoning and to probe these details the reader is encouraged to examine chap. 16 of Callen (1985) or the first sections of Feynman (1972). In the following section we will examine how this machinery may be brought to bear on some model problems.

3.3.4 Information Theoretic Approach to Statistical Mechanics

Until now, our treatment of statistical mechanics has been founded upon the assumption that this theory reflects the average treatment of the coupled microscopic degrees of freedom that make up a material. However, in this section we follow an alternative view in which it is held that statistical mechanics is not a physical theory at all, but is rather an example of a calculus of uncertainty in which inference in the face of incomplete information is put on a rigorous footing. Indeed, the Boltzmann factor introduced above has been used in problems ranging from image reconstruction to the analysis of time series data lending support to the claim that the fundamental ideas transcend their historical origins in physics.

The central idea is that the algorithm of statistical mechanics is but one example of how reasoning should proceed when we are confronted with incomplete knowledge. The ideas associated with the information theoretic approach to statistical mechanics may be sketched as follows. We begin with the observation that our task is to infer the probability of the various microstates that the system is capable of adopting. From the macroscopic perspective, all that we may know about a system might be its mean energy, $\langle E \rangle$. Given so little information about the distribution of microstates it seems brash to assume that anything definitive may be argued for. However, such problems form the province of probabilistic reasoning, and in conjunction with the ideas of information theory, much progress can be made. We follow Shannon in attaching a measure to the quality of our knowledge of a system's disposition as revealed by the probability distribution that describes it. For an overview of these ideas, we refer the reader to Jaynes (1983). For example, if we consider the rolling of a die, the best we might do is to enumerate the six possible outcomes and to attribute to each an equal likelihood. This is low-quality information. On the other hand, if the die were loaded and we could say with certainty that only the face with a one would be revealed, our state of knowledge is of the highest quality.

Associated with a discrete probability distribution with N outcomes Shannon identifies an entropy, $H(p_1, p_2, \ldots, p_N)$, which lends quantitative weight to our intuitive notions about the quality of information associated with that distribution. The derivation of the particular functional form taken by $H(\{p_i\})$ is based on a few fundamental *desiderata*. First, given N possible outcomes, the case in which all such outcomes are equally likely will maximize the function H. This statement corresponds to our intuition that the case of equal likelihood of all outcomes corresponds to maximal ignorance. The probability distribution is featureless and there is nothing about it that inclines us to favor one outcome over any other. Secondly, if we now imagine increasing the number n of equally likely outcomes, the function $H(\{1/n\})$ should be a monotonically increasing function of n. This too jibes with our intuition that as the number of outcomes is increased, our knowledge is degraded. The final requirement of the function H is based upon its additivity. If we specify the state of the system in two steps (i.e. there are n rooms, each of which contains m boxes), the reckoning of the information entropy must be consistent with the result of specifying all of the information at once (i.e. there are $n \times m$ outcomes). On the basis of these few basic requirements, the claim is that the function $H(p_1, p_2, \ldots, p_N)$ is uniquely determined and that it provides a measure of the information content of the probability distribution it characterizes.

The outcome of the reasoning hinted at above is the claim that the function H is given by

$$H(p_1, p_2, \ldots, p_N) = -k \sum_i p_i \ln p_i. \qquad (3.102)$$

That is, if we are given a discrete probability distribution $\{p_i\}$, the function $H(\{p_i\})$ provides a unique measure of the information content of that distribution. The beauty of this result is that it provides a foundation upon which to do inferential reasoning. In particular, when faced with the question of determining a probability distribution (such as, for example, in the context of constructing a statistical ensemble which gives the probabilities of various microstates), it gives us a way to determine the *best* distribution. As yet, we have said nothing of statistical mechanics, and it is just this fact that makes the information theoretic approach to inference of utility to X-ray crystallographers, radio astronomers attempting to decipher noisy data and statistical mechanicians alike. In light of the developments described above, the Gibbsian variational principle may be seen in a different light. The idea is that the maximization of entropy may be alternatively viewed as a statement of unbiased inference: choose that probability distribution that assumes nothing that is not forced on us by the data themselves.

To see the workings of the maximum entropy approach, we begin with the trivial example of N outcomes with our only knowledge being the enumeration of the

states themselves. In this case, our intuition anticipates the probability distribution
in which all outcomes are equally likely. To see how this arises from the principle
of maximum entropy, we maximize the augmented function

$$S' = -\sum_i p_i \ln p_i + \lambda \left(\sum_i p_i - 1 \right), \tag{3.103}$$

where λ is a Lagrange multiplier used to enforce the constraint that the distribution
be normalized. Also, we have absorbed the constant k into the Lagrange multiplier.
Maximizing this function with respect to the p_is results in $p_i = e^{\lambda-1}$, independent
of i, revealing that all outcomes are equally likely. The determination of the
constant is based on the recognition that $\sum_i p_i = 1$, and yields $p_i = e^{\lambda-1} = 1/N$.
Note that we have provided theoretical justification for the principle of equal *a
priori* probabilities that was adopted earlier. We see that this is the best guess that
can be made on our very limited data. The key point is that rather than asserting
equal *a priori* probabilities as an article of faith we have shown that it is the
best distribution that we can possibly assign in the sense that it most faithfully
represents our degree of ignorance concerning the various outcomes.

A more enlightening example is that in which not only is the constraint of
normalization of the probability distribution imposed, but it is assumed that we
know the average value of some quantity, for example, $\langle E \rangle = \sum_i E_i p_i(E_i)$. In
this case, we again consider an augmented function, which involves two Lagrange
multipliers, one tied to each constraint, and given by

$$S' = -\sum_i p_i \ln p_i - \lambda \left(\sum_i p_i - 1 \right) - \beta \left(\sum_i p_i E_i - \langle E \rangle \right). \tag{3.104}$$

Note that once again the first constraint imposes normalization of the distribution.
Maximizing this augmented function leads to the conclusion that $p_j = e^{-1-\lambda-\beta E_j}$.
Imposition of the normalization requirement, $\sum_j p_j = 1$, implies that

$$e^{-1-\lambda} = \frac{1}{\sum\limits_j} e^{-\beta E_j}, \tag{3.105}$$

which is consonant with our expectation that we should recover the partition
function as the relevant normalization factor for the canonical distribution. A
deeper question concerns the Lagrange multiplier β which is formally determined
via the nonlinear equation

$$\langle E \rangle = -\frac{\partial}{\partial \beta} \ln Z. \tag{3.106}$$

3.3.5 Statistical Mechanics Models for Materials

Statistical Mechanics of the Harmonic Oscillator. As has already been argued in this chapter, the harmonic oscillator often serves as the basis for the construction of various phenomena in materials. For example, it will serve as the basis for our analysis of vibrations in solids, and, in turn, of our analysis of the vibrational entropy which will be seen to dictate the onset of certain structural phase transformations in solids. We will also see that the harmonic oscillator provides the foundation for consideration of the jumps between adjacent sites that are the microscopic basis of the process of diffusion.

In light of these claims, it is useful to commence our study of the thermodynamic properties of the harmonic oscillator from the statistical mechanical perspective. In particular, our task is to consider a single harmonic oscillator in presumed contact with a heat reservoir and to seek the various properties of this system, such as its mean energy and specific heat. As we found in the section on quantum mechanics, such an oscillator is characterized by a series of equally spaced energy levels of the form $E_n = (n + \frac{1}{2})\hbar\omega$. From the point of view of the previous section on the formalism tied to the canonical distribution, we see that the consideration of this problem amounts to deriving the partition function. In this case it is given by

$$Z = \sum_n e^{-[(n+\frac{1}{2})\hbar\omega/kT]}.$$
(3.107)

The contribution of the zero point energy can be factored out of the sum leaving us with

$$Z = e^{-\hbar\omega/2kT} \sum_n e^{-n\hbar\omega/kT},$$
(3.108)

which upon the substitution $x = e^{-\hbar\omega/kT}$ is seen to be a simple geometric series and can be immediately summed to yield

$$Z = \frac{e^{-\hbar\omega/2kT}}{1 - e^{-\hbar\omega/kT}}.$$
(3.109)

As a result, we are left with the insight that the free energy of such an oscillator is given by

$$F = \frac{\hbar\omega}{2} + kT \ln(1 - e^{-\hbar\omega/kT}).$$
(3.110)

This equation will arise in chap. 6 in the context of structural transformations in solids when it comes time to evaluate the contribution made by vibrations to the total free energy of a solid.

As shown earlier, the thermodynamic properties of this system can be reckoned as derivatives of the logarithm of the partition function with respect to the inverse temperature. For example, the mean energy is given by

$$\langle E \rangle = -\frac{\partial \ln Z}{\partial \beta} = \frac{\hbar \omega}{2} + \frac{\hbar \omega}{e^{\hbar \omega / kT} - 1}. \tag{3.111}$$

This result should be contrasted with expectations founded on the classical equipartition theorem which asserts that an oscillator should have a mean energy of kT. Here we note two ideas. First, in the low-temperature limit, the mean energy approaches a constant value that is just the zero point energy observed earlier in our quantum analysis of the same problem. Secondly, it is seen that in the high-temperature limit, the mean energy approaches the classical equipartition result $\langle E \rangle = kT$. This can be seen by expanding the exponential in the denominator of the second term to leading order in the inverse temperature (i.e. $e^{\hbar \omega / kT} - 1 \simeq \hbar \omega / kT$). As a result, the mean energy is thus $\langle E \rangle \simeq (\hbar \omega / 2) + kT$. With the presumed largeness of the temperature as revealed in the inequality $\hbar \omega / kT \ll 1$, this yields the classical equipartition result. These results will all make their appearance later in several different contexts including in our discussion of phase diagrams in chap. 6.

Statistical Mechanics of the Ising Model. A tremendously important model within statistical mechanics at large, and for materials in particular, is the Ising model. This model is built around the notion that the various sites on a lattice admit of two distinct states. The disposition of a given site is represented through a spin variable σ_i which can take the values ± 1, with the two values corresponding to the two states. With these 'kinematic' preliminaries settled, it is possible to write the energy (i.e. the Hamiltonian) as a function of the occupations of these spin variables as

$$E_{Ising} = \sum_{\langle ij \rangle} J_{ij} \sigma_i \sigma_j. \tag{3.112}$$

In its simplest form, we can take the coupling constants J_{ij} as being equal so that the energy may be rewritten as $E_{Ising} = J \sum_{ij} \sigma_i \sigma_j$ and further we restrict our attention to the near-neighbor variant of this model in which only spins on neighboring sites interact. At this point, we see that the physics of this model is determined in large measure by the sign of J, with $J > 0$ favoring those states in which adjacent sites are unlike, whereas when $J < 0$, the energy favors alignment of neighboring spins. The statistical mechanics of this model surrounds the question of the competition between the energy which favors an ordering of the various spins and the entropy term in the free energy which favors a disordering of the arrangement of spins.

Before entering into questions concerning the 'solution' of this model, we should first recount some of the uses to which it has been put and to foreshadow its significance to our endeavor. The Ising model has enormous significance beyond the limited range of applications it has in modeling materials. Nevertheless, even for what we will want to say, the Ising model will often serve as a paradigm for the analysis of a wide range of problems in materials modeling. In chronological terms, the Ising model will see duty as the basis for our analysis of phase diagrams, the various stacking arrangements that arise in the so-called polytypes, some of our discussion of surface reconstructions and will provide a foundation for the use of the Potts model in the analysis of grain growth.

Though the Ising model is amenable to full analytic solution in one and two dimensions, we will make use of it in the more general three-dimensional setting and will therefore have to resort to either analytic approximations or full numerics to really probe its implications. On the other hand, examination of this model in the reduced dimensionality setting can be very revealing to illustrate the various tools of statistical mechanics in action. In addition, the reduced dimensionality problems will also help foreshadow the difficulties we shall face in confronting the three-dimensional problem that will arise in the consideration of phase diagrams, for example. As has already been noted throughout this chapter, the engine for our analysis of the thermodynamic properties of a given model is the partition function, which for the Ising model may be written as

$$Z = \sum_{\sigma_1 = \pm 1} \cdots \sum_{\sigma_N = \pm 1} \exp(-\beta \sum_{ij} J_{ij} \sigma_i \sigma_j). \tag{3.113}$$

The sums in front of the exponential are an instruction to consider all possible 2^N configurations of the N-spin system.

For the one-dimensional case, there are a variety of possible schemes for evaluating the partition function, and thereby the thermodynamics. One appealing scheme that illustrates one of the themes we will belabor repeatedly is that of degree of freedom thinning in which degrees of freedom are eliminated by integrating over particular subsets of the full set of degrees of freedom. Indeed, in chap. 12, we will attempt to make generic comments on strategies for eliminating degrees of freedom. In the present case, the idea is to evaluate the partition sum explicitly for half the spins. For example, we can elect to integrate over the even-numbered spins, resulting in an effective Hamiltonian in which only the odd-numbered spins appear explicitly, though at the price of introduction of new coupling parameters in the model. An instructive description of this procedure may be found in Maris and Kadanoff (1978), whom we follow below.

To see the idea in practice, the partition function of eqn (3.113) is reorganized as

$$Z = \sum_{\sigma_1 = \pm 1} \cdots \sum_{\sigma_N = \pm 1} e^{K(\sigma_1 \sigma_2 + \sigma_2 \sigma_3)} e^{K(\sigma_3 \sigma_4 + \sigma_4 \sigma_5)} \cdots, \tag{3.114}$$

where the constant $K = J/kT$ has been introduced. We may now sum over the spins with even labels, resulting in

$$Z = \sum_{\sigma_1=\pm1} \sum_{\sigma_3=\pm1} \cdots (e^{K(\sigma_1+\sigma_3)} + e^{-K(\sigma_1+\sigma_3)})(e^{K(\sigma_3+\sigma_5)} + e^{-K(\sigma_3+\sigma_5)}) \cdots . \quad (3.115)$$

The goal is to seek a scheme to rewrite this result that takes precisely the same form as the original partition function. In concrete terms, this implies that we seek an $f(K)$ such that

$$e^{K(\sigma_i+\sigma_{i+2})} + e^{-K(\sigma_i+\sigma_{i+2})} = f(K)e^{K'\sigma_i\sigma_{i+2}}. \quad (3.116)$$

Note that this equation involving the unknown function $f(K)$ must be true whatever our choices for $\sigma_i = \pm1$ and $\sigma_{i+2} = \pm1$ may be. In particular, for $\sigma_i = \sigma_{i+2} = 1$, this becomes

$$e^{2K} + e^{-2K} = f(K)e^{K'}, \quad (3.117)$$

while for $\sigma_i = -\sigma_{i+2}$, eqn (3.116) becomes

$$2 = f(K)e^{-K'}. \quad (3.118)$$

Eqns (3.117) and (3.118) may now be solved for $f(K)$ and K' with the result that $f(K) = 2\cosh^{\frac{1}{2}} 2K$ and $K' = \frac{1}{2} \ln \cosh 2K$. Note that in the more formidable setting in higher dimensions, the degree of freedom thinning procedure leads to equations for the 'renormalized' coupling constant (i.e. K') that may be treated only approximately. The result obtained above leads now to the key insight of the analysis, namely, the relation between the partition functions of the system with N degrees of freedom and that of the thinned down model involving only $N/2$ degrees of freedom. Using our new found results for the constant K' and the function $f(K)$, we may relate these partition functions as

$$Z(N, K) = [f(K)]^{\frac{N}{2}} Z(N/2, K'), \quad (3.119)$$

where the dependence of the partition function on both the number of degrees of freedom and the the coupling constant has been made explicit.

The next step in the argument is the recognition that the free energy (i.e. $F = -kT \ln Z$) scales with the system size and thus we may write, $\ln Z = N\zeta(K)$. Taking the logarithm of eqn (3.119) and using the fact that $\ln Z(N, K) = N\zeta(K)$, results in

$$\zeta(K') = 2\zeta(K) - \ln[2\cosh^{\frac{1}{2}}(2K)]. \quad (3.120)$$

What is the good of this set of results? The key point is the identification of a recursive strategy for obtaining the partition function. In particular, in the present setting, we note that in the high-temperature limit, $Z = 2^N$ resulting in the observation that $\zeta(K \to 0) = \ln 2$. On the basis of the knowledge of $\zeta(K')$ at this particular point, we can now iterate to determine the functions of interest at other points in parameter space.

Our discussion of the Ising model was primarily intended to introduce the physics of the model itself. However, because of its *conceptual* relevance to many of the strategies to be adopted later in the book, we have also illustrated the use of real space renormalization group ideas for degree of freedom thinning. We note again that our later uses of the Ising model will be for problems in which exact analytic progress is presently unavailable, and for which either approximate or numerical procedures will be resorted to.

Statistical Mechanics of Electrons. Later in the book, we will repeatedly require insights into the generalization of the arguments made earlier in the context of the electron gas for filling up energy states at finite temperatures. Recall our discussion of the electron gas given earlier in which *all* states with wavevectors $|\mathbf{k}| \leq k_F$ were unequivocally occupied and those with $|\mathbf{k}| > k_F$ were unequivocally not occupied. This picture must be amended at finite temperatures, since in this case, by virtue of thermal excitation, some fraction of electrons will fill states of higher energy. The distribution which governs the finite temperature distribution of electrons is the so-called Fermi–Dirac distribution, and it is that distribution (both its derivation and significance) that occupies our attention presently.

The first question we pose is of wider generality than is needed to consider the question of electron occupancies. In particular, we ask for the relevant probability distribution for those problems in which not only is the average value of the energy prescribed, but so too is the average number of particles, $\langle N \rangle$. In this case, using the maximum entropy ideas introduced above, the entropy and its associated constraints may be written as

$$S' = -\sum_i p_i \ln p_i - \lambda \left(\sum_i p_i - 1 \right) - \beta \left(\sum_i p_i E_i - \langle E \rangle \right) - \alpha \left(\sum_i p_i N_i - \langle N \rangle \right).$$
$$(3.121)$$

The index i is a superindex that refers to a particular realization of both E and N. If we now follow the same procedure used earlier to derive the canonical distribution by maximizing the entropy, it is found that the resulting probabilities are of the form

$$p_i = P(E, N) = \frac{1}{Z} e^{-\beta(E - \mu N)}. \qquad (3.122)$$

In this case, we have introduced the Gibbs sum defined as

$$Z = \sum_{states} e^{-(E - \mu N)/kT}. \qquad (3.123)$$

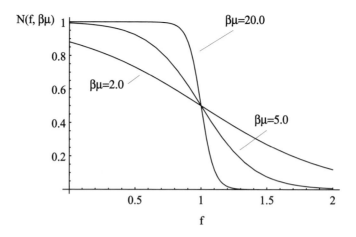

Fig. 3.20. Fermi–Dirac distribution for several different temperatures.

Note that the passage from the formal result based on information theoretic reasoning to the physical setting of interest here involved the recognition of the Lagrange multiplier α being related to the chemical potential through $\alpha = \mu/kT$.

The results obtained above are of such general importance when applied to the case of electrons (more generally to fermions) that the resulting distribution has its own nomenclature and is referred to as the Fermi–Dirac distribution. To see how eqn (3.122) is applied to determine the average number of particles in a given quantum state when the particles themselves obey the Pauli principle, we consider a particular electronic state with energy ϵ. In this case, the Gibbs sum is of the form

$$Z = 1 + e^{-\beta(\epsilon-\mu)}. \tag{3.124}$$

The reasoning behind this assertion is the statement that there are only two allowed states to consider: the state is unoccupied ($E = 0$, $N = 0$) and the state is occupied by a single particle ($E = \epsilon$, $N = 1$). As a result, the mean occupancy of this state is obtained as

$$\langle N(\epsilon)\rangle = \frac{e^{-\beta(\epsilon-\mu)}}{1 + e^{-\beta(\epsilon-\mu)}} = \frac{1}{e^{\beta(\epsilon-\mu)+1}}, \tag{3.125}$$

which is the sought after Fermi–Dirac distribution which presides over questions of the occupancy of electron states at finite temperatures.

In fig. 3.20, we show this distribution for several different temperatures. For ease of plotting, we have rescaled variables so that the Fermi–Dirac distribution is written in the form

$$N(f, \beta\mu) = \frac{1}{e^{\beta\mu(f-1)} + 1}, \tag{3.126}$$

where f has been introduced as a way of characterizing what fraction the energy of interest is of the energy μ. It is evident from the figure that at low temperatures (i.e. $\beta\mu \gg 1$) we recover the results of our earlier analysis, namely, that all states with energy lower than ϵ_F (note that ϵ_F is the zero temperature μ, i.e. $\epsilon_F = \mu(0)$) are occupied while those above this energy are not. On the other hand, we also see that at higher temperatures there is a reduction in the mean occupancy of states with energy below ϵ_F and a corresponding enhancement in the occupancy of those states just above ϵ_F.

3.3.6 Bounds and Inequalities: The Bogoliubov Inequality

There are many instances in which an exact analytic solution for the partition function (or the free energy) will be beyond our reach. One notable example that will arise in chap. 6 is that of the free energy of a liquid. In such instances, there are several related strategies that may be adopted. It is possible to adopt perturbative approaches in which the statistical mechanics of the unperturbed state is in hand, and approximations for the problem of interest are arrived at through a series of systematic corrections. A related approach is the attempt to bound the exact free energy through the use of robust inequalities such as the Bogoliubov inequality to be introduced below. With this principle in hand, the bounds can be variationally improved by introducing free parameters in the 'reference' Hamiltonian, which are then chosen to yield the best possible bound.

The basic idea is as follows: we assume that the free energy of real interest is that associated with a Hamiltonian H, for which the exact calculation of the free energy is intractable. We then imagine an allied Hamiltonian, denoted H_0, for which the partition function and free energy may be evaluated. The inequality that will serve us in the present context asserts that

$$F_{exact} \leq F_0 + \langle (H - H_0) \rangle_0, \qquad (3.127)$$

where the average on the right hand side is evaluated with respect to the distribution that derives from the Hamiltonian H_0. What this inequality teaches us is that if we can find an approximate Hamiltonian for which we can evaluate the partition function, then we can provide bounds on the free energy of the system described by the full $H = H_0 + H_1$, where we have introduced the notation H_1 to describe the perturbation that is added to the original H_0.

Presently, we consider a few of the details associated with the derivation of the Bogoliubov inequality. Our treatment mirrors that of Callen (1985) who, once again, we find gets to the point in the cleanest and most immediate fashion. As said above, we imagine that the problem of real interest is characterized by a Hamiltonian H which may be written in terms of some reference Hamiltonian H_0

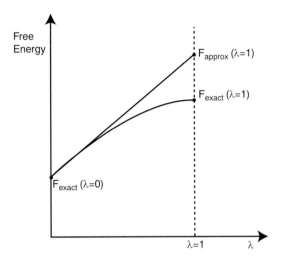

Fig. 3.21. Schematic of the free energy as a function of the parameter λ discussed in the context of the Bogoliubov inequality.

as

$$H(\lambda) = H_0 + \lambda H_1, \tag{3.128}$$

where λ is a tunable parameter which weights the contribution of the perturbation to the reference Hamiltonian. We introduce the notation $Z = \mathrm{tr}\, e^{-\beta H}$, with the symbol tr referring to the trace operation, though it should be read as 'sum over all configurations'. Evaluation of the derivatives of the exact free energy $F_{exact}(\lambda) = -kT \ln \mathrm{tr}\, e^{-\beta H}$ with respect to λ reveal that

$$\frac{d F_{exact}}{d\lambda} = \langle H_1 \rangle, \tag{3.129}$$

$$\frac{d^2 F_{exact}}{d\lambda^2} = -\beta \langle (H_1 - \langle H_1 \rangle)^2 \rangle. \tag{3.130}$$

The significance of the latter equation is that it demonstrates that for all λ, the curvature of the function $F_{exact}(\lambda)$ is negative. Why should this observation fill us with enthusiasm? The relevance of this insight is revealed graphically in fig. 3.21, where it is seen that we may construct an approximate free energy (a first-order Taylor expansion in powers of λ) of the form

$$F_{approx}(\lambda) = F_{exact}(\lambda = 0) + \left(\frac{d F_{exact}}{d\lambda} \right)_{\lambda=0} \lambda, \tag{3.131}$$

which is guaranteed to satisfy the inequality $F_{exact}(\lambda = 1) \leq F_{approx}(\lambda = 1)$, by virtue of what we have learned about the curvature of the function $F_{exact}(\lambda)$.

What has been done, in essence, is to exploit our certainty about the sign of the curvature of the free energy with respect to λ to carry out a Taylor expansion of that free energy with respect to λ on the hope that the leading order expansion will be sufficiently accurate. Note that when the slope of $F_{exact}(\lambda)$ is evaluated explicitly, we find

$$\frac{d F_{exact}}{d\lambda} = \frac{\operatorname{tr} H_1 e^{-\beta(H_0 + \lambda H_1)}}{\operatorname{tr} e^{-\beta(H_0 + \lambda H_1)}}, \qquad (3.132)$$

which for the special case of $\lambda = 0$ yields

$$\frac{d F_{exact}}{d\lambda} = \langle H_1 \rangle_0, \qquad (3.133)$$

where $\langle H_1 \rangle_0$ signifies the evaluation of the average value of H_1 (or $H - H_0$) with respect to the distribution function arising only from the reference Hamiltonian. The implication of these results is that with only a knowledge of the distribution function for the reference state, we are able to put bounds on the exact free energy. To see this mathematically, we plug the result of eqn (3.133) into eqn (3.131) with the result that

$$F_{exact} \leq F_{exact}(\lambda = 0) + \langle H - H_0 \rangle_0. \qquad (3.134)$$

Recall that $F_{exact}(\lambda = 0)$ is nothing more than the free energy associated with the Hamiltonian H_0 which, by hypothesis, we know how to compute. As a result, we have demonstrated the existence of the sought after bound on the exact free energy. The finishing touches are usually put on the analysis by allowing the reference Hamiltonian to have some dependence on adjustable parameters that are determined to minimize the right hand side of eqn (3.134). We note that as a result of this scheme, the better the choice of H_0, the higher will be the quality of the resulting variational estimate of the free energy.

3.3.7 Correlation Functions: The Kinematics of Order

Conventional kinematic notions are built around the idea that the system of interest can be characterized by a series of simple states without any reference to microscopic fluctuations or disorder. On the other hand, in the analysis of systems at finite temperature or in the presence of the effects of disorder, it is essential to build up measures that characterize the disposition of the system while at the same time acknowledging the possibility of an incomplete state of order within that system. Just as strain measures served as the primary kinematic engine used in our analysis of continua, the notion of a correlation function will serve to characterize the disposition of those systems that are better thought of statistically. Indeed, we will find that in problems ranging from the 'spins' introduced earlier in conjunction with the Ising model, to the density (and its fluctuations), and even to the heights on fracture surfaces, correlation functions provide a quantitative language for describing the statistical state of systems with many degrees of freedom.

As is already evident from the discussion of this chapter, many-particle systems are often described in terms of averages over microscopic variables. For example, the net magnetization M of a spin system can be thought of as an average of the form $M = N\mu\langle\sigma_i\rangle$, where N is the total number of spins per unit volume, μ is the magnetic moment per spin and σ_i is the spin variable at site i. The notion of the correlation function is introduced in light of the knowledge that by virtue of the dynamics of the microscopic variables, the macroscopic variables themselves are subject to fluctuations. The extent of these fluctuations is conveniently quantified using the notion of the correlation function which achieves a greater significance because of its relation to the response of a system to external probes, including the scattering of X-rays, electrons or neutrons. The basic idea is to quantify the extent to which the behavior of a particular microscopic quantity such as the spin at site i, σ_i, is correlated with that of the spin σ_j at site j.

By definition, the pair correlation function of σ_i and σ_j is defined by

$$C_{ij} = \langle\sigma_i\sigma_j\rangle - \langle\sigma_i\rangle\langle\sigma_j\rangle. \tag{3.135}$$

What we note is that if the joint probability distribution $P(\sigma_i, \sigma_j)$ is of the form $P(\sigma_i, \sigma_j) = P(\sigma_i)P(\sigma_j)$, then $\langle\sigma_i\sigma_j\rangle = \langle\sigma_i\rangle\langle\sigma_j\rangle$, the spins are uncorrelated, and correspondingly, the correlation function C_{ij} vanishes. The importance of correlation functions as a way of characterizing systems with many degrees of freedom can be argued for from a number of different angles, but here we simply note that a knowledge of such functions serves as a basis for computing 'susceptibilities' of various types in a number of different systems, and can be thought of as providing a kinematics of order.

From a structural perspective, these ideas can be couched in slightly different terms by considering the two-particle distribution function defined by

$$\rho^{(2)}(\mathbf{r}_1, \mathbf{r}_2) = \left\langle \sum_{i \neq j} \delta(\mathbf{r}_1 - \mathbf{r}_i)\delta(\mathbf{r}_2 - \mathbf{r}_j) \right\rangle. \tag{3.136}$$

The two-particle distribution tells us the probability that particles will be found at $\mathbf{r}_1$ and $\mathbf{r}_2$ and will thereby serve as the basis for determining the energy of interaction between different pairs of particles when the best statements we can make about the system are statistical. For example, when describing the energetics of liquids our statements about the structure of that liquid will be couched in the language of distribution functions. In preparation for our later work on the free energy of liquids (see chap. 6), we find it convenient to introduce an auxiliary quantity, the pair correlation function $g(\mathbf{r}_1, \mathbf{r}_2)$, which is defined through

$$\rho^{(2)}(\mathbf{r}_1, \mathbf{r}_2) = \langle \rho(\mathbf{r}_1) \rangle \langle \rho(\mathbf{r}_2) \rangle g(\mathbf{r}_1, \mathbf{r}_2). \tag{3.137}$$

This definition simplifes considerably in the case of an isotropic liquid for which the single particle density is a constant (i.e. $\rho(\mathbf{r}) = N/V$), resulting in the radial distribution function $g(r)$. As said above, these definitions will form a part of our later thinking.

3.3.8 Computational Statistical Mechanics

As yet, we have described some of the analytic tools that are available to forge a statistical connection between macroscopic observables and the microstates that lead to them. The advent of high-speed numerical computation opened up alternative strategies for effecting this connection which permitted a literal interpretation of the notion of an average. The key idea is to allow the system to visit a sequence of microscopic 'realizations' of the macroscopic state of interest. By effecting an average for the quantity of interest over the various members in this set of realizations, the statistical average of interest is performed explicitly.

On the one hand, molecular dynamics obtains average quantities by repeatedly evaluating the dynamical variable of interest as it takes on a succession of values as the system winds through phase space. Alternatively, Monte Carlo methods allow for construction of copies of a particular system that one envisions in speaking of the 'ensemble averages' of Gibbs. In this case, rather than explicitly tracking the system along some trajectory through phase space, we imagine an ensemble of copies of the system of interest, each corresponding to some different microstate and which are representative members of the set being averaged over. In both the molecular dynamics and Monte Carlo methods, the mentality is to construct explicit microscopic realizations of the system of interest in which once the microscopic degrees of freedom have been identified and the nature of their interaction understood, one explicitly constructs some subset of the allowable microstates and performs averages over them.

In this part of the chapter, our aim is to describe a few of the highlights of both the molecular dynamics and Monte Carlo approaches to computational statistical mechanics. We begin with a review of the notion of microstate, this time with the idea of identifying how one might treat such states from the perspective of the computer. Having identified the nature of such states, we will take up in turn the alternatives offered by the molecular dynamics and Monte Carlo schemes for effecting the averages over these microstates.

As noted in the preface, the availability of computers has in many cases changed the conduct of theoretical analysis. One of the fields to benefit from the emergence

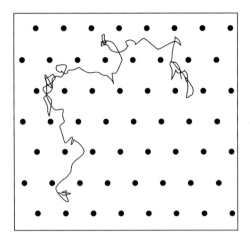

Fig. 3.22. Wandering of a Ag adatom on the Ag(111) surface as revealed using molecular dynamics (adapted from Boisvert and Lewis (1996)).

of powerful computational alternatives to traditional analytic approaches is that of statistical mechanics. If we follow the line of reasoning suggested by molecular dynamics, the argument is made that the averages of interest can be realized by following the system in its trajectory in phase space via the expression

$$\langle A \rangle_t = \frac{1}{T} \int_0^T A(\{p(t), q(t)\}) dt. \tag{3.138}$$

In this case, our notation indicates that we are considering the time average ($\langle \ \rangle_t$) of the quantity A which itself is a function of the dynamical variables. Those variables are represented here generically by $q(t)$ and $p(t)$, with q a configurational coordinate and p the associated momentum. The averaging process is tied to the type of trajectory indicated schematically in fig. 3.22, where we have artificially reduced the problem to a two-dimensional scenario that allows for visualization. Molecular dynamics provides a scheme for explicitly evaluating these averages by computing A at various time intervals while allowing the particles to range over the trajectories resulting from their respective equations of motion.

In the simplest classical terms, carrying out a molecular dynamics simulation involves a few key steps. First, it is necessary to identify some force law that governs the interactions between the particles which make up the system. The microscopic origins of the force laws themselves will be taken up in chap. 4, and for the moment we merely presuppose their existence. It is then imagined that these particles evolve under their mutual interactions according to Newton's second law of motion. If we adopt the most naive picture in which all of the atoms are placed in a closed box at fixed energy, we find a set of $3N$ coupled second-order ordinary differential equations of the form

$$m_i \frac{d^2\mathbf{r}_i}{dt^2} = \mathbf{F}_i. \tag{3.139}$$

A host of clever schemes have been devised for integrating these equations of motion in a manner that is at once efficient and stable. One well-known integration scheme is that of Verlet (though we note that more accurate methods are generally used in practice). One notes that for a sufficiently small time step, the position at times $t \pm \delta t$ may be expressed as

$$\mathbf{r}_i(t \pm \delta t) \approx \mathbf{r}_i(t) \pm \frac{d\mathbf{r}_i}{dt}\delta t + \frac{1}{2}\frac{d^2\mathbf{r}_i}{dt^2}\delta t^2. \tag{3.140}$$

By adding the two resulting expressions and rearranging terms, we find

$$\mathbf{r}_i(t + \delta t) = -\mathbf{r}_i(t - \delta t) + 2\mathbf{r}_i(t) + \frac{d^2\mathbf{r}_i}{dt^2}\delta t^2. \tag{3.141}$$

Though there are a variety of sophisticated schemes for carrying out the integrations required here, we choose to emphasize the physical insights gained thus far. The choice of a time step δt strikes to the heart of one of the key difficulties in carrying out the types of simulations advocated here. The processes of interest to the mechanics of materials involve a broad range of time scales. At the smallest end of that spectrum, we have the time scale associated with atomic vibrations. Clearly, if we are to simulate the thermal properties of solids, the time step must be a small fraction of the vibrational period. By way of contrast, if our aim is the investigation of deformation processes that are slaved to mass transport, the time scale can be many orders of magnitude larger than the picosecond scale associated with thermal vibrations, resulting in a gross mismatch between the time step required for carrying out temporal integrations, and the time scale associated with the process of interest. This question will be addressed in chap. 12 when we describe the hyperdynamics method.

In addition to questions surrounding the treatment of multiple time scales, another intriguing problem is posed by simulating a system in contact with a thermal reservoir. Until now, we have envisaged a system that is maintained at fixed total energy. In many cases, one would rather mimic a system at fixed temperature. One of the more interesting aspects of molecular dynamics is the role of effective theories for mimicking the presence of either thermal or pressure reservoirs. The point is that through the trick of incorporating a single extra set of conjugate variables, the effect of a thermal reservoir may be accounted for rigorously while maintaining the Hamiltonian description of the underlying dynamics. The approach we will follow here was pioneered by Nosé (1984) and Hoover (1985).

Nosé argues that the presence of the thermal reservoir may be accounted for in the Hamiltonian context by incorporating an additional set of conjugate variables, $\{s, p_s\}$. The Hamiltonian for the extended system is given by

$$H = \sum_i \frac{\mathbf{p}_i^2}{2ms^2} + V_{eff}(\{\mathbf{q}_i\}) + \frac{p_s^2}{2Q} + gkT \ln s. \tag{3.142}$$

For the present purposes, all that we really wish to note is that in light of this Hamiltonian, it becomes possible to carry out averages that are effectively those associated with a canonical ensemble, or said differently, with a system maintained at a temperature T. These ideas are described very nicely in Frenkel and Smit (1996).

As already mentioned, an alternative to the explicit contemplation of microscopic trajectories is the construction of a stochastic approach in which various microscopic realizations of a system are constructed so as to respect the underlying distribution. Much has been made of the necessity of ergodic assumptions in statistical mechanics which allow for a replacement of the type of temporal average described above by the alternative ensemble averages immortalized by Gibbs. Here the idea is that rather than following the system on its excursions through phase space, one generates a series of macroscopically equivalent copies of the system, and averages over the members of this 'ensemble'. For example, if our interest is in the thermodynamic properties of an alloy with constituents A and B, we can imagine generating alternative occupations of the various crystal lattice sites on the basis of random occupancy. Two snapshots corresponding to different members of the ensemble under consideration here were shown in fig. 3.16. The underlying physical contention is the equivalency of the averaging process when effected either via the ensemble trick introduced above or by following the system as it ranges over its dynamic trajectory. Besides providing the foundation on which Monte Carlo methods are built, we have seen that the ensemble picture serves as the basis for much of the analytic machinery which is used to conduct the investigations of statistical mechanics. if one generates a sequence of states based upon the rule that a given state is 'accepted' with probability 1 if the new state lowers the energy and with probability $e^{-\beta\Delta E}$ is the new state increases the energy, then the resulting distribution will be canonical. For further details, I am especially fond of the discussion of Frenkel and Smit (1996).

3.4 Further Reading

Quantum Mechanics

Principles of Quantum Mechanics by R. Shankar, Plenum Press, New York: New York, 1980. A personal favorite on the subject of quantum mechanics. Shankar does a fine job of integrating the mathematical formalism and physical principles of quantum mechanics with their roots in many of the historical ideas of classical mechanics. His discussion of path integral methods is enlightening and readable.

Quantum Theory by David Bohm, Prentice-Hall, Inc., Englewood Cliffs: New Jersey, 1951. Bohm's book brings both clarity and completeness to many of the features of quantum mechanics that are of importance for the work carried out in the present book.

Quantum States of Atoms, Molecules and Solids by M. A. Morrison, T. L. Estle and N. F. Lane, Prentice-Hall, Inc., Englewood Cliffs: New Jersey, 1976. Not a well-known book, but one I like very much as it has much that is practical to offer concerning the application of quantum mechanics to understanding bonding in solids.

Thermodynamics and Statistical Mechanics

E. T. Jaynes: Papers on Probability and Statistics edited by R. D. Rosenkranz, D. Reidel Publishing Company, Dordrecht: The Netherlands, 1983. The presence of this work on the list reflects my enormous prejudice in favor of the information theoretic approach to statistical mechanics as pioneered by Edwin T. Jaynes. This book is one of my prize possessions and at the time of this writing has no trace of its original hardback cover as a result of the many trips the book has made.

Statistical Mechanics by R. P. Feynman, The Benjamin/Cummings Publishing Company, Inc., Reading: Massachusetts, 1972. This book bears the usual stamp of Feynman's idiosyncratic approach to physics. There are many perspectives to be found here that will not be found elsewhere.

States of Matter by D. L. Goodstein, Prentice-Hall, Inc., Englewood Cliffs: New Jersey, 1975. This often humorous book is full of insights into everything from the meaning of the partition function to the origins of Bragg scattering.

Principles of Statistical Mechanics by A. Katz, W. H. Freeman and Company, San Francisco: California, 1967. This book remains virtually unknown, but again reflects my own taste for the information theoretic perspective. This book offers some hints as to the possible uses of this perspective to build more complete theories of nonequilibrium processes.

Thermodynamics and an Introduction to Thermostatics by H. B. Callen, John Wiley and Sons, New York: New York, 1985. The perceptive reader will recognize on short order my own affinity for the book of Callen. His treatment of the central ideas of both thermodynamics and statistical mechanics is clean and enlightening, and I share the author's view that classical thermodynamics is primarily a theory of thermostatics.

Fundamentals of Statistical Mechanics, Manuscript and Notes of Felix Bloch, prepared by John Dirk Walecka, Stanford University Press, Stanford: California, 1989. I have found this book eminently readable and interesting.

'The Expanding Scope of Thermodynamics in Physical Metallurgy' by J. W. Cahn, *Materials Transactions, JIM*, **35**, 377 (1994). Cahn's article is full of interesting ideas concerning both the significance of previous thermodynamic achievements and the possible expansion of thermodynamic ideas to novel ways of treating interfaces and lattice effects.

Computational Statistical Mechanics

Understanding Molecular Simulation by Daan Frenkel and Berend Smit, Academic Press, San Diego: California, 1996. An extremely serious book written at a high level and full of detailed insights.

The Art of Molecular Dynamics Simulation by D. C. Rapaport, Cambridge University Press, Cambridge: England, 1995. This book gives a modern introduction to the subject of computer simulation, with special reference to molecular dynamics.

'The Virtual Matter Laboratory', M. J. Gillan, *Contemporary Physics*, **38**, 115 (1997) provides an overview of some of the key ideas associated with the use of the ideas presented in this chapter as the basis of simulations of the properties of matter.

3.5 Problems

1 Particle in a Box by the Finite Element Method

Flesh out the details of the discussion of the use of the finite element method to solve the problem of a particle in a box. In particular, derive eqns (3.44), (3.46), (3.48), (3.50) and table 3.1.

2 Finite Elements and the Schrödinger equation

In this chapter, we discussed the use of the finite element method to solve the Schrödinger equation. Use the same formalism to solve the Schrödinger equation for a quantum corral of elliptical shape.

Fig. 3.23. Two three-atom configurations to be considered in problem 4.

3 Two-Dimensional Electron Gas

Construct a two-dimensional analog of the electron gas model developed in this chapter. Begin by solving the relevant Schrödinger equation and obtaining the set of discrete energy eigenvalues. Assume that in the square box of area A there are N free electrons. Obtain the electronic energy by filling up the various energy states. How does the resulting energy depend upon the density?

4 Tight-Binding Model of Three-Atom Molecules

Using the arguments introduced earlier for considering the H_2 molecule, consider two separate three-atom configurations and their associated binding energies. In particular, consider both a three-atom chain with near-neighbor interactions as well as an equilateral triangle geometry as shown in fig. 3.23. Use the same basis (i.e. one s-orbital per atom) and maintain the notation introduced earlier, namely, that $\langle i|H|j \rangle = h$. Construct the appropriate 3×3 Hamiltonian matrices for these two different geometries and find their corresponding eigenvalues. Fill up the lowest lying energy states and thereby find the electronic contribution to the binding energies of both configurations in the case in which there is one electron per atom.

5 The Stirling Approximation Revisited

To obtain an explicit analytic expression for the entropy of mixing as shown in eqn (3.89), we exploited the Stirling approximation. Plot both the exact and the approximate entropy (as gotten from the Stirling approximation) for systems of size $N = 10$, 100 and 1000.

6 Statistical Mechanics of a Two-Level System

Consider a system characterized by two distinct energy levels, ϵ_0 and $\epsilon_0 + \Delta$. Write down the partition function and compute the average energy and specific heat for this system.

7 Entropy Maximization and the Dishonest Die

In this problem we will examine the way in which maximum entropy can be used to confront a problem in incomplete information. The basic idea is that we are told that the mean value that emerges after many rolls of a die is 4.5. On this information, we are urged to make a best guess as to the probability of rolling a 1, 2, 3, 4, 5 and 6. Apply the principle of maximum entropy in order to deduce the probability $p(n)$ by applying the constraint $4.5 = \sum_n np(n)$.

8 Partition Function for the One-Dimensional Ising Model

For the one-dimensional Ising model with near-neighbor interactions with a Hamiltonian of the form (i.e. zero field)

$$H = \sum_{\langle ij \rangle} J\sigma_i\sigma_j, \qquad (3.143)$$

compute the partition function, the free energy and the average energy.

9 Statistical Mechanics of the Harmonic Oscillator

In the text we sketched the treatment of the statistical mechanics of the harmonic oscillator. Flesh out that discussion by explicitly obtaining the partition function, the free energy, the average energy, the entropy and specific heat for such an oscillator. Plot these quantities as a function of temperature.

10 Monte Carlo Treatment of the Ising Model

Write a computer program that allows for the Monte Carlo simulation of the two-dimensional Ising model. The program should use periodic boundary conditions. Generate representative configurations at a number of different temperatures and attempt to deduce the nature of the phase transition found as a function of temperature. To monitor the state of the system, evaluate the average spin on each site as a function of temperature. In addition you may wish to monitor the values of spin correlation functions such as were described in section 3.3.7.

Part two

Energetics of Crystalline Solids

FOUR

Energetic Description of Cohesion in Solids

4.1 The Role of the Total Energy in Modeling Materials

A recurring theme in the study of materials is the connection between structure and properties. Whether our description of structure is made at the level of the crystal lattice or the defect arrangements that populate the material or even at the level of continuum deformation fields, a crucial prerequisite which precedes the connection of structure and properties is the ability to describe the total energy of the system of interest. In each case, we seek a function (or functional) such that given a description of the system's geometry, the energy of that system can be obtained on the basis of the kinematic measures that have been used to characterize that geometry. As we have discussed earlier (see chap. 2), the geometry of deformation may be captured in a number of different ways. Microscopic theories are based upon explicitly accounting for each and every atomic coordinate. Alternatively, a continuum description envisions a sort of kinematic slavery in which the motion of a continuum material particle implies the associated evolution of large numbers of microscopic coordinates.

From the standpoint of microscopic theories, the computation of the total energy requires a description in terms of the atomic positions. In this instance, we seek a reliable function $E_{tot}(\{\mathbf{R}_i\})$, where $\{\mathbf{R}_i\}$ refers to the set of all nuclear coordinates and serves to describe the geometry at the microscale. For a more complete analysis, this picture should reflect the electronic degrees of freedom as well. By way of contrast, a description in terms of a continuum model often results in a phenomenological treatment of the total energy in which the energy is postulated to vary in accordance with some functional of the relevant strain measures. Here we seek a functional $E_{tot}[\mathbf{F}]$ which relates the spatially varying deformation field and the attendant energy. Note that in accordance with the distinction made between interesting and uninteresting motions in chap. 2, the energy of deformation should depend only on relative motions of material particles. As will become evident in this and subsequent chapters, there are a variety of means for describing the energetics of deformation, each with its own merits and advantages. It is of special interest to uncover the way in which insights at the microscopic scale can be bootstrapped for use at scales normally reserved for continuum approaches.

As mentioned above, the total energy often serves as the point of entry for the analysis of material behavior. To whet our appetite for subsequent developments and to illustrate how the total energy impacts material properties we consider three different examples; the elastic moduli, the diffusion constant and the so-called gamma surface. As was already made clear in chap. 2, the elastic moduli are the phenomenological cornerstone of the linear theory of elasticity. Given these parameters, one is positioned to evaluate the stored energy of an arbitrary state of deformation so long as the displacement gradients are not too large. In chap. 5, we will show how by exploiting an analogy between the microscopic and macroscopic descriptions of the total energy the elastic moduli themselves may be computed strictly on the basis of microscopic considerations such as those that form the backbone of this chapter. The basic idea is founded on the recognition that for small deformations, the stored energy is quadratic in the appropriate strain measure. In addition to their ability to probe the energy cost for small excursions from some reference state, microscopic calculations can also be used to examine the energetics of large homogeneous deformations. This type of behavior is indicated schematically in fig. 4.1. By exploiting a series of total energy calculations at the microscopic scale, each associated with a different overall state of strain, the curvature of the energy about the minimum may be determined and thus the moduli.

Just as the elastic moduli characterize the properties of a material in the context of elastic boundary value problems, the diffusion coefficient serves as the phenomenological basis for solving boundary value problems involving mass transport. In chap. 7, we will explicitly evaluate the diffusion constant on the basis of microscopic arguments. To do so, we will require an energetic analysis of certain saddle point configurations on a multi-dimensional potential energy surface. To compute them one must have solved the problem posed in this chapter: given a certain geometric configuration, what is the total energy? For the problem of diffusion, it is the energy as a function of the diffusing atom's position, also known as the 'reaction coordinate', that determines the diffusion rate. This type of analysis is indicated in fig. 4.2 where the energy of a Pb adatom on a Ge surface is computed for a series of different adatom positions. These calculations were carried out using the types of first-principles techniques to be described at the end of this chapter.

In section 12.3.4, we will discuss a wide class of constitutive models that are broadly known as cohesive surface models. These models postulate the existence of an elastic potential that can be used to describe sliding and opening of interfaces.

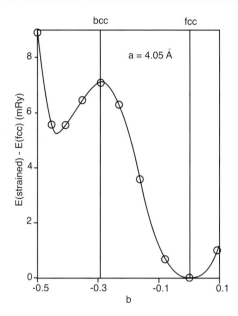

Fig. 4.1. Energy as a function of shear deformation in Al (adapted from Mehl and Boyer (1991)). The lattice parameter *a* is fixed during the deformation and hence the energy characterizes a one-parameter family of deformations of the fcc lattice, with the members of the family parameterized by *b*.

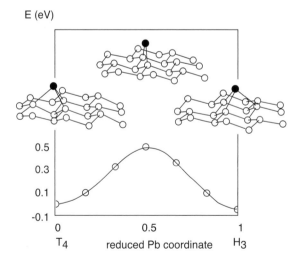

Fig. 4.2. Energy of a diffusing adatom as a function of reaction coordinate (after Kaxiras and Erlebacher (1994)).

In the context of the mechanics of materials, such analyses are of value in describing both plasticity and fracture. What will be required in the determination of these elastic potentials, like for the problems described above, is the relative energies of a number of different configurations. Our reason for calling attention to these models at this point is to drive home the significance of the total energy, whether that energy is being considered in the context of elasticity, mass transport or fracture. The aim of the present chapter is to give examples of the broad classes of microscopic energy functionals that are available to make such calculations.

4.2 Conceptual Backdrop for Characterizing the Total Energy

4.2.1 Atomistic and Continuum Descriptions Contrasted

One of the key questions posed in this chapter is how best to reconcile microscopic descriptions, which are presumed to represent the true mechanistic underpinnings of a given process, and phenomenological descriptions, whose aim is to capture the essential physics of a particular process, and to use this reduced description as a predictive tool. The philosophies of these two types of calculations are strikingly different. The microscopic viewpoint champions a complete description of the 'relevant' degrees of freedom, with an explicit reckoning of the interactions and evolution of all of them. The total energy may then be written in the form $E_{tot}(\{\mathbf{R}_i, \mathbf{r}_n\})$ which highlights the dependence of the total energy on each and every microscopic degree of freedom. Note that here we have explicitly acknowledged the presence not only of the nuclei $\{\mathbf{R}_i\}$, but also the electrons which are characterized in a classical description by coordinates $\{\mathbf{r}_n\}$. We note that in some instances, such as in density functional theory, the dependence on the electronic degrees of freedom is effected via a dependence of the total energy on the electron density, $\rho(r)$, which may be written as $E_{tot}[\rho(\mathbf{r}); \{\mathbf{R}_i\}]$. Continuum descriptions are built around energy functionals which depend upon kinematic measures such as displacement fields as opposed to atomic coordinates, and the neglected atomic-level degrees of freedom make their appearance only implicitly through the presence of material parameters such as the elastic moduli.

Whether the description of the total energy is based upon microscopic or continuum notions, ultimately, all such energy functionals represent a reduction in complexity relative to the full microscopic Hamiltonian. The type of degree of freedom reduction we have in mind can be illustrated on the basis of an analogy borrowed from statistical mechanics. If we begin with a system of spins on a lattice which interact according to a near-neighbor Ising model, the total energy may be written $E_{tot}(\{\sigma_i\}) = \sum_i J\sigma_i\sigma_{i+1}$. If we now use the dodge of constructing block spins by adding up the spins in small groups, the total energy may be rewritten in terms of the new block spin variables mediated by some new effective coupling constant J_{eff}. That is, the reduced description of the total energy is $E_{tot}(\{\sigma_i^{block}\}) = \sum_i J_{eff}\sigma_i^{block}\sigma_{i+1}^{block}$. In this case, the missing degrees of freedom only make their presence known through the effective coupling constant J_{eff}.

The simplest example of this type of degree of freedom as pertains to the energy of solids emerges through consideration of the meaning of the elastic moduli. From the continuum perspective, the elastic moduli are material parameters whose values cannot be determined by the continuum theory itself. Within such a framework, the moduli represent the energy cost associated with special affine deformations of the solid. For example, a pure breathing mode has an energy that scales with the bulk modulus B. Thus, we write the energy change upon altering the volume by an amount ΔV as

$$\Delta E = \frac{1}{2} B \frac{(\Delta V)^2}{V}. \tag{4.1}$$

One then makes a bolder assertion. It is hypothesized that even in the presence of some inhomogeneous deformation, the total energy may be obtained as an integral over the strain energy density which is itself obtained by appealing only to the local state of deformation. This leads to the form described in chap. 2:

$$E_{tot} = \frac{1}{2} \int_{\Omega} C_{ijkl} \epsilon_{ij}(\mathbf{x}) \epsilon_{kl}(\mathbf{x}) d\mathbf{x}. \tag{4.2}$$

On the other hand, the microscopic view of such elastic distortions reflects the modification of a multitude of atomic bonds. The energy cost associated with a given deformation is nothing more than the outcome of evaluating the energies of the distinct atomic constituents. Remarkably, these two perspectives merge in the form of the elastic moduli. That is, the energetics of the complex arrangements of atomic bonds that result from some complicated inhomogeneous deformation can in the end be represented (at least in the case of small deformations) via the relevant elastic moduli. In subsequent chapters, we will refine this description by demonstrating the nature of the coupling between microscopic and continuum descriptions of elasticity by explicitly computing a microscopic expression for the moduli using a simple model potential.

However, bonding in solids is an inherently quantum mechanical process and thereby, at least in principle, requires a consideration of the electronic degrees of freedom. In the remainder of this chapter we will investigate the ways in which the total energy can be evaluated from the microscopic perspective. First, we will examine those methods in which the electronic degrees of freedom are subsumed in some effective description. The result of adopting this strategy is that the total energies we will write can only be considered as approximations. We begin by considering pair potential models of bonding in solids. Such models provide a transparent scheme for evaluating the total energy, though they are often of limited quantitative value. In this case, the idea is that the pair potential is an effective representation of the underlying electronic degrees of freedom which enter only through a parametric dependence of the potentials on the electron density. A more thorough analysis of the electronic degrees of freedom suggests a class of total energy functionals known as effective medium and angular force schemes in which additional features of the physics of bonding are incorporated. These methods are evaluated in turn, with a summarizing commentary on the limitations inherent in all of these methods.

Subsequently, we will highlight the key ideas involved in extending the microscopic analysis to permit incorporation of the electronic degrees of freedom. The crowning achievement here is the Hohenberg–Kohn theorem which shows how the total energy may be evaluated as a functional of the electronic density. Further, this scheme allows for a variational treatment of the total energy which amazingly leads to a set of effective single particle equations that are an exact description of the complex many-body system encountered when thinking about electrons in solids.

4.2.2 The Many-Particle Hamiltonian and Degree of Freedom Reduction

From a quantum mechanical perspective, the starting point for any analysis of the total energy is the relevant Hamiltonian for the system of interest. In the present setting, it is cohesion in solids that is our concern and hence it is the Hamiltonian characterizing the motions and interactions of all of the nuclei and electrons in the system that must be considered. On qualitative grounds, our intuition coaches us to expect not only the kinetic energy terms for both the electrons and nuclei, but also their mutual interactions via the Coulomb potential. In particular, the Hamiltonian may be written as

$$\sum_i \frac{\mathbf{P}_i^2}{2M} + \sum_j \frac{\mathbf{p}_j^2}{2m} + \frac{1}{2}\sum_{ij} \frac{e^2}{|\mathbf{r}_i - \mathbf{r}_j|} - \sum_{ij} \frac{Ze^2}{|\mathbf{r}_i - \mathbf{R}_j|} + \sum_{ij} \frac{Z^2 e^2}{|\mathbf{R}_i - \mathbf{R}_j|}. \quad (4.3)$$

The terms in this Hamiltonian are successively, the kinetic energy associated with the nuclei, the kinetic energy associated with the electrons, the Coulomb interaction between the electrons, the electron–nucleus Coulomb interaction and finally, the nuclear–nuclear Coulomb interaction. Note that since we have assumed only a single mass M, our attention is momentarily restricted to the case of a one-component system.

As an antidote to the nearly hopeless complexity presented by the problem of the coupling of degrees of freedom revealed in eqn (4.3), a number of different strategies have been constructed. Such strategies can be divided along two main lines, although this division is in the end an oversimplification. One set of ideas surrounds the attempt to remove the electronic degrees of freedom altogether, electing to bury them in effective interactions between the ions. Note that in these approaches, the exact energy is replaced by an approximate surrogate via

$$E_{exact}(\{\mathbf{R}_i, \mathbf{r}_n\}) \rightarrow E_{approx}(\{\mathbf{R}_i\}). \tag{4.4}$$

As will be shown in coming sections, wide classes of total energy descriptions fall under this general heading including traditional pair potential descriptions, pair functionals, angular force schemes and cluster functionals.

The second general class of schemes, which almost always imply a higher level of faithfulness to the full Hamiltonian, are those in which reference to the electrons is maintained explicitly. In this case, the mapping between the exact and approximate descriptions is of the form,

$$E_{exact}(\{\mathbf{R}_i, \mathbf{r}_n\}) \rightarrow E_{approx}(\{\mathbf{R}_i, \mathbf{r}_n\}), \tag{4.5}$$

or as was said already

$$E_{exact}(\{\mathbf{R}_i, \mathbf{r}_n\}) \rightarrow E_{approx}[\rho(\mathbf{r}), \{\mathbf{R}_i\}], \tag{4.6}$$

where the second expression highlights the use of a function $\rho(\mathbf{r})$ to characterize the disposition of the electronic degrees of freedom. Later in the present chapter, we will discuss both tight-binding models (section 4.5) and the use of density functional theory (section 4.6) as ways to effect this type of approximate mapping between the full Hamiltonian and more tractable approximations. The ideas embodied in eqns (4.5), (4.6) and (4.7) constitute profound exercises in effective theory construction. Though considerable hype surrounds the use of the term 'multiscale modeling', efforts to confront the difficulties inherent in the many-body problem that is cohesion in solids provide a perfect example of the philosophy behind systematic degree of freedom elimination which is itself at the heart of multiscale modeling.

The first step taken in the attempt to pare down the problem is to exploit the so-called Born–Oppenheimer approximation. Though this approximation will not enter into our calculations in any material way, it is crucial to our enterprise from a philosophic perspective since it illustrates the way in which degrees of freedom can be unlinked, and thereby, problems can be rendered more tractable. The central physical idea is to separate the nuclear and electronic motions on the basis of the smallness of the ratio m/M, where m is the electron mass and M is that of the relevant nuclei. Indeed, from a multiscale perspective, the Born–Oppenheimer approximation represents the simplest of cases in which there is a natural separation of scales, upon which may be founded the scale-bridging *ansatz*. The physical argument that is made is that the electronic motions are so fast relative to the nuclear motions that we may imagine the electrons to respond to an essentially static set of nuclear positions. The outcome of adopting this view is that the ionic coordinates only appear parametrically in the Schrödinger equation for the electrons.

The key observations may be summarized as follows. The Schrödinger equation we are faced with solving in the context of the Hamiltonian of eqn (4.3) is of the form

$$H\psi(\{\mathbf{R}_i, \mathbf{r}_n\}) = E\psi(\{\mathbf{R}_i, \mathbf{r}_n\}), \tag{4.7}$$

where our notation highlights the dependence of the wave function on all of the nuclear and electronic coordinates. The strategy at this point is to attempt a solution of the form

$$\psi(\{\mathbf{R}_i, \mathbf{r}_n\}) = \psi_e(\{\mathbf{r}_n\}; \{\mathbf{R}_i\})\psi_n(\{\mathbf{R}_i\}), \tag{4.8}$$

where ψ_e and ψ_n refer to the solutions to the electron and nuclear problems, respectively. To be more precise, $\psi_e(\{\mathbf{r}_n\}; \{\mathbf{R}_i\})$ is a solution to the problem

$$(T_e + V_{ee} + V_{en})\psi_e(\{\mathbf{r}_n\}; \{\mathbf{R}_i\}) = E_e\psi_e(\{\mathbf{r}_n\}; \{\mathbf{R}_i\}), \tag{4.9}$$

where we have introduced the notation T_e to describe the electron kinetic energy, V_{ee} is the electron–electron interaction and V_{en} is the interaction between the electrons and the nuclei. In addition, this problem is solved with respect to *fixed* nuclear coordinates $\{\mathbf{R}_i\}$ on the grounds that the nuclei are effectively frozen on the time scale of electronic motions. The final step in the approximation is to solve the Schrödinger equation for the nuclei of the form

$$(T_n + V_{nn} + E_e)\psi_n(\{\mathbf{R}_i\}) = E_n\psi_n(\{\mathbf{R}_i\}), \tag{4.10}$$

where it is to be noted that the energy eigenvalue from the electron problem ends up as part of the effective potential for the nuclear problem.

4.3 Pair Potentials

4.3.1 Generic Pair Potentials

As said earlier, almost the entire business of the present chapter will be to examine the various ways in which the overwhelming complexity presented by the full Hamiltonian of eqn (4.3) has been tamed. Perhaps the simplest example of simplifying the total energy is via an approximation of the form

$$E_{exact}(\{\mathbf{R}_i, \mathbf{r}_n\}) \rightarrow E_{approx}(\{\mathbf{R}_i\}) = \frac{1}{2}\sum_{ij} V(R_{ij}), \tag{4.11}$$

where $V(R_{ij})$ is the pair potential, and R_{ij} is the distance between the ith and j^{th} particles. Examples of the pairwise decomposition of the total energy are as old as Newtonian physics itself. Both the gravitational and Coulomb problems exhibit

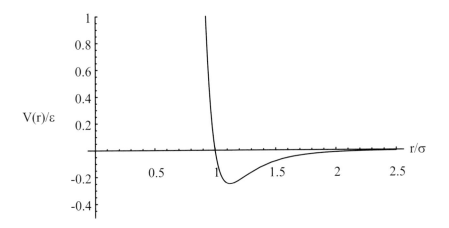

Fig. 4.3. Lennard–Jones pair potential plotted in dimensionless units.

such additivity. Indeed, the ability to write the gravitational potential of a given mass distribution with density $\rho(\mathbf{r})$ as

$$\phi(\mathbf{r}) = G \int \frac{\rho(\mathbf{r}')d^3\mathbf{r}'}{|\mathbf{r} - \mathbf{r}'|} \tag{4.12}$$

may be traced directly to such pairwise additivity.

To understand the origins of pair potentials, we begin by reflecting on their nature in a generic sense, and then turn to an explicit evaluation of quantum mechanically based pair potentials from the standpoint of perturbation theory. As noted above, the total energy within a pair potential setting may be written as $E_{tot} = (1/2) \sum_{ij} V_{eff}(R_{ij})$, where R_{ij} is the distance between the ith and j^{th} atoms. The logic in this case is that the total energy may be decomposed into a number of distinct contributions, one from each of the bonds that make up the system. Perhaps the most familiar example of such potentials is that of the Lennard–Jones potentials which are intended to reflect the polarization interactions that occur in closed-shell systems in which the charge distribution on a given atom may be rearranged as a result of the presence of the neighboring atoms. Within the context of the Lennard–Jones potentials, the energy of interaction between atoms at sites separated by a distance r is

$$V(r) = V_0 \left[\left(\frac{\sigma}{r} \right)^{12} - \left(\frac{\sigma}{r} \right)^6 \right], \tag{4.13}$$

where V_0 is a parameter determining the depth of the potential well and σ is a length scale that determines the position of the minimum. An example of a Lennard–Jones potential, plotted in dimensionless units, is shown in fig. 4.3.

Once the pair potential has been specified, it is a straightforward exercise given the atomic positions to compute the total energy. For simplicity, consider a pair potential $V_{eff}(r)$ whose form is left unspecified. Further, consider the simplifying assumption that this potential reflects a coupling only between 'nearest neighbors'. To compute the total energy of a given configuration, all that is required is that we sum up the independent contributions of all distinct bonds. For example, if we consider an fcc crystal with lattice parameter a_0, the total energy within the context of a generic potential in which only near-neighbor atoms are presumed to interact is given by

$$E_{tot} = \frac{N}{2} \sum_{neighbors} V_{eff}(r_{nn}) = 6N V_{eff}\left(\frac{a_0}{\sqrt{2}}\right). \qquad (4.14)$$

In the previous expression we have exploited the fact that in the fcc crystal, each atom is surrounded by an identical environment that is composed of twelve near neighbors at a distance of $a_0/\sqrt{2}$. Because of the high symmetry of the perfect fcc geometry, it was possible to carry out the total energy computation analytically. The basic idea associated with a pair potential calculation of the total energy is to sum over the entirety of bonds between each atom and its 'neighbors'. When confronted with a more complex geometry, it is likely that we will be reduced to numerical computation, but the concept remains unchanged.

4.3.2 Free Electron Pair Potentials

Though simple analytic expressions such as the Lennard–Jones form adopted above (at least the attractive part) may be constructed on the basis of a sound quantum mechanical plausibility argument, the derivation of pair potentials for free electron systems is even more enlightening and strikes to the heart of bonding in metallic systems. An even more important reason for the current discussion is the explicit way in which the effective description of the total energy referred to in eqn (4.4) is effected. As will be shown below, the electronic degrees of freedom are literally integrated out with the result that

$$E_{exact}(\{\mathbf{R}_i, \mathbf{r}_n\}) \rightarrow E_{approx}(\{\mathbf{R}_i\}; \bar{\rho}) = \frac{1}{2} \sum_{ij} V(R_{ij}; \bar{\rho}). \qquad (4.15)$$

Note, however, that there is a remnant of the electronic degrees of freedom in the form of a dependence of the potentials on the mean electron density.

The starting point for our free electron analysis is the assertion that to a first approximation, simple metallic systems may be thought of as consisting of a uniform electron gas with a density determined by how many electrons each atom has donated to the 'electron sea'. The underlying conceptual model is that described in chap. 3 for electrons in a three-dimensional box. That model is now supplemented by imagining the box to contain not only the free electrons, but also the ions that have donated them. Our task is to compute the total energy of such a system. The presence of the ion cores which have donated the free electrons to the electron gas is then imagined to impose a weak periodic perturbation on the quiescent electron gas. For the present purposes, the key result is that within the confines of perturbation theory, if the effect of these ion cores is carried to second order in the weak perturbing potential, the total energy of the system may be written in real space in exactly the form of a pair potential. We have now advanced the view that the effect of the ionic cores may be treated as a weak perturbating potential $V_p(\mathbf{r})$ on the free electron states discussed earlier. The presence of the perturbing potentials will have the effect of altering the free electron energies, $E_k = \hbar^2 k^2/2m$, and the free electron wave functions, $\psi_{\mathbf{k}}^{(0)}(r) = (1/\sqrt{\Omega})e^{i\mathbf{k}\cdot\mathbf{r}}$.

We begin by recalling that within quantum mechanical perturbation theory, the first- and second-order corrections to the energy of the nth eigenstate, $E_n^{(1)}$ and $E_n^{(2)}$, may be written as

$$E_n^{(1)} = \int \psi_n^{(0)*}(\mathbf{r}) V_p(\mathbf{r}) \psi_n^{(0)}(\mathbf{r}) d^3\mathbf{r} \tag{4.16}$$

and

$$E_n^{(2)} = \sum_{k \neq n} \frac{|\int \psi_n^{(0)*}(\mathbf{r}) V_p(\mathbf{r}) \psi_k^{(0)}(\mathbf{r}) d^3\mathbf{r}|^2}{E_n^{(0)} - E_k^{(0)}}, \tag{4.17}$$

where $V_p(\mathbf{r})$ is the perturbing potential derived from the ionic potentials, $\psi_n^{(0)}(\mathbf{r})$ is the nth wave function of the unperturbed system and $E_n^{(0)}$ is the nth energy eigenvalue in the absence of the perturbation. The results written above should be thought of in the following context. Our preliminary investigation of the electrons in a metal was restricted to the assumption that the presence of the ionic cores is completely irrelevant. We solved this problem in the introductory sections on quantum mechanics in the last chapter. We have now introduced a periodic array of ionic potentials and have included their contribution to the total energy perturbatively.

The picture at this point is that the total potential experienced by an electron in the box as a result of the presence of the ionic cores may be written as

$$V(\mathbf{r}) = \sum_n V_{ps}(|\mathbf{r} - \mathbf{R_n}|). \tag{4.18}$$

We have introduced the notation V_{ps} to denote the atomic pseudopotentials. These pseudopotentials are meant to characterize the interaction of the valence (i.e. outer) electrons with the nucleus and its associated tightly bound core electrons. Our aim now is to assess the corrections to the total energy of the electron gas as a result of the perturbation due to the ionic cores. To begin with, we consider the alteration of the energy in first-order perturbation theory. The change in energy of a plane wave state with wavevector $\mathbf{k}$ is obtained using eqn (4.16) with wave functions $\psi_{\mathbf{k}}^{(0)}(\mathbf{r}) = (1/\sqrt{\Omega})e^{i\mathbf{k}\cdot\mathbf{r}}$ and is given by

$$E_k^{(1)} = \frac{1}{\Omega}\int_\Omega e^{-i\mathbf{k}\cdot\mathbf{r}}V(\mathbf{r})e^{i\mathbf{k}\cdot\mathbf{r}}d^3\mathbf{r}. \tag{4.19}$$

We see immediately that this results in an energy shift $\Delta E = (1/\Omega)\int_\Omega V(\mathbf{r})d^3\mathbf{r}$ that is independent of the wavevector. As a result, the conclusion is that to first order, the presence of the ionic cores does nothing more than impose a uniform shift on the energies of all of the free electron states. It is thus uninteresting as concerns different structural arrangements but contributes to the overall cohesive energy.

We now look to second-order pertubation theory for a determination of a nontrivial change in the energy. Our strategy will be to first compute the energy change associated with a state of a particular wavevector and, subsequently, to sum the change over all occupied states. The net result is the second-order energy change for the electron gas in the presence of a particular ionic configuration. The second-order shift to a state with wavevector $\mathbf{k}$ is obtained using eqn (4.17), again with wave functions $\psi_{\mathbf{k}}^{(0)}(\mathbf{r}) = (1/\sqrt{\Omega})e^{i\mathbf{k}\cdot\mathbf{r}}$ and energies $E_k^{(0)} = \hbar^2 k^2/2m$ and is given by

$$E_k^{(2)} = \sum_{\mathbf{q}\neq 0}\frac{\left|\frac{1}{\Omega}\int_\Omega e^{-i\mathbf{k}\cdot\mathbf{r}}V(\mathbf{r})e^{i(\mathbf{k}+\mathbf{q})\cdot\mathbf{r}}d^3\mathbf{r}\right|^2}{\frac{\hbar^2}{2m}(k^2 - |\mathbf{k}+\mathbf{q}|^2)}. \tag{4.20}$$

We now use eqn (4.18) to obtain

$$E_k^{(2)} = \sum_{\mathbf{q}\neq 0}\frac{\left|\frac{1}{N}\sum_n e^{i\mathbf{q}\cdot\mathbf{R}_n}\right|^2 \left|\frac{1}{\Omega_a}\int_\Omega V_{ps}(\mathbf{r})e^{i\mathbf{q}\cdot\mathbf{r}}d^3\mathbf{r}\right|^2}{\frac{\hbar^2}{2m}(k^2 - |\mathbf{k}+\mathbf{q}|^2)}. \tag{4.21}$$

We have used a simple relation between the total volume Ω and the atomic volume Ω_a, namely, $\Omega = N\Omega_a$. Note that the factors in the numerator are both of particularly simple forms, with the term involving V_{ps} clearly revealed as the Fourier transform of the atomic pseudopotential. It is convenient to invoke the notational convention

$$S(\mathbf{q}) = \frac{1}{N}\sum_n e^{i\mathbf{q}\cdot\mathbf{R}_n}, \tag{4.22}$$

which is the structure factor and is only dependent upon the *geometry* of the ionic positions. If we replace the sum over **q** with the corresponding integral, using the well-worn identity $\sum_{\mathbf{q}} \rightarrow (V/2\pi)^3 \int d^3\mathbf{q}$, our expression for the energy change to second order for the state with wavevector **k** is

$$E_k^{(2)} = \frac{V}{(2\pi)^3} \int d^3\mathbf{q} \frac{|S(\mathbf{q})|^2 |V_{ps}(\mathbf{q})|^2}{\frac{\hbar^2}{2m}(k^2 - |\mathbf{k} + \mathbf{q}|^2)}. \tag{4.23}$$

Our claim is that this expression may now be manipulated in such a way that when it is time to sum the energy change over all of the occupied states, the total energy change will be seen as a sum over effective pair potentials.

With the Fourier transform of the atomic pseudopotential in hand, it is possible to evaluate the energy change explicitly. However, rather than concerning ourselves with the particulars associated with the specific choice of pseudopotential, we will work in the abstract, holding a course that will reveal the emergence of the pair potential as an effective theory for the electronic degrees of freedom as quickly as possible. To demonstrate that the energy shift described above may be written as a sum over pair terms, we proceed as follows. Note that what we have done above is to identify the energy change to second order for the free electron state of wavevector **k**. In order to assess the total energy change of the free electron gas in the presence of the potentials due to the ionic cores, we must now sum over all of the occupied free electron states. This results in a second-order change in the band structure part of the total energy per atom of

$$\Delta E_{tot}^{band} = \frac{2}{N} \sum_{|\mathbf{k}| \le k_F} \frac{V}{(2\pi)^3} \int d^3\mathbf{q} \frac{|S(\mathbf{q})|^2 |V_{ps}(\mathbf{q})|^2}{\frac{\hbar^2}{2m}(k^2 - |\mathbf{k} + \mathbf{q}|^2)}. \tag{4.24}$$

Note that everything except for $S(\mathbf{q})$ is structure-*independent*. By expanding the structure factor explicitly in terms of its definition given in eqn (4.22), it is evident that the energy has decomposed into a series of pair terms as

$$\Delta E_{tot}^{band} = \frac{1}{2N} \sum_{m,n} \phi_{eff}(R_m - R_n), \tag{4.25}$$

with the effective pair interaction given by

$$\phi_{eff}(R_m - R_n) = \frac{4V}{(2\pi)^3} \sum_{|\mathbf{k}| \le k_F} \int d^3\mathbf{q} \frac{e^{i\mathbf{q}\cdot(\mathbf{R_m} - \mathbf{R_n})} |V_{ps}(\mathbf{q})|^2}{\frac{\hbar^2}{2m}(k^2 - |\mathbf{k} + \mathbf{q}|^2)}. \tag{4.26}$$

We now note that in addition to the electronic contribution to the structural part of the energy, the term computed above must be supplemented by the repulsion between the ions. We should also note that if our emphasis had been on computing

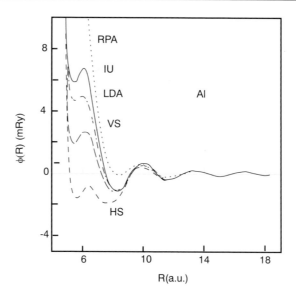

Fig. 4.4. Pair potentials for Al as computed using perturbation analysis (adapted from Hafner (1987)). The four separate potentials shown here reflect different treatments of the electron–electron interaction, where RPA, IU, LDA and VS refer to the names of different approximations used in the construction of potentials.

the potential itself, rather than on demonstrating the way in which the electronic degrees of freedom are subsumed in an effective potential, we could have worked out expressions in which the Fourier transform of the pseudopotential is used explicitly and in which the sums over k and q are performed explicitly leaving in the end a single one-dimensional integral. As a final word of caution, it should be noted that we have been lax in our treatment of screening effects which would modify the quantitative details of what we have said, though not the overall argument concerning the integrating out of the electronic degrees of freedom. For further details, the reader is encouraged to consult Harrison (1980) or Hafner (1987).

The outcome of a calculation such as that given above is not entirely unambiguous. In particular, the choice of atomic pseudopotential and the treatment of screening effects can lead to potentials with different qualitative features. For the sake of discussion, in fig. 4.4 we show a series of pair potentials for Al developed using different schemes for handling these effects. For our purposes, the key insight to emerge from this analysis is that we have seen how, on the basis of a free electron picture of metals, one may systematically construct effective pair potentials.

Though our primary emphasis will center on the calculation of energies, it is also worth remembering that the total energy serves as the basis for the determination of forces as well. In many instances (i.e. static relaxation or molecular dynamics), the

calculation of forces is a prerequisite to the performance of structural relaxation. If
we are to consider a generic pair potential of the form

$$E_{tot} = \frac{1}{2} \sum_{mn} \phi(R_{mn}), \tag{4.27}$$

then the force in the αth Cartesian direction on particle i is given by

$$f_{i\alpha} = -\frac{\partial E_{tot}}{\partial R_{i\alpha}}, \tag{4.28}$$

where we have introduced the notation $R_{i\alpha}$ to characterize the αth component of
the position vector $\mathbf{R}_i$. Explicit evaluation of the derivative we are instructed to
perform results in the force

$$f_{i\alpha} = -\sum_{m} \phi(R_{im}) \frac{(R_{i\alpha} - R_{m\alpha})}{R_{im}}. \tag{4.29}$$

Despite the virtues inherent in using pair potentials, they also exhibit certain
physical features that are troubling. Once the total energy has been written in the
pair potential form, we note that the bond strength is explicitly independent of the
environment. What this means is that no matter how many neighbors atom i may
have, its bond with atom j has energy $\phi_{eff}(R_{ij})$. In the next section, we will see
that *ab initio* calculations suggest that the environmental independence presumed
in pair potential models is a faulty assumption. In addition to this fundamental
qualitative shortcoming, there are other quantitative problems as well. When we
undertake an analysis of elastic properties in ensuing chapters, the conceptual
foundations for the microscopic evaluation of the moduli will be laid in terms of
pair potentials. This development will be made not so much out of necessity, but
rather because the pair potential description lays bare the central idea behind the
evaluation of the moduli. One of the outcomes of such an analysis will be the
realization of just how restrictive is a pair potential description of the total energy
(especially in those cases in which volume-dependent terms are neglected). We
will show, for example, that for cubic crystals, the elastic moduli C_{12} and C_{44} are
equal, in contradiction with observation. Though there are ways to soften this blow,
fundamentally, the pair description is left wanting.

A second difficulty with the pair potential description is manifest when we
attempt to consider defect formation energies. Specifically, we will also show that
the pair potential description implies a formation energy for vacancies that is a vast
overestimate relative to the numbers found in experiment. As is true for the elastic
moduli, there are modifications that can be considered, but not without a sense that
the crucial physics is being ignored. In response to these difficulties, and others,
concerted efforts have been made to extend our description of the energetics of

solids without paying the full price of *ab initio* calculation. We undertake such analyses now.

4.4 Potentials with Environmental and Angular Dependence

As we have noted above and will see in more detail later, the pair potential description is littered with difficulties. There are a few directions which can be taken that will be seen to cure some of them. We begin with a qualitative discussion of these issues, and turn in what follows to the development of both angular force schemes and effective medium theories that patch some of these troubles, primarily through the device of incorporating more of what we know about the nature of bonding in solids.

4.4.1 Diagnostics for Evaluating Potentials

We begin with a few qualitative insights into where the pair potential description might be lacking. One question we might pose that could expose a particularly problematic assumption of a pair potential description would be: how does the strength of a given bond depend upon whether or not the atom is participating in other bonds? Using *first-principles* calculations one can address this question systematically by repeatedly evaluating the strength of bonds for a range of different geometries. By first-principles calculations, we mean fully quantum mechanical calculations that are, at least in principle, parameter-free. An example of this strategy is revealed in fig. 4.5 where it is seen that the strength of the bond goes roughly inversely as the square root of the coordination number. That is, as the number of neighbors to a given atom increases, the energy associated with each bond goes down. The first-principles calculations were performed by considering a number of different structures, some observed, others hypothetical, which have near-neighbor environments that differ considerably in terms of coordination number. For example, the diamond cubic structure features only four near-neighbor bonds while the fcc structure is characterized by twelve. The energies of each of these structures were computed using the type of first-principles techniques to be described later in the chapter, and then the resulting energies were plotted in a normalized form that revealed the energy per bond. In the context of pair potentials, what do these results teach us? Stated simply, the conclusion is that the bond strength has a certain environmental dependence. As will become evident in what follows, this environmental dependence can be mimicked by a class of energy functionals that are known variously as effective medium theories, Finnis–Sinclair potentials, glue models, the embedded atom method and pair functionals.

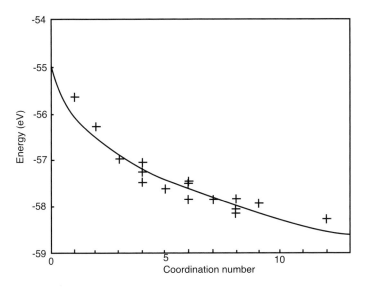

Fig. 4.5. Energy per atom for a variety of different structures in Al having different coordination numbers (after Payne *et al.* (1996)). The curve is a fit to the functional form $E = E_0 + \alpha Z^{\frac{1}{2}} + \beta Z$, where Z refers to the coordination number of a given atom.

A second perspective on the weakness of pair potentials comes from a consideration of bond angles. For example, if we are to consider a simple pair potential of the Morse or Lennard–Jones form in which there is a minimum associated with the near-neighbor separation, it is difficult to imagine how open structures such as those found in the covalent semiconductors could be stabilized. The argument in this case might go as follows: twelve bonds taking advantage of the near-neighbor separation is better than only four. As a result, we are inclined to the view that there must be something particularly energetically important about the existence of certain bond angles.

4.4.2 Pair Functionals

At the next level of sophistication beyond pair potentials, a variety of methods which can broadly be classified as pair functionals have been developed. In keeping with the various surrogates for the full Hamiltonian introduced in eqns (4.4), (4.5) and (4.30), the total energy in the pair functional setting is given by

$$E_{exact}(\{\mathbf{R}_i, \mathbf{r}_n\}) \rightarrow E_{approx}[\rho(\mathbf{r}), \{\mathbf{R}_i\}] = \frac{1}{2} \sum_{ij} \phi(R_{ij}) + \sum_i F(\rho_i), \qquad (4.30)$$

as will be seen below. Pair functionals can be motivated in a number of different ways, a few of which will be taken up here. One route to pair functionals is to

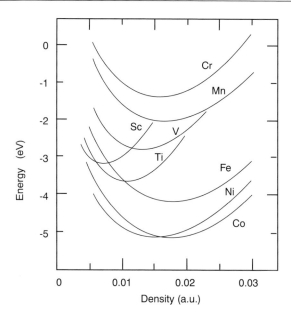

Fig. 4.6. Energy cost to embed ion in electron gas of particular density (from Jacobsen *et al.* (1987)).

revisit the electron gas, seen in a different light from that discussed in the context of free electron pair potentials. The physical picture that we will adopt here may be summarized as follows. The question that is posed is what is the energy cost to deposit an ion into a homogeneous electron gas of density ρ? What is found and will be discussed at some level of detail below is that there is an optimum electronic density (i.e. an energy minimizing density), and that optimal density depends upon the identity of the ion being placed in the electron gas. Consideration of fig. 4.6 reveals the outcome of a series of numerical calculations made using density functional theory for a host of elements. What is learned from these calculations is that an ion of a particular type has a taste as to what the optimal local electron density should be. This insight provides the opportunity to construct a new class of energy functional which overcomes some of the limitations of the pair potential description.

The philosophy of the pair functional approach is perhaps best illustrated via a concrete example. Consider fcc Ni. In this case, we know on the basis of the results shown above, that the Ni atom prefers to live in a certain equilibrium electron density. As yet, we have not seen how to exploit this insight in the context of particular geometric arrangements of a series of Ni atoms. We may see our way through this difficulty as follows. If we center our attention on a particular Ni atom, we may think of it as being deposited in an electron gas, the density of which is dictated by the number and proximity of its neighbors. The hypothesized energy functional that captures that dependence may be written as

$$E_{embedding} = F(\rho),$$ (4.31)

where we further suppose that the electron density of the electron gas that the Ni atom finds itself in is given by

$$\rho_i = \sum_{j \neq i} f(R_{ij}).$$ (4.32)

This equation tells us how to determine the 'density' of the electron gas at the site of interest which we denote ρ_i in terms of the geometric disposition of the neighboring atoms. In particular, $f(R)$ is a pairwise function that characterizes the decay of electronic density from a given site. Hence, if we wish to find the contribution of atom j to the density at site i, the function $f(R)$ evaluated at the distance $R = |\mathbf{R}_i - \mathbf{R}_j|$ tells us exactly how much electronic density bleeds off from site j onto its neighbors.

To bring the development of the pair functional approach to its completion, we assume that the total energy may be divided into two parts; one part whose aim is to mimic the type of electron gas embedding terms envisaged above and another whose intent is to reflect the Coulombic (possibly screened) interactions between the various nuclei. As yet, we have examined the embedding energy on the basis of the quantum mechanics of the electron gas. We have argued that on the basis of legitimate numerical evidence from quantum calculations one can construct the functional $F(\rho)$. On the other hand, a large body of work has been built around the idea of taking one's cue from the quantum mechanical foundations laid above, but instead to assume a pragmatist's position and to *fit* the various functional forms on the basis of known data. The embedded atom method is predicated on the latter alternative, and posits a total energy of the form,

$$E_{tot} = \frac{1}{2} \sum_{ij} \phi(R_{ij}) + \sum_i F(\rho_i),$$ (4.33)

where the density on site i is given in turn by an equation of the form of eqn (4.32) above.

Our appeal that pair potentials were inadequate for capturing many of the energetic features of real materials was founded in part upon the environmental dependence of the bond strengths, captured in fig. 4.5. It is of interest to revisit this question in light of the simple embedding form adopted above. As with our discussion of the free electron pair potentials, our aim at the moment is not to discuss the optimal choice of embedding functions but rather to identify the extra physics that makes such energy functionals an improvement over pair potentials. For the sake of argument, consider the same near-neighbor model discussed earlier

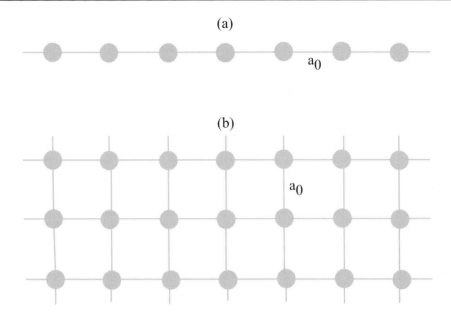

Fig. 4.7. Model geometries used to illustrate computation of total energy within pair functional formalism: (a) linear chain, (b) square lattice.

in which the density due to a single neighbor at the equilibrium distance R_{nn} is given by f_0. Our aim now is to consider the energy/atom for a few distinct structures in order to explicitly exhibit the environmental dependence that was claimed to be of interest above, and to build facility in using this functional form.

For the linear chain of atoms (see fig. 4.7), the total energy per atom within the embedded atom context is given by

$$E_{chain} = \phi(R_{nn}) + F(2f_0). \tag{4.34}$$

This result reflects the fact that every atom on the chain has an identical environment with two near neighbors at the equilibrium distance. If we continue along the lines set above and consider the total energy of the square lattice depicted in fig. 4.7, it is seen to have a total energy of

$$E_{square} = 2\phi(R_{nn}) + F(4f_0). \tag{4.35}$$

The analysis precedes similarly for other cases of interest, such as, for example, the fcc structure, which has a total energy of

$$E_{fcc} = 6\phi(R_{nn}) + F(12f_0). \tag{4.36}$$

Note that in the absence of the embedding term the total energy per bond is constant as expected. However, if we adopt the form $F(\rho) = \sqrt{\rho}$ which will be motivated

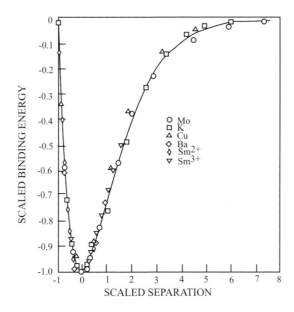

Fig. 4.8. Universal binding energy relation (adapted from Rose *et al.* (1984)).

after we have introduced the tight-binding method, the energy per bond is clearly *not* constant.

One of the cleanest routes to determining embedding functions is by invoking the universal binding energy relation (UBER) (Rose *et al.* 1984). On the basis of a vast collection of experience with first-principles calculations of a range of materials, it was found that the total energy as a function of lattice parameter may be fitted to

$$E_{tot}(r) = -E_c \left[1 + \alpha \left(\frac{r}{r_0} - 1 \right) \right] e^{-\alpha(\frac{r}{r_0} - 1)}, \qquad (4.37)$$

where r is a measure of the interatomic separation and α and r_0 are scaling parameters which collapse the specific material data onto the universal curve. An example of this scaling behavior as a function of the lattice parameter for a range of materials is shown in fig. 4.8. This form for the total energy suggests a strategy for constructing embedding functions via the following argument. We assume the pair potential is of some specified form. Once this form is adopted, the embedding function may be deduced as the difference between the total energy as given by the universal binding energy relation and the total contribution to the pair potential.

To illustrate this procedure, we follow Johnson (1988) by assuming a near-neighbor only interaction. Hence, as was shown in eqn (4.36), the energy of a given atom is $E_i = F(\rho_i) + 6\phi(R_{nn})$. If we now assert that the left hand side of this equation is determined by the universal binding energy relation, then the embedding function is

$$F[\rho(R)] = E_{UBER}(R) - 6\phi(R). \tag{4.38}$$

Johnson makes the explicit choice of density function

$$\rho(r) = fe^{-\beta(\frac{r}{r_0}-1)}, \tag{4.39}$$

and defines the pair potential as

$$\phi(r) = \phi_0 e^{-\gamma(\frac{r}{r_0}-1)}. \tag{4.40}$$

Note that we have imitated Johnson's notation in which the quantity f is a constant factor that appears in the determination of the density ρ, and the parameters r_0, β, γ and ϕ_0 are other parameters that are determined through the fitting procedure. Once the density function and the pair potential have been specified, the determination of the embedding function is complete in principle. In this case, because of the choice of functions, the relevant equations can be solved analytically with the result that the embedding energy is given by

$$F(\rho) = -E_c\left(1 - \frac{\alpha}{\beta}\ln\frac{\rho}{\rho_0}\right)\left(\frac{\rho}{\rho_0}\right)^{\frac{\alpha}{\beta}} - 6\phi_0\left(\frac{\rho}{\rho_0}\right)^{\frac{\gamma}{\beta}}. \tag{4.41}$$

The basic strategy used in deriving this form for the embedding function is the recognition that the density function $\rho(r)$ may be inverted to yield

$$\frac{r}{r_0} - 1 = -\frac{1}{\beta}\ln\frac{\rho(r)}{\rho_0}, \tag{4.42}$$

where ρ_0 is the density when the interatomic separation is r_0. This form for the embedding energy is contrasted with others in fig. 4.9. The central importance of this particular embedding strategy is that now that the analytic form is in hand, it is possible to examine a range of secondary quantities such as the elastic moduli, surface energies and point defect formation energies *analytically*. Of course, there exist significantly more sophisticated (and accurate) schemes for carrying out pair functional calculations. While such functions may be preferable in the attempt to calculate real material properties, the Johnson potential serves as an ideal pedagogic vehicle for discussing embedding energies. From the standpoint of 'back of the envelope' exercises, one may use analytic near-neighbor models such as that proposed by Johnson to estimate quantities ranging from the (111) surface energy to the vacancy formation energy.

In addition to the importance of energies, a number of situations demand a knowledge of the forces between atoms. We have already seen in eqn (4.29) how forces are computed within the pair potential setting. The presence of the

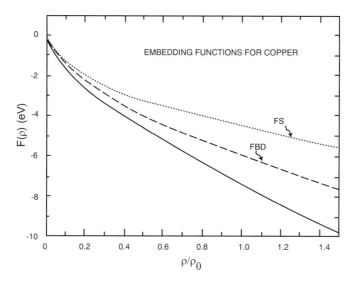

Fig. 4.9. Various representations of the embedding energy for Cu (adapted from Johnson (1988)). The solid curve is the embedding function for Cu derived from UBER as described in text, the dashed curve is a fit for the embedding function and the dotted curve corresponds to an embedding function of the Finnis–Sinclair form in which $F(\rho) \propto \sqrt{\rho}$.

embedding term in the pair functional setting leads to additional subtlety in the computation of forces. In particular, the expression for the force on the ith atom due to the embedding term is given by

$$\mathbf{f}_i = -\sum_{j \neq i}(F'(\rho_j)\rho_i' + F'(\rho_i)\rho_j')\hat{\mathbf{R}}_{ji}, \qquad (4.43)$$

where $\hat{\mathbf{R}}_{ji}$ is the unit vector pointing from the jth to the ith atom. It is left to the reader to derive this result explicitly in the problems at the end of the chapter. A final intriguing point about energy functionals like those described in this section is that they may be transformed into an effective pair potential description of the total energy. This claim is an interesting example of the various layers in effective theory construction and reflects a transformation of the form

$$E_{approx}[\rho(\mathbf{r}), \{\mathbf{R}_i\}] \rightarrow E_{approx}(\{\mathbf{R}_i\}; \rho_{ref}) = \frac{1}{2}\sum_{ij} V_{eff}(R_{ij}; \rho_{ref}), \qquad (4.44)$$

where ρ_{ref} denotes the dependence of the resulting pair potential on a reference density. Thus, the effective pair potential developed to describe bulk environments will differ considerably from that derived to account for free surfaces. Indeed, these potentials will reflect exactly the environment-dependent bond strengthening that is the hallmark of the pair functional approach. The reader is asked to fill in the details of this development in the problem section at the end of the chapter.

4.4.3 Angular Forces: A First Look

There are some circumstances in which the inclusion of the 'many-body' terms present in pair functionals is insufficient and recourse must be made to angular terms. As yet, our discussion has been dominated by metals. On the other hand, more than fifty different potentials have grown up around the question of how best to model covalent materials, with Si as the dominant example. One of the remarkable features of Si in the solid state is its open structure, characterized by only four neighboring atoms in the first neighbor shell in contrast with the twelve neighbors found in close-packed structures. As a result of this structural clue, it seems that the insistence on the coupling between bond lengths and energy must be supplemented by consideration of the energetics of bond angles.

As will be seen below, both angular potentials and cluster functionals have been developed to characterize materials such as Si. Our aim here is to give a flavor of these developments by recourse to two of the most celebrated total energy schemes used in semiconductors: the Stillinger–Weber angular potentials and the Tersoff cluster functional. The Stillinger–Weber potentials were developed with the mind of explaining the condensed phases of Si, including the fact that in the liquid state, the coordination number of Si atoms increases. The basic structure of the Stillinger–Weber potential is revealed in the expression for the total energy which is

$$E_{tot}(\{\mathbf{R}_i\}) = \frac{1}{2} \sum_{i,j} V_2(R_{ij}) + \frac{1}{3!} \sum_{i,j,k} V_3(\mathbf{R}_i, \mathbf{R}_j, \mathbf{R}_k), \tag{4.45}$$

with the first term corresponding to a pair potential of the type with which we are already familiar, and the second term is an angular term which imposes penalties when the angles in three-atom triangles deviate from particularly stable values. The potentials themselves are represented as

$$V_2(R_{ij}) = \epsilon f_2\left(\frac{R_{ij}}{\sigma}\right), \tag{4.46}$$

and

$$V_3(\mathbf{R}_i, \mathbf{R}_j, \mathbf{R}_k) = \epsilon f_3\left(\frac{\mathbf{R}_i}{\sigma}, \frac{\mathbf{R}_j}{\sigma}, \frac{\mathbf{R}_k}{\sigma}\right). \tag{4.47}$$

The functions $f_2(R_{ij})$ and $f_3(\mathbf{R}_i, \mathbf{R}_j, \mathbf{R}_k)$ are defined by

$$f_2(R) = \begin{cases} A(BR^{-p} - R^{-q})e^{[(r-a)^{-1}]} & \text{if } r < a \\ 0 & \text{otherwise} \end{cases} \tag{4.48}$$

and

$$f_3(\mathbf{R}_i, \mathbf{R}_j, \mathbf{R}_k) = h(R_{ij}, R_{ik}, \theta_{jik}) + h(R_{ji}, R_{jk}, \theta_{ijk}) + h(R_{ki}, R_{kj}, \theta_{ikj}). \tag{4.49}$$

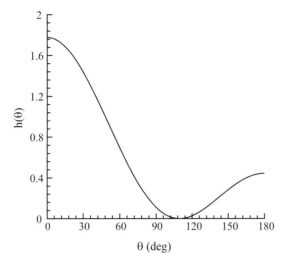

Fig. 4.10. Stillinger–Weber angular term.

The novel feature associated with the Stillinger–Weber potentials is this angular term. To be more specific, the three-body terms have an angular dependence of the form

$$h(R_{ij}, R_{ik}, \theta_{jik}) = \lambda e^{[\gamma(R_{ij}-a)^{-1}+\gamma(R_{ik}-a)^{-1}]}\left(\cos\theta_{jik} + \frac{1}{3}\right)^2, \qquad (4.50)$$

where only the angle-dependent part of this function is shown in fig. 4.10.

Recall that in our discussion of pair potentials, structural stability was attributed to the fact that particular structures have bond lengths that take advantage of minima in the pair potential. On the other hand, the existence of angular potentials paves the way for a second form of correspondence between structure and energetics. In particular, now not only must bond length constraints be satisfied, but bond angle constraints as well. In this case, we see that the angular term is minimized in the case that $\theta \approx 109.47°$, corresponding to the favorability of the tetrahedral angles that characterize the sp^3 bonding in many covalent materials.

The computation of the total energy in the context of a total energy scheme involving angular forces is quite analogous to that considered in the pair potential setting. Indeed, that part of the energy due to the pair potential terms is identical to those considered earlier. However, the angular terms involve a more careful analysis of the neighbor environment for a given atom. In particular, these terms require that a three-body contribution to the total energy be made for each and every triplet of atoms in which the distances between neighbors are smaller than the range of V_3. To illustrate the reckoning of these three-body contributions

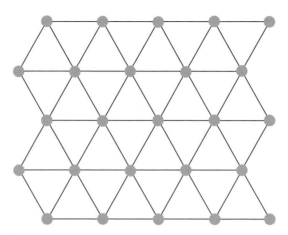

Fig. 4.11. Schematic of the triangular lattice used in computing the total energy using angular forces.

to the total energy, we consider the simple case of the perfect two-dimensional triangular lattice shown in fig. 4.11, and we further assume that the range of V_3 is chosen such that only near-neighbor atoms are considered. Given these simplifying assumptions, the total energy per atom may be computed as

$$E_{tot} = 3V_2(R_{nn}) + 6h(R_{nn}, 60°) + 6h(R_{nn}, 120°) + 3h(R_{nn}, 180°), \qquad (4.51)$$

where we have used the notation $h(R_{nn}, \theta) = h(R_{nn}, R_{nn}, \theta)$. The contribution of the angular term was arrived at by noting that by translational invariance, each and every atom has an identical local environment. In more complicated geometrical situations, the strategy required is similar. Essentially, it is necessary to find all three-atom clusters that are within the required proximity and to sum up their contributions separately.

Note that our example is especially artificial in the context of the Stillinger–Weber potential since for the triangular lattice it is $60°$ angles, and not the tetrahedral angles, that form the central angular motif. However, our reason for including this example is not to argue for the stability of the triangular lattice in the Stillinger–Weber context, but rather to show how calculations are made using this potential. We also note that in the remainder of the book, the Stillinger–Weber potential will often be invoked as the model of choice for discussing empirical treatments of Si in much the same way that the Johnson potential will serve as one of our key paradigms for pair functionals.

The Stillinger–Weber scheme lacks the explicit environmental dependence that was argued to be of importance in our discussion of pair functionals. However, environmental dependence has been accounted for in covalent materials as well. Tersoff (1988) was one of the first to suggest such a scheme, and his total energy takes the form

$$E_{tot} = \frac{1}{2} \sum_{ij} [V^r(R_{ij}) + b_{ij} V^a(R_{ij})], \tag{4.52}$$

where $V^r(R_{ij})$ is a repulsive potential of the form $Ae^{-\lambda_1 R_{ij}}$ and, similarly, $V^a(R_{ij})$ is an attractive potential of the form $Be^{-\lambda_2 R_{ij}}$. As Tersoff notes, the key novelty is buried in the function b_{ij}. Indeed, it is via this function that both the environmental *and* angular dependences enter the potential. The form adopted by Tersoff is

$$b_{ij} = (1 + \beta^n \zeta_{ij}^n)^{(-1/2n)}, \tag{4.53}$$

where the function ζ_{ij} is in turn given by

$$\zeta_{ij} = \sum_{k \neq ij} f_c(R_{ik}) g(\theta_{ijk}). \tag{4.54}$$

The function $f_c(R)$ is a cutoff function and passes quickly but smoothly from unity to zero as $R \to R_c$, where R_c is a cutoff distance. The function $g(\theta_{ijk})$ is an angular term of the form

$$g(\theta) = 1 + \frac{c^2}{d^2} - \frac{c^2}{d^2 + (h - \cos\theta)^2}. \tag{4.55}$$

While these potentials are notationally complex, they are conceptually simple in terms of the basic physics they incorporate. In particular, potentials of this type invoke both of the new physical ideas presented in this section: environmental dependence and angular forces. In particular, the quantity denoted b_{ij} guarantees that the interaction between two given atoms will depend upon both the number and disposition of neighboring atoms. For our purposes, the key advantage of the Tersoff potential is its attempt to capture effects which are outside of the purview of more traditional potentials.

The Stillinger–Weber and Tersoff schemes as introduced above are essentially *ad hoc*. In forthcoming chapters, we will describe some of the successes and shortcomings of models like these. These examples will show that it is important to build intuition concerning the bonding in solids in order to formulate a clear idea as to when a particular empirical description of the energy is suitable and when it can be expected to fail. One way to service this ambition is to develop a more systematic scheme for deriving such potentials, or if not deriving them, then at least developing plausibility arguments for their functional form. The tack adopted here is to appeal to an understanding of the electronic structure of a material with special reference to the electronic density of states and its moments (which will be defined shortly).

4.5 Tight-Binding Calculations of the Total Energy

Thus far, our emphasis has been on methods in which no explicit reference is made to the electronic degrees of freedom. In the present section we make our first concession in the direction of incorporating electronic structure information into the total energy in the form of the tight-binding method. From the perspective of the general discussion given at the beginning of the chapter, the tight-binding total energy can be thought of in terms of the replacement

$$E_{exact}(\{\mathbf{R}_i, \mathbf{r}_n\}) \rightarrow E_{approx}(\{\mathbf{R}_i, \alpha_i\}), \tag{4.56}$$

where the coefficients $\{\alpha_i\}$ are expansion coefficients that characterize the electronic wave function. That is, the tight-binding scheme represents an approximate treatment of the electronic structure of the material. The rudiments of the tight-binding method can be spelled out as follows. First, we take our cue from the discussion of the H_2 molecule given in chap. 3, where we noted that the total molecular wave function could be written as a linear combination of atomic wave functions centered on the various sites making up the molecule. In the case of a solid, we will adopt the same strategy, this time with the electronic wave function of the entire solid being built up as a linear combination of the atomic wave functions centered on the various sites that make up the solid. The coefficients in the linear combination of these basis functions will then be thought of as variational parameters, chosen so as to obtain the lowest possible energy. Finally, the resulting Hamiltonian matrix is diagonalized with the resulting eigenvalues serving as the basis for the calculation of the electronic contribution to the total energy.

The tight-binding method will be exploited in two distinct senses. The first main thrust of this section will be to show how the tight-binding method can be used to compute the total energy directly without recourse to some reduced description in terms of pair functionals or angular forces. Secondly, tight-binding arguments will be invoked in order to provide a plausibility argument for the form of both the pair functionals and angular potentials introduced in previous sections. Ultimately, the potentials will be seen to inherit the angular features of the moments of the electronic density of states. Prior to undertaking either of these objectives, we must first describe the tight-binding method itself.

4.5.1 The Tight-Binding Method

The tight-binding method is based upon the same set of observations that inspired the approach to molecular bonding introduced in chap. 3. We imagine that in the act of assembling the solid, the wave functions that characterize the isolated atoms are viable building blocks for describing even the complex wave function appropriate to the solid. When atoms are brought into one another's proximity, the atomic energy levels associated with the outer valence electrons are disturbed as a result of the overlap of wave functions on adjacent atoms. This simple observation leads to a naive picture of the formation of the chemical bond. In particular, by summing up the occupied one-electron eigenvalues, one determines the electronic part of the total energy. As was shown in fig. 3.13, as the number of atoms participating in the molecule increases, the resulting electronic spectrum becomes increasingly complex. Our discussion of the tight-binding method is intended to set forth in quantitative detail the recipe for computing the one-electron eigenvalues for solids.

As noted already, the starting point of the analysis is an approximation to the total electronic wave function for the solid. In particular, we seek a wave function of the form

$$\psi(\mathbf{r}) = \sum_{i=1}^{N} \sum_{\alpha=1}^{n} a_{i\alpha} \phi_{\alpha}(\mathbf{r} - \mathbf{R}_i), \tag{4.57}$$

which we will write more compactly as

$$|\psi\rangle = \sum_{i=1}^{N} \sum_{\alpha=1}^{n} a_{i\alpha} |i, \alpha\rangle. \tag{4.58}$$

The sum over i in eqns (4.57) and (4.58) runs over the number of atoms N in the solid, while the sum on α is over the number of orbitals per site. The atomic-like orbital of type α centered on site i is written as $|i, \alpha\rangle = \phi_{\alpha}(\mathbf{r} - \mathbf{R}_i)$. Indeed, the nature of these orbitals and the size of the parameter n strikes right to the heart of the minimalist character of the semiempirical tight-binding method.

As an example of the characteristics of the type of basis $|i, \alpha\rangle$ that might be used in a tight-binding calculation, we consider two distinct cases, that of Si, a covalently bonded semiconductor with an energy gap, and W, a transition metal with an electronic structure that is dominated by its d-electrons. We consider both examples, since each sheds different light on the questions that arise in the tight-binding context. As mentioned above, the basic spirit of tight-binding theory is minimalist in the sense that one searches for the smallest possible basis for expanding the total wave function. For the case of W, this might mean considering only five d-orbitals per atom. Recall from the discussion of chap. 3 that the solution of the atomic hydrogen problem resulted in a series of energy eigenvalues and their associated wave functions. These solutions could be organized along the lines of their angular momenta, with the possibility of s-orbitals ($l = 0$), p-orbitals ($l = 1$), d-orbitals ($l = 2$) and so on. For the case at hand, it is the 5d orbitals of W that serve as our basis. That is, each W atom has electrons from the 5d shell which contribute to the overall electronic wave function for the solid. We also observe that if we are interested in Si, for example, rather than

W, our expansion will center on the outer 3s and 3p electrons. In this case, the total wave function is built up out of the four such basis functions at each site which for the ith site we label as $|i, 3s\rangle$, $|i, 3p_x\rangle$, $|i, 3p_y\rangle$ and $|i, 3p_z\rangle$. Note that in keeping with our aim to shed unnecessary degrees of freedom, all reference to core electrons has been omitted; they are assumed to be unchanged by the proximity of neighboring atoms. In addition, the basis itself is *incomplete* in the mathematical sense, since we have not included higher energy atomic states such as $|i, 4s\rangle$, etc. in the basis.

Having identified the nature of our approximation to the wave function, the next step in the argument concerns the evaluation of the energy. In particular, we aim to determine the series of coefficients $a_{i\alpha}$ which state with what weight the αth orbital on site i (i.e. $|i, \alpha\rangle$) contributes to the total wave function $|\psi\rangle$. The scheme is to construct an energy function that depends upon the set of coefficients $a_{i\alpha}$ and to find the values of those coefficients that minimize that energy. Just as was done in our finite element treatment of the Schrödinger equation in chap. 3, we have replaced the problem of finding unknown functions (i.e. the wave functions) with that of finding unknown coefficients through the device of writing the energy function

$$\Pi(\{a_{i\alpha}\}) = \langle\psi|\hat{H}|\psi\rangle - E\langle\psi|\psi\rangle, \tag{4.59}$$

where it is understood that the wave function ψ depends explicitly on the as-yet undetermined coefficients $a_{i\alpha}$. Substitution of the total wave function given in eqn (4.58) into eqn (4.59) results in

$$\Pi(\{a_{i\alpha}\}) = \sum_{i,j=1}^{N} \sum_{\alpha,\beta=1}^{n} (h_{ij}^{\alpha\beta} - E S_{ij}^{\alpha\beta}) a_{i\alpha}^* a_{j\beta}, \tag{4.60}$$

where we have introduced the notation

$$h_{ij}^{\alpha\beta} = \langle i, \alpha|\hat{H}|j, \beta\rangle = \int_{\Omega} \phi_\alpha^*(\mathbf{r} - \mathbf{R}_i)\hat{H}\phi_\beta(\mathbf{r} - \mathbf{R}_j)d^3\mathbf{r}, \tag{4.61}$$

and

$$S_{ij}^{\alpha\beta} = \langle i, \alpha|j, \beta\rangle = \int_{\Omega} \phi_\alpha^*(\mathbf{r} - \mathbf{R}_i)\phi_\beta(\mathbf{r} - \mathbf{R}_j)d^3\mathbf{r}. \tag{4.62}$$

The nature of the matrices $h_{ij}^{\alpha\beta}$ and $S_{ij}^{\alpha\beta}$ is subtle and as a result has been tackled from a number of different perspectives. Since our aim is to present ideas rather than to give an exhaustive description of each and every alternative approach to a given problem, the present discussion will try to highlight some of the distinguishing features of these quantities. The matrix $S_{ij}^{\alpha\beta}$ provides a measure of the overlap of wave functions on adjacent sites, and is often described as the 'overlap matrix'. For the purposes of the present discussion, we adopt the simplest approximation of orthonormality which is characterized mathematically as

$$S_{ij}^{\alpha\beta} = \delta_{ij}\delta_{\alpha\beta}, \tag{4.63}$$

with the proviso that nonorthogonal tight-binding methods have much to recommend them and hence, the interested reader should pursue this matter further. We make equally strong approximations in the context of the matrix $h_{ij}^{\alpha\beta}$ as well by resorting to the so-called two-center approximation. The basic idea of this approximation may be understood in qualitative terms by noting that the total Hamiltonian is constructed as a sum of potentials arising from the different atoms (i.e. $V(\mathbf{r}) = \sum_{i=1}^{N} V_{atom}(\mathbf{r} - \mathbf{R}_i)$). The essence of the two-center approximation is to say that the only terms from $V(\mathbf{r})$ contributing to $h_{ij}^{\alpha\beta}$ are those arising from $V_{atom}(\mathbf{r} - \mathbf{R}_i)$ or $V_{atom}(\mathbf{r} - \mathbf{R}_j)$. Within this approximation, a given matrix element depends upon both the separation and relative orientation of atoms i and j. The interactions between orbitals on different sites can be parameterized in terms of certain energetic factors $V_{ss\sigma}$, $V_{pp\sigma}$, etc. and the direction cosines l, m, n of the vector connecting the ith and j^{th} atoms. As yet, however, our discussion has been generic and has not revealed insights into the nature of the tight-binding matrix elements themselves. To best see in qualitative terms where these matrix elements come from, we return to the simpler molecular setting. Recall that the orbitals of different angular momentum character have wave functions characterized by different numbers of lobes. An s-orbital is spherically symmetric, while p-orbitals have single lobes. As a result, the matrix elements which parameterize the interactions between orbitals on different sites depend exclusively on geometric/symmetry characteristics of the participating orbitals and, in particular, on the symmetry of these orbitals about the axis joining the two sites of interest. For example, for s-orbitals, the symmetry about the bond axis is such that we can rotate the orbitals about the bond axis with impunity (i.e. there is axial symmetry). Such bonds are labeled as σ bonds, and lead to the coupling $V_{ss\sigma}$. However, axial symmetry bonds can also occur for p-orbitals. For example, if we have two atoms aligned along the z-axis, and consider the interaction between the p_z-orbital on the first atom and the p_z-orbital on the second atom, this too has azimuthal symmetry and is parameterized by the coupling $V_{pp\sigma}$. On the other hand, if we consider the coupling of p_z-orbitals between two atoms that are neighbors along the x-axis, then the symmetry of their coupling is denoted by π, and leads to couplings of the form $V_{pp\pi}$. The geometry of orbitals associated with these couplings is depicted in fig. 4.12. We take up the question of how precisely parameters such as $V_{ss\sigma}$, $V_{pp\sigma}$, $V_{pp\pi}$, $V_{dd\sigma}$, etc. are determined later in this section.

For example, for two p_y-orbitals, the matrix element $h_{ij}^{p_y p_y}$ can be split up into a term involving $V_{pp\sigma}$ and a second term involving $V_{pp\pi}$. As a result of these

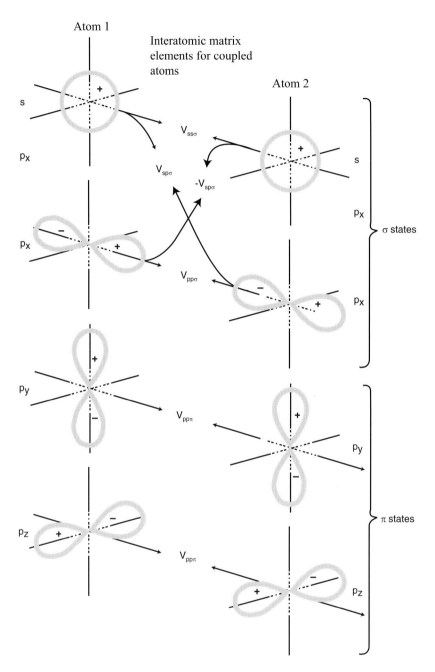

Fig. 4.12. Schematic indication of the different symmetry types associated with couplings between orbitals on different sites (adapted from Harrison (1980)).

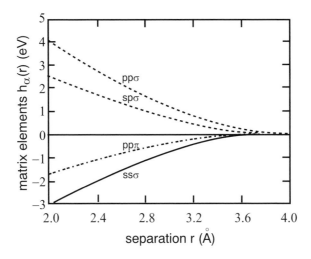

Fig. 4.13. Radial dependence of the tight-binding coupling parameters for Si (adapted from Kwon *et al.* (1994)).

arguments, the calculation of the Hamiltonian matrix is reduced to a series of simple geometric operations (i.e. the determination of the direction cosines of the vector joining the two atoms of interest) in conjunction with a table known as the Slater–Koster table which illustrates how two orbitals are coupled. In addition to the dependence of the coupling between two sites on the relative orientation of the vector joining them, there is also a strong dependence of these couplings on the distance between the two sites. An example of the radial dependence of these couplings is shown in fig. 4.13.

Once the Hamiltonian matrix is in hand, the next step in the procedure is to seek the energy minimizing coefficients $\{a_{i\alpha}\}$. This minimization is effected by computing $\partial \Pi / \partial a_{i\alpha}^* = 0$ for all i and α, and results in a series of coupled algebraic equations that may be written in matrix form as

$$\sum_{p'} (H_{pp'} - E\delta_{pp'})a_p = 0, \tag{4.64}$$

where we have introduced the superindex p that runs from 1 to nN, with each p corresponding to a different one of the nN choices of i and α. This implies that we will have nN equations in the unknowns $\{a_{i\alpha}\}$. The result of solving these equations will be nN different energy eigenvalues and their corresponding set of expansion coefficients which in turn determine the wave function itself.

What has been learned thus far? We have noted that in the context of our truncated basis, built from the atomic orbitals characterizing the free atoms, the solution of the Schrödinger equation is mapped onto the equivalent problem of finding that set of expansion coefficients that minimize the total energy. Like in the finite element context discussed earlier, minimization problems of this type lead to coupled equations such as that found in eqn (4.64). From the point of view of that set of equations, our problem has been reduced to a series of coupled linear algebraic equations and to find the undetermined coefficients we must diagonalize the Hamiltonian matrix. The result of this exercise is a series of energy eigenvalues that represent, within the confines of the present model, the allowed energy states of the solid.

The spectrum of energy eigenvalues that emerge from the type of calculation spelled out above can be complex in the sense that there can be a large number of such eigenvalues spread across a wide range of energies. One way to tame this complexity is by virtue of the electronic density of states (DOS). The density of states is essentially a histogram that characterizes the distribution of eigenvalues which arise from solving problems concerning either the vibrational or electronic states of solids. In particular, the electronic density of states is defined as

$$\rho(E) = \sum_i \delta(E - E_i), \qquad (4.65)$$

where the sum i is over the energy eigenvalues. Our present objective is to develop an understanding of the physical origins of these eigenvalues in the context of the electronic structure of a material. We already saw a schematic example of such a density of states in fig. 3.13 which shows the densities of states for a series of model problems treated in the one-electron framework adopted in chap. 3. The vertical axis in that figure specifies the energy of a given eigenstate. The more complex tight-binding analysis introduced here results in nN eigenvalues which may be used within the context of eqn (4.65) to obtain the density of states.

Once the density of states has been determined, the electronic contribution to the energy may be obtained by summing the energies of all the occupied states. In terms of the density of states itself, this amounts to an evaluation of the integral

$$E_{el} = 2 \int_{-\infty}^{\epsilon_F} \rho(E) E \, dE, \qquad (4.66)$$

where the factor of 2 accounts for the fact that each state may be occupied by an electron of two different spins. The equation written above is the continuum limit of the discrete sum over energy eigenvalues (i.e. $E_{el} = 2 \sum_i \epsilon_i$) that we adopted earlier. In order to compute the *total* energy, the electronic term introduced above must be supplemented by a phenomenological treatment of those terms left out of the tight-binding description including the interactions between the various nuclei.

This extra piece of the energy is often represented as a pair potential resulting in a total energy of the form

$$E_{tot}(\{\mathbf{R}_i\}) = \frac{1}{2} \sum_{ij} V_{eff}(R_{ij}) + 2 \int_{-\infty}^{\epsilon_F} E\rho(E)dE. \qquad (4.67)$$

The effective pair potential can be determined by insisting that certain properties such as the lattice parameter, the cohesive energy and the bulk modulus are accurately reproduced. Of course, more sophisticated fitting schemes are also possible.

From a more concrete perspective, the application of eqn (4.67) is carried out as follows. A computational cell is considered in which the set of nuclear coordinates $\{\mathbf{R}_i\}$ is known. Given these coordinates, it is then possible to determine the various matrix elements $h_{ij}^{\alpha\beta}$ which characterize the coupling of the αth orbital on site i to the βth orbital on site j. By assembling all such matrix elements, the full Hamiltonian matrix is thus known, and one can resort to either k-space or real space methods to determine the all-important density of states, $\rho(E)$. Given this density of states, the electronic contribution to the total energy is determined by carrying out the integration indicated in eqn (4.67). The Fermi level itself (i.e. the upper limit of the integral) is determined by insisting that the atoms have the appropriate number of electrons. By tuning the Fermi energy, one tunes between the empty band and full band limits. In addition to the electronic term, we are also required to consider the pair term which demands lattice sums no different than those undertaken in our discussion of pair potentials. This procedure, therefore, gives an unambiguous transcription between a set of atomic positions and a corresponding total energy, thus paving the way for the investigation of questions relating to the coupling of structure and material properties.

In chap. 6, we will return to the types of total energy analysis discussed here with special reference to the ability of tight-binding models to capture the relative energetics of competing crystal structures. In addition, in subsequent chapters, we will repeatedly make reference to the application of these methods for examining the energetics of a wide range of processes and mechanisms. The tight-binding method is a convenient entry level way to undertake calculations of the total energy in which reference to the electronic degrees of freedom is made explicitly. In addition, it has the added virtue that the calculations are computationally less demanding than those involving a more faithful treatment of the electronic degrees of freedom.

Despite our ability to now write down with some level of confidence the tight-binding description, our treatment has been overly cavalier in that we have as yet not stated how one might go about determining the Hamiltonian matrix elements, $h_{ij}^{\alpha\beta}$. Note that in precise terms these matrix elements describe the coupling between orbitals of character α and β on sites i and j, respectively. The basic thrust of the type of semiempirical tight-binding theories that it is our aim to discuss here is to assume that the interactions between orbitals on different sites may be empirically parameterized (through the parameters $V_{ss\sigma}$, $V_{sp\sigma}$, etc.) without ever really entering the debate as to the fundamental interactions between nuclei and electrons. The question of how to go about determining these fitting parameters will be taken up once we have seen how the tight-binding description is carried out in the context of periodic solids.

4.5.2 An Aside on Periodic Solids: k-space Methods

The description of tight binding introduced above advanced no assumptions about the symmetries present in the atomic-level geometry. As a result, the Hamiltonian matrix and the symmetries of the eigenfunctions defied any generic classification of the states along lines of their symmetries. On the other hand, in crystalline solids, the existence of translational symmetry leads to the expectation that the wave functions will have very special properties that guarantee that the charge density is the same in every unit cell. In particular, the existence of translational symmetries gives rise to so-called k-space methods which are the thrust of the present discussion. We specialize our discussion to periodic solids with the recognition that there are a wide variety of problems for which translational symmetry is lacking and that will therefore demand the more general perspective introduced above.

Because of the presence of the regularity associated with a crystal with periodicity, we may invoke Bloch's theorem which asserts that the wave function in one cell of the crystal differs from that in another by a phase factor. In particular, using the notation from earlier in this section, the total wave function for a periodic solid with one atom per unit cell is

$$|\psi_k\rangle = \frac{1}{\sqrt{nN}} \sum_{i=1}^{N} \sum_{\alpha=1}^{n} a_\alpha e^{i\mathbf{k}\cdot\mathbf{R}_i} |i, \alpha\rangle. \tag{4.68}$$

The first sum is over cells, the ith of which is specified by the Bravais lattice vector $\mathbf{R}_i$, while the second sum is over the set of orbitals α which are assumed to be centered on each site and participating in the formation of the solid. If there were more than one atom per unit cell, we would be required to introduce an additional index in eqn (4.68) and an attendant sum over the atoms within the unit cell. We leave these elaborations to the reader. Because of the translational periodicity, we know much more about the solution than we did in the case represented by eqn (4.58). This knowledge reveals itself when we imitate the procedure described earlier in that instead of having nN algebraic equations in the nN unknown coefficients $a_{i\alpha}$, translational symmetry collapses the problem into a series of n coupled algebraic equations in the n unknowns a_α, where n is the number of orbitals on each site.

Imitating the treatment given in eqn (4.60), we can once again elucidate our variational strategy, but this time with the added knowledge that the wave function takes the special form given in eqn (4.68). In particular, the function we aim to minimize is

$$\Pi_{periodic}(\{a_\alpha\}) = \langle \psi_{\mathbf{k}} | \hat{H} | \psi_{\mathbf{k}} \rangle - E_{\mathbf{k}} \langle \psi_{\mathbf{k}} | \psi_{\mathbf{k}} \rangle, \tag{4.69}$$

which in light of eqn (4.68) may be rewritten as

$$\Pi_{periodic}(\{a_\alpha\}) = \sum_{j=1}^{N} \sum_{\alpha=1}^{n} \sum_{\beta=1}^{n} e^{i\mathbf{k}\cdot\mathbf{R}_j} a_\alpha^* a_\beta h_{ij}^{\alpha\beta} - E_{\mathbf{k}} \sum_{\alpha=1}^{n} a_\alpha^* a_\alpha. \tag{4.70}$$

The energy eigenstates in the periodic case are labeled by the wave vector $\mathbf{k}$. If we now seek those a_αs that minimize eqn (4.70), and introduce the notation

$$H_{\alpha\beta}(\mathbf{k}) = \sum_{j=1}^{N} h_{ij}^{\alpha\beta} e^{i\mathbf{k}\cdot\mathbf{R}_j}, \tag{4.71}$$

then the resulting problem may be rewritten as

$$\sum_{\beta} (H_{\alpha\beta}(\mathbf{k}) - E_{\mathbf{k}} \delta_{\alpha\beta}) a_\beta = 0. \tag{4.72}$$

The structure of our result is identical to that obtained in eqn (4.64) in the absence of any consideration of crystal symmetry. On the other hand, it should be remarked that the dimensionality of the matrix in this case is equal to the number of orbitals per atom, n, rather than the product nN (our statements are based on the assumption that there is only one atom per cell). As a result of the derivation given above, in order to compute the eigenvalues for a periodic solid, we must first construct the Hamiltonian matrix $\mathbf{H}(\mathbf{k})$ and then repeatedly find its eigenvalues for each k-point. Thus, rather than carrying out one matrix diagonalization on a matrix of dimension $nN \times nN$, in principle, we carry out N diagonalizations of an $n \times n$ matrix.

To complete the description, it is also necessary to understand the nature of the $\mathbf{k}$-vectors themselves. For the purposes of the present discussion, we consider a one-dimensional periodic chain along the x-direction of N atoms with a spacing a between the atoms on the chain. One statement of Bloch's theorem, which we set forth in one-dimensional language, is the assertion that the wave function in the nth cell ψ_n and that in the $(n+1)^{th}$ cell, ψ_{n+1} is given by

$$\psi_{n+1} = e^{ika} \psi_n. \tag{4.73}$$

The periodicity of this one-dimensional crystal implies that $\psi_1 = \psi_{N+1}$, or

$$e^{ikNa} = 1,\tag{4.74}$$

which illustrates that the allowed **k**-vectors are of the form

$$k_n = \frac{2\pi n}{Na},\tag{4.75}$$

where the integer n can assume the values $0, 1, \ldots, N - 1$. This collection of vectors defines a finite region in k-space known as the first Brillouin zone. For the one-dimensional case being discussed here, it is convenient to consider the N vectors that run between $-\pi/a$ and π/a. The three-dimensional analogs of the concepts introduced here involve no particular subtlety and are left to the reader as an exercise. In addition, further elaboration of these concepts can be found in Ashcroft and Mermin (1976).

To illustrate the ideas presented above in concrete form, we now consider two separate one-dimensional examples of increasing sophistication. In the first case, we consider a one-dimensional periodic solid and examine a tight-binding basis that consists of one s-orbital per site. As seen from the definition of the matrix **H(k)** given in eqn (4.71), this case is elementary since the Hamiltonian matrix itself is one-dimensional and may be written as

$$H(k) = he^{ika} + he^{-ika},\tag{4.76}$$

where we have introduced the parameter h to characterize the coupling between orbitals on neighboring sites. All interactions beyond those at the near-neighbor level are neglected and we have assumed that the 'on-site' coupling (i.e. $\langle i|H|i\rangle$) is zero. Hence, the outcome of this calculation is a dispersion relation for the electronic states of the form

$$E(k) = 2h\cos ka.\tag{4.77}$$

This dispersion relation is plotted in fig. 4.14.

A more revealing example is provided by that of a linear chain in which each atom has four-basis functions, namely, s, p_x, p_y and p_z. This basis is of the same variety as that which is often used in the analysis of Si. In the present setting, the Hamiltonian matrix **H**(k) is obtained by recourse to the definition of eqn (4.71). In particular, we note that in the one-dimensional setting considered here, there are only six nonvanishing matrix elements, namely, $h_{i,i}^{ss} = \epsilon_s$, $h_{i,i}^{p_x p_x} = h_{i,i}^{p_y p_y} = h_{i,i}^{p_z p_z} = \epsilon_p$, $h_{i,i\pm 1}^{ss} = V_{ss\sigma}$, $h_{i,i\pm 1}^{p_x p_x} = V_{pp\sigma}$, $h_{i,i\pm 1}^{sp_x} = V_{sp\sigma}$ and $h_{i,i\pm 1}^{p_y p_y} = V_{pp\pi}$. As a

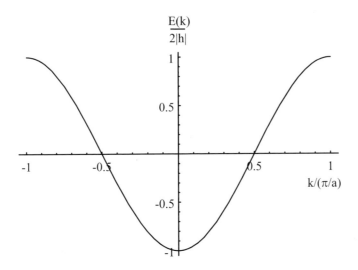

Fig. 4.14. Dispersion relation for a one-dimensional s-band tight-binding model.

result, the Hamiltonian matrix can be written down immediately as

$$
\mathbf{H}(k) =
\begin{pmatrix}
\epsilon_s + 2V_{ss\sigma}\cos ka & 2V_{sp\sigma}\cos ka & 0 & 0 \\
2V_{sp\sigma}\cos ka & \epsilon_p + 2V_{pp\sigma}\cos ka & 0 & 0 \\
0 & 0 & \epsilon_p + 2V_{pp\pi}\cos ka & 0 \\
0 & 0 & 0 & \epsilon_p + 2V_{pp\pi}\cos ka
\end{pmatrix}.
$$

$$(4.78)$$

Because the structure of this matrix involves only a 2×2 subblock that is nondiagonal, two of the four energy bands are already determined and are degenerate with the form

$$E(k) = \epsilon_p + 2V_{pp\pi}\cos ka. \tag{4.79}$$

To find the remaining two energy bands, we diagonalize the remaining 2×2 piece with the result that

$$
E_\pm(k) = \frac{\epsilon_s + \epsilon_p}{2} + (V_{ss\sigma} + V_{pp\sigma})\cos ka
$$

$$
\pm \frac{1}{2}[(\Delta\epsilon)^2 + 4(\Delta V_\sigma)(\Delta\epsilon)\cos ka + (4(\Delta V_\sigma)^2 + 16V_{sp\sigma}^2)\cos^2 ka]^{\frac{1}{2}},
$$

$$(4.80)$$

where we have introduced the notation $\Delta\epsilon = \epsilon_s - \epsilon_p$ and $\Delta V_\sigma = V_{ss\sigma} - V_{pp\sigma}$.

Given our discussion of the band structures that arise on the basis of tight-binding arguments, we are now prepared to look a little more deeply into the logic behind the fitting of tight-binding parameters. It is not really possible to provide any definitive statement which encompasses the diversity of ideas that have been brought to bear on the question of how best to determine tight-binding parameters. Evidently, there are any of a number of different schemes, many of which are founded on a sensible rationale. On the one hand, it is possible to imagine constructing parameterizations on the basis of the goodness of the fit to the band structure itself. In these cases, one notes that the values of the energy at certain k-points within the Brillouin zone can be determined simply in terms of the unknown tight-binding parameters, with the result that a fit is possible. However, in another sense, it is often the total energy that is of concern, so one can imagine determining the tight-binding parameters on the basis of energetic criteria. In either case, the central idea is more or less to determine the tight-binding expressions for a series of properties of interest $a_1, a_2, \ldots, a_n$. In each case, from the tight-binding perspective, these properties are functions of the unknown parameters, and one's task is to determine these parameters on the basis of these constraints. In mathematical terms, we are noting that

$$\left. \begin{aligned} f_1(V_{ss\sigma}, V_{sp\sigma}, \ldots) &= a_1, \\ f_2(V_{ss\sigma}, V_{sp\sigma}, \ldots) &= a_2, \\ &\vdots \\ f_n(V_{ss\sigma}, V_{sp\sigma}, \ldots) &= a_n. \end{aligned} \right\}$$

(4.81)

Depending upon the strategy that has been adopted, it is possible to have as many equations as there are unknown coefficients, in which case one is reduced to solving n equations in n unknowns. On the other hand, the number of equations could conceivably outnumber the number of unknowns in which case one resorts to solving these equations in the least-squares sense.

An example of the type of success that can be attained by exploiting the tight-binding framework in the context of periodic solids is shown in fig. 4.15, where the result of a tight-binding characterization of the band structure of Mo is contrasted with that determined from a full first-principles calculation. For the present purposes, the key recognition is that once this type of fitting procedure has been undertaken, it is possible to carry out calculations on both the electronic and structural properties of essentially arbitrary geometric configurations as long as the number of atoms is not so large as to be beyond the capacity to calculate.

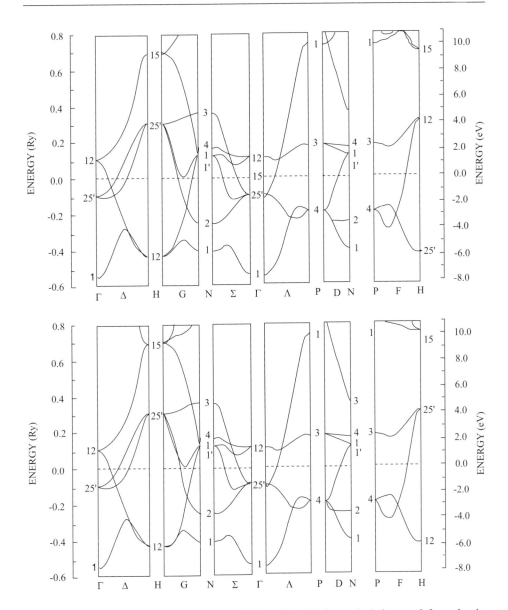

Fig. 4.15. Band structure for Mo as obtained from tight-binding calculations and from density functional theory (from Mehl and Papaconstantopoulos (1996)). The notation Γ, N, H, etc. refers to particular *k*-points within the Brillouin zone.

4.5.3 Real Space Tight-Binding Methods

Earlier in this chapter, we introduced pair and cluster functionals as approximate effective theories in which explicit reference to the electronic degrees of freedom had been removed. In each case, we introduced these total energy schemes on heuristic grounds without giving any strong arguments as to the origins of their particular functional forms. In the present section, we will see how tight-binding theory may be used to motivate both pair and cluster functionals, yielding both of the essential pieces of new physics represented by these schemes: environmental and angular dependence.

As a preliminary to our discussion of the emergence of both pair functionals and angular forces from tight-binding theory, we first discuss the moments of the density of states which serve as a useful basis for these developments. We begin with the recognition that every distribution has its moments. In particular, the nth moment of the density of states is defined by

$$\mu_n = \int_{-\infty}^{\infty} E^n \rho(E) dE. \tag{4.82}$$

The moments of the density of states can be seen to be related to the electronic structure itself through the following argument. The shape of the electronic density of states determines the outcome of the one-electron sum embodied in eqn (4.66). The shape of the density of states is in turn determined by its moments.

The connection between geometry and the electronic structure in the moments' setting is revealed when we note that the moments may be written as products of matrix elements over directed paths starting and ending on the same atom. For the purposes of the present discussion we will consider an s-band tight-binding method in which our notation is collapsed in the form $|i, s\rangle \to |i\rangle$. The elaboration to the more general situation in which there is more than one orbital per site can be carried out relatively painlessly. If we are interested in the second moment associated with a particular site, a more useful but equivalent definition to that given in eqn (4.82) is $\mu_2(i) = \langle i|H^2|i\rangle$. Making use of the identity $\mathbf{I} = \sum_j |j\rangle\langle j|$, we see that the second moment may be written in the form

$$\mu_2(i) = \sum_j h(R_{ij})h(R_{ji}), \tag{4.83}$$

where $h(R_{ij}) = \langle i|H|j\rangle$. What this means is that to find the second moment associated with a given site, we sum up all the products called for in eqn (4.83). For example, for an fcc crystal in which the matrix elements are assumed to vanish for distances greater than the near-neighbor distance, the second moment is $12h(R_{nn})^2$. For the higher-order moments, we have

$$\mu_3(i) = \sum_{jk} h(R_{ij})h(R_{jk})h(R_{ki}) \tag{4.84}$$

and

$$\mu_4(i) = \sum_{jkl} h(R_{ij})h(R_{jk})h(R_{kl})h(R_{li}). \tag{4.85}$$

These are the two lowest-order moments beyond the second moment itself. In this expression, the quantities $h(R)$ characterize the strength of interaction between orbitals that are separated by a distance R. Yet higher-order moments can be obtained by recourse to the same basic ideas.

Rectangular Band Model and Pair Functionals. As stated earlier, the 'derivation' of the pair functional ansatz may be approached from a variety of different viewpoints. One particularly revealing avenue which is relevant to understanding the emergence of pair functional descriptions of cohesion in transition metals such as Pd or Cu is to adopt the rectangular band model . This model density of states is shown in fig. 4.16(a), and is contrasted with a rigorous density functional calculation of a transition metal density of states in fig. 4.16(b). Despite the presence of a range of detail in the first-principles density of states, the rectangular band model is very useful in elucidating qualitative trends in quantities such as the electronic contribution to the cohesive energy that depend upon *integrated* properties of the density of states.

The rectangular band model posits that the density of states is constant over some range of energies and zero otherwise. The constant value associated with the density of states is chosen such that

$$N_{full} = 2 \int_{-W/2}^{W/2} \rho(E) dE, \tag{4.86}$$

where N_{full} is the total number of electrons that can be accommodated in the band (including spin). This implies that the density of states within our rectangular band ansatz is given by $\rho(E) = N_{full}/2W$. We may use this density of states to evaluate the electronic contribution to the total energy as a function of the filling of the band (i.e. the number of electrons, N_e). The electronic contribution to the energy is given by

$$E_{el} = \frac{N_{full}}{W} \int_{-W/2}^{\epsilon_F} E dE, \tag{4.87}$$

where the Fermi energy, ϵ_F, is dependent upon the band filling N_e through

$$N_e = \frac{N_{full}}{W} \int_{-W/2}^{\epsilon_F} dE, \tag{4.88}$$

yielding

$$\epsilon_F = W \left(\frac{N_e}{N_{full}} - \frac{1}{2} \right). \tag{4.89}$$

In light of these insights, it is seen then that the electronic contribution to the

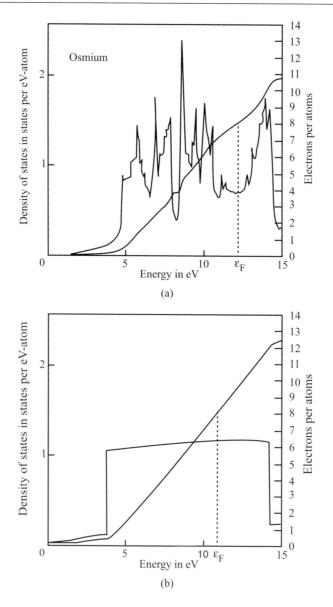

Fig. 4.16. Comparison of 'exact' and rectangular band densities of states for osmium (adapted from Harrison (1980)): (a) density of states for osmium as computed using density functional theory and (b) density of states constructed by superposing a rectangular band model with a free-electron-like density of states. Rising lines in the two curves are a measure of the integrated density of states.

bonding energy is given by

$$E_{el} = -\frac{N_{full}W}{2}\left(\frac{N_e}{N_{full}}\right)\left(1 - \frac{N_e}{N_{full}}\right). \tag{4.90}$$

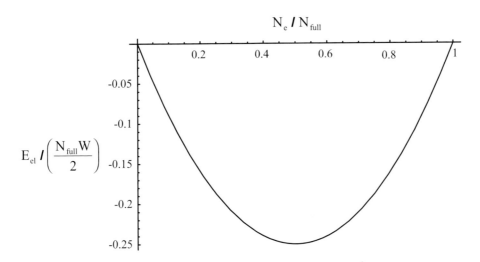

$$E_{el} \left/ \left(\frac{N_{full} W}{2} \right) \right.$$

$$N_e / N_{full}$$

Fig. 4.17. Band energy associated with rectangular band density of states. Energy is plotted in dimensionless units, scaled by the width of the band, W. Similarly, the band filling is plotted as a fraction of the full band.

The parabolic bonding profile revealed above is shown in fig. 4.17 and may be contrasted with the measured cohesive energies shown in fig. 4.18, which in the case of the transition metals reveal just the same trends. The picture that we have advanced here is one of rigid bands in which it is imagined that in passing from one element to the next as we traverse the transition series, all that changes is the filling of the d-band, with no change in the density of states from one element to the next. It is also important to note that the bonding energy scales with the bandwidth, an observation that will makes its appearance shortly in the context of pair functionals.

The rectangular band model developed above may be used as a tool to gain insights into the type of functional form that should be presumed in pair functionals such as those introduced earlier. In particular, if we consider a particular choice of band filling (i.e. fix our attention on describing the energetics of a particular element), then the electronic contribution to the total energy of eqn (4.90) scales with the second moment of the density of states, since eqn (4.90) can be rewritten on the basis of the observation that $W \propto \sqrt{\mu_2}$. To see that the bandwidth and the second moment of the density of states are indeed related, a plausibility argument can be made by appealing to a calculation of the second moment of the rectangular band density of states. In this case, the second moment is $\mu_2 = 5W^2/12$, showing that the bandwidth scales with the square root of the second moment. Thus, if we consider as an example the case of a half-filled d-band (i.e. $N_e/N_{full} = 1/2$) then eqn (4.90) tells us that $E_{el} = -\sqrt{15\mu_2}/4$. On the basis of the reasoning given

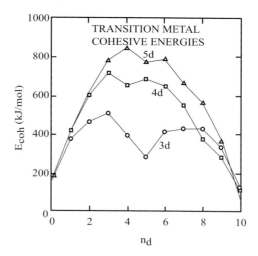

Fig. 4.18. Experimental energies of cohesion as a function of the number of d-electrons (n_d) for the three transition metal series (adapted from Young (1991)).

above, the rectangular band model suggests a total energy that is written as the sum of the electronic contribution and a pair potential term that accounts for the interactions between nuclei, resulting in

$$E_{tot} = \frac{1}{2} \sum_{ij} \phi(R_{ij}) - \sum_i \alpha \sqrt{\mu_2(i)}, \qquad (4.91)$$

where the constant α is a parameter that depends upon the band filling and can be deduced on fundamental grounds or used as a fitting parameter. To finish the argument, we need to know the way in which the second moment itself depends upon the geometric arrangement of atoms. At this point, it suffices to note that if we are interested in the second moment associated with a particular site, it may be determined using eqn (4.83). The fact that the second moment can be written in pairwise form immediately suggests the correspondence between the pair functionals presented earlier and our present argument. Because the second moment is itself of the pairwise additive form, it is now seen that, like in the embedded atom description explored earlier, the energy per bond depends upon the coordination number.

Moments and Angular Forces. Our aim in the present setting is to exploit the geometry of the densities of states in order to construct effective potentials for describing the bonding in solids which go beyond the pairwise forms considered thus far. We follow Carlsson (1991) in formulating that description. First, we reiterate that our aim is *not* the construction of the most up-to-date potentials that may be imagined. Rather, we wish to understand the microscopic origins of the angular terms which might appear in a total energy functional. For example, the total energy can take the form

$$E_{tot} = \frac{1}{2} \sum_{ij} V_2^{eff}(\mathbf{R}_i, \mathbf{R}_j) + \frac{1}{3!} \sum_{ijk} V_3^{eff}(\mathbf{R}_i, \mathbf{R}_j, \mathbf{R}_k)$$

$$+ \frac{1}{4!} \sum_{ijkl} V_4^{eff}(\mathbf{R}_i, \mathbf{R}_j, \mathbf{R}_k, \mathbf{R}_l) + \cdots. \tag{4.92}$$

Our aim at this point is an analysis of the quantum mechanical underpinnings for the angular terms introduced above. We begin by assuming the existence of some reference state, the energy of which shall serve as our zero and from which all other energies are measured. In this state, if we assume that only n moments of the density of states are known, the electronic contribution to the total energy is given by

$$E_{el} = E(\mu_2^{ref}, \mu_3^{ref}, \ldots, \mu_n^{ref}). \tag{4.93}$$

By way of contrast, if we interest ourselves in a particular configuration and aim to determine its energy *relative* to the reference state, that energy may be written

$$E_{el} = E(\mu_2^{ref} + \delta\mu_2, \mu_3^{ref} + \delta\mu_3, \ldots, \mu_n^{ref} + \delta\mu_n), \tag{4.94}$$

where the $\delta\mu_i$s reflect the difference between the moments of the reference state and the structure of interest. We now adopt the usual dodge of Taylor expanding in the presumed small deviations from the reference environment, resulting in an expression for the total energy change of the form

$$\Delta E_{el} = \sum_n \left(\frac{\partial E}{\partial \mu_n}\right)_{ref} (\mu_n - \mu_n^{ref}). \tag{4.95}$$

This expression is enlightening in that it reveals that the effective interatomic potentials have inherited the explicit angular dependence of the moments and, further, that the potentials have an environmental dependence that is dictated by the 'susceptibility' $\partial E/\partial \mu_n$. By absorbing the constant terms involving the reference state into our definition of the potentials, the energy of a given state may now be written as

$$E_{el} = \sum_n \left(\frac{\partial E}{\partial \mu_n}\right)_{ref} \sum_{k=1}^n \mu_n^{(k)}. \tag{4.96}$$

We have adopted the notation $\mu_n^{(k)}$ to signify the contribution to the nth moment arising from paths connecting k atoms. We now seek a correspondence between eqns (4.92) and (4.96). In particular, $V_k^{eff}(\mathbf{R}_1, \ldots, \mathbf{R}_k)$ is obtained by adding up all the terms in eqn (4.96) involving atoms $1, 2, \ldots, k$. For example, the effective pair potential is given by

$$V_2^{eff}(\mathbf{R}_i, \mathbf{R}_j) = 2 \sum_{n=2}^{N_m} \left(\frac{\partial E}{\partial \mu_n} \right)_{ref} \mu_n^{(2)}(\mathbf{R}_i, \mathbf{R}_j), \qquad (4.97)$$

where we have defined $\mu_n^{(2)}(\mathbf{R}_i, \mathbf{R}_j)$ as the two-atom contribution to the nth moment involving atoms i and j. Note that we are considering two-atom contributions to all moments between the 2nd moment and the N_m^{th} moment. Similar expressions are obtained in the case of the higher-order potentials. As said before, the key observation is that the potentials inherit the angular dependence of the moments.

Though the notation used above is admittedly cumbersome, the basic idea associated with the development of these potentials is relatively straightforward. The idea was to expand the energy about a reference environment and to adopt a functional form for the potentials that is a direct inheritance from the moments of the density of states which themselves are defined as directed paths involving the atoms in a given cluster. We must maintain the insight that the energies are measured relative to a reference environment. In particular, the susceptibilities are evaluated for the reference environment. As a result, depending upon the choice of reference environment, the relative strengths of the potentials will vary. To see this explicitly, consider the effective pair potential tied to the rectangular band model discussed above. Since E_{el} scales as $\sqrt{\mu_2}$, we see that $\partial E_{el}/\partial \mu_2 \propto 1/\sqrt{\mu_2}$. The implication of this result is that for a reference environment with a small μ_2 (i.e. the atoms have fewer neighbors, such as at a surface) the effective potential is stronger.

An example of the type of potentials that result upon adopting a strategy like that given above is shown in fig. 4.19. We focus attention on the effective interactions involving clusters of four atoms The four-body terms are based upon a slightly more sophisticated scheme than that used here involving a basis more complex than the simple tight-binding basis considered here, but the basic concepts are identical. In this case, we see that for a metal such as Mo, four-body planar clusters with 90° angles are energetically costly and point to a substantial lowering of the energy of the bcc structure (which has no such clusters based on first neighbors) and the fcc structure.

As we have seen in the preceding sections, the development of empirical potentials is a mixture of plausibility arguments and sound guesswork. Despite their successes in accounting for the features of defects in a range of materials, even the best of potentials at times suffer from fatal weaknesses. Our aim now is to give a flavor of the many attempts that have been put forth to build satisfactory interatomic descriptions for the direct simulation of problems of interest concerning real materials. One perspective has been offered by Heine et al. (1991) who note

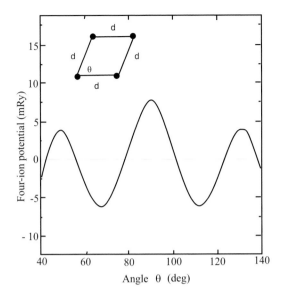

Fig. 4.19. Four-body potential for Mo (adapted from Moriarty (1990)).

"The success of the 'glue' models, which take many-atom forces specifically into account, has been impressive particularly in accounting for subtle effects such as the reconstructions of the Au surface. By way of contrast, one can point to the sad history of attempts to find a satisfactory force model for Si. There has been a whole succession of models, each fitted to experimental data plus the computed energies of a range of unstable structures covering quite a variety of bonding geometries. Yet each model failed when confronted a year or so later by the energy of some new structure." At the time of this writing it remains an open question whether or not any satisfactory scheme can be developed for treating covalent materials. For the moment, the most reliable scheme is to use the full quantum mechanical description described in the next section.

4.6 First-Principles Calculations of the Total Energy

As yet, our discussion has focused on the attempt to construct energy functionals in which there is some level of empiricism that mimics to a greater or lesser extent the presumed exact results of a full quantum mechanics calculation. Our reason for shying away from the full quantum mechanical calculations themselves was founded upon the fact that these methods are computationally intensive, and hence must be undertaken with the foresight to insure that they are used only when really necessary. It is the purpose of the present section to sketch the way in which first-principles calculations are constructed.

As our starting point, we begin by reminding ourselves of the basic facts that make the full many-electron problem so difficult. A useful starting point is the Hamiltonian that describes both the electrons and the nuclei from which they are derived. The full many-particle Hamiltonian made its appearance in eqn (4.3), where it was shown that the various particle coordinates are linked in an exceedingly complicated way. If we adopt the purists's perspective, we demand the full wave function which is a function not only of the electronic degrees of freedom, but of the nuclear degrees of freedom as well. As is clear from the full Hamiltonian, the Schrödinger equation corresponding to this many-body description is a coupled problem of enormous proportions. Nevertheless, through a combination of path-breaking theorems and numerical inventiveness, the calculation of electronic structure and associated total energies has still become routine.

4.6.1 Managing the Many-Particle Hamiltonian

Density Functional Theory and the Local Density Approximation

Even in light of the insights afforded by the Born–Oppenheimer approximation, our problem remains hopelessly complex. The true wave function of the system may be written as $\psi(\mathbf{r}_1, \mathbf{r}_2, \mathbf{r}_3, \ldots, \mathbf{r}_N)$, where we must bear in mind, N can be a number of Avogadrian proportions. Furthermore, if we attempt the separation of variables ansatz, what is found is that the equation for the ith electron depends in a nonlinear way upon the single particle wave functions of all of the other electrons. Though there is a colorful history of attempts to cope with these difficulties, we skip forth to the major conceptual breakthrough that made possible a systematic approach to these problems.

In the language of the beginning of the chapter, the basic idea used in the density functional setting is to effect a replacement

$$E_{exact}(\{\mathbf{R}_i, \mathbf{r}_n\}) \rightarrow E_{approx}[\rho(\mathbf{r}), \{\mathbf{R}_i\}]. \tag{4.98}$$

In particular, Hohenberg and Kohn advanced the view that the total energy of the many-electron system may be written as a unique functional of the electron density of the form,

$$E[\rho(r)] = T[\rho(\mathbf{r})] + E_{xc}[\rho(\mathbf{r})] + \frac{1}{2} \int \int \frac{\rho(r)\rho(r')}{|\mathbf{r} - \mathbf{r}'|} d^3\mathbf{r} d^3\mathbf{r}' + \int V_{ext}(\mathbf{r})\rho(\mathbf{r})d^3\mathbf{r}, \tag{4.99}$$

and that the ground state energy of the system is attained on minimizing this functional. The various terms in this equation shape up as follows. $T[\rho(\mathbf{r})]$ is the kinetic energy term, while the term $E_{xc}[\rho(\mathbf{r})]$ is the so-called exchange-correlation energy and refers to parts of the many-electron interaction whose origin is quantum mechanical. The remaining terms correspond to the direct electron–electron interaction and the interaction of electrons with an external potential (which can include the ions). In the problem section at the end of the chapter, we take up the question of how a simple model for the kinetic energy functional can be built up using what we have already learned about the electron gas. As it stands, the statement given above is exact. The profundity of the insight buried in the density functional given above is related to the fact that it allows for an exact mapping of our perceived hopeless many-electron problem onto a set of effective single particle equations. The only difficulty is that the functional itself is unknown. However, the strategy suggested in eqn (4.98) is to use a local approximation to the exchange-correlation functional which results in a series of approximate, but highly accurate, one-electron equations. We also note that local approximations like those introduced here will reveal themselves of relevance in other continuum settings.

As said above, the key assertion is that the ground state energy corresponds to that density that minimizes the functional of eqn (4.99). Of course, until something can be said about the form of $T[\rho(\mathbf{r})]$ and $E_{xc}[\rho(\mathbf{r})]$, we don't have a practical scheme for obtaining the total energy. The next key step in the analysis was the introduction of an approximation which converted the original problem into a corresponding problem for a fictitious system involving noninteracting electrons in the presence of an effective potential. The key approximation is known as the local density approximation and consists in the statement

$$E_{xc}[\rho(\mathbf{r})] = \int \epsilon_{xc}(\rho(\mathbf{r}))d^3\mathbf{r}. \qquad (4.100)$$

As we said above, in the absence of knowledge of the exchange-correlation functional $E_{xc}[\rho(\mathbf{r})]$ itself, our results were only suggestive. The key additional insight that makes this formulation a viable calculational scheme is the local density approximation in which, once again, appeal is made to the properties of the electron gas. In this instance, the idea is to presume that the exchange-correlation functional may be approximated by the electron gas exchange-correlation functional $\epsilon_{xc}(\rho)$, which is known. The use of this functional in the solid is based upon the assumption that the local behavior of the electrons in the solid mimic those of the corresponding homogeneous electron gas at the same density. In light of this approximation, we are now prepared to seek the energy minimizing density $\rho(\mathbf{r})$ which is obtained by minimizing this functional, that is by effecting the functional derivative

$$\delta E[\rho(\mathbf{r})]/\delta\rho(\mathbf{r}) = 0. \qquad (4.101)$$

This results in a set of single particle equations, known as the Kohn–Sham equations given by

$$-\frac{\hbar^2}{2m}\nabla^2\psi_i(\mathbf{r}) + V_{eff}(\mathbf{r})\psi_i(\mathbf{r}) = \epsilon_i\psi_i(\mathbf{r}). \qquad (4.102)$$

The effective potential is given in turn as

$$V_{eff}(\mathbf{r}) = V_{ext}(\mathbf{r}) + \int\frac{\rho(\mathbf{r}')d\mathbf{r}'}{|\mathbf{r}-\mathbf{r}'|} + \frac{\delta E_{xc}}{\delta\rho(\mathbf{r})}. \qquad (4.103)$$

Note that at this point we have turned the original (hopeless) many-body problem into a series of effective single particle Schrödinger equations.

As will become increasingly evident below, there are a number of different particular realizations of the general ideas being set forth here. For example, in the context of the exchange-correlation energy itself, there are several popular parameterizations of the density dependence of the exchange-correlation energy that are in use. Recall that the exchange-correlation energy is representative of those parts of the electron–electron interaction which are purely quantum in nature. Part of the variety that attends the various computer implementations of density functional theory in the local density approximation surrounds the different parameterizations for this term. The key point is that from an analytic perspective, even after making the simplifying assumption that it is the exchange-correlation energy of the *uniform* electron gas that is needed, $\epsilon_{xc}(\rho)$ cannot be described fully. Often, one resorts to parameterizations which satisfy known constraints in the limits that the density of the electron gas is either very high or very low, and then searches for interpolation formulae in the intervening density regime which can be informed on the basis of quantum Monte Carlo calculations of this property. A detailed discussion of many of the popular parameterizations for this term can be found in table 7.1 of Dreizler and Gross (1990) and the attendant prose.

4.6.2 Total Energies in the Local Density Approximation

Once the effective single particle equations have been derived, there are almost limitless possibilities in terms of the different schemes that have been invented to solve them. By way of contrast, the assertion is often made that if the calculations in these different realizations of the first-principles methodology are made correctly, then the predictions of these different implementations should be the same.

Interactions Between Nuclei and Electrons

Note that in our treatment of the Kohn–Sham equations, we assume the existence of an external potential $V_{ext}(\mathbf{r})$ which characterizes the interactions of the nuclei and the electrons. Given our interest in modeling the properties of solid materials,

V_{ext} will arise as a superposition of the potentials associated with each of the nuclei making up the solid. In the most direct of pictures, one can imagine treating the solid by superposing the 'bare' nuclear potentials for each atom, and subsequently solving for the disposition of each and every electron. This approach is the basis of the various 'all-electron' methods.

As an alternative to the explicit treatment of each and every electron, a variety of methods have grown up around the use of effective potentials in which V_{ext} is meant to mimic the nucleus and its complement of core electrons which are presumed to be indifferent to the changes in environment in going from the free atom to the solid. Indeed, much of the spirit of this book centers on the idea of reduced descriptions in which degrees of freedom are eliminated in the name of both conceptual simplicity and computational tractability and the use of pseudopotentials like those described above is a compelling example of this strategy. It is crucial to realize that in the process of formation of the solid, many of the low-lying atomic states (i.e. those electrons that are tightly bound to the nucleus in inner shells) are disturbed very little by the formation of the solid. As a result, pseudopotential schemes attempt to exploit this observation by creating an effective potential in which the nucleus and the core electrons together serve to influence the outer electrons. The key idea is a choice of pseudopotential that reproduces known properties such as atomic wave functions or the scattering properties associated with the full potential. An example of a pseudopotential is given in fig. 4.20. For our purposes our interest in this idea is as observers rather than practitioners, with the key insight gleaned from the observation being the formal way in which degrees of freedom are eliminated.

Choosing a Basis

In our statement of the Kohn–Sham equations, nothing has been said about the way in which $\psi(\mathbf{r})$ itself will be determined. Many methods share the common feature that the wave function is presumed to exist as a linear combination of some set of basis functions which might be written generically as

$$\psi(\mathbf{r}) = \sum_n \alpha_n b_n(\mathbf{r}), \qquad (4.104)$$

where α_n is the weight associated with the nth basis function $b_n(\mathbf{r})$. Just as with the earlier statement of tight-binding theory, solving the relevant equations becomes a search for unknown coefficients rather than unknown functions.

Part of the variety associated with the many different incarnations of density functional codes is associated with this choice of basis functions. For example, in the spirit of the earlier tight-binding discussion, one can imagine a basis constructed out of functions localized about each atomic site. Alternatively, we can

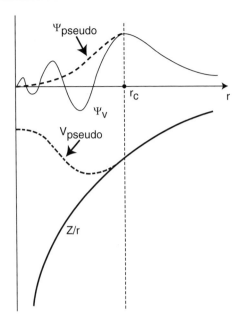

Fig. 4.20. Schematic of pseudopotential and associated pseudowave function (adapted from Payne *et al.* (1992)).

take our cue from the physics of the electron gas and attempt to expand the total wave function in terms of a free-electron-like basis. If one is to use the so-called plane wave basis set, then the generic wave function for a state characterized by wave vector $\mathbf{k}$ is represented as

$$\psi_{\mathbf{k}}(\mathbf{r}) = \sum_{\mathbf{G}} a_{\mathbf{k}+\mathbf{G}} e^{i(\mathbf{k}+\mathbf{G})\cdot\mathbf{r}}. \tag{4.105}$$

In this case, the vectors $\mathbf{G}$ are members of the set of reciprocal lattice vectors and eqn (4.105) really amounts to a three-dimensional Fourier series. The task of solving the Kohn–Sham equations is then reduced to solving for the unknown coefficients $a_{\mathbf{k}+\mathbf{G}}$. The only difficulty is that the potential that arises in the equation depends upon the solution. As a result, one resorts to iterative solution strategies.

Computing the Electronic Structure and the Total Energy

Once the Kohn–Sham equations have been solved, we are in a position to evaluate energies, forces and the electronic structure of a material itself. In particular, with the eigenvalues ϵ_i and corresponding wave functions $\psi_i(\mathbf{r})$ we can compute the energy of the system explicitly. As can be seen from the discussion given above, and as has been true with each of the total energy methods introduced in this chapter, in the end we are left with a scheme such that once the nuclear

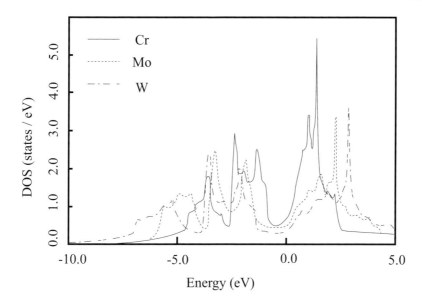

Fig. 4.21. Densities of states for several bcc transition metals as computed using density functional theory (adapted from Eriksson *et al.* (1992)).

coordinates have been provided, one can compute the corresponding energy. From the perspective of computational materials science, this step is of primary importance since it is the key to first examining the structure and energetics of the various defects that populate materials, and ultimately to explicating how those structures conspire to give rise to the properties of materials.

Part of the added value of using methods in which the electronic degrees of freedom are treated explicitly is the insights that may be garnered concerning the electronic properties of a material. To provide some feel for the types of results that emerge, an example of the densities of states for several materials is shown in fig. 4.21.

4.7 Choosing a Description of the Total Energy: Challenges and Conundrums

This chapter has emphasized the microscopic approach to computing the total energy of solids. We have seen that there are a number of different schemes for computing the total energy, each with its own advantages and difficulties. In simplest terms, the largest conundrum faced in the selection of a total energy description concerns the compromise that must usually be made between computational tractability and accuracy. Treatments featuring the electronic degrees of freedom explicitly are often plagued by the inability to treat system sizes capable of resolving the problem of interest without the contaminating influences of nearby boundaries. On the other hand, empirical schemes, while able to handle much larger system sizes, suffer a shortcoming of similar magnitude in that the results can often be charged with lacking physical realism.

In subsequent chapters, we will show how the types of energy functionals described here may be used as a tool for computing the properties of materials. Part of our approach will be to attempt to tackle particular case studies from the perspective of several different total energy methods. Our intention in adopting this approach is to illustrate the way in which some methods provide a clearer physical picture of the problem at hand, while others are to be noted for their physical accuracy.

4.8 Further Reading

'Beyond Pair Potentials' by A. E. Carlsson, in *Solid State Physics* edited by H. Ehrenreich and D. Turnbull, Academic Press, San Diego: California, Vol. 43, pg. 1, 1990. This article gives a thorough discussion of the conceptual underpinnings associated with efforts to build effective potentials as a way to eliminate explicit reference to electronic degrees of freedom.

'Many-atom interactions in solids' by V. Heine, I. J. Robertson and M. C. Payne, *Phil. Trans. Roy. Soc. Lond.*, **A334**, 393 (1991) gives a thoroughly thoughtful overview of the difficulties that must be faced in trying to create effective theories of interatomic interactions in which explicit reference to the electronic degrees of freedom has been renounced.

Electronic Structure and the Properties of Solids by W. A. Harrison, W. H. Freeman and Company, San Francisco: California, 1980. Harrison's book espouses the tight-binding perspective in a way that reveals how this method may be applied to consider materials ranging from Si to Ta. I have learned much from this book.

'Tight-binding modelling of materials' by G. M. Goringe, D. R. Bowler and E. Hernández, *Rep. Prog. Phys.*, **60**, 1447 (1997). This article gives an up-to-date review of the continued uses of the tight-binding formalism.

Solids and Surfaces: A Chemist's View of Bonding in Extended Structures by R. Hoffmann, VCH Publishers, New York: New York, 1988. This book reflects the chemist's intuition as regards bonding in solids. Hoffmann has a special talent for constructing plausible models of the electronic structure of solids on the basis of insights strictly in terms of the crystalline geometry and the character of the underlying atomic orbitals.

Another book that espouses the chemical perspective on bonding in solids is

Chemical Bonding in Solids by J. K. Burdett, Oxford University Press, Oxford: England, 1995. One of the key features that I enjoy in looking at the work of both Hoffmann and Burdett is their willingness to confront systems not only with a high level of structural complexity, but chemical complexity as well.

Bonding and Structure of Molecules and Solids by D. Pettifor, Clarendon Press, Oxford: England, 1995. A superb contribution aimed at explaining the fundamental *physical* reasoning associated with constructing models of bonding in solids. Primary emphasis is placed on metals and covalent materials.

The Electronic Structure and Chemistry of Solids by P. A. Cox, Oxford University Press, Oxford: England, 1987. Like the book by Hoffmann, this book takes a more chemical perspective of bonding in solids. Though the mathematical sophistication demanded of the reader is fairly low, this does not imply that there is not much to be learned here. A book I have returned to repeatedly.

'Iterative minimization techniques for *ab initio* total energy calculations: molecular dynamics and conjugate gradients', by M. C. Payne, M. P. Teter, D. C. Allan, T. A. Arias and J. D. Joannopoulos, *Rev. Mod. Phys.*, **64**, 1045 (1992). Despite the title which makes it appear highly specialized, this article has much to offer on everything from the LDA formalism itself, to issues of convergence and the use of LDA calculations (rather than classical potentials) as the basis of molecular dynamics.

'Modern Electron Theory' by M. W. Finnis in *Electron Theory in Alloy Design* edited by D. G. Pettifor and A. H. Cottrell, Institute of Materials, London: England, 1992. This is my favorite article on first-principles calculations. Finnis spells out all of the key points involved in carrying out first-principles calculations of the total energy.

'Understanding quasi-particles: observable many-body effects in metals and ^{3}He' by J. W. Wilkins in *Fermi Surface* edited by M. Springford, Cambridge University Press, New York: New York, 1980. Wilkins' article describes many of the issues that arise in thinking about the many-body aspects of the total energy in solids.

Density Functional Theory by R. M. Dreizler and E. K. U. Gross, Springer-Verlag, Berlin: Germany, 1990. This book concerns itself primarily with the formalism of density functional theory and less with the specifics of a given computer implementation. It has much to offer on the origins of the Kohn–Sham equations, the treatment of the exchange-correlation energy and on refinements such as gradient corrections.

4.9 Problems

---

1 Fitting parameters in the Lennard–Jones potential

In this chapter, we introduced the Lennard–Jones potential given by

$$V(r) = V_0 \left[\left(\frac{\sigma}{r} \right)^{12} - \left(\frac{\sigma}{r} \right)^6 \right]. \tag{4.106}$$

Determine the constants V_0 and σ for fcc Xe by using the fact that the cohesive energy of Xe is 0.17 eV/atom and the lattice parameter is 6.13 Å. Energy units for the potential should be in eV and length units in Å. Carry out the solution to this problem both analytically (by fitting these energies only for a near-neighbor model) and numerically, by using six-neighbor shells. For further details see Kittel (1976).

---

2 Morse Potential and Cohesion in Cu

The Morse potential is given by $V(r) = V_0(e^{-2a(r-r_0)} - 2e^{-a(r-r_0)})$, where V_0, a and r_0 are parameters to be determined by fitting to experimental data. Using the experimental data for the cohesive energy of fcc Cu ($E_{coh} = 336$ kJ/mole), the fcc Cu equilibrium lattice parameter ($a_0 = 3.61$ Å) and the bulk modulus ($B = 134.3$ GPa), determine the parameters V_0 (in eV), a (in Å^{-1}) and r_0 (in Å) under the assumption that only near-neighbor atoms interact.

---

3 Forces within the Pair Functional Formalism

Given the prescription for the total energy as defined within the pair functional formalism, compute the expression for the force on atom i as a result of its interactions with its neighbors. In particular, compute $F_i = -\partial E_{tot}/\partial x_i$, remembering that there is implicit dependence on x_i through the density. Use your result to demonstrate that if the cutoff range of the interactions is denoted as R_{cut}, then the motion of an atom at a distance of $2R_{cut}$ from the atom of interest can result in a force on that atom. The result of this calculation is given in eqn (4.43).

---

4 Effective Pair Interactions from Pair Functionals

The pair functional formalism may be recast in the language of pair potentials by computing an *effective* pair potential which is valid for environments close to some reference environment. In this problem, starting with eqn (4.33), consider small deviations about some reference environment characterized by an electron density ρ_{ref}. By writing the total density at a given site as $\rho_{ref} + \delta\rho$ and Taylor expanding the embedding term, show that the resulting expression for the total energy takes the form of a pair potential. Using Johnson's analytic embedded-atom functions introduced in this chapter, obtain and plot the effective pair potentials valid for both the bulk environment and for the (100) surface of an fcc crystal.

---

5 Stillinger–Weber Energy for Si in Diamond Cubic Structure

In eqn (4.51) we derived an expression for the total energy of a two-dimensional triangular lattice. That expression was for a geometrically unrealistic structure for a covalently bonded material. Derive an analogous expression for the total energy of Si when in the diamond cubic structure.

---

6 Forces Associated with Stillinger–Weber Potentials

Obtain an expression for the forces between atoms in a system characterized by the Stillinger–Weber potential. An explicit expression may be found in Wilson *et al.* (1990).

---

7 Electronically Driven Structural Transformations

Consider an s-band tight-binding Hamiltonian with matrix elements $\langle i|H|j \rangle = h_0/r$, where r is the distance between atoms i and j. The goal of this problem is to examine the electronic driving force for dimerization of a linear chain of atoms by examining the electronic structure and resulting cohesive energy for a chain of atoms with nominal lattice parameter a_0 as shown in fig. 4.22. Consider a dimerization reaction in which the new lattice is characterized by alternating spacings $a_0 - \delta$ and $a_0 + \delta$ with the result that there are two atoms per unit cell with a lattice parameter of $2a_0$. Assume that each atom has one electron (i.e. the case of half-filling) and obtain the electronic contribution to the energy as a function of the parameter δ. Note that each atom is coupled only to its near neighbors on the left and right, though the strength of that coupling is scaled by $1/r$ as described above. Plot the electronic energy as a function of δ and comment on the implications of your result.

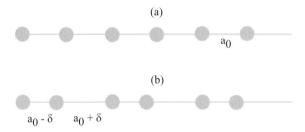

Fig. 4.22. Chain geometries to be considered in problem 7 corresponding to: (a) the undimerized state and (b) the dimerized state.

---

8 d-Band Tight-Binding Model for fcc Metal

In this problem you are asked to obtain an expression for the dispersion relation along the (100) direction in k-space for fcc Ni. The calculation should be done using a basis of five d-orbitals per atom, with the parameters $V_{dd\sigma}$, $V_{dd\pi}$ and $V_{dd\delta}$ chosen in the ratio $V_{dd\sigma} : V_{dd\pi} : V_{dd\delta} = -6 : 4 : -1$.

---

9 Tight-Binding for Crystals with Multiple Atoms per Unit Cell

In eqn (4.68), we developed the tight-binding formalism for periodic systems in which there was only one atom per unit cell, with each such atom donating n basis functions to the total electronic wave function. Generalize the discussion presented there to allow for the presence of more than one atom per cell. In concrete terms, this means that the basis functions acquire a new index, $|i, J, \alpha\rangle$, where the index i specifies the unit cell, J is a label for the atoms within a unit cell and α remains the label of the various orbitals on each site.

---

10 Density of States for Electron Gas

In the previous chapter we described the physics of the electron gas. Using the ideas described there, deduce the density of states, $\rho(E)$ for the one-, two- and three-dimensional electron gas. In particular, show how the density of states scales with the energy and plot your results. Recover the expression for the cohesive energy of the electron gas by evaluating integrals of the form

$$E_{el} = 2 \int_0^{\epsilon_F} \rho(E) E \, dE. \tag{4.107}$$

11 *Simple Model for the Kinetic Energy Functional*

The density functional introduced in eqn (4.99) featured a kinetic energy term denoted $T[\rho(\mathbf{r})]$. In this problem, derive an expression for a kinetic energy functional based upon the simple electron gas physics described in chap. 3. In particular, if the total kinetic energy per atom is of the form

$$\frac{E_{eg}}{N} = \frac{3}{5}\frac{\hbar^2}{2m}(3\pi^2\rho)^{\frac{2}{3}}, \tag{4.108}$$

then what form should the kinetic energy density $t(\rho(\mathbf{r}))$ take in writing a kinetic energy functional of the form $T[\rho(\mathbf{r})] = \int t(\rho(\mathbf{r}))d^3\mathbf{r}$? Now that this kinetic energy functional is in hand, write an approximate functional for thetotal energy of the form

$$E_{TF}[\rho(\mathbf{r})] = \int t(\rho(\mathbf{r}))d^3\mathbf{r} + \frac{1}{2}\int\int\frac{\rho(r)\rho(r')}{|\mathbf{r}-\mathbf{r}'|}d^3\mathbf{r}d^3\mathbf{r}' + \int V_{ext}(\mathbf{r})\rho(\mathbf{r})d^3\mathbf{r}, \tag{4.109}$$

subject to the constraint that

$$\int\rho(\mathbf{r})d^3\mathbf{r} = N, \tag{4.110}$$

where N is the total number of electrons. Note that we have introduced the notation $E_{TF}[\rho(\mathbf{r})]$ to refer to the fact that we are making the Thomas–Fermi approximation. By computing the functional derivative of the Thomas–Fermi functional subject to the constraint given above, obtain the equation satisfied by the ground state density.

FIVE

Thermal and Elastic Properties of Crystals

5.1 Thermal and Elastic Material Response

As yet, our discussion of modeling the energetics of crystalline solids has centered on the analysis of atomic configurations that are undisturbed by deformation, by the thermal vibrations that are inevitably present at finite temperatures or by defects. Before embarking on an analysis of the full-fledged defected solid, we begin with a consideration of geometric states that may be thought of as relatively gentle perturbations about the perfect crystal. In particular, the aim of the present chapter is to assess the energetics of both thermal vibrations and elastic deformations, with a treatment that is unified by the fact that in both cases the energy can be expanded about that of the perfect crystal which serves as the reference state.

Prior to examining the ways in which the methods of chap. 4 can be used to examine the energetics of both thermal vibrations and elastic deformations, we consider the significance of these excitations to the properties of materials at large. The methods of the present chapter feed into our quest to understand the properties of materials in several distinct ways. First, as will be described in the following paragraphs, a variety of important material properties involve the energetics of thermal vibration or elastic deformation directly, including the specific heat, the thermal conductivity, the thermal expansion and the elastic moduli themselves. However, in addition to the direct significance of the ideas to be presented here to the analysis of thermal properties, much of what we will have to say will form the methodological backdrop for our discussion of the thermodynamic ideas invoked to understand phase diagrams, diffusion and microstructural evolution.

In chap. 4 we argued that much can be learned by comparing the energetics of different competing states of a crystalline solid. One set of material parameters that can be understood on the basis of total energy methods like those presented in the previous chapter in conjunction with the methods of statistical mechanics are those related to the thermal response of materials. Indeed, one of the workhorses

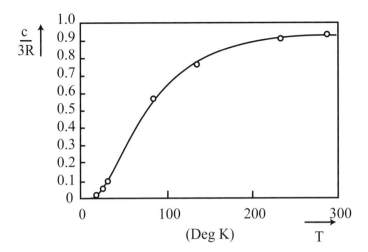

Fig. 5.1. Specific heat for Cu as obtained from Debye model contrasted with experimental results (adaptation of Debye's original results by Reif (1965)).

of the statistical mechanics curriculum is the evaluation of the specific heat of crystalline solids as a function of the temperature. A representative example of the experimentally observed specific heat of solids is shown in fig. 5.1. Our discussion of the theoretical apparatus that is needed to understand these data will provide a variety of lessons in the use of total energy methods in conjunction with statistical mechanics.

A second class of material parameters associated with the thermal behavior of solids is more traditionally thought of as those associated with 'anharmonic effects' in solids and includes the thermal conductivity and the coefficient of thermal expansion. As already shown in fig. 1.2, the thermal conductivity of different material classes ranges over a number of orders of magnitude and involves a variety of different mechanisms including the scattering of phonons by point defects as will be briefly described in chap. 7. The coefficient of thermal expansion also looms as an important material parameter, with an increasingly important technological role as a result of the reliance on complex heterostructures in which many different materials are joined together. In these cases, it is the mismatch in thermal expansion coefficient which is of particular significance since this mismatch results in thermal stresses that can lead to mechanical failure. The use of thermal expansion mismatch as an engineering tool in the form of the bimetallic strip dates at least as far back as the creation of the famed clocks of John Harrison described in Dava Sobel's *Longitude* (1995). It is amusing to note that the application of thermal mismatch preceded its understanding on the basis of statistical mechanics by nearly 250 years.

In addition to the thermal properties mentioned above, the present chapter also has as its charter the investigation of the elastic (both linear and nonlinear) properties of solids. We begin by noting that the linear theory of elasticity is one of several prototypical continuum theories and is at the heart of analyses of problems ranging from the formation of dislocations in Si_xGe_{1-x} to the deformation of bridges subjected to loads from wind and earthquakes. As noted earlier, continuum theories such as the linear theory of elasticity are silent on the values of the material parameters used to characterize solids. As a result, it is the business of subscale analysis, such as is offered by the types of microscopic energy functionals described in the previous chapter, to evaluate the elastic response of different materials. An example of the diversity of elastic properties as measured by the Young modulus is shown in fig. 1.1.

In the preceding paragraphs, we have given a sketch of some of the material parameters that are explicitly tied to the thermal and elastic response of solids. However, as we already noted, the tools developed to examine these material parameters are of much broader importance to the generic aim of modeling materials and their response. Indeed, the ideas to be presented in the present chapter serve as the backbone for an analysis of thermostatics (i.e. the study of equilibria) and of kinetics in solids. The discussion above was meant to highlight the various material parameters related to the thermal properties of solids. In the following paragraphs we also show the origins of the methodological relevance of the material in the present chapter to our later examination of both equilibria and kinetics in solids.

One of the central points of contact between the methods of the present chapter and an analysis of material response will be the application of these methods to the construction of free energies. Whether our aim is the investigation of phase diagrams or the driving forces on interfaces, one of the elemental challenges that will be faced will be that of the construction of the relevant free energy. For example, in the context of phase diagrams, we will see that a central part in the competition between different phases is played by the vibrational entropy of these competing phases.

In addition to the consideration of equilibrium problems, the treatment of thermal excitations is also important to the investigation of kinetics. An intriguing insight into kinetics is that in many cases, the tools for treating thermal excitations to be described in this chapter can be used to consider driven processes such as diffusion. As will be noted in chap. 7, models of kinetics are tied to the idea that as a result of the random excursion of atoms from their equilibrium positions, a system may occasionally pass from one well in the multi-dimensional potential energy surface to another. Indeed, this eventuality has already been rendered pictorially in fig. 3.22 where we saw an adatom hopping from one energy well to the next in its excursion across a crystal surface. As will be seen below, the treatment of these excursions, at least in the case of small displacements, can be put on a formal basis that allows for the construction of explicit models of thermal activation.

A less explicit way in which the methods of the present chapter are of paramount importance is the way in which they can instruct us in our aim of building effective theories based on some reduced set of degrees of freedom. As we have emphasized repeatedly, one of our central tasks is the construction of effective theories in which the degrees of freedom used in the model subsume the full set of microscopic degrees of freedom. For instance, in chap. 8, we will see that despite the complicated atomic-level motions which attend plastic deformation, it is possible to construct models based on defect-level entities, namely, dislocations, without mandating the consideration of atomic positions or motions explicitly. These dislocations represent the collective motion of huge numbers of atoms which have suffered relative displacements across a common slip plane. An interesting side effect of our construction of such effective models is that they require that we institute some notion of dissipation. For example, as the dislocation traverses the crystal some of the energy associated with its center of mass motion is transferred to the microscopic degrees of freedom which are disguised from the effective theory that interests itself only in the overall dislocation motion. The methods of the present chapter begin to teach us how to think of the coupling of the center of mass motions of defects to atomic-level motions, and thereby to construct models of dissipation.

The plan of the following sections is as follows. First, we undertake an analysis of the energetics of a solid when the atoms are presumed to make *small* excursions about their equilibrium positions. This is followed by a classical analysis of the normal modes of vibration that attend such small excursions. We will see in this case that an arbitrary state of motion may be specified and its subsequent evolution traced. The quantum theory of the same problem is then undertaken with the result that we can consider the problem of the specific heat of solids. This analysis will lead to an explicit calculation of thermodynamic quantities as weighted integrals over the frequencies of vibration.

5.2 Mechanics of the Harmonic Solid

As noted in the introduction to the present chapter, the contemplation of thermal and elastic excitations necessitates that we go beyond the uninterrupted monotony of the perfect crystal. As a first concession in this direction, we now spell out the way in which the energy of a weakly excited solid can be thought of as resulting from excursions about the perfect crystal reference state. Once the total energy of such excursions is in hand, we will be in a position to write down equations of motion for these excursions, and to probe their implications for the thermal and elastic properties of solids.

5.2.1 Total Energy of the Thermally Fluctuating Solid

Our starting point for the analysis of the thermal and elastic properties of crystals is an approximation. We begin with the assertion that the motions of the ith atom, will be captured kinematically through a displacement field $\mathbf{u}_i$ which is a measure of the deviations from the atomic positions of the perfect lattice. It is presumed that this displacement is small in comparison with the lattice parameter, $|\mathbf{u}_i| \ll a_0$. Though within the context of the true total energy surface of our crystal (i.e. $E_{tot}(\mathbf{R}_1, \mathbf{R}_2, \ldots, \mathbf{R}_N)$) this approximation is unnecessary, we will see that by assuming the smallness of the excursions taken by the atoms of interest, a well-defined and tractable formulation of the thermal properties of the solid may be made.

As in chap. 4, we posit the existence of a total energy function of the form $E_{tot}(\mathbf{R}_1, \mathbf{R}_2, \ldots, \mathbf{R}_N) = E_{tot}(\{\mathbf{R}_i\})$, where $\{\mathbf{R}_i\}$ denotes the entire set of ionic coordinates. For simplicity, the total energy is allowed to depend only upon the ionic positions (i.e. we have ignored the explicit dependence of the total energy function on the electronic degrees of freedom thus assuming that the system remains on the Born–Oppenheimer energy surface). Since our interest is in excursions about the equilibrium positions $\mathbf{R}_i^{ref}$, we now advance an expansion of the total energy about the reference configuration in powers of the small displacements $\mathbf{u}_i$. In particular, the total energy of the instantaneous fluctuating snapshot of the solid characterized by atomic positions $\{\mathbf{R}_i^{ref} + \mathbf{u}_i\}$ is written as

$$E_{tot}(\{\mathbf{R}_i^{ref} + \mathbf{u}_i\}) = E_{tot}(\{\mathbf{R}_i^{ref}\}) + \sum_{i,\alpha} \frac{\partial E_{tot}}{\partial R_{i\alpha}} u_{i\alpha}$$

$$+ \frac{1}{2} \sum_{i\alpha} \sum_{j\beta} \frac{\partial^2 E_{tot}}{\partial R_{i\alpha} R_{j\beta}} u_{i\alpha} u_{j\beta} + \cdots. \qquad (5.1)$$

The αth component of displacement of the ith ion is denoted $u_{i\alpha}$. The rationale for our strategy is cast in geometric terms in fig. 5.2 which shows the quadratic approximation to a fully nonlinear energy surface in the neighborhood of the minimum for an idealized two-dimensional configuration space.

This figure allows us to contrast the energy in both its exact and approximate guises. It is clear that in the vicinity of the minimum, the total energy is well characterized by what will shortly be revealed as a far-reaching but tractable model. Further progress can be made in trimming down the various terms embodied in the equation above by recognizing that the expansion is built around

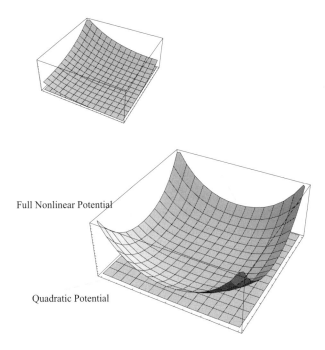

Fig. 5.2. Nonlinear potential energy $V(x_1, x_2)$ and corresponding quadratic approximation to that potential. The cutaway in the upper left hand corner shows one quadrant of the two potentials to illustrate their correspondence in the limit of small displacements.

the equilibrium configuration and hence the terms linear in $u_{i\alpha}$ can be eliminated since at equilibrium we have $\partial E_{tot}/\partial R_{i\alpha} = 0$. Furthermore, since the first term is an uninteresting constant it is found convenient to define our zero of energy as $E_{tot}(\{\mathbf{R}_i^{ref}\}) = 0$ leaving us with the result that

$$E_{tot}(\{\mathbf{R}_i^{ref} + \mathbf{u}_i\}) = \frac{1}{2} \sum_{i\alpha} \sum_{j\beta} K_{ij}^{\alpha\beta} u_{i\alpha} u_{j\beta}, \qquad (5.2)$$

where the stiffness matrix (more conventionally known as the force constant matrix) is defined by

$$K_{ij}^{\alpha\beta} = \frac{\partial^2 E_{tot}}{\partial R_{i\alpha} \partial R_{j\beta}}. \qquad (5.3)$$

Note that the expressions we have written thus far are entirely generic. All that we have assumed is the existence of a total energy function that is dependent upon the set of ionic positions. Our analysis is thus indifferent to the microscopic energy functional used to compute E_{tot} and hence the results are equally valid for total energy methods ranging from pair potentials to fully quantum mechanical E_{tot}s. On the other hand, the ease with which we can expect to determine the stiffness matrix is significantly different, with pair potentials allowing for analytic determination of **K** and density functional calculations demanding a significant computational effort. From the standpoint of our consideration of effective theories, it is especially interesting to note that regardless of our choice of microscopic energy functional, the result embodied in eqn (5.2) is nothing but a sophisticated model of atoms coupled with springs.

5.2.2 Atomic Motion and Normal Modes

Having identified our approximation to the total energy associated with small vibrations about equilibrium, we are now in a position to construct the classical equations of motion which dictate the ionic motions. If we make the simplifying assumption that we are concerned with an elemental material (i.e. all the masses are identical, etc.), thus surrendering some of the possible generality of our treatment, the equation of motion for the αth component of displacement associated with the ith ion is given by

$$m\frac{d^2 u_{i\alpha}}{dt^2} = -\frac{\partial E_{tot}}{\partial R_{i\alpha}} = -\sum_{j\beta} K_{ij}^{\alpha\beta} u_{j\beta}. \tag{5.4}$$

We note, however, that the added complication of including different masses at different sites does not change any of the key physical ideas. For further details the reader is urged to consult Maradudin *et al.* (1971). What has emerged from our analysis is a series of coupled linear differential equations in which the accelerations at one site are coupled to the displacements at the other sites. Though clearly our ambition at this point is to solve these coupled differential equations, we may recast this objective in different terms. In particular, we are after the appropriate linear combinations of the atomic-level displacements which will render the matrix **K** diagonal and lead to independent 'modes' of vibration. We anticipate the nature of the desired result by insisting that the solution take the form of a normal mode solution (which in the quantum setting we will refer to as phonons) in that each and every atom is imagined to vibrate with the same frequency ω but with different amplitudes. This conjectured behavior calls for the trial solution

$$u_{i\alpha}(t) = u_{i\alpha} e^{i\omega t}, \tag{5.5}$$

where $u_{i\alpha}$ is the amplitude of the displacement and where the physical motion is obtained by considering the real part of this expression. We hope the reader will forgive our notation in which the letter i serves both as a label of atomic positions as well as the unit imaginary number. Note that as yet we have presumed nothing

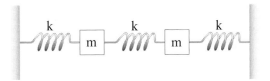

Fig. 5.3. Mass–spring system consisting of two point masses used in normal mode analysis considered in eqn (5.9).

about the possible geometric symmetries of the atomic positions. On the assumed solution given above, we find that the equations of motion may be rewritten as

$$-m\omega^2 u_{i\alpha} = -\sum_{j\beta} K_{ij}^{\alpha\beta} u_{j\beta}. \qquad (5.6)$$

This series of coupled linear equations in the unknowns $u_{i\alpha}$ may be seen in a different light if written in direct form as the matrix equation,

$$[\mathbf{K} - m\omega^2 \mathbf{I}]\mathbf{u} = 0, \qquad (5.7)$$

where $\mathbf{I}$ is the identity matrix. Here we use the compact notation in which the displacements $u_{i\alpha}$ are contained in the vector

$$\mathbf{u} = \begin{pmatrix} u_{11} \\ u_{12} \\ u_{13} \\ u_{21} \\ \vdots \\ u_{N3} \end{pmatrix}. \qquad (5.8)$$

When our problem is recast in this form it is clear that we are reconsidering it as one of matrix diagonalization. Our task is to find that change of variables built around linear combinations of the $u_{i\alpha}$s that results in a diagonal $\mathbf{K}$. Each such linear combination of atomic displacements (the eigenvector) will be seen to act as an *independent* harmonic oscillator of a particular frequency, and will be denoted as 'normal coordinates'.

Before undertaking the formal analysis of vibrations in solids, it is worthwhile to see the machinery described above in action. For concreteness and utter simplicity, we elect to begin with a two-degrees-of-freedom system such as that depicted in fig. 5.3. Displacements in all but the x-direction are forbidden, and we assume that both masses m are equal and that all three springs are characterized by a stiffness k. By inspection, the total potential energy in this case may be written as

$$V(u_1, u_2) = \frac{1}{2}ku_1^2 + \frac{1}{2}ku_2^2 + \frac{1}{2}k(u_2 - u_1)^2, \qquad (5.9)$$

where u_i refers to the x-component of displacement of mass i. Following the prescription outlined in eqn (5.4), the two equations of motion become

$$m\ddot{u}_1 + 2ku_1 - ku_2 = 0, \tag{5.10}$$

and

$$m\ddot{u}_2 + 2ku_2 - ku_1 = 0. \tag{5.11}$$

These equations can be rewritten in the more sympathetic matrix form as

$$\begin{pmatrix} 2k & -k \\ -k & 2k \end{pmatrix} \begin{pmatrix} u_1 \\ u_2 \end{pmatrix} = m\omega^2 \begin{pmatrix} u_1 \\ u_2 \end{pmatrix}. \tag{5.12}$$

This equation may be rewritten as the requirement

$$\det \begin{pmatrix} 2k - m\omega^2 & -k \\ -k & 2k - m\omega^2 \end{pmatrix} = 0, \tag{5.13}$$

which results in the vibrational frequencies $\omega_- = \sqrt{k/m}$ and $\omega_+ = \sqrt{3k/m}$. In fact, we could have arrived at the lower of these frequencies without calculation by noting that this mode corresponds to the motion of the two masses in unison without any stretching of the spring that couples them. In effect, they are vibrating as if they are only attached to the springs that are tied to the walls. The eigenvectors that are allied with our two frequencies correspond precisely to the motion in unison described above and the opposite mode in which the two masses are exactly out of phase. The general solution in this instance may then be written as

$$\begin{pmatrix} u_1(t) \\ u_2(t) \end{pmatrix} = A_1 \begin{pmatrix} 1 \\ 1 \end{pmatrix} \cos \omega_- t + B_1 \begin{pmatrix} 1 \\ 1 \end{pmatrix} \sin \omega_- t$$
$$+ A_2 \begin{pmatrix} 1 \\ -1 \end{pmatrix} \cos \omega_+ t + B_2 \begin{pmatrix} 1 \\ -1 \end{pmatrix} \sin \omega_+ t. \tag{5.14}$$

The undetermined coefficients A_i and B_i allow for the specification of the relevant initial data in the form of the initial positions and velocities of the two masses.

In the more general setting in which we imagine $3N$ degrees of freedom, our analysis goes through in the same way. As mentioned earlier, the strategy remains one of finding independent normal modes in which every ion is presumed to vibrate at the same frequency. As with our two-masses problem, in principle all that remains is a diagonalization of the stiffness matrix $\mathbf{K}$. If we consider the generalization to a fully three-dimensional situation, then the αth component of displacement of the ith ion may be represented in general terms as

$$u_{i\alpha}(t) = \sum_s u_{i\alpha}^{(s)} (A_s \cos \omega_s t + B_s \sin \omega_s t), \tag{5.15}$$

where A_s and B_s are as-yet undetermined constants that insure that our solution has the full generality needed to capture an arbitrary state of displacement, $u_{i\alpha}^{(s)}$ is that part of the eigenvector associated with the sth normal mode that is tied to the αth component of displacement of atom i and the sum over s is a sum over the various normal modes.

As yet, we have illustrated that the mechanics of harmonic vibrations can be reduced to the consideration of a series of independent normal modes. This normal mode decomposition is an especially powerful insight in that it points the way to the statistical mechanical treatment of the problem of the thermal vibrations of a crystal. The Hamiltonian for the problem of harmonic vibrations can be written as

$$H = \sum_{i\alpha} \frac{p_{i\alpha}^2}{2m} + \frac{1}{2} \sum_{ij} \sum_{\alpha\beta} K_{ij}^{\alpha\beta} u_{i\alpha} u_{j\beta}, \tag{5.16}$$

where $p_{i\alpha}$ is the αth component of momentum associated with atom i and the second term is the harmonic part of the potential energy of the system as introduced in eqn (5.2). The key result we will highlight presently is that when this Hamiltonian is rewritten in terms of the normal modes, it reduces to a series of uncoupled harmonic oscillators which from the point of view of both quantum and statistical mechanics may each be treated independently. As it stands, the Hamiltonian as written in eqn (5.16) represents a *coupled* problem since the equation of motion for the ith atom depends upon the coordinates of the remaining atoms. Again in keeping with our contention that much of the business of effective theory construction is about finding the *right* degrees of freedom, we note that in the present context by adopting the correct linear combinations of the original coordinates $\{u_{i\alpha}\}$, the original problem can be recast in much simpler terms. To simplify the analysis, we introduce the notation $\eta_r(t) = A_r \cos \omega_r t + B_r \sin \omega_r t$. In terms of this notational convention, eqn (5.15) may be rewritten as

$$u_{i\alpha}(t) = \sum_s u_{i\alpha}^{(s)} \eta_s(t), \tag{5.17}$$

and hence the original Hamiltonian may similarly be rewritten as

$$H = \frac{1}{2} m \sum_{i\alpha} \sum_{rs} u_{i\alpha}^{(r)} u_{i\alpha}^{(s)} \dot{\eta}_r(t) \dot{\eta}_s(t) + \frac{1}{2} \sum_{ij} \sum_{\alpha\beta} \sum_{rs} K_{ij}^{\alpha\beta} u_{i\alpha}^{(r)} u_{i\beta}^{(s)} \eta_r(t) \eta_s(t). \tag{5.18}$$

This expression may now be cast in its most useful form by recourse to certain key orthogonality relations satisfied by the eigenvectors $u_{i\alpha}^{(s)}$. In particular, we assert without proof (see Marion (1970)) that $\sum_{i\alpha} u_{i\alpha}^{(r)} u_{i\alpha}^{(s)} = \delta_{rs}$ and that $\sum_i K_{ij}^{\alpha\beta} u_{i\alpha}^{(r)} u_{ibeta}^{(s)} = \omega_r^2 \delta_{rs}$. In light of these orthogonality relations, our Hamiltonian reduces to

$$H = \sum_r \left(\frac{1}{2} \dot{\eta}_r^2(t) + \frac{1}{2} \omega_r^2 \eta_r^2(t) \right). \tag{5.19}$$

What we have learned is that our change to normal coordinates yields a series of independent harmonic oscillators. From the statistical mechanical viewpoint, this signifies that the statistical mechanics of the collective vibrations of the harmonic solid can be evaluated on a mode by mode basis using nothing more than the simple ideas associated with the one-dimensional oscillator that were reviewed in chap. 3.

We have already learned a great deal about the atomic-level treatment of vibrations in solids and in the paragraphs that follow intend to consolidate our gains by invoking the known translational periodicity of crystals. Let us examine what more can be said about the vibrational spectrum without actually doing any calculations. First, we note that for a generic configuration of N atoms, there will be $3N$ vibrational modes, corresponding to the dimensionality of the stiffness matrix itself. We have found that the eigenvectors associated with the stiffness matrix may be constructed to form an orthonormal set, and that we can determine coefficients such that arbitrary initial data may be matched. However, as yet there has been a serious oversight in our reasoning. We have forgotten that in the case of crystalline solids, the repetitive nature of the geometry implies special symmetry relations that lead to vast simplifications in the analysis of both their electronic and vibrational states. At this point we are poised to replace our problem of diagonalization of a $3N \times 3N$ matrix by N diagonalizations of a 3×3 matrix. The crucial insight that makes this further simplification of our problem possible is known as either Bloch's theorem or Floquet analysis. As was already shown in chap. 4 (see eqn (4.68)), the basic idea is the assertion that on account of the translational periodicity, the solution in one unit cell can differ from that in other cells by at most a phase factor.

The physical content of the statement above can be translated into mathematical terms by recognizing that this extra piece of information about the solution allows us to go beyond the trial solution of eqn (5.5), to suggest

$$u_{i\alpha}(t) = u_\alpha e^{i\mathbf{q}\cdot\mathbf{R}_i - i\omega t}. \tag{5.20}$$

Note that we have presumed something further about the solution than we did in our earlier case which ignored the translational periodicity. Here we impose the idea that the solutions in different cells differ by a phase factor tied to the Bravais lattice vector $\mathbf{R}_i$. The result is that the eigenvector u_α has no dependence upon the site i; we need only find the eigenvectors associated with the contents of a single unit cell, and the motions of atoms within all remaining cells are unequivocally determined. Substitution of the expression given above into the equations of motion yields

$$-m\omega_\mathbf{q}^2 u_\alpha(\mathbf{q})e^{i\mathbf{q}\cdot\mathbf{R}_i} + \sum_{j\beta} K_{ij}^{\alpha\beta} u_\beta(\mathbf{q})e^{i\mathbf{q}\cdot\mathbf{R}_j} = 0. \tag{5.21}$$

By virtue of the crystal's periodicity we can, without loss of generality, take $\mathbf{R}_i =$

0. The equation given above may consequently be rewritten

$$-m\omega_{\mathbf{q}}^2 u_\alpha(\mathbf{q}) + \sum_\beta D_{\alpha\beta}(\mathbf{q})u_\beta(\mathbf{q}) = 0, \tag{5.22}$$

where we have defined the 'dynamical matrix' as

$$D_{\alpha\beta}(\mathbf{q}) = \sum_j K_{0j}^{\alpha\beta} e^{i\mathbf{q}\cdot\mathbf{R}_j}. \tag{5.23}$$

As a result of these definitions, eqn (5.22) may be rewritten in matrix form as

$$[\mathbf{D}(\mathbf{q}) - m\omega^2\mathbf{I}]\mathbf{u}(\mathbf{q}) = 0. \tag{5.24}$$

Just as was found in the discussion prior to eqn (4.75), the set of allowed $\mathbf{q}$ vectors is restricted because of the condition that the crystal is periodic. Indeed, because of this restriction, the $\mathbf{q}$ vectors are of the form $q_x = 2\pi m_1/Na$, $q_y = 2\pi m_2/Na$ and $q_z = 2\pi m_3/Na$, where m_1, m_2 and m_3 take only integer values. We have thus reduced our original problem which demanded the diagonalization of a $3N \times 3N$ matrix to a problem in which we diagonalize a 3×3 matrix N times, where $\mathbf{q}$ adopts N different values and labels the different vibrational states. Here we have trimmed our analysis down such that we consider only the case in which there is one atom per unit cell. Aside from index gymnastics, the extension to more atoms in each cell is relatively uneventful. Note also that the present discussion parallels, even in the details, that already given in the context of the electronic structure of periodic solids in chap. 4.

What we have learned is that our solutions may be labeled in terms of the vector $\mathbf{q}$, which is known as the wavevector of the mode of interest. Each such mode corresponds to a displacement wave with wavelength $\lambda = 2\pi/|\mathbf{q}|$. This interpretation makes it clear that there is a maximum $|\mathbf{q}|$ above which the oscillations are associated with lengths smaller than the lattice parameter and hence imply no additional information. This insight leads to an alternative interpretation of the significance of the first Brillouin zone relative to that presented in chap. 4. As an aside, we also note that had we elected to ignore the translational periodicity, we could just as well have solved the problem using the normal mode idea developed earlier. If we had followed this route, we would have determined $3N$ vibrational frequencies (as we have by invoking the translational periodicity), but our classification of the solutions would have been severely hindered.

The standard textbook exercise for illuminating the ideas discussed above in their full generality is that of the linear chain of atoms. The simplest variant of this exercise assumes that all the masses are the same as are all of the springs joining them. In this case, the dynamical matrix itself is one-dimensional and hence its eigenvalues are immediate. From above, we note that

$$D_{11}(q) = \sum_j K_{0j}^{11} e^{iqR_j}, \tag{5.25}$$

where in our case we have $K_{00} = 2k$, and $K_{01} = K_{0,-1} = -k$. To see this, recall that the total potential energy of our one-dimensional chain can be written in simplest terms as

$$E_{tot} = \frac{1}{2} \sum_{ij} k_{ij} (u_i - u_j)^2. \tag{5.26}$$

If we make the simplifying assumption of only nearest-neighbor interactions via springs of stiffness k, then the only terms involving atom 0 are

$$E_0 = \frac{1}{2} k \left[(u_0 - u_{-1})^2 + (u_1 - u_0)^2 \right]. \tag{5.27}$$

In light of this insight, it follows that the elements of the stiffness matrix are those assigned above. Explicitly evaluating the sums demanded by the dynamical matrix, yields the spectrum

$$\omega(q) = \sqrt{\frac{4k}{m}} \left| \sin \frac{qa}{2} \right|. \tag{5.28}$$

This spectrum of vibrational frequencies is shown in fig. 5.4. We note again that the significance of this result is that it tells us that if we are interested in a particular displacement wave characterized by the wavevector q, the corresponding frequency associated with that wave will be $\omega(q)$. One of the key outcomes of this calculation is that it shows that in the limit of long wavelengths (i.e. small qs) the conventional elastic wave solution is recovered (i.e. $\omega = Cq$, where C is the relevant sound velocity), while at the zone edges the waves are clearly dispersive. The use of the word dispersive is intended to convey the idea that the wave speed, $d\omega/dq$, depends explicitly on q.

An additional way in which the outcome of the normal mode analysis may be analyzed is by recourse to the distribution of vibrational frequencies. This distribution of modes is captured most succinctly via the vibrational density of states defined by

$$\rho(\omega) = \sum_{n=1}^{N_{modes}} \delta(\omega - \omega_n). \tag{5.29}$$

The idea here is that the N_{modes} vibrational frequencies may be bunched up over certain frequencies, and will have some maximum frequency. Just as the electronic density of states introduced in chap. 4 succinctly encompassed much of the key information about the electronic spectrum, the vibrational density of states will

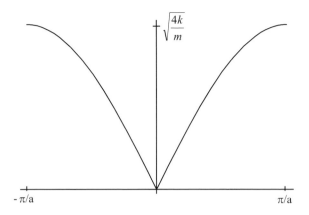

Fig. 5.4. Dispersion relation, ω vs q, for a linear chain of atoms.

play an analogous role in our analysis of the thermodynamics of solids. In the case of the linear chain, we may arrive at the vibrational density of states analytically. Before venturing our analytic solution for the vibrational density of states, we examine the ideas pictorially. In fig. 5.4 we examine the dispersion relation. Note that each region of q-space of width $2\pi/Na$ corresponds to a single mode. (In fact, we need to be slightly cautious on this point since in addition to the mode with wavevector q there is a partner mode with wavevector $-q$.) However, we have interested ourselves in the distribution of states as a function of frequency ω. The correspondence is made in the figure where we note that in regions where $\partial\omega/\partial q \approx 0$, there are many q-vectors per unit interval in ω while for those regions where $\partial\omega/\partial q$ is large, there are but a few.

To compute the vibrational density of states explicitly, we make the correspondence $\rho(\omega)d\omega = (Na/\pi)dq$. What we have noted is that in a region of q-space of width dq, there are $2dq/(2\pi/Na)$ states since the uniform spacing between q-points is $2\pi/Na$ and the factor of 2 in the numerator accounts for the q and $-q$ degeneracy. We may reorganize this statement as

$$\rho(\omega) = \frac{Na}{\pi} \frac{1}{|d\omega/dq|}. \tag{5.30}$$

For the dispersion relation given above for the linear chain, we have

$$\frac{d\omega}{dq} = \frac{\omega_0 a}{2} \cos \frac{qa}{2}, \tag{5.31}$$

where we have defined $\omega_0 = \sqrt{4K/m}$. We must now rewrite the right hand side of this equation in terms of ω. We note that

$$\frac{qa}{2} = \sin^{-1}\left(\frac{\omega}{\omega_0}\right). \tag{5.32}$$

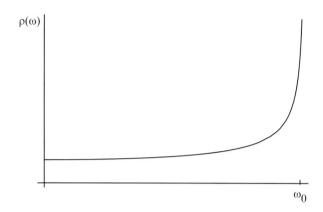

Fig. 5.5. Vibrational density of states associated with linear chain.

This is nothing more than an inversion of the dispersion relation. In geometric terms, what we have learned is that $\cos(qa/2) = \sqrt{\omega_0^2 - \omega^2}/\omega_0$ and thus

$$\rho(\omega) = \frac{2N}{\pi} \frac{1}{\sqrt{\omega_0^2 - \omega^2}}, \tag{5.33}$$

which is shown in fig. 5.5. As noted earlier, the vibrational density of states provides an alternative representation of the information content present in the dispersion relation, though there is a loss of information concerning the q vector associated with each frequency. As will be seen in subsequent discussions, a knowledge of the vibrational density of states will allow us to compute key thermodynamic properties such as the entropy associated with thermal vibrations.

What has been lost in our treatment of the restricted geometry of the one-dimensional chain is the possibility for different wave polarizations to be associated with each wavevector. In particular, for a three-dimensional crystal, we expect to recover modes in which the vibrations are either parallel to the wavevector (longitudinal) or perpendicular to it (transverse). On the other hand, the formalism outlined above already makes the outcome of this analysis abundantly clear. In particular, for a simple three-dimensional problem in which we imagine only one atom per unit cell, we see that there are three distinct solutions that emerge from our matrix diagonalization, each of which corresponds to a different polarization.

To make these ideas concrete, we consider the case of a solid characterized by a pair potential and with a total energy of the form

$$E_{tot} = \frac{1}{2} \sum_{ij} \phi(R_{ij}). \tag{5.34}$$

Our charter is to first compute the matrix $K_{ij}^{\alpha\beta}$ introduced earlier and then to evaluate its Fourier transform in preparation for the computation of the spectrum of vibrational frequencies. By definition, the matrix $K_{ij}^{\alpha\beta}$ is obtained as

$$K_{ij}^{\alpha\beta} = \frac{\partial^2 E_{tot}}{\partial R_{i\alpha} \partial R_{j\beta}}. \tag{5.35}$$

Within the pair potential formalism under consideration presently, this involves a series of exercises in the use of the chain rule since we have to evaluate terms of the form

$$\frac{\partial E_{tot}}{\partial R_{i\alpha}} = \frac{1}{2} \sum_{lm} \phi'(R_{lm}) \frac{\partial R_{lm}}{\partial R_{i\alpha}}. \tag{5.36}$$

Note that there is a slight burden of notational baggage which we will try to dispel by observing that we have defined $R_{lm} = |\mathbf{R}_l - \mathbf{R}_m|$, and at the same time have used $R_{i\alpha}$ to denote the αth Cartesian component of the position vector of atom i. Thus we have adopted a convention in which a pair of Latin indices implies consideration of a pair of atoms, while a mixed pair of a single Latin and a single Greek index refers to a particular component of the position vector of a given atom. Carrying out the differentiations demanded by eqn (5.35), the force constant matrix may be written as

$$\begin{aligned}
K_{ij}^{\alpha\beta} = &\sum_m \phi''(R_{im}) \frac{(R_{i\alpha} - R_{m\alpha})(R_{i\beta} - R_{m\beta})}{R_{im}^2} \delta_{ij} \\
&- \phi''(R_{ij}) \frac{(R_{i\alpha} - R_{j\alpha})(R_{i\beta} - R_{j\beta})}{R_{ij}^2} \\
&+ \sum_m \phi'(R_{im}) \frac{\delta_{ij}\delta_{\alpha\beta}}{R_{im}} - \sum_m \phi'(R_{im}) \frac{(R_{i\alpha} - R_{m\alpha})(R_{i\beta} - R_{m\beta})}{R_{im}^3} \delta_{ij} \\
&+ \phi'(R_{ij}) \frac{(R_{i\alpha} - R_{j\alpha})(R_{i\beta} - R_{j\beta})}{R_{ij}^3} - \phi'(R_{ij}) \frac{\delta_{\alpha\beta}}{R_{ij}}.
\end{aligned} \tag{5.37}$$

We show this result primarily to illustrate that what are really required to determine the force constant matrix are the derivatives of the relevant interatomic potential, and then the consideration of a series of lattice sums as embodied in the various $\sum_m$s that appear above. Once the force constant matrix is in hand, the next step in the procedure for obtaining the spectrum of vibrational frequencies is the determination of the dynamical matrix. For the simple case in which there is only one atom per unit cell, this amounts to carrying out the sums illustrated in eqn (5.23).

It is worth exploring this question a little further. We note that we require the sum $D_{\alpha\beta}(\mathbf{q}) = \sum_j K_{0j}^{\alpha\beta} e^{i\mathbf{q}\cdot\mathbf{R}_j}$. Analytic progress can be made by exploiting several key symmetries of the force constant matrix $K_{ij}^{\alpha\beta}$. First, we claim that there is a sum rule of the form $\sum_j K_{0j}^{\alpha\beta} = 0$ which arises from the fact that if every atom is displaced by the same amount, there is no net force on any atom (for further details see the problems at the end of the chapter). This then allows us to note that we can rewrite the self-term in the force constant matrix as

$$K_{00}^{\alpha\beta} = -\sum_{j\neq 0} K_{0j}^{\alpha\beta}. \tag{5.38}$$

As a result of the observation implicit in eqn (5.38), we can now write the dynamical matrix as

$$D_{\alpha\beta}(\mathbf{q}) = \sum_{j\neq 0} K_{0j}^{\alpha\beta} (e^{i\mathbf{q}\cdot\mathbf{R}_j} - 1). \tag{5.39}$$

This may be further simplified by noting that for every term $\mathbf{R}_j$ in the sum we will have a partner term $-\mathbf{R}_j$, and also by using the fact that the force constant matrix is the same for these partner terms. As a result, we have

$$D_{\alpha\beta}(\mathbf{q}) = \frac{1}{2} \sum_{j\neq 0} K_{0j}^{\alpha\beta} (e^{i\mathbf{q}\cdot\mathbf{R}_j} + e^{-i\mathbf{q}\cdot\mathbf{R}_j} - 2), \tag{5.40}$$

which after simplification leaves us with

$$D_{\alpha\beta}(\mathbf{q}) = -2 \sum_{j\neq 0} K_{0j}^{\alpha\beta} \sin^2 \frac{\mathbf{q}\cdot\mathbf{R}_j}{2}. \tag{5.41}$$

The conclusion of this analysis is that we can obtain the dynamical matrix by performing lattice sums involving only those parts of the force constant matrix that do not involve the self-terms (i.e. $K_{00}^{\alpha\beta}$). We note that the analysis we have made until now has been performed in the somewhat sterile setting of a pair potential description of the total energy. On the other hand, the use of more complex energy functionals does not introduce any new conceptual features.

As a case study in the calculation of the relation between ω and $\mathbf{q}$ (also known as phonon dispersion relations) for three-dimensional crystals, we consider the analysis of normal modes of vibration in fcc Al. As will be evident repeatedly as coming chapters unfold, one of our principal themes will be to examine particular problems from the perspective of several different total energy schemes simultaneously. In the present context, our plan is to consider the dispersion relations in Al as computed using both empirical pair functional calculations as well as first-principles calculations.

As the analysis given above illustrated for the case of a pair potential, the determination of the dynamical matrix demands the evaluation of lattice sums involving the products of geometric factors and the derivatives of the various

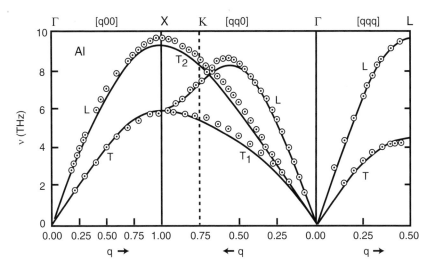

Fig. 5.6. Phonon dispersion relation for Al as computed using pair functional approach (adapted from Mishin *et al.* (1999)). The lines correspond to the theoretical results and the points are experimental data obtained from neutron scattering.

energy terms. The basic procedure is unchanged in the case of the pair functional scheme although the presence of the embedding term introduces a whole new set of terms in the force constant matrix and hence in the dynamical matrix as well. A particularly simple description of these additional terms may be found in Finnis and Sinclair (1984). Once the lattice sums required in the computation of the force constant matrix have been carried out, all that remains is the calculation and subsequent diagonalization of the dynamical matrix with results like those shown in fig. 5.6.

The results given above can be measured not only against those resulting from experiments, but also against those deduced on the basis of first-principles calculations. An example of the outcome of a first-principles analysis of phonons in fcc Al is shown in fig. 5.7 (Quong and Klein 1992). These calculations were carried out using pseudopotentials to represent the ion–electron interaction and with a plane wave basis for expanding the wave function. We note in passing that the use of first-principles techniques to compute the phonon spectrum involves further subtleties than were hinted at in our discussion above, and we content ourselves only with showing the results of such calculation. We wish to emphasize that our primary ambition in the present setting has not been to debate the vices and virtues of any particular treatment of the lattice vibrations of a particular system. Rather, our aim has been to illustrate what is involved in carrying out such calculations and to reveal the way in which the outcome of these calculations is communicated,

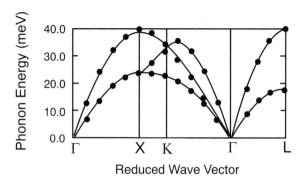

Fig. 5.7. Phonon dispersion relation for Al as computed using first-principles methods (adapted from Quong and Klein (1992)).

namely in terms of the types of phonon dispersion relations shown in figs. 5.6 and 5.7.

5.2.3 Phonons

Until now, our treatment has been built in exactly the same terms that might have been used in work on normal modes of vibration in the latter part of the nineteenth century. However, it is incumbent upon us to revisit these same ideas within the quantum mechanical setting. The starting point of our analysis is the observation embodied in eqn (5.19), namely, that our harmonic Hamiltonian admits of a decomposition into a series of independent one-dimensional harmonic oscillators. We may build upon this observation by treating each and every such oscillator on the basis of the quantum mechanics discussed in chap. 3. In light of this observation, for example, we may write the total energy of the harmonic solid as

$$E = \sum_{\mathbf{q},\alpha} \left[n(\mathbf{q}, \alpha) + \frac{1}{2} \right] \hbar \omega_{\mathbf{q},\alpha}. \tag{5.42}$$

The interpretation that we adopt in this setting is that the various modes are each characterized by a different occupation number, $n(\mathbf{q}, \alpha)$, where $\mathbf{q}$ characterizes the wavevector of the mode of interest and α labels the various polarizations. In this context, the occupation of each normal mode is associated with a discrete entity known as a phonon. Each such phonon carries an energy $\hbar \omega(\mathbf{q}, s)$ and a momentum $\hbar \mathbf{q}$ and may be assigned properties traditionally thought of as being tied to particles. For example, phonons may scatter off of one another. However, we also note that in order to develop a full picture of the phonons and their interactions,

C_{60}

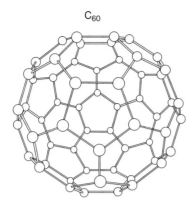

Fig. 5.8. Schematic of the C_{60} molecule (adapted from Woo *et al.* (1993)).

it is necessary to extend the harmonic treatment of vibrations entertained above to include the anharmonic effects which give rise to such scattering.

5.2.4 *Buckminsterfullerene and Nanotubes: A Case Study in Vibration*

One of the approaches favored in the present work is the idea of attempting to give a thorough description of robust physical principles followed by more speculative application of those principles to problems of current interest. One class of materials that has garnered particular attention in recent years is the class of fullerenes in which by virtue of either five- or seven-membered rings, curvature is induced in the normally flat packings of carbon based upon six-membered rings. The most celebrated such example is perhaps the C_{60} molecule in which a total of 60 carbon atoms form a molecular structure with vertices reminiscent of those on a soccer ball. Beyond the interest that has attached to C_{60} itself, there has been a surge of interest in carbon nanotubes. The aim of the present subsection is to give a sense of how the physics of phonons introduced above plays out in the context of relatively complex molecules and solids. The C_{60} molecule is shown in fig. 5.8. The distinctive five-fold rings are clearly evident in the figure. In this instance, we have $3N = 180$ independent degrees of freedom, although six of them are associated with uninteresting rigid body translations and rotations.

Our aim in the present section is to examine the nature of vibrations in several exotic carbon structures, and we begin with the C_{60} itself. The calculations to be described here were based on first-principles techniques in which the electron–ion interaction was handled using pseudopotentials in conjunction with a mixed basis including both plane wave states and localized s- and p-orbitals associated with the carbon atoms (Bohnen *et al.* 1995). As has already been made clear, to

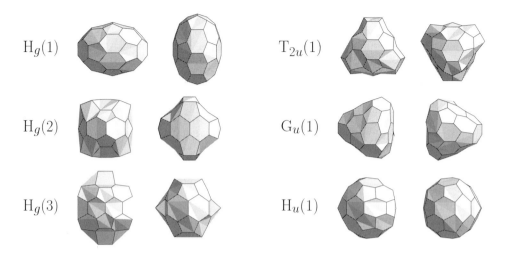

$H_g(1)$ $T_{2u}(1)$

$H_g(2)$ $G_u(1)$

$H_g(3)$ $H_u(1)$

Fig. 5.9. Vibrational displacement patterns associated with five different modes of the C_{60} molecule (adapted from Heid *et al.* (1997)). The left and right hand panels correspond to opposite extremes in motion associated with a given mode. Labels associated with each mode characterize the symmetry of the displacement pattern.

obtain the phonon frequencies, it is necessary to construct the dynamical matrix. Diagonalization of this matrix results in a series of vibrational frequencies ω_s and associated eigenvectors which give the relative amplitudes of vibration for the different degrees of freedom. The disposition of the vibrational frequencies may be summarized pictorially via the vibrational density of states. Note that in a highly symmetric object like the C_{60} molecule the occurrence of certain degeneracies in this spectrum is not accidental and can be predicted on the grounds of this symmetry without doing any explicit matrix diagonalizations. In addition to our interest in the spectrum of vibrational frequencies that emerge from the type of analysis being described here, it is also of interest to characterize the nature of the associated eigenvectors. Recall that these eigenvectors yield the nature of the displacement patterns that will be associated with a given mode of vibration. In the case of C_{60}, some of the different displacement patterns are revealed in fig. 5.9, though the amplitudes of the displacements are exaggerated to more easily reveal the symmetry of the various modes.

As noted above, in addition to the interest that has been attached to the C_{60} molecule and its crystalline incarnations, a torrent of interest has also developed around the existence of carbon nanotubes in which planar graphite-like packings of carbon atoms are folded onto themselves to form cylinders. We note that there is great variety in the structures that can be created since the line along which the original planar sheet is cut prior to wrapping it to form a tube can be chosen in many different ways. In addition, different tube radii can be selected. In fig. 5.10

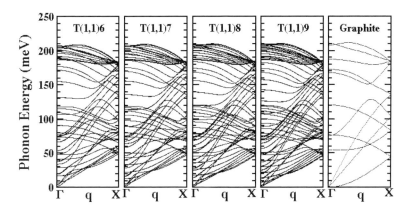

Fig. 5.10. Dispersion relations for several different carbon nanotube structures (adapted from Yu *et al.* (1995)).

we show the results of tight-binding calculations of the vibrational spectrum of several different nanotube variants. Part of the value in presenting these results is the reminder that calculations of quantities such as the vibrational spectrum of a material can be carried out in any of a number of different ways. Already in this chapter we have seen such dispersion relations computed using pair functionals (for Al), first principles (for fcc Al and C_{60}) and tight-binding analysis (the present case). Just as was possible with the analysis revealed of the displacement patterns in C_{60}, it is possible to examine the nature of the vibrations associated with nanotubes as well. An example of the various eigenvectors associated with a particular nanotube is shown in fig. 5.11.

5.3 Thermodynamics of Solids

5.3.1 Harmonic Approximation

As we saw above, what emerges from our detailed analysis of the vibrational spectrum of a solid can be neatly captured in terms of the vibrational density of states, $\rho(\omega)$. The point of this exercise will be seen more clearly shortly as we will observe that the thermodynamic functions such as the Helmholtz free energy can be written as integrals over the various allowed frequencies, appropriately weighted by the vibrational density of states. In chap. 3 it was noted that upon consideration of a single oscillator of natural frequency ω, the associated Helmholtz free energy is

$$F = \frac{\hbar\omega}{2} + kT \ln(1 - e^{-(\hbar\omega/kT)}). \qquad (5.43)$$

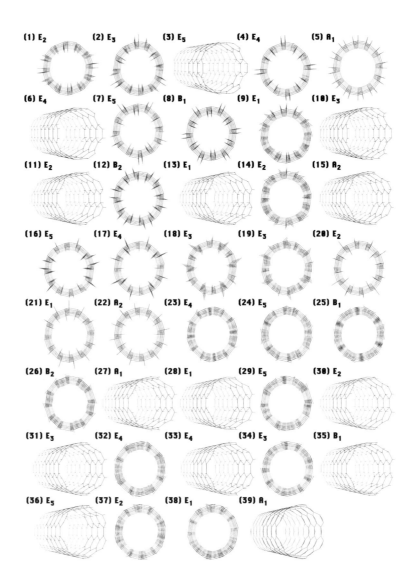

Fig. 5.11. Illustration of a variety of different vibrational modes of one of the broad class of carbon nanotubes (adapted from Yu *et al.* (1995)).

In light of our observations from above, namely that the vibrational contribution to the energy of the crystal may be written as a sum of independent harmonic oscillators, this result for the Helmholtz free energy may be immediately generalized. In particular, we note that once the vibrational density of states has been determined it is immediate that

$$F = \int_0^\infty \frac{\hbar\omega}{2} \rho(\omega)d\omega + kT \int_0^\infty \rho(\omega)\ln(1 - e^{-(\hbar\omega/kT)})d\omega. \qquad (5.44)$$

Note that all reference to the wavevector has been eliminated as a result of the fact that $\rho(\omega)$ is obtained by averaging over *all* wave vectors within the Brillouin zone, and is hence indifferent to the polarization and propagation vector of a given normal mode.

In sections 5.22 and 5.24, we have made schematic evaluations of the nature of the vibrational spectrum (see fig. 5.5). At this point, it is convenient to construct approximate model representations of the vibrational spectrum with the aim of gleaning some insight into how the vibrational free energy affects material properties such as the specific heat, the thermal expansion coefficient and processes such as structural phase transformations. One useful tool for characterizing a distribution such as the vibrational density of states is through its moments, defined by

$$\mu_n = \int_0^\infty \omega^n \rho(\omega)d\omega. \qquad (5.45)$$

The most naive model we can imagine is to insist that whatever model we select for the density of states faithfully reproduces the first moment of the exact vibrational density of states. That is,

$$\int_0^\infty \omega\rho_{exact}(\omega)d\omega = \int_0^\infty \omega\rho_{model}(\omega)d\omega. \qquad (5.46)$$

We can guarantee this by following Einstein and asserting that all of the spectral weight is concentrated in a single vibrational frequency, ω_E. That is, we approximate the full density of states with the replacement

$$\rho_{exact}(\omega) \to \rho_E(\omega) = 3N\delta(\omega - \omega_E). \qquad (5.47)$$

The partition function for the model presented above is given by

$$Z = \left(\frac{e^{-\frac{\beta\hbar\omega_E}{2}}}{1 - e^{-\beta\hbar\omega_E}} \right)^{3N}. \qquad (5.48)$$

This may be seen by noting that the partition function may be written as a series of sums of the form

$$Z = \sum_{n_1=0}^\infty \cdots \sum_{n_{3N}}^\infty \exp\left[-\beta \sum_i \left(n_i + \frac{1}{2} \right)\hbar\omega_i \right]. \qquad (5.49)$$

However, in the case of the Einstein model $\omega_i = \omega_E$ for all ω and hence the sum indicated above may be rewritten as

$$Z = \sum_{n_1=0}^{\infty} \exp\left[-\beta\left(n_1 + \frac{1}{2}\right)\hbar\omega_E\right] \cdots \sum_{n_{3N}=0}^{\infty} \exp\left[-\beta\left(n_{3N} + \frac{1}{2}\right)\hbar\omega_E\right], \quad (5.50)$$

which is equivalent to the product written in eqn (5.48). As noted already in chap. 3, once the partition function is in hand, we are in a position to evaluate properties of thermodynamic interest such as the average energy which is given by eqn (3.98) and in the present context may be written as

$$\langle U \rangle = -\frac{\partial}{\partial\beta}\ln Z = \frac{3N\hbar\omega_E}{2} + \frac{3N\hbar\omega_E}{e^{\beta\hbar\omega_E} - 1}. \quad (5.51)$$

Recalling the definition of the specific heat as the temperature derivative of the average energy (see eqn (3.99)), it is found that for the Einstein solid

$$C_V = 3Nk(\beta\hbar\omega_E)^2 \frac{e^{\beta\hbar\omega_E}}{(e^{\beta\hbar\omega_E} - 1)^2}. \quad (5.52)$$

The average energy and the associated specific heat are plotted in fig. 5.12. These are precisely the same results derived in our treatment of the statistical mechanics of a single oscillator in chap. 3, but now imbued with the extra insights that come from knowing *how* the vibrational frequencies emerge from microscopic analysis. This model represents one of the first (if not the first) examples of the use of reasoning based on quantum theory in order to deduce the properties of materials. One of the key features of this result is the fact that at high temperatures it reduces to the known high-temperature limit, $C_V = 3Nk$, while vanishing at low temperature (though not with the proper scaling with T as we shall see in our discussion of the Debye model).

As yet, we have made a very crude approximation to the exact density of states. In particular, we have replaced the full vibrational density of states by a single delta function according to the prescription

$$\rho(\omega) = \sum_n \delta(\omega - \omega_n) \rightarrow 3N\delta(\omega - \omega_E). \quad (5.53)$$

We now consider a more quantitative model of the vibrational density of states which makes a remarkable linkage between continuum and discrete lattice descriptions. In particular, we undertake the Debye model in which the vibrational density of states is built in terms of an isotropic linear elastic reckoning of the phonon dispersions. Recall from above that in order to effect an accurate calculation of the true phonon dispersion relation, one must consider the dynamical matrix. Our approach here, on the other hand, is to produce a model representation of the phonon dispersions which is valid for long wavelengths and breaks down at

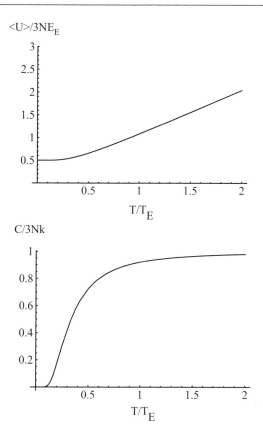

Fig. 5.12. Plots of the dimensionless average energy and specific heat resulting from the Einstein model. The average energy is scaled by $3NE_E$, where $E_E = \hbar\omega_E$, while the specific heat is reported in units of $3Nk$. The temperature is also plotted in units of the Einstein temperature given by $T_E = \hbar\omega_E/k$.

shorter wavelengths. One approach to this analysis is to linearize the dynamical matrix. An alternative route is to compute the linear elastic dispersion relation explicitly. In order to make this calculation it is well to recall the equations of linear elastodynamics, namely,

$$C_{ijkl}u_{k,lj} = \rho\frac{\partial^2 u_i}{\partial t^2}. \tag{5.54}$$

In the present setting ρ is the mass density while the subscript i identifies a particular Cartesian component of the displacement field. In this equation recall that C_{ijkl} is the elastic modulus tensor which in the case of an isotropic linear elastic solid is given by $C_{ijkl} = \lambda\delta_{ij}\delta_{kl} + \mu(\delta_{ik}\delta_{jl} + \delta_{il}\delta_{jk})$. Following our earlier footsteps from chap. 2 this leads in turn to the Navier equations (see eqn (2.55)) which it is to be remembered are given by

$$(\lambda + \mu)\nabla(\nabla \cdot \mathbf{u}) + \mu\nabla^2\mathbf{u} = \rho\frac{\partial^2\mathbf{u}}{\partial t^2}, \tag{5.55}$$

in the dynamical case.

These equations admit of wave solutions of the form $u_i = u_{i0}e^{i(\mathbf{q}\cdot\mathbf{r}-\omega t)}$ like those advanced in the treatment of vibrations in terms of the dynamical matrix. Substitution of this trial solution into the Navier equations yields their q-space analog, namely,

$$[(\lambda + \mu)q_iq_j + \mu q^2\delta_{ij} - \rho\omega^2\delta_{ij}]u_{j0} = 0. \tag{5.56}$$

Once again, our problem of determining the wave solutions associated with our medium has been reduced to a problem of matrix diagonalization. The eigenvalues of the acoustic tensor or Christoffel tensor, $A_{ij} = (\lambda + \mu)q_iq_j + \mu q^2\delta_{ij}$, are found to be

$$\omega_T = \sqrt{\frac{\mu}{\rho}}q \tag{5.57}$$

and

$$\omega_L = \sqrt{\frac{\lambda + 2\mu}{\rho}}q, \tag{5.58}$$

corresponding to transverse (ω_T) and longitudinal (ω_L) waves, in which the displacements are either perpendicular or parallel to the propagation direction, respectively. The solution associated with transverse waves is doubly degenerate in that there are two distinct solutions with this same eigenfrequency.

The solution arrived at in our linear elastic model may be contrasted with those determined earlier in the lattice treatment of the same problem. In fig. 5.13 the dispersion relation along an arbitrary direction in q-space is shown for our elastic model of vibrations. Note that as a result of the presumed isotropy of the medium, no q-directions are singled out and the dispersion relation is the same in every direction in q-space. Though our elastic model of the vibrations of solids is of more far reaching significance, at present our main interest in it is as the basis for a deeper analysis of the specific heats of solids. From the standpoint of the contribution of the thermal vibrations to the specific heat, we now need to determine the density of states associated with this dispersion relation.

As was noted earlier in the context of phonons, the density of states in q-space is constant, with one state per unit volume $V/(2\pi)^3$ of q-space. That is, the set of allowed q-vectors form a lattice in q-space with each and every lattice point corresponding to a different allowed state. Our intention now is to recast this result in frequency space. We know that in a volume d^3q of q-space there

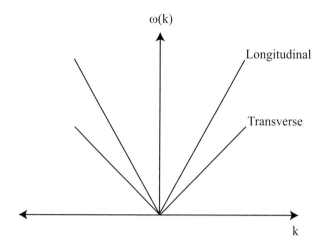

Fig. 5.13. Dispersion relation obtained if solid is characterized by linear elastic constitutive model.

are $V/(2\pi)^3 d^3q$ states. We now interest ourselves in the total number of states between two constant ω surfaces at ω and $\omega + d\omega$. The total number of states in this shell is

$$N = \frac{V}{(2\pi)^3} dq_\perp \int dA,$$ (5.59)

where $dq_\perp$ is the height of the volume element at a given position on the constant frequency surface being integrated over and dA is the area element on the constant frequency surface. The logic behind this equation is that we have asked how many states are in the shell of thickness $dq_\perp$. Now we note that the increment in frequency over that same height is given by $d\omega = |\nabla_q \omega| dq_\perp$, suggesting the alternative and more useful expression

$$\rho(\omega)d\omega = \frac{V}{(2\pi)^3} \int dA \frac{d\omega}{|\nabla_q \omega|},$$ (5.60)

for the vibrational density of states. Use of this expression in the context of the dispersion relations obtained above yields

$$\rho(\omega) = \frac{V\omega^2}{2\pi^2}\left(\frac{1}{c_L^3} + \frac{2}{c_T^3}\right),$$ (5.61)

where we have introduced the elastic wave speeds, $c_L = d\omega_L/dq$ and $c_T = d\omega_T/dq$.

We are now in a position to make an evaluation of the specific heat within this model. Recall from earlier that the average energy of an oscillator of frequency ω

is given by

$$\langle U \rangle = \frac{\hbar\omega}{e^{\beta\hbar\omega} - 1},\tag{5.62}$$

where we have dropped reference to the zero point energy for simplicity. For our collection of oscillators described by the vibrational density of states determined above, we have then

$$\langle U \rangle = \frac{\hbar V}{2\pi^2}\left(\frac{1}{c_L^3} + \frac{2}{c_T^3}\right)\int_0^{\omega_D} \frac{\omega^3 d\omega}{e^{\beta\hbar\omega} - 1},\tag{5.63}$$

where as yet all that remains undetermined is the upper cutoff frequency ω_D, known as the Debye frequency. However, we know that the total number of modes of the system is $3N$, where N is the number of atoms. As a result, we know that the density of states when integrated up to ω_D must yield $3N$. That is,

$$\int_0^{\omega_D} \rho(\omega)d\omega = 3N.\tag{5.64}$$

The reasoning invoked here is really no different than that associated with the Fermi level in electronic problems where we insisted that upon integrating the electronic density of states up to this level, the correct number of electrons was obtained (see eqn (4.88), for example). If we define the mean wave speed via

$$\frac{3}{\bar{c}^3} = \frac{1}{c_L^3} + \frac{2}{c_T^3},\tag{5.65}$$

then the density of states may be rewritten as $\rho(\omega) = 3V\omega^2/2\pi^2\bar{c}^3$, which when substituted into the equation constraining the number of modes to $3N$ yields $\omega_D = \bar{c}(6\pi^2 N/V)^{\frac{1}{3}}$. If we further define the Debye temperature θ_D via $\hbar\omega_D = k\theta_D$, our expressions for the average energy may be differentiated to yield the specific heat which may itself be written in the considerably simplified form

$$C_v = 9kN\left(\frac{T}{\theta_D}\right)^3 \int_0^{\frac{\theta_D}{T}} \frac{u^4 e^u du}{(e^u - 1)^2}.\tag{5.66}$$

The Debye result for the specific heat of a solid is of profound significance in elucidating the key physical features that attend the thermal properties of solids. One key observation that is immediate is that in the limit when the temperature is low in comparison with the Debye temperature, we have $\theta_D/T \rightarrow \infty$, which implies that the upper limit in the integral given in eqn (5.66) may be taken as infinity. The consequence of this insight is that the low-temperature specific heat scales with the temperature as T^3. This result emerges in light of the fact that the integral is a pure number which has no dependence on the temperature. In particular, we note the characteristic T^3 scaling that is compared with experimental values in fig. 5.1. In the other extreme, namely at very high temperatures, we note that the Debye result may be rearranged to recover the classical result $C_V = 3Nk$ of Dulong and Petit. Physically, the success of the Debye model may be attributed to the fact that at low temperatures it is only the low-energy (or low-frequency) parts of the phonon spectrum which are excited. This part of the spectrum corresponds in part to the long-wavelength part of the spectrum which is precisely that part of the dispersion relation that is captured correctly by the Debye model. By way of contrast, the high-temperature result recovers the equipartition theorem and is predicated upon a correct assessment of the total number of modes to be shared. The Debye result then smoothly interpolates between these two limits.

As a reminder of what we have accomplished and where it fits into our broader program, we summarize the arguments of the chapter up until this point. We begin by noting that in the limit in which it is imagined that the excursions of atoms as they exercise thermal vibrations are sufficiently small, the potential energy of all solids reduces to a sophisticated model of coupled springs. This is a fascinating example of the type of effective theory construction that we have harped on repeatedly and note that in this case it is an especially intriguing conclusion since this result is indifferent to the type of bonding in the solid, whether it be metallic, covalent, ionic or something else. Once the energetics of the harmonic solid was in hand, we were then able to demonstrate that there exists a natural set of coordinates (i.e. normal coordinates) for characterizing the vibrations of a solid as a set of independent oscillators. Finally, as a result of our knowledge of the vibrational spectrum of a solid, the tools of statistical mechanics were invoked in order to determine the mean energy and specific heat, and as will be shown in the next chapter, these same arguments can be exploited to examine the vibrational entropy and to determine its role in the free energy balance separating competing different crystal structures. We now turn to an examination of some of the *anharmonic* effects in solids such as the thermal expansion and the thermal conductivity.

5.3.2 Beyond the Harmonic Approximation

Thus far, our analysis has been entirely predicated upon the harmonic approximation advanced in eqn (5.2). On the other hand, real crystals are characterized by nonlinear force laws and the resulting anharmonic effects are part of the story from the outset. If we confine our interest to the specific heat, we see that much may be made of the harmonic approximation. However, if we aim to press our modeling efforts further, for example, to consider the thermal expansion or the thermal conductivity, we will see that the harmonic approximation does not suffice. Let us flesh out the logic in the case of the coefficient of thermal expansion. Our first ambition will be to demonstrate that the harmonic approximation predicts zero thermal expansion. This will be followed by an explicit incorporation of anharmonicity.

We have seen that as a consequence of the harmonic Hamiltonian that has been set up thus far, our oscillators decouple and in cases which attempt to capture the transport of energy via heating, there is no mechanism whereby energy may be communicated from one mode to the other. This shortcoming of the model may be amended by including coupling between the modes, which as we will show below, arises naturally if we go beyond the harmonic approximation. The simplest route to visualizing the physics of this problem is to assume that our harmonic model is supplemented by anharmonic terms, such as

$$H^{anh} = \sum_{i\alpha} \sum_{j\beta} \sum_{k\gamma} \Gamma_{ijk}^{\alpha\beta\gamma} u_{i\alpha} u_{j\beta} u_{k\gamma} + \cdots . \tag{5.67}$$

This result urges us to consider the nonlinear terms which must supplement the quadratic model set forth in eqn (5.2). The coefficients $\Gamma_{ijk}^{\alpha\beta\gamma}$ are defined in analogy with the $K_{ij}^{\alpha\beta}$ defined in eqn (5.3) and are given by

$$\Gamma_{ijk}^{\alpha\beta\gamma} = \frac{\partial^3 E_{tot}}{\partial R_{i\alpha} \partial R_{j\beta} \partial R_{k\gamma}} . \tag{5.68}$$

There are certain subtleties associated with the order to which the expansion of eqn (5.67) is carried and for a discussion of which we refer the perspicacious reader to chap. 25 of Ashcroft and Mermin (1976). If we maintain the use of normal coordinates, but now including the anharmonic terms which are represented via these coordinates, it is seen that the various modes are no longer independent. For the purposes of the present discussion, we see that these anharmonic terms have the effect of insuring that the equation of motion for the nth mode is altered by coupling to the rest of the mode variables. This result bears formal resemblance to the Navier–Stokes equations which when written in terms of the Fourier variables yield a series of mode coupled equations. We have already noted that the physics of both thermal expansion and thermal conductivity demand the inclusion of these higher-order terms.

Recall that the coefficient of thermal expansion is defined as

$$\alpha = \frac{1}{3B} \left(\frac{\partial P}{\partial T} \right)_V , \tag{5.69}$$

where B is the bulk modulus, itself defined as $B = -V(\partial p/\partial V)_T$. We may evaluate the pressure, needed to compute the thermal expansion coefficient, as $p = -(\partial F/\partial V)_T$. Using the fact that the Helmholtz free energy is obtained as a sum over all modes as

$$F = \sum_{\mathbf{q},\alpha} \frac{\hbar\omega(\mathbf{q},\alpha)}{2} + kT \sum_{\mathbf{q},\alpha} \ln(1 - e^{-\beta\hbar\omega(\mathbf{q},\alpha)}), \qquad (5.70)$$

the pressure may be written in turn as

$$p = \sum_{\mathbf{q},\alpha} \frac{\hbar}{2}\left[\frac{-\partial\omega(\mathbf{q},\alpha)}{\partial V}\right] + \sum_{\mathbf{q},\alpha} \hbar\left[\frac{-\partial\omega(\mathbf{q},\alpha)}{\partial V}\right]\frac{1}{e^{\beta\hbar\omega(\mathbf{q},\alpha)} - 1}. \qquad (5.71)$$

In both of these sums, we are summing over the wavevector $\mathbf{q}$ and the polarization α of the various phonons.

What we have found is that all of the putative volume dependence of the free energy is tied to the vibrational frequencies themselves. Once we have determined p, then the thermal expansion coefficient may be computed as

$$\alpha = \frac{1}{3B}\sum_{\mathbf{q},\alpha}\left[-\hbar\frac{\partial\omega(\mathbf{q},\alpha)}{\partial V}\right]\left[\frac{\partial n(\omega,T)}{\partial T}\right], \qquad (5.72)$$

where we have defined

$$n(\omega,T) = \frac{1}{e^{(\hbar\omega/kT)} - 1}. \qquad (5.73)$$

Though we have determined the coefficient of thermal expansion in principle, the entirety of our analysis is focused on how the vibrational frequencies are altered under a uniform dilatational strain. However, in the absence of some form of anharmonicity, this model is as yet plagued with a fatal flaw; the vibrational frequencies within the harmonic approximation have no such volume dependence. At this point, our idea is to revisit the expansion of eqn (5.2) with the ambition of determining how an effective volume dependence to the phonon frequencies can be acquired. In particular, we return to the expansion given in eqn (5.2), but now adjusted to reflect the presence of a homogeneous strain. If we imagine that the equilibrium positions of the undeformed crystal are given by $\mathbf{R}_i^{ref}$, then in the presence of a state of deformation characterized by the deformation gradient $\mathbf{F}$, these positions are amended to $\mathbf{F}\mathbf{R}_i^{ref}$. The thermal vibrations about these positions follow by expanding $E_{tot}(\{\mathbf{F}\mathbf{R}_i^{ref} + \mathbf{u}_i\})$. Terms will now arise that couple the homogeneous strain to the thermal vibrations. In particular, if we keep terms only to second order in the expansion we have,

$$E_{tot}(\{\mathbf{F}\mathbf{R}_i^{ref} + \mathbf{u}_i\}) = E_{tot}(\{\mathbf{F}\mathbf{R}_i^{ref}\}) + \sum_{i,\alpha}\left(\frac{\partial E_{tot}}{\partial R_{i,\alpha}}\right)_{\mathbf{F}} u_{i\alpha}$$

$$+ \frac{1}{2}\sum_{i,\alpha}\sum_{j,\beta}\left(\frac{\partial^2 E_{tot}}{\partial R_{i\alpha}\partial R_{j\beta}}\right)_{\mathbf{F}} u_{i\alpha}u_{j\beta}. \qquad (5.74)$$

Note that the stiffness coefficients are now evaluated at the strained reference state (as signified by the notation $(\partial E_{tot}/\partial R_{i\alpha})_{\mathbf{F}}$) rather than for the state of zero strain considered in our earlier treatment of the harmonic approximation. To make further progress, we specialize the discussion to the case of a crystal subject to a homogeneous strain for which the deformation gradient is a multiple of the identity (i.e. strict volume change given by $\mathbf{F} = \lambda\mathbf{I}$). We now reconsider the stiffness coefficients, but with the aim of evaluating them about the zero strain reference state. For example, we may rewrite the first-order term via Taylor expansion as

$$\left(\frac{\partial E_{tot}}{\partial R_{i,\alpha}}\right)_{\mathbf{F}} = \left(\frac{\partial E_{tot}}{\partial R_{i,\alpha}}\right)_0 + \left(\frac{\partial^2 E_{tot}}{\partial R_{i,\alpha}\partial\lambda}\right)_0 \lambda. \tag{5.75}$$

Note that when this expansion is reinserted into eqn (5.74), the first term will clearly vanish on account of the equilibrium condition, namely, $(\partial E_{tot}/\partial R_{i\alpha})_0 = 0$. Through use of the chain rule, we rewrite the term that is linear in λ as

$$\left(\frac{\partial^2 E_{tot}}{\partial R_{i\alpha}\partial\lambda}\right)_0 \lambda = \sum_{k\gamma}\left(\frac{\partial^2 E_{tot}}{\partial R_{i\alpha}\partial R_{k\gamma}}\right)_0 \lambda R_{k\gamma}. \tag{5.76}$$

Substituting this result back into the expansion for the total energy, it is noted that for crystals with centrosymmetry, this term will vanish because every term involving $\lambda R_{k\gamma}$ will have a compensating partner $-\lambda R_{k\gamma}$.

We now adopt precisely the same strategy, this time with respect to the terms in the total energy that are quadratic in the displacements. In particular, we have

$$\left(\frac{\partial^2 E_{tot}}{\partial R_{i\alpha}\partial R_{j\beta}}\right)_{\mathbf{F}} = \left(\frac{\partial^2 E_{tot}}{\partial R_{i\alpha}\partial R_{j\beta}}\right)_0 + \left(\frac{\partial^3 E_{tot}}{\partial R_{i\alpha}\partial R_{j\beta}\partial\lambda}\right)_0 \lambda. \tag{5.77}$$

The first term on the right hand side is recognized precisely as the matrix $K_{ij}^{\alpha\beta}$ introduced earlier. The term that is linear in λ may be evaluated via the chain rule as

$$\left(\frac{\partial^3 E_{tot}}{\partial R_{i\alpha}\partial R_{j\beta}\partial\lambda}\right) = \sum_{k\gamma}\left(\frac{\partial^3 E_{tot}}{\partial R_{i\alpha}\partial R_{j\beta}\partial R_{k\gamma}}\right)\lambda R_{k\gamma}. \tag{5.78}$$

As a result of these manipulations, eqn (5.74) may be rewritten as

$$E_{tot}(\{\mathbf{FR}_i^{ref} + \mathbf{u}_i\}) = \frac{1}{2}\sum_{i,\alpha}\sum_{j,\beta}K_{ij}^{\alpha\beta}u_{i\alpha}u_{j\beta} + \frac{1}{2}\sum_{i\alpha}\sum_{j\beta}\sum_{k\gamma}\Gamma_{ijk}^{\alpha\beta\gamma}\lambda R_{k\gamma}u_{i\alpha}u_{j\beta}, \tag{5.79}$$

where we have thrown away uninteresting constant terms. This result may be rewritten more provocatively as

$$E_{tot}(\{\mathbf{FR}_i^{ref} + \mathbf{u}_i\}) = \frac{1}{2}\sum_{i,\alpha}\sum_{j,\beta}(K_{ij}^{\alpha\beta} + \delta K_{ij}^{\alpha\beta})u_{i\alpha}u_{j\beta}, \tag{5.80}$$

where we have made the definition

$$\delta K_{ij}^{\alpha\beta} = \sum_{k\gamma} \Gamma_{ijk}^{\alpha\beta\gamma} \lambda R_{k\gamma}. \tag{5.81}$$

The consequence of this analysis is that the original stiffness matrix has been amended by terms that are linear in the strain λ and that therefore adjust the vibrational frequencies under a volume change. In particular, the 'renormalized' frequencies arise as the eigenvalues of the stiffness matrix

$$\mathbf{K}' = \mathbf{K} + \delta\mathbf{K}. \tag{5.82}$$

We are now in a position to reconsider the vibrational frequencies in light of the imposition of a small homogeneous strain. Recall that our strategy was to use the quadratic potential energy to produce equations of motion. Inspection of the potential energy developed above makes it clear that the term that is quadratic in the displacements has acquired additional complexity. In particular, if we derive the equations of motion on the basis of both our earlier quadratic terms and those considered here, then we find that they may be written as

$$m\ddot{u}_{i\alpha} = -\sum_{j\beta} (K_{ij}^{\alpha\beta} + \delta K_{ij}^{\alpha\beta}) u_{j\beta}. \tag{5.83}$$

Because of the term $\delta\mathbf{K}$, when we evaluate $\partial\omega(\mathbf{q}, \alpha)/\partial V$, as demanded by eqn (5.72) in our deduction of the thermal expansion coefficient, it will no longer vanish as it would if we were only to keep $\mathbf{K}$. The phonon frequencies have acquired a volume dependence by virtue of the quasiharmonic approximation which amounts to a volume-dependent renormalization of the force constant matrix.

There are a few lessons learned from this analysis. First, we note that as was claimed above, keeping only the harmonic terms in the expansion of the potential energy results in no volume dependence to the vibrational frequencies, and hence to no thermal expansion. This is seen because the parameter λ, which carries the information about such volume dependence, makes its first appearance in conjunction with the manifestly anharmonic term, $\Gamma_{ijk}^{\alpha\beta\gamma}$. This insight prepares us for the quasiharmonic approximation in which the volume dependence of the vibrational frequencies appears in a way that masquerades as a harmonic term.

We are now in a position to revisit the analysis of the thermal expansion presented earlier. In particular, we note that the thermal expansion coefficient may now be evaluated by attributing a particular form to the volume dependence of the phonon frequencies. In order to reveal the application of the developments advanced above, we undertake an analysis of the thermal expansion in the relatively sterile setting of a one-dimensional chain of atoms interacting through a near-neighbor pair potential model for the total energy. As usual, the elaborations of the model to either higher dimensions or other total energy methods do not involve any new questions of logic, though they may imply severe notational and computational demands.

It is one of my largest regrets to have foregone a discussion of the thermal conductivity. Such a discussion would have been opportune at this point since it involves much of the same logic that we invoked in our discussion of the thermal expansion, though reworked along the lines of examining the way in which mode coupling leads to phonon scattering. The interested reader is referred to several of the books described in the Further Reading section at the end of the present chapter.

5.4 Modeling the Elastic Properties of Materials

As was discussed in some detail in chap. 2, the notion of an elastic solid is a powerful idealization in which the action of the entirety of microscopic degrees of freedom are subsumed into but a few material parameters known as the elastic constants. Depending upon the material symmetry, the number of independent elastic constants can vary. For example, as is well known, a cubic crystal has three independent elastic moduli. For crystals with lower symmetry, the number of elastic constants is larger. The aim of the present section is first to examine the physical origins of the elastic moduli and how they can be obtained on the basis of microscopic reasoning, and then to consider the nonlinear generalization of the ideas of linear elasticity for the consideration of nonlinear stored energy functions.

5.4.1 Linear Elastic Moduli

If a crystal is subjected to small strain elastic deformation it is convenient to imagine the energetics of the strained solid in terms of the linear theory of elasticity. As we noted in chap. 2, the stored strain energy may be captured via the elastic strain energy density which in this context is a strictly local quantity of the form

$$W(\epsilon) = \frac{1}{2} C_{ijkl} \epsilon_{ij} \epsilon_{kl}. \tag{5.84}$$

As usual, we are employing the summation convention which demands that we sum over all repeated indices. This relation is posited to apply pointwise, meaning that even if the state of deformation is spatially varying, the energy stored in the body by virtue of elastic deformation is obtained by adding up the energy of each volume element *as though* it had suffered homogeneous deformation. Despite our replacement of the entirety of the set of atomistic degrees of freedom with an effective description in terms of elastic strains, we may restore contact with the microscopic degrees of freedom via the elastic moduli in a way that will be made evident below.

For our present purposes, we note that the correspondence of interest is given by

$$\frac{1}{2}C_{ijkl}\epsilon_{ij}\epsilon_{kl} = \frac{1}{\Omega}[E_{tot}(\{\mathbf{R}_i^{def}\}) - E_{tot}(\{\mathbf{R}_i^{undef}\})]. \tag{5.85}$$

The logic embodied in this expression is built around the recognition that from the continuum perspective the elastic strain energy density is the energy per unit volume above and beyond that of the perfect crystal and is given by the left hand side of eqn (5.85). The microscopic evaluation of this same quantity is obtained by computing the energy of the deformed state (i.e. $E_{tot}(\{\mathbf{R}_i^{def}\})$) and subtracting off the energy of the perfect crystal which we have referred to as $E_{tot}(\{\mathbf{R}_i^{undef}\})$, and then appropriately normalizing that energy with a volume factor so as to produce an energy density. To make the connection given in eqn (5.85) computationally useful, we expand the total energy $E_{tot}(\{\mathbf{R}_i^{def}\})$ about the unstrained reference state, keeping terms only to second order in the strains. In particular, we write

$$E_{tot}(\{\mathbf{R}_i^{def}\}) \approx E_{tot}(\{\mathbf{R}_i^{undef}\}) + \left(\frac{\partial E_{tot}}{\partial \epsilon_{ij}}\right)_{ref}\epsilon_{ij} + \frac{1}{2}\left(\frac{\partial E_{tot}^2}{\partial \epsilon_{ij}\partial \epsilon_{kl}}\right)_{ref}\epsilon_{ij}\epsilon_{kl}. \tag{5.86}$$

On the recognition that the expansion is about the reference state and hence that the linear term in the expansion vanishes (i.e. $(\partial E_{tot}/\partial \epsilon_{ij})_{ref} = 0$), eqn (5.85) reduces to

$$\frac{1}{2}C_{ijkl}\epsilon_{ij}\epsilon_{kl} = \frac{1}{2\Omega}\frac{\partial^2 E_{tot}}{\partial \epsilon_{ij}\partial \epsilon_{kl}}\epsilon_{ij}\epsilon_{kl}, \tag{5.87}$$

where as usual we sum over all repeated indices, which illustrates that the elastic moduli are obtained as

$$C_{ijkl} = \frac{1}{\Omega}\frac{\partial^2 E_{tot}}{\partial \epsilon_{ij}\partial \epsilon_{kl}}. \tag{5.88}$$

It is worth noting that this expression is generic as a means of extracting the elastic moduli from microscopic calculations. We have not, as yet, specialized to any particular choice of microscopic energy functional. The physical interpretation of this result is recalled in fig. 5.2. The basic observations may be summarized as follows. First, we note that the actual energy landscape is nonlinear. One next adopts the view that even in the presence of finite deformations, there is a strain energy density of the form $W(\mathbf{F})$ which is a function of the macroscopic deformation gradient. In the neighborhoods of the minima of this nonconvex energy surface, it is convenient to parameterize the potential as a series of parabolic wells, the curvatures of which may be identified with the elastic moduli as we have done in eqn (5.88).

In light of these gains we are now in a position to explicitly determine the elastic properties. As a first step in this direction we exploit a pair potential description of the total energy. As promised in chap. 4, it will emerge shortly that such a description has serious shortcomings as concerns the elastic properties of materials. On the other hand, the pair potential picture has the salutary effect of demonstrating just how the atomic-level interactions manifest themselves in macroscopic observables such as the elastic moduli. Once we have seen just how these ideas emerge, the way will be paved for a more quantitative analysis of elasticity using more sophisticated total energy functionals.

Within the pair potential setting where the total energy is given by $E_{tot} = \frac{1}{2}\sum_{mn} V_{eff}(r_{mn})$, the elastic moduli are given by

$$C_{ijkl} = \frac{1}{2\Omega}\frac{\partial}{\partial\epsilon_{ij}}\frac{\partial}{\partial\epsilon_{kl}}\sum_{mn} V_{eff}(r_{mn}). \qquad (5.89)$$

Through judicious use of the chain rule, we end up with the generic expression

$$C_{ijkl} = \frac{1}{2\Omega}\sum_{mn}\left[V''(r_{mn})\frac{\partial r_{mn}}{\partial\epsilon_{ij}}\frac{\partial r_{mn}}{\partial\epsilon_{kl}} + V'(r_{mn})\frac{\partial^2 r_{mn}}{\partial\epsilon_{ij}\partial\epsilon_{kl}}\right]. \qquad (5.90)$$

In this case, r_{mn} refers to the distance between the mth and n^{th} atoms in the deformed state in which the crystal is subjected to some homogeneous strain. If we assume that the deformed and reference positions are related by $\mathbf{r}_i = \mathbf{F}\mathbf{R}_i$, then r_{ij} may be rewritten as

$$r_{ij} = \sqrt{\mathbf{R}_{ij}\cdot\mathbf{R}_{ij} + 2\mathbf{R}_{ij}\cdot\mathbf{E}\cdot\mathbf{R}_{ij}}, \qquad (5.91)$$

where we have used the fact that the deformation gradient $\mathbf{F}$ and the Lagrangian strain tensor $\mathbf{E}$ are related by $\mathbf{E} = \frac{1}{2}(\mathbf{F}^T\mathbf{F} - \mathbf{I})$ as shown in eqn (2.6), and we have used the definition $\mathbf{R}_{ij} = \mathbf{R}_i - \mathbf{R}_j$. In the limit of small displacement gradients we may replace $\mathbf{E}$ with ϵ in eqn (5.91). As a result of these geometric insights, the geometric derivatives written above may be evaluated with the result that

$$C_{ijkl} = \frac{1}{2\Omega_{at}}\sum_{m}\left[V''_{eff}(R_m) - \frac{V'_{eff}(R_m)}{R_m}\right]\frac{R_m^i R_m^j R_m^k R_m^l}{R_m^2}. \qquad (5.92)$$

We have added another layer of notation in the form of the quantity Ω_{at} which is the volume per atom associated with the crystal of interest as well as the quantities R_m^i, which refer to the ith Cartesian component of the position vector of the mth atom, with the sum over m referring to the various neighbors that the atom at the origin is coupled to via the potential $V_{eff}(R_m)$. The notation is admittedly unpleasant, but I am too entrenched in the habit of writing C_{ijkl} rather than the more consistent $C_{\alpha\beta\gamma\delta}$, where Greek indices are associated with Cartesian components.

The result obtained in eqn (5.92) is of interest for a number of reasons including what it has to teach us about the energetics associated with pair potentials. For concreteness, we consider the case of an fcc crystal and specialize to the case of a near-neighbor pair potential for which $V'_{eff}(R_m) = 0$ as a result of the equilibrium condition. We are interested in obtaining the three cubic elastic moduli, C_{11}, C_{12} and C_{44}. Note that all neighbors of a given atom are characterized by position vectors of the form $\mathbf{R} = (a_0/2)(110)$. If we are to assess the sums called for in eqn (5.92), what emerges is that $C_{11} = 2V''_{eff}(a_0/\sqrt{2})/a_0$ and $C_{12} = C_{44} = C_{11}/2$, a result which is generally at odds with experimental findings. In fairness to our pair potential calculations, our results could have been improved by the inclusion of terms in the energy which are explicitly volume-dependent and structure-independent such as the electron gas terms described in chap. 3 (see eqn (3.64)). In addition, our reason for introducing this calculation was not so much to criticize pair potentials as to illustrate the conceptual ideas associated with a real calculation of the elastic moduli.

An alternative to the analytic approach adopted above is to take a computational strategy in which the energies of the various deformed states in the neighborhood of the reference state are computed explicitly using whatever total energy method is under consideration. The set of total energies obtained in this way is then fit as a quadratic function of the relevant strain measure, with the elastic modulus of interest (or some linear combination of the moduli) being related to the curvature of this function about the minimum. For example, as shown in fig. 5.14, one can carry out a series of calculations of the energy of a crystal as a function of the volume. As long as the departures from the equilibrium volume are sufficiently small, the energy can be written in the form

$$E_{tot} = \frac{1}{2}\alpha(V - V_0)^2, \tag{5.93}$$

where α is a stiffness parameter related to the bulk modulus and V_0 is the equilibrium volume. Similar results are obtained for other types of deformation (for example, shearing deformations) resulting in the possibilty of obtaining a complete set of linear elastic moduli directly on the basis of microscopic calculations. Note also that in describing this strategy we have not made any particular choice about the nature of the underlying microscopic energy functional. Indeed, this strategy is robust and can be carried out with methods ranging all the way from pair potentials to first-principles techniques.

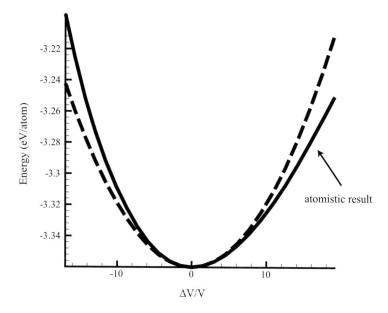

Fig. 5.14. Energy as a function of volumetric strain as computed using atomic-scale analysis in terms of embedded-atom potentials (courtesy of D. Pawaskar). The atomistic result is compared with the quadratic fit like that suggested in eqn (5.93).

5.4.2 Nonlinear Elastic Material Response: Cauchy–Born Elasticity

As yet, our reflections on the elastic properties of solids have ventured only so far as the small-strain regime. On the other hand, one of the powerful inheritances of our use of microscopic methods for computing the total energy is the ease with which we may compute the energetics of states of arbitrarily large homogeneous deformations. Indeed, this was already hinted at in fig. 4.1.

As one of our central missions is to uncover the relation between microscopic and continuum perspectives, it is of interest to further examine the correspondence between kinematic notions such as the deformation gradient and conventional ideas from crystallography. One useful point of contact between these two sets of ideas is provided by the Cauchy–Born rule. The idea here is that the rearrangement of a crystalline material by virtue of some deformation mapping may be interpreted via its effect on the Bravais lattice vectors themselves. In particular, the Cauchy–Born rule asserts that if the Bravais lattice vectors before deformation are denoted by $\mathbf{E}_i$, then the deformed Bravais lattice vectors are determined by $\mathbf{e}_i = \mathbf{F}\mathbf{E}_i$. As will become evident below, this rule can be used as the basis for determining the stored energy function $W(\mathbf{F})$ associated with nonlinear deformations $\mathbf{F}$.

The Cauchy–Born strategy yields interesting insights in a number of contexts one of which is in the kinematic description of structural transformations. A

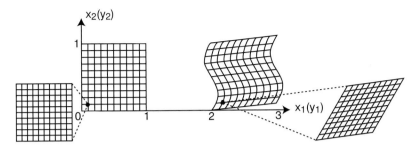

Fig. 5.15. Schematic of the assumption inherent in the Cauchy–Born rule which assumes that the crystal lattice may be thought of as locally having suffered an affine deformation (courtesy of K. Bhattacharya).

schematic of the fundamental geometric idea associated with the Cauchy–Born rule is shown in fig. 5.15 in which it is seen that even though the overall state of deformation may be spatially varying, it is possible to advance the assumption of a deformation that is *locally* affine. As seen in the figure, despite the fact that the overall body has been subjected to an inhomogeneous deformation, locally, the crystal behaves as though it is of infinite extent and has been subjected to homogeneous deformation.

From the standpoint of the microscopic total energy methods presented in the previous chapter, the Cauchy–Born rule may be used to compute the stored energy function $W(\mathbf{F})$. Indeed, we have already shown two different examples of this thinking in figs. 2.14 and 4.1. The key message of both of these figures is not only the nonlinearity of the energy landscape once the displacements become too large, but even further, its nonconvexity. The idea invoked in carrying out such calculations is that we can evaluate the energy of a given configuration, characterized by Bravais lattice vectors $\mathbf{e}_i = \mathbf{F}\mathbf{E}_i$ as $W(\{\mathbf{e}_i\})$, via

$$W(\{\mathbf{e}_i\}) = \frac{E_{tot}(\{\mathbf{e}_i\}) - E_{tot}(\{\mathbf{E}_i\})}{\Omega}. \qquad (5.94)$$

In a word, the problem is reduced to one of computing the energy of two different homogeneous configurations, and as usual, the concept is indifferent to choice of underlying microscopic energy functional. A particularly important example of this strategy concerns the energetics of shearing deformations. In particular, as shown in fig. 2.14, the energetics of shearing deformations immediately reveals the presence of lattice invariant shears. Further, the presence of zero-energy wells corresponding to specific values of the shear is precisely the physics that gives rise to the existence of dislocations. We will later see (see section 10.3) that similar ideas are relevant to the emergence of microstructures in systems having multiple degenerate energy wells. It is also important to note that a strain energy density function like that given in eqn (5.94) can be used as the basis of a hyperelastic constitutive model which permits the calculation of stresses via

$$\mathbf{P} = \frac{\partial W(\mathbf{F})}{\partial \mathbf{F}}. \tag{5.95}$$

Further details of this idea can be found in Tadmor *et al.* (1996).

5.5 Further Reading

Introduction to Lattice Dynamics by M. Dove, Cambridge University Press, Cambridge: England, 1993. I really like Dove's book. Unlike the present work, it has the virtue of being short and to the point. The discussion of anharmonic effects is instructive.

'Group Theory and Normal Modes' by A. Nussbaum, *Am. J. Phys.*, **36**, 529 (1968) provides an interesting tutorial introduction to the use of symmetry to reduce the computational overhead associated with a given calculation of the normal modes.

Theory of Lattice Dynamics in the Harmonic Approximation by A. A. Maradudin, E. W. Montroll, G. H. Weiss and I. P. Ipatova, Academic Press, New York: New York, 1971. A formidable volume with not only a full accounting for the fundamentals of vibrations in solids, but also including issues such as the effects of defects on vibrations in solids.

Imaging Phonons: Acoustic Wave Propagation in Solids by James P. Wolfe, Cambridge University Press, Cambridge: England, 1998. Wolfe's book has a number of different insights into the concepts discussed in the present chapter as well as a wealth of material which goes well beyond the present discussion.

Thermodynamics of Crystals by D. C. Wallace, Dover Publications, Inc., Mineola: New York, 1998. Wallace's chaps. 2–4 give a thorough discussion of the material that has been sketched in the present chapter.

Thermal Conduction in Solids by R. Berman, Clarendon Press, Oxford: England, 1976. A useful compendium of insights into the problem of thermal conduction.

'Thermal Conductivity and Lattice Vibrational Modes' by P. G. Klemens, in *Solid State Physics* edited by F. Seitz and D. Turnbull, Academic Press, San Diego: California, Vol. 7, pg. 1, 1958. I find the sections on scattering due to defects in this article to be quite useful.

'Theory of Thermal Conductivity of Solids at Low Temperatures' by P. Carruthers, *Rev. Mod. Phys.*, **33**, 92 (1961). This article is probably severely dated, but I still find it to be instructive reading.

5.6 Problems

1 Normal Modes for Chain with Three Masses

Consider a three-mass generalization of the system shown in fig. 5.3. Compute the vibrational frequencies and associated eigenvectors on the assumption that each mass has only one degree of freedom, namely, motions along the line joining the masses. Assume that the masses are all identical as are the spring constants.

2 Properties of the Force Constants and the Dynamical Matrix

In this problem, we revisit the formalism associated with the force constants and the dynamical matrix. Recall that the force on the ith atom in the αth direction is given by

$$F_{i\alpha} = -\sum_{j\beta} K_{ij}^{\alpha\beta} u_{j\beta}, \qquad (5.96)$$

where $u_{j\beta}$ denotes the displacement of the jth atom in the βth Cartesian direction. Using this expression for the force, show that if all atoms are displaced by the same constant vector, that this implies the sum rule

$$\sum_j K_{ij}^{\alpha\beta} = 0. \qquad (5.97)$$

3 Phonons in Morse Potential Cu

Obtain an analytic expression for the zone edge phonons in fcc Cu using the Morse potential derived in the previous chapter. To do so, begin by deriving eqn (5.37) and then carry out the appropriate lattice sums explicitly for the Morse potential to obtain the force constant matrix. In addition, obtain a numerical solution for the phonon dispersion relation along the (100) direction.

4 Phonons using the Pair Functional Formalism

Generalize the discussion given in this chapter based on pair potentials in order to deduce the force constant matrix for a total energy based on pair functionals. The result of this analysis can be found in Finnis and Sinclair (1984). Having obtained this expression for the force constant matrix, use the Johnson potential presented in the previous chapter and compute the phonon dispersion relations in Cu. In addition, compute the phonon density of states for Cu and compare the resulting structure of this function to the parabolic shape expected on the basis of the Debye model.

---

5 Elaboration of the Debye Model

In this problem, deduce the eigenvalues of the Christoffel tensor and also recreate the vibrational density of states for the Debye model given in eqn (5.61). In light of these results, fill in the various steps culminating in eqn (5.66).

---

6 Elastic Moduli for fcc Cu from Morse Potential

Using the Morse pair potential for Cu derived in the problems of the previous chapter, derive C_{11}, C_{12} and C_{44}. Carry out this analysis both analytically and numerically. For the numerical results, you will need to compute the energetics of three independent deformations and then to fit the resulting energy curves to obtain these moduli.

---

7 Expression for Elastic Moduli of Cubic Crystals from Pair Functionals

Using the generic definition of the total energy as defined within the pair functional formalism, derive an expression for the elastic moduli. Once this result is in hand, use the explicit expressions associated with the Johnson potential to obtain the elastic moduli for Cu.

---

8 Cauchy–Born Elasticity

Using the Cauchy–Born rule in conjunction with the Johnson embedded-atom potential, compute the energy of a shearing deformation on (111) planes in the [110] direction. Show that the energy has the form given in fig. 2.14.

SIX

Structural Energies and Phase Diagrams

6.1 Structures in Solids

A central tenet of materials science is the intimate relation between structure and properties, an idea that has been elevated to what one might call the structure–properties paradigm. However, we must proceed with caution, even on this seemingly innocent point, since materials exhibit geometrical structures on many different scales. As a result, when making reference to the connection between structure and properties, we must ask ourselves: structure at what scale? This chapter is the first in a long series that will culminate in a discussion of microstructure, in which we will confront the various geometric structures found in solids. The bottom rung in the ladder of structures that make up this hierarchy is that of the atomic-scale geometry of the perfect crystal. It is geometry at this scale that concerns us in the present chapter.

In crystalline solids, the atomic-scale geometry is tied to the regular arrangements of atoms and is described in terms of the Bravais lattice. At this scale, the modeling of structure refers to our ability to appropriately decipher the phase diagram of a particular material. What is the equilibrium crystal structure? What happens to the stable phase as a function of an applied stress, changes in temperature or changes in composition? And a more subtle question, what are the favored *metastable* arrangements that can be reached by the system? At the next level of description, there are a host of issues related still to the atomic-level geometry, but now as concerns the various defects that disturb the perfect crystal. For example, as we raise the temperature, what will be the distribution of vacancies within a crystal, and what governs the dislocation structures that are revealed in electron microscopy images? These defects will occupy much of our attention in coming chapters. At larger scales yet, an investigation of the relation between structure and properties must by necessity also acknowledge structure at the scale of the microstructure itself. Here, it is evident that we are faced again with questions related to metastability and that our theory must answer to these demands. Our first step towards answering these questions will show how microscopic analysis can yield meaningful insights into the nature of atomic-level geometry.

6.2 Atomic-Level Geometry in Materials

As noted above and hinted at in the sections on microscopic approaches to the total energy, the ability to correctly capture the energetics of competing crystal structures and their various defects is a central mission faced in the modeling of materials. Before embarking on an analysis of the structure and energetics of the many defects that populate solids and lend them their fascinating properties, we must first ensure our ability to differentiate the various competing crystal structures found in the elements and their compounds that serve as the backbone of materials science.

Casual inspection of the periodic table reveals that already at the level of elemental materials, a whole range of atomic-level structures may occur. The room temperature structures of elemental materials are summarized in fig. 6.1. A fascinating resource concerning the structure of the elements is Young's *Phase Diagrams of the Elements* (1991), a book that rewards its reader with insights into the startling complexity found in crystalline solids, even without the complicating influence of the compositional degrees of freedom that are present in compounds.

Fig. 6.1 reveals a number of important and intriguing structural trends, all of which serve as a challenge to our analysis. For example, it is seen that a wide assortment of elements adopt the close-packed fcc and hcp structures. In particular, the rare-gas solids (aside from He) all adopt the fcc structure. We also note that the transition metals (i.e. the rows running from Sc to Cu, Y to Ag and La to Au) adopt a nearly regular progression starting with the hcp structure, passing to the bcc structure at the middle of the transition series and ending with the fcc structure at the noble metals (Cu, Ag, Au). These structural trends can be traced in turn to the filling of the transition metal d-band as the number of d-electrons per atom increases as the transition series is traversed from left to right. The bcc structure also plays a crucial role in the description of elemental materials, with the most regular trend being the occurrence of this structure in the alkali metals such as Li, Na and their partners in the same column of the periodic table. Covalent materials, such as Si and Ge, are also cubic but take on the much more open diamond structure. A more exotic cubic structure is adopted by elemental Mn which is built up of 58 atoms. In this case, it is the complicating effects of magnetism that can be blamed for the structural complexity.

In addition to the high-symmetry cubic phases, there are a collection of bad actors which adopt structures of much lower symmetry. For example, the group

Periodic Table of the Elements

Elements given as Symbol (atomic number, crystal structure, atomic weight):

Group IA / VIIIA and main blocks

Row 1: H (1, h, 1.008) ; He (2, h, 4.003)

Row 2: Li (3, b, 6.94) ; Be (4, h, 9.01) ; B (5, h, 10.81) ; C (6, c, 12.01) ; N (7, h, 14.01) ; O (8, c, 16.00) ; F (9, 19.00) ; Ne (10, f, 20.13)

Row 3: Na (11, b, 22.99) ; Mg (12, h, 24.31) ; Al (13, f, 26.98) ; Si (14, d, 28.09) ; P (15, c, 30.97) ; S (16, o, 32.06) ; Cl (17, t, 35.45) ; Ar (18, f, 39.95)

Row 4: K (19, b, 39.10) ; Ca (20, b, 40.08) ; Sc (21, h, 44.96) ; Ti (22, h, 47.90) ; V (23, b, 50.94) ; Cr (24, b, 52.00) ; Mn (25, c, 54.94) ; Fe (26, b, 55.85) ; Co (27, h, 58.93) ; Ni (28, f, 58.71) ; Cu (29, f, 63.54) ; Zn (30, h, 65.37) ; Ga (31, o, 69.72) ; Ge (32, d, 72.59) ; As (33, r, 74.92) ; Se (34, h, 78.96) ; Br (35, o, 78.91) ; Kr (36, f, 83.80)

Row 5: Rb (37, b, 85.47) ; Sr (38, f, 87.62) ; Y (39, h, 88.91) ; Zr (40, h, 91.22) ; Nb (41, b, 92.91) ; Mo (42, b, 95.94) ; Tc (43, 98) ; Ru (44, h, 101.07) ; Rh (45, f, 102.91) ; Pd (46, f, 106.4) ; Ag (47, f, 107.87) ; Cd (48, h, 112.40) ; In (49, t, 114.82) ; Sn (50, t, 118.69) ; Sb (51, r, 121.75) ; Te (52, h, 127.60) ; I (53, o, 126.90) ; Xe (54, f, 131.30)

Row 6: Cs (55, b, 132.91) ; Ba (56, b, 137.34) ; La (57, h, 138.91) ; Hf (72, h, 178.49) ; Ta (73, b, 180.95) ; W (74, b, 183.85) ; Re (75, h, 186.2) ; Os (76, h, 190.2) ; Ir (77, f, 192.2) ; Pt (78, f, 195.09) ; Au (79, f, 196.97) ; Hg (80, r, 200.59) ; Tl (81, h, 204.37) ; Pb (82, f, 207.19) ; Bi (83, r, 208.98) ; Po (84, m, 210) ; At (85, 210) ; Rn (86, f, 222)

Row 7: Fr (87, b, 223) ; Ra (88, 226) ; Ac (89, 227)

Column headers shown: IA, IIA, IIIB, IVB, VB, VIB, VIIB, VIII, IB, IIB, IIIA, IVA, VA, VIA, VIIA, VIIIA

Lanthanides: Ce (58, f, 140.12) ; Pr (59, h, 140.91) ; Nd (60, h, 144.24) ; Pm (61, h, 147) ; Sm (62, h, 150.35) ; Eu (63, b, 151.96) ; Gd (64, h, 157.25) ; Tb (65, h, 158.92) ; Dy (66, h, 162.50) ; Ho (67, h, 164.93) ; Er (68, h, 167.26) ; Tm (69, h, 168.93) ; Yb (70, f, 173.04) ; Lu (71, h, 174.97)

Actinides: Th (90, f, 232.04) ; Pa (91, 231) ; U (92, o, 238.03) ; Np (93, 237) ; Pu (94, 242) ; Am (95, 243) ; Cm (96, 247) ; Bk (97, 247) ; Cf (98, 249) ; Es (99, 254) ; Fm (100, 253) ; Md (101, 256) ; No (102, 254) ; Lw (103, 257)

Legend:
c = cubic
f = face-centered cubic
b = body-centered cubic
r = rhombohedral
d = diamond cubic
h = hexagonal
m = monoclinic
t = tetragonal
o = other

Fig. 6.1. Periodic table with identification of the crystal structures adopted by the various elements (adapted from McMahon and Graham (1992)).

V elements (N, P, As, Sb and Bi) adopt several orthorhombic and rhombohedral phases. Additional structural complexity is ushered in as a result of the fact that in certain cases, the basic building blocks are no longer individual atoms, but rather molecules. The most stunning example of this effect is revealed in S which has a room temperature orthorhombic crystal structure built up of 16 S_8 molecules in each unit cell.

To drive home the complexity revealed at the structural level even in the case of elemental materials, fig. 6.2 shows the temperature–pressure phase diagrams of a few representative elements. There are a few trends to be remarked. First, we note that for several of the close-packed metals that at high temperatures they undergo a transformation to the bcc structure, as seen in the figure for the representative example of Ti, and as is true in addition for Zr and Hf. A more complex situation featuring a series of cubic structures is that of Mn which undergoes a number of different structural transformations with increasing temperature. Oxygen is also characterized by a phase diagram of remarkable complexity, in this case with the fundamental structural building blocks being O_2 molecules. In addition to the role of the temperature as a control parameter which serves to tune the relative stability of competing phases, the application of a hydrostatic stress can also induce structural change. This claim is substantiated in the particular case of Si shown in fig. 6.2 which undergoes a transition from the diamond cubic to the centered

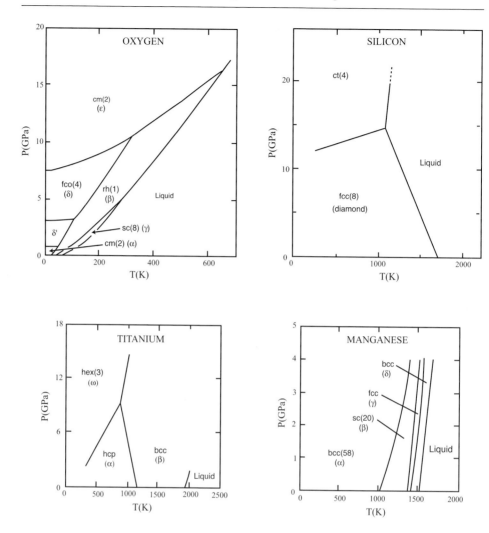

Fig. 6.2. Pressure–temperature phase diagrams of a few representative elements (adapted from Young (1991)).

tetragonal β-Sn phase at a pressure of roughly 11.3 GPa, and more generically in fig. 6.3. Indeed, an important open question is how such phase diagrams will generalize when the material is subjected to more general states of stress. In the case of a material like sulfur (see fig. 1.4), we have already seen that the equilibrium structure is a delicate function of the particular region of parameter space under consideration. From the standpoint of equilibrium thermodynamics, despite the unarguable complexity of the phase diagrams described above, we must conclude that these various structures are nothing more than different realizations of the minima in the appropriate free energy. Our present challenge is to see if

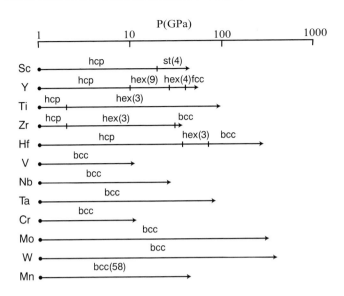

Fig. 6.3. Dependence of the equilibrium structure on the pressure for several representative transition metals (adapted from Young (1991)).

we can learn what competing features give rise to the stability of the different competitors.

Beyond the simple elemental structures considered above, we are faced with the more serious task of deducing the structures that arise once we allow for more than a single element and the attendant chemical complexity. Here again, the challenges to the materials modeler are seemingly limitless. To reveal this complexity, we consider a few representative binary systems as shown in fig. 6.4. For intermetallics, the Cu–Au system may be the most celebrated of all examples. In this case, one of the key qualitative features to note is the existence of both low-temperature ordered phases and a high-temperature disordered phase. The low-temperature ordered phases are particularly appealing from a modeling perspective since they can all be thought of as superstructures on the fcc lattice which are built up as different decorations of the fcc lattice on the basis of the various Cu and Au atoms. A second example of an intermetallic phase diagram is that of the TiAl intermetallic system shown in fig. 6.4(b), a material that has been the subject of intense interest in recent years. In addition to the α (hcp) and β (bcc) solid solutions, we note the presence of the ordered Ti_3Al hexagonal phase, the TiAl $L1_0$ phase and the ordered fcc $TiAl_2$ phase.

Phase diagrams associated with systems other than intermetallics are also of interest. Our third example, shown in fig. 6.4(c), the Si–Ge system, is interesting primarily for what it is not. In particular, we note the marked lack of complexity

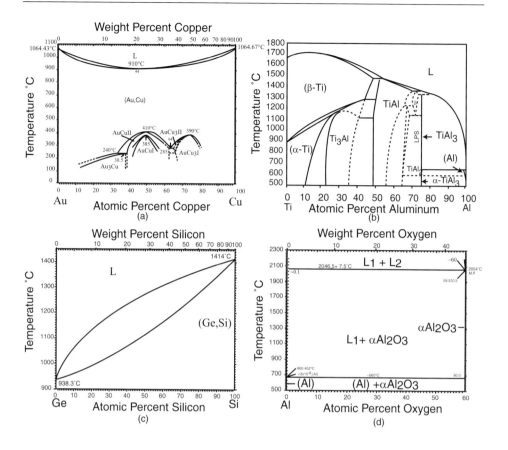

Fig. 6.4. Experimental temperature–composition phase diagrams for several representative binary systems: (a) Cu–Au (adapted from Massalski (1990)), (b) Ti–Al (adapted from Asta *et al.* (1992)), (c) Si–Ge (adapted from Massalski (1990)), (d) Al–O (adapted from Massalski (1990)).

in this phase diagram which reveals that Ge is completely soluble in Si and vice versa. This fact is of interest from the standpoint of the various thin film systems that can be built around different compositions of Si and Ge. Oxides are another important class of compounds. An example of the Al–O phase diagram is shown in fig. 6.4(d). This system is interesting since it reveals the coexistence of an Al solid solution with very small O concentration and hexagonal Al_2O_3 (alumina). This type of two-phase mixture will reveal itself of interest later when we consider the implications of second-phase particles for mechanical properties. We should note that in the context of the materials of everyday life, even binary systems are a facile oversimplification of most materials of technological relevance.

From the standpoint of constructing plausible models of the type of structural complexity discussed above, we must find a way to assess the free energy of the

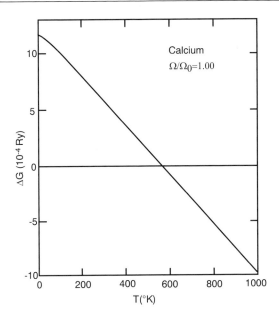

Fig. 6.5. Difference in free energy between bcc and fcc Ca. Plot is of $G_{bcc} - G_{fcc}$ (adapted from Moriarty (1973)).

relevant competing phases. To foreshadow what we are after, fig. 6.5 shows the free energy as a function of temperature for Ca in both the fcc and bcc phases. At low temperatures, we see that the fcc phase is more stable (i.e. it has lower internal energy). On the other hand, at sufficiently high temperatures, the entropic contribution to the free energy swings the balance in favor of the bcc structure. Indeed, Ca is characterized by a relatively simple phase diagram featuring a progression with increasing temperature from fcc to bcc, followed by melting at yet higher temperatures to the liquid phase, although the picture has been complicated somewhat by the presence of a high-pressure simple cubic structure (see, Ahuja *et al.* (1995) for example). Our present ambition is to review the ways in which the types of total energy methods presented in chap. 4 may be exercised in conjunction with statistical mechanics for the evaluation of phase diagrams.

   In response to the types of structural questions posed above, the present chapter will be developed along the following lines. We begin with an analysis of the zero-temperature internal energies associated with different structural competitors. These analyses are carried out on the basis of the entirety of different schemes for computing the total energy that were introduced in chap. 4 and should be thought of as both the logical extension and culmination of the efforts begun there. Once the zero-temperature contribution to the free energy is in hand, we will turn to an analysis of the entropic contributions to the free energy. These ideas will be played

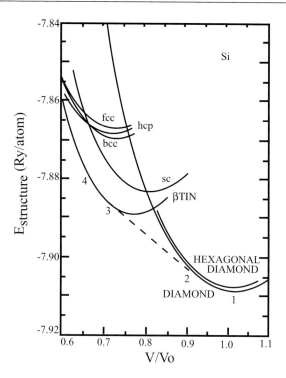

Fig. 6.6. Structural energies for various competing crystal structures in Si as computed using density functional theory (adapted from Yin and Cohen (1982)).

out first in the context of elemental phase diagrams, with special reference to the case of Mg, and later with respect to alloy systems, culminating in the example of oxygen disorder in high-temperature superconductors.

6.3 Structural energies of solids

Our discussion begins with an assessment of the zero-temperature part of the free energy difference separating different competing structures. The intention of this section is twofold. On the one hand, we wish to show how the various schemes for computing the total energy already introduced in chap. 4 can be used to characterize the energetics of competing structures. On the other hand, this section will further serve to illustrate the inadequacy of certain total energy functions, emphasizing the need for caution in off-the-shelf application of a given energy function without some knowledge of its limitations.

As a paradigm for the type of work that has been done to determine the internal energy contribution to the free energy, fig. 6.6 shows the classic work of Yin and Cohen (1982) in which some of the possible competing crystal structures for Si were assessed from the standpoint of density functional theory. Indeed, these calculations have served as the gold standard against which the many total energy schemes used to model the structure and energetics of Si have been measured. These calculations examined the energy vs volume relations, $E(V)$, for Si in seven distinct crystal structures with the calculations done using a pseudopotential to describe the ion–electron interaction and a plane wave basis for describing the wave function. What may be learned from these curves is first, that at zero temperature and pressure of all of the competing phases examined, the diamond cubic structure is the most stable. It is worth noting, however, that the search over the space of energy minimizers is discrete, in contrast with the type of analysis that is the mainstay of the calculus of variations and in which an entire continuum of solutions are compared. Here, one is reduced to identifying possible competitors and querying their respective energies. The structure with the lowest energy is the winner among the limited list of competitors. A second observation that emerges from the Yin and Cohen analysis may be gleaned from considering the pressure dependence of the total energy. What is remarked is that by using the common tangent construction in which the common tangent between the two relevant $E(V)$ curves is constructed, it is possible to deduce a transition pressure. In the case of Si, Yin and Cohen argued that a structural transformation to the β-Sn structure is expected at a pressure of 99 kbar which is to be compared with the experimentally observed value of 125 kbar.

Our present aim is to take a hierarchical view of structural energies, beginning with the type of insights into structural stability that have emerged from pair potentials and working our way to the full first-principles treatment of the type embodied in the work of Yin and Cohen. Our task in this instance is to discriminate between alternative structures in the same way as done by Yin and Cohen. Our purpose will be not only to elucidate the ways in which such zero-temperature calculations are performed and what may be learned from them, but also to acknowledge the serious limitations that are confronted in attempting to use empirical descriptions of the total energy such as were described in the first part of chap. 4. Having understood the zero-temperature part of the elemental phase diagram, we will then turn to the inclusion of finite-temperature effects. Once the free energy is known, our analysis will have reached the point that it may be used to discriminate between competing phases and, in fact, to compute the phase diagram itself.

6.3.1 Pair Potentials and Structural Stability

In chap. 4, we discussed the different ways in which the total energy of a collection of atoms may be approximated. Our intention now is to examine the success with which these different approximations can characterize the structural energies of competing crystal structures. The pair potential approach was justified in earlier sections (see section 4.3.2) on the basis of a perturbative treatment in which the free electron gas is considered as the reference state, and the ionic potentials are seen as a perturbing influence on this medium. What emerged from that discussion was the recognition that the total energy may be written as the sum of a structure-independent term which depends upon the mean electron density (or volume) and a structure-dependent pair interaction. In particular, we adopt the notation

$$E_{tot}(\{\mathbf{R}_i\}) = E_0(\Omega) + \frac{1}{2} \sum_{ij} \phi_{eff}(R_{ij}; \bar{\rho}), \qquad (6.1)$$

where the first term is the structure-independent volume term and the second term involves the effective pair potentials which themselves depend upon the mean electron density, $\bar{\rho}$.

   With the pair potential scheme noted above as the starting point, a variety of systematic studies of structural stability have been effected. The basic picture that emerges in making such calculations is one in which the origins of the structural energies are traced to the distributions of distances between neighboring atoms. As a rule of thumb, those structures with bond lengths that are able to take advantage of the minima in the effective pair interaction are more stable than those that do not. To make this analysis concrete, we examine the work of Hafner and Heine (1983) in which a systematic analysis of a range of elements was carried out. In fig. 6.7, we show a series of effective pair interactions that are appropriate to a representative cross section of the elements they considered. The plots are so constructed that one may also discern the relation between the minima in the pair potential and the distribution of bond lengths in the various competing crystal structures. The distribution of such bond lengths is in some ways the simplest geometric descriptor of the underlying structure. Stability of a particular lattice has been examined in this context in settings ranging from the structure of elemental metals to the formation of intermetallics and related defect structures. To examine how such arguments transpire, consider fig. 6.7 in more detail. The basic idea behind rationalizing the existence of a given structural type is that the minima in the potential are aligned with the bond lengths that occur in the structure. In this figure, we see that the near neighbors in fcc Mg and Al, for example, take full advantage of the minimum in the pair potential. Note that in contrast to the Lennard–Jones potential, the free electron pair potentials shown in the figure have a series of oscillations.

   Before embarking upon an analysis of structural energies from the standpoint of methods constructed at the next level of sophistication, we would do well to reflect on the significance and limitations of those things said so far. First, recall

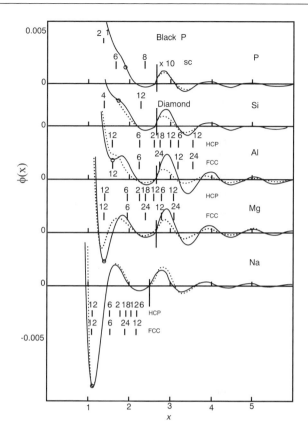

Fig. 6.7. Relation between structure and energetics as characterized by pair potentials (adapted from Hafner and Heine (1983)). Note that when x reaches the vertical bar between 2 and 3 the value of the potential is multiplied by 10. The various numbers associated with vertical lines for a given structure (e.g. 12, 6, 24, etc. for Al) reveal the number of neighbors a given atom has at that distance. Dotted and solid lines correspond to different schemes for computing the pair potential.

the expected range of validity for potentials of this type. Our derivation of pair potentials was constructed on the grounds that the material of interest could be thought of, to lowest order, as an electron gas. Viewed from the perspective of the periodic table, this conjures images of those elements that easily surrender their outer electrons such as those found in columns I and II of the periodic table. In addition to the free electron pair potentials described here, we might also imagine that some form of pair potential description might suffice for ionic materials in which the dominant contribution to the cohesive energy may be traced to long-range interactions between the charged ions. On the other hand, it is unlikely that an analysis of structural stability in transition metals could be built around similar reasoning.

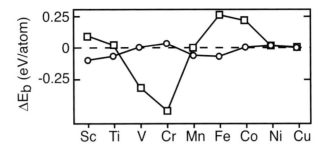

Fig. 6.8. Structural energy difference between fcc and both bcc and hcp structures as a function of the filling of the d-band (adapted from Paxton *et al.* (1990)). The hcp–fcc energy difference is labeled by circles while the bcc–fcc energy difference is labeled by squares.

6.3.2 Structural Stability in Transition Metals

Thus far, we have how pair potentials may be used to assess the structural energy contribution to the free energy. The present discussion is aimed at examining the ways in which questions of structural stability have been investigated in metals from the standpoint of total energy functionals more sophisticated than the pair potentials described above. We reserve for the next section an analogous discussion which highlights the question of structural stability in Si.

From the standpoint of structural stability in metals, one of the intriguing structural trends considered in fig. 6.1 was the progression of structures exhibited by the transition metals with increasing filling of the d-band. This progression has now been explained on the basis of models of cohesion ranging from angular force schemes to tight-binding theory to full-fledged density functional calculations. From the angular force perspective, these trends in structural stability have been attributed to the fact that in the vicinity of half-filling of the d-band, there is a strong energy penalty associated with the presence of four-atom clusters with 90° angles. The absence of such near-neighbor clusters in the bcc structures is argued to give it an energetic advantage at this band filling. This result was already foreshadowed in fig. 4.19 which shows the energy penalty associated with 90° angles such as are characteristic of the fcc structure.

While structural stability can be rationalized on the basis of the type of real space arguments given above in which either special bond lengths or bond angles are tied to structural stability, first-principles calculations are often the court of final appeal. The energy differences between both the hcp and bcc structures relative to the fcc structure are plotted as a function of the filling of the d-band in fig. 6.8. These calculations were carried out using the full-potential LMTO method (linear muffin tin orbitals – one of the many varieties of first-principles techniques) by Paxton *et al.* (1990). The key observation to be made with regard to these results is the stability of the hcp structure for the left hand side of the transition series and a preference for the bcc structure near half-filling. In addition, it is important to note how small the energy differences are to the right hand side of the diagram, posing a challenge not only to empirical schemes, but even to first-principles methods themselves.

6.3.3 Structural Stability Reconsidered: The Case of Elemental Si

The challenge of structural stability in metals has been greeted successfully from the standpoint of an entire spectrum of total energy functionals. The question of such stability in covalent semiconductors is another story altogether. In these cases, we know that the crystal structures are open and the atoms are highly undercoordinated relative to the close-packed limit seen in fcc and hcp materials. Note that from the type of neighbor distance perspective adopted thus far, the diamond cubic structure has only four neighbors at the near-neighbor separation. The logic of the present section is to examine structural stability in Si from the perspective of at least five different total energy methods, and with the use of the Yin–Cohen curve introduced in fig. 6.6 as the standard of comparison. The goal of this exercise is multifaced with the hope that this discussion will illustrate how such total energy calculations are done and to what extent empirical schemes are effective in describing covalent systems.

In chap. 4, we remarked that there have been over 100 different empirical schemes developed to describe the energetics of covalent materials, with special reference to Si. A particularly interesting assessment of many of these total energy schemes can be found in Balamane *et al.* (1992). One of the benchmarks by which such potentials are judged is their ability to account for structural stability in a number of different settings: clusters, bulk crystalline structures, surfaces, liquid and amorphous phases. One of the abiding conclusions of these efforts has been the inability of any such scheme to really successfully account for Si over the entire range of possible structures for which one might wish to compute properties.

The simplest energetic scenario within which to explore structural stability in Si is via models which are built around pairwise summations. Fig. 6.9(b) shows a comparison of the different energy vs volume curves for Si as computed using such a model as developed by Ackland (1989). Surprisingly, the energetics produced by this energy functional is in reasonable accord with the full density functional results. At the next level of sophistication in the hierarchy of total energy methods are angular force schemes. In chap. 4 we noted how angular force schemes could be seen as a natural outgrowth of models of bonding in systems with electronic orbitals exhibiting strong angular character. For the moment, we pose the question of how two such schemes, the Stillinger–Weber and Tersoff potentials, fare in the

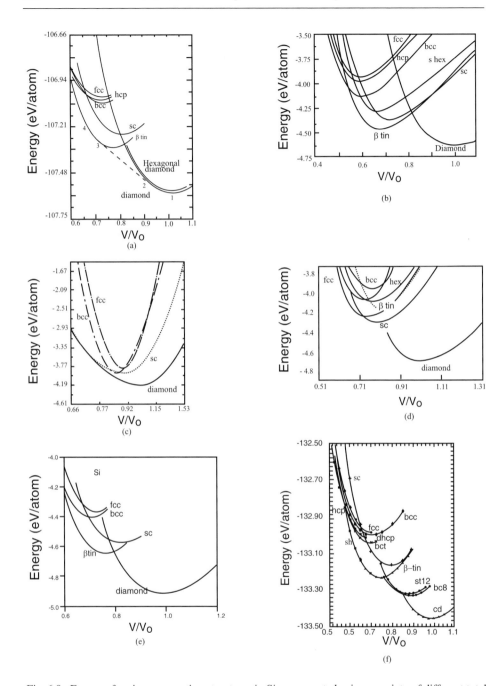

Fig. 6.9. Energy of various competing structures in Si as computed using a variety of different total energy methods: (a) density functional results of Yin and Cohen (1982), (b) pair potential results of Ackland (1989), (c) angular force results of Stillinger and Weber (1985), (d) empirical results of Tersoff (1988), (e) tight-binding results of Mercer and Chou (1993) and (f) density functional results of Needs and Mujica (1995).

attempt to reproduce the Yin–Cohen curve of fig. 6.6. As shown in chap. 4, the Stillinger–Weber potentials acknowledge the presence of strong angular preferences in covalent materials by supplementing the standard pair potential term with a three-body term which penalizes not only bond stretching distortions, but bond bending distortions as well. The alternative energetic description that we will use in the present analysis is that of the Tersoff potentials discussed already in chap. 4. Again, our logic is that these calculations will reveal the strengths and limitations of empirical approaches while at the same time elaborating how calculations of phase diagrams can really be effected. Figs. 6.9(c) and (d) reveal the Stillinger–Weber and Tersoff versions of the Yin–Cohen curve. Note that the values of the energy on the energy axis associated with the Stillinger–Weber potential have a bizarre spacing since I converted the original numbers which were computed in different units. In addition, the original curve was plotted with respect to density rather than volume and, as a result, the x-axis in this case is not even linear. In the end, it is not entirely clear how to assess the success (or failure) of the Stillinger–Weber results, and we note that via a different choice of parameters the curves can be altered. The energetics associated with the Tersoff potential is more evidently flawed. In particular, the energy vs volume curve for the β-Sn structure using the Tersoff potential is shown as the dotted line in fig. 6.9(d) and reveals that contrary to experiment, this structure is not one of the viable structural competitors according to this version of the Tersoff potential. However, like with the case of the Stillinger–Weber potential, it has been shown that by changing the cutoff parameters describing the range of the potentials, the β-Sn results can be brought more in line with the density functional results. One of the compelling conclusions of this part of our discussion must be the insight that results can depend sensitively on the choice of fitting parameters.

As we increase the level of sophistication invoked to explain structural stability, we now turn to the use of tight-binding models to compute structural energy curves such as that of Yin and Cohen. The tight-binding Hamiltonian used in the calculations to be reported here was based upon the parameters shown already in fig. 4.13. In fig. 6.6(e), we see the hierarchy of structural energies as obtained via this tight-binding parameterization. Just as the Stillinger–Weber results had to be recast in terms of volume (as opposed to density), I had to convert the values on the x-axis from near-neighbor distance into volume with the result that the values on this axis are nonlinear. Note that at this level of description, the correspondence between the full density functional calculations and the approximate methods is beginning to appear tolerable. Indeed, fig. 6.6(e) puts the tight-binding results (indicated by lines) in juxtaposition with the Yin–Cohen results (discrete points) and the level of quantitative agreement is apparent. As seen above, in the time since the path-breaking calculations of Yin and Cohen, not only have a host of empirical schemes been put forth to imitate the energetics found there, but more recent density functional calculations have also been made which may well supersede their original work. An example of such recent calculations are those shown in fig. 6.9(f).

We have so far only set up the tools that will be needed in the zero-temperature analysis of the total energy. To span the set of possible structural outcomes for a range of different temperatures requires a knowledge of the entropy of the competing phases. In order to effect such a calculation, we must first revisit the finite-temperature properties of crystals discussed in the previous chapter with an eye to how the entropy of the competing phases (including the liquid) may be determined.

6.4 Elemental Phase Diagrams

Recall that we started this chapter by noting the challenge posed by both elemental and alloy phase diagrams. There we claimed that the first step in modeling materials from the microscopic level up was to insure that the correct equilibrium phases can be accounted for. Failure to satisfy this primary criterion destroys the credibility of analyses aimed at uncovering the structure and energetics of the defects which are at the heart of many of the intriguing behaviors of materials. Armed with our knowledge of the tools of statistical mechanics (chap. 3), how to go about computing the structural energies of solids (chap. 4 and the earlier part of the present chapter) and how to account for the thermal properties of crystals (chap. 5), we are now ready to revisit the question of structural stability at finite temperatures. At this point, our task is to assemble the various contributions to the overall free energy of the competing phases. What this means in concrete terms is to bring together what we have learned about structural energies and an assessment of the vibrational and electronic contributions to the free energy.

6.4.1 Free Energy of the Crystalline Solid

Our aim is a proper reckoning of the free energies of all of the competing phases, be they solid or liquid. In this section, we undertake a systematic analysis of the various terms that need to be accounted for in the free energy of an elemental crystalline solid. An important simplification results if we are willing to neglect the possibility of the electron–phonon interaction. In this case, the total free energy admits of an additive decomposition in which the various contributions to the overall free energy can be reckoned separately. The basic structure of our results may be decomposed as follows,

$$F_{tot} = F_{el} + F_{vib}^{h} + F_{vib}^{anh}. \tag{6.2}$$

The first term on the right hand side, F_{el}, refers to the contribution of the electronic degrees of freedom to the overall free energy. In particular, we require a knowledge of

$$F_{el}(T, V) = U_{el}(T, V) - T S_{el}(T, V), \tag{6.3}$$

where $U_{el}(T, V)$ may be written as

$$U_{el}(T, V) = 2 \int_{\infty}^{\infty} \rho(E) f(E, T) E dE, \tag{6.4}$$

where $\rho(E)$ is the electronic density of states and $f(E, T)$ is the Fermi–Dirac distribution. We will return to the question of the electronic contribution to the entropy shortly. The second term, F_{vib}^h, refers to the harmonic contribution of thermal vibrations to the free energy of the solid. Recall from our discussion of chap. 3 that, roughly speaking, the entropy is a measure of the number of states available to the system. In the present context, the vibrational entropy reflects the number of ways in which the atoms can exercise small excursions about their equilibrium positions, a quantity that will itself depend upon the particular crystal structure in question. However, this contribution to the free energy is computed on the assumption that the lattice is appropriately described in the harmonic approximation. On the other hand, at higher temperatures, anharmonic terms in the description of the lattice vibrations must be taken into account and it is these contributions that are represented in the term F_{vib}^{anh}. Wallace (1992) notes that in the context of 'nearly free-electron metals', the relative magnitudes of the various contributions to the entropy are of the order $S_{harmonic} \approx 10k/$atom, $S_{anh} \approx \pm 0.1k/$atom and $S_{el} \approx 0.1k/$atom. At least in the context of these metals, this suggests an approach to the analysis of phase diagrams predicated entirely on the treatment of the internal energy of the structural competitors and the contributions of the harmonic phonons to the free energy, though the veracity of this assumption must await an assessment of the *differences* in these various contributions to the free energy for different phases. We should also note that we have explicitly ignored any contribution to the overall free energy budget resulting from electron–phonon interactions. It is now our aim to give each of these terms an examination in turn, with particular reference to the details of how one might go about calculating these terms in the context of a real material.

The Contribution of Harmonic Lattice Vibrations to the Free Energy. Recall the treatment of lattice vibrations in chap. 5. In that discussion, we noted that in the harmonic approximation, the Hamiltonian for harmonic vibrations of a crystalline solid (see eqn (5.19)) may be written as

$$H_{vib} = \sum_{k,s} \left(\frac{p_{k,s}^2}{2m} + \frac{1}{2} m\omega_{k,s}^2 q_{k,s}^2 \right),$$ (6.5)

where by virtue of the transformation to normal coordinates, the Hamiltonian is diagonal. Note that we have used slightly different notation than we did there. In particular, here we have introduced the variables $\{p_{k,s}, q_{k,s}\}$ which represent the generalized momenta and generalized coordinates, with the label $\mathbf{k}$ referring to a particular wavevector from the phonon spectrum and the label s referring to the polarization of the mode in question. The corresponding quantum partner of eqn (6.5) is

$$E_{vib} = \sum_{k,s} \left(n_{k,s} + \frac{1}{2} \right) \hbar\omega_{k,s}.$$ (6.6)

Once this result is in hand, the statistical mechanics of this system reduces to repeated application of the statistical mechanics of a simple one-dimensional quantum harmonic oscillator already discussed in chap. 3. This can be seen by noting that the partition function may be written as

$$Z_{vib} = \sum_{n_{k_1,1}} \cdots \sum_{n_{k_N,3}} \exp\left[-\sum_{k,s} \left(n_{k,s} + \frac{1}{2} \right) \hbar\omega_{k,s} \right].$$ (6.7)

This sum instructs us to sum over the occupation numbers of each and every mode, with each of the N possible $\mathbf{k}$-vectors having three distinct polarizations to consider. This expression can be factorized, in turn, into a product of the partition functions of the individual oscillators with the result

$$Z_{vib} = \prod_{k,s} \frac{e^{-\beta\hbar\omega_{k,s}/2}}{(1 - e^{-\beta\hbar\omega_{k,s}})}.$$ (6.8)

Once we have evaluated the partition function, it is a simple matter (as shown in the previous chapter) to compute the free energy of harmonic vibrations which may be written

$$F_{vib}^h = \sum_{k,s} \left[\frac{1}{2} \hbar\omega_{k,s} + kT \ln(1 - e^{-\beta\hbar\omega_{k,s}}) \right].$$ (6.9)

We will show below the way in which this result influences the outcome of the competition between different phases.

One point that was omitted in the discussion given above was the evaluation of the $\omega_{k,s}$s themselves. In order to perform the sums denoted above to evaluate the free energy we require a knowledge of the $3N$ vibrational frequencies available to the lattice. How does one determine such frequencies? In the previous chapter, we made it clear that the fundamental idea is the diagonalization of the dynamical matrix. However, rather than really attempting to compute each and every such eigenvalue, one may resort to clever or not-so-clever schemes in which some subset of k points within the Brillouin zone are visited and their eigenvalues computed. Knowledge of such eigenfrequencies gives the harmonic contribution to the free energy by the summation indicated in eqn (6.9). As we have also already described, an alternative way of collecting the information associated with the vibrational spectrum is via the vibrational density of states. Once we have determined the vibrational density of states, the contribution of harmonic vibrations to the free energy may be found by performing the integral

$$F_{vib}^h = kT \int_0^\infty \rho(\omega) \left[\frac{1}{2}\hbar\omega + \ln(1 - e^{-\hbar\omega/kT}) \right] d\omega. \tag{6.10}$$

Hence, once the vibrational densities of states of the competing phases are known, we may begin to explore the origins of structural transformations.

The Anharmonic Contribution to the Free Energy. Recall from our discussion in the previous chapter that the harmonic approximation represents but a first term in the expansion of the potential energy in powers of the displacements the ions may exercise about their equilibrium positions. As the amplitudes of the vibrations increase with increasing temperature, it is reasonable to expect that there will be a corresponding contribution to the entropy arising from anharmonicities. Probably the simplest approach to the treatment of such anharmonicities is perturbative, with the idea being that one constructs the anharmonic corrections to the free energy as a series, with the hope that only the lowest-order terms will contribute. On the other hand, a number of experimental efforts in which the anharmonic entropy is deduced from measurements suggest that such perturbative approaches don't jibe particularly well with experiment.

An example of the measured anharmonic contributions to the entropy for a number of different transition metals is shown in fig. 6.10. The entropy reported in these curves is measured in units of k/atom. For our present purposes, we note that the theoretical description of the anharmonic contributions to the entropy remains incomplete. A direct approach to the computation of such terms based on molecular dynamics is possible, but does not leave us with any particular analytic understanding. In the case study of interest to us later in the chapter, we will seize on the smallness of this contribution with the result that it can be ignored.

The Electronic Contribution to the Free Energy. Of course, as the physics of semiconductors instructs us, the distribution of electrons is also affected by the presence of finite temperatures, and this effect can make its presence known in

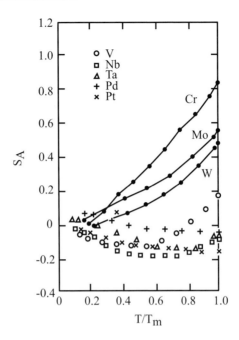

Fig. 6.10. Measured anharmonic contribution to the entropy for a series of transition metals (adapted from Eriksson *et al.* (1992)).

the context of structural transformations as well. Rearrangements of electrons with energies near that of the Fermi level provide an additional contribution to the entropy which are reflected in the free energy. Our picture of the electronic states is that of the independent-electron approximation in which the electrons are assumed to occupy a series of electronic states obtained from solving the one-electron Schrödinger equation. To see the origins and form of the electronic contribution to the free energy, we need to recall certain basic notions from chap. 3.

The first point is to recall eqn (3.102) in which it was seen that a system characterized by a series of discrete states labeled by the integer i with associated probabilities p_i has an entropy $S = -k \sum_i p_i \ln p_i$. In the context of the electronic states resulting from the Schrödinger equation, the key realization is that a state with energy E_i has a probability $f(E_i, T)$ of being occupied and a probability $1 - f(E_i, T)$ of being unoccupied. Again, drawing from what we have already described in chap. 3, $f(E, T)$ is the Fermi–Dirac distribution. By virtue of the reasoning presented above, the entropy associated with the ith electronic state is given by $S_{el}^i = -k\{f(E_i, T) \ln f(E_i, T) + [1 - f(E_i, T)] \ln[1 - f(E_i, T)]\}$.

If we now consider the entire electronic spectrum as characterized by the density of states $\rho(E)$, the total electronic entropy is gotten by adding up the contributions energy level by energy level. In particular, this results in the expression

$$S_{el} = -2k \int_{-\infty}^{\infty} \rho(E)\{f(E,T)\ln f(E,T) + [1 - f(E,T)]\ln[1 - f(E,T)]\}dE,$$

$$(6.11)$$

where $f(E,T)$ is the Fermi–Dirac distribution and the factor of 2 accounts for spin degeneracy. Just as with the vibrational contribution to the entropy (or free energy), once the spectrum is known, the entropy can be computed. We see, therefore, that in order to evaluate the electronic entropy, it is incumbent on the modeler to compute the electronic density of states. Once this function is known, the evaluation of the entropy itself becomes a matter of simple integration. In addition, we have already shown with eqn (6.4) that the electronic contribution to the internal energy may similarly be evaluated by integration over the density of states. As we have continually reiterated, our main purpose here is not to fly the flag of any particular microscopic modeling scheme. As a result, we note that whether one has elected the computationally less-intensive tight-binding approach or a more rigorous density functional approach, the conceptual steps raised in computing the electronic contribution to the free energy are identical.

The central result of eqn (6.11) can be played out both at the model level and using full quantum mechanical treatments of the electron spectrum. The most immediate route to a concrete result is to evaluate the electronic contribution to the free energy directly on the basis of the physics of the electron gas model. Recall that in this case, the electron density of states is of the form $\rho(E) \propto \sqrt{E}$. As a result, we are faced with the task of evaluating eqn (6.11) for the square-root density of states. As will be spelled out in the problems at the end of this chapter, the result of carrying out this analysis is

$$S_{el} \approx \frac{2\pi^2}{3} \rho(\epsilon_F) k^2 T. \qquad (6.12)$$

As was noted above, in addition to the attempt to get model results such as that of eqn (6.12), considerable effort has also been put forth into the first-principles evaluation of the electronic contribution to the free energy. One interesting group of results concerns the question of structural transformations in the transition metals, where we recall that elements such as Ti and Zr undergo a transition with increasing temperature from the hcp to the bcc structures. A few representative examples of the relevant electronic densities of states are shown in fig. 6.11. These results were obtained using the so-called LMTO method which is one of the density functional schemes that is especially popular. As an indication of the type of effort demanded in order to obtain tolerable accuracy, this set of calculations was done using 140 k points for the bcc calculations, and 240 k points in the fcc case.

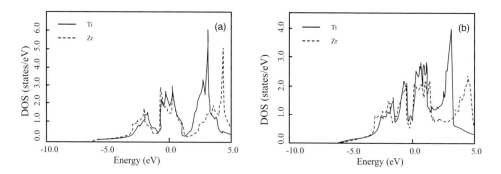

Fig. 6.11. Representative examples of electronic densities of states for (a) bcc and (b) hcp Ti and Zr (adapted from Eriksson *et al.* (1992)). The Fermi level is at the zero of energy.

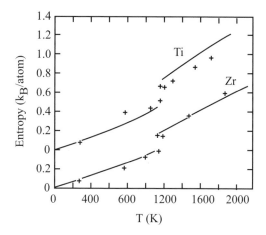

Fig. 6.12. Comparison of theoretical and experimental electronic contributions to the entropy (adapted from Eriksson *et al.* (1992)). The solid lines are the results of the theoretical analysis and the symbols correspond to experimental values.

As a result of a knowledge of the electronic density of states, it is then possible to assess the electronic contribution to the entropy. An example of these results is given in fig. 6.12 which shows not only the computed electronic entropies for Ti and Zr (the solid lines) but also the experimental values for these entropies as obtained by taking the measured total entropy and subtracting off the contributions due to vibrations. In addition to their ability to shed light on the entropies themselves, these results were also used to examine the relative importance of vibrations and electronic excitations to the hcp–bcc structural transformation in these metals. Unlike in the case of simple metals, it is found that the electronic contribution to the free energy difference is every bit as important as the vibrational terms. In particular, in the case of Ti, it is claimed that the measured entropy change is $0.43k$/atom, with a predicted $0.29k$/atom due to vibrations (primarily harmonic) and an electronic contribution of $0.27k$/atom. Though the role of finite-temperature electronic excitations is important in transition metals, we will see in the case study in Mg given below that the electronic entropy can be neglected altogether.

As yet, our analysis has illustrated some of the key contributions to the free energy of crystalline solids. We have noted that in the absence of electron–phonon coupling, there is an additive decomposition that allows for a decoupling of the vibrational and electronic contributions. The evaluation of these various contributions to the free energy can be effected using what we have learned about computing total energies (chap. 4) and the thermal properties of solids (chap. 5). To complete our analysis of the various competitors available to elemental materials we now turn to an evaluation of the free energy of liquids.

6.4.2 Free Energy of the Liquid

If our ambition is the determination of the equilibrium phase diagram for an elemental material, one of the competing states is that of the liquid. At high enough temperatures we expect the crystalline solid to melt, and the point at which such melting occurs is an important part of our phase diagram calculation. Unlike in the case of the crystalline solid, in which the high degree of structural order may be exploited to simplify the evaluation of the free energy, the calculation of the free energy of liquids requires a different approach. In this case, a variational theorem in conjunction with enlightened guessing allow us to put bounds on the free energy of the liquid, which may in turn be validated using explicit calculations such as those offered by molecular dynamics.

In previous chapters, we have already made use of variational ideas on a few different occasions. In the present setting, our aim is to identify variational bounds that will insure that the free energy we calculate is an upper bound on the exact free energy that it is our true aim to deduce. To make this calculation, we use the Bogoliubov inequality introduced in chap. 3. The basic idea is to identify some reference state for which the free energy of the liquid can be computed and then to use this result as the reference state in the context of the Bogoliubov inequality. Recall that our objective is to place bounds on the exact free energy through an appeal to the statement that $F_{exact} \leq F_{model} + \langle (H - H_0) \rangle_0$.

Our starting point is a restatement of the energy as a function of atomic positions, now stated in terms of the type of statistical distribution functions introduced in chap. 3. The purpose of the first part of our discussion is to establish a means for evaluating the quantity $\langle (H - H_0) \rangle$. For simplicity, imagine that the potential energy of interaction between the various atoms is characterized by a pair potential of the form $V(|\mathbf{r} - \mathbf{r}'|)$. Our claim is that the average value of the energy of

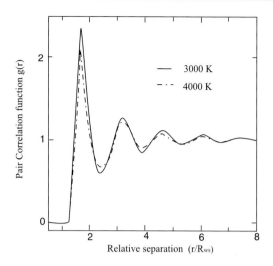

Fig. 6.13. Pair correlation function for liquid Mg (adapted from Moriarty and Althoff (1995)).

interaction is given by

$$\langle V \rangle = \frac{1}{2} \int \int \rho^{(2)}(\mathbf{r}, \mathbf{r}') V(|\mathbf{r} - \mathbf{r}'|) d^3\mathbf{r} d^3\mathbf{r}', \qquad (6.13)$$

where $\rho^{(2)}(\mathbf{r}, \mathbf{r}')$ is the two-particle density defined as

$$\rho^{(2)}(\mathbf{r}, \mathbf{r}') = \sum_{i \neq j} \delta(\mathbf{r} - \mathbf{r}_i)\delta(\mathbf{r}' - \mathbf{r}_j).$$

In the context of an isotropic liquid, this expression can be collapsed by recourse to the definition of the pair correlation function and the recognition that the density assumes a constant value, $\rho(\mathbf{r}) = N/V$. In particular, we introduce the pair correlation function $g(|\mathbf{r} - \mathbf{r}'|)$ which is defined through the relation

$$\rho^{(2)}(|\mathbf{r} - \mathbf{r}'|) = \rho(\mathbf{r})\rho(\mathbf{r}')g(|\mathbf{r} - \mathbf{r}'|). \qquad (6.14)$$

As a result of this observation, the energy may be rewritten as

$$\langle V \rangle = \frac{N^2}{2V} \int d^3\mathbf{r} g(r) V(r). \qquad (6.15)$$

The physical idea being represented here is that the liquid has an isotropic structure characterized by the radial distribution function $g(r)$. The averaging being performed may be visualized with reference to fig. 6.13, where it is seen that the structure factor for a liquid is characterized by peaks corresponding to the various shells of neighbors surrounding a given atom. The average taken in eqn (6.15) builds up the energy, roughly speaking, by weighting the interaction

$V(r)$ by the number of neighbors at a given distance. Hence, we are now in a position to be able to evaluate the liquid state free energy.

As mentioned above, in light of eqn (6.15) we now aim to exploit the Bogoliubov inequality to bound the free energy. Our bound on the liquid state free energy may be written as

$$F_{exact} \leq F_0(z) + 2\pi N\rho \int_0^\infty g_0(r, z)[V(r) - V_0(r, z)]r^2 dr. \qquad (6.16)$$

$V(r)$ refers to the true interatomic interactions of interest (i.e. the real interaction we wish to compute the free energy of) while $V_0(r, z)$ is the reference state interactions. We also note that $g_0(r, z)$ is the radial distribution function of the *reference* system, and it is by virtue of this function that we effect the average $\langle (H - H_0) \rangle_0$ demanded by the Bogoliubov inequality. We follow Moriarty in denoting the adjustable parameter z within the description of the reference state. The argument that is made now is that for each volume and temperature of interest, the parameter z is tuned so as to minimize the right hand side of the equation, taking full advantage of the variational power of the Bogoliubov inequality.

6.4.3 Putting It All Together

In this section we undertake the business of assembling the phase diagram on the basis of the various insights that were developed above. Let us review what we have learned. In order to compute the phase diagram, we must make a series of comparisons. We note again that, unfortunately, the space over which we compare the competing alternatives is discrete and allows us to consider only those competitors that we have been clever enough to imagine in advance. This state of affairs should be contrasted with the wide sweeping searches that are made possible within the formalism of the calculus of variations where it is possible to search over a space of functions in order to determine the winning candidate.

The set of comparisons we make in determining the equilibrium phase diagram for an elemental material is between the liquid state and any of a number of possible crystalline phases that the material might adopt for the temperatures and pressures of interest. For the case study given below, that of magnesium, the fcc, bcc and hcp structures were considered as viable alternatives. The calculation begins by computing the zero-temperature structural energy of each of the competing alternatives. This determines the free energy zero for the various structures. The next task is to evaluate the contribution of the vibrational entropy to the free energy balance. To do so, one must compute the vibrational spectrum using the methods described in chap. 5, from which the vibrational entropy is computed using eqn (6.9). Prior to embarking on a detailed study of the phase diagram of

Mg we first reconsider some of the salient points in our analysis by appealing to a model system which undergoes a structural change with increasing temperature.

6.4.4 An Einstein Model for Structural Change

Recall that for many elemental metals, there is a structural transformation between a close-packed structure at low temperatures and the bcc structure at high temperatures. If we assume that this transformation is driven by the harmonic part of the vibrational entropy (later we will see that this is exactly what happens in some cases), then the simplest such model is that in which the entire vibrational spectrum of each phase is collapsed into a delta function. One might dub this model for the structural transformation between the fcc (low-temperature) and bcc (high-temperature) phases, an 'Einstein' model of structural change since it shares precisely the same density of states as the Einstein model originally introduced to remedy the low-temperature paradox associated with the specific heats of solids. In essence, we replace the function $\rho(\omega)$, with all of its structure, by a single delta function at the appropriate value of ω. If we further assume that the frequency associated with that delta function is lower for the bcc phase than for the fcc phase, the free energy balance is thereby guaranteed to turn in favor of the bcc phase at high temperatures. A visual reckoning of this model is given in fig. 6.14.

To be more quantitative, we assign an internal energy E_{fcc} to the fcc phase, E_{bcc} to the bcc phase, and associated Einstein frequencies ω_{fcc} and ω_{bcc}. As a result, the free energies (neglecting anharmonicity and the electronic contribution) may be written as

$$F_{fcc} = E_{fcc} + \frac{\hbar\omega_{fcc}}{2} + kT \ln(1 - e^{-\beta\hbar\omega_{fcc}}) \tag{6.17}$$

and

$$F_{bcc} = E_{bcc} + \frac{\hbar\omega_{bcc}}{2} + kT \ln(1 - e^{-\beta\hbar\omega_{bcc}}). \tag{6.18}$$

The resulting free energy difference $\Delta F = F_{bcc} - F_{fcc}$ may be written in turn as

$$\Delta F = E_{bcc} - E_{fcc} + \frac{\hbar}{2}(\omega_{bcc} - \omega_{fcc}) + kT \ln\left(\frac{1 - e^{-\beta\hbar\omega_{bcc}}}{1 - e^{-\beta\hbar\omega_{fcc}}}\right). \tag{6.19}$$

This simple model exhibits many of the qualitative features that govern the actual competition between different phases. First, by virtue of the fact that $E_{fcc} < E_{bcc}$, the low-temperature winner is the fcc structure. On the other hand, because the bcc phonon density of states has more weight at lower frequencies (here we put this in by hand through the assumption $\omega_{bcc} < \omega_{fcc}$), at high temperatures, the bcc phase has higher entropy and thereby wins the competition. These arguments are illustrated pictorially in fig. 6.14 where we see that at a critical temperature T_c,

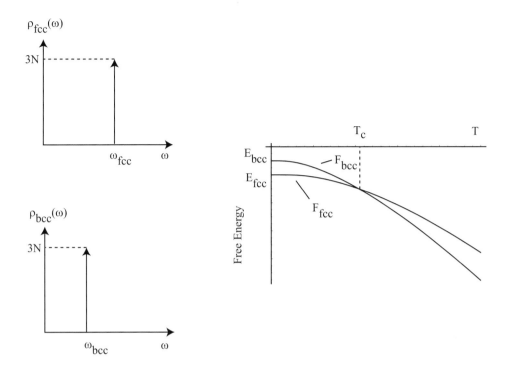

Fig. 6.14. Densities of states and free energies associated with Einstein model of structural change.

there is a crossing point at which F_{bcc} falls below F_{fcc}. Further elaborations of this model are possible such as replacing the delta function density of states used here with the rectangular band analog familiar already from our study of cohesion in solids, now tailored to the case of the phonon density of states rather than the electronic density of states. This elaboration is taken up in the problems.

To make the elements of this model more concrete, we consider the way in which the transition temperature depends upon the difference in the Einstein frequencies of the two structural competitors. To do so, it is convenient to choose the variables $\bar{\omega} = (\omega_B + \omega_A)/2$ and $\Delta\omega = (\omega_B - \omega_A)/2$, where ω_A and ω_B are the Einstein frequencies of the two structural competitors. If we now further measure the difference in the two frequencies, $\Delta\omega$, in units of the mean frequency $\bar{\omega}$ according to the relation $\Delta\omega = f\bar{\omega}$, then the difference in free energy between the two phases may be written as

$$\Delta F = F_B - F_A = E_B - E_A + \frac{1}{\beta} \ln \left(\frac{1 - z^{\beta(1+f)}}{1 - z^{\beta(1-f)}} \right), \qquad (6.20)$$

where we have introduced the variable $z = e^{-\hbar\bar{\omega}}$. In terms of this analysis, what we now seek for each value of f (i.e. for each value of $\Delta\omega = f\bar{\omega}$) is the value of

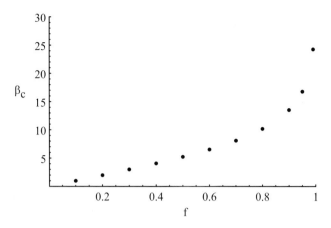

Fig. 6.15. Dependence of the critical temperature, here reported as $\beta_c = 1/kT_c$, on the difference in Einstein frequencies of the competing phases.

β for which $\Delta F = 0$, corresponding to the critical temperature for the structural transformation. A numerical solution to the problem for the case in which $\Delta E = E_B - E_A = -0.2$ eV and $\hbar\bar{\omega} = 0.02$ eV is shown in fig. 6.15.

The outstanding issue in our simplified model of structural change is the physical origins of the different spectral weights in the phonon density of states associated with different structures. In the context of our Einstein model, what we are asking is why should one structure have a lower Einstein frequency than another? Though we don't have the space to enter into such details here, we note that in an interesting analysis Friedel (1974) argues for the excess entropy of the bcc phase in terms of a smaller Einstein frequency than that of close-packed structures.

6.4.5 A Case Study in Elemental Mg

In this section we will sketch the work of Moriarty and Althoff (1995) which reveals the details of how elemental phase diagrams may be computed for the special case of Mg. In order to make the total energy calculations that are necessary for evaluating the free energy, they use volume-dependent pair potentials of the type described in section 4.3.2. The potentials for Mg are shown in fig. 6.16 for a range of different atomic volumes, indicating the volume dependence of both the strength of the interactions and the positions of the various minima. Once the potentials have been derived, the calculations of the structural energies, while tedious, are a straightforward application of ideas sketched earlier. These same potentials can be used to evaluate the dynamical matrix and thereby the phonon spectrum for Mg which can provide the contribution of vibrations to the free energy of the solid.

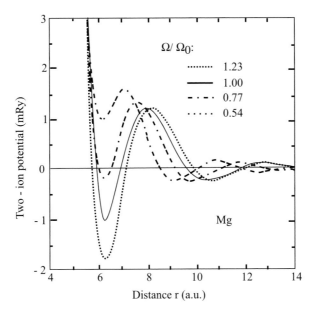

Fig. 6.16. Effective pair interactions for Mg at different electron densities (adapted from Moriarty and Althoff (1995)).

In addition to requiring the free energy of the solid, we also demand a knowledge of the liquid state free energy. For the purposes of evaluating the liquid state free energy within the context of the Bogoliubov inequality, a r^{-12} reference state is used in eqn (6.16). In particular, within this setting,

$$V_0(r, z) = \epsilon \left(\frac{\sigma}{r} \right)^{12} \tag{6.21}$$

is the appropriate reference potential, where the adjustable parameter is defined as $z = (\sigma^3 / \sqrt{2}\Omega)(\epsilon/kT)^{1/4}$. The key point is that the key functions $F_0(z)$ and $g_0(r, z)$ are known and may thus be exploited within the Bogoliubov inequality context described above.

The phase diagram that results from their analysis is shown in fig. 6.17. One striking feature of the phase diagram is the paucity of experimental data. We note that the solid/liquid melting line is known at very low pressures, and a single data point is also available from diamond-anvil cell experiments at high pressures. Nevertheless, given the essentially parameter-free manner with which this phase diagram was computed, it should be seen as a remarkable success for both the total energy and statistical mechanical features of the calculation.

This section has attempted an overview of the key conceptual features involved in computing elemental phase diagrams on the basis of microscopic insights. The

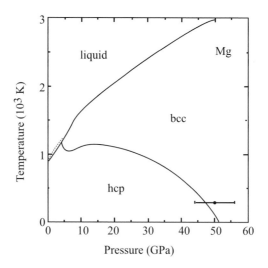

Fig. 6.17. Computed Mg phase diagram (adapted from Moriarty and Althoff (1995)).

general principles have been illustrated through application to the special case of Mg where we have seen that the computation of the phase diagram is a culmination of all of the ideas developed thus far in the book.

6.5 Alloy Phase Diagrams

As stated earlier, in many ways alloy phase diagrams are the road map of the materials scientist. In effect, the phase diagram is a diagrammatic representation of the various thermodynamic equilibria for a given system as control parameters such as the temperature or alloy composition are varied. From a modeling perspective, phase diagrams are especially fascinating since they represent the confluence of a variety of different modeling paradigms, and serve as a highly instructive and successful example of the type of effective theory construction being advocated in this book.

In previous sections we have shown how the structures of elemental materials may be rationalized on the basis of microscopic considerations. Our idea was to build up the free energy of a discrete set of competitors and then to make a free energy comparison between these competitors as a function of temperature and perhaps pressure. We next turn to the analysis of alloy phase diagrams. Here we will have to expand the scope of our previous analysis in order to account for the fact that the presence of more than one chemical constituent will at the very least alter the configurational and vibrational entropy. We define alloy in the present context to include any system in which there is more than one fundamental entity from which the structure of interest is built up. Our definition of alloy is broad enough to encompass systems such as the high-temperature superconductors in which oxygen vacancies serve as the second 'component' in the system. The logical structure of our discussion of alloy phase diagrams will unfold as follows. First, we examine the key ideas associated with the construction of an effective energy function for describing the energetics of the various competitors of relevance to the construction of alloy phase diagrams. These ideas will be illustrated with special reference to the Cu–Au system. Once the alloy Hamiltonian is in hand, we turn to an analysis of the use of statistical mechanics to determine the thermodynamics of this Hamiltonian. There are a number of subtleties associated with the effective Hamiltonian itself such as the role of structural relaxations and thermal vibrations. We will continue with a discussion of these subtleties, illustrated by means of the example of the Cd–Mg system. Finally, we round out the discussion through an analysis of oxygen ordering in high-temperature superconductors, the intention being to use this example to illustrate the various ideas introduced in the course of the preceding discussion.

6.5.1 Constructing the Effective Energy: Cluster Expansions

The issue of alloy phase stability poses challenges at both the philosophical and practical level. On the philosophical side, we note that in principle we must conduct a search over not only the space of possible structures, but also over a space of configurations at fixed structure (i.e. all the different ways of arranging atoms of more than one type on a given lattice). Because of the presence of more than a single chemical constituent, it is possible that different chemical occupancies can be found on a given lattice. We have already registered the complaint that often when confronted with questions of structure in solids, one is reduced to a set of guesses based upon intuition (a strategy which Zunger (1994) aptly describes as 'rounding up the usual suspects'), in stark contrast with the types of continuous searches that are possible when the function being minimized depends continuously on some parameter. One partial remedy to this problem is to find a way to enlarge the discrete space of competitors through the use of an effective Hamiltonian. The cluster expansion technique to be described here is just such a model.

On the practical side, the difficulties involved in constructing the alloy free energy are related to the fact that somehow the effective Hamiltonian must be informed on the basis of a relatively small set of calculations. For concreteness, we consider a binary alloy in which there are two different chemical constituents. For a lattice built up of N sites, the freedom to distribute the two constituents on these N sites results in 2^N configurations (i.e. each site may be assigned one of two values). We wish to construct an effective Hamiltonian that is cognizant of the various configurational competitors. On the other hand, we do not wish to determine the parameters in this effective Hamiltonian by recourse to calculations on some large subset of these 2^N configurations.

As noted above, our charter is to map the energetics of the various alloy configurations onto an effective Hamiltonian. Before embarking on this path we must first settle the question of configurational kinematics in a way that allows us to specify the configurational geometry using a scheme that is both transparent and will serve as a viable basis for the description of alloy energetics. One noted example of this strategy is the set of lattice gas models familiar from statistical mechanics in which the various configurational states of the alloy are labeled by occupation numbers σ_i, with $\sigma_i = 1$ signifying the occupation of the ith site by an A atom, and $\sigma_i = -1$ signifying the occupation of this site by a B atom. From the point of view of the kinematics of configuration, the disposition of the alloy is characterized by the complete set of occupation numbers $\{\sigma_i\}$, where the label i runs over all N sites of the crystal. Since each site is allowed one of two occupancies, as noted above, the total number of configurations in question is 2^N. The effective Hamiltonian amounts to a total energy of the form $E_{tot} = E_{tot}(\{\sigma_i\})$ which is then amenable to analysis in terms of the standard tools of statistical mechanics.

Our first quest is to examine the ways in which the types of total energy calculations presented at the beginning of this chapter can be used to produce a total energy in which only the configurational degrees of freedom (i.e. the $\{\sigma_i\}$) remain. The objective in the construction of an effective Hamiltonian is to be able to write the total free energy in such a way that the space of structural competitors has been collapsed onto the space of configurations $\{\sigma_i\}$. This means that the thermodynamics of alloys then becomes a problem in the statistical mechanics of lattice models. The particular idea taken up in the present setting is to write the total Hamiltonian in extended Ising form, such as

$$H = J_0 + \sum_i J_i \sigma_i + \frac{1}{2} \sum_{ij} J_{ij} \sigma_i \sigma_j + \frac{1}{3!} \sum_{ijk} J_{ijk} \sigma_i \sigma_j \sigma_k + \cdots . \qquad (6.22)$$

In a landmark paper, Sanchez *et al.* (1984), examined the extent to which the types of cluster functions introduced above can serve as a basis for representing any function of configuration. For our present purposes, the key result to take away from their analysis is that this set of cluster functions is complete in the mathematical sense, implying that any function of configuration can be written as a linear combination of these cluster functions. On the other hand, for the whole exercise to be useful, we seek a minimal basis, that is to say, we truncate at some order with the expectation that a low-order expansion in such cluster functions will be sufficiently accurate. The strategy we will outline below is to truncate the cluster expansion at some relatively low order, to perform a series of atomic-level calculations that will serve to *inform* the cluster expansion, and then to carry out the appropriate statistical mechanics analysis to investigate the phase diagrammatics.

Eqn (6.22) may be written more formally, as

$$H = \sum_{\alpha} J_{\alpha} \Phi_{\alpha}(\{\sigma\}), \tag{6.23}$$

where the sum on α is organized in terms of real space clusters of increasing size, with $\Phi_{\alpha}(\{\sigma\})$ being a shorthand representation for an average over the αth cluster type which itself can be written as

$$\Phi_{\alpha}(\{\sigma\}) = \frac{1}{N_{\alpha}} \sum_{clusters} \sigma_1 \sigma_2 \cdots \sigma_{n_{\alpha}}. \tag{6.24}$$

What this equation tells us is to find all N_{α} clusters of type α and to evaluate the product $\sigma_1 \sigma_2 \cdots \sigma_n$ for each such cluster and to add up the contributions from them all. In the particular case written above, our notation has been chosen such that there are n sites involved in the αth cluster type. The entirety of these averaged correlation functions provides a measure of the configurational order associated with the structure of interest. What these equations tell us physically is that the energetics is expanded in terms of various correlation functions that characterize the extent to which adjacent atoms are like or unlike.

Though there are several strategies for informing effective Hamiltonians like that introduced above, we will sketch here in generic terms their common features. The basic idea is to choose some group of representative structures, the energies of which are to be determined 'exactly' on the basis of an underlying atomistic total energy. At this point, a database of structural energies is in hand and the next step is to find a choice of parameters within the effective Hamiltonian that reflects the energies found in this database. Note that the database consists of a set of energies of different configurations of the various A and B atoms making up the alloy on a given lattice. Of course, in order to turn this idea into a practical calculational scheme, it is necessary to first truncate the cluster expansion at some order. An example of the types of terms that are retained in these cluster expansions is shown in fig. 6.18. In particular, the figure makes reference to the various cluster types associated with structures built around the fcc lattice. The set $J2$, $K2$, $L2$ and $M2$ refer to the pair interactions between first, second, third and fourth neighbors on the fcc lattice, as is evident from the figure. $J3$ is is the coupling constant associated with a three-atom near-neighbor cluster, and $J4$ refers to a four-atom near-neighbor cluster.

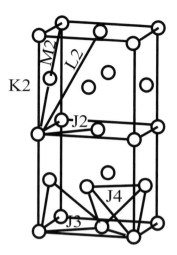

Fig. 6.18. Schematic of the low-order clusters considered in the treatment of the energetics of alloys built on a host fcc lattice (adapted from Lu *et al.* (1991)).

In the approach which has come to be known as the Connolly–Williams method (Connolly and Williams, 1983), a systematic inversion of the cluster expansion is effected with the result that the parameters in the effective Hamiltonian are determined explicitly. In this case, the number of energies in the database is the same as the number of undetermined parameters in the cluster expansion. Other strategies include the use of least-squares analysis (Lu *et al.* 1991) and linear programming methods (Garbulsky and Ceder, 1995) in order to obtain some 'best' choice of parameters. For example, in the least-squares analyses, the number of energies determined in the database exceeds the number of undetermined parameters in the effective Hamiltonian, which are then determined by minimizing a cost function of the form

$$C(J_0, J_1, J_2, \ldots, J_N) = \sum_{j=1}^{p} w_j \left[E_j - \sum_{\alpha} J_\alpha \Phi_\alpha(\{\sigma\})_j \right]^2. \tag{6.25}$$

The subscript j refers to the jth member of the database, and thus E_j is the energy of that member as obtained from the relevant atomistic calculation and $\Phi_\alpha(\{\sigma\})_j$ is the averaged correlation function for the αth cluster type in the jth structure. In addition, the factor w_j is a weighting factor that quantifies the importance of various contributions to the least squares fit.

The basic analysis of the Connolly–Williams method may be spelled out as follows. If the effective Hamiltonian is characterized by $M + 1$ distinct coupling constants $J_0, J_1, \ldots, J_M$, then we wish to perform $M + 1$ distinct total energy

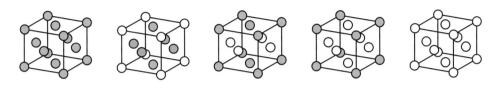

Fig. 6.19. Schematic of the five structures used in the evaluation of the parameters J_0–J_5.

calculations on different configurations which will serve as the basis of the fit for these coupling constants. The set of resulting total energies form a vector $\mathbf{E}_{tot} = (E_0, E_1, \ldots, E_M)$. On the other hand, these same energies may be written in terms of our effective Hamiltonian as

$$E_0 = \Phi_0(\{\sigma\})_0 J_0 + \Phi_1(\{\sigma\})_0 J_1 + \cdots + \Phi_M(\{\sigma\})_0 J_M,$$

$$E_1 = \Phi_0(\{\sigma\})_1 J_0 + \Phi_1(\{\sigma\})_1 J_1 + \cdots + \Phi_M(\{\sigma\})_1 J_M,$$

$$\vdots$$

$$E_M = \Phi_0(\{\sigma\})_M J_0 + \Phi_1(\{\sigma\})_M J_1 + \cdots + \Phi_M(\{\sigma\})_M J_M.$$

With apologies for the clumsy notation, we note that the coefficients $\Phi_i(\{\sigma\})_j$ are averages over all of the clusters of type i associated with the jth structure. In matrix form, what we have said above may be restated as

$$\mathbf{E}_{tot} = \Phi\mathbf{J}. \qquad (6.26)$$

Hence, for a determination of the unknown parameters $\mathbf{J}$ our problem is reduced to that of matrix inversion, with the result that the vector of interaction strengths is given by

$$\mathbf{J} = \Phi^{-1}\mathbf{E}_{tot}. \qquad (6.27)$$

To make the ideas presented above more concrete, we follow Connolly and Williams in considering a five-parameter model of ordering on the fcc lattice. We consider a model characterized by the parameters J_0, J_1 (single atom clusters), J_2 (near-neighbor pair interactions), J_3 (a triplet of atoms that are all near neighbors) and J_4 (a tetrahedron of atoms that are all near neighbors). Corresponding to these five unknown parameters, we require a knowledge of the energies of five distinct configurations which are chosen to be fcc A, A_3B in the $L1_2$ structure, AB in the $L1_0$ structure, AB_3 in the $L1_2$ structure and fcc B. As a reminder, these five structures are all shown in fig. 6.19. The outcome of performing the calculations dictated by eqn (6.24) is shown in table 6.1. The values in the table were arrived at by considering the various sums demanded by eqn (6.24). To be concrete in our example, we analyze Φ_2 for the AB structure. In this structure, the key point

Table 6.1. *Averaged correlation functions required in evaluation of effective Hamiltonian for various fcc-based structures.*

Structure	Φ_0	Φ_1	Φ_2	Φ_3	Φ_4
fcc A	1	1	1	1	1
L1$_2$ A$_3$B	1	$\frac{1}{2}$	0	$-\frac{1}{2}$	-1
L1$_0$ AB	1	0	$-\frac{1}{3}$	0	1
L1$_2$ AB$_3$	1	$-\frac{1}{2}$	0	$\frac{1}{2}$	-1
fcc B	1	-1	1	-1	1

to realize is that every atom has four like near neighbors and eight unlike near neighbors. In order to effect the sum of eqn (6.24) we note that for the four atoms per unit cell there are a total of 48 two-atom clusters to be considered. For an unlike near-neighbor pair, $\sigma_i \sigma_j = -1$ while for a like pair the product is positive. As a result, we are left with

$$\Phi_2^{AB} = \frac{1}{48}[32 \times (+1)(-1) + 8 \times (+1)(+1) + 8 \times (-1)(-1)], \qquad (6.28)$$

resulting in $\Phi_2^{AB} = -1/3$. The remainder of the analysis is taken up in the problems at the end of the chapter.

Armed with our knowledge of the averaged correlation functions as embodied in table 6.1, we can now write down eqn (6.26) for the particular case of present interest. To do so, we recall that the energy of the jth structure is given by $E_j = \sum_\alpha J_\alpha \Phi_\alpha^{(j)}$, where we denote the correlation function for the αth cluster type in the jth structure as $\Phi_\alpha^{(j)}$. The result in the case of interest here is

$$\left. \begin{aligned} E_A &= J_0 + J_1 + J_2 + J_3 + J_4, \\ E_{A_3 B} &= J_0 + \frac{J_1}{2} - \frac{J_3}{2} - J_4, \\ E_{AB} &= J_0 - \frac{J_2}{3} + J_4, \\ E_{AB_3} &= J_0 - \frac{J_1}{2} + \frac{J_3}{2} - J_4, \\ E_B &= J_0 - J_1 + J_2 - J_3 + J_4. \end{aligned} \right\} \qquad (6.29)$$

The determination of the J_is now amounts to an inversion of the matrix of coefficients. In the present case, this results in

$$
\left.\begin{aligned}
J_0 &= \frac{1}{16}E_A + \frac{1}{4}E_{A_3B} + \frac{3}{8}E_{AB} + \frac{1}{4}E_{AB_3} + \frac{1}{16}E_B, \\
J_1 &= \frac{1}{4}E_A + \frac{1}{2}E_{A_3B} - \frac{1}{2}E_{AB_3} + \frac{1}{4}E_B, \\
J_2 &= \frac{3}{8}E_A - \frac{3}{4}E_{AB} + \frac{3}{8}E_B, \\
J_3 &= \frac{1}{4}E_A - \frac{1}{2}E_{A_3B} + \frac{1}{2}E_{AB_3} - \frac{1}{4}E_B, \\
J_4 &= \frac{1}{16}E_A - \frac{1}{4}E_{A_3B} + \frac{3}{8}E_{AB} - \frac{1}{4}E_{AB_3} + \frac{1}{16}E_B.
\end{aligned}\right\}
\tag{6.30}
$$

The details of this simple implementation of the Connolly–Williams method can be found in Terakura *et al.* (1987). Our main point in carrying out this analysis has been to illustrate in full detail the *conceptual* underpinnings associated with the determination of effective cluster interactions. The main point is that the quantities on the right hand side of eqn (6.30) can be computed by recourse to atomic-level analysis. As a result of the inversion shown above, we are then in possession of a set of parameters $\{J_i\}$ which can then be used to generate the energy of *any* configuration using eqn (6.23), and thereby, we are poised to invoke the tools of statistical mechanics to examine the phase diagrammatics of the resulting effective Hamiltonian.

As a second foray into the detailed implementation of cluster expansions, we consider efforts that have been put forth in the quest to better understand intermetallic phase diagrams with special reference to the Cu–Au system. In this case, the results of the analysis are almost entirely numerical and hence we will resort to a graphical depiction of the outcome of the relevant calculations. Thus far, we have emphasized the conceptual features associated with determining the coefficients in the effective Hamiltonian. The routine calculation of total energies using density functional theory provides the opportunity to determine the interaction parameters on the basis of a full quantum mechanical treatment of the electronic degrees of freedom. The idea is to use electronic structure calculations to generate the set of energies in the vector $\mathbf{E}_{tot}$ discussed above in the context of the Connolly–Williams method. One of the most studied of alloy phase diagrams is that of the Cu–Au system which will serve as our case study for the calculation of effective cluster interactions using first-principles techniques.

Unlike in the direct inversion scheme presented above, we follow the analysis of Ozoliņš *et al.* (1998) in which the determination of the parameters for the Cu–Au system was carried out by recourse to least-squares fitting. The advantage of this approach is that it allows for a systematic analysis of the errors incurred by carrying

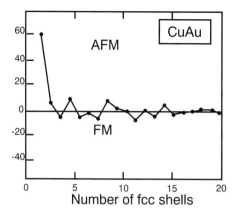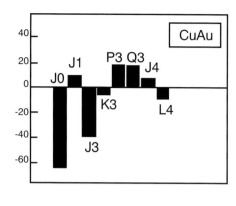

Fig. 6.20. Cluster interaction results for the Cu–Au system (adapted from Ozoliņš *et al.* (1998)).

the cluster expansion to a given order and allows for strategies to improve these fits. For the Cu–Au interaction parameters of interest here, the pair interactions were assumed to extend for a number of neighbor shells, and similarly, the set of three- and four-atom clusters was extended to include neighbors beyond the first neighbor shell. The cluster interactions resulting from this procedure are shown in fig. 6.20. As an aside, we also refer the reader to the original paper of Ozoliņš *et al.* (1998) in the hope that he or she will contrast the Cu–Au results with those for Ag–Au as shown in their figs. 5 and 6. One observation to be made concerning the pair interactions in this case is that the short-range contributions have a sign for the near-neighbor interaction corresponding to 'antiferromagnetic' ordering (i.e. a preference for unlike near neighbors) consistent with the low-temperature ordering tendency in this system. In fact, we have taken the present calculations out of context. In the paper of Ozoliņš *et al.* (1998) the analysis of Cu–Au is but one of a series of alloy systems which includes Ag–Au, Cu–Ag and Ni–Au, which when viewed collectively reveal intriguing phenomenological distinctions which are explained by the type of calculations described above and which serve to convincingly demonstrate the power of the use of effective Hamiltonians. Our main ambition in the present section has been to illustrate the way in which information can be passed from computer intensive atomic-level calculations of the energetics of alloys to the construction of effective alloy Hamiltonians which permit the modeler to consider more than just 'the usual suspects' noted by Zunger.

Before proceeding, we pause to note both the elegance and power of the ideas introduced in this section. The main thesis that has been argued for is the necessity of enlarging the space of structural competitors considered in questions of alloy phase stability. The brute force calculation of the energies of a series of intuitive guesses is intrinsically unsatisfying, and computationally prohibitive to boot. On the other hand, by invoking an effective Hamiltonian, it has been seen that an entire set of configurations can be considered. By way of contrast, criticisms can still be leveled on the grounds that this approach presumes that the structural winners are members of the class of lattices that are used in constructing the lattice Hamiltonian.

6.5.2 Statistical Mechanics for the Effective Hamiltonian

To the extent that we have successfully mapped the alloy energetics onto an effective Hamiltonian like those described above, the next step in determining the phase diagram is to carry out a statistical mechanical analysis of this Hamiltonian. However, as we noted in chap. 3, three-dimensional lattice gas models (and in particular the three-dimensional Ising model) have resisted analytic solution and so we must turn to other techniques for finding the free energy minimizing statistical distribution. As a result, we take up two schemes in turn. First, we examine a series of approximations to the lattice gas partition function which are systematized generically as the cluster variation method (CVM). Our second primary approach will be to consider the direct numerical evaluation of the thermodynamics of our alloy Hamiltonian through Monte Carlo simulation.

Historically, the statistical mechanics of the class of lattice-gas Hamiltonians like those introduced above has been attacked using approximations of increasing sophistication. The first such scheme and the pioneering effort on alloy phase stability and the origins of order–disorder phenomena is the Bragg–Williams approximation. This scheme represents the simplest approach to such problems by treating each and every site independently and insisting only that the overall concentration is preserved site by site. At the next level of approximation, the Bethe–Peierls scheme acknowledges correlations between the occupancies on neighboring sites by insisting that the overall number of AA, AB and BB bonds is respected. The cluster variation method continues with this systematic imposition of correlation by insisting that larger and larger clusters are reproduced with the correct statistical weight.

To note the nature of the problem, we begin with a reminder concerning the entropy in the case in which no correlation between the occupancies of adjacent sites is assumed. In this limit, the entropy reduces to that of the ideal entropy already revealed in eqn (3.89). In preparation for the notation that will emerge in our discussion of the cluster variation method, we revisit the analysis culminating in eqn (3.89). Recall from chap. 3 that the entropy of a system characterized by a series of discrete states with probabilities p_i is given by

$$S = -k \sum_i p_i \ln p_i. \qquad (6.31)$$

In the present case, there are 2^N such states (i.e. the various occupancies of the N sites of the lattice by A and B atoms), each of which has a probability $P(\sigma_1, \sigma_2, \ldots, \sigma_N)$. This notation refers to the probability that site 1 has occupation number σ_1, site 2 has occupation number σ_2 and so on. We introduce the alternative notation $P(\{\sigma_i\})$, which again signifies the probability of a given state characterized by the set of occupation numbers $\{\sigma_i\}$. In light of this more explicit notation, we can rewrite eqn (6.31) as

$$S = -k \sum_{\{\sigma_i\}} P(\{\sigma_i\}) \ln P(\{\sigma_i\}), \tag{6.32}$$

in which the sum $\sum_{\{\sigma_i\}}$ instructs us to sum over all 2^N configurations. The problem is that we don't know the $P(\{\sigma_i\})$s.

Point Approximation. The simplest approximation is to advance the assumption that the probability may be written as $P(\{\sigma_i\}) = P_1(\sigma_1) P_2(\sigma_2) \cdots P_N(\sigma_N)$. Conceptually, what this means is that we have abandoned the treatment of correlations between the occupancies of different sites. As a result, certain configurations are given much higher statistical weight than they really deserve. On the other hand, once this form for the probabilities has been accepted and using the fact that $P_i(1) + P_i(-1) = 1$, the sum in eqn (6.32) can be simplified to the form

$$S = -Nk[P_i(1) \ln P_i(1) + P_i(-1) \ln P_i(-1)]. \tag{6.33}$$

In addition, because we know the overall concentration x of A atoms and $1 - x$ of B atoms, we can write down the probabilities as $P_i(1) = x$ and $P_i(-1) = 1 - x$, resulting in an approximation to the entropy of the form

$$S = -Nk[x \ln x + (1 - x) \ln(1 - x)]. \tag{6.34}$$

This is no great surprise. We have recovered the result of eqn (3.89), but now seen as the first in a series of possible approximations. We will denote the present approximation as the 'point' approximation.

Pair Approximation. At the next level of approximation, we can insist that our probabilities $P(\{\sigma_i\})$ reproduce not only site occupancies with correct statistical weight, but also the distribution of AA, AB and BB bonds. We begin from a kinematic perspective by describing the relations between the number of A and B atoms (N_A and N_B), the number of AA, AB and BB bonds (N_{AA}, N_{AB} and N_{BB}) and the total number of sites, N. Note that we have not distinguished AB and BA bonds. These ideas are depicted in fig. 6.21. What we note is that for our one-dimensional model, every A atom is associated with two bonds, the identity of which must be either AA or AB. Similar remarks can be made for the B atoms.

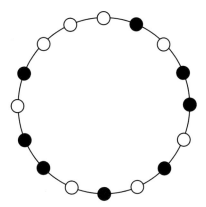

Fig. 6.21. Schematic of a one-dimensional chain of atoms (with periodic boundary conditions) illustrating the connection between the number of A and B atoms, and the number of AA, AB and BB bonds.

As a result, we see that the the numbers N_A, N_B, N_{AA}, N_{AB}, N_{BB} and N are not independent and in particular,

$$\left. \begin{array}{l} 2N_A = 2N_{AA} + N_{AB}, \\ 2N_B = 2N_{BB} + N_{AB}. \end{array} \right\} \tag{6.35}$$

Our reason for carrying out this discussion is to note that there are further overall constraints on the distribution of A and B atoms that should be respected. In particular, the probabilities $P(\{\sigma_i\})$ should be selected so that in a statistical sense, the constraints of eqn (6.35) are reproduced.

To close our discussion of the pair approximation, we must now discover how to write the probabilities $P(\{\sigma_i\})$. Schematically, we know that what we are after is a probability distribution of the form

$$P(\{\sigma_i\}) \propto \prod_{ij} P_{ij}(\sigma_i, \sigma_j), \tag{6.36}$$

where the total probability has been decomposed into a product over pair probabilities. In addition, we know that a crucial *desideratum* for this distribution is that it reduce to the uncorrelated point approximation in the high-temperature limit. In that limit, we see that $P_{ij}(\sigma_i, \sigma_j) = P_i(\sigma_i) P_j(\sigma_j)$. When the product of eqn (6.36) is evaluated in this limit, we see that there is an overcounting with each $P_i(\sigma_i)$ appearing as many times as there are bonds that it participates in. If we follow Finel (1994) in using the label p to characterize the number of bonds that each site participates in, then the appropriate probability distribution can be written as

$$P(\{\sigma_i\}) = \frac{\prod_{ij} P_{ij}(\sigma_i, \sigma_j)}{\prod_i (P_i(\sigma_i))^{p-1}}. \tag{6.37}$$

General Description of the Cluster Variation Method. In the previous paragraphs we have described several approximation schemes for treating the statistical mechanics of lattice-gas Hamiltonians like those introduced in the previous section. These approximations and systematic improvements to them are afforded a unifying description when viewed from the perspective of the cluster variation method. The evaluation of the entropy associated with the alloy can be carried out approximately but systematically by recourse to this method. The idea of the cluster variation method is to introduce an increasingly refined description of the correlations that are present in the system, and with it, to produce a series of increasingly improved estimates for the entropy.

As shown above, in the point approximation to the cluster variation method (essentially the Bragg–Williams approach), we carry out the counting of configurations by finding all the ways of distributing A and B atoms that guarantee an overall concentration x of A atoms and $1 - x$ of B atoms. The number of such configurations is

$$\Omega = \frac{N!}{(xN)!((1-x)N)!},$$ (6.38)

resulting in an entropy $S = -kN[x \ln x + (1 - x) \ln(1 - x)]$. The Bethe–Peierls approximation can be thought of as a pair approximation in the sense that we count up all the ways of distributing the A and B atoms that simultaneously guarantee an overall concentration x of A atoms and $1 - x$ of B atoms that at the same time guarantees that the number of AA, AB and BB bonds is reproduced statistically.

In general, once both the effective cluster interactions and the statistical treatment resulting from the cluster variation method are in hand, the alloy free energy can be written as

$$F = \sum_\alpha J_\alpha \Phi_\alpha(\{\sigma\}) - kT \sum_{k=1}^{K} m_k \gamma_k \sum_{\alpha_k} P_k(\alpha_k) \ln P_k(\alpha_k).$$ (6.39)

The first term is familiar from the previous section and represents the energy of a given configuration. The second term is more complex, and reflects the contributions of various order to the entropy arising from the inclusion of clusters starting from the point and going up to the Kth cluster.

Again harking back to our use of the Cu–Au system to serve as a case study, many of the ideas associated with the approximation schemes given above can be elucidated pictorially. For the present discussion, we consider a model Hamiltonian in which only pair interactions between neighboring atoms are considered. That is, the cluster expansion of the total energy (as given in eqn (6.22)) is carried only to the level of terms of the form $\sigma_i \sigma_j$. Even within the confines of this trimmed down effective Hamiltonian, the treatment of the statistical mechanics can be carried out

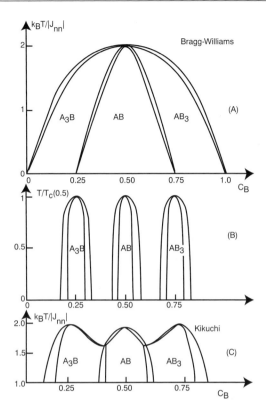

Fig. 6.22. Phase diagram for the Cu–Au system with energetics described by a near-neighbor pair approximation and with thermodynamics treated to various levels of approximation (adapted from Binder (1980)).

to various levels of approximation. In particular, in fig. 6.22, the Cu–Au phase diagram is represented at three different levels of approximation, namely at the level of the Bragg–Williams approximation (top panel), at the level of the Bethe–Peierls approximation (middle panel) and finally at the level of the tetrahedron approximation to the cluster variation method. Though certain critical features of the experimental phase diagram are missing even from the most accurate of these phase diagrams, our current interest is with the fidelity of our treatment of the statistical mechanics.

A more direct brute force approach to examining the statistical implications of a given model is to resort to Monte Carlo simulation. In simplest terms, the idea is to sample the free energy and to effect the relevant averages by passing through a series of configurations that are selected randomly from the full ensemble of possibilities but which are accepted in a fashion that is consistent with their statistical weights. In the context of the phase diagram problem of interest here,

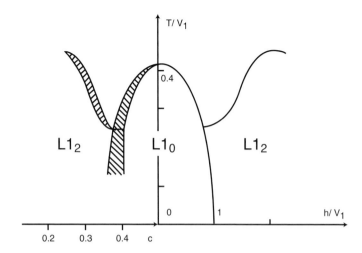

Fig. 6.23. Phase diagram for the Cu–Au system obtained via Monte Carlo treatment of model Hamiltonian (adapted from Ducastelle (1991)).

the simplest algorithm amounts to choosing two atoms at random within the computational cell and swapping them. The effective Hamiltonian is evaluated with the new configuration resulting from the atom swap and the new configuration is kept or rejected on the basis of the usual rules of Monte Carlo analysis. This procedure is equilibrated (i.e. each occupation number is changed a large number of times so that a great variety of configurations are sampled) and then statistical averages are evaluated that signal features of the system such as the onset of ordering.

To continue with our evaluation of the Cu–Au system and as a concrete realization of the Monte Carlo method with respect to a particular system, fig. 6.23 shows the results of carrying out a direct Monte Carlo evaluation of the phase diagram, again in the case of a pair interaction model. These calculations are detailed in Diep *et al.* (1986), Gahn (1986), Ackermann *et al.* (1986) and Ducastelle (1991). What is noted in particular about the Monte Carlo results is their strong correspondence with the results obtained using the cluster variation method as shown in fig. 6.22(c). On the other hand, the agreement with experiment leaves much to be desired. In this case, it is a failure not of the treatment of the statistical mechanics of the Hamiltonian, but rather of the Hamiltonian itself. In particular, the assumption of near-neighbor pair interactions is clearly insufficient to characterize the energetics of this system.

Extension of the model Hamiltonian to include higher-order interactions results in the Cu–Au phase diagram shown in fig. 6.24. In particular, the cluster expansion for the total energy in this case includes four-site interactions between those sites

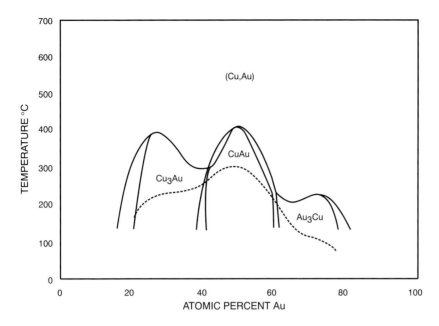

Fig. 6.24. Phase diagram for the Cu–Au system with energetics obtained by cluster expansion including tetrahedral clusters and with the thermodynamics treated in the cluster variation method in the tetrahedron approximation approximation (adapted from de Fontaine (1979)).

occupying the vertices of a tetrahedron. The calculation is described in more detail in de Fontaine (1979). What is noted here is that by virtue of the inclusion of the higher-order terms, the characteristic asymmetry about the point $x = 1/2$ is reproduced.

6.5.3 The Effective Hamiltonian Revisited: Relaxations and Vibrations

In contemplating the total free energy of an alloy, it is clear that in addition to the dependence of the free energy on configuration, there are terms arising from the various excitations about static states that can be considered. In particular, just as we found earlier in the context of elemental phase diagrams, lattice vibrations contribute to the overall free energy budget.

In an attempt to formalize the true meaning of the effective Hamiltonian in which ostensibly only configurational degrees of freedom are featured, Ceder (1993) has shown how the cluster Hamiltonian may be interpreted as a model in which excitations such as thermal vibrations have been integrated out. The starting point of the analysis is the assertion that the full partition function may be decomposed into two sets of sums, one over configurations, the other over excitations associated with a given configuration. This decomposition is shown schematically in fig. 6.25

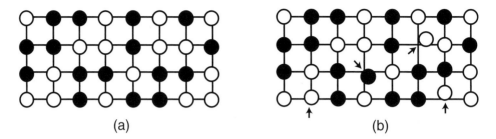

Fig. 6.25. Schematic illustrating both (a) the kinematic description of various configurations and (b) the types of excitations that are possible for a given configuration (adapted from Ceder (1993)).

and is represented mathematically as

$$Z(N, V, T) = \sum_{\substack{configs.}} \sum_{\substack{excitations}} e^{-\beta E_\alpha}. \tag{6.40}$$

The label α is a generalized index to indicate the energy of a particular excitation associated with a given configuration. We note that the present discussion emphasizes the local atomic-level relaxations and how they can be handled within the context of the Ising-like model introduced earlier. However, there are additional effects due to long-range strain fields that require further care (they are usually handled in reciprocal space) which are described in Ozoliņš *et al.* (1998).

To make the decoupling of the configurations and excitations about them explicit, we can rewrite eqn (6.40) as

$$Z(N, V, T) = \sum_{\substack{configs.}} \Lambda(N, V, T, \{\sigma\}), \tag{6.41}$$

where $\Lambda(N, V, T, \{\sigma\})$ is defined as

$$\Lambda(N, V, T, \{\sigma\}) = \sum_{\substack{excitations}} e^{-\beta E_i} \tag{6.42}$$

and E_i is the energy of the ith excitation associated with configuration $\{\sigma\}$. Note that while we have found it convenient to attribute discrete labels to the excitations about a given configuration, in fact, these excitations often refer to a continuous set of states.

We are now prepared to define an effective Hamiltonian in which the excitations have been 'integrated out' through the relation

$$H_{eff}(\{\sigma\}, N, V, T) = -\frac{1}{\beta} \ln \Lambda(N, V, T, \{\sigma\}), \tag{6.43}$$

resulting in the ability to write the alloy partition function as a sum *only* over configurational states as

$$Z(N, V, T) = \sum_{\{\sigma\}} e^{-\beta H_{eff}(\{\sigma\}, N, V, T)}. \qquad (6.44)$$

In the discussion given above, we have shown how increasing levels of sophistication may be achieved in the nature of the effective cluster interactions. In the most simple-minded approach, the cluster interactions describe configurational energetics on a fixed lattice. However, we have seen that the effective Hamiltonian can in fact be built up in such a way that various relaxations have been taken into account and the vibrational contributions to the free energy can also be considered. As an illustration of the successive layers of sophistication that can be included in these models, we consider the Cd–Mg phase diagram which has been calculated by Asta *et al.* (1993) from the perspective of the various refinements that can be added to the model. An example of the type of phase diagram that emerges from the calculations sketched here is shown in fig. 6.26. The key point to be made about these calculations is that as the effective Hamiltonian has become more sophisticated, namely, the inclusion of relaxation and vibrational effects, the structure of the phase diagram has moved closer to that from experiment.

6.5.4 The Alloy Free Energy

In the previous subsections, we have described some of the key contributions to the free energy of an alloy. In principle, one can imagine a calculational scheme in which these various contributions to the free energy are painstakingly assembled on the basis of standard microscopic techniques like those described in this and earlier chapters. Once the free energies of the various structural competitors are in hand, structural selectivity becomes nothing more than a matter of effecting the relevant comparisons at each temperature and composition, for example. On the other hand, this program can be criticized on the grounds that it depends entirely on the foresight of the modeler with the hope being that intuition is sufficiently refined as to have identified all of the relevant structural competitors. We have already belabored the shortcomings of intuitive discrete searches of this type and will return to this theme repeatedly with an additional difficult example being that of identifying the relevant transition pathways for thermally activated processes. However, in the present section, our main point is the observation that discrete searches based on intuition over the space of possible minimizers of some function are difficult and largely unsatisfying. Within the alloy community, recourse to variational methods has been made possible in the context of the phase diagram problem through the construction of an appropriate effective alloy Hamiltonian.

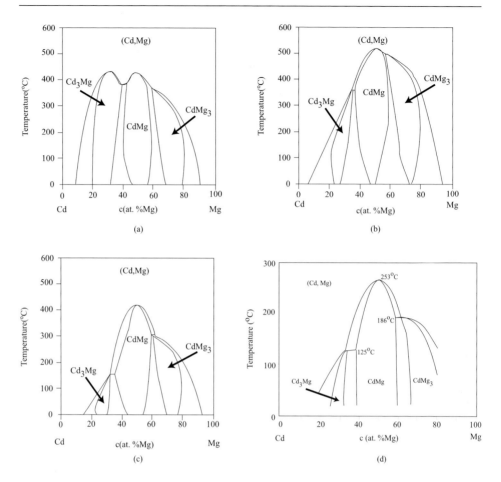

Fig. 6.26. Phase diagram for the Cd–Mg system (adapted from Asta *et al.* (1993)): (a) configurational contributions to free energy included, (b) configurational and relaxational contributions to the free energy included, (c) all contributions to the free energy included including both vibrational and electronic effects and (d) experimental phase diagram.

6.5.5 Case Study: Oxygen Ordering in High T_C Superconductors

We have now seen that by systematic exploitation of our ability to evaluate the relative structural energies of different competing phases at zero temperature as well as the entropy balance separating them, it is possible to compute the free energies themselves and thereby construct the phase diagram. Though we will not have resolved the problem of metastable phases, our work will have done much to enlighten us as concerns the equilibrium phases favored by a given material. Later, we will use the same type of thinking rounded out here to evaluate the structure and energetics of defects.

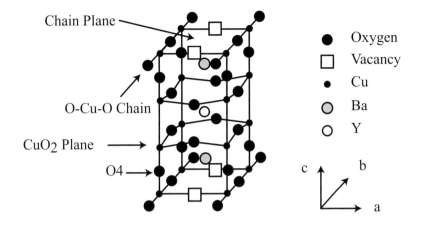

Fig. 6.27. Structure of the YBCO superconductor (adapted from Ceder (1994)).

As a final case study, we consider the case of oxygen ordering in the context of the YBCO class of high-temperature superconductor. We note that by the late 1980s, oxide superconductors were among the most studied of materials. The surge of interest in this class of materials was inspired largely by the fact that they have superconducting transition temperatures which exceed liquid nitrogen temperatures. One class of materials that has attracted particular attention with reference to the character of its different phases are the $YBa_2Cu_3O_z$ materials, where the stoichiometric parameter z falls between 6 and 7. The structure of these materials is indicated in fig. 6.27, and for our present purposes, the key feature of this structure that we wish to acknowledge is the presence of Cu–O planes in which the distribution of oxygen atoms depends upon the concentration. What makes this material especially interesting for the present discussion is the fact that as the parameter z is varied, several different structures are observed. In particular, a transition from an orthorhombic to tetragonal structure can be induced by reducing the oxygen content. In fact, the nature of the oxygen ordering in the Cu–O planes can be more subtle with longer period orthorhombic structures observed as well with different oxygen concentrations.

An intriguing suggestion to emerge from the dominance of two-dimensional features in the structure of this material is the possibility of modeling not only the electronic structure from a two-dimensional perspective, but the phase diagrammatics as well. Because of the predominant planar character of the oxygen part of the atomic-level geometry, we follow de Fontaine *et al.* (1987) in considering a two-dimensional model of this system. The cluster expansion used in the consideration of this system is known as the asymmetric next-nearest-neighbor

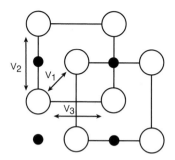

Fig. 6.28. Schematic illustrating the various effective interactions that are invoked in order to capture the energetics of oxygen ordering (adapted from Ceder *et al.* (1991)). Open circles correspond to O atoms and filled circles are Cu atoms.

Ising model (ASYNNNI) and may be written explicitly as

$$H = -h \sum_i \sigma_i + V_1 \sum_{nn} \sigma_i \sigma_j + V_2 \sum_{2nd} \sigma_i \sigma_j + V_3 \sum_{2nd} \sigma_i \sigma_j. \tag{6.45}$$

Note that the two different sums over second neighbors refer to the two classes of second neighbor interactions, both those mediated by copper intermediaries (i.e. V_2) and those with no intermediate copper atom. The significance of the various terms is shown in fig. 6.28. The motivation behind this particular choice of cluster expansion is the recognition that the phase diagram for this generic class of model is rich enough to account for the structural diversity noted above. A standard tool for informing such models (i.e. determining the coefficients) is to make quantum mechanical calculations on several representative structures and to use the resulting energies as the basis of a fit for the parameters. In the present setting, this program was carried out through an appeal to calculations using the LMTO method, with the result that the parameters are $V_1 = -6.71$ mRy, $V_2 = -5.58$ mRy and $V_3 = 0.73$ mRy (Liu *et al.* 1995). Note that there is nothing sacred about the parameter set we present here and that other calculations have been done with different outcomes. On the other hand, the most notable comment to be made on this and the other sets of parameters determined from first principles is that they satisfy certain inequalities.

A number of different approaches have been put forth in order to investigate the phase diagram for oxygen ordering in light of the class of ASYNNNI models presented in eqn (6.45). One class of investigations has been built around examining the consequences of this model for generic values of the parameters $x = V_2/V_1$ and $y = V_3/V_1$. The analysis of the thermodynamics of this model has been carried out using the cluster variation method, Monte Carlo methods, transfer matrix methods and others. In addition to the use of model parameters, first-principles parameters (including the set presented above) have also been used

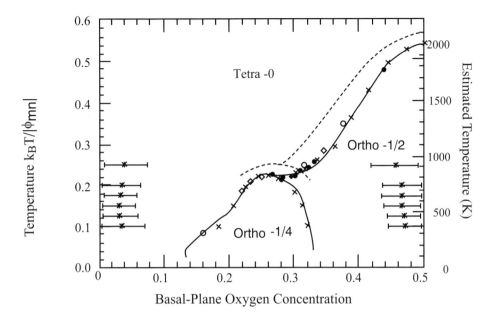

Fig. 6.29. Phase diagram for oxygen ordering in $YBa_2Cu_3O_z$ as obtained using Monte Carlo calculations (adapted from Aukrust *et al.* (1990)). Tetra refers to the disordered tetragonal phase, while ortho-1/4 and ortho-1/2 are ordered orthorhombic phases. The filled circles are experimental data points, and diamonds and crosses correspond to different Monte Carlo schemes. The dotted lines correspond to cluster variation calculations of the same phase diagram.

as the basis for the analysis of the phase diagram. We note that we have already presented the outcome of an analysis of this phase diagram using the cluster variation method in fig. 1.7. To illustrate the outcome of several other approaches to the same problem, figs. 6.29 and 6.30 show the results of an analysis of the phase diagram using Monte Carlo techniques and the transfer matrix, respectively.

There is much to be admired in the analyses presented here. In my opinion, this serves as a shining example of many different elements in the quest to build effective theories of material response. First, it is to be noted again that the key structural features of these complex materials have been mapped onto a corresponding two-dimensional model, itself already an intriguing and useful idea. In addition, the construction of the cluster expansion illustrates the way in which first-principles calculations for a relatively complex material can be used to inform a more tractable (and enlightening) model. Finally, the tools of statistical mechanics can be used to explore the phase diagrams of these systems which helps explain observed features of these problems but also suggests further complexity as can be seen in fig. 3 of de Fontaine *et al.* (1990).

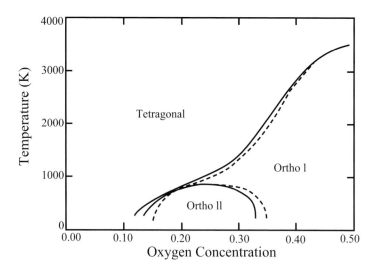

Fig. 6.30. Phase diagram for oxygen as obtained using the transfer matrix method in conjunction with effective Hamiltonian parameters from first-principles calculations (adapted from Liu *et al.* (1995)).

6.6 Summary

This chapter has shown how the zero-temperature analyses presented earlier in the book may be extended to incorporate finite-temperature effects. By advancing the harmonic approximation we have been able to construct classical and quantum mechanical models of thermal vibrations that are tractable. These models have been used in the present chapter to examine simple models of both the specific heat and thermal expansion. In later chapters we will see how these same concepts emerge in the setting of diffusion in solids and the description of the vibrational entropies that lead to an important class of structural phase transformations.

6.7 Further Reading

Thermodynamics of Crystals by D. C. Wallace, Dover Publications, Inc., Mineola: New York, 1998. This book has already appeared in the further reading list for chap. 5, but it is pertinent to the subject of the present chapter as well, especially his chap. 4.

Thermophysical Properties of Materials by G. Grimvall, North-Holland Publishing Company, Amsterdam: The Netherlands, 1986. Grimvall's book is full of insights into the vibrational and electronic contributions to the entropy, in addition to much else.

Order and Phase Stability in Alloys by F. Ducastelle, North Holland, Amsterdam:

The Netherlands, 1991. Ducastelle's book is full of insights into the tools needed to compute phase diagrams, including a very detailed treatment of the Ising model and its relation to phase diagrams. I wish I had more time to spend reading this one.

From Hamiltonians to Phase Diagrams by J. Hafner, Springer-Verlag, Berlin: Germany, 1987. An excellent account of the microscopic tools that are required in constructing phase diagrams on the basis of microscopic analysis. Special emphasis is given to the case of metals and alloys.

'Configurational Thermodynamics of Solid Solutions' by D. de Fontaine, in *Solid State Physics* edited by H. Ehrenreich and D. Turnbull, Academic Press, San Diego: California, 1979, Vol. 34, pg. 73.

'The Cluster Variation Method and Some Applications' by A. Finel in *Statics and Dynamics of Alloy Phase Transformations* edited by P. E. A. Turchi and A. Gonis, Plenum Press, New York: New York, 1994. This article gives an enlightening description of the cluster variation method.

Phase Equilibria, Phase Diagrams and Phase Transformations by M. Hillert, Cambridge University Press, Cambridge: England, 1998. The present chapter has neglected enormous amounts of knowledge about phase diagrams constructed on the basis of macroscopic thermodynamics. Hillert's book provides a thorough and insightful treatment of phase diagrams from this perspective.

Materials Fundamentals of Molecular Beam Epitaxy by J. Y. Tsao, Academic Press, Inc., San Diego: California, 1993. Tsao's book has the best pedagogical treatment of the cluster variation method that I have seen anywhere. In fact, the title provides a false impression since this book is full of deep insights into many *generic* issues from materials science.

6.8 Problems

---

1 Stillinger–Weber Version of the Yin–Cohen Curve

Use the Stillinger–Weber potential to determine the energy vs volume curves for Si in the diamond cubic, simple cubic, fcc, bcc and β-Sn structures.

---

2 Rectangular Density of States Model for Electronic Entropy

In this problem we imagine two competing structures, both characterized by electronic densities of states that are of the rectangular-band-type. In particular, the electronic density of states is given by

$$\rho(E) = \begin{cases} \dfrac{1}{w} & \text{if } 0 < E < w \\ 0 & \text{otherwise} \end{cases}, \tag{6.46}$$

where w is the width of the density of states and is w_1 or w_2 for structures 1 and 2, respectively. Compute the electronic entropy of each in terms of the bandwidth and evaluate the difference in electronic entropy of the two phases as a function of temperature. In addition, compute the electronic contribution to the free energy and use the results to plot the electronic contribution to the free energy difference of the two phases.

3 Free Electron Model for Electronic Entropy

Using the free electron model of chap. 3, estimate the entropy of metals. Begin by obtaining an expression for the electronic density of states for the three-dimensional electron gas. Then, make a low-temperature expansion for the free energy of the electron gas. Use your results to derive eqn (6.12).

4 Rectangular Band Model for Structural Change

In this problem, elaborate on the Einstein model for structural change given in the chapter by replacing the delta function densities of states by rectangular densities of states. Assume that the phase with the lower internal energy has the broader vibrational density of states. Compute the transformation temperature as a function of the difference in the widths of the two rectangular bands.

5 Debye Model of Vibrational Entropy

Consider two different structures, each with a vibrational density of states of the type employed in the Debye model. Assuming that the densities of states are of the form $\rho_1(\omega) = \alpha_1\omega^2$ and $\rho_2(\omega) = \alpha_2\omega^2$. Compute the Debye frequency for each of these structures and the corresponding vibrational entropy and finally, evaluate the entropy difference between them.

6 High-Temperature Expansion for Vibrational Free Energy

Obtain a high-temperature expansion for the free energy due to thermal vibrations given in eqn (3.110).

7 Pair Correlations and Total Energies for Crystalline Solids

In describing the energy of a liquid, we exploited the result

$$E = \int g(r)V(r)d^3r. \tag{6.47}$$

In the present problem, write down the function $g(r)$ for a crystalline solid (say, an fcc crystal) and show that the expression for the total energy given above reduces to the the conventional pair potential expression.

8 Cluster Counting on the fcc Lattice

Our discussion of cluster expansions allowed us to determine the Ising parameters $J_0, \ldots, J_5$ using a knowledge of the energies of the structures shown in fig. 6.19. Deduce the entries in table 6.1 and use these results to deduce eqn (6.30).

9 Cluster Interactions for an fcc Lattice Using a Pair Potential Description of E_{tot}

Consider a cluster expansion only up to pair terms, and use your knowledge of the energy of pure A in the fcc structure, pure B in the fcc structure and AB in the L1$_0$ structure to deduce values of J_0, J_1 and J_2 for an alloy described by a near-neighbor pair potential. The interactions between atoms are of the form V_{AA} when the near-neighbor atoms are both of type A, V_{AB} when the near-neighbor pair are of different types and V_{BB} when the neighboring atoms are both of type B. Express your results for the J coefficients in terms of the interactions V_{AA}, V_{AB} and V_{BB}.

10 Monte Carlo Analysis of Ordering on the fcc Lattice

The analysis of phase diagrams in systems such as Cu–Au has been undertaken from the perspective of Monte Carlo simulation. In this problem consider an fcc lattice characterized by an effective Hamiltonian of the form

$$H = J \sum_{ij} \sigma_i \sigma_j. \tag{6.48}$$

Perform a Monte Carlo simulation of this system at fixed composition for a series of different temperatures and determine the transition temperature. To determine the thermodynamic state of the system it is necessary to monitor the average 'spin' on each sublattice. For further details, see Ackermann *et al.* (1986), Diep *et al.* (1986) and Gahn (1986).

Part three

Geometric Structures in Solids: Defects and Microstructures

SEVEN

Point Defects in Solids

7.1 Point Defects and Material Response

This chapter is the first in a series that will make the case that many of the important features of real materials are dictated in large measure by the presence of defects. Whether one's interest is the electronic and optical behavior of semiconductors or the creep resistance of alloys at high temperatures, it is largely the nature of the defects that populate the material that will determine both its subsequent temporal evolution and response to external stimuli of all sorts (e.g. stresses, electric fields, etc.). For the most part, we will not undertake an analysis of the widespread electronic implications of such defects. Rather, our primary charter will be to investigate the ways in which point, line and wall defects impact the thermomechanical properties of materials.

Though we will later make the case (see chap. 11) that many properties of materials involve a complex interplay of point, line and wall defects, our initial foray into considering such defects will be carried out using dimensionality as the central organizing principle. In many instances, it is convenient to organize defects on the basis of their underlying dimensionality, with point defects characterized mathematically through their vanishing spatial extent, line defects such as dislocations sharing some of the mathematical features of space curves, and surfaces, grain boundaries and other wall defects being thought of as mathematical surfaces. Indeed, one of the interesting *physics* questions we will ask about such defects is the extent to which such dimensional classifications jibe with the properties of the defects themselves.

The condensed matter physicist has traditionally built up a picture of a solid which is measured against the perfect crystal of infinite extent. The structure is imagined as a monotonous continuation of perfectly repeating units. Far reaching theorems such as that of Bloch emanate from this periodicity and allow for the determination of the types of insights we constructed for the vibrational motions and electronic states of periodic solids in previous chapters. In this case, deviations from this regularity can be thought of perturbatively, with defects seen as disturbances in the reference configuration. To a materials scientist, on the other hand, microstructure is the handle whereby one manipulates a material to give it properties such as high strength or corrosion resistance. For our present purposes, microstructure represents the sum total of all deviations from the type of perfect crystallinity described above. In this setting, we have given broad compass to the notion of microstructure to emphasize just what we are up against. In broad terms, we know that most crystals are polycrystalline, with grains of many different orientations meeting to form an array of grain boundaries. In addition, a material may be riddled with precipitates that may also have a damning influence on its mechanical properties. At a lower level of description, any successful theory of a materials microstructure must account for the tangled web of dislocations that populate the grains themselves. Finally, at an even smaller scale, point defects form the backdrop against which the rest of these microstructural features play out.

The present chapter will undertake an analysis of point defects from both a microscopic and continuum perspective with the aim of understanding the factors that give rise to such defects and their motion within a solid, as well as how they affect the ultimate macroscopic properties of materials. The chapter begins with an overview of the phenomenology of point defects: how often are they encountered, what properties do they influence, what is their relation to diffusion? After introducing the broad range of issues encountered in thinking about point defects, our analysis will center primarily on a few key ideas. First, we will examine the factors that determine the equilibrium concentration of point defects. This analysis will demand an evaluation not only of the structure and energetics of isolated point defects, but also of the configurational entropy that arises when many such defects have the freedom to occupy different lattice sites. We will then undertake an analysis of the kinetics of point defect motion with an eye to a microscopic prediction of the diffusion coefficient. This section will be in keeping with our quest to determine material properties from appropriate microscopic analysis.

7.1.1 Material Properties Related to Point Disorder

Point Defects and Phase Diagrams. As will become more evident in subsequent parts of this chapter, substitutional impurities are one of the key types of point disorder. These defects correspond to foreign atoms that are taken into the lattice and which occupy sites normally reserved for the host atoms. For example, in the case of fcc Al some small fraction of the host lattice sites can be occupied by Cu

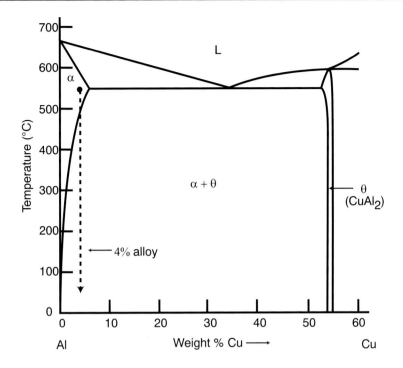

Fig. 7.1. Al-rich end of the Al–Cu phase diagram as an illustration of the concept of the solubility limit (adapted from Ashby and Jones (1986)).

atoms. As long as the concentration of such Cu atoms is not too large, the Cu atoms will assume positions in the underlying fcc structure without further ado. The extent to which such substitution is allowed is an important part of the alloy phase diagram. At a given temperature, only so many substitutional impurities may be incorporated into the host lattice, with the boundary separating full solubility from not known as the solubility limit. The specific properties of the Al–Cu system are shown in fig. 7.1. Note that on the Al-rich side of the phase diagram (i.e. the left hand side), there is a line sloping up and to the right that represents the solubility limit described above.

The solubility of foreign atoms within a given parent lattice is more than a subject of idle curiosity. The capacity of a lattice to take up such foreign atoms will reveal itself both in the propensity for subsequent microstructural evolution, as well as in partially determining the mechanisms for hardening within a given material. For example, in the case in which the foreign atoms are distributed randomly throughout the sites of the host lattice, these impurities hinder the motion of dislocations and are responsible for solid solution hardening. By way of contrast, if the concentration of impurities is too high, precipitation reactions can occur which have their own associated hardening mechanism.

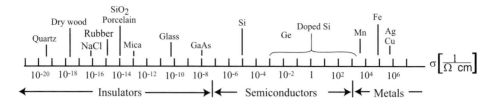

Fig. 7.2. Schematic of the range of room temperature electrical conductivities observed in a broad class of materials (adapted from Hummel (1998)).

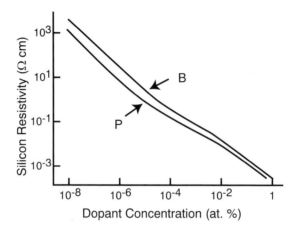

Fig. 7.3. Electrical resistivity of Si as a function of concentration of boron (B) and phosphorus (P) (adapted from McMahon and Graham (1994)).

Doping in Semiconductors. The electrical conductivity is one of the most diverse of material parameters. In passing from the best insulators to the best conductors (excluding the case of superconductivity), measured conductivities vary by more than 20 orders of magnitude, as illustrated in fig. 7.2.

As was remarked as early as chap. 1, many of the properties of materials are not intrinsic in the sense that they are highly dependent upon the structure, purity and history of the material in question. The electrical conductivity is one of the most striking examples of this truth. Nowhere is this more evident than in considering the role of doping in semiconductors. This is hinted at in fig. 7.2 where it is seen that the conductivity of Si ranges over more than 6 orders of magnitude as the concentration of impurities is varied. In fig. 7.3 this effect is illustrated concretely with the variation in resistivity of Si as a function of the concentration of impurities.

Though we will not undertake any systematic analysis of the electronic implications of the presence of boron (or other dopants) in Si, it is evident that questions surrounding the energetics of impurities in Si are an important part of the whole picture of defects in semiconductors. One question of obvious significance is that

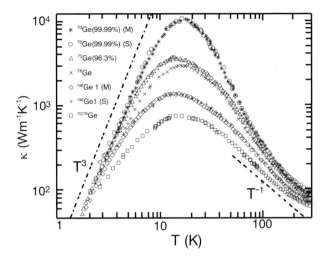

Fig. 7.4. Thermal conductivity for different isotopic compositions (taken from Asen-Palmer *et al.* (1997)).

of the diffusivity of such defects in the host Si lattice since the rate of such diffusion will largely govern the stability of the microstructure. From the standpoint of modeling, the additional subtlety that attends the presence of point defects of this type in Si is that their electronic structure is entirely coupled to their energetics.

Thermal Conductivity. Another example of the way in which point disorder can alter the underlying properties of materials is seen in the case of the thermal conductivity. By virtue of the presence of more than a single isotope of the same underlying chemical element, it is found that the thermal conductivity can be altered significantly as indicated in fig. 7.4. Our discussion in chap. 5 made it clear that the thermal conductivity is tied in part to the vibrations that characterize the solid. In the present setting, it is clear that the presence of a foreign atom (characterized by a mass that is different than that of the atoms of the host lattice) will alter the vibrational spectrum of the material, and thereby the thermal conductivity. In particular, when referenced to the normal modes of the perfect crystal, the point defects act as scattering centers that couple these modes. The resulting scattering of phonons degrades the thermal current and thus alters the thermal conductivity.

Internal Friction. The application of a stress to a material containing point defects leads to measurable consequences that differ from those of an imaginary solid which is free of such defects. In particular, through the application of an alternating stress field, it is possible to induce microscopic rearrangements involving such

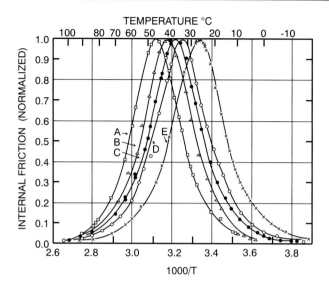

Fig. 7.5. Internal friction due to substitutional impurities in Fe (adapted from A. S. Nowick and B. S. Berry (1972)).

point defects. Stated differently, the coupling of acoustic waves to the motions of point defects is one of many dissipative mechanisms that are found in solids. These microscopic rearrangements are typified by a characteristic signature when examining the acoustic absorption as a function of the driving frequency or the temperature. An example of the type of behavior suggested above is shown in fig. 7.5.

Hardening in Metals. As was already made evident in our consideration of electrical conduction in semiconductors, doping a material with chemical impurities is one of the primary ways in which the materials scientist can exercise control over the properties of that material. This observation is as relevant for mechanical properties as it is for those tied to the electronic structure of a material. The incorporation of foreign atoms in metals leads to two of the key hardening mechanisms observed in such materials, namely, solid solution and precipitate hardening. An example of the phenomenology of hardening due to impurities is given in fig. 7.6 in which we show the increase in the flow stress as the number of impurities is increased. As will be discussed in more detail later, the underlying mechanism is that dislocations are inhibited in their motion through the crystal because of the barriers to their motion provided by foreign atoms or precipitates. The ideas set forth in the present chapter serve as a backdrop for our eventual discussion of these hardening mechanisms.

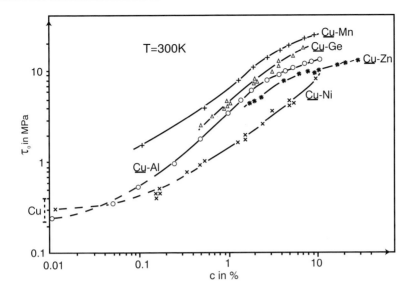

Fig. 7.6. Increase in flow stress as a function of the concentration of impurities illustrating solid solution hardening (adapted from Neuhäuser and Schwink (1993)).

Creep. Point defects are in many cases the lubricants for a range of more insidious and subtle mechanisms for permanent deformation than the glide of dislocations that characterizes simple yielding. The qualitative insights afforded by deformation mechanism maps make them a good place to start in our quest to unearth one of the roles of point defects in the mechanics of materials. In fig. 7.7, we show a deformation mechanism map for Ni. Such maps attempt to convey the dominant mechanism mediating plastic deformation at a given stress and temperature. One of the key observations to be gleaned from the figure is the onset of permanent deformation at stresses well below those necessary to commence conventional plastic deformation (i.e. that due to the motion of dislocations). The stress to induce conventional plastic deformation via dislocation motion begins at a stress on the order of 0.01μ, where μ is the shear modulus. On the other hand, it is clear that there are a host of mechanisms in which plastic deformation takes place at stresses well below this level. In one way or another, all of these mechanisms involve mass transport which is mediated by the presence of point defects such as vacancies. The creep mechanisms evident in the map are related to processes such as dislocation climb or mass transport along grain boundaries that is biased by the presence of a stress and leads to a macroscopic change of shape. The evidence mounted above in the context of deformation mechanism maps provides ample motivation for the need to give more than a passing eye to point defects and their motion, even in the context of the mechanics of materials.

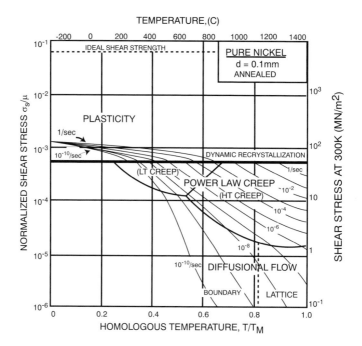

Fig. 7.7. Deformation mechanism map for Ni (adapted from Frost and Ashby (1982)).

7.2 Diffusion

In the previous section, we put forth evidence for the claim that the presence of point defects can completely alter the observed properties of materials. However, this discussion was set forth without entering into the question of how materials are brought to a state in which there is a given distribution of point defects. Many processes in materials are mediated by diffusion in which mass is transported from one part of the material to another. When viewed from the atomic scale, these diffusive processes may be seen as a conspiratorial effect resulting from the repeated microscopic hopping of particles over time. Our aim in the present section is to examine several different perspectives on diffusion, and to illustrate how a 'fundamental solution' may be constructed for diffusion problems in much the same way we constructed the fundamental solution for elasticity problems in section 2.5.2.

7.2.1 Effective Theories of Diffusion

Ultimately, diffusion is an atomic-level process characterized by the jiggling motions of atoms within their local energy wells, now and then punctuated by substantive motions from one well on the potential energy surface to another. The sum total of these motions constitutes what we observe macroscopically as diffusion. Part of the ambition around which much of the work in this book is built is to construct effective theories in which complex atomic-level processes are replaced by continuum surrogates that obviate the need to consider each and every atomic-level degree of freedom. The problem of diffusion has been considered using a number of different effective theories, some discrete and others continuous. In this subsection, we consider several such theories.

The Random Walk. The most compelling discrete effective theory of diffusion is that provided by the random walk model. This picture of diffusion is built around nothing more than the idea that the diffusing entities of interest exercise a series of uncorrelated hops. The key analytic properties of this process can be exposed without too much difficulty and will serve as the basis of an interesting comparison with the Fourier methods we will undertake in the context of the diffusion equation.

If we are to consider the accumulated excursion made by a random walker in a succession of N hops the ith of which is characterized by the hop vector $\mathbf{r}_i$, this excursion may be written as the vector sum

$$\mathbf{r} = \sum_{i=1}^{N} \mathbf{r}_i. \tag{7.1}$$

We imagine that each hop is of a magnitude a (i.e. the lattice parameter) and for the purposes of simplicity, will further constrain our analysis to a simple cubic lattice in three dimensions. We are interested in the mean excursion made by the walker in N steps, namely,

$$\langle \mathbf{r} \rangle = \left\langle \sum_{i=1}^{N} \mathbf{r}_i \right\rangle. \tag{7.2}$$

As was noted above, each hop is assumed to be independent of its predecessor and is just as likely to be in the forward direction as it is along the reverse direction and hence $\langle \mathbf{r}_i \rangle = 0$. For the three-dimensional simple cubic lattice (spanned by the three basis vectors $\{\mathbf{e}_1, \mathbf{e}_2, \mathbf{e}_3\}$) of interest here, what we are claiming is that each of the six possible hops that may emanate from a given site has an equal probability of $1/6$. Hence, the average $\langle \mathbf{r}_i \rangle$ is given by $\frac{1}{6}(\mathbf{e}_1 - \mathbf{e}_1 + \mathbf{e}_2 - \mathbf{e}_2 + \mathbf{e}_3 - \mathbf{e}_3) = 0$. As a result, we must seek a measure of the mean excursion which is indifferent to the forward–backward symmetry. In particular, we note that the mean-square displacement

$$\langle r^2 \rangle = \sum_i \langle \mathbf{r}_i \cdot \mathbf{r}_i \rangle + \sum_{i \neq j} \langle \mathbf{r}_i \cdot \mathbf{r}_j \rangle \tag{7.3}$$

only measures the mean-square distance the walker has advanced from the origin without any concern for direction. Since $\mathbf{r}_i$ and $\mathbf{r}_j$ are uncorrelated, the second term vanishes. Further, because we have confined our attention to walks on a lattice with lattice parameter a, each step satisfies $\langle \mathbf{r}_i \cdot \mathbf{r}_i \rangle = a^2$, with the result that $\langle r^2 \rangle = Na^2$. Essentially, what we have done is to compute the second moment of the probability distribution for a diffusing species given the certainty that it began its life at the origin. What more can be said about the nature of the probability distribution itself?

For simplicity, we now contract the discussion so as to only consider the one-dimensional version of the arguments put forth above. In addition, we maintain the assumption that the probability of steps to the left and right are equal (i.e. $p_{left} = p_{right} = 1/2$). The question we pose is what is the probability that in a total of N steps the particle will have suffered a net wandering of m steps to the right. The probability of a given sequence of N steps in which there were n_r steps to the right and $N - n_r$ to the left is $\frac{1}{2}^N$. Note that the total travel distance to the right is $d = ma$, where $m = n_r - n_l = 2n_r - N$. Since we are indifferent to the particular way in which this net excursion took place, we must add up the contributions from all sequences of hops that result in a net number of m steps to the right. The total number of ways of accomplishing this net motion is

$$\text{number of equivalent hopping sequences} = \frac{N!}{\left(\dfrac{N+m}{2}\right)!\left(\dfrac{N-m}{2}\right)!} \qquad (7.4)$$

with the result that the probability distribution for a net motion of m steps to the right given a total of N hops is

$$P(m, N) = \frac{N!}{\left(\dfrac{N+m}{2}\right)!\left(\dfrac{N-m}{2}\right)!}\left(\dfrac{1}{2}\right)^N. \qquad (7.5)$$

We note that this distribution is symmetrically disposed about the origin (i.e. $P(m, N) = P(-m, N)$) in keeping with the equal likelihood of hops to the left and right. Though the formal connection between this distribution and $\langle x^2 \rangle$ can be examined at this point, we find it more convenient to explore the large-N behavior of this result. Our reason for pursuing this point is for the insight it provides when compared to the Green function that will later emerge in the context of the continuum linear diffusion equation.

The starting point for our discussion of the large-N behavior of this distribution is Stirling's expansion which we invoke here in order to examine $\log P(m, N)$. In particular, we use the Stirling formula in a slightly amended form relative to that of chap. 3, namely,

$$\log N! \approx \left(N + \frac{1}{2} \right) \log N - N + \frac{1}{2} \log 2\pi. \tag{7.6}$$

If we now evaluate $\log P(m, N)$, using the Stirling formula in the form given above, and using the expansion

$$\log \left(1 \pm \frac{m}{N} \right) \approx \pm \frac{m}{N} - \frac{m^2}{2N^2}, \tag{7.7}$$

on the basis of the assumption that $m \ll N$, then we find that

$$\log P(m, N) \approx -\frac{1}{2} \log N - \frac{1}{2} \log 2\pi + \log 2 - \frac{m^2}{2N}. \tag{7.8}$$

This statement is equivalent to

$$P(m, N) = \sqrt{\frac{2}{\pi N}} e^{-m^2/2N}, \tag{7.9}$$

which is the sought after probability distribution.

We may now reinterpret this result using the more familiar notions of space and time by recognizing that the total distance traveled by the walker is $x = ma$, while the total elapsed time is given by $t = N/\Gamma$, where we should think of Γ as the rate (i.e. number of hops per second) at which the walker makes hops from one site to the next. In light of these arguments, eqn (7.9) may be rewritten by noting that

$$P(x, t)\Delta x = P(m, N)\frac{\Delta x}{2a}, \tag{7.10}$$

where the factor $\Delta x/2a$ comes from the recognition that m can either be even (if N is even) or odd (if N is odd), and in either case this means that the distance between successive hops is measured in units of $2a$. Hence, we have found that the continuum limit of our random walker model implies a probability distribution for a walker starting at the origin at time $t = 0$ of the form

$$P(x, t) = \frac{1}{\sqrt{2\pi \Gamma a^2 t}} e^{-x^2/2\Gamma a^2 t}. \tag{7.11}$$

We will later recognize this as the kernel for the continuum linear diffusion equation once we make the observation that the jump rate and the diffusion constant are related by $D = \frac{1}{2}\Gamma a^2$. For now we content ourselves by noting that the model of diffusion put forth here is invested with relatively little complexity. All we postulated was the uncorrelated hopping of the particles of interest. Despite the simplicity of the model, we have been led to relatively sophisticated predictions for what one might call the kinematics (which is statistical) of the diffusion field that results once the diffusive process has been set in motion. Indeed, the spreading

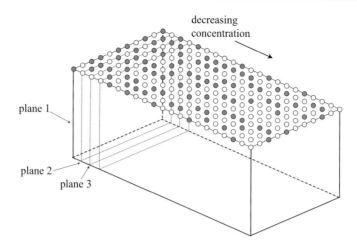

Fig. 7.8. Series of atomic planes illustrating one-dimensional concentration profile.

of the concentration field embodied in eqn (7.11) has significant implications for the processing of materials.

Master Equation Approach. The random walk arguments made above can be put on an alternative mathematical footing by writing down the rate of change of concentration directly. In particular, we adopt the so-called master equation approach in which the probability that the nth site is occupied at time t is characterized by a governing equation of the form

$$\frac{\partial P(n, t)}{\partial t} = \sum_{n'} [\Gamma_{n',n} P(n', t) - \Gamma_{n,n'} P(n, t)]. \tag{7.12}$$

As noted above, $P(n, t)$ is the probability that the nth site is occupied at time t, while $\Gamma_{n',n}$ is the rate at which jumps are made from the n'^{th} site to the nth site. The essence of this equation is the idea of accounting for the flow into and out of the nth site. In the context of diffusion, n labels different lattice sites, or in the simpler one-dimensional setting to be explored below, different lattice planes. The first term on the right hand side accounts for the accumulation of particles on site n as a result of jumps from the set of sites $\{n'\}$. The second term reflects a reduction in the net occupation of site n since particles are leaving this site and arriving at sites among the set $\{n'\}$.

As a concrete incarnation of this idea, imagine a geometry in which the diffusion is one-dimensional involving a series of equally spaced planes such as those shown in fig. 7.8. Each of these planes is characterized by an associated concentration $c(x, t)$, which gives the population of the diffusing species per unit volume. If we consider the plane with coordinate x, the statement of mass conservation is built around the idea that the net change in the population on this plane is given by the difference between the flow of diffusing species onto that plane from its neighbors, and the flux away from that plane of those atoms that initially occupied it. If we assign a jump probability $\Gamma_{i,i+1}$ to describe the likelihood that a particle in the ith plane (with position $x = ia$) will jump to the $(i+1)^{th}$ plane, the rate of change in the occupancy of the plane of interest may be written

$$\frac{\partial c(x,t)}{\partial t} = -\Gamma_{i,i+1}c(x,t) - \Gamma_{i,i-1}c(x,t) + \Gamma_{i+1,i}c(x+a,t) + \Gamma_{i-1,i}c(x-a,t).$$
(7.13)

In this expression, the first two terms on the right hand side represent the transfer of material from the plane of interest to its neighbors, while the second two terms reflect the influx of material onto the plane of interest.

For the purposes of simplification, we now assume that the jump rates are independent of position, which amounts to assuming that the potential energy surface describing the atomic motions is symmetrically disposed and that there are no applied fields. This assumption tells us that all of the jump rates may be replaced by $\Gamma/2$, where the factor of 2 derives from the fact that Γ is the total jump rate (i.e. the *total* number of atoms leaving a given site per unit time). If we now expand the concentrations $c(x \pm a, t)$ as

$$c(x \pm a, t) = c(x,t) \pm \frac{\partial c}{\partial x}a + \frac{1}{2}\frac{\partial^2 c}{\partial x^2}a^2,$$
(7.14)

we can rewrite the equation given above as

$$\frac{\partial c}{\partial t} = \frac{1}{2}\Gamma a^2 \frac{\partial^2 c}{\partial x^2}.$$
(7.15)

As will be seen below, we have recovered the diffusion equation.

Phenomenological Continuum Arguments. Though we have arrived at the one-dimensional diffusion equation on the basis of arguments concerning microscopic hopping, an alternative treatment is to consider the balance laws of continuum mechanics supplemented by constitutive assumptions connecting the mass current and the relevant gradient in chemical potential. The phenomenological starting point of most analyses of diffusion is the statement of mass conservation in conjunction with Fick's first law which states that the mass current (in fact, since we are using the concentration field, the current is the number of particles crossing unit area per unit time) is linear in the relevant compositional gradient. That is,

$$\mathbf{j}_{mass} = -D\nabla c,$$
(7.16)

where D is the appropriate diffusion coefficient and c gives the concentration of the diffusing species. We note that as was found in the case of linear elasticity and will be found again later, the introduction of a phenomenological statement such as that embodied in Fick's law requires the attendant introduction of a material parameter, the job of which is to mask the subscale processes which are not treated explicitly. In local form, the statement of mass conservation tells us that the rate of change of density and the current are related by

$$\frac{\partial c}{\partial t} + \nabla \cdot \mathbf{j}_{mass} = 0, \tag{7.17}$$

which in light of Fick's law and after evaluating the divergence of this expression yields the standard diffusion equation

$$\frac{\partial c}{\partial t} = D\nabla^2 c. \tag{7.18}$$

Note that we have assumed that D itself is not spatially varying.

As noted above, our continuum arguments concerning mass transport have resulted in the introduction of a new material parameter, namely, the diffusion constant. One interesting point of contact between the continuum arguments and the more microscopic arguments described in connection with the random walk and the master equation is the diffusion constant itself. In particular, these arguments provide a plausible connection between quantities of microscopic origin, namely, the atomic-level jump rates Γ and the macroscopic diffusion constant D. Comparison of the expressions that we have described thus far suggests that for the simple one-dimensional example under consideration here, the diffusion constant and the jump rate are related by $D = \frac{1}{2}\Gamma a^2$. Our discrete arguments using both the random walk and the master equation approach have both culminated in ideas familiar in the analysis of diffusion. Part of the significance of the developments we have made thus far is the light they shed on the connection between the discrete atomic-level processes that really characterize diffusion and the continuum arguments that are often more convenient.

In preparation for the exploitation of continuum models of diffusion in later chapters, we presently make a brief foray into solving the diffusion equation itself in the case in which the initial concentration is localized. As with our treatment of many linear elastic problems, the examination of diffusion is immediate if we first learn how to solve the problem for the case of a unit source (i.e. exploit Green functions). Our strategy for working out this solution will be to exploit Fourier transform methods. For simplicity, we begin with the one-dimensional problem. Our Fourier transform conventions are such that the transformed concentration, $\tilde{c}(k, t)$ is given by

$$\tilde{c}(k, t) = \int_{-\infty}^{\infty} c(x, t)e^{-ikx}dx. \tag{7.19}$$

If we Fourier transform the one-dimensional version of eqn (7.18), the resulting differential equation in the Fourier transformed concentration is

$$\frac{\partial \tilde{c}(k, t)}{\partial t} = -Dk^2 \tilde{c}(k, t).$$ (7.20)

The solution to this ordinary differential equation may be written down by separation of variables as

$$\tilde{c}(k, t) = \tilde{c}_0 e^{-Dk^2 t},$$ (7.21)

where the constant $\tilde{c}_0$ is to be determined on the basis of initial conditions.

Recall that our intention was to solve the diffusion problem for a unit source. What we have in mind in this case is an initial concentration profile of the form

$$c(x, 0) = \delta(x).$$ (7.22)

This represents an initial concentration field in which all of the diffusing species is concentrated at the origin. To find c_0 of eqn (7.21), we require $\tilde{c}(k, 0)$. If we Fourier tranform the initial concentration profile given in eqn (7.22), the result is $\tilde{c}(k, 0) = 1$. We are now prepared to invert the transformed concentration profile to find its real space representation given by

$$c(x, t) = \frac{1}{2\pi} \int_{-\infty}^{\infty} e^{-Dk^2 t} e^{ikx} dk.$$ (7.23)

This integral is easily evaluated by completing the square and the resulting real space concentration field is

$$c(x, t) = \frac{1}{2\sqrt{\pi Dt}} e^{-x^2/4Dt}.$$ (7.24)

Note that this solution yields exactly the same physics arrived at in eqn (7.11) as a result of our analysis of the random walk. In addition, the correspondence between the diffusion constant D and the jump rate Γ also becomes evident.

This solution is the basis of a wide range of insights into diffusion phenomena, several of which we will consider presently. One insight provided by our fundamental solution to the diffusion problem concerns the spreading of the initial unit source concentration field given by $c(x, 0) = \delta(x)$. The spreading of this profile is of interest since we will find that the result depends critically on the diffusion constant, the calculation of which will stand out as one of the primary missions of the remainder of the chapter. Our measure of the spreading of the concentration will be undertaken essentially through reference to the second moment, $\langle x^2 \rangle$. The perspective that we adopt now is that eqn (7.24) is the probability distribution for finding a particle at position x at time t. In light of this assertion, the width of the diffusion profile is given by

$$\langle x^2 \rangle = \int_{-\infty}^{\infty} x'^2 \frac{1}{2\sqrt{\pi Dt}} e^{-x'^2/4Dt} dx', \tag{7.25}$$

and yields the result

$$\langle x^2 \rangle = 2Dt. \tag{7.26}$$

We have essentially recovered the physics of the random walker as derived earlier where we noted that $\langle r^2 \rangle = Na^2$ which is equivalent to eqn (7.26) if we recall that $N = \Gamma t$ and that $D = \frac{1}{2}\Gamma a^2$ in the one-dimensional setting considered here.

The nature of the concentration profile resulting from some imposed boundary conditions can be of significance in the contemplation of material processing. For example, in the carburization process, a steel is exposed to an atmosphere which maintains a fixed concentration of carbon at the steel surface. Consequently, these carbon atoms bleed into the solid with a profile that is essentially dictated by the formula derived above.

A second insight offered by our determination of the fundamental solution for diffusion is afforded by exploiting the linearity of the governing equation (in this case the diffusion equation). In particular, we know that the sum of a number of solutions is still a solution. This observation provides the basis for constructing solutions to arbitrary initial concentration profiles, $n(x, 0) = c_0(x)$. In particular, the solution to this problem is given by

$$c(x, t) = \frac{1}{\sqrt{4\pi Dt}} \int_{-\infty}^{\infty} c_0(x') e^{-(x-x')^2/4Dt} dx'. \tag{7.27}$$

We note that this is one of many examples within the book wherein a 'fundamental solution' is derived and used as the basis of solution for more complex problems using superposition. This same viewpoint will rear its head again in the context of dislocations when we show that the solution for the single straight dislocation may be repeatedly superposed to construct extremely complex solutions such as those for cracks and dislocation pileups.

7.3 Geometries and Energies of Point Defects

As we have already remarked, point defects mediate an entire range of phenomena in materials that are especially important at high temperatures. The starting point of our investigation of such defects will center on the idea of understanding what determines their equilibrium disposition. A primitive model of the formation energy of a point defect can be constructed on the basis of linear elasticity theory. In this case, the point defect can be thought of as carrying different elastic properties than the parent lattice, either by virtue of a difference in elastic moduli,

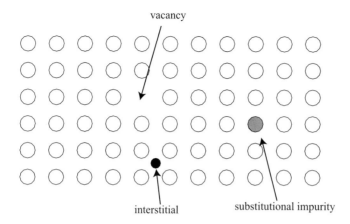

Fig. 7.9. Schematic illustrating a few of the key types of point defects.

or of size. Another conceptually simple approach is to attempt to glean microscopic insights by resorting to bond counting in which one tallies just how many bonds have been broken in the act of creating the point defect. Prior to embarking on these energetic questions, we must first settle even more fundamental geometric questions about the nature of point defects themselves.

7.3.1 Crystallographic Preliminaries

For the purposes of the discussion given here, we will only consider point defects in crystalline solids. Our main objective in this section is to provide the semantic backdrop for the remainder of the chapter. Our starting point is the perfect crystal since this is the reference state against which the defected crystal is measured. In fig. 7.9, we provide a visual catalog of some of the key types of point defects that can perturb the uninterrupted regularity of the perfect crystal.

We begin with an emphasis on three of the simplest point defects, namely, the vacancy, the interstitial and the substitional impurity. These defects are indicated schematically in fig. 7.9 where it is seen that a vacancy is nothing more than the failure of a given site to be occupied. Similarly, an interstitial refers to the presence of an atom at a position within the crystal that is normally unoccupied. In the case of interstitials, our intuition suggests that for close-packed lattices, only with foreign species in which the size of the foreign atoms is smaller than those of the host do we expect any appreciable fraction of interstitials. In those cases, in which there is more than a single chemical species present, it is also possible for the foreign atoms to occupy sites on the parent lattice, defects that we will denote as substitutional impurities, such as that of Cu in Al below the solubility limit as discussed in section 7.1.1 and illustrated in fig. 7.1. What we will learn in the pages that follow is how one may go about estimating the energy of formation of such defects, and in turn, what is their equilibrium concentration. We also note that our treatment of point defects is woefully incomplete since for reasons of space limitations, we have ignored other classes of point defects.

7.3.2 A Continuum Perspective on Point Defects

We begin by examining what continuum mechanics might tell us about the structure and energetics of point defects. In this context, the point defect is seen as an elastic disturbance in the otherwise unperturbed elastic continuum. The properties of this disturbance can be rather easily evaluated by treating the medium within the setting of isotropic linear elasticity. Once we have determined the fields of the point defect we may in turn evaluate its energy and thereby the thermodynamic likelihood of its existence.

The art in building a plausible elastic model of the point defect arises in determining the displacement fields that attend the presence of that defect. Different boundary conditions yield different solutions, each with its own particular virtues. For example, one might imagine the point defect as a spherical hole in which either the displacements or the pressure at the surface of the hole are specified. Alternatively, following the work of Eshelby (1975a), one could imagine a point defect as a misfitting sphere which is stuffed into a hole of the 'wrong' size and which induces an appropriate elastic relaxation. In any case, these models are just that – fictions, invented to provide understanding, and presently we will consider a few different possibilities.

We begin with the special case in which it is assumed that the displacements resulting from the presence of the point defect are spherically symmetric. We know that the fields must satisfy the Navier equations derived in section 2.4.2, namely

$$(\lambda + \mu)\nabla(\nabla \cdot \mathbf{u}) + \mu\nabla^2\mathbf{u} = 0. \tag{7.28}$$

Because of the presumed spherical symmetry of the solution, we expect the displacement field to have only a nonvanishing radial component, u_r which implies the need to consider only the equation in the radial degree of freedom which is given by

$$r^2\frac{d^2u_r}{dr^2} + 2r\frac{du_r}{dr} - 2u_r = 0. \tag{7.29}$$

The general solution to this problem is given by

$$u_r = ar + \frac{b}{r^2}. \tag{7.30}$$

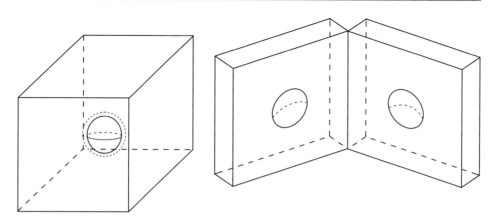

Fig. 7.10. Representation of point defect as spherical hole of radius a_0 with displacement δ. The left hand frame shows the spherical hole while the right hand one shows a view of the body after being cut down the middle.

In addition to satisfying the differential equation itself, there are boundary conditions that must be respected as well. Indeed, this is when the 'modeling' begins for it is through the application of these boundary conditions that we are forced to posit the elastic implications of such a defect.

Our analysis thus far has been built around the argument that a point defect may be thought of from the continuum perspective as a source of elastic distortion. We found ourselves (on the assumption of a spherically symmetric displacement field) able to write the solution to the equilibrium equations without much ado. We note that the business of solving the equilibrium equations themselves is in this case not a source of great difficulty. However, the real trick is figuring out how, with a judicious choice of boundary conditions, we may get our continuum model to mimic the presumed properties of point defects. For the moment, we idealize the point defect as a spherical hole of radius a_0 with prescribed displacement δ on the surface of that hole as indicated schematically in fig. 7.10. The parameter δ is a measure of the deviation of the size of the point defect from that associated with the underlying lattice. For a large substitutional impurity, δ will be positive, while for an impurity atom smaller than the host, or a vacancy itself, we might speculate that δ would be negative. In the far field (i.e. $r \to \infty$) it is assumed that the displacements vanish. These two conditions determine the constants a and b in eqn (7.30). In particular, if we insist that $u_r(r \to \infty) = 0$ and $u_r(a_0) = \delta$ then we find $a = 0$ and $b = \delta a_0^2$, resulting in the observation that the displacement fields for point defects decay as $1/r^2$.

In light of this solution, we may now compute the continuum estimate for the elastic energy stored in the displacement fields as a result of the presence of the point defect. This estimate will provide our first insight into the types of energies that arise for point defect formation. The energy is evaluated using

$$E_{strain} = \int_\Omega W(\epsilon) dV, \qquad (7.31)$$

where Ω is the volume extending from the surface $r = a_0$ to infinity, and $W(\epsilon)$ is the strain energy density. To make contact with two familiar elastic moduli, namely Young's modulus E and Poisson's ratio v, we write the strain energy density as

$$W = \frac{E}{2(1+v)} \left[\epsilon_{ij}\epsilon_{ij} + \frac{v}{1-2v} \operatorname{tr}(\epsilon)^2 \right]. \qquad (7.32)$$

For the problem solved above, we have

$$\epsilon_{rr} = -\frac{2b}{r^3} = -\frac{2\delta a_0^2}{r^3}, \qquad (7.33)$$

$$\epsilon_{\theta\theta} = \epsilon_{\phi\phi} = \frac{b}{r^3} = \frac{\delta a_0^2}{r^3} \qquad (7.34)$$

and hence

$$W = \frac{E}{2(1+v)} \frac{6\delta^2 a_0^4}{r^6}. \qquad (7.35)$$

Performing the integration suggested in eqn (7.31) results in

$$E_{strain} = \frac{4\pi E \delta^2 a_0}{1+v}, \qquad (7.36)$$

which is an estimate for the point defect formation energy for this particular choice of boundary conditions. In order to foreshadow results that will emerge later, we note that the continuum theory of the formation energy for point defects results in an energy that is bounded, even as the size of the system grows without bound, in contrast with what will be seen in conjunction with dislocations for which the energy of a single dislocation diverges as the system size increases.

For the choice of parameters relevant to a typical metal like Cu, we may estimate this energy as follows. Consider a relaxation due to the point defect of $\delta = 0.1$ Å. Assume that the spherical hole has radius $a_0 = 4.0$ Å and elastic moduli $v = 1/3$ and $E = 0.8$ eV/Å^3 (i.e. roughly 125 GPa). If we use these numbers in eqn (7.36), the resulting energy is $E_{strain} \approx 0.3$ eV. This simple model gives an estimate of the energy scale associated with the relaxations around a point defect. If we really wish to finesse the problem, we could also attempt to account for the bond breaking that attends the formation of the point defect by attributing an interfacial energy penalty for the interface separating the spherical hole from the rest of the body. This argument is taken up in the problems at the end of the chapter.

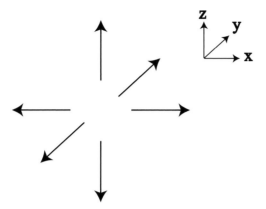

Fig. 7.11. Collection of force dipoles used to mimic the presence of a point defect as idealized within an elastic continuum.

An alternative treatment of the elastic consequences of the presence of point defects can be built up using the elastic Green function technology described in chap. 2. In this case, one imagines that the point defect elastic fields arise as the superposition of elastic dipoles of various strengths. Recall that the elastic Green function characterizes the displacements induced by a unit force. The physical idea in the present setting is to create three sets of force dipoles as shown in fig. 7.11 in order to mimic the elastic distortions induced by the point defect. In particular, if we assume that the point sources are placed at $\pm a/2$ along the three Cartesian directions, then the total displacement induced by such dipoles is given by using the superposition principle to obtain

$$\mathbf{u}(\mathbf{r}) = \sum_{n=1}^{3} \left[\mathbf{G}\left(\mathbf{r} - \frac{a}{2}\mathbf{e_n}\right) - \mathbf{G}\left(\mathbf{r} + \frac{a}{2}\mathbf{e_n}\right) \right] f_0 \mathbf{e_n}. \qquad (7.37)$$

The strength of the forces is measured by the parameter f_0. What we have said is that a collection of three force dipoles, one along each Cartesian direction suffices to characterize the types of displacements associated with a point defect. All that we have done is to add up the separate contribution to the displacement from each such dipole. If we now advance the argument that the separations of the elements in these dipoles are sufficiently small, then the displacement field may be written simply in terms of gradients in the elastic Green function. That is, the displacement from a force applied at a position slightly displaced from the origin can be found by Taylor expanding the Green function about the origin. In particular, the displacements are given by

$$u_i(\mathbf{r}) = -f_0 a G_{ij,j}(\mathbf{r}), \qquad (7.38)$$

which yields in Cartesian coordinates the expression

$$u_i = \frac{f_0 a}{4\pi (\lambda + 2\mu)} \frac{x_i}{r^3}.$$ (7.39)

If we rewrite this result in spherical coordinates, we recover the spherically symmetric displacement field deduced earlier which written in terms of the parameters used here takes the form

$$u_r = \frac{f_0 a}{4\pi (\lambda + 2\mu)} \frac{1}{r^2}.$$ (7.40)

The use of elastic Green functions has not as yet revealed a different mathematical result for the distortions about a point defect than that obtained by other means. However, the alternative interpretation afforded by this approach puts us in a position to be able to evaluate the displacements around defects with symmetry other than the simple spherical symmetry favored here. Thus far, we have interested ourselves in defects whose effects are completely isotropic. This description may be appropriate in some instances, but if we turn our attention to interstitial impurities in bcc materials, for example, with the canonical example being that of C in Fe, the resulting point defect geometries exhibit tetragonal symmetry. As we will see later, the symmetry of a given point defect can have deep significance for the resulting interactions between that point defect and dislocations. The particulars of the displacements associated with point defects of tetragonal symmetry are taken up in the problems at the end of the chapter.

The solutions that we have worked out above all teach us that from the standpoint of linear elasticity, there is a relaxation around the point defect that carries with it a certain energy penalty. The energy stored in the elastic fields associated with these defects may be reckoned in the usual way by integrating over the strain energy density, and yields the conclusion that typical point defect energies hover around the 1 eV scale. From the microscopic perspective, it is the rearrangements of atomic bonds that gives rise to point defect energies, and we take up such estimates in the following section.

7.3.3 *Microscopic Theories of Point Defects*

An alternative perspective on the subject of point defects to the continuum analysis advanced above is offered by atomic-level analysis. Perhaps the simplest microscopic model of point defect formation is that of the formation energy for vacancies within a pair potential description of the total energy. This calculation is revealing in two respects; first, it illustrates the conceptual basis for evaluating the vacancy formation energy, even within schemes that are energetically more accurate. Secondly, it reveals additional conceptual shortcomings associated with pair potential models. In the absence of a 'volume term', pair potential models imply a set of very restrictive conditions on the elastic moduli. In the present context, we will see that the pair potential also implies a very serious restriction on the form of the vacancy formation energy.

If a vacancy is created in an otherwise perfect lattice, the implication is that an additional atom must now occupy the surface of the crystal. The basic idea behind the vacancy formation energy is that it is a measure of the difference in energy between two states: one being the perfect crystal and the other being that in which an atom has been plucked from the bulk of the crystal and attached to the surface. From a computational perspective, the vacancy formation energy may be defined as

$$\epsilon_{vac}^f = E_{tot}(vacancy, N) - E_{tot}(perfect, N). \qquad (7.41)$$

This notation is meant to imply that we are interested in evaluating the difference in energy for an N-atom configuration without a defect and an N-atom configuration with the defect. From this point of view, the vacancy formation energy is perhaps not entirely well defined since different choices of binding site on the surface will result in different net changes in the total energy of the N atoms that make up the crystal. As an antidote to this conundrum, it is convenient to construct a theorist's artifice in which the vacancy formation energy is defined as the difference in energy between a periodic supercell with N atoms and a vacancy and the perfect crystal with N atoms.

From the standpoint of the simple pair potential analyses advocated in earlier chapters, we may evaluate this energy difference for the case of the *unrelaxed* vacancy as follows. For simplicity, consider an fcc crystal described via a near-neighbor pair potential model. In this case, the total energy of the N atoms in the perfect crystal is given by $E_{tot} = \frac{1}{2} \sum_{ij} \phi(R_{ij}) = 6N\phi(R_{nn})$. To compute the total energy of the crystal with a single vacancy we note that all but the twelve atoms adjacent to the vacancy see near-neighbor environments that are identical to those in the perfect crystal. For the twelve atoms neighboring the vacancy, a reduced coordination number of eleven is obtained. Thus,

$$E_{tot}(vacancy, N) = \frac{1}{2}[(N - 12)12\phi(R_{nn}) + 12(12 - 1)\phi(R_{nn})]. \qquad (7.42)$$

If we now form the difference between the energy of the perfect lattice and that of the defected lattice, we are left with

$$\epsilon_{vac}^f = -6\phi(R_{nn}). \qquad (7.43)$$

What is remarkable about this result is that we have found that the vacancy formation energy is with opposite sign equal to the cohesive energy per atom. An assessment of the accuracy of this conclusion is given in table 7.1. We note

Table 7.1. *Cohesive energies and measured vacancy formation energies (in eV) for several representative metals (adapted from Carlsson (1990)).*

Material	$|E_{coh}|$	ϵ_{vac}^f
V	5.31	2.1 ± 0.2
Nb	7.57	2.6 ± 0.3
W	8.90	4.0 ± 0.2

that in general the vacancy formation energy is but a fraction of the cohesive energy, suggesting that the pair potential lacks a significant element of realism in its treatment of point defects. One might object, however, that our failure to account for the relaxation of the atoms about the defect may be responsible for the discrepancy. Recall that from the continuum perspective explored earlier, our estimate for the vacancy formation energy was predicated entirely on such relaxation. It is clear that the relaxation will *reduce* the energy difference between the perfect crystal and the defected crystal. This question is easily examined within the context of the Morse potential described in chapters 4 and 5 and is taken up in the problems at the end of the chapter. Suffice it to note that, as we found in our continuum model, the energy scale associated with structural relaxations is on the order of a tenth of an electron volt rather than something like the full electron volt needed to reconcile the differences between the model and experiment.

We must look for a more fundamental explanation for the discrepancy between the computed and observed vacancy formation energies. As noted earlier, the key piece of new physics that arises as a consequence of adopting more sophisticated energetic descriptions such as are offered by pair functionals is that the bond strength is environmentally dependent. In the context of vacancies, this means that the bonds between the atoms adjacent to the vacancy and their neighbors are strengthened because they are missing part of their full complement of neighbors due to the presence of the vacancy. This feature of the pair functional treatment of the vacancy formation energy is shown in fig. 7.12.

We may again adopt a simple near-neighbor model to elucidate the fundamental ideas associated with a pair functional treatment of the vacancy. As noted earlier, within a pair functional description of the total energy, the energy of the perfect fcc crystal is given by

$$E_{tot} = 6N\phi(R_{nn}) + NF(12\rho_{nn}). \qquad (7.44)$$

In the absence of any relaxation, the total energy of the N-atom configuration in

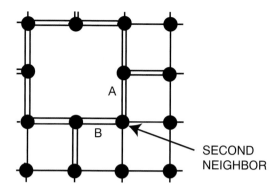

Fig. 7.12. Schematic illustrating the strengthening of bonds in the vicinity of a vacancy in a model square lattice (adapted from Carlsson (1990)). Double lines indicate bonds that are stronger than those in the bulk as a result of the reduced coordination of the atoms that form them.

the presence of the vacancy is given by

$$E_{tot} = 6(N-12)\phi(R_{nn}) + \frac{1}{2}12 \times 11\phi(R_{nn}) + (N-12)F(12\rho_{nn}) + 12F(11\rho_{nn}).$$
$$(7.45)$$

In this expression, we have made the simplifying assumption that the model is strictly reckoned on the basis of near-neighbor interactions. The outcome of the calculation is that the vacancy formation energy is given by

$$\epsilon_{vac}^{f} = -6\phi(R_{nn}) + 12[F(11\rho_{nn}) - F(12\rho_{nn})].$$
$$(7.46)$$

This result illustrates the amendment to the vacancy formation energy due to the embedding term. For Johnson's simple analytic model discussed in chap. 4, recall that the embedding function is given by

$$F(\rho) = -E_c\left(1 - \frac{\alpha}{\beta}\ln\frac{\rho}{\rho_0}\right)\left(\frac{\rho}{\rho_0}\right)^{\frac{\alpha}{\beta}} - 6\phi_0\left(\frac{\rho}{\rho_0}\right)^{\frac{\gamma}{\beta}}.$$
$$(7.47)$$

As a result, we find that within this model,

$$\epsilon_{vac}^{f} = 12E_c\left[1 - \left(1 - \frac{\alpha}{\beta}\ln\frac{11}{12}\right)\left(\frac{11}{12}\right)^{\frac{\alpha}{\beta}}\right] + 12\left[6\phi_0\left(1 - \left(\frac{11}{12}\right)^{\frac{\gamma}{\beta}}\right)\right] - 6\phi_0. \quad (7.48)$$

For the particular choice of parameters adopted by Johnson in his treatment of Cu, this result implies a vacancy formation energy of $\epsilon_{vac}^{f} \approx 1.3$ eV. Note that this result refers to the *unrelaxed* vacancy formation energy, and hence we should anticipate a further reduction in the energy to form such a vacancy once the atoms in the vicinity of the vacancy are allowed to adjust their positions.

Of course, the results given above are advanced not so much as the final word on the subject of point defect formation energies, but rather to illustrate the

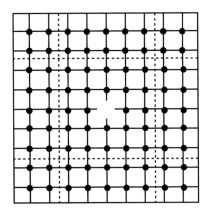

Fig. 7.13. Schematic illustrating the geometry of a supercell of the type used to evaluate the vacancy formation energy (adapted from Payne *et al.* (1992)). Dotted lines denote the limits of the computational cell which is then repeated periodically.

physical ideas that attend the examination of these defects. With the conceptual understanding in hand, we now turn to the description of point defect formation energies from the standpoint of first-principles calculation. Care must be taken with such calculations in the sense that we must concern ourselves with the question of whether or not the supercell used to make the calculation is large enough to preclude defect interactions which will contaminate the assessment of the formation energies. As a first example, we consider the case of vacancies in Al since this case has served as something of a proving ground for the evaluation of energetics of defects in metals.

Perusal of the literature on the subject of vacancies in Al (Gillan 1989, Mehl and Klein 1991, Benedek *et al.* 1992, Chetty *et al.* 1995, Turner *et al.* 1997) reveals some of the pitfalls encountered in the first-principles investigation of problems such as that of the zero-temperature vacancy formation energy. Two of the most immediate difficulties that must be owned up to are that of the system size and the level of refinement exploited in evaluating integrals over the Brillouin zone. A schematic example of such a supercell is shown in fig. 7.13. The key idea of the supercell approach is to construct an N-atom configuration in which a single atom is missing (i.e. the vacancy). This cell is then repeated periodically, as shown in the figure, so that the real calculation is that for a periodic array of vacancies. Once the strategy of adopting a periodic representation of the defected state has been taken, the full machinery associated with the existence of translational periodicity may be brought to bear on the problem.

Once the geometry has been identified, the calculation proceeds in much the same way described in earlier chapters. Depending upon the choice of basis (i.e. plane waves, localized orbitals, etc.) and how the potentials are represented (i.e.

pseudopotentials, full-potential, etc.), specific details of the implementation vary from one case to the next, but the basic idea is to compute the relevant energy eigenvalues. Once these eigenvalues are known, the internal energy of the system may then be computed. If the aim is to obtain the relaxed vacancy formation energy, then the forces on the various atoms within the supercell must be computed, and the atoms are then moved until an equilibrium state, in which the force on all atoms vanishes, is obtained. The work on vacancies in Al represented above all acknowledges the difficulties that must be faced, both in terms of system size and in terms of the size of the k-point set used to do the sampling. However, there seems to be reasonable accord on the point that the 'converged' vacancy formation energy (including relaxation) is on the order of 0.66 eV.

In earlier sections, we have already drawn attention to the issue of the relaxations that attend the presence of a point defect. We have already examined this question both from the perspective of elasticity theory and atomistics. In fig. 7.14 we show the computed internal relaxations in the neighborhood of the vacancy as obtained by Turner *et al.* (1997). We should reemphasize that despite the fact that these results were obtained using first-principles calculations, they should be seen as provisional in the sense that it is likely that with increased computational power, both larger system sizes and larger k-point sets will lead to a refinement in these conclusions, though perhaps not in the fundamental arguments used to obtain them. A particularly important feature of this figure is the recognition that at a distance of roughly 5.5 Å, a given atom finds itself just as close to the vacancy in question as it does to the vacancy in the next cell. This can be understood by noting that if our computational cell is cubic and of edge length L, then an atom at a distance of $L/2$ from the vacancy in the host cell has another vacancy as a neighbor at roughly the same distance. As a result, the full equilibrium (i.e. long-range) part of the relaxation cannot be present in this solution because a given atom is responding to the competing demands of the central vacancy and its periodic partners.

A final point of interest concerning the relaxations is the magnitude of the energy that they imply. Note that there are two contributions to the relaxation energy. First, there is a relaxation associated with the overall volume change, a feature that can be exploited to determine the concentration of such vacancies. This effect is captured in the calculations by allowing for the supercell lattice parameter to enter the calculation as a dynamical degree of freedom. However, as was noted above, there are also local atomic relaxations in the vicinity of the vacancy. Both of these effects together conspire to yield the overall relaxation energy. Recall from our discussion of pair potential approaches that we speculated that one possible source of the serious discrepancy between the measured vacancy formation energies and those obtained from pair potentials was the relaxation effect. We argued against this scenario on the grounds that the relaxation energies are much too small to

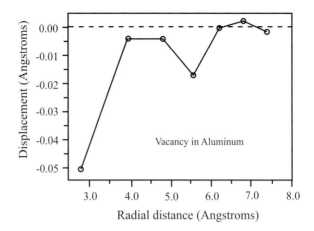

Fig. 7.14. Relaxations of atoms in various neighbor shells around a vacancy in Al (adapted from Turner *et al.* (1997)).

account for the large difference between the measured results and those emerging from pair potentials. This argument is borne out in the first-principles calculations in the sense that $\Delta E_{relax}/\epsilon_{vac}^{f} \approx 0.14$, revealing that the relaxation energy is a minor fraction of the overall vacancy formation energy.

A more meaningful analysis of the energetics of point defects in metals is afforded by a systematic examination of the trends in the formation energies for these defects from one material to the next. As will become evident as our analysis of various types of defects unfolds, the group of Skriver has repeatedly carried out systematic evaluations of the energies of different defects for a number of different materials. In addition, these analyses are carried out using basically the same method (first-principles LMTO method) from one calculation to the next, making an honest assessment of the trends more likely. In the context of the vacancy formation energy, their results for all three transition metal series are presented in fig. 7.15. These calculations were carried out without allowing for local relaxations of the atoms in the vicinity of the vacancy. In addition to the theoretical values, the figure also reveals the experimental data for the same set of metals, though it is clear from the figure that there is significant scatter for a given element from one experiment to the next. The most reliable data are those associated with the open squares, and for which the calculations are in quite good accord with the experiments.

A second key observation associated with the trends presented in fig. 7.15 is the variation in the vacancy formation energy as a function of band filling (i.e. the filling up of the transition metal d-band in passing from the left to right of each figure). In particular, there is the same trend in the vacancy formation energy

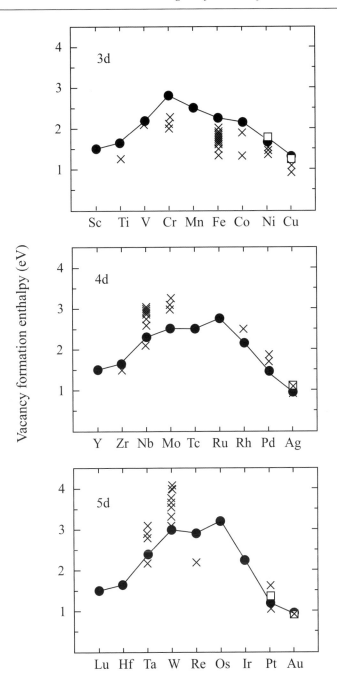

Fig. 7.15. Vacancy formation energies for a series of elements in both the fcc and bcc structures (adapted from Korzhavyi *et al.* (1999)).

that was noted earlier (see fig. 4.18) with respect to the energy of cohesion across the transition metal series. Indeed, a noted rule of thumb is the assertion that the vacancy formation energy and the cohesive energy (when reckoned on a per atom basis) are related by

$$\epsilon^f_{vac} \approx -\frac{1}{3}|E_{coh}|. \tag{7.49}$$

A knowledge of the point defect formation energies constitutes only a partial victory over the question of observed point defect concentrations. It is well known that in addition to the energetic contribution to the free energy, we must similarly examine the entropic origins of equilibrium defect configurations. As has been noted previously, configurational entropy terms may be accounted for (to a first approximation) by the simple entropy of mixing discussed in section 3.3.2. If we further assume that the presence of the point defect amends the vibrational state of the crystal, there is clearly an additional term in the entropy that accounts for this vibrational entropy change.

Now that we have seen how in principle the free energies of point defects may be determined microscopically, our task is to find what concentration of point defects minimizes the free energy. As with many such exercises, this minimization reflects a competition between the energy penalty associated with increasing the defect population and the entropy gain that attends it. To the level of approximation at which we are working, the total free energy per atom as a function of the defect concentration is given by

$$F(c) = c(\epsilon^f_{vac} - T\Delta S_{vib}) + kT[c\ln c + (1-c)\ln(1-c)]. \tag{7.50}$$

This expression is predicated upon a few key approximations. First, we assert that the energy to form a vacancy is given by ϵ^f_{vac} and takes no account of the possible alterations in this energy that would be present if interactions between the vacancies were considered. This amounts to a low-concentration approximation, and we will see that for low temperatures, this approximation is borne out by the results. The quantity ΔS_{vib} corresponds to the change in the entropy of vibration as a result of the presence of the vacancy. In addition, the third term is the configurational entropy associated with the presence of the vacancies. Our treatment of this term assumes that there is no correlation in the positions of the vacancies, again a reflection of the presumed diluteness of the vacancies.

By differentiating the previous expression with respect to the as-yet undetermined concentration (i.e. $\partial F/\partial c = 0$), we are left with an algebraic equation in the equilibrium concentration which may be solved analytically. The resulting equilibrium concentration is given by

$$\frac{c}{1-c} = e^{-(\epsilon^f_{vac} - T\Delta S_{vib})/kT}, \tag{7.51}$$

which in the limit of low concentrations reduces to the more familiar form

$$c = e^{\Delta S_{vib}/k} e^{-\epsilon_{vac}^f/kT}. \tag{7.52}$$

Note that via the potent combination of microscopic evaluation of relevant material parameters such as the vacancy formation energy and statistical mechanical reasoning to treat the entropic effects of the presence of vacancies we have arrived at a prediction for the equilibrium concentration of point defects.

7.3.4 Point Defects in Si: A Case Study

One of our recurring themes has been the capacity to examine a particular question from a number of different energetic perspectives. In broadest terms, we have divided our theoretical perspectives along the lines of whether or not they reflect a vision of the defected solid as a continuum or as a discrete collection of atoms. However, as we have discussed in detail already, even within the picture of a crystalline solid as a collection of atoms, it is possible to consider widely different frameworks for the total energy. In the present section, we use Si as a case study in order to contrast empirical and first-principles approaches to point defects as was done already for metals.

Semiempirical descriptions of the energetics of point defects in Si. The energetics of covalent materials is notoriously tricky when viewed from the perspective of empirical potentials. As was already evident when we considered the litany of empirical approaches to the Yin–Cohen-type energy curves (see section 6.3.3) for Si and its various structural variants, there can be blatant disagreement as to the relative energies of competing structural alternatives. This subtlety extends beyond the relatively simple question of structural energies and becomes even more damning once we attempt to examine the energetics of defects.

For the present purposes, our plan is to examine several competing empirical alternatives for considering the energetics of defects in Si. Our primary emphasis will center on vacancies. From a structural perspective, there is nothing particularly novel in switching from the geometries of point defects in metals to those of covalent materials such as diamond-cubic semiconductors. On the other hand, the energetics of bond-breaking is considerably more subtle in this case and poses serious challenges for any viable empirical scheme. As was already made evident in chap. 4, there are any number of different schemes for computing the energetics of Si, and thereby, for determining the energetics of point defects. For the moment, we restrict our attention to the two classes of empirical schemes represented by the Stillinger–Weber potential and the Tersoff potentials and their many variants. As said above, to compute the energetics of a vacancy using these potentials is usually

Table 7.2. *Computed and experimental values of the vacancy formation energy in Si.*
Units are eV. Values obtained from (a) Maroudas and Brown (1993), (b) Ungar et al.
(1994), (c) Tang et al. (1997), (d) Mercer et al. (1998).

	Stillinger–Weber	Tersoff	Tight-Binding	LDA	Expt.
ϵ_{vac}^f	2.66[a]	3.70[b]	3.97[c]	3.63[d]	3.6 ± 0.2[d]

something that must be done numerically (at least if relaxations are to be included),
and involves the creation of a supercell and the evaluation of an expression like that
of eqn (7.41). Representative examples of the outcome of calculations of this type,
in addition to those from more sophisticated total energy schemes, are shown in
table 7.2.

Because of the difficulties that have already been noted with respect to the
transferability of empirical potentials for Si, there has been tremendous pressure
to seek alternative schemes which are at once more accurate and still represent a
reduction in the computational burden associated with first-principles calculations.
In the context of Si in particular, a powerful alternative is offered by tight-binding
analysis. Like in the case of the empirical potentials, there are a number of different
schemes and fits, each with its own particular patriots and warriors. On the other
hand, the basic ideas don't really differ from what has been spelled out in chaps. 3
and 4.

One popular parameterization has been brought to bear on the question of point
defects in Si by Wang *et al.* (1991). They use tight-binding molecular dynamics to
simultaneously relax the nuclear positions and to find the ground state electronic
configuration. An example of the equilibrium structure of the vacancy in Si
emerging from these calculations is shown in fig. 7.16. One immediate conclusion
of this analysis is the tetragonal nature of the symmetry breaking that attends these
atomic-level distortions. The computed vacancy formation energy associated with
their largest (512 atoms) computational cell is $\epsilon_{vac}^f = 4.12$ eV.

An interesting question that may be posed about the relaxations around a
vacancy is the extent to which they admit of description in terms of the ideas
of linear elasticity. In fig. 7.17, we show the magnitude of the displacements
of various shells of atoms around the vacancy in Si, again as computed using
tight-binding molecular dynamics. In the problems at the end of the chapter,
the reader is asked to use the results of eqn (7.30) to estimate the magnitude of
these displacements by fitting the first shell displacements to those obtained in the
calculation.

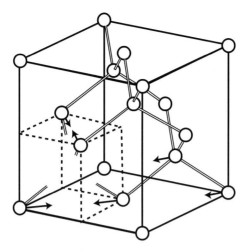

Fig. 7.16. Relaxations around a vacancy in Si (adapted from Wang *et al.* (1991)).

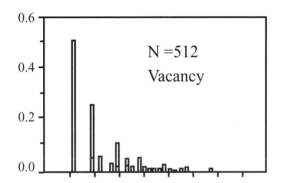

Fig. 7.17. Relaxation magnitudes around a vacancy in Si, plotted as a function of distance from the vacancy itself (adapted from Wang *et al.* (1991)).

First-Principles Descriptions of the Energetics of Point Defects in Si. Just as was possible with the question of vacancies in metals, similar first-principles calculations have been set forth for the energetics of defects in semiconductors. In the present case, we examine the outcome of analyses of the energetics of vacancies in these materials. From a structural perspective, the vacancy in Si appears to admit of two important geometric competitors, one with a tetragonal distortion, the other with a more symmetric tetrahedral distortion. Note that once again, the space of possible outcomes is searched over discretely. What this means in practice is that informed guessing must be set forth and each such guess is tried on a case by case basis. In this particular case, the two competing structures are found to lie within 20 meV of one another with the tetragonal distorted state, already discussed in terms of tight-binding calculations, slightly lower in energy. So as to collect all of our insights in one place for further inspection, table 7.2 shows the computed vacancy formation energy in Si for a few different classes of total energy scheme. It should be remarked that not only are the energies a challenge for empirical schemes, but in some instances even the symmetry of the relaxed defect structure is not correct.

Recall our earlier commentary of the type of work that has been done thus far on the question of the energetics of point defects in simple metals such as that of Al. We noted that one of the pitfalls of first-principles calculations is the difficulty of guaranteeing convergence of the results with respect to system size, energy cutoff and k-point sampling. From a philosophical perspective these issues arise as a result of the fact that those who make these calculations are almost always in the process of pushing their calculations to the extreme. Revisiting the Yin–Cohen curve described in chap. 6 in light of the huge computational advances that have taken place in the intervening years is hardly as exciting as trying to tackle new and unexplored areas. On the other hand, in the name of careful calculation, such calculations have much to recommend them.

Mercer and coworkers have systematically studied the convergence of the vacancy formation energy in Si using first-principles calculations. Indeed, in fig. 7.18 we show the convergence of both the relaxations around the vacancy and the vacancy energy with respect to system size. An additional insight into convergence questions is provided by fig. 7.19 which in addition to the dependence on system size shows the way in which the vacancy formation energy depends upon the k-space integrations required to compute the electronic energy of a periodic supercell. Our main reason for showing these terms is to provide graphical weight to the assertion that while studying convergence may be boring, it can spell the difference between results that are correct and those that are not.

7.4 Point Defect Motions

As yet, we have concerned ourselves with the static structure of point defects. On the other hand, part of what makes such defects so important is their mobility. In the present section, we undertake an analysis of the various schemes that have been set forth to model the motion of point defects in solids, though only from the perspective of classical physics without consideration of the role of quantum effects. In particular, it is the nature of diffusion that occupies us. However, whereas the plan of section 7.2 was the phenomenology of diffusion, the plan of the present section is to show how a combination of microscopic calculations of the total energy and arguments from statistical mechanics may be assembled to construct a general description of the rates of defect migration.

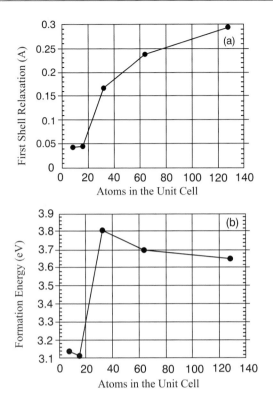

Fig. 7.18. Illustration of convergence with system size of both relaxation of the first shell of atoms in vicinity of vacancy and the vacancy formation energy itself (adapted from Mercer *et al.* (1998)).

7.4.1 *Material Parameters for Mass Transport*

As we have noted repeatedly, one of the features of effective theories is that they often exploit phenomenological parameters that capture the material specificity and reflect the subscale processes that are not treated explicitly in that effective theory. Our arguments concerning diffusion illustrate this thinking through the diffusion constant which is linked in turn to microscopic hopping. It is useful to construct a plausibility argument for the form adopted by Γ and thereby D. For concreteness, we consider the self-diffusion of atoms within a crystal via a vacancy mechanism in which the hopping of a vacancy from one site to the next can be interpreted alternatively as the motion of atoms themselves. What we will learn is that a proper description of the rate of microscopic hopping requires that we characterize the probability of occurrence of the point defects themselves, as well as the likelihood that the diffusing species will traverse the barriers separating adjacent wells. The conceptual picture is that the atom under consideration is bound to a particular potential energy well. This is indicated schematically in

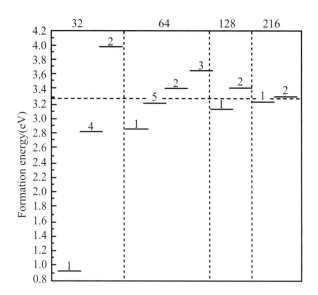

Fig. 7.19. Illustration of convergence of vacancy formation energy in Si with respect to both system size and k-point sampling scheme (adapted from Puska et $al.$ (1998)). In going from left to right the results are shown for supercells containing 32, 64, 128 and 216 atomic sites. Within each group the labels 1, 2, 4, etc. are a measure of the size of the set of k-points used to perform the Brillouin zone sampling.

fig. 7.20 which shows two adjacent wells and the saddle point separating them. As a result of thermal vibrations, the particles rattle about within the well of interest, and may occasionally drum up sufficient energy involving motions in the right direction to successfully traverse the saddle point configuration. The remainder of the chapter will adopt as one of its key objectives the elucidation of the qualitative arguments given here though it is clear that the differences from one material to the next are dictated both by geometric considerations emanating from the underlying crystallography as well as the energetics associated with atomic-level bonding.

7.4.2 Diffusion via Transition State Theory

Unlike the analysis of the structure and equilibrium concentrations of point defects, questions of defect motion take us directly to the heart of some of the most important unsolved questions in nonequilibrium physics. In particular, while there are a host of useful empirical constructs built around the idea of Arrhenius activated processes with rates determined generically by expressions of the form

$$D = D_0 e^{-E/kT},$$ (7.53)

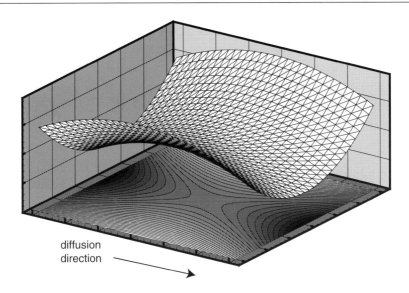

diffusion
direction

Fig. 7.20. Potential energy surface near the saddle point configuration. Contour plot shows lines of constant potential energy.

and similarly, there are well-defined theoretical ideas for thinking of nonequilibrium processes as weak perturbations about privileged equilibrium states, there is no convincing framework for treating generic questions about kinetics. As a fallback position, many treatments (ours included) resort to the use of transition state theory in which arguments are borrowed from equilibrium statistical mechanics which is used to describe the character not only of the various wells which diffusing species normally occupy, but the saddle points which they cross as well.

To illustrate the conceptual underpinnings of transition state theory, we begin with a simplified one-dimensional model to be followed later by the higher-dimensional analog. For ease of visualization, we imagine surface diffusion. Our picture is one of a perfectly flat geometry which is nothing more than the truncation of the underlying crystal on a particular plane which becomes the surface of interest. We then imagine the addition of a single atom on the surface of this crystal (the adatom), the diffusion of which it is our aim to study. We consider a crystal surface with a structure that is like a series of channels, and because of the corrugations in the surface due to the atomic-level geometry, the adatom is confined to diffuse only along these channels. We temporarily simplify our theoretical construction by assuming that the substrate is entirely rigid and serves only to provide the potential energy landscape upon which the diffusing atom surfs. The one-dimensional potential energy profile encountered by the diffusing atom is indicated schematically in fig. 7.21. We will adopt a local equilibrium hypothesis in which we assume that even when the diffusing particle is in the process of

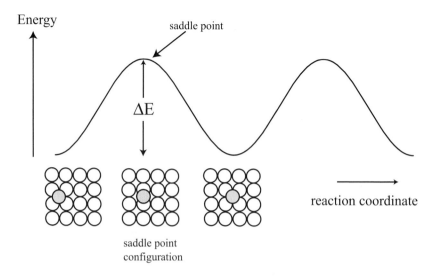

Fig. 7.21. One-dimensional potential energy profile for a diffusing atom. This schematic illustrates the type of energy that would be encountered by an adatom wandering on a *rigid* substrate.

traversing the saddle point, its statistical properties may be reckoned on the basis of the arguments that emanate from *equilibrium* statistical mechanics.

In anticipation of the work we will do quantitatively below, let's first work up a qualitative understanding of the process. We begin by imagining the diffusing atom to occupy one of the wells as depicted in fig. 7.21. By virtue of thermal excitation, it is known that the atom is vibrating about within that well. The question we pose is how likely is it that the diffusing atom will reach the saddle point between the two wells? The answer to that question allows us to determine the flux across the saddle point, and thereby, the rate of traversal of the saddle configuration which can then be tied to the diffusion constant.

The physical interpretation of these arguments may be summarized as follows. The traversal of the saddle point separating adjacent minima occurs by virtue of thermal fluctuations. The theoretical treatment of this phenomenon is to contract the full microscopic Hamiltonian, projecting out those parts involving the 'reaction coordinate'. The flux is computed by appealing to a hypothesis of local equilibrium which leads to a net rate that depends upon an activation energy (i.e. the energetic barrier height separating the different wells). Thus, if one is to make use of these ideas from the microscopic perspective, it is necessary to calculate two key quantities. First, we seek the activation energy which is the energy difference between the wells and the saddle point configuration. In addition, the curvature at the well of departure is required so as to obtain the prefactor in the transition rate expression.

Our one-dimensional example imitates the presentation given by Eshelby (1975a) of the Vineyard (1957) rate theory. The transition rate may be computed as the ratio of two quantities that may themselves be evaluated on the basis of notions familiar from equilibrium statistical mechanics. The numerator of the expression of interest is given by the total number of particles crossing the saddle per unit time, while the denominator reflects the number of particles available to make this transition in the well from which the particles depart. In particular, the transition rate is

$$\Gamma = \frac{\text{number crossing saddle point per unit time}}{\text{number in well}}. \tag{7.54}$$

In the context of fig. 7.21 if we consider the process of atomic hopping from left to right, the numerator requires us to evaluate properties at the saddle point, while the denominator demands an analysis of the properties of the well to the left of the saddle point.

Our intention is to invoke the tools of statistical mechanics developed earlier to quantify these ideas. From the standpoint of our one-dimensional example, the transition rate of eqn (7.54) may be rewritten mathematically as

$$\Gamma = \frac{\langle \delta(q)\dot{q}\theta(\dot{q}) \rangle}{\langle \theta(-q) \rangle}. \tag{7.55}$$

Despite the cumbersome notation, the idea is rather simple. The numerator expresses the average current passing across the saddle point, where the reaction coordinate has been identified with the variable q. The delta function ensures that the current is evaluated at the saddle point itself while the use of the θ function in the velocity $\dot{q}$ instructs us only to consider right moving particles. The use of the θ function in the denominator indicates that only those points to the left of the saddle point as indicated in fig. 7.20 can contribute particles to the right going current.

However, until the averages represented in the equation given above are effected explicitly, the treatment of the transition rate remains abstract. We know from statistical mechanics that to evaluate averages like those demanded above, we must sum over all allowed values of the quantity being averaged, with the contribution of each term modulated by a Boltzmann weight of the form $e^{-H(\langle p,q \rangle)/kT}$. For the present problem, the Hamiltonian is of the form

$$H(p, q) = \frac{p^2}{2m} + \phi(q), \tag{7.56}$$

where $\phi(q)$ reflects the profile of the energy landscape itself. The average called for in eqn (7.55) is

$$\Gamma = \frac{\int_{-\infty}^{\infty} dp \int_{-\infty}^{\infty} dq\, \delta(q)\dot{q}\theta(p)e^{-\beta H(p,q)}}{\int_{-\infty}^{\infty} dp \int_{-\infty}^{\infty} dq\, \theta(-q)e^{-\beta H(p,q)}}. \tag{7.57}$$

In writing the transition rate in this way, we have eliminated constant factors that are shared by numerator and denominator alike. If we now perform the instructions inherent in the presence of both the δ and θ functions, the rate may be written more concretely as

$$\Gamma = e^{-\beta\phi(q_s)} \frac{\int_0^\infty dp \frac{p}{m} e^{-\frac{\beta p^2}{2m}}}{\int_{-\infty}^\infty dp \int_{well} dq e^{-\beta H(p,q)}}. \tag{7.58}$$

We have introduced the notation q_s to signify the value of the coordinate q at the saddle point, and will similarly find it convenient to introduce q_w to denote the coordinate for the well under consideration. As yet, our statements are general and involve no approximations other than those present in transition state theory itself.

A standard approximation at this point is to note that in the vicinity of the well it is appropriate to represent the energy surface as a quadratic function in the variable q, and in addition it is asserted that one may make the transcription

$$\int_{-\infty}^\infty dp \int_{well} dq e^{-\beta(\frac{p^2}{2m}+\phi(q_w)+\frac{1}{2}\phi''(q_w)q^2)}$$
$$\rightarrow \int_{-\infty}^\infty dp \int_{-\infty}^\infty dq e^{-\beta(\frac{p^2}{2m}+\phi(q_w)+\frac{1}{2}\phi''(q_w)q^2)}. \tag{7.59}$$

The significance of this assertion is that we contend that by virtue of the rapid vanishing of the integrand even in those parts of the well over which the well is harmonic, there is no harm done in replacing the limits of integration such that all of space is considered as opposed to only the well itself. Once this approximation has been made, all remaining integrals are Gaussian integrals and may be handled analytically with the result that the numerator in eqn (7.58) is $kT e^{-\beta\phi(q_s)}$, while the denominator is $2\pi kT \sqrt{m/\phi''(q_w)}e^{-\beta\phi(q_w)}$. If we now define the frequency

$$\nu = \frac{1}{2\pi}\sqrt{\frac{\phi''(q_w)}{m}}, \tag{7.60}$$

interpreted as the frequency with which the adatom rattles about in the well (and sometimes called the 'attempt frequency'), then the the transition rate may be expressed finally as

$$\Gamma = \nu e^{-(\phi(q_s)-\phi(q_w))/kT}. \tag{7.61}$$

Note that if the kinetics of the process under consideration is dominated by a single mechanism, then the rate obtained in eqn (7.61) has a temperature dependence of exactly the Arrhenius form mentioned at the beginning of this section. An additional comment to be made about our assessment of the transition rate is that it presupposed a knowledge of the pathway connecting the various wells.

This, in turn, depends upon the intuition of the one carrying out the analysis and, as will be seen later, there are many instances in which the best of intuition could not have guessed the complex *collective* motions associated with some diffusion mechanisms.

The saddle point analysis described above can serve as the foundation for the more general multi-dimensional setting that attends a more complete assessment of diffusion. In this case, one imagines the multi-dimensional potential energy surface in its entirety. We assess the transition rate on the basis of the same conceptual argument: compare the number of saddle crossings to the total possible number of such crossings as evaluated on the initial well as required by eqn (7.54). Following the full multi-dimensional analysis to its conclusion results in

$$\Gamma = \left(\frac{\prod_{i=1}^{N} \nu_i^w}{\prod_{i=1}^{N-1} \nu_i^s} \right) e^{-[\phi(q_s) - \phi(q_w)]/kT}. \tag{7.62}$$

Here we have adopted the notation ν_i^w to signify the ith eigenvalue of the dynamical matrix evaluated at the well, while ν_i^s signifies the ith eigenvalue of the dynamical matrix at the saddle point. Further, N is the dimensionality of the energy landscape and determines the number of eigenvalues of the dynamical matrix. Note that because the curvature of the energy surface along the reaction coordinate direction is negative at the saddle point, there is one less real frequency there.

With the machinery of transition state theory in place, it is now possible to examine the predicted diffusion rates associated with a host of different important situations ranging from the bulk diffusion of impurities to the motion of adatoms on surfaces to the short-circuit diffusion of atoms along the cores of dislocations. It is evident that besides being of academic interest (which they definitely are), diffusive processes such as those mentioned above are a key part of the processing steps that take place in both the growth and subsequent microstructural evolution of the materials around which modern technology is built.

7.4.3 Diffusion via Molecular Dynamics

The transition state analysis favored above examines diffusion from the standpoint of a number of essentially static configurations without recourse to any explicit reckoning of atomic motions. Though this type of analysis has a certain elegance and instructive power, it relies heavily on an intuition as to the underlying mechanistic details of the transition state itself. However, as we will see in detail in chap. 11, there is substantial evidence that in certain cases the underlying mechanism is completely different than that suggested by intuition. As a result, there is a certain advantage to setting forth numerical schemes in which the system is allowed to undergo the entirety of allowed transitions without the artificial constraints imposed in the transition state setting. In this section, we consider molecular dynamics approaches to diffusion and the allied theoretical machinery used to understand the results of such simulations.

Molecular dynamics examines the temporal evolution of a collection of atoms on the basis of an explicit integration of the equations of motion. From the point of view of diffusion, this poses grave problems. The time step demanded in the consideration of atomic motions in solids is dictated by the periods associated with lattice vibrations. Recall our analysis from chap. 5 in which we found that a typical period for such vibrations is smaller than a picosecond. Hence, without recourse to clever acceleration schemes, explicit integration of the equations of motion demands time steps yet smaller than these vibrational periods.

By way of contrast, the time scale associated with diffusive processes can be estimated by recourse to ideas developed earlier in the chapter. Recall that one measure of the distance that an atom has wandered during the diffusion process is given by $\sqrt{\langle r^2 \rangle} = \sqrt{Dt}$, where D is the diffusion constant. As a result, the time scale associated with wandering a distance l is of the order of $t_{wander} = l^2/D$. We now pose the question: what is the time scale associated with the wandering of an atom by a distance characteristic of typical interatomic spacings, say, 0.5 nm? We obtain an estimate by using generic numbers for self-diffusion and through appeal to the definition of the diffusion coefficient as $D = D_0 e^{-E_{act}/kT}$, where the prefactor $D_0 \approx 10^{-5}$ m²/s and the activation energy is given by $E_{act} \approx 2.0$ eV. If we further assume a temperature (pretty high for many materials) such as $T = 4T_{room}$, then we find

$$t_{wander} = \frac{l^2}{D_0 e^{-E_{act}/kT}} \approx 10^{-5} \text{ s.} \qquad (7.63)$$

An essentially equivalent argument can be made by noting that the hopping rate is given by $\Gamma = \nu e^{-E_{act}/kT}$, and hence if we use $\nu = 10^{13}$ s⁻¹ and the activation energy and temperatures given above, the mean time between successful hops is $t_{wander} = 1/\Gamma \approx 10^{-5}$ s. Note that if viewed from the perspective of the minimum time step demanded in the molecular dynamics setting, this is an eternity. Part of the conclusion to be drawn from this argument is the rarity of diffusion events themselves. In addition, these conclusions pose a modeling conundrum that we will take up in turn.

Despite the reservations set down above, to carry out a molecular dynamics study of the diffusion process itself one resorts to a computational cell of the type described earlier. The temperature is assigned and maintained via some scheme such as the Nosé thermostat (Frenkel and Smit, 1996), and the atomic-level trajectories are obtained via a direct integration of the equations of motion. In fig. 3.22, we showed the type of resulting trajectories in the case of surface diffusion. For the purposes of reconciling the results of a molecular dynamics simulation of diffusion with the traditional empirical machinery used to describe mass transport, there are several interesting possibilities. One such possibility is based on the insight encompassed in eqn (7.26). In particular, we can take advantage of the fact that the average distance wandered by a diffusing atom scales with the diffusion constant as $\sqrt{D}$. The formula that allows us to examine the wandering of a particle during diffusion is

$$\langle r^2(t) \rangle = \frac{1}{N} \sum_{i=1}^{N} [r_i(t) - r_i(0)]^2. \tag{7.64}$$

What this formula tells us is to evaluate the excursion undertaken by each atom during the time t and to average over all N particles. As a result of this expression, by 'measuring' the mean excursion taken by a diffusing atom in a known time, the diffusion constant can be inferred.

A more sophisticated version of basically this same idea is associated with the Green–Kubo formalism in which the relevant transport coefficient is seen as an average of the appropriate velocity autocorrelation function. In particular, the diffusion coefficient may be evaluated as

$$D = \frac{1}{3N} \int_0^\infty \left\langle \sum_{i=1}^{N} \mathbf{v}_i(t) \cdot \mathbf{v}_i(0) \right\rangle dt. \tag{7.65}$$

For details on the derivation of this formula the reader is referred to section 4.5.1 of Frenkel and Smit (1996). The physical message to be taken away from this discussion is that by carefully observing the statistics of particle trajectories it is possible to reconcile the relevant microscopic motions seen in a molecular dynamics simulation with the macroscopic reflection of these motions, namely, the existence of a diffusion constant. On the other hand, we have not fully owned up to the rarity of diffusion events as measured on the time scale of atomic vibrations. This topic will be taken up again in chap. 12.

7.4.4 A Case Study in Diffusion: Interstitials in Si

Our goal in this section is to examine diffusion in Si from several of the per-spectives introduced in previous sections. For the purposes of illuminating the connection between transition state theory and microscopic analysis as presented in earlier chapters, we consider the diffusion of a self-interstitial in Si. As with many of the case studies to be presented in this book, the particulars of the implementation described here may well not stand the test of time. On the other hand, these implementations will more than suffice to illustrate *how* to apply these ideas in the context of a concrete problem.

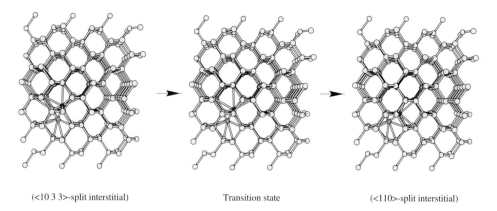

(<10 3 3>-split interstitial)   Transition state   (<110>-split interstitial)

Fig. 7.22. Illustration of the transition path connecting two different wells for an interstitial in Si (adapted from Munro and Wales *et al.* (1999)).

Tight-Binding Transition State Analysis of Interstitial Diffusion. The tight-binding method often represents a reasonable compromise between accuracy and computational tractability, especially in those problems for which the evaluation of a number of different configurations is necessary. In the present context, a systematic tight-binding analysis of the problem of interstitial diffusion in Si has been undertaken by Munro and Wales (1999) using the types of transition state arguments introduced above. To carry out such calculations they used a tight-binding Hamiltonian which is a nonorthogonal generalization of those introduced in chap. 4. Their calculations are made on both 64- and 216-atom supercells with periodic boundary conditions. As part of their analysis they considered a variety of different starting configurations (i.e. different wells) and then searched for various low-energy pathways connecting these wells. As noted above, one of the demands placed on the modeler in exploiting the transition state philosophy is that of finding the relevant transition pathway. In the present case, such pathways were obtained by recourse to several hybrid computational schemes, resulting in reaction pathways like that shown in fig. 7.22.

Tight-Binding Molecular Dynamics Approach to Interstitial Diffusion. Again for the purposes of illuminating the basic theoretical constructs, we now turn to a specific implementation of the molecular dynamics approach to diffusion. Our earlier treatment of interstitial diffusion in Si was predicated upon the identification of a series of suspected diffusion pathways, and the evaluation of their implications for diffusion via transition state theory. These same problems have been examined from the point of view of tight-binding molecular dynamics by Tang *et al.* (1997). By 'measuring' the wandering of a given interstitial as a function of time it is possible to deduce the diffusion constant in accord with the arguments of

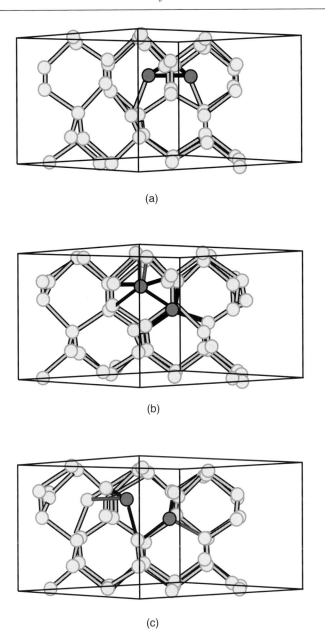

(a)

(b)

(c)

Fig. 7.23. Illustration of the transition path connecting two different wells for an interstitial in Si (adapted from Tang *et al.* (1997)).

section 7.4.3, and they were able to evaluate both the vibrational prefactor and the migration energy for interstitial diffusion. Several snapshots from a trajectory involving interstitial migration are revealed in fig. 7.23.

7.5 Defect Clustering

Our analysis thus far has been built on a defect by defect basis. On the other hand, given the presence of mass transport via diffusion, it is possible for adjacent point defects to find each other and form complexes with yet lower free energies of formation. Two relevant examples of this phenomenon are that of vacancy–interstital pairs and divacancies.

Microscopic Considerations. Our earlier analysis of point defect concentrations was predicated at the outset on the presumed diluteness of such defects. The necessity for this assumption was related to the fact that we had explicitly neglected any possible interaction effects between different point defects. On the other hand, it is clear that a vacancy cluster, for example, is a viable structural alternative that must be considered. As a starting point, we consider the case of a divacancy. A divacancy refers to a defect complex in which two vacancies fall under each others influence and form a partnership which it requires energy to dissolve. In energetic terms, the question we are posing is whether or not the energy difference $\Delta E = 2\epsilon_{vac}^{f} - \epsilon_{divac}$ is positive or negative. In the case in which this energy is positive, the conclusion is that it is energetically more favorable for two vacancies to join forces and bind than it is for them to remain in isolation.

The formation of larger vacancy clusters is an important question as well. In a series of systematic tight-binding analyses, the structure and energetics of such clusters in Si have been investigated with the emergence of a picture of 'magic numbers' in which vacancy clusters involving particular numbers of vacancies are especially stable. As with the divacancy considered above, the metric used for evaluating structural stability is the energy difference

$$\Delta E = n\epsilon_{vac}^{f} - \epsilon_{nvac}^{f}. \tag{7.66}$$

Here we have introduced the notation ϵ_{nvac}^{f} to denote the energy associated with a n-vacancy cluster. An example of the energy as a function of vacancy cluster size is shown in fig. 7.24. Associated with the various clusters described above are a series of favored shapes for the vacancy clusters. An example of several such structures is shown in fig. 7.25.

7.6 Further Reading

Point Defects and Diffusion by C. P. Flynn, Clarendon Press, Oxford: England, 1972. A definitive treatise on the subject of point defects and diffusion as seen from the perspective which predates the routine use of first-principles simulations to inform models of diffusion.

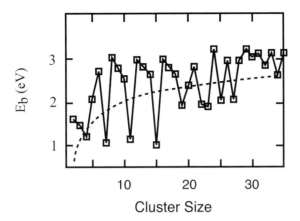

Fig. 7.24. Energetics of vacancy clusters for various sizes (adapted from La Manga *et al.* (1999)).

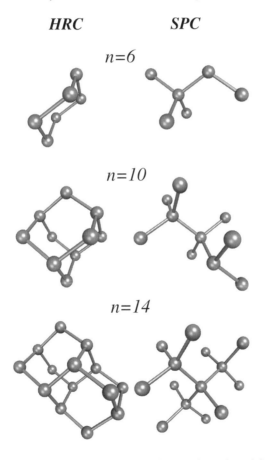

Fig. 7.25. Cluster shapes for various sized clusters of vacancies (adapted from Bongiorno *et al.* (1998)).

Atomic Transport in Solids by A. R. Allnatt and A. B. Lidiard, Cambridge University Press, Cambridge: England, 1993. This book is an exhaustive modern treatment of point defects and their motion. I turn to this book often.

Atom Movements by Jean Philibert, Les Editions de Physique, Les Ulis Cedex A: France, 1991. My personal favorite on the subject of diffusion. This book is filled with both descriptive and quantitative accounts of point defects and their motion.

Engineering Materials 2 by Michael F. Ashby and David R. H. Jones, Pergamon Press, Oxford: England, 1986. As with many other topics of relevance in materials, Ashby and Jones have insightful remarks to make on diffusion.

Dislocations by J. Friedel, Pergamon Press, Oxford: England, 1964. Though Friedel's book will show up again in the Further Reading section of chap. 8, his discussion of point defects is also enlightening and deserves to have attention called to it in the present setting as well.

'Point Defects' by J. D. Eshelby in *The Physics of Metals 2. Defects* edited by P. B. Hirsch, Cambridge University Press, London: England, 1975. Eshelby's article provides an interesting perspective on the structure, energetics and mobility of point defects.

A wonderful article on random walks that gives a broad coverage of this important subject is 'On the wonderful world of random walks' by E. W. Montroll and M. F. Shlesinger, in Studies in Statistical Mechanics, Volume XI, Elsevier Science Publishers, Amsterdam: The Netherlands, 1984. This book goes under the title *Nonequilibrium Phenomena II From Stochastics to Hydrodynamics*, edited by J. L. Lebowitz and E. W. Montroll.

A second outstanding source for discussions of some of the statistical issues raised in this chapter is 'Stochastic Problems in Physics and Astronomy' by S.Chandrasekhar, *Rev. Mod. Phys.*, **15**, 1 (1943). Our discussion of the large-N behavior of the random walk closely follows the discussion in Chandrasekhar.

An excellent article with broad scope on the analysis of transition rates may be found in 'Reaction-rate theory: fifty years after Kramers' by P. Hängi, P. Talkner and M. Borkovec, *Rev. Mod. Phys.*, **62** 251 (1990). Our discussion of transition state theory borrows the compact notation used in this article.

Phase Transformations in Metals and Alloys by D. A. Porter and K. E. Easterling, Chapman and Hall, London: England, 1992. Chap. 2 of this book has much of interest to say concerning point defects and their motion.

Diffusion in Solids by P. G. Shewmon, McGraw Hill, New York: New York, 1963. Shewmon's book is one of the classic texts on diffusion.

Diffusion in the Condensed State by J. S. Kirkaldy and D. J. Young, Institute of Metals, London: England, 1987. A detailed account of diffusion in a wide variety of settings ranging from ternary alloys to colonies of *Penicillium*.

7.7 Problems

---

1 Fourier transforms and the Diffusion Equation

In this problem, we take up the details of the Fourier transform approach to the diffusion equation, generalizing the one-dimensional treatment given in the text. Evaluate the Fourier transform in space of the three-dimensional version of the diffusion equation,

$$\frac{\partial c(\mathbf{x}, t)}{\partial t} = D\nabla^2 c(\mathbf{x}, t). \tag{7.67}$$

Solve the resulting ordinary differential equation in time and deduce the three-dimensional analog of eqn (7.24).

---

2 Continuum Model of Point Defect Energetics

In this chapter, we showed how to estimate the energetics of the structural relaxations around a point defect using continuum mechanics. Our arguments culminated in eqn (7.36). In this problem, we 'refine' that model by adding an energy penalty for the surface of the tiny sphere of radius a_0 shown in fig. 7.10. The total formation energy of the point defect is now the sum of two terms, one relating to elastic relaxations and the other to broken bonds. Complete the estimate by using characteristic numbers from real materials.

---

3 Continuum Models of Point Defects with Tetragonal Symmetry

One of the approaches we used to examine the structure and energetics of point defects from the continuum perspective was to represent the elastic consequences of the point defect by a collection of force dipoles. The treatment given in the chapter assumed that the point defect was isotropic. In this problem, derive an equation like that given in eqn (7.38). Assume that the forces at $\pm(a/2)\mathbf{e}_1$ and $\pm(a/2)\mathbf{e}_2$ have strength f_0 while those at $\pm(a/2)\mathbf{e}_3$ have strength f_1. Evaluate the displacement fields explicitly and note how they differ from the isotropic case done earlier.

4 Relaxation Energy for Point Defects

Using the Morse potential that was fit to the cohesive energy, lattice parameter and bulk modulus of Cu in problem 2 of chap. 4, compute the relaxation energy associated with a vacancy in Cu. In light of this relaxation energy, compute the relaxed vacancy formation energy in Cu and compare it to experimental values. How do the structure and energy of the vacancy depend upon the size of the computational cell?

5 Vacancy Formation Energy within Embedded-Atom Formalism

In this chapter, we argued that the pair functional approach incorporates additional features of the physics of bonding that improve the treatment of lattice defects such as vacancies. Use the explicit embedding function of Johnson given in eqn (7.47) in order to deduce the expression for the vacancy formation energy given in eqn (7.48).

6 Vacancy Formation Energy using Angular Forces

Use the Stillinger–Weber potential to compute the unrelaxed vacancy formation energy in Si.

7 Comparison between Atomistic and Continuum Relaxations Near a Vacancy

In fig. 7.17 we showed the relaxations in the vicinity of a vacancy in Si as computed using tight-binding theory. Compute your own estimate for the elastic relaxations using eqn (7.30) with the constants determined by insisting that the displacement of the first shell agrees with the tight-binding results. Note that although the relaxations are actually tetragonal, we are assuming they are spherically symmetric. Compare your results with those obtained in the tight-binding analysis.

8 Transition State Theory in the Multi-dimensional Context

Our discussion of transition state theory in this chapter was laid out in detail in only the one-dimensional setting. Provide the detailed arguments for the multi-dimensional version of transition state theory that culminates in eqn (7.62).

9 Surface Diffusion on a Rigid Surface via Transition State Theory

Using the Cu Morse potential developed in problem 2 of chap. 4, compute the diffusion constant for diffusion on the (100) and (111) faces of Cu. Assume that the surface is unreconstructed and rigid. As a result, the only free degrees of freedom in the problem are those of the diffusing adatom.

10 Surface Diffusion on a Rigid Surface via Molecular Dynamics

Using the Cu Morse potential developed in problem 2 of chap. 4, compute the diffusion constant for diffusion on the (100) and (111) faces of Cu using molecular dynamics. Assume that the surface is unreconstructed and rigid. As a result, the only free degrees of freedom in the problem are those associated with the diffusing adatom. To maintain the appropriate temperature, use the dynamics introduced in eqn (12.74) and explained in Allen and Tildesley (1987).

EIGHT

Line Defects in Solids

8.1 Permanent Deformation of Materials

Steel conjures up images of an absolute rigidity. Yet, as a visit to the most ordinary of steel mills quickly demonstrates, at high enough temperatures a steel bar can be elongated by many orders of magnitude with respect to its initial length. Indeed, it is staggering to stand near a rolling mill, one's face aglow, as a giant block of steel is flattened and stretched into sheet or wire. Similarly, the tungsten filaments which illuminate our homes have had a history rich in permanent deformation with the length of a given bar undergoing extension by factors of as much as 500 000. The physical mechanisms that make such counterintuitive processes possible pose a puzzle that leads us to the consideration of one of the dominant lattice defects, the dislocation.

Dislocations occupy centerstage in discussions of the permanent deformation of crystalline solids largely because of their role as the primary agents of plastic change. The attempt to model such plasticity can be based upon an entire spectrum of strategies. At the smallest scales, appeal can be made to the atomic structure of dislocation cores which can suggest features such as the fact that in certain materials the flow stress increases with increasing temperature. At the other extreme, continuum models of single crystal plasticity are purely phenomenological and make no reference to dislocations except to the extent that they motivate the choice of slip systems that are taken into account.

The task of this chapter is to introduce the key concepts (both continuum and discrete) used in thinking about dislocations with the aim of explaining a range of observations concerning plastic deformation in crystalline solids. In the opinion of the author, one of the primary conclusions to emerge from the discussions to be made here is that despite the fact that the theory of a single dislocation has reached a high level of sophistication, the promise of dislocation-based models of plasticity with predictive power remains elusive. One reason for this is the fact that true plasticity represents a conspiracy of multiple dislocations in concert with other defects such as substitutional impurities, grain boundaries and inclusions. As a result, plasticity is a 'many-body' problem and poses huge statistical challenges in that what is sought are suitable averages over the entire distribution of dislocations as they traverse a disordered medium. From a broader perspective, line defects have a more far reaching significance than that presented by the plastic deformation of solids. The motion of vortices in fluids, of magnetic flux lines in type-II superconductors and the slithering arrangements of the long molecules making up polymeric materials all place demands on our ability to characterize the motion and entanglement of extended defects. From our present perspective, we take up one such example (that of dislocations), and examine how far we can go in our aim of building theories of line defect motion.

8.1.1 Yield and Hardening

The simplest window on the processes that operate when a crystal is subject to plastic deformation is the stress–strain curve. As shown in fig. 2.10, once the stress exceeds the elastic limit, not only is there a marked deviation from linearity, but the crystal is also permanently deformed. If such stresses are reached and the crystal is subsequently unloaded, there will be a residual plastic strain. One conventional measure of the yield point is to consider that stress for which there is a 0.2% residual plastic strain upon unloading. If the sample is reloaded, it will be found that the yield point has moved to higher stresses – the crystal has hardened.

One of the unifying themes in this book has been the notion of material parameters. We have repeatedly interested ourselves in the existence of quantities that serve to characterize a material under particular temperatures, loading conditions, and past history. The discussion of the stress–strain curve suggests further candidates. The most immediate suggestion in the present context is that of the yield stress. Despite the presence of some arbitrariness in the definition of the yield stress, if we adopt the convention of the 0.2% offset point, a case can be made for the idea that different materials have reproducible yield stresses. A schematic rendering of the typical scales for the yield stress is given in fig. 8.1 in which it is seen that, for the vast majority of materials, the yield stress is a small fraction of the elastic moduli.

Our attempt to define a unique yield stress for a given material is flawed, however, on the grounds that there is a large dependence of that stress on a material's past history. In particular, the mechanisms of solid solution hardening, precipitation hardening, grain size strengthening and work hardening can each greatly alter the putative yield stress. As a result, we must be mindful of the material's past. A collage of the type of data that provides evidence for these

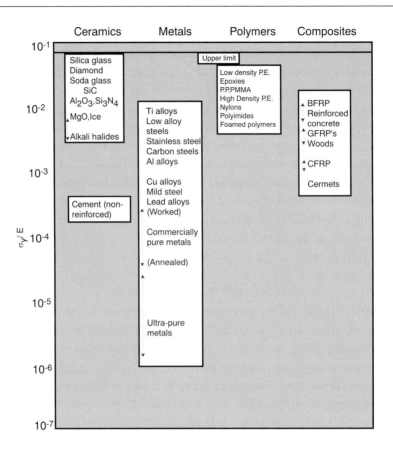

Fig. 8.1. Representation of data for yield stress in a broad range of materials (adapted from Ashby and Jones (1996)). The upper limit is associated with the stress to induce plastic flow in the *absence* of defects and will be taken up in detail in our discussion of the ideal shear strength.

various mechanisms is shown in fig. 8.2. The solid solution hardening data illustrate that as the concentration of substitutional impurities is increased, so too is the stress to initiate plastic deformation. If there is a supersaturation of substitutional impurities, subsequent annealing will lead to the formation of precipitates which also influence the stress to initiate plastic deformation. The data shown in the figure illustrate that for the purposes of strengthening, there is an optimal particle size. The dependence of the yield stress on a material's grain size is revealed in the celebrated Hall–Petch relation which notes that the yield stress scales with the grain size d as $\sigma_y \propto d^{-\frac{1}{2}}$. Finally, depending upon the extent of prior working of a material, the preexisting dislocation density can vary by orders of magnitude, with the result that the yield stress can be all over the map. Each of these mechanisms will be taken up in turn in this and coming chapters.

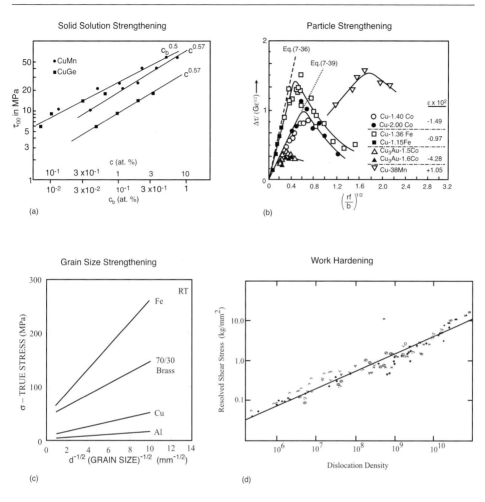

Fig. 8.2. Representative data associated with four of the key strengthening mechanisms within a material. (a) Increase in flow stress as a function of concentration of substitutional impurities. (b) Dependence of flow stress on mean particle size for material in which there are second-phase particles. (c) Dependence of yield stress on mean grain size of material. (d) Relation between yield stress and mean dislocation density. (Adapted from (a) Neuhäuser and Schwink (1993), (b) Reppich (1993), (c) Hansen (1985), (d) Basinski and Basinski (1979).)

8.1.2 *Structural Consequences of Plastic Deformation*

The arrival at the yield point on the stress–strain curve is revealed at the most gross structural level by a permanent change in the body's shape. However, there are more subtle structural changes as well. One of the intriguing features that arise with the onset of plastic flow and a hint in favor of the claim that dislocations are the carriers of plastic change is the occurrence of micron-sized steps on the crystal faces. In particular, if an optical microscope is used to examine the crystal surface

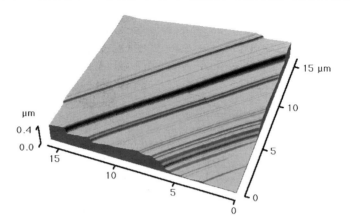

Fig. 8.3. Slip traces on crystal surface. Slip traces are a microstructural consequence of the onset of plastic deformation (courtesy of C. Coupeau).

a series of fine lines are revealed. With increasing strain, the number of such lines is increased and it becomes evident that their orientation is not haphazard. An example of the emergence of slip steps as a result of plastic deformation as revealed using atomic force microscopy is shown in fig. 8.3. As will become more evident below, these lines are the point of exit of dislocations after their tortuous journey through the crystal's interior. Each dislocation carries with it an elementary unit of lattice translation (the so-called Burgers vector) which implies a relative displacement across a given plane by a length of order of the lattice parameter. If many such dislocations cross the surface in the same vicinity, the surface has an appearance like that shown in fig. 8.3. The immediate and abiding conclusion of this structural clue is the role of shearing processes in plastic deformation, with the dominant mechanism being the sliding of different parts of the crystal with respect to one another.

Higher resolution reveals surprises as well. Indeed, if a thin foil of the sample is examined via the electron microscope, it is found that the sample is seething with activity. Consider the sequence of snapshots shown in fig. 8.4. What is revealed is a series of line objects in motion, objects which ultimately it can be concluded are dislocations.

As will be seen below, the presence of dislocations has the effect of producing local disturbances of the atomic-level geometry that are so severe as to produce scattering that is not consonant with that from the remainder of the crystal, giving rise to the contrast revealed in the figure. Evidently, the onset of plastic deformation reveals its presence at multiple scales simultaneously. Indeed, as was seen in fig. 1.9, the properties of individual dislocation cores can be observed using high-resolution transmission electron microscopy.

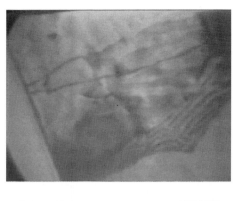

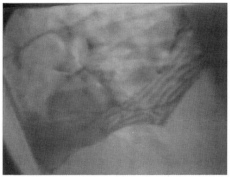

Fig. 8.4. Series of snapshots from *in situ* experiments using the transmission electron microscope (courtesy of M. Legros and K. Hemker).

8.1.3 Single Crystal Slip and the Schmid Law

We have claimed that it is via the process of crystalline slip that the majority of crystal plasticity takes place. In particular, we marshalled the evidence from slip steps at surfaces as an argument in favor of shearing motions as the basis of plastic deformation. It is now of interest to examine the coupling of these geometric features to the state of applied stress. Typical structural materials are invariably polycrystalline. As a result, it is difficult to precisely determine characterizations of the conditions leading to plastic deformation in this setting. In particular, while certain grains may be favorably oriented with respect to the loading axis for the commencement of crystalline slip, other grains may experience much lower resolved shear stresses. The interpretation of yield in this case is confused by questions of how slip is transmitted from favorably oriented grains to their less favorably oriented partners. Consequently, the ideal forum within which to probe the factors leading to the motion of dislocations is that of single crystals. Here, the relation between the geometry of the crystal and that of the loading apparatus can be carefully constructed so that it is clear what resolved shear stresses competing

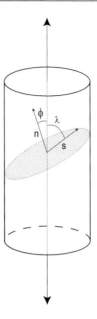

Fig. 8.5. Geometry illustrating concept of resolved shear stress.

slip systems are faced with. Experiments of this form culminate in the elucidation of the Schmid law for crystalline slip.

In single crystal tensile tests it is imagined that the loading axis is oriented with respect to the crystal in the way depicted in fig. 8.5. The angle ϕ denotes the relative orientation of the loading axis and the normal to the slip plane of interest. Similarly, the angle λ characterizes the relative orientation of the loading axis and the slip direction within the slip plane. Arduous experiments in which these relative orientations are changed incrementally lead to the assertion that slip commences when the resolved shear stress (see section 2.3.1) projected along the slip direction reaches a certain critical value. The resolved shear stress can be found by projecting the traction vector $\mathbf{t}^{(\mathbf{n})}$ on the plane with normal $\mathbf{n}$ onto the slip direction. In particular, this leads to the result that

$$\tau_{rss} = \mathbf{s} \cdot \boldsymbol{\sigma} \mathbf{n} = \sigma_0 \cos \phi \cos \lambda, \tag{8.1}$$

where $\mathbf{s}$ is a unit vector in the slip direction and σ_0 is the magnitude of the externally applied tensile load. This result calls for a very definite scaling of the stress to induce plastic flow with the geometric parameters that define the relative orientation of the loading axis and the slip system. An example of this scaling in cadmium and zinc is shown in fig. 8.6.

The hypothesis of the existence of a parameter such as τ_{rss} raises serious questions for the modeler. From a single crystal perspective, this is our entry point

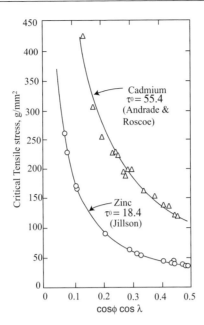

Fig. 8.6. Scaling of the stress to induce plastic flow with the geometric parameters characterizing the relative orientations of the loading axis and the relevant slip system (adapted from Gilman (1969)). Data are for both Cd and Zn.

into the consideration of calculating material parameters such as the yield stress. The most naive model that we might imagine constructing is one founded upon the critical resolved shear stress to induce motion in a single straight dislocation. We will later encounter this idea in the guise of the Peierls stress. However, it is not clear that this model is really what we are after. Once the problem of glide of a single dislocation is tackled, we are still left with the 'many-body problem' in which a statistical average must be taken over the distribution of all such dislocations in the fields of their mutual interactions and those with other defects such as precipitates.

8.2 The Ideal Strength Concept and the Need for Dislocations

The simplest model of plastic flow was historically founded on the idea of a shearing process in which entire crystal planes underwent motion with respect to one another in a uniform fashion. This model envisions a certain energy cost for relative motion of adjacent crystal planes. The picture that emerges is that of uniform translation of the upper half of the crystal with respect to the lower half by multiples of the appropriate lattice translation vector. For example, if we imagine a perfect crystal as indicated schematically in the left hand panel of fig. 8.7, the

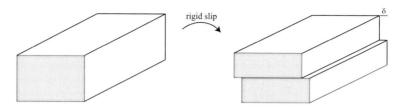

Fig. 8.7. Schematic illustrating outcome of crystal slip with one part of the crystal translated with respect to another.

outcome of plastic deformation is that indicated schematically in the right hand panel. The important issue is the mechanism whereby the state shown in the right hand panel is attained.

An estimate of the energy cost for such uniform deformations can be built on the basis of the crystal periodicity. In particular, a simple model of this energy can be written in the form

$$E(\delta) = \frac{\mu b^2}{4\pi^2 d} \left(1 - \cos \frac{2\pi\delta}{b} \right), \tag{8.2}$$

where δ is the relative translation of the two half-spaces and b is the periodicity associated with lattice translations in the slip direction. To determine the prefactor $\mu b^2/4\pi^2 d$, appeal has been made to the behavior of the crystal when it suffers small deformations. In particular, it is assumed that the energy cost of relative translation of adjacent planes is captured by appealing to linear elasticity. For a relative translation of adjacent planes by small δ, the energy cost within isotropic linear elasticity is $W = (\mu/2)(\delta/d)^2$, where d is the distance between these planes. By equating this representation for the strain energy density with that obtained from eqn (8.2) in the small-δ limit we can determine the prefactor. Since it is stresses, not energies, that are typically measured in a mechanical test, it is of interest to cast this model of the ideal strength in those terms. To compute the stress associated with the shearing motion described above we evaluate the change in energy with respect to the parameter δ with the result that

$$\tau = \frac{\mu b}{2\pi d} \sin\left(\frac{2\pi\delta}{b} \right). \tag{8.3}$$

The ultimate resistance of a crystal, and hence the ideal shear strength, is the maximum value of τ given by

$$\tau_{ideal} = \frac{\mu b}{2\pi d} \approx \frac{\mu}{10}. \tag{8.4}$$

For a typical metal, the shear modulus is measured in gigapascals, implying an ideal strength of the same order. This value is to be contrasted with typical

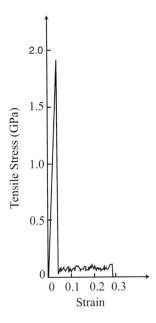

Fig. 8.8. Stress–strain curve for Cu whisker (adapted from Boyer (1987)).

measured yield stresses which are found in the megapascal range (see fig. 8.1), resulting in the observation that typically $\sigma_y \ll \tau_{ideal}$. Something is amiss with the hypothesized deformation mechanism involving uniform slip. In fact, the presence of dislocations reduces the stress for the onset of plastic flow well below that of the ideal shear strength. On the other hand, for small perfect crystals that can be produced virtually defect-free (known as whiskers), the ideal shear strength can be realized.

A stress–strain curve for a Cu whisker is shown in fig. 8.8. It is seen that the stress scale associated with whiskers is in the gigapascal range in support of our hypothesis that in whiskers the ideal strength may be more approximately realized. It is not surprising that the ideal strength estimate is not reproduced exactly since our argument assumed a contrived sinusoidal dependence to the energy associated with uniform slip.

8.3 Geometry of Slip

We have claimed that dislocations are often the primary agents of plasticity. Our present intention is to consider the geometric features implied by their presence. First, we consider the generic topological features that are shared by all dislocations, and then turn to the particulars induced by crystallography.

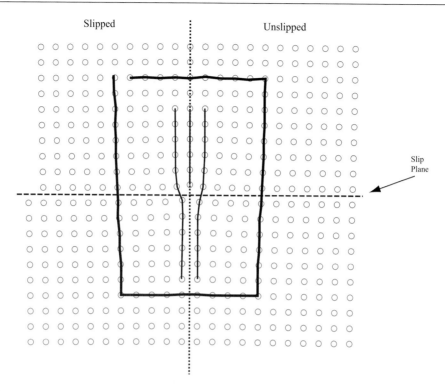

Fig. 8.9. Schematic illustrating process whereby crystalline slip is induced by the passage of a dislocation (courtesy of Ron Miller).

8.3.1 Topological Signature of Dislocations

The net result of plastic deformation is indicated schematically in the right hand panel of fig. 8.7. Our first instinct was to attribute this process to a uniform translation thought to take place across the interfacial plane (the so-called slip plane) all at once. However, this relative translation can occur sequentially via the presence of dislocations at much lower stresses. Formally, a dislocation is a line object that is the boundary that separates those regions on the slip plane which have undergone slip from those that have not. As a result, rather than the wholesale relative translation of different parts of the crystal shown in fig. 8.7, slip takes place as indicated schematically in fig. 8.9. The extent to which one part of the crystal has slipped relative to the other is determined by the periodicity of the underlying lattice, and corresponds to shifting the registry of adjacent atoms across the slip plane such that atoms that were formerly partners across the slip plane are now shifted relative to each other by a lattice translation vector dubbed the Burgers vector. Indeed, the Burgers vector emerges from the existence of a series of lattice invariant shears that leave the overall lattice unchanged. Indeed, as seen on the left

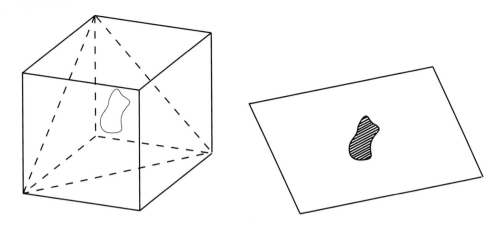

Fig. 8.10. Dislocation loop on a (111) plane in a cubic crystal.

hand side of the slip plane the atoms, though shifted, are in registry with their new partners across the slip plane.

As will be seen below, in addition to the discrete perspective we have adopted here, a dislocation can also be thought of from a continuum perspective as the result of cutting the crystal over some part of the slip plane, imposing a relative translation and rejoining the cut region. In particular, this suggests the definition of a dislocation loop as the boundary between some finite region of the slip plane which has been slipped and the remainder of the slip plane. A schematic example of this idea is given in fig. 8.10 which shows a (111) plane in a cubic crystal and a dislocation loop.

Historically, one of the most ingenious venues within which to view lattice defects was through the use of bubble rafts. The basic idea is that a series of soap bubbles arrange themselves in a crystalline form which is occasionally disturbed by defects including dislocations. A more modern analog is provided by colloidal crystals in which colloidal particles crystallize and may be viewed via optical microscopy. As a supplement to the schematic of fig. 8.9, fig. 8.11 shows a dislocation in a colloidal crystal.

A dislocated crystal has a distinct geometric character from one that is dislocation-free. From both the atomistic and continuum perspectives, the boundary between slipped and unslipped parts of the crystal has a unique signature. Whether we choose to view the material from the detailed atomic-level perspective of the crystal lattice or the macroscopic perspective offered by smeared out displacement fields, this geometric signature is evidenced by the presence of the so-called Burgers vector. After the passage of a lattice dislocation, atoms across the slip plane assume new partnerships. Atoms which were formerly across from

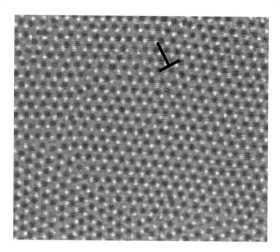

Fig. 8.11. Lattice defects as revealed in a colloidal crystal (courtesy of X. Ling).

one another on the slip plane are no longer in registry and find that they have experienced a relative translation characterized by the Burgers vector. The idea is to march around a dislocation line, always insuring that the number of northbound steps matches the number of southbound steps and similarly for the east- and westbound steps. These ideas are illustrated in fig. 8.9. The key point is that in the slipped part of the crystal, as mentioned above, the atoms across the slip plane have formed new partnerships, while those parts of the slip plane that the dislocation has not yet reached maintain the integrity of their initial neighbors. As a result, the act of traversing the Burgers circuit reveals the distinct topological disturbance of the initially perfect lattice and is sometimes referred to as the closure failure of the Burgers circuit.

The Burgers circuit concept introduced above from the discrete perspective has a continuum analog. Mathematically, the origin of this analog is the fact that there is a jump in the continuum displacement fields used to characterize the geometric state of the body. Recall that above we described the Volterra procedure in which the body is cut and rejoined after a relative translation operation. The resulting displacement jump reveals itself upon consideration of the integral

$$b_i = \oint du_i = \oint u_{i,j} dx_j, \tag{8.5}$$

which instructs us to add up the displacements in traversing a closed contour surrounding the dislocation line. The outcome of this analysis is the emergence of a distinct 'topological charge' associated with the presence of a dislocation and signaling the presence of the jump in the displacement fields associated with this defect. As will be seen below, the elastic solution for a dislocation arises from insisting on this displacement discontinuity across the slip plane. Furthermore, it should be noted that the presence of the nonvanishing circulation around the dislocation line is an invariant and will be revealed whatever path we choose to traverse, so long as it encircles the dislocation.

8.3.2 Crystallography of Slip

The arguments given above suggest that dislocations have a deep geometric significance that is independent of the particulars of a given crystal structure type. On the other hand, the response of such dislocations to the stresses present in a solid can be widely different depending upon the crystallographic features of a particular material. The goal of the present section is to describe the basic crystallographic features of dislocations with special reference to fcc crystals.

There are a few primary clues that suggest the correspondence between crystallography and plastic deformation. First, we noted above that concomitant with the onset of plastic deformation was the development of definite surface structure (see fig. 8.3) which we attributed to the relative translation of different halves of the crystal on certain crystallographic planes. These experiments reveal a privileged status for certain sets of vectors associated with the slip plane normal $\mathbf{n}$ and the slip direction $\mathbf{s}$. The linkage between crystallography and plastic deformation is further strengthened by appealing to experiments like those described above, in which these same privileged directions are revealed in tensile tests (see figs. 8.5 and 8.6).

From the perspective of the present discussion, our key objective is to uncover the detailed crystallographic underpinnings of single crystal plasticity. From the standpoint of a finite element model of such plasticity, what distinguishes the kinematic treatment of fcc Al from that of hcp Zn? In a word, the primary distinction is the presence of different *slip systems* in these different materials. The notion of a slip system refers to a partnership between two directions, the slip plane normal and the slip direction.

In our earlier description of the geometry of deformation (viz. chap. 2), we found that the strain associated with shear in a direction $\mathbf{s}$ for a plane with normal $\mathbf{n}$ could be written as

$$\epsilon = \frac{\gamma}{2}(\mathbf{s} \otimes \mathbf{n} + \mathbf{n} \otimes \mathbf{s}). \tag{8.6}$$

As a result of these arguments, within the confines of continuum models of single crystal plastic deformation, the total plastic strain rate can be written as

$$\dot{\epsilon}_P = \frac{1}{2}\sum_i \dot{\gamma}_i(\mathbf{s}_i \otimes \mathbf{n}_i + \mathbf{n}_i \otimes \mathbf{s}_i), \tag{8.7}$$

where the subscript i labels the various slip systems which are each characterized by their own slip plane and slip direction. The basic idea is that the net plastic strain may be thought of as an accumulation of the contributions arising on all of the relevant slip systems. Thus far we have seen that dislocation mediated plastic deformation is characterized by two crystallographic features, namely, the slip plane of interest characterized by a vector $\mathbf{n}$ and the Burgers vector $\mathbf{b}$ which characterizes the direction of slip ($\mathbf{s} = \mathbf{b}/|\mathbf{b}|$) associated with the passage of a given dislocation.

Though we have not yet made reference to it, a third vector of interest is associated with the direction of the dislocation line itself. That is, we consider the local tangent vector $\boldsymbol{\xi}$ and note that the angle between the Burgers vector and this local tangent is another key descriptor of the character of a given dislocation. The simplest dislocations are those in which the dislocation line is either parallel to, $\mathbf{b} \parallel \boldsymbol{\xi}$, (a screw dislocation) or perpendicular to, $\mathbf{b} \perp \boldsymbol{\xi}$, (an edge dislocation) the slip direction. A generic dislocation will be neither screw nor edge in character and will have a Burgers vector that has some other orientation with respect to the local tangent to the dislocation line. Such dislocations are known as mixed dislocations. We now take up the particulars of these ideas in the context of several different crystal classes, with special emphasis on fcc materials.

fcc Materials

The kinematic distinction of one crystal type from another as regards their propensity for plastic deformation is determined by the particular slip systems that are available to that material. One of the simplest settings within which to examine crystalline slip is that of fcc materials. In fcc metals, the dominant slip system is that in which slip occurs along $\langle 110 \rangle$-type directions with corresponding Burgers vectors $\mathbf{b} = (a_0/2)\langle 110 \rangle$ and is associated with $\{111\}$-type planes. This example reveals a few generic hints about the nature of crystalline slip. First, in this case it is seen that the slip planes are those of closest packing. Further, the direction of slip is that along which the packing is again closest. Later, we will see that there are energetic arguments (Frank's rule) that support the empirical observation about the role of close-packed planes and slip directions. In particular, we will find that the energy cost associated with a given dislocation scales with the square of the Burgers vector magnitude implicating the close-packed directions as those that will have associated dislocations of lowest energy. Similarly, we will find (at least within a simplified model of the stress needed to move a dislocation) that the minimum stress to induce plastic flow at a fixed Burgers vector is associated with the close-packed planes.

Because of the high symmetry of cubic crystals, there are a wide variety of equivalent slip systems. In particular, in the fcc case, each of the four $\{111\}$ planes

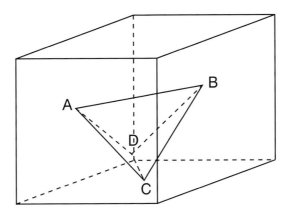

Fig. 8.12. Schematic of the Thompson tetrahedron illustrating the four {111} planes (adapted from Hull and Bacon (1995)).

supports three distinct Burgers vectors, making for a total of twelve distinct slip systems. A particularly useful device for classifying the various slip systems is provided by the Thompson tetrahedron which shows the geometric relation between all twelve of these slip systems. The Thompson tetrahedron is shown in fig. 8.12. Each of the faces of the Thompson tetrahedron coincides with a particular {111} plane, while each of its edges corresponds to an allowable Burgers vector direction for either of the planes that it shares. Hence, the tetrahedron is a compact way of characterizing all of the crystallographic choices available to an fcc material during plastic deformation. As will be seen in more detail later, in addition to the 'perfect' dislocations we have discussed until now, it will be seen that there is the possibility of dislocation chemistry in which different dislocations react to form others. The Thompson tetrahedron serves as something of a geometric phase diagram in that it encompasses in simple terms the various reactions that can be imagined and can be used for easy reference in uncovering the possible reaction pathways.

One example of dislocation chemistry that is revealed by the Thompson tetrahedron is the possibility of dissociation into Shockley partial dislocations. The key point is that a given Burgers vector can be decomposed into two shorter vectors associated with the face centers of the Thompson tetrahedron. The Burgers vector associated with these partial dislocations is no longer the full lattice translation of the perfect crystal, but rather a smaller translation. Recall that an fcc crystal may be thought of as the regular ABC stacking of spheres. The consequence of the presence of partial dislocations is the presence of a mistake in the regular ABC packing of the fcc crystal, the stacking fault. From a different point of view, the emergence of stacking faults may be pictured as follows. Once a perfect

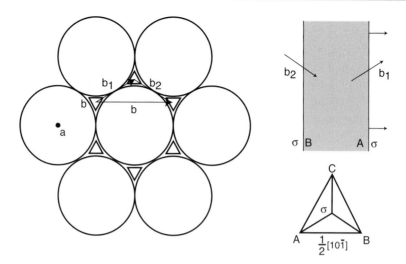

Fig. 8.13. Schematic of slip plane region of a crystal with an edge dislocation (adapted from Amelinckx (1979)). The left hand figure shows the way in which the full Burgers vector **b** may be decomposed into two smaller vectors **b**₁ and **b**₂. The upper right panel illustrates the stacking fault region (in gray) bounding the two partials.

dislocation has passed a certain point on the slip plane, the crystal may be thought of as having suffered a lattice invariant shear which restores the atoms in the wake of the dislocation into perfect crystal registry. This is indicated schematically in fig. 8.13. On the other hand, if the unit of slip carried by a particular dislocation does *not* correspond to a full lattice translation, the outcome of the passage of such a dislocation will be that in its wake, the crystal will not be restored into perfect registry. The entire slip plane is thus occupied by an interfacial defect, precisely the stacking fault of which we have made mention above.

The relevant slip vectors associated with the presence of partial dislocations are depicted schematically in fig. 8.13 and may be associated with a reaction of the type

$$\frac{a_0}{2}[110] \rightarrow \frac{a_0}{6}[211] + \frac{a_0}{6}[12\bar{1}]. \tag{8.8}$$

The point is that if we consider the regular ABC stacking associated with the perfect fcc crystal, and imagine dividing this crystal into two half-spaces with the dividing plane itself being a (111) plane, and then shift the upper half-space with respect to the lower half-space by some vector, only for certain choices of the slip vector will the crystal be restored into perfect registry. For the class of vectors of the form $(a_0/2)\langle 110 \rangle$, indeed, the crystal is restored without defect. On the other hand, if we elect to slide the upper half-space with respect to the lower half-space by a vector connecting the B sites to the C sites (i.e. the Burgers

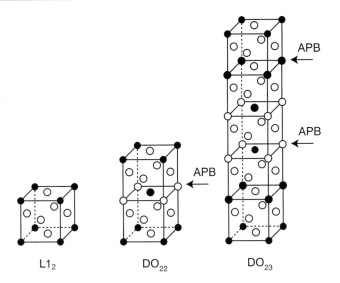

L1$_2$ DO$_{22}$ DO$_{23}$

Fig. 8.14. A few representative examples of structures adopted by intermetallics (adapted from Yamaguchi and Umakoshi (1990)).

vector of a Shockley partial given above as of the type $(a_0/6)\langle 211 \rangle$), the perfect ABC stacking sequence is spoiled and we are left with a packing sequence such as ABCBCABC.... Such disturbances in the regular fcc stacking result in a certain energy penalty, and the energy is most conveniently partitioned by asserting that the energy cost for such disruptions is associated with an interfacial energy, such as the stacking fault energy. The energetics of stacking faults will be taken up in the next chapter.

Intermetallics

Intermetallics pose the next level of complexity in considering the geometry of slip. Some of the key structural types such as the $L1_2$, DO_{22} and DO_{23} structures built around fcc-like packings are shown in fig. 8.14. Recall from our earlier discussion that the Burgers vector is usually associated with the smallest lattice translation vector supported by the particular crystal structure. In the intermetallic context, we note that for structures such as the $L1_2$ structure which are fcc-like, the fundamental lattice translations are of the form $a_0\langle 110 \rangle$ since glide within these materials occurs on {111} planes. Note that because of the presence of more than a single chemical species, the fundamental unit of translation has been doubled relative to the fcc counterpart of this structure. The significance of this observation is that the number of possible dissociations is even larger in this context. In particular, the stacking fault concept introduced in the context of fcc materials

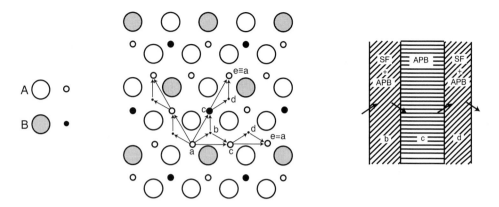

Fig. 8.15. Extended dislocation core resulting from dissociation reaction in $L1_2$ structure (adapted from Amelinckx (1979)). The left hand figure shows the various slip vectors that are possible in intermetallics. The right hand figure shows in schematic form the splitting of the dislocation into four distinct partials bounding several different types of faulted region.

must now be extended to include a series of planar faults: complex stacking faults, antiphase boundaries (APB), superintrinsic stacking faults.

Our present perspective will be to make a geometric assessment of such faults only to the extent that it serves us in our aim to understand crystalline slip in intermetallics. In chap. 9, we will return to the question of the energetics of such faults, since it is not only the existence of such faults, but also their extent which really governs the observed mechanical properties. It will be seen that the extent of such faults is a direct consequence of their energies. In fig. 8.15, we show a series of possible slip vectors that can be associated with the {111} planes in $L1_2$ crystals. What we note first of all is that the perfect dislocation is that associated with the vector $a_0\langle 110\rangle$. This vector is twice as large as that associated with the elemental fcc structure since because of the presence of more than a single species the lattice translation vectors are twice as long. One possible dissociation reaction that is available for such crystals is given by

$$a_0[\bar{1}10] = \frac{a_0}{6}[\bar{2}11] + \frac{a_0}{6}[\bar{1}2\bar{1}] + \frac{a_0}{6}[\bar{2}11] + \frac{a_0}{6}[\bar{1}2\bar{1}]. \tag{8.9}$$

In the figure, the sequence of vectors on the right hand side of this equation are those connecting a to b, b to c, c to d and finally, d to e which then represent a full lattice translation. As has already been noted in the section on fcc crystals, failure of the slip vector to be a full lattice translation implies the presence of a mistake in the regular packing along the [111] direction. In the present context, there are three distinct faulted regions associated with the complex core described in the reaction above. A schematic of the dissociation given above is shown in fig. 8.15.

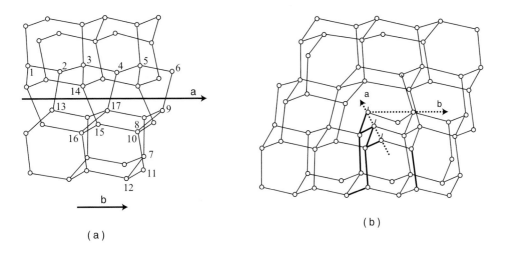

Fig. 8.16. Screw and 60° dislocations for Si (adapted from Amelinckx (1979)).

Semiconductors

Silicon will serve as the paradigmatic example of slip in covalent materials. Recall that Si adopts the diamond cubic crystal structure, and like in the case of fcc materials, the relevant slip system in Si is associated with {111} planes and ⟨110⟩ slip directions. However, because of the fact that the diamond cubic structure is an fcc lattice with a basis (or it may be thought of as two interpenetrating fcc lattices), the geometric character of such slip is more complex just as we found that, in the case of intermetallics, the presence of more than one atom per unit cell enriches the sequence of possible slip mechanisms.

In general, dislocation lines in Si run along ⟨110⟩ directions. Hence, a generic perfect dislocation will have a Burgers vector that is either parallel to the dislocation line itself, or at 60° with respect to the line. The geometry of such dislocations is beautifully illustrated with ball and stick models. Here we resort to the imperfect device of pictorial representations of these core geometries. Fig. 8.16 shows the core geometries relevant for both the screw and 60° dislocations, purely on the basis of reconnecting the appropriate bonds so as to take the presence of slip into account.

The structure of the perfect 60° dislocation is especially provocative. What we notice is that by virtue of the passage of such a dislocation, an entire series of threefold coordinated Si atoms has been exposed. As discussed in earlier chapters in some detail, one of the distinctive characteristics of covalent materials is the strongly directional preferences adopted as a result of the covalent bonds and similarly, there is an abhorrence for the presence of dangling bonds (i.e. Si atoms prefer to maintain their conventional fourfold coordination). Because of this, the structure introduced above provides an immediate hint that we should expect that once atomic-level relaxations are allowed to take place, the arrangements in the immediate core region will change. Insights into core geometries and the consequences of such geometries for the motion of dislocations are probably the largest single achievement in the atomic theory of dislocations.

8.4 Elastic Models of Single Dislocations

In the previous section we have highlighted some of the key ambitions in the modeling of dislocations. There we argued that a first step in the chain of reasoning that begins at the microscopic scale and whose aim is the derivation of constitutive models for plasticity is the development of precise insights into the behavior of single dislocations. Indeed, the linear elastic theory we will elucidate in the following pages must be seen as the proper foundation for much of our attempt to construct such models.

Some of the main areas in which we will query our models of dislocations and their interactions surround the parallel notions of yield and hardening. Our discussion of yielding behavior will first focus on the response of single dislocations to an applied stress, and will lead to a discussion of the Peach–Koehler force which tells us how to compute the force on a dislocation given a knowledge of the state of stress. These arguments will culminate in the identification of the Peierls stress which is the critical stress needed to induce dislocation motion. We will then be forced on experimental grounds to refine our picture even in this highly idealized setting to reflect the fact that dislocations are not generally straight (i.e. kink mechanisms must be considered) and even in the case of straight dislocations, spreading of the core geometry can lead to marked differences from the predictions based upon elastic models of straight dislocations. Once these ideas are in hand, our models must be refined to consider the role of interactions of dislocations with point defects, other dislocations and interfaces.

8.4.1 The Screw Dislocation

The very existence of dislocations within the context of crystalline solids is predicated upon the intrinsic discreteness of the crystal lattice. From the standpoint of the eigenstrain concept introduced in chap. 2, the stress-free strains of significance for the emergence of dislocations are those corresponding to lattice invariant shears along the slip directions. Nevertheless, a host of useful and intriguing insights can be obtained from an elastic theory of such defects which, with the possible exception of elastic anisotropy and the nature of the Burgers vector itself, is entirely blind to the underlying lattice. As is the case with many of the defects that make for the fascinating properties of materials (i.e. cracks in solids, vortices in superconductors, etc.), the continuum theory of dislocations results in singular elastic fields, a result that was apparently first established in the work of Volterra, for whom the elastic theory of straight dislocations is named.

For many purposes, we will find that antiplane shear problems in which there is only one nonzero component of the displacement field are the most mathematically transparent. In the context of dislocations, this leads us to first undertake an analysis of the straight screw dislocation in which the slip direction is parallel to the dislocation line itself. In particular, we consider a dislocation along the x_3-direction (i.e. $\xi = (001)$) characterized by a displacement field $u_3(x_1, x_2)$. The Burgers vector is of the form $\mathbf{b} = (0, 0, \pm b)$. Our present aim is to deduce the equilibrium fields associated with such a dislocation which we seek by recourse to the Navier equations. For the situation of interest here, the Navier equations given in eqn (2.55) simplify to the Laplace equation ($\nabla^2 u_3 = 0$) in the unknown three-component of displacement. Our statement of equilibrium is supplemented by the boundary condition that for $x_1 > 0$, the jump in the displacement field be equal to the Burgers vector (i.e. $u_3(x_1, 0^+) - u_3(x_1, 0^-) = b$). Our notation $u_3(x_1, 0^+)$ means that the field u_3 is to be evaluated just above the slip plane (i.e. $x_2 = \epsilon$).

This geometry of this problem inspires a description in terms of cylindrical coordinates, within which a plausible form for the displacements is the assumption that, like a helical ramp in a parking garage, the displacements increase linearly in the winding angle, θ, yielding the solution

$$u_3(r, \theta, z) = \frac{b\theta}{2\pi}. \tag{8.10}$$

This displacement field may also be represented simply in terms of Cartesian coordinates, resulting in the expression

$$u_3(x_1, x_2) = \frac{b}{2\pi} \tan^{-1}\left(\frac{x_2}{x_1}\right), \tag{8.11}$$

and the associated strains

$$\epsilon_{31}(x_1, x_2) = -\frac{b}{4\pi} \frac{x_2}{x_1^2 + x_2^2}, \tag{8.12}$$

$$\epsilon_{32}(x_1, x_2) = \frac{b}{4\pi} \frac{x_1}{x_1^2 + x_2^2}. \tag{8.13}$$

The stress tensor corresponding to these strains may be written as

$$\sigma = \frac{\mu b}{2\pi(x_1^2 + x_2^2)} \begin{pmatrix} 0 & 0 & -x_2 \\ 0 & 0 & x_1 \\ -x_2 & x_1 & 0 \end{pmatrix}. \tag{8.14}$$

Note that care must be taken with the choice of sign of the Burgers vector to actually use these equations in practice.

The significance of the results given above may be seen in a number of different contexts. Indeed, the solution for the perfect screw dislocation in an isotropic linear elastic medium provides a window on much of what we will wish to know even in the more generic setting of an arbitrary curved dislocation. Further, this simple result also serves as the basis for our realization that all is not well with the continuum solution. One insight that emerges immediately from our consideration of the screw dislocation is the nature of the energies associated with dislocations.

Cursory inspection of the dislocation solution determined above reveals that the strain fields are singular as the dislocation core region is approached; a feature that highlights the ultimate need for calculations that respect the discrete lattice scale, either through explicit resolution of the atomic degrees of freedom or through use of suitable constitutive and kinematic nonlinearity within a continuum framework. If we are willing to suspend momentarily the question of the behavior in the near-core region, it is of great interest to determine that part of the total energy of a dislocation that arises by virtue of the elastic energy stored in its deformation fields. The simplest statement of the energy associated with a dislocation by virtue of its deformation fields is to integrate the strain energy density over the volume. Recall that for a linear elastic solid, the strain energy density may be written as

$$W(\epsilon) = \frac{1}{2} C_{ijkl} \epsilon_{ij} \epsilon_{kl}. \tag{8.15}$$

Alternatively, if we assume that there are no body forces, then the divergence theorem in conjunction with the equilibrium equations asserts

$$\int_{\Omega} \frac{1}{2} C_{ijkl} \epsilon_{ij} \epsilon_{kl} = \frac{1}{2} \int_{\partial\Omega} t_k u_k dA. \tag{8.16}$$

Hence, we find that the total elastic energy can be found by performing surface integrations associated with the tractions. It is this device that we will follow here. Though in principle we must evaluate the surface integral not only on the slipped surface, but also on both an outer and an inner circular surface, the only surface that must be considered in $\partial\Omega$ is that of the cut along the slip plane as shown in fig. 8.17. In fact, it can be shown that the contribution on both the outer and inner surfaces vanish. In terms of the coordinate system specified above, we are left with the necessity of evaluating the traction vector $t_i = \sigma_{i2} n_2$, which yields $t_3 = \sigma_{32}$. From our assessment of the displacement fields, we note that the relevant component of the stress tensor is given as

$$\sigma_{32} = \mu u_{3,2} = \frac{\mu b}{2\pi} \frac{x_1}{x_1^2 + x_2^2}. \tag{8.17}$$

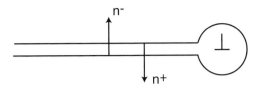

Fig. 8.17. Schematic of slip plane as used within a continuum setting.

In light of this result, we are prepared to evaluate the strain energy which is obtained by integrating

$$E = 2\frac{1}{2} \int_0^L dz \int_{r_0}^R dx \frac{\mu b^2}{4\pi x},$$

(8.18)

where we have exploited the fact that along the slip plane, $x_2 = 0$. The factor of 2 in front of the integral reflects the presence of both symmetry equivalent terms, σ_{32} and σ_{23}. The limits of integration for the integral along the line direction make it clear that we are reckoning the energy per unit length of dislocation, while the x-integral reflects the existence of a large distance cutoff corresponding to the dimensions of the body and a lower scale cutoff, corresponding loosely with the scale at which elasticity theory breaks down.

The result of this analysis is that the energy per unit length along the dislocation line is

$$E = \frac{\mu b^2}{4\pi} \ln\left(\frac{R}{r_0}\right).$$

(8.19)

This result is at once enlightening and troubling. On the one hand, on the basis of relatively little effort we have constructed an evaluation of the elastic energy associated with a screw dislocation, and while our analysis depended upon the simplicity of the screw dislocation solution, we will find that the qualitative features of our result are unchanged by the introduction of subtleties such as mixed character to the dislocation's Burgers vector and elastic anisotropy. On the other hand, to avoid a divergence in the energy we have been forced to introduce a phenomenological parameter r_0 known as the core radius, which can only be guessed at within the confines of our continuum model. Further, we note that our result for the energy is logarithmically divergent not only for the short-range part of the fields, but for the long-range parts as well. Clearly, there is more to be discovered.

An interesting estimate of the significance of the core parameter can be constructed by harking back to our discussion of the ideal strength concept. There, we noted that when shear stresses reach a value on the order of $\mu/2\pi$, the forces

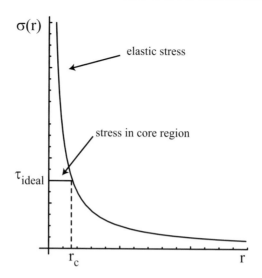

Fig. 8.18. Model of stresses in core region of dislocation based upon the ideal shear strength. The elastically diverging stresses are cutoff for $r < r_c$ on the assumption that the stresses within the core region have a constant value τ_{ideal}.

are sufficiently high to begin the shearing of adjacent crystal planes. Hence, we take the view here that the core radius may be defined as that region over which the stresses have reached values consonant with the ideal strength. One estimate of the core radius is to take that distance along the x-axis at which the shear stress reaches the ideal strength, which yields $r_0 = b$. As noted above, it is standard to idealize the core as a cylindrical region, although the locus of points of constant shear stress is certainly not circular as such a model would suppose.

This simple model of the core radius can be exploited to make an estimate for the energy associated with the core itself. The basic idea is to make the pessimistic assumption that within the region of radius r_0, the stresses are constant at the level of the ideal shear strength. This situation is depicted schematically in fig. 8.18. Our estimate of the core energy amounts to nothing more than computing the strain energy stored in this region on the assumption of a constant state of strain corresponding to that at the ideal strength limit. In particular, we note that the only terms appearing in the assessment of the strain energy density are of the form $\sigma_{31}\epsilon_{31}$ and $\sigma_{32}\epsilon_{32}$. The total strain energy within the core region is given as

$$E_{core} = \frac{1}{2}(\sigma_{31}\epsilon_{31} + \sigma_{32}\epsilon_{32})(\pi r_0^2)L, \qquad (8.20)$$

where the last two factors are geometric and characterize the cylindrical volume element under consideration. We note that for the special case of the screw dislocation, $\sigma_{31}\epsilon_{31} + \sigma_{32}\epsilon_{32} = \mu b^2/8\pi^2 r_0^2$, with the result that $E_{core} = \mu b^2 L/8\pi$.

Hence, within this simple model, the energy of the dislocation per unit length may be written as

$$\frac{E_{disloc}}{L} = \frac{\mu b^2}{4\pi}\left(\ln\frac{R}{r_0} + \frac{1}{2}\right). \tag{8.21}$$

For concreteness, it is of interest to examine the typical energy scale associated with a straight dislocation line. Consider a hypothetical material for which the shear modulus is 50 GPa, the Burgers vector is 3 Å and take as our outer cutoff and inner cutoff radii, 1 μm and 3 Å respectively. In this case, we see that the energy associated with each segment with a length comparable to the lattice spacing is of order 6 eV. Though our analysis can be refined in a number of ways, there is no escaping the fact that the energy scale here is huge on the scale of that associated with room temperature. As a result, we are momentarily left to wonder whether dislocations can arise purely by virtue of thermal fluctuations.

We have still not exhausted the insights that can be gleaned from this simplest of dislocation models. Before taking up the question of the displacement fields associated with a generic dislocation loop, we first examine the question of whether or not it is plausible that entropic effects can compensate for the huge energy associated with the dislocation, thereby raising the possibility of the existence of an equilibrium distribution of such defects. There are a few different entropic mechanisms that can be imagined as a consequence of the presence of dislocations. On the one hand, it is clear that the presence of such dislocations will alter the vibrational spectrum of the material with an attendant change in its vibrational entropy. Another contribution to the entropy is that due to the wandering of the dislocation line itself and corresponds to a configurational term. A crude estimate of this latter effect is to assume that on passing from one atomic layer to the next, the dislocation can change directions towards any of the neighboring atoms. Hence, the entropy per layer spacing is of the order

$$S = k \ln z, \tag{8.22}$$

where z represents the number of neighbors along the line direction and we will make the simplifying assumption that $\ln z \approx 1$. Hence, the contribution of configurational terms to the free energy of a dislocation is of order $TS = kT$, which is negligible in comparison with the enormous internal energy cost of the dislocation line itself.

Our initial foray into the elastic theory of dislocations has revealed much about both the structure and energetics of dislocations. From the continuum standpoint, the determination of the displacement fields (in this case u_3) is equivalent to solving for the 'structure' of the dislocation. We have determined a generic feature of such fields, namely, the presence of long-range strains that decay as r^{-1}. Another generic feature of the elastic solution is the fact that in the immediate vicinity of the dislocation core, the elastic theory breaks down. It has also become evident from our analysis that unlike point defects, dislocations have energies well above those for which an equilibrium distribution can be expected.

8.4.2 The Volterra Formula

As yet, our treatment of the elastic theory of single dislocations has depended upon the simplicity of the equilibrium equations that emerge for the particular geometry of a straight screw dislocation. While further progress can be made for specialized geometries, it is perhaps more instructive to seek generic insights into the deformation fields that attend the presence of dislocations of arbitrary geometry. In particular, we will concern ourselves with a closed dislocation loop. One approach for the elastic analysis of dislocations that is especially appealing is built around the use of the elastic Green function. As is customary, we will initially restrict our analysis to isotropic materials, though extension to an anisotropic setting, while not trivial, can be managed, at least formally. The aim of our elastic analysis will be a full characterization of the elastic fields (i.e. stresses, strains and displacements) associated with the dislocation of interest, as well as a description of its energy. Though the elastic fields are of interest in their own right, we will see in section 8.5 that it is the coupling of these fields, either to applied stresses or to internal stresses due to other defects, that makes for the beginnings of a picture of the origins of phenomena such as strain and solution hardening.

The reciprocal theorem points the way to an idea that will allow for the calculation of the displacements associated with a dislocation loop of arbitrary shape. As discussed in chap. 2, the reciprocal theorem asserts that two sets of equilibrium fields $(\sigma^{(1)}, \mathbf{u}^{(1)}, \mathbf{f}^{(1)})$ and $(\sigma^{(2)}, \mathbf{u}^{(2)}, \mathbf{f}^{(2)})$, associated with the same body (i.e. they satisfy the equilibrium equations), are related by the expression

$$\int_{\partial\Omega} \mathbf{t}^{(1)} \cdot \mathbf{u}^{(2)} dA + \int_{\Omega} \mathbf{f}^{(1)} \cdot \mathbf{u}^{(2)} dV = \int_{\partial\Omega} \mathbf{t}^{(2)} \cdot \mathbf{u}^{(1)} dA + \int_{\Omega} \mathbf{f}^{(2)} \cdot \mathbf{u}^{(1)} dV. \quad (8.23)$$

Within the context of the elastic Green function, the reciprocal theorem serves as a jumping off point for the construction of fundamental solutions to a number of different problems. For example, we will first show how the reciprocal theorem may be used to construct the solution for an arbitrary dislocation loop via consideration of a distribution of point forces. Later, the fundamental dislocation solution will be bootstrapped to construct solutions associated with the problem of a cracked solid.

We consider the generic problem of a dislocated solid with a loop such as that shown in fig. 8.10. Here we imagine two different configurations associated with the same infinite body that is described appropriately as an isotropic linear elastic solid. In the first body, we imagine a dislocation loop as shown with fields labeled by the superscript (d), while in the second, we consider a point force at position $\mathbf{r}'$ and the associated fields which are labeled with a superscript (p). For the dislocation case there are no body forces and hence $\mathbf{f}^{(d)} = 0$. By the reciprocal theorem we note that

$$\int_{\partial\Omega} \mathbf{t}^{(d)} \cdot \mathbf{u}^{(p)} dA = \int_{\partial\Omega} \mathbf{t}^{(p)} \cdot \mathbf{u}^{(d)} dA + \int_{\Omega} \mathbf{f}^{(p)} \cdot \mathbf{u}^{(d)} dV. \qquad (8.24)$$

The basic idea is to exploit the fact that the displacements associated with the point force are already known, and correspond to the elastic Green function. Further, the displacements associated with the dislocation are *prescribed* on the slip plane. We can rewrite the surface integral that spans the slipped region as

$$\int_{\partial\Omega} \mathbf{t}^{(d)} \cdot \mathbf{u}^{(p)} dA = \int_{\partial\Omega_-} \mathbf{t}^{(d)}(-) \cdot \mathbf{u}^{(p)}(-) dA + \int_{\partial\Omega_+} \mathbf{t}^{(d)}(+) \cdot \mathbf{u}^{(p)}(+) dA. \qquad (8.25)$$

In this case $\mathbf{t}^{(d)}(+)$ refers to the traction associated with the dislocation just above the slip plane while $\mathbf{t}^{(d)}(-)$ refers to the value of the traction just below the slip plane. However, since the displacement fields $(\mathbf{u}^{(p)})$ associated with the point force are continuous across the slip plane, $\mathbf{u}^{(p)}(-) = \mathbf{u}^{(p)}(+)$. Similarly, the tractions across the slip plane must be equal and opposite (i.e. $\mathbf{t}^{(d)}(+) = -\mathbf{t}^{(d)}(-)$). As a result, we are left with

$$\int_{\partial\Omega} \mathbf{t}^{(d)} \cdot \mathbf{u}^{(p)} dA = \int_{\partial\Omega_-} \mathbf{t}^{(d)}(+) \cdot \mathbf{u}^{(p)}(+) dA - \int_{\partial\Omega_+} \mathbf{t}^{(d)}(+) \cdot \mathbf{u}^{(p)}(+) dA = 0. \qquad (8.26)$$

Hence, the only terms remaining from our implementation of the reciprocal theorem yield the rather simpler expression

$$\int_{\partial\Omega} \mathbf{t}^{(p)} \cdot \mathbf{u}^{(d)} dA = -\int_{\Omega} \mathbf{f}^{(p)} \cdot \mathbf{u}^{(d)} dV \qquad (8.27)$$

Following the same decomposition of the surface integrals suggested above, we may rewrite the term involving the tractions as

$$\int_{\partial\Omega} \mathbf{t}^{(p)} \cdot \mathbf{u}^{(d)} dA = \int_{\partial\Omega} \mathbf{t}^{(p)}(+) \cdot [\mathbf{u}^{(d)}(+) - \mathbf{u}^{(d)}(-)] dA. \qquad (8.28)$$

If we now recall that the jump in the displacement field across the slip plane is nothing more than the Burgers vector and we exploit the fact that the traction vector is related to the stress tensor via $\mathbf{t}^{(p)} = \boldsymbol{\sigma}^{(p)}\mathbf{n}$, then the left hand side of eqn (8.28) may be rewritten as

$$\int_{\partial\Omega} \mathbf{t}^{(p)} \cdot \mathbf{u}^{(d)} dA = -\int_{\partial\Omega} \sigma_{ij}^{(p)} n_j b_i dA. \qquad (8.29)$$

Recalling our linear elastic constitutive assumption, we may further simplify this result to read

$$\int_{\partial\Omega} \mathbf{t}^{(p)} \cdot \mathbf{u}^{(d)} dA = -\int_{\partial\Omega} C_{ijmn} u^{(p)}_{m,n} n_j b_i dA. \tag{8.30}$$

Thus far, we have treated only one side of eqn (8.27). Our next task is to probe the significance of the term involving the body force. Through the artifice of choosing the fields associated with a point force as our complementary fields, the dislocation solutions can be obtained directly. In particular, we note that if $f_j^{(p)}(\mathbf{r}') = f_j \delta(\mathbf{r} - \mathbf{r}')$, then we have

$$\int_{\Omega} \mathbf{f}^{(p)} \cdot \mathbf{u}^{(d)} dV = u_k^{(d)}(\mathbf{r}) f_k, \tag{8.31}$$

which involves the unknown dislocation displacement field. We may tidy up this result by recalling that $u_m^{(p)}$ that appears in eqn (8.30) is the solution for the point force problem and hence is given by $u_m^{(p)} = -G_{mk} f_k$. As a result, our efforts culminate in the expression,

$$u_k(\mathbf{r}) = -\int_{\partial\Omega} C_{ijmn} G_{mk,n}(\mathbf{r} - \mathbf{r}') n_j b_i dA', \tag{8.32}$$

a result known as Volterra's formula. Note that this result is a special case of the formula we wrote earlier in the eigenstrain context as eqn (2.96). The physical content of this result is that we have found a way by integrating only on the slipped surface to determine the displacement fields (in linear elasticity) throughout space. In practice, this method is unwieldy for consideration of the standard textbook examples concerning straight dislocations of either pure edge or screw character. On the other hand, for determining the fields resulting from dislocation loops of arbitrary shape, this formula and its variants are considerably more enlightening.

Despite the heavy-handedness of the Volterra formula in the context of straight dislocations, it is instructive to see it in action. As an example of this line of thought, we now use the Volterra formula to reconstruct the displacement fields associated with a screw dislocation. As was shown earlier, the displacement fields that are carried by a screw dislocation in an isotropic linear elastic medium may be easily arrived at through the insight that the equilibrium equations reduce to Laplace's equation in the single unknown field, u_3. Here we instead consider these fields as the result of our Green function analysis. We orient our coordinate system such that the Burgers vector is given by $\mathbf{b} = (0, 0, b)$ and the normal to the slip plane is $\mathbf{n} = (0, 1, 0)$. Further, we recall eqn (2.54) which gives the elastic modulus tensor in terms of the Lamé constants. Substitution of these quantities into the

Volterra formula yields

$$u_m(\mathbf{r}') = -\mu b \int_{\partial \Omega} (G_{33,2} + G_{23,3}) dA. \tag{8.33}$$

The derivatives of the Green function that we are instructed to evaluate can be carried out through reference to eqn (2.91) resulting in

$$u_3(\mathbf{r}') = \frac{-\mu b y'}{8\pi \mu (\lambda + 2\mu)} \int_{-\infty}^{0} dx \int_{-\infty}^{\infty} dz \left\{ \frac{(\lambda + 3\mu)}{[(x - x')^2 + y'^2 + z^2]^{\frac{3}{2}}} \right.$$
$$\left. + \frac{3(\lambda + \mu)z^2}{[(x - x')^2 + y'^2 + z^2]^{\frac{5}{2}}} \right\}. \tag{8.34}$$

Not surprisingly, the resulting integrals are entirely tractable and lead to the result

$$u_3(\mathbf{r}') = -\frac{b}{2\pi} \tan^{-1} \left(\frac{y'}{x'} \right), \tag{8.35}$$

in exact coincidence with the displacements that emerge in the simpler, more standard treatment as already revealed in eqn (8.11). A similar analysis can be performed in the context of the conventional edge dislocation within the isotropic elasticity context (see Problems).

8.4.3 The Edge Dislocation

We have now succeeded in identifying the deformations induced by a screw dislocation, and have also found a generic formula for the fields due to an arbitrary dislocation. Our present task is to return to the edge dislocation for the purposes of completeness and because this solution will make its way into our future analysis. This is particularly evident in the case of isotropic linear elasticity where we will find that even for dislocations of mixed character, their geometries may be thought of as a superposition of pure edge and pure screw dislocations.

Recall from our discussion in chap. 2 that the solution of elasticity problems of the two-dimensional variety presented by the edge dislocation is often amenable to a treatment in terms of the Airy stress function. Consultation of Hirth and Lothe (1992), for example, reveals a well-defined prescription for determining the Airy stress function. The outcome of this analysis is the recognition that the stresses in the case of an edge dislocation are given by

$$\sigma_{xx}(x, y) = -\frac{\mu b}{2\pi (1 - \nu)} \frac{y(3x^2 + y^2)}{(x^2 + y^2)^2}, \tag{8.36}$$

$$\sigma_{yy}(x, y) = \frac{\mu b}{2\pi (1 - \nu)} \frac{y(x^2 - y^2)}{(x^2 + y^2)^2}, \tag{8.37}$$

$$\sigma_{xy}(x, y) = \frac{\mu b}{2\pi(1 - \nu)} \frac{x(x^2 - y^2)}{(x^2 + y^2)^2}.$$ (8.38)

Earlier we made the promise that the energy stored in the elastic fields had a generic logarithmic character, regardless of the type of the dislocation. To compute the elastic strain energy associated with the edge dislocation we use eqn (8.16). In the context of the stress fields given above,

$$\frac{E_{strain}}{L} = \frac{b}{2} \frac{\mu b}{2\pi(1 - \nu)} \int_{r_0}^{R} \frac{dx}{x} = \frac{\mu b^2}{4\pi(1 - \nu)} \ln \frac{R}{r_0}.$$ (8.39)

The integrand is determined by computing the tractions *on the slip plane* using the stresses for the edge dislocation described above, and as claimed, the structure of the energy is exactly the same as that found earlier for the screw dislocation.

8.4.4 Mixed Dislocations

As noted earlier, generic dislocations have neither pure screw nor pure edge character. If we consider a straight dislocation with line direction $\boldsymbol{\xi} = (0, 1, 0)$, then the dislocation has a Burgers vector of the form $\mathbf{b} = b(\sin \theta, \cos \theta, 0)$, where θ is the angle between the line direction and the Burgers vector. We see that using this notation, a screw dislocation is characterized by $\theta = 0$ and an edge dislocation similarly by $\theta = 90°$. For the special case in which we consider an isotropic linear elastic material, the total stress field due to a mixed dislocation is a superposition of the fields due to its edge and screw components treated separately. What this means in particular is that the stress tensor can be written as

$$\sigma = \begin{pmatrix} \sigma_{11}^e & \sigma_{12}^e & \sigma_{13}^s \\ \sigma_{12}^e & \sigma_{22}^e & \sigma_{23}^s \\ \sigma_{13}^s & \sigma_{23}^s & \sigma_{33}^e \end{pmatrix}.$$ (8.40)

The superscripts e and s refer to edge and screw and serve as an instruction to use the isotropic linear elastic stress fields for the edge and screw dislocation, respectively, but with the Burgers vector adjusted to account for relevant trigonometric weighting factors.

Elastic isotropy considerably simplifies the analyses that we are forced to undertake in our goal of characterizing the deformation fields associated with a dislocation. On the other hand, there are some instances in which it is desirable to make the extra effort to include the effects of elastic anisotropy. On the other hand, because the present work has already grown well beyond original intentions and because the addition of anisotropy is for the most part an elaboration of the physical ideas already set forth above, we refer the reader to the outstanding work of Bacon *et al.* (1979).

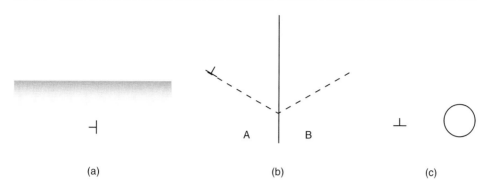

Fig. 8.19. Schematic illustration of breaking of translational invariance by (a) the presence of a free surfaces, (b) an internal interface between two dissimilar materials, and (c) a precipitate particle.

8.5 Interaction Energies and Forces

Though the elastic theory of single dislocations is interesting in its own right, our elucidation of the elastic fields of such dislocations leaves us in a position to begin to explore what can be learned about the plastic deformation of crystalline solids on the basis of dislocations. At the very heart of the questions of both yield and hardening is the issue of how a given dislocation responds to the presence of a stress field. Whether that stress field has its origins in some remotely applied external load or from the presence of inhomogeneities such as point defects, surfaces and interfaces, voids or other dislocations themselves, ultimately we require a quantitative theory of how the dislocation will respond to such forces.

We follow the lead of Eshelby already introduced in chap. 2 in our discussion of configurational forces. Recall that a small excursion $\delta\xi_i$ of the defect of interest will result in a change of the total energy of the system and attendant loading devices of the form $\delta E_{tot} = -F_i\delta\xi_i$, where the configurational force F_i is the work-conjugate partner of the excursion $\delta\xi_i$. As we noted earlier, the origin of such configurational forces is the breaking of translational symmetry. In the case of a single dislocation in an infinite body that is treated within the confines of the theory of linear elasticity there is no such symmetry breaking, and this is manifested in the fact that translation of the dislocation core results in no energy change. This situation is to be contrasted with those shown in fig. 8.19 in which we examine a dislocation in the presence of other defects. In these cases, it is clear that a small translation of the dislocation will result in a change in the total energy, signaling the presence of a nonzero configurational force on the dislocation.

For the moment, we will restrict our attention to the general statements that can be made concerning such forces in the context of the linear theory of elasticity. In this case, the results simplify considerably in that we can think of such energetic

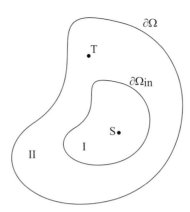

Fig. 8.20. Organization of surface integrals for computing the interaction energy between the fields of the defect of interest and all other fields.

forces as arising from the interaction terms between the separate stress systems. Let us parcel the stresses and strains into two groups, those of the singularity of interest which we will label σ_{ij}^S and those from the remaining singularities and external fields, σ_{ij}^T. Eshelby's idea is to note that if we write the total elastic energy as

$$E_{elastic} = \frac{1}{2} \int_\Omega (\sigma_{ij}^S + \sigma_{ij}^T)(\epsilon_{ij}^S + \epsilon_{ij}^T)dV, \tag{8.41}$$

then only the cross terms will be affected by the small excursions we intend to impose on the singularity S. Hence, we may rewrite the part of the total energy that is not translationally invariant as

$$E_{interaction} = \frac{1}{2} \int_\Omega (\sigma_{ij}^S \epsilon_{ij}^T + \sigma_{ij}^T \epsilon_{ij}^S)dV. \tag{8.42}$$

Consideration of fig. 8.20 suggests a strategy for rewriting the interaction energy. Note that within region I, the fields T are jump free while for region II, the same may be said of the fields S. Hence, we rewrite the interaction energy as

$$E_{interaction} = \int_I \sigma_{ij}^S u_{i,j}^T dV + \int_{II} \sigma_{ij}^T u_{i,j}^S dV. \tag{8.43}$$

This expression insures that our integrations are organized such that the integrands are well behaved in the regions of interest. If we use the fact that the stress fields satisfy the equilibrium equations (i.e. $\sigma_{ij,j} = 0$) in conjunction with the divergence theorem, this expression may be rewritten as

$$E_{interaction} = \int_{\partial \Omega_{in}} (\sigma_{ij}^S u_i^T - \sigma_{ij}^T u_i^S)n_j dS, \tag{8.44}$$

where the surface integral is evaluated on the dividing surface $\partial \Omega_{in}$ separating the defect of interest from the rest of the crystal. This interaction energy is independent of the choice of $\partial\Omega_{in}$ so long as it encloses the singularity S. Though eqn (8.44) is abstract, it forms the basis of our calculation of the configurational forces on dislocations. In particular, on the basis of this expression we are now in a position to derive the Peach–Koehler formula which relates the force on a dislocation to the local stress field and the geometric descriptors of the dislocation itself such as the Burgers vector $\mathbf{b}$ and the line direction $\boldsymbol{\xi}$.

8.5.1 The Peach–Koehler Formula

The expression derived above lays the foundation for much of our work on the forces on dislocations. In particular, we shall find it convenient to direct our attention to the way in which an externally applied stress couples to the motion of a dislocation. This will culminate in the elucidation of the so-called Peach–Koehler force which explicitly reckons the force on a dislocation. Before invoking the Peach–Koehler formula, it is perhaps useful to first examine the force between two screw dislocations from the perspective of the interaction energy. Our interest is in determining the force of interaction between these dislocations using the Eshelby philosophy espoused above. In particular, we must evaluate the interaction energy when the dislocations are separated by a distance d and when they are separated by a distance $d + \xi$, which reflects the existence of the small excursion which serves as the basis of our calculation of the configurational force. In this case, we note that the total energy is given by

$$E_{elastic} = \frac{1}{2}\int_{\Omega}\sigma_{ij}^{(1)}\epsilon_{ij}^{(1)}dV + \frac{1}{2}\int_{\Omega}\sigma_{ij}^{(2)}\epsilon_{ij}^{(2)}dV + \frac{1}{2}\int_{\Omega}\sigma_{ij}^{(1)}\epsilon_{ij}^{(2)}dV + \frac{1}{2}\int_{\Omega}\sigma_{ij}^{(2)}\epsilon_{ij}^{(1)}dV.$$

$$(8.45)$$

The last two terms, which are equal, represent the interaction energies, and are the only terms that will be affected by the excursion of the dislocation. Hence, we may write

$$E_{int} = \int_{\Omega}\sigma_{ij}^{(1)}\epsilon_{ij}^{(2)}dV,$$

$$(8.46)$$

which in terms of the particular dislocation fields for this case is

$$E_{int} = \frac{\mu b_1 b_2}{4\pi^2}\int_{\Omega}\left\{\frac{y^2}{(x^2+y^2)[(x-d)^2+y^2]} + \frac{x(x-d)}{(x^2+y^2)[(x-d)^2+y^2]}\right\}dV.$$

$$(8.47)$$

The result of this integration may be derived after some calculations demanding algebraic care, with the result that

$$E_{int}(d) = \frac{\mu b_1 b_2}{2\pi}\ln\left(\frac{R}{d}\right),$$

$$(8.48)$$

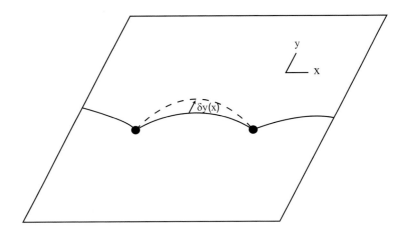

Fig. 8.21. Small excursion of dislocation segment.

where R is a large distance cutoff. In order to compute the configurational force, we require the difference between $E_{int}(d + \xi)$ and $E_{int}(d)$ which to lowest order in ξ is given by

$$\Delta E_{int} = -\frac{\mu b_1 b_2 \xi}{2\pi d},$$

(8.49)

which implies that the interaction force is given by

$$F = \frac{\mu b_1 b_2}{2\pi d}.$$

(8.50)

We see that in the case in which the Burgers vectors are equal, the force of interaction is repulsive, whereas in the case in which they are equal and opposite, the resulting force is attractive.

In reality, all of this was vastly more work than was really needed to derive this elementary result. Our reason for pursuing this heavy-handed approach was not so much in the name of the particular result as to illustrate how the broken translational invariance induced by the presence of *two* dislocations manifests itself in the emergence of an interaction energy. The whole enterprise of computing the configurational forces can be worked out in explicit and gory detail in this case, although a more intelligent approach to the integrations would have been to compute the work done in bringing the second dislocation up to the first one.

The result derived above is but a special case of a more general formula which can be derived from eqn (8.44). In order to compute the more general result, we imagine a small excursion of the segment of dislocation as shown in fig. 8.21. In light of eqn (8.44), we must evaluate

$$E_{int}(S, T) = \int_{\partial\Omega_{in}^+} \sigma_{ij}^S(+)u_i^T(+)n_j(+)dS - \int_{\partial\Omega_{in}^+} \sigma_{ij}^T(+)u_i^S(+)n_j(+)dS$$

$$+ \int_{\partial\Omega_{in}^-} \sigma_{ij}^S(-)u_i^T(-)n_j(-)dS - \int_{\partial\Omega_{in}^-} \sigma_{ij}^T(-)u_i^S(-)n_j(-)dS, \quad (8.51)$$

where the label S refers to the fields associated with the dislocation that is bowing out and the label T refers to all other contributions to the stress. The labels $+$ and $-$ refer to the integration surfaces which are just above and just below the slip plane. We now use the fact that for the case in question, the stresses σ_{ij}^S associated with the dislocation are continuous (i.e. $\sigma_{ij}^S(+) = \sigma_{ij}^S(-)$) and that the displacements associated with system T are continuous (i.e. $u_i^T(+) = u_i^T(-)$). Finally, upon noting that $n_i(+) = -n_i(-)$, we may rewrite the expression above as

$$E_{int}(S, T) = \int_{\partial\Omega_{in}} b_i \sigma_{ij}^T n_j(-)dS. \quad (8.52)$$

Recall that the definition of the Burgers vector is $b_i = u_i(+) - u_i(-)$, which has been used in writing the above expression in this final form.

For the special case of the infinitesimal bow-out of a dislocation segment under the action of the stress σ_{ij}^T, we note that

$$\delta E_{int} = b_i \sigma_{ij}^T (d\mathbf{y} \times d\mathbf{l})_j. \quad (8.53)$$

This equation is written on the basis of the observation that the infinitesimal element of area swept out by the dislocation is characterized by the vector product $d\mathbf{y} \times d\mathbf{l}$. Note that $d\mathbf{y}$ is the local excursion of the segment, while $d\mathbf{l}$ is a vector along the line at the point of interest. If we now recall that the configurational force is given as $\delta E_{int} = -F_m \delta\xi_m$, then we may evidently write the configurational force in the present context as

$$\frac{F_m}{dl} = \epsilon_{mjn} b_i \sigma_{ij} \zeta_n, \quad (8.54)$$

where we have used $d\mathbf{l} = dl\boldsymbol{\zeta}$. We apologize for temporarily denoting the local tangent vector as $\boldsymbol{\zeta}$, but note that had we not done so, the symbol ξ would be doing double time, serving to denote both the excursion of the dislocation as well as its local tangent vector. In vector form, we may write this as

$$\frac{\mathbf{F}}{dl} = (\boldsymbol{\sigma}\mathbf{b}) \times \boldsymbol{\zeta}. \quad (8.55)$$

The physical interpretation of this result is that it is the force per unit length of dislocation acting on a segment by virtue of the presence of the stress field $\boldsymbol{\sigma}$. This expression is the famed Peach–Koehler formula and will be seen to yield a host of interesting insights into the interactions between dislocations.

8.5.2 Interactions and Images: Peach–Koehler Applied

In the discussion given above, we have resurrected the notion of an energetic force as first described in chap. 2. The key point is the existence of a configurational force associated with energy changes that attend small excursions of the defect of interest. In the dislocation setting, we have shown that such arguments lead naturally to the Peach–Koehler formula. Presently, we use this formula to investigate the interactions between dislocations in different contexts. Part of our interest in this topic arises from our ultimate interest in explicating hardening in crystals as a consequence of dislocation interactions. Though the elastic theory will be found wanting once the dislocations in question are in too close a proximity, for larger distances of separation, the elastic theory offers an enlightening description of the types of dislocation interactions that are important.

Interactions of Two Straight Edge Dislocations

We begin with an analysis of the simple case of two parallel edge dislocations. Our aim in the present context is to compute the component of the force which will lead to gliding motion if it is sufficiently large as to overcome the intrinsic lattice resistance. The line directions of both dislocations are given by $\boldsymbol{\xi}_1 = \boldsymbol{\xi}_2 = (0, 0, 1)$, and their Burgers vectors are $\mathbf{b}_1 = \mathbf{b}_2 = (b, 0, 0)$. The Peach–Koehler formula instructs us to compute $\mathbf{F} = \boldsymbol{\sigma}\mathbf{b} \times \boldsymbol{\xi}$. Further, we consider a geometry in which dislocation 1 is at the origin while dislocation 2 is to be found at (x, d). The state of stress at dislocation 2 due to the presence of dislocation 1 is given by

$$\boldsymbol{\sigma} = \frac{\mu b}{2\pi(1-\nu)} \begin{pmatrix} -\dfrac{d(3x^2+d^2)}{(x^2+d^2)^2} & \dfrac{x(x^2-d^2)}{(x^2+d^2)^2} & 0 \\ \dfrac{x(x^2-d^2)}{(x^2+d^2)^2} & \dfrac{d(x^2-d^2)}{(x^2+d^2)^2} & 0 \\ 0 & 0 & -\dfrac{d\nu}{x^2+d^2} \end{pmatrix}. \tag{8.56}$$

By exercising the instructions dictated by the Peach–Koehler formula using the state of stress given above, the force on dislocation 2 is given by

$$\mathbf{F}_{1on2} = \frac{\mu b^2}{2\pi(1-\nu)(x^2+d^2)^2}\{\mathbf{i}[x(x^2-d^2)] + \mathbf{j}[d(3x^2+d^2)]\}. \tag{8.57}$$

If we work in units in which x is measured as a fraction f of d (i.e. $x = fd$) then the glide force (i.e. the force in the x-direction) may be rewritten in dimensionless form as

$$\frac{2\pi d(1-\nu)F}{\mu b^2} = \frac{f(f^2-1)}{(f^2+1)^2}. \tag{8.58}$$

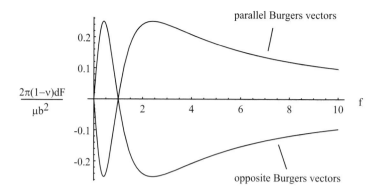

Fig. 8.22. Force of interaction between two dislocations on adjacent slip planes.

The normalized glide force is plotted in fig. 8.22 where it is seen that there are two equilibrium points for the dislocations one of which is stable and the other of which is unstable. In fact, we have plotted the interaction force for the alternative case in which the dislocations have opposite Burgers vectors as well so as to contrast the resulting solutions.

In preparation for our discussion of dislocation junctions, we now consider the case of two dislocations that are not parallel. This analysis is important as a first step in evaluating the origins of 'forest hardening' in which it is presumed that dislocations gliding on one plane are blocked as a result of dislocations that are piercing their slip plane. The suggestive imagery of a forest of trees serves as a mnemonic for this blocking action and is taken up presently. For simplicity, consider the case depicted in fig. 8.23 in which two screw dislocations characterized by $\boldsymbol{\xi}_1 = (0, 0, 1)$, $\boldsymbol{\xi}_2 = (0, 1, 0)$, $\mathbf{b}_1 = (0, 0, b)$ and $\mathbf{b}_2 = (0, b, 0)$, interact according to

$$\boldsymbol{\sigma}^{(1)}\mathbf{b}^{(2)} = \frac{\mu b}{2\pi(s^2 + y^2)} \begin{pmatrix} 0 & 0 & -y \\ 0 & 0 & s \\ -y & s & 0 \end{pmatrix} \begin{pmatrix} 0 \\ b \\ 0 \end{pmatrix}. \qquad (8.59)$$

Note that we denote the distance of closest approach of the dislocations (i.e. along the x_1-axis) as s. Carrying out the relevant algebra in conjunction with the Peach–Koehler formula leads to the result that

$$\mathbf{F} = -\frac{\mu b^2 s}{2\pi(s^2 + y^2)}\mathbf{i}. \qquad (8.60)$$

This result yields the insight that everywhere along the dislocation line the force of interaction is attractive. However, we note that as a function of the perpendicular

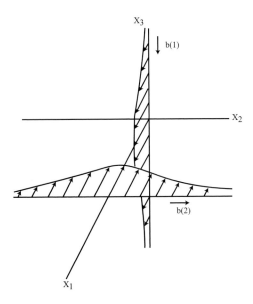

Fig. 8.23. Interaction force between two perpendicular screw dislocations (adapted from Hartley and Hirth (1965)).

distance from the line of closest interaction, the interaction becomes weaker. The force profile is indicated schematically in fig. 8.23.

Image Forces

One of the key manifestations of the broken translational invariance resulting from the presence of a free surface is that of an image force. The presence of a free surface (or an internal interface) implies a breaking of the translational symmetry so that a small excursion of a dislocation will have an attendant energy cost and thereby an attendant configurational force. In the present section, we undertake an analysis of the image forces that emerge in the simplest of cases, namely that of a screw dislocation near a free surface. The geometry of interest is depicted in fig. 8.24 which shows a screw dislocation a distance d below a free surface. From the standpoint of the elastic boundary value problem involving the dislocation and the free surface, the crucial additional requirement beyond the normal jump condition on the dislocations slip plane is the requirement that the free surface be traction free. What this means in particular is that $\mathbf{t_n} = \boldsymbol{\sigma}\mathbf{n} = 0$.

If we begin our analysis by considering the fields due to the dislocation beneath the surface, pretending for the moment that the appropriate fields are those for a dislocation in an infinite body, we have

$$\sigma_{31}(x, y) = -\frac{\mu b}{2\pi} \frac{y + d}{x^2 + (y + d)^2}, \qquad (8.61)$$

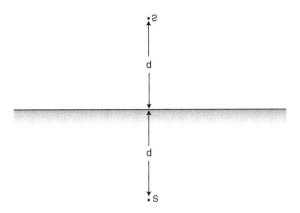

Fig. 8.24. Subsurface screw dislocation and its image partner.

$$\sigma_{32}(x, y) = \frac{\mu b}{2\pi} \frac{x}{x^2 + (y + d)^2}. \tag{8.62}$$

These fields imply a surface traction of the form $\mathbf{t_n} = (0, 0, \sigma_{32})$. This observation inspires the conclusion that if we could arrange for a compensating stress $-\sigma_{32}$ at the free surface, then the boundary conditions would be satisfied. In particular, for this simple geometry it is well known that a parallel dislocation of Burgers vector $-b$ at a distance d *above* the surface (as depicted in fig. 8.24) accomplishes this end.

The full fields for this problem are

$$\sigma_{31}(x, y) = -\frac{\mu b}{2\pi} \frac{y + d}{x^2 + (y + d)^2} + \frac{\mu b}{2\pi} \frac{y - d}{x^2 + (y - d)^2}, \tag{8.63}$$

$$\sigma_{32}(x, y) = \frac{\mu b}{2\pi} \frac{x}{x^2 + (y + d)^2} - \frac{\mu b}{2\pi} \frac{x}{x^2 + (y - d)^2}. \tag{8.64}$$

A few remarks on the significance of our assumptions are in order. First, we note that the linearity of our elasticity problem is central to the ability to write this solution. Because the sum of two solutions is itself a solution, we know that the elastic fields themselves are a solution to the problem. Furthermore, by construction, the fields satisfy the free surface boundary conditions, and hence these fields are the solution. In light of this result, we are now ready to pose the question concerning the configurational force induced by the presence of the free surface. We ask the question just how does the energy change if the subsurface dislocation makes a small excursion towards the free surface. However, we have already successfully carried out this calculation in an earlier section and the resulting force was given in eqn (8.50), although in the present case the signs of the dislocations in question are opposite and hence there is an attractive force

$$F_{image} = -\frac{\mu b^2}{4\pi d}.$$ (8.65)

Note that in contrast to the result in eqn (8.50), there is a factor of 4 rather than 2 in the denominator. This arises because the separation of the image dislocations themselves is $2d$, not d. We have now seen that by virtue of the presence of a free surface, an image force is induced on a subsurface dislocation. In addition, the idea of an image force has further reinforced our use of configurational forces and their origins in the breaking of translational symmetry.

8.5.3 The Line Tension Approximation

In many instances, the evaluation of the complete set of interactions between several dislocations can be prohibitive, at least at the level of the kind of analysis that one might wish to make with a stick in the sand. Indeed, though in the vast majority of this chapter we have been engaged in the development of the elastic theory of dislocations, we now undertake a different framework in which the complex interactions between various infinitesimal segments of a given dislocation are entirely neglected.

   An alternative to the full machinery of elasticity that is especially useful in attempting to make sense of the complex properties of three-dimensional dislocation configurations is the so-called line tension approximation. The line tension idea borrows an analogy from what is known about the perturbations of strings and surfaces when they are disturbed from some reference configuration. For example, we know that if a string is stretched, the energetics of this situation can be described via

$$\text{energy change} = T\delta l.$$ (8.66)

Similarly, if the surface area of a particular interface is altered, the energetics may be handled through a similar idea, namely,

$$\text{energy change} = \gamma \Delta A,$$ (8.67)

where γ is a surface energy.

   The analogy with the interfacial energy is more appropriate since dislocations, like free surfaces, have an anisotropy which leads to the conclusion that the energy cost for a small excursion from some flat reference state depends upon the orientation of that reference state. However, the analogy between dislocations and interfaces is imperfect since models of this type have an inherent locality assumption which is exceedingly fragile. To explore this further, we note that in the line tension approximation, the total energy of a dislocation line may be written as

$$E_{disloc} = \int_{\Gamma} E[\theta(s)]ds, \qquad (8.68)$$

where we have written the dislocation line parametrically in terms of the parameter s, Γ reminds us to integrate along the length of the dislocation of interest and θ is the angle between the local tangent vector and the Burgers vector. The subtlety of the assertion embodied in this equation is that of locality. The basic premise is that the contribution of a given segment to the total energy of the dislocation line depends only on one thing, namely, its local orientation, and is *indifferent* to the disposition of the remaining segments on the dislocation line. In the context of surfaces, this locality assumption may well be appropriate. On the other hand, because of the long-range elastic interactions between different segments, the energy of a given segment is patently not independent of the disposition of the remaining segments on the line.

Nevertheless, as an approximate procedure for reckoning the total energy of some complex dislocation configuration, we can make this locality assumption and use what we have learned earlier about the energetics of straight dislocations to build up a local line energy $E(\theta)$ which depends only on the misorientation θ between the line direction and the Burgers vector. The most naive of all line tension models is that in which a single constant energy is assigned to all orientations. Based on what has been learned earlier (see eqns (8.19) and (8.39)), the assertion is made that $E(\theta) = T = \frac{1}{2}\mu b^2$, independent of the angle θ. In this case, the line tension becomes a strictly geometric approximation and results in comparisons between different geometries predicated on the overall dislocation line length. In particular, in this approximation the energy is given by

$$E_{disloc} = T \int_{\Gamma} \sqrt{\frac{d\mathbf{x}(s)}{ds} \cdot \frac{d\mathbf{x}(s)}{ds}} ds. \qquad (8.69)$$

Note that we have exploited a parameteric representation of the dislocation line in which the position of the line is given by $\mathbf{x}(s)$.

A more sophisticated version of the line tension approximation is based upon the idea of using the energy per unit length that is obtained for a mixed dislocation in which the angle between the Burgers vector and the line direction is denoted by θ. In this case, the isotropic linear elastic results of section 8.4.4 can be used to obtain the energy per unit length of a mixed dislocation which is given by

$$\frac{E(\theta)}{L} = \frac{\mu b^2}{4\pi}\left(\frac{\sin^2\theta}{1-\nu} + \cos^2\theta\right)\ln\left(\frac{R}{r_0}\right). \qquad (8.70)$$

Once we are in possession of this result, we are then prepared to examine the energetics of a given dislocation configuration by exploiting it in conjunction with eqn (8.68).

We will resort to the line tension approximation repeatedly since as was noted above, much may be learned about the key features of a given problem on the basis of such arguments, which are largely geometrical. Again, what is especially appealing about the line tension approach is the prospect for making analytic headway on fully three-dimensional problems.

8.6 Modeling the Dislocation Core: Beyond Linearity

Our discussion of dislocations has thus far made little reference to the significant role that can be played by the dislocation core. Whether one thinks of the important effects that arise from the presence of stacking faults (and their analogs in the alloy setting) or of the full scale core reconstructions that occur in materials such as covalent semiconductors or the bcc transition metals, the dislocation core can manifest itself in macroscopic plastic response. Our aim in this section is to take stock of how structural insights into the dislocation core can be obtained.

Perhaps the simplest of those features of dislocations that may be thought of as core effects are those associated with the splitting of dislocations into partial dislocations which have associated Burgers vectors that do not correspond to lattice invariant shears. These ideas were already introduced in section 8.3.2. Though the slip associated with partial dislocations does not restore the lattice completely it is nevertheless energetically tolerable with an energy that scales with the stacking fault energy. A dislocation may undergo a splitting reaction into two dislocations that bound a stacking fault. In simple terms, we would describe the resulting dislocation as having a complex core structure.

Our treatment in this section will cover three primary thrusts in modeling dislocation core phenomena. Our first calculations will consider the simplest elastic models of dislocation dissociation. This will be followed by our first foray into mixed atomistic/continuum models in the form of the Peierls–Nabarro cohesive zone model. This hybrid model divides the dislocation into two parts, one of which is treated using linear elasticity and the other of which is considered in light of a continuum model of the atomic-level forces acting across the slip plane of the dislocation. Our analysis will finish with an assessment of the gains made in direct atomistic simulation of dislocation cores.

8.6.1 Dislocation Dissociation

In this section, our aim is to compute the core geometry associated with a dissociated dislocation in an fcc material. Our model will be founded upon a

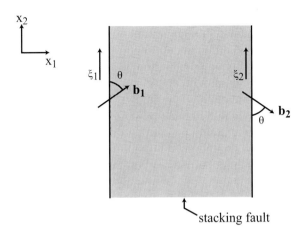

Fig. 8.25. Geometry of dislocation splitting.

linear elastic analysis of the interactions between the partial dislocations, but supplemented with a constant energy cost γ_{sf} per unit area of stacking fault. To make the analysis eminently tractable, we restrict our attention to an isotropic material. The total energy of the system composed of two partials and the stacking fault they bound is

$$E_{tot}(d) = E_{self} + E_{int}(d) + \gamma_{sf}d. \qquad (8.71)$$

What we have noted here is that the self-energies of the two Shockley partials are independent of their spacing. It is only the interaction energy between the partials and the stacking fault area itself that involves the splitting distance d. The equilibrium spacing is to be determined by minimizing this total energy with respect to d.

We note that the interaction energy may be written as

$$E_{int} = \int_{\Omega} \sigma_{ij}^{(1)} \epsilon_{ij}^{(2)} dV, \qquad (8.72)$$

where the labels 1 and 2 refer to the two partners in the dissociation reaction as shown in fig. 8.25. The dislocation lines are characterized by $\boldsymbol{\xi}^{(1)} = \boldsymbol{\xi}^{(2)} = (0, 1, 0)$, $\mathbf{b}^{(1)} = (b\sin\theta, b\cos\theta, 0)$ and $\mathbf{b}^{(2)} = (b\sin\theta, -b\cos\theta, 0)$. For the special case of elastic isotropy we note that the stress tensor decomposes as shown in eqn (8.40) where we have invoked the observation that the partial dislocations themselves are mixed dislocations. The significance of this observation is that it implies that the interaction energy may be written as the sum of two interactions, $E_{int} = E_{int}^{edge} + E_{int}^{screw}$. Using the interaction energies derived earlier, this result may be rewritten as

$$E_{int} = -\frac{\mu b^2 \sin^2 \theta}{2\pi(1-\nu)} \ln\left(\frac{d}{R}\right) + \frac{\mu b^2 \cos^2 \theta}{2\pi} \ln\left(\frac{d}{R}\right). \qquad (8.73)$$

The minimization of eqn (8.71) is effected by differentiating the total interaction energy with respect to d and leads to

$$d_{eq} = \frac{\mu b^2}{2\pi \gamma_{sf}} \left[\frac{\sin^2 \theta}{(1-\nu)} - \cos^2 \theta\right]. \qquad (8.74)$$

Thus we see that the splitting of dislocations into partials is characterized by a stacking fault width that is set by the magnitude of the stacking fault energy, with low stacking fault energy materials having larger splittings. As said above, the splitting of dislocations into partials can be thought of as the most elementary of core effects.

8.6.2 The Peierls–Nabarro Model

Despite the significant gains that have been made with atomistic models of the dislocation core (which we will take up in the next section), it remains of interest to produce analytical insights into the role of the core. One compelling strategy in this regard is that known as the Peierls–Nabarro model which is but one example of the general class of cohesive surface strategies that we will take up in more depth in chap. 12. The cohesive surface strategy in this context is an artificial mixture of conventional continuum thinking in the form of a linear elastic treatment of the vast majority of the body with a planar constitutive model whose aim it is to mimic the atomic-level forces that act across the slip plane of the dislocated body. This example serves as yet another reminder of the way in which information gleaned at one scale (in this case the interplanar potential) can be used to inform models of a different type.

The basic idea of such thinking is embodied in fig. 8.26, where the separation of the dislocated body into two elastic half-spaces joined by atomic-level forces across their common interface is shown. The formulation is aimed at determining the energy minimizing slip distribution on the slip plane. From a kinematic perspective the dislocation is characterized by the slip distribution $\delta(x) = u(x, 0^+) - u(x, 0^-)$ which is a measure of the disregistry across the slip plane. In order to provide a proper variational setting within which to pose this problem we must first identify the total energy itself as a functional of this unknown slip distribution. The total energy is built up of two distinct contributions that impose competing influences on the equilibrium slip distribution. One of the terms in the total energy accounts for the atomic-level interactions across the slip plane which reflect the fact that there is

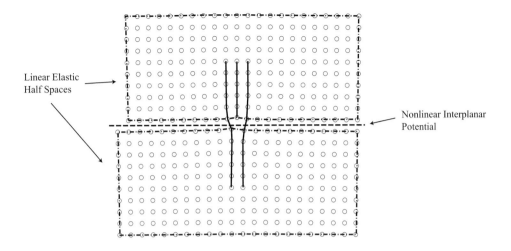

Fig. 8.26. Schematic of key elements in Peierls–Nabarro dislocation model (courtesy of R. Miller). The key idea is the use of a nonlinear constitutive model on the relevant slip plane which is intended to mimic the atomic-level forces resulting from slip.

an energetic penalty for the disregistry across the slip plane. In particular, we may write

$$E_{misfit} = \int_{\infty}^{-\infty} \phi(\delta(x))dx, \qquad (8.75)$$

where ϕ is the interplanar potential which accounts for the disregistry mentioned above. This interplanar potential is the essence of the cohesive surface approach in this context. The most distinctive characteristic of the interplanar potential is the fact that it is nonconvex, implying the existence of multiple wells, and thereby, the possibility for the emergence of complex microstructures such as dislocations. The particular significance of this feature in the present setting is that the multiple well character of the interplanar potential allows for a disregistry across the slip plane of a Burgers vector without any attendant energy cost. For the present purposes, the interplanar potential should be thought of as the energy cost associated with sliding two half-spaces with respect to one another by some slip vector. We have already encountered this idea in our discussion of the ideal strength, and in fact, will once again invoke the energetics of rigid sliding given in eqn (8.2). Note that the interplanar potential has the same periodicity as the lattice and that whenever atoms across the slip plane have suffered a relative translation equal to a lattice translation, there is no attendant energy penalty.

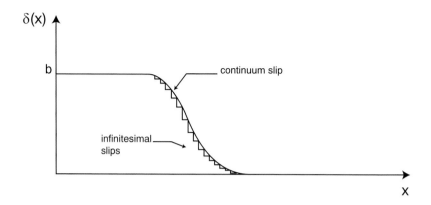

Fig. 8.27. Decomposition of slip distribution into a superposition of infinitesimal dislocations.

The second contribution to the total energy, which also depends upon the as-yet undetermined slip distribution $\delta(x)$ is the elastic energy associated with the dislocation. As indicated schematically in fig. 8.27, the Peierls–Nabarro dislocation may be thought of as a continuous distribution of infinitesimal dislocations. These dislocations in turn are imagined to suffer the usual elastic interactions between dislocations that were described above. Because of our use of linear elasticity, the dislocation fields are obtained through the superposition principle by adding up the contribution from each infinitesimal dislocation separately. Recall that for the case of two edge dislocations at positions x and x' within the confines of isotropic linear elasticity, the interaction energy is given by

$$E_{int} = \frac{\mu b^2}{2\pi(1-\nu)} \frac{1}{x-x'}.$$
(8.76)

This result may be borrowed for the continuous distribution imagined here once we recognize that the Burgers vector of each such infinitesimal dislocation is related to the local gradient in the slip distribution as is seen in fig. 8.27. Hence, the elastic energy due to all such dislocations is given by

$$E_{elastic} = \frac{\mu}{2\pi(1-\nu)} \int dx \int dx' \ln\left(\frac{R}{x-x'}\right) \frac{d\delta(x)}{dx} \frac{d\delta(x')}{dx'}.$$
(8.77)

In the equation given above, R is an inconsequential constant introduced as a large distance cutoff for the computation of the logarithmic interaction energy.

Given the energy functional which depends upon the unknown slip distribution we are now set up to use standard techniques to find the energy minimizing slip distribution. Evaluating the functional derivative of the total energy with respect to the function $\delta(x)$ (which amounts to balancing the atomic-level forces with those due to the elastic interactions) results in the Peierls–Nabarro integro-differential equation which may be written as

$$-\frac{\partial \phi(\delta)}{\partial \delta} = \frac{\mu b}{2\pi(1-\nu)} \int \frac{1}{x-x'} \frac{d\delta(x')}{dx'} dx'.$$ (8.78)

For the special case in which we make the simplifying assumption that the interplanar potential takes the simple cosine form given in eqn (8.2) and advocated earlier in the context of the ideal strength model, this problem allows for direct analytic solution. In particular, the solution is

$$\delta(x) = \frac{b}{2} - \frac{b}{\pi} \tan^{-1}\left(\frac{x}{a}\right).$$ (8.79)

One of the key features that emerges from this solution, and a crucial hint for the type of analyses we will aim for in the future, is the fact that the stresses implied by the Peierls–Nabarro solution do not suffer from the same singularities that plague the linear elastic solution. A detailed examination of this point may be found in Hirth and Lothe (1992) and is illustrated in their eqn (8-13). By introducing an element of constitutive realism in the form of a nonlinear (and in fact nonconvex) interplanar potential, the solution is seen to be well behaved.

One of the outcomes we require of our analysis of dislocations is an insight into the value of the stress needed to move a dislocation. The Peierls–Nabarro model offers the possibility of examining this question. The key qualitative feature that will emerge from our calculations will be the idea that a straight dislocation may be thought of as living in an effective potential that is periodic. The periodicity of this potential reflects the fact that there are many degenerate wells within which we may place the dislocation, each separated by a fundamental length determined by the crystal symmetry. In order to induce dislocation motion, the applied stress must be sufficiently high so as to overcome the *lattice resistance* provided by the periodic Peierls potential. As usual, we hark back to the role of symmetry breaking terms in inducing configurational forces, of which the Peierls stress is but one example.

One of the dominant objectives posed in the context of the theory of dislocations is how such models may assist us in the understanding of crystal plasticity. As noted earlier, one measure of our success is our ability to shed light on the tandem questions of yield and hardening. As yet, our focus has primarily centered on the behavior of single dislocations, and is hence at best only an initial foray into these type of problems. From the single dislocation perspective, one of the first questions we might hope to answer is that of how a single dislocation responds to the presence of an applied stress. Our earlier efforts were based upon the linear elastic theory of dislocations, and revealed that even in the presence of a very small stress, a dislocation will experience a force compelling it to motion. On the other hand, the phenomenology of plasticity teaches us that dislocation motion is one of many examples of a phenomenon that exhibits threshold behavior. Until a certain critical stress is reached, the dislocation remains sessile. Our task is to uncover the microscopic origins of this threshold response.

One immediate route that allows for the examination of this question is offered by the Peierls–Nabarro model discussed above. Recall that the total energy of the Peierls dislocation was seen as arising from two distinct sources. On the one hand, there is an elastic contribution that arises from the two half-spaces that are adjacent to the slip plane. Secondly, we argued that there is an energy cost associated with the disregistry across the slip plane (the so-called misfit energy). Within the confines of the continuum description offered above, the energy of the Peierls dislocation is invariant under translations of the dislocation coordinate. This result emerges from the fact that we treat the misfit energy as an integral over the slip plane disregistry. To remedy this flaw, it is necessary to make further concessions in the direction of atomistic realism. In particular, we must sample the interplanar potential at those discrete positions that are consonant with the underlying crystal lattice.

In particular, to compute the Peierls stress, we make the transcription

$$E_{misfit}(x_c) = \int_{-\infty}^{\infty} \phi(\delta(x - x_c))dx \rightarrow \sum_{n=-\infty}^{\infty} \phi(\delta(nb - x_c))\Delta x, \qquad (8.80)$$

where we have introduced the parameter x_c to denote the position of the dislocation as measured by the position on the slip distribution where the slip is $b/2$, for example (i.e. $\delta(x_c) = b/2$). The significance of this replacement is that rather than evaluating the misfit energy at a continuum of points along the slip plane, the energy of disregistry is computed at a series of equally spaced points meant to mimic the atomic positions themselves. In this context, it is imagined that the dislocation core is translated with respect to the atomic positions. For the particular choice of interplanar potential suggested by Frenkel, this sum may be written as

$$E_{misfit}(x_c) = \frac{\mu b^2}{4\pi^2 d} \sum_{n=-\infty}^{\infty} \left(1 - \cos\left\{\frac{2\pi}{b}\left[\frac{b}{2} - \frac{b}{\pi}\tan^{-1}\left(\frac{x_c + nb}{a}\right)\right]\right\}\right). \qquad (8.81)$$

All that we have done to arrive at this equation is to use eqn (8.2) to evaluate the energy at various points on the slip plane which have undergone slip characterized by eqn (8.79). After appropriate massaging using obvious trigonometric identities, this sum may be rewritten as

$$E_{misfit}(x_c) = \frac{\mu b^2 a^2}{2\pi^2 d} \sum_{n=-\infty}^{\infty} \frac{1}{a^2 + (x_c + nb)^2}, \qquad (8.82)$$

which has the look of a sum that might be handled by the Poisson summation formula. The Poisson summation formula states

$$\sum_{n=-\infty}^{\infty} f(n) = \sum_{k=-\infty}^{\infty} \int_{-\infty}^{\infty} f(x)e^{2\pi i k x}\,dx, \tag{8.83}$$

and is predicated on the hope that the transformed sum in the variable k will be more directly amenable than was the sum in n. If we introduce the dimensionless variables $\gamma = a/b$ and $\delta = x/b$, the sum may be rewritten as

$$E_{misfit} = \frac{\mu a^2}{2\pi^2 d} \sum_{k=-\infty}^{\infty} e^{-2\pi i k \delta} \int_{-\infty}^{\infty} \frac{d^{2\pi i k u}\,du}{\gamma^2 + u^2}, \tag{8.84}$$

which after performing the relevant integrations leads in turn to

$$E_{misfit}(x_c) \approx \frac{\mu a b}{2\pi d}\left\{1 + 2\cos\frac{2\pi x_c}{b}e^{-\frac{2\pi a}{b}}\right\}. \tag{8.85}$$

In performing this sum, we have kept only the terms corresponding to $k = 0$ and $k = \pm 1$.

What has been achieved thus far? Our claim is that the calculation just performed tells us the energy profile for a straight dislocation as that dislocation is moved against the background of the fixed lattice. That is, the straight dislocation is now seen to have a series of wells in which it has a minimum energy and in the act of passing from one well to the next it must cross a barrier known as the Peierls barrier which gives rise to the intrinsic lattice resistance to dislocation motion.

Our next task is to use our newly found Peierls potential to estimate the stress to move a straight dislocation. The stress is essentially provided as the derivative of the energy profile with respect to the coordinate x, and is given by

$$\frac{\partial E(x)}{\partial x} = -\frac{\mu}{1-\nu}\sin\left(\frac{2\pi x}{b}\right)e^{-\pi d/(1-\nu)b}, \tag{8.86}$$

where we have used the recollection that the parameter a is a measure of the core width and is given by $a = d/2(1 - \nu)$. By evaluating the derivative of this expression in turn, we find that the *maximum* value of the stress occurs when $x = b/4$, yielding an estimate of the Peierls stress of

$$\tau_P = \frac{\mu}{1-\nu}e^{-\pi d/(1-\nu)b}. \tag{8.87}$$

Our analysis calls for a number of observations. First, we note that the formula for the Peierls stress leads to the expectation that the stress to move a dislocation will be lowest for those planes that have the largest interplanar spacing at fixed b. In the case of fcc crystals, we note that the spacing between (001) planes is given by $a_0/2$, while for (110) planes the spacing is $\sqrt{2}a_0/4$, and finally for the (111) planes this value is $a_0/\sqrt{3}$. A second observation to be made concerns the magnitude of the Peierls stress. What we note is that the stress to move a dislocation is down by an exponential factor in comparison with the shear modulus. Recall that in our discussion of the ideal strength, the expectation was that aside from numerical factors, the ideal strength goes as the shear modulus itself. By way of contrast, we now see that the presence of the dislocation modulates this estimate. A final observation that should be made is that even this case is too idealized. Real dislocation motion often takes place by kink mechanisms in which a bulge on the dislocation line brings a segment of the dislocation into adjacent Peierls wells, and the resulting kinks propagate with the end result being a net forward motion of the dislocation.

8.6.3 Structural Details of the Dislocation Core

Thus far, our discussion of dislocation core effects has been built around those tools admitting of direct analytic progress such as is offered by the Peierls–Nabarro dislocation core model. This model has many virtues. However, it is also strictly limited, with one of its primary limitations being that without huge machinations, it may only be applied to dislocations with planar cores. To explore the full atomic-level complexity that arises in dislocation cores, we must resort to numerical techniques, and direct atomistic simulation in particular. The remainder of our discussion on dislocation cores will take up the question of how one goes about carrying out such simulations and what may be learned from them.

One of the central conclusions derived from the Peierls–Nabarro analysis is the role of nonlinear effects in the dislocation core. From an atomistic perspective, the far field atomic displacements could be derived just as well from linear elasticity as the full nonlinear function that results from direct atomistic calculation. By way of contrast, in the core region it is the nonlinear terms that give rise to some of the complex core rearrangements that we take up now. Our discussion will be built around two key examples: cores in fcc metals and the core reconstructions found in covalent semiconductors such as in Si.

Before embarking on the specifics of these cores, we begin with an assessment of the basic ideas needed to effect an atomistic simulation of the dislocation core. We begin with a picture of lattice statics in which the core geometry is determined by energy minimization. To proceed, what is needed is an energy function of the type discussed in chap. 4 which may be written generically as

$$E_{tot} = E(\{\mathbf{R}_i\}), \tag{8.88}$$

which gives the total energy once the atomic positions have been specified. For the moment, we are indifferent to the particulars of the choice of total energy scheme. Later we will see that the detailed core geometry *can* depend critically on this choice, and that not all choices are equally satisfying. On the other hand, since our present ambition is the description of how to perform such a simulation we leave these details aside. From the standpoint of energy minimization, we have said all there is to say except for the all important issue of boundary conditions.

For the moment, we describe the most naive approach to this problem and will reserve for later chapters (e.g. chap. 12) more sophisticated matching schemes which allow for the nonlinear calculations demanded in the core to be matched to linear calculations in the far fields. As we have repeatedly belabored, the objective is to allow the core geometry to emerge as a result of the full nonlinearity that accompanies that use of an atomistic approach to the total energy. However, in order to accomplish this aim, some form of boundary condition must be instituted. One of the most common such schemes is to assume that the far field atomic positions are dictated entirely by the linear elastic fields. In particular, that is

$$\mathbf{x}_i = \mathbf{X}_i + \mathbf{u}(\mathbf{X}_i), \tag{8.89}$$

where $\mathbf{x}_i$ is the position of the far field atom i in the dislocated crystal, $\mathbf{X}_i$ is the position of the same atom in the perfect crystal and $\mathbf{u}(\mathbf{X}_i)$ is the displacement of that atom in the dislocated crystal as obtained using Volterra fields for a single straight dislocation like those described in section 8.4.

To perform the energy minimization, the atoms in the far field are rigidly fixed to their linear elastic values. As an initial guess, it is often elected to assume that the initial positions even for the atoms that are being relaxed is well described by the linear elastic solution. The forces are computed using the total energy scheme and the local minimum is found by any of a number of schemes including conjugate gradient, Monte Carlo, Newton–Raphson, etc.

Core structures in fcc materials

A key example of the strategy for direct atomistic simulation of dislocation cores advocated above is that of fcc metals. In this case, we consider the atomic-level geometry of a perfect dislocation (which by virtue of the relaxation will split into partials) with line direction $(1\bar{1}2)$ and Burgers vector $(a_0/2)(\bar{1}10)$. The Volterra fields associated with this geometry are introduced using the scheme suggested in eqn (8.89). For the case shown here, the atomic-level relaxation has been performed using the conjugate gradient method on an arrangement of Al atoms described by the embedded-atom potentials of Ercolessi and Adams. These potentials are of the traditional pair functional form. As noted above, the key structural rearrangement that occurs for this type of dislocation is the splitting of the perfect dislocation into Shockley partials. This result is shown in fig. 8.28. We note that the resulting atomic-level geometry is a concrete realization of the arguments made earlier in this section concerning dislocation dissociation.

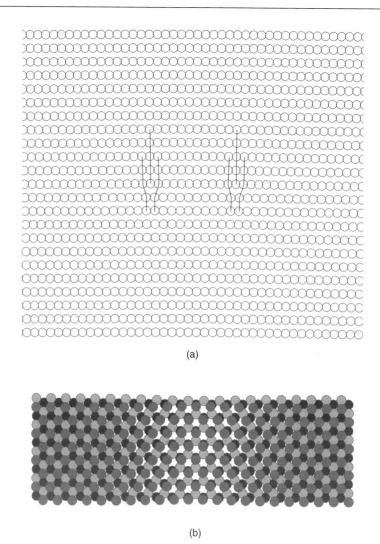

(a)

(b)

Fig. 8.28. Atomic-level geometry of the dislocation core for the conventional edge dislocation in Al (courtesy of Kedar Hardikar): (a) a view along the line direction of the dislocation showing two partials, and (b) a view looking down on the slip plane, revealing the stacking fault region bounded by the two partials.

Core reconstructions in Si

A more substantial rearrangement in the dislocation core can be found in the case of a covalent material. Probably the single greatest influence on the structures adopted by defects in covalent materials is the severe energy penalty that attends dangling bonds. Free surfaces, grain boundaries and dislocation cores all have geometries that reconstruct in a manner that preserves (albeit in a distorted way)

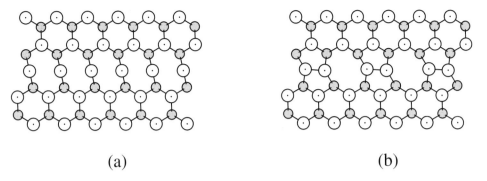

(a)   (b)

Fig. 8.29. Atomic-level geometry of the dislocation core for the 30° partial dislocation in Si (adapted from Bulatov *et al.* (1995)). (a) Unreconstructed core structure and (b) reconstructed core structure with dimerization along line direction.

the fourfold coordination of the atoms in question. In the present discussion, we examine a manifestation of this antipathy for dangling bonds in the context of dislocation core structures. For concreteness, we consider the core geometry of the 30° partial in Si as computed using Stillinger–Weber potentials. Recall from the discussion of chap. 4 that the Stillinger–Weber potentials incorporate an angular term which penalizes deviations from the ideal tetrahedral angles found in the cubic diamond structure. For the present purposes, the key feature of the atomic-level relaxations is the presence of a dimerization reaction as seen in fig. 8.29. Prior to relaxation, the Volterra fields leave each atom in the core region missing one of its full complement of four neighbors. The relaxation remedies this loss by pairing up the atoms along the core direction, restoring fourfold coordination and lowering the energy.

8.7 Three-Dimensional Dislocation Configurations

To have any hope of honestly addressing the dislocation-level processes that take place in plastic deformation, we must consider fully three-dimensional geometries. Though we earlier made disparaging remarks about the replacement of understanding by simulation, the evaluation of problems with full three-dimensional complexity almost always demands a recourse to numerics. Just as the use of analytic techniques culminated in a compendium of solutions to a variety of two-dimensional problems, numerical analysis now makes possible the development of catalogs of three-dimensional problems. In this section, we consider several very important examples of three-dimensional problems involving dislocations, namely, the operation of dislocation sources and dislocation junctions.

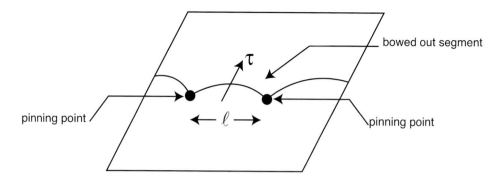

Fig. 8.30. Geometry associated with dislocation bow-out between two pinning points.

8.7.1 Dislocation Bow-Out

A recurring theme in a number of different contexts is that of dislocation bow-out. In a number of different circumstances, a dislocation segment of finite extent, pinned at its two ends, is subjected to a resolved shear stress which has the effect of bowing out the segment in its slip plane. The simplest setting within which to view this process is the line tension approximation set forth earlier in this chapter. Dislocation pinning is a ubiquitous phenomenon. Pinning arguments arise in the context of solution and precipitate hardening as well as the discussion of dislocation sources. Our present discussion has as its aim the development of some of the theoretical apparatus needed to discuss the response of dislocations in this context, without giving any attention to the attributes of the pinning centers themselves. We simply imagine a geometry like that shown in fig. 8.30 in which the dislocation is pinned at the two endpoints, leaving a segment of length l that is free to glide.

The energetics of the bowed-out segment may be constructed simply on the basis of line tension arguments. We begin by adopting the simplest possible model of line tension, namely, that in which the angular dependence is ignored altogether. In this case, the energy of the bowed segment as a functional of its displacement profile $u(x)$ is given by

$$E[u(x)] = \int_{-\frac{l}{2}}^{\frac{l}{2}} T\sqrt{1 + u'^2}\,dx - \int_{-\frac{l}{2}}^{\frac{l}{2}} \sigma b u(x)\,dx. \qquad (8.90)$$

Our interest is in determining the energy minimizing configuration of the bowed-out segment. To do so, we note that this has become a simple problem in variational calculus, with the relevant Euler–Lagrange equation being

$$\frac{d}{dx}\left[\frac{T u'}{(1 + u'^2)^{\frac{1}{2}}}\right] = -\sigma b. \qquad (8.91)$$

Even at the level of the line tension approximation, this equation may be solved to various levels of approximation. In the limit in which the bow-out is small (i.e. $|u'| \ll 1$), the denominator in the expression given above can be neglected with the result that the equilibrium equation takes the form,

$$\frac{d^2 u}{dx^2} = -\frac{\sigma b}{T}. \tag{8.92}$$

The solution to this equation can be written down by inspection as

$$u(x) = -\frac{\sigma b}{2T} x^2 + cx + d, \tag{8.93}$$

where c and d are as-yet undetermined constants. Note that the solution we seek satisfies the boundary conditions $u(-l/2) = u(l/2) = 0$. Application of these boundary conditions results in the solution

$$u(x) = \frac{\sigma b}{2T} \left(\frac{l^2}{4} - x^2 \right), \tag{8.94}$$

which may be recast in dimensionless terms if the stress is measured in units of $\sigma = f(2T/bl)$, and f is what fraction of $2T/bl$ the stress has attained, and similarly, we take $x = \delta(l/2)$. In light of these notational simplifications, our expression for the bowed-out profile may be rewritten as

$$u(x) = \frac{fl}{4} (1 - \delta^2). \tag{8.95}$$

The equation for bow-out may also be solved exactly without resorting to the approximation in which $|u'(x)| \ll 1$. In this case, the equation of equilibrium may be written (after performing the first integration)

$$\frac{u'^2}{1 + u'^2} = \left(-\frac{\sigma bx}{T} + c \right)^2, \tag{8.96}$$

where c is a constant of integration to be determined in light of the boundary conditions. This equation is separable and leads immediately to the solution

$$u(x) = \sqrt{\frac{T^2}{\sigma^2 b^2} - x^2} - \sqrt{\frac{T^2}{\sigma^2 b^2} - \frac{l^2}{4}}. \tag{8.97}$$

Like its approximate partner derived above, this equation can be rewritten in dimensionless variables as

$$u(\delta) = \frac{l}{2} \left(\sqrt{\frac{1}{f^2} - \delta^2} - \sqrt{\frac{1}{f^2} - 1} \right). \tag{8.98}$$

In the treatment above, we have used the line tension approximation to deduce both an exact and an approximate treatment of bow-out. The comparison between

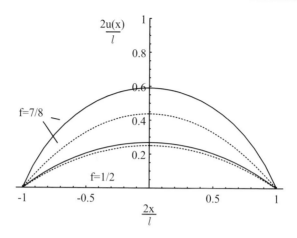

Fig. 8.31. Comparison of the bow-out geometry resulting from exact (full line) and approximate (dashed line) treatment of the line tension approximation. f is a dimensionless stress such that $f = \sigma b l / 2T$.

the resulting solutions is shown in fig. 8.31. As expected, we note that in the small bow-out limit, the exact and approximate treatments are in very good accord, while for larger stresses, their correspondence is completely unsatisfactory. We have deemed it worthwhile to examine this comparison since in many instances, the treatment of curved dislocations is made not only in the constant line tension approximation, but also with the assumption that $|u'| \ll 1$, an assumption we are now in a better position to view skeptically.

8.7.2 Kinks and Jogs

Our treatment thus far has centered on idealized geometries in which the dislocation is presumed to adopt highly symmetric configurations which allow for immediate insights from the linear elastic perspective. From the phenomenological standpoint, it is clear that we must go beyond such idealized geometries and our first such example will be the consideration of kinks. The formation of kink–antikink pairs plays a role in the interpretation of phenomena ranging from plastic deformation itself, to the analysis of internal friction.

For the purposes of our modeling efforts here, we will treat the formation energy of kinks via a hybrid scheme in which the elastic energy of the dislocation line that has adopted a configuration like that shown in fig. 8.32 is treated in the line tension approximation. The line tension picture supposes that the energy associated with perturbing a dislocation line can be accounted for in much the same way that one describes the potential energy of a stretched string. If we ignore the dependence of the line energy on the angle between the line and the Burgers vector, then the line

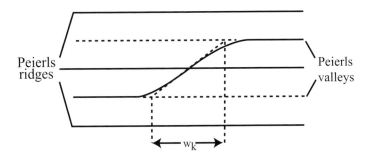

Fig. 8.32. Schematic of the geometry associated with the formation of a kink in the dislocation line (adapted from Schöck (1980)).

tension may be treated as a scalar T and the energy associated with distorting the line is given by

$$E_{elastic} = \int T\sqrt{1 + u'^2} dx. \qquad (8.99)$$

The kink configuration emerges as the result of a competition between the elastic energy as characterized by the line tension which favors a straight dislocation line and a nonlinear term whose origins are crystallographic. By virtue of the translational periodicity of the underlying lattice the dislocation line may be thought of as riding in a periodic potential that we will call $W_p(u)$ and which is really nothing more than the result obtained in eqn (8.85). The idea here is that there are a series of wells that the line can occupy and the kink is a mechanism whereby the line can pass from one such well into another.

As noted above, the nature of the kink reflects a competition between two different factors. Consideration of fig. 8.32 illustrates that in the limit that the energy scale associated with the line tension is much larger than that of the multiple welled Peierls potential, the dislocation line will vary slowly with a resulting wide kink. Alternatively, in the limit that the peaks in the Peierls potential are high, the lowest-energy configuration will correspond to a kink that has as little of its length on that peak as possible. The mathematical setting for examining this competition is the functional minimization of the energy

$$E[u(x)] = \int T\sqrt{1 + u'^2} dx + \int W_p(u) dx. \qquad (8.100)$$

As seen in the figure, the function $u(x)$ characterizes the displacement of the line from the well at the origin. The boundary condition of interest here is that $u(-\infty) = 0$ and $u(\infty) = b$.

Our ambition is to find the energy minimizing disposition of the dislocation as described by the function $u(x)$. This function in the present context is a complete characterization of the kink geometry. Evaluation of the variational derivative of the total energy given in eqn (8.100) leads to an Euler–Lagrange equation of the form

$$\frac{d}{dx}\left(T\frac{u'}{\sqrt{1+u'^2}}\right) - W_p'(u) = 0. \tag{8.101}$$

If we further assume that $|u'| \ll 1$, then the equation simplifies to

$$Tu'' - W_p'(u) = 0. \tag{8.102}$$

Upon multiplication by u', we are immediately led to a first integral of the problem, which implies in turn that the profile satisfies

$$\frac{du}{dx} = \sqrt{\frac{2}{T}W_p(u)}. \tag{8.103}$$

This equation may be integrated to yield a kink geometry of the form

$$u(x) = \frac{2b}{\pi}\tan^{-1}\left(e^{\frac{2\pi x}{b}\sqrt{\frac{W_p}{T}}}\right). \tag{8.104}$$

Furthermore, we may determine the energy of the kinked configuration in light of our solution in conjunction with the first integral of the motion found above. The definition of the kink formation energy is

$$E_k^f = E_{kink} - E_{straight}. \tag{8.105}$$

This may be rewritten using our knowledge of the line energy and the periodic potential $W_p(u)$ as

$$E_k^f = \int_{-\infty}^{\infty} T\sqrt{1+u'^2}dx + \int_{-\infty}^{\infty} W_p\left(1 - \cos\frac{2\pi u}{b}\right)dx - \int_{-\infty}^{\infty} Tdx. \tag{8.106}$$

If we now expand the square root and recall from our first integral that $Tu'^2/2 = W_p[1 - \cos(2\pi u/b)]$, the resulting kink energy is given by

$$E_k^f = 2W_p\int_{-\infty}^{\infty} W_p\left(1 - \cos\frac{2\pi u}{b}\right)dx. \tag{8.107}$$

If we now substitute the solution for the kink profile (i.e. $u(x)$) obtained in eqn (8.104), the resulting integral is elementary and results in a kink energy

$$E_k^f = \frac{4b}{\pi}\sqrt{TW_p}. \tag{8.108}$$

In addition to the insights that can be gleaned from continuum arguments, atomistic analysis has been used to shed light on some of the structural complexity that attends the motion of kinks. The basis for our case study will be the partial dislocations that are known to exist in covalent materials such as Si. In particular,

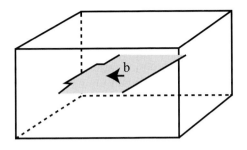

Fig. 8.33. Schematic of the computational cell used to model the structure and energetics of kinks using atomic-level analysis (adapted from Bulatov *et al.* (1995)).

there have been a variety of atomic-level calculations of the structure and energetics of kinks in Si using both empirical potentials such as the Stillinger–Weber potentials introduced in chap. 4, but also on the basis of tight-binding theory and density functional theory. The key insight that will emerge from our discussion of atomic-level calculations of the structure and energetics of kinks is that the kinks themselves are capable of supporting a number of defects within a defect which can alter the properties of the kinks and which may invalidate some of the ideas on kinks advanced in the continuum setting.

As was shown in fig. 8.29, partial dislocations in Si admit of a core reconstruction that includes dimerization along the line direction. It is that reconstructed geometry that will serve as our jumping off point for the calculation of the structure and energetics of kinks in Si. The calculations to be considered here were carried out using a computational cell of the form shown in fig. 8.33 in which a kink–antikink pair are imposed on the dislocation line. The energy minimizing geometry was determined in this case by resorting to a simulated annealing algorithm. Since the kinked configuration, and even the dislocation configuration itself, is metastable, care had to be taken to find only local minima and to guard against falling into the deeper well corresponding to a configuration free of any vestiges of the dislocation. Though the issues that were confronted in achieving this are interesting in themselves, we leave them here for our more immediate concern, namely, the kink geometries themselves. As can be seen in fig. 8.34,

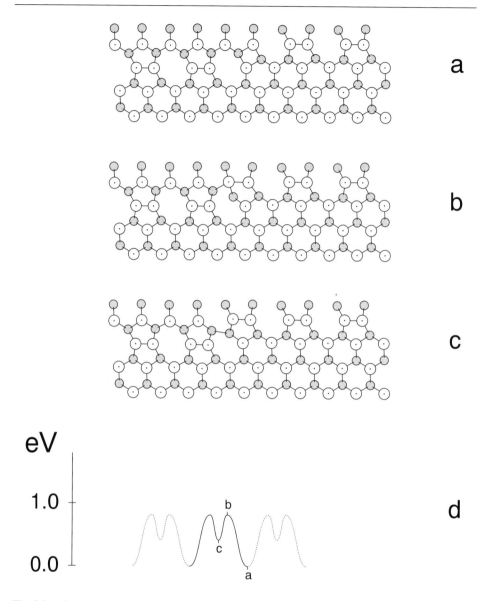

Fig. 8.34. Series of configurations associated with motion of kink on the 30° partial in Si (adapted from Bulatov *et al.* (1995)).

in addition to the ability to determine the energy wells corresponding kinks, it is also possible to find the transition pathways connecting one kink configuration to another.

In addition to the questions that can be raised about kink motion, the type of atomistic analysis being described here has also been used to examine the propensity for nucleation of kinks themselves. Though plausible mechanisms for homogeneous nucleation of kink–antikink pairs exist, a more compelling argument can be made for heterogeneous nucleation off of in-core defects known as antiphase defects (APD). These antiphase defects arise from the fact that when the core of the partial dislocation like that under consideration here reconstructs, a given atom can form a partnership (i.e. dimerize) with its partners either to the left or to the right. If a sequence of partnerships is struck to the right and then changes to a partnership to the left, the point at which this transition occurs will bear the imprint of this transition in the form of a dangling bond. An especially interesting conclusion deriving from the atomistic analysis is that these antiphase defects can serve as the hosts for the nucleation of kink–antikink pairs. The mechanism for this nucleation is shown in fig. 8.35.

The atomic-level calculations described above were carried out using the Stillinger–Weber potentials. A step toward a higher level of realism in the description of the total energy can be made by accounting for the electronic degrees of freedom. As has already been made evident throughout the book, an intermediate position which strikes a compromise between computational tractability and accuracy is the tight-binding method. A series of tight-binding calculations have been carried out on the same systems described above (Nunes *et al.* 1998) resulting in the same sort of structural information. An example of these results is shown in fig. 8.36. Our main objective in this section has been to illustrate the way in which continuum and several different atomistic schemes can be brought to bear on the same underlying problem.

8.7.3 Cross Slip

Another important class of three-dimensional dislocation configurations are those associated with the cross slip process in which a screw dislocation passes from one glide plane to another. The most familiar mechanism for such cross slip is probably the Friedel–Escaig mechanism, which is illustrated schematically in fig. 8.37. The basic idea is that an extended dislocation suffers a local constriction at some point along the line. This dislocation segment, which after constriction is a pure screw dislocation, can then glide in a different slip plane than that on which is gliding the parent dislocation. This mechanism, like those considered already, is amenable to treatment from both continuum and atomistic perspectives, and we take them each up in turn.

Since it is presumed that the process of cross slip is thermally activated, to the extent that we are trying to explicate the unit processes that make up plasticity it is of great interest to come up with an estimate for the activation energy for cross slip. In broad terms, what this implies is a comparison of the energies between the unconstricted and constricted states. The simplest continuum estimate

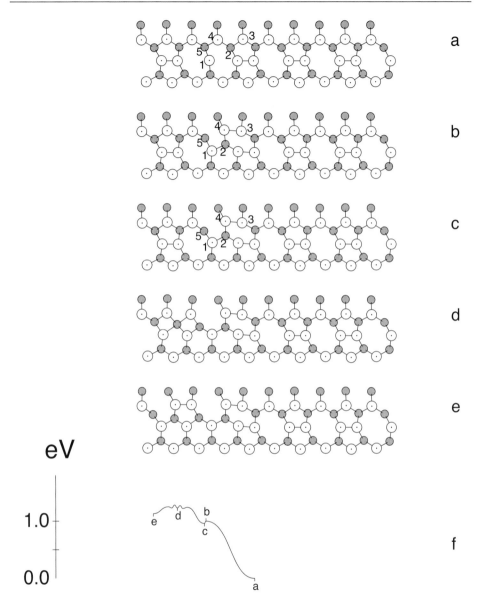

Fig. 8.35. Nucleation of a kink–antikink pair from an antiphase defect (adapted from Bulatov *et al.* (1995)).

of this process derives from the line tension approximation. At a higher level of sophistication it is also possible to investigate this process from the point of view of full three-dimensional elasticity. An overview of some of the continuum thinking on these problems can be found in Saada (1991). On the other hand, the continuum treatment of cross slip is attended by a number of uncertainties. An

(a)

(b)

(c)

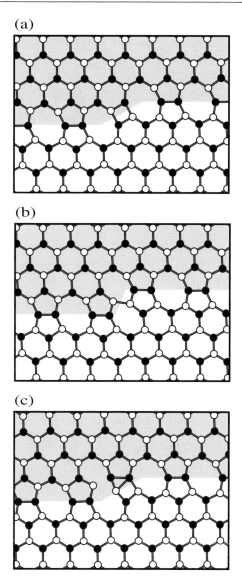

Fig. 8.36. Series of configurations associated with motion of kink on the 30° partial in Si as computed using the tight-binding model (adapted from Nunes *et al.* (1998)).

attractive feature of atomic-level calculations is that they allow for a systematic investigation of some of the complex questions that have earlier been attacked from the line tension or linear elastic perspective. The cross slip process has attracted attention with a resulting series of impressive calculations of the various intermediate configurations connecting the initial and final states of a dislocation that has undergone cross slip as well as the energy barrier that separates them

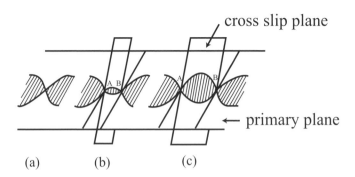

Fig. 8.37. Schematic of the Friedel–Escaig cross slip mechanism (adapted from Bonneville *et al.* (1988)).

(Rasmussen *et al.* 1997a).   These calculations were undertaken using a pair functional description of the total energy.

   Though there are a number of technical details associated with the calculations to be presented here, we will concentrate primarily on what these calculations have to say about cross slip itself without entering too much into the fine points of the calculation. Suffice it to say that like in any problem in which one is in search of an activation pathway, it is not always true that this pathway is that which presents itself to intuition, and hence a number of clever schemes have been set forth for finding the transition state. In the present setting, the series of steps taking a dislocation from one glide plane to another are shown in fig. 8.38. From a visual perspective, what we wish to take away from this calculation is that the mechanism of cross slip in this case is largely similar to that indicated schematically above in fig. 8.37. The energy of the intermediate state is $\approx$ 2.7 eV higher than that of the starting configuration, which raises serious questions as to whether the homogeneous nucleation of a constriction is really plausible. Future work will have to explore the possible catalytic effect of heterogeneities which can induce cross slip. An interesting continuum model of cross slip induced by the presence of slip plane heterogeneities may be found in Li (1961). As an illustration of the type of observations that can be made on cross slip, we show a set of beautiful experimental pictures of the cross slip process in fig. 8.39.

8.7.4 Dislocation Sources

Frank–Read Sources. It is well known that with increasing deformation, the density of dislocations increases. This effect is illustrated in fig. 8.40 in which the number of dislocations as a function of the strain is indicated schematically. The immediate conclusion from this observation is the fact that there are sources

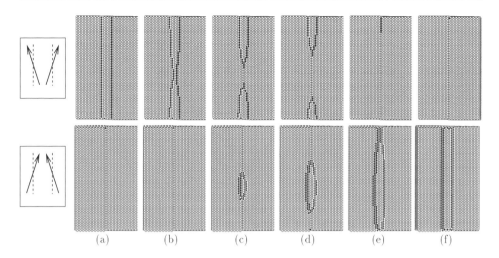

Fig. 8.38. Series of intermediate configurations during the cross slip process as computed using atomistics (adapted from Rasmussen *et al.* (1997b)).

Fig. 8.39. Experimental image of the cross slip process (courtesy of I. Robertson).

of dislocations within solids which serve to increase the number of dislocations present. Though there have been a number of mechanisms postulated for the multiplication of dislocations, the most celebrated mechanism is known as the Frank–Read source and will serve as the cornerstone of the present discussion.

The increase in the net dislocation density with increasing deformation illustrates that even with the passage of some fraction of the dislocations out of the crystal at the surface, there must be sources operating within the crystal. From an experimental perspective, the nature of these sources at the statistical level remains largely unexplored. On the other hand, specific mechanisms have been suggested, one of which we take up in quantitative detail here. An overview on the subject of dislocation sources can be found in Li (1981). The Frank–Read mechanism postulates the existence of some finite segment that can be bowed out in the same sense as was discussed earlier in this section. With increasing stress, the segment

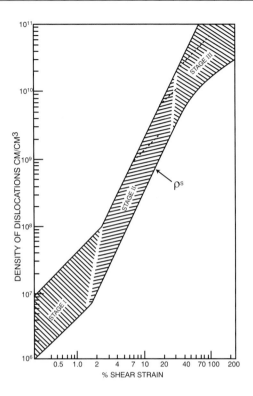

Fig. 8.40. Illustration of the increase in dislocation density with deformation (adapted from Asaro (1983)).

continues to bow out until a critical stress is reached beyond which the dislocation's line tension is no longer able to resist the bow-out and the segment continues to expand indefinitely.

The specific analytic ideas associated with the bowing-out of a pinned segment were already described earlier in this section. As a result of these arguments, we find that the applied stress and the radius of curvature of the bowed-out segment are related by

$$R = \frac{T}{\tau b}. \tag{8.109}$$

An additional insight may be garnered from this discussion which applies to the Frank–Read source. For a segment that is pinned between two points by a spacing d, and that will operate as a Frank–Read source, we now see that with increasing stress the radius of curvature will reduce. For a given pinning geometry, the maximum stress will occur when the radius of curvature is half the spacing between obstacles or

$$\tau_{source} = \frac{2T}{bd}. \tag{8.110}$$

If we now invoke an approximate expression for the line energy, namely, $T = \frac{1}{2}\mu b^2$, then the critical stress may be rewritten as

$$\tau_{source} = \mu\left(\frac{b}{L}\right). \tag{8.111}$$

Thus we note that we have argued for a dislocation source which operates at a stress that is smaller than the shear modulus by a factor of b/L.

Though the argument given above was offered in the spirit of the line tension approximation, the Frank–Read mechanism can be examined from the linear elastic perspective as well. The segment which is to serve as the Frank–Read source has a geometry that is determined by two competing effects. On the one hand, the applied stress has the tendency to induce glide in the various segments that make up the dislocation. On the other hand, the mutual interactions of the various segments (the effect giving rise to the line tension in the first place) tend to maintain the dislocation line in its straight configuration. The playing out of this balance can be carried out dynamically by supposing that a given dislocation segment glides with a velocity that is proportional to the force it experiences. That is, the nth segment on the dislocation line is attributed a velocity

$$v_n = M F_n, \tag{8.112}$$

where M is a parameter characterizing the mobility of dislocations in the over-damped limit considered here. The main point behind a dynamics like that described here is that we are considering the limit in which dissipative effects are dominant and the dislocation is maintained in a steady state of motion with the work of the applied stress all going into feeding the relevant dissipative mechanisms such as phonon drag which results from the coupling of the dislocations center of mass motion to the vibrations of the atoms around it. The sequence of states reached by a dislocation subjected to a sufficiently large applied stress is shown in fig. 8.41. A second view of the dislocation dynamics approach to Frank–Read sources is shown in fig. 8.42. A further elaboration of this so-called 'dislocation dynamics' approach will be set forth in chap. 12. One of the more intriguing features of figure 8.42 is the realization that the full elasticity treatment confirms our intuition about those parts of the bowing-out process which occur at stresses larger than the critical stress.

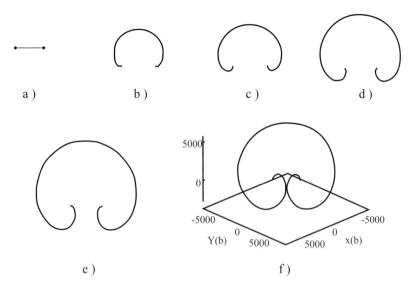

Fig. 8.41. Frank–Read source as obtained using dislocation dynamics (adapted from Zbib *et al.* (1998)).

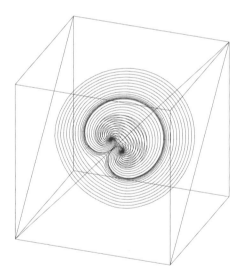

Fig. 8.42. Three-dimensional rendering of a Frank–Read source as obtained using dislocation dynamics (adapted from Ghoniem *et al.* (1999)).

8.7.5 Dislocation Junctions

From the standpoint of the phenomenology of plastic deformation, one of the most important classes of three-dimensional configuration is that associated with dislocation intersections and junctions. As will become more evident in our

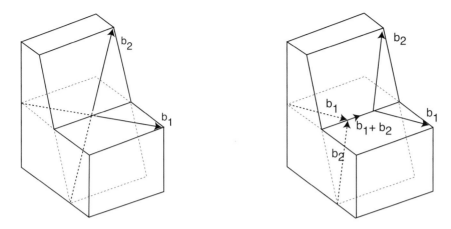

Fig. 8.43. Two different competing geometries (courtesy of D. Rodney): (a) Two dislocations with Burgers vectors $\mathbf{b}_1$ and $\mathbf{b}_2$ and (b) junction case in which the junction segment has Burgers vector $\mathbf{b}_1 + \mathbf{b}_2$.

discussion of work hardening (see chap. 11), dislocations on one glide plane can pierce the glide plane of another slip system, with the result that the gliding dislocation must pass through a 'forest' of obstacles. From our present perspective, we are interested in examining the geometric and energetic characteristics of the intersection of two such dislocations. Our arguments will begin in generic terms with the aim being an overall assessment of the structure and energetics of dislocation junctions with special reference to the stress needed to break them.

The simplest picture of junction formation is largely geometric and relies on the line tension approximation introduced earlier. Perhaps the most fundamental insight to emerge from these studies is the conclusion that the stress to break such junctions scales as (arm length)$^{-1}$. Important discussions of the line tension approach may be found in Saada (1960) and Hirth (1961). The basic idea concerning the structure of junctions is the assessment of the energetics of two competing states, one in which the two parent dislocations (with Burgers vectors $\mathbf{b}_1$ and $\mathbf{b}_2$) remain unattached and the other in which the two parent segments join along the line of intersection of the relevant slip planes to form a new dislocation segment with Burgers vector $\mathbf{b}_1 + \mathbf{b}_2$. The competition being played out here in three dimensions is much like its two-dimensional analog in which a perfect dislocation dissociates to form two partial dislocations bounding a stacking fault.

The two competing geometries are illustrated in fig. 8.43. For the purposes of examining the energetics of junction formation, we will concentrate on a highly symmetric geometry such that the arm lengths l_1 and l_2 are equal, and similarly, such that the angles ϕ_1 and ϕ_2 between the two dislocations and the line of intersection of the two relevant glide planes are equal. Within the confines of this

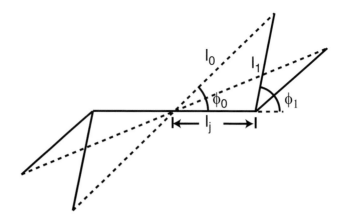

Fig. 8.44. Schematic of the behavior of the arms of a dislocation before and after junction formation.

geometric setup the energy of the state in which the two lines have not formed a junction is given by

$$E_{lines} = 2\frac{1}{2}\mu b_1^2 l_0 + 2\frac{1}{2}\mu b_2^2 l_0. \tag{8.113}$$

Similarly, within the line tension approximation, the energy of the state in which a junction of length l_j has formed is given by

$$E_{junction} = 2\frac{1}{2}\mu b_1^2 l_1 + 2\frac{1}{2}\mu b_2^2 l_1 + 2\frac{1}{2}\mu |\mathbf{b}_1 + \mathbf{b}_2|^2 l_j. \tag{8.114}$$

Because of the choice of the highly symmetric geometry, the junction geometry is characterized by a single parameter. In particular, the parameters l_j and l_1 may be rewritten purely in terms of the angle ϕ_1 and the initial arm length l_0. These geometric relations are illustrated in fig. 8.44. What we note is that by the law of sines, l_j and l_1 are given by

$$l_j = \frac{\sin(\phi_1 - \phi_0)}{\sin\phi_1}l_0, \tag{8.115}$$

and

$$l_1 = \frac{\sin\phi_0}{\sin\phi_1}l_0. \tag{8.116}$$

Further, we choose a configuration in which $|\mathbf{b}_1| = |\mathbf{b}_2| = |\mathbf{b}_1 + \mathbf{b}_2|$, corresponding to the three edges of one of the faces of the Thompson tetrahedron. The significance of this geometric choice is that it guarantees that the line tensions of all three dislocation segments of interest will be equal.

To examine the structure and energetics of junction formation from the line tension perspective, we now seek the equilibrium angle ϕ_1 corresponding to a given

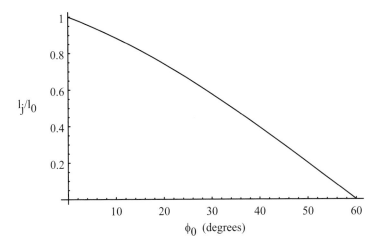

Fig. 8.45. Dimensionless length of junction within the line tension approximation as a function of the initial angle.

choice of initial angle ϕ_0 and arm length l_0. The equilibrium condition is equivalent to finding the energy minimizer, which for the symmetric geometry considered here results in a one-parameter optimization problem. From above, we note that the energy change relative to the junction-free state is given by

$$E = E_{junction} - E_{lines} = 2\mu b^2 \frac{\sin \phi_0}{\sin \phi_1} l_0 + \mu b^2 \frac{\sin(\phi_1 - \phi_0)}{\sin \phi_1} l_0 - 2\mu b^2 l_0, \quad (8.117)$$

where l_j and l_1 have been eliminated in favor of ϕ_1 and we have used the fact that the magnitude of all relevant Burgers vectors are equal. If we now seek equilibrium with respect to the angle ϕ_1, we are led to the equilibrium condition

$$\frac{1}{\mu b^2 l_0} \frac{\partial E}{\partial \phi_1} = \frac{\sin \phi_0}{\sin \phi_1} (-2 \cot \phi_1 + \operatorname{cosec} \phi_1) = 0. \quad (8.118)$$

Solution of this equation leads to the conclusion that for this particular choice of Burgers vectors (i.e. the Lomer–Cottrell junction) the equilibrium angle is $60°$. Using this result, and in light of eqn (8.115), we can now plot the junction length as a function of initial angle, namely,

$$\frac{l_j}{l_0} = \frac{\sin\left(\frac{\pi}{3} - \phi_0\right)}{\sin \frac{\pi}{3}}. \quad (8.119)$$

The result is plotted in fig. 8.45.

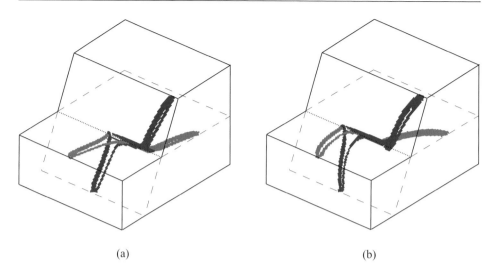

(a)   (b)

Fig. 8.46. Junction results from atomistics as computed using atomic-level analysis (adapted from Rodney and Phillips (1999)).

Atomistic Models of Dislocation Junctions. As noted above, the line tension approximation is largely geometrical and it is clear that an assessment of the accuracy of these ideas can be addressed from an atomistic perspective. One of the questions that one hopes to address from the atomistic perspective is the extent to which dislocation core effects (including the presence of dislocation splittings) alter the basic picture set forth in the line tension approximation. In keeping with the arguments given above, we continue to consider the geometry of the junction in which $|\mathbf{b}_1| = |\mathbf{b}_2| = |\mathbf{b}_1 + \mathbf{b}_2|$. When carrying out an atomistic calculation of a problem of this type, it is obviously necessary to impose some choice of boundary conditions. Often, the imposition of such boundary conditions has unwanted side effects, and that is the case thus far with regard to the structure and energetics of junctions.

Fig. 8.46 shows a sequence of snapshots in the deformation history of a junction in fcc Al as computed using pair functionals of the embedded-atom variety. Fig 8.46(a) illustrates the junction geometry in the presence of zero applied stress, while fig. 8.46(b) illustrates the evolution of the junction in the presence of an applied stress.

Linear Elastic Models of Junction Formation. Yet another scheme for addressing the structure and energetics of dislocation junctions is using the machinery of full three-dimensional elasticity. Calculations of this type are especially provocative since it is not clear *a priori* to what extent elastic models of junction formation will be hampered by their inability to handle the details of the shortest range part

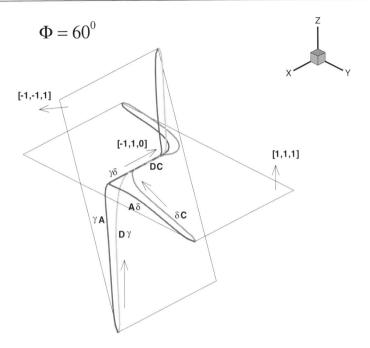

Fig. 8.47. Lomer–Cottrell junction as computed using three-dimensional elasticity representation of dislocation dynamics (adapted from Shenoy *et al.* (2000)).

of the dislocation–dislocation interaction. It is important to emphasize that these calculations include far more physics than do their line tension counterparts since within the elasticity framework the interactions between the various segments are taken into account. The calculations to be considered here, shown in fig. 8.47, are built around a linkage of the type of dislocation dynamics models set forth earlier in the context of Frank–Read sources and the finite element method which is used to discretize the dislocation lines themselves. In addition, this elastic model explicitly treated the interaction energies (and corresponding forces) that arise from the presence of stacking faults.

8.8 Further Reading

The entire series *Dislocations in Solids* edited by F. R. N. Nabarro, North Holland Publishing Company, Amsterdam: The Netherlands is replete with articles pertinent to all aspects of dislocations (and many other topics as well). Many of those articles are relevant to the discussion advanced here. In particular, with reference to the crystallographic aspects of dislocations and the complex dislocation arrangements that occur in three dimensions I strongly recommend the article by S. Amelinckx in volume 2.

Dislocations by J. Friedel, Pergamon Press, Oxford: England, 1964. One of the great classics in the field, what makes Friedel's book particularly appealing is that almost every time a new theoretical construct is introduced, it is done so with reference to a particular experimental question in mind.

'Dislocations and Walls in Crystals' by J. Friedel in *Physique des Défauts/Physics of Defects* edited by R. Balian *et al.*, North Holland Publishing Company, Amsterdam: The Netherlands, 1980 makes for very interesting reading as Friedel revisits the theory of dislocations after a twenty-year interlude. I recommend this article very highly.

Dislocation Based Fracture Mechanics by Johannes Weertman, World Scientific, Singapore: Singapore, 1996. What I find especially engaging about this book is its idiosyncratic treatment of a number of problems. Weertman shows repeatedly how in adopting the notion of dislocation solutions as being fundamental, many interesting concepts may be explained concerning cracks and dislocations.

Introduction to Dislocations by D. Hull and D. J. Bacon, Butterworth-Heinemann Ltd, Oxford: England, 1984. Ostensibly, this book is aimed at an audience uninitiated in the subject of dislocations. However, this may be deceptive since there are a variety of interesting discussions that are of interest even to seasoned veterans.

Theory of Dislocations by J. P. Hirth and J. Lothe, Krieger Publishing Company, Malabar: Florida, 1992. The classic treatise on the mathematical theory of dislocations. This is always the *first* place I look when I want to find out something about dislocations.

Mathematical Theory of Dislocations and Fracture by R. W. Lardner, University of Toronto Press, Toronto: Canada, 1974. This book treats a variety of interesting problems in dislocation theory without shying away from mathematically sophisticated treatments using the elastic Green function. A reference that I return to repeatedly.

Thermodynamics and Kinetics of Slip by U. F. Kocks, A. S. Argon and M. F. Ashby, Pergamon Press, Oxford: England. This book is volume 19 of the series Progress in Materials Science, edited by B. Chalmers, J. W. Christian and T. B. Massalski. This book reflects the considerable knowledge of three leaders in the field. One of the key features that makes the book so useful is that rather than concentrating on elastic solutions it considers how dislocations give rise to observed macroscopic behavior.

Micromechanics of Defects in Solids by T. Mura, Kluwer Academic Publishers, Dordrecht: The Netherlands, 1993. A highly nonconventional treatment of dislocations built around the idea of eigenstrains. This perspective is quite valuable in that it calls special attention to the lattice invariant shears that make such defects possible.

'Dislocations in Particular Structures' by S. Amelinckx from the 1979 volume of *Dislocations in Solids* is an incredible achievement in which Amelinckx considers the peculiarities of dislocations that attend particular crystal structures. A truly incredible article!

8.9 Problems

1 Microscopic description of the ideal shear strength

Use the Morse pair potential constructed in chap. 4 to compute the periodic energy profile for sliding one half-space relative to another for the fcc crystal. In particular, consider the sliding of adjacent (111) planes along the [$\bar{1}$10]-direction. Make a plot of this periodic energy profile and include for comparison the Frenkel sinusoid for the same slip system. You will need to be careful to insure that your program computes the distribution of neighbors of an atom of interest to large enough distances. Also, multiply your potential by some rapidly decaying function at large distances so as to eliminate the artifact of atoms going in and out of the cutoff radius.

2 Displacements associated with an edge dislocation

In this chapter we derived the strain fields associated with an edge dislocation. In this problem your goal is to derive the displacement fields implied by those strains and to use them to generate a projected atomic view of an fcc lattice with such a dislocation. (a) Using the strains for an edge dislocation, derive an expression for the *displacements* associated with such a dislocation. Hint: You will want to integrate the strains with respect to specific variables, but bear in mind that when you integrate with respect to x, for example, you still have an unknown function of y that is unspecified. (b) Consider the fcc crystal with a conventional perfect edge dislocation with a line direction of the type [112]. Use the displacement fields to generate a plot of the displaced atoms resulting from the Volterra solution. Explain why the picture results in the addition of two additional planes of atoms.

3 Asymptotic fields for dislocation dipoles

In this problem, our aim is to investigate the far fields produced by a dislocation dipole. For simplicity, begin by considering two screw dislocations, both situated on the x-axis at positions $(\delta, 0)$ and $(-\delta, 0)$. (a) Using linear superposition of the fields due to these two screw dislocations, write an analytic expression for the fields at an arbitrary position (x, y). Comment on how the solution depends upon the relative signs of the Burgers vectors at these two dislocations. (b) Consider both the case in which the dislocations have opposite signs and that in which they have the same signs, and find the leading order expansion for the far fields for both components of the stress tensor (i.e. σ_{xz} and σ_{yz}) in terms of the parameter δ. Note the enhanced decay rate of the stresses with distance in the case in which the dislocations have opposite signs. Further, observe that for the case in which the dislocations have the same sign, the far field behavior is equivalent to that of a single dislocation with Burgers vector $2b$. (c) Make your observations from the end of the previous part of the problem concrete by making a plot of the far fields for the case in which the signs of the two dislocations are the same. In particular, plot the exact stress field for this dipole configuration (which you know from part (a)) and that of a dislocation with Burgers vector $2b$. Make your plot with respect to the dimensionless parameter r/δ and for a line at $45°$ with respect to the x-axis.

4 Edge dislocation fields using the elastic Green function

This problem is a bit nasty, but will illustrate the way in which the Volterra formula can be used to derive dislocation displacement fields. (a) Imitate the derivation we performed for the screw dislocation, this time to obtain the displacement fields for an edge dislocation. Use the Volterra formula (as we did for the screw dislocation) to obtain these displacements. Make sure you are careful to explain the logic of your various steps. In particular, make sure at the outset that you make a decent sketch that explains what surface integral you will perform and why.

5 Isotropic Line Tension

We advanced the idea that the energy of a dislocation line could be written in a *local* approximation as

$$E_{line} = \int_\Gamma E(\theta)ds. \qquad (8.120)$$

(a) Explain the meaning of 'locality' and reproduce the argument leading to

$E(\theta)$ given as eqn (8.70). (b) By averaging over all θ, derive a result for a θ-independent line energy. Compare this with the conventional use of $\frac{1}{2}\mu b^2$.

6 Bow-out in the Line Tension Approximation

In this chapter, the bow-out of a dislocation segment of length a was derived on the basis of an isotropic line tension approximation. This derivation culminated in eqn (8.97). Derive this equation and its dimensionless counterpart by first writing down the functional $E[u(x)]$ and then deducing the corresponding Euler–Lagrange equation which will then be solved via separation of the resulting differential equation.

7 Energy of Curved Dislocation in Line Tension Approximation

We wrote down the energy of a curved dislocation as

$$E_{line} = \int_0^L T\sqrt{1 + u'^2(x)}dx. \qquad (8.121)$$

The problem with this expression is that it assumes that the line can be parameterized in terms of x. In this problem I want you to write down a fully general version of this result such that the line can wander in any direction and we can still write the energy as an integral over the line. Hint: Consider a parameteric representation of the dislocation and write the arc length in terms of this parameter. Show that your new expression reduces to the expression given above in the case in which your curve is parameterized by x. In addition, sketch a winding line for which more than one part of the line corresponds to a given x (i.e. it has overhangs) and discuss why the form derived in this problem is still applicable while that given in the equation above is not.

8 Elastic Forces for Three-Dimensional Dislocation Configurations

Use isotropic linear elasticity to compute the elastic forces of interaction between two dislocations oriented as: (a) two perpendicular screw dislocations, (b) two perpendicular edge dislocations.

9 Line Tension Model for Dislocation Junctions

Reproduce the arguments culminating in eqn (8.119).

---

10 fcc Dislocation Core Structure using Pair Functional Description of Total Energy

Using the Johnson potential, determine the structure of the dislocation core in an fcc material. Consider a conventional edge dislocation on the (111) plane with a $[\bar{1}\bar{1}2]$ line direction. For a computational cell use a large cylinder in which the atoms in an annular region at the edge of the cell are frozen in correspondence with the Volterra solution.

---

11 Displacements due to a Dislocation Loop

In eqn (2.96) we obtained the displacements due to a distribution of eigenstrains. In this problem, show that the eigenstrain associated with a dislocation loop with Burgers vector **b** on a plane with normal **n** is given by

$$\epsilon_{ij}^* = \frac{1}{2}(b_i n_j + b_j n_i)\delta(\mathbf{r}' \cdot \mathbf{n}).\qquad(8.122)$$

By integrating eqn (2.96) by parts and substituting this experession for the eigenstrain into the resulting formula, show that the expression given in eqn (8.32) is recovered.

NINE

Wall Defects in Solids

9.1 Interfaces in Materials

Much of materials science and engineering is played out at the level of a material's microstructure. From a more refined perspective, the quest to tune or create new microstructures may also be thought of as a series of exercises in interfacial control. Indeed, interfaces in materials are the seat of an extremely wide variety of processes ranging from the confinement of electrons to the fracture of polycrystals. The character of the interfaces that populate a given material leaves an indelible imprint on that material's properties. Whether we consider the susceptibility of a given material to chemical attack at its surfaces, the transport properties of complex semiconductor heterostructures, or the fracture of steels, ushered in by the debonding of internal interfaces, ultimately, interfaces will be seen as one of the critical elements yielding a particular material response. In some cases, the presence of such interfaces is desirable, in others, it is not. In either case, much effort attends the ambition of controlling such interfaces and the properties that they control.

Until this point in the book we have concerned ourselves with defects of reduced dimensionality. The present chapter aims to expand that coverage by turning to two-dimensional wall defects. The analysis of these defects will serve as our jumping off point for the consideration of microstructures within materials. From the start, the notion of material interface will be broadly construed so as to encompass free surfaces which are the terminal interface separating a material from the rest of the world, stacking faults which are planar defects in the stacking arrangement of a given crystal, grain boundaries which separate regions with different crystalline orientations and the interfaces between chemically or structurally distinct phases. The plan of the chapter is to begin with an introductory commentary on the role of interfaces in materials. This will be followed by three primary sections which treat surfaces, stacking faults and grain boundaries in turn.

441

9.1.1 Interfacial Confinement

One organizing paradigm for examining the significance of material interfaces is to think of them in terms of their role in producing confinement. For the purposes of the present discussion, confinement refers to the localization of some process or object either to the vicinity of the interface itself, or to a region that is bounded by such interfaces. In this section, we will consider examples of such confinement in the context of elastic waves, which in certain cases are confined to the vicinity of interfaces (i.e. Rayleigh and Stoneley waves), mass transport, in which interfaces serve as short-circuit diffusion paths and sinks for chemical impurities, and as impediments to dislocation motion, thus confining plastic action to certain favorably oriented grains. Finally, we will note the significance of internal interfaces in their role in providing a localizing pathway for fracture.

Interfacial Confinement of Mass and Charge. One of the key forms of confinement induced by interfaces is associated with mass transport. What we have in mind is the localization of the diffusion process to the immediate vicinity of the interface in question. In particular, the presence of interfaces can lead to wholesale alteration of the rate at which processes involving mass transport transpire with a huge enhancement in the rate of diffusion along interfaces relative to bulk values. The macroscopic evidence for this claim is presented in fig. 9.1 in which the rate of diffusion is shown not only for the bulk processes considered in chap. 7, but also for the short-circuit diffusion pathways presented by dislocation cores, free surfaces and internal interfaces. The origins of such phenomena are the confining properties of interfaces with respect to point defects such as vacancies and substitutional impurities. As will be seen in more detail later, the structure of interfaces (whether surfaces or internal interfaces) reduces the barrier to the activated processes that occur during diffusion. From the standpoint of the analytic machinery set up in chap. 7, these short-circuit diffusion pathways are characterized by an activation energy for diffusion that is smaller than that found in the bulk.

In addition to the role of interfaces in serving as a conduit for easy diffusion, the chemical potential profile near such interfaces also makes them an ideal sink for point defects. The existence of such sinks leads in turn to the emergence of precipitate-free zones, local regions near grain boundaries characterized by the conspicuous absence of second-phase particles such as those that riddle the remainder of the material which is more distant from boundaries. An example of this phenomenon is depicted in fig. 9.2. Again, the explanation for this behavior is the argument that because of the strong driving force which encourages foreign atoms in the vicinity of boundaries to diffuse to those boundaries, these foreign atoms are unavailable to participate in the nucleation and growth of second-phase particles.

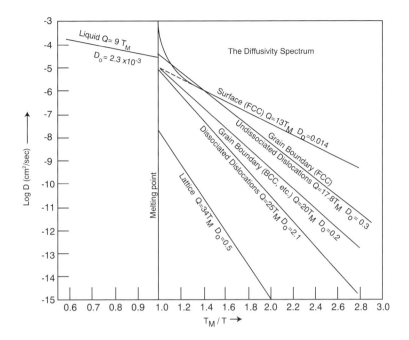

Fig. 9.1. Diffusion data revealing the alteration in diffusion rates as a result of short-circuit pathways provided by interfaces and other defects such as dislocations (adapted from Gjostein (1972)).

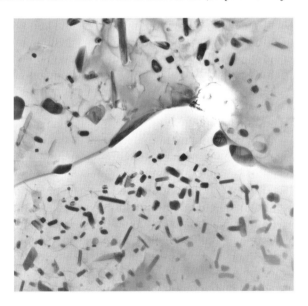

Fig. 9.2. Illustration of a precipitate-free zone in a Cr-3 at.% Nb alloy (courtesy of Sharvan Kumar). The grain boundary runs from left to right through the middle of the picture, and reveals the existence of Cr_2Nb precipitates along the boundary itself and in the Cr matrix and the absence of such precipitates in a band near the boundary.

In addition to their role in confining mass, in some cases interfaces also have the effect of confining charge to certain regions. As will be discussed in greater detail below, one of the most significant types of internal interface is that representing a boundary between chemically distinct phases. Such interfaces play a pivotal role in modern microelectronics, and can be understood in part as a result of their role in serving as potential barriers for electronic properties. In chap. 3, the problem of confinement in this context was evident in our solution of both the particle in the box and the quantum corral. Our interest at this juncture is to call attention to three-dimensional analogs of those systems, with special emphasis on the ways in which the internal interface serves as such a potential barrier which can have the effect of confining electrons.

Interfacial Confinement of Waves. A knowledge of the complex interactions between waves and interfaces is at least as old as the first explanation of Snell's law of refraction on the basis of the wave theory of light. Indeed, with the further insight that upon passing from a medium with larger index of refraction to one with a smaller index of refraction, light could suffer total internal reflection, we see the roots of the ideas of interest to the current discussion. Not only do interfaces serve as the basis of confinement in the context of electromagnetic waves, but they similarly alter the behavior of elastic waves. As was already discussed in chap. 5, the theory of linear elasticity may be used in order to consider dynamics in solids, illustrating the possibility of elastic waves within a medium. From the perspective of bulk excitations, we have already seen (see section 5.3.1) the formalism involving the acoustic tensor, and the subsequent propagating solutions. This earlier work may be generalized to consider waves at interfaces. Such waves have been dubbed Rayleigh waves in the context of those elastic waves confined to surfaces, and Stoneley waves for the case in which such propagation is confined to internal interfaces. The basic argument is that the equations of elastodynamics already highlighted in eqn (5.55), for example, admit of solutions not only in the bulk context, but also those localized to a region near the interface of interest. To see this, a trial solution to the equations of elastodynamics of the form

$$u_i(\mathbf{x}, t) = A e^{-\gamma z} e^{i(\mathbf{k} \cdot \mathbf{r} - \omega t)}, \tag{9.1}$$

is considered. The key feature of this trial solution is its property that the displacement fields decay in the direction z perpendicular to the interface substantiating the claim that the wave amplitude is nonzero only in the vicinity of the interface.

Interfacial Confinement of Plasticity and Fracture. From the perspective of the mechanical properties of materials, interfaces are key players as a result of their unequivocal role in dictating mechanical response. The role of such interfaces in

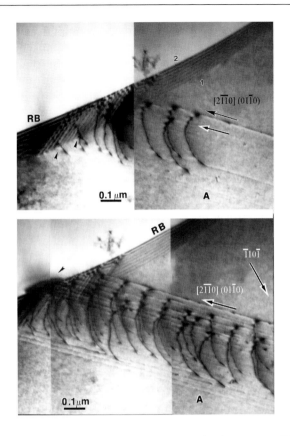

Fig. 9.3. Pile up of dislocations against a grain boundary in Ti (courtesy of Ian Robertson).

mechanical response is perhaps most evident in the context of intergranular fracture in which the crack path follows the grain boundaries. As the crack propagates, it traverses the medium in a manner such that it is internal interfaces themselves that are the victims of bond breaking. Such fracture processes jibe with our contention that by virtue of interfaces, there is a confinement of processes to a spatial region localized about these interfaces.

However, internal interfaces can manifest their presence in the mechanical properties of materials in more subtle ways. In particular, in the context of plasticity in polycrystals, grain boundaries serve as obstacles to dislocation motion, leading to either dislocation pile-ups or grain boundary absorption of these dislocations. An example of the piling up of dislocations against a grain boundary in Ti as revealed by transmission electron microscopy is shown in fig. 9.3. The detailed accounting for the elastic fields associated with such pile-ups as well as their implications for plastic deformation is given in chap. 11, whereas at present we only wish to remark upon the obstacle to dislocation motion offered by the presence of interfaces.

In the discussion given above, we have illustrated the sense in which interfaces can alter material properties. The entirety of all interfaces found within a given material forms one of the key backbones of that material's microstructure. Indeed, much of the actual business of microstructural control takes place at interfaces, either by eliminating them through processes such as grain growth, or by intentionally creating them such as in processes like the carburization of steel.

Having motivated the idea of the importance of material interfaces, we will begin our analysis of interfaces in much the same way we treated point and line defects. Our analysis will begin by examining interfacial phenomena from a geometric perspective with an attempt to consider not only the geometry at the level of atomic positions, but also with regard to the macroscopic geometries associated with interfaces. Particular attention will be directed first towards surfaces which are largely the point of communication between a crystal and the external world. Once the geometric and energetic preliminaries have been dispensed with, the discussion will lead in the direction of increasing geometric complexity as we confront first, the reconstructions that attend many surfaces, and then the step features and roughening that are present on most real surfaces.

Surfaces will be followed by a discussion of stacking faults and twin boundaries. The point of this part of the discussion will be to examine how the structure and energetics of such interfaces may be determined, and once such features are known, how they may be used to provide insights into observed material behavior. Stacking faults and twins may perhaps be thought of as the least severe of internal interfaces in that they represent the smallest departure from the monotony of the perfect crystal. After having examined the fascinating structural polytypes that can occur as a result of repeated 'mistakes' in the packing sequence of various crystal planes, we will take up the geometric and energetic features of grain boundaries. Our discussion on grain boundaries will commence with bicrystallography. However, the ultimate aim of these conceptual notions will never be far from the fore: an analysis of grain boundaries should leave us poised to tackle the more difficult problem of the structure and energetics of polycrystals.

9.2 Free Surfaces

The crystal surface is the seat of the vast majority of interactions that take place between a material and the rest of the world. Catalysis, crystal growth, and degradation via corrosion are but a few of the key processes that take place at surfaces. In each case, it is the interaction between a crystal on the one hand and the atoms arriving at the surface on the other that leads to the properties of interest. Another way in which the crystal surface has emerged as one of the key players in the study of solids is by virtue of the recent development of a number of experimental probes which allow the direct observation of the behavior in the vicinity of the surface.

In addition to their importance to the enterprise of materials science itself, surfaces have been one of the central proving grounds in the quest to construct a viable *computational materials science*. Because of the experimental capacity for explicit determination of surface structure, many calculations have been aimed at establishing the nature of the atomic-level geometries that occur on a given surface. With the increasing sophistication of experimental and theoretical tools alike, not only have the geometries of an increasing number of metal, semiconductor and ceramic surfaces been determined unambiguously, but also a rich variety of new phases associated with adsorbates, islands and steps have been uncovered. Our starting point for the analysis of this class of problems will be to consider the structure and energetics of ideal crystal surfaces.

9.2.1 Crystallography and Energetics of Ideal Surfaces

From a geometrical perspective, if one imagines terminating a crystal on some particular crystal plane, there is a well-defined and unambiguous idealized geometry. By this we mean that by virtue of the simple conceptual act of terminating the repeated packing of crystal planes along a particular direction, we are left with an exposed surface whose structure is determined unequivocally by the particulars of the underlying crystal lattice. We adopt the standard characterization of the crystallographic features of such a surface by exploiting the Miller indices. Stated simply, the Miller indices of a given surface are determined by first noting the points of intersection of the surface of interest and the Cartesian axes. These numbers are then inverted and multiplied by an appropriate integer such that all three fractions have been converted into integers. These integers are the Miller indices.

For the purposes of the present argument, our intention is to convey a sense of the structure and energetics of ideal surfaces. The single most obvious feature of these geometries is the loss of neighbors to those atoms at the surface. In the case of a covalent material, such as Si, this loss will be most significant as a result of the formation of dangling bonds in which the Si atom is left with fewer neighbors than is energetically favorable. Clearly, the creation of such surfaces has energetic consequences. As a starting point for our foray into the energetics and motion of surfaces, we will consider the energetics of such idealized geometries. These calculations will serve as our point of departure for more elaborate analyses involving reconstructions, surface steps and islands. For the sake of physical transparency, our calculations will begin on the basis of an idealized pair potential model. Once these calculations are in hand, a more general

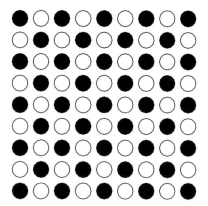

Fig. 9.4. Schematic of the (100) surface of an fcc crystal. Filled circles correspond to atoms on the top layer, open circles are those one layer beneath the surface.

discussion of computing surface energies will be undertaken, first with respect to pair functionals, and then through consideration of the general concepts appropriate to all energetic descriptions.

Our endeavor to uncover the energetic features of surfaces begins with a pair potential description. For convenience, we leave the specific details of the pair potential unspecified, noting only that the total energy of the solid is written as

$$E_{tot} = \frac{1}{2} \sum_{ij} V(R_{ij}).$$ (9.2)

Further, we restrict our attention to the simplified analysis in which it is assumed that only near-neighbor interactions contribute to the energy. Consider a slab of fcc material with periodic boundary conditions in two of the three Cartesian directions, and exposed (100) surfaces in the third direction with a geometry like that shown in fig. 9.4. For concreteness, we imagine a total of N atoms in the slab with m atoms per layer. Once we have adopted the pair potential mindset, which will be our temporary preferred mode of thought, the determination of the surface energy is reduced to little more than a proper accounting of the number of bonds lost by those atoms in the vicinity of the surface. In this case, it is only those atoms on the surface itself that suffer any reduction in coordination number and hence have an altered energy. In particular, consideration of fig. 9.4 reveals that surface atoms on the fcc (100) surface are deprived of four neighbors. As a result of this argument, the total energy of the slab of atoms is given by

$$E_{tot} = 6(N - 2m)V(R_{nn}) + 8mV(R_{nn}),$$ (9.3)

where the first term on the right hand side corresponds to bulk atoms that are not disturbed by the presence of the free surface, while the second term corresponds to the $2m$ atoms participating in the two free surfaces.

One of the key linkages between continuum and microscopic descriptions of interfaces is the notion of an interfacial energy. Though the use of such energies is by now entirely routine, nevertheless, this exemplifies the type of information passage that is one of the hallmarks of multiscale modeling. Note that from the perspective of the continuum mechanics of materials with interfaces, the total energy is often written in the form

$$E_{tot} = \int_\Omega W(\mathbf{F}) dV + \int_{\partial\Omega} \gamma(\mathbf{n}) dA, \qquad (9.4)$$

where $\mathbf{F}$ is the deformation gradient and $W(\mathbf{F})$ is the strain energy density accounting for the energy of bulk deformation and where we have included the possibility of an anisotropic surface energy through the dependence of the surface energy γ on the surface normal $\mathbf{n}$. The question of current interest is the microscopic origins of such a surface energy. On the basis of the total energy for the slab considered above, we are in a position to evaluate the surface energy. The key point is that the surface energy is computed by evaluating the difference between the N-atom system including the free surfaces and that of N atoms in the perfect crystal, and then dividing by an area factor that reflects the amount of exposed surface.

For the case of the (100) surface, this amounts to

$$\gamma_{100} = \frac{E_N(surface) - E_N(bulk)}{2A}, \qquad (9.5)$$

where the factor of 2 in the denominator accounts for the fact that there are two exposed faces on the slab under consideration. Our notation calls for the energy of the N-atom system in the presence of free surface (i.e. $E_N(surface)$) and the energy of an N-atom bulk solid (i.e. $E_N(bulk)$). In this case, $E_N(surface)$ is given by eqn (9.3), while $E_N(bulk) = 6NV(R_{nn})$, our result for the energy of an fcc crystal in a near-neighbor pair potential approximation. Hence, we find that the surface energy is given by

$$\gamma_{100} = -\frac{4V(a_0/\sqrt{2})}{a_0^2}. \qquad (9.6)$$

Note that in deducing this result we have not allowed for the possibility of structural relaxations that might occur for surface and near-surface atoms.

The type of calculation illustrated above can be elaborated by appeal to more sophisticated descriptions of the total energy. Regardless of choice of energy function, the fundamental concept is embodied in eqn (9.5). In preparation for our discussion of surface reconstructions, we will generalize the description given above to the case of pair functionals such as the embedded-atom method. We adopt the Johnson-type embedded-atom functions already described in chap. 4 which permit analytic progress in the context of questions such as that raised here. Recall that in this case, our task is to supplement the pair potential description of the surface energy (already deduced above) with the contribution due to the embedding energy.

Following the discussion from chap. 4, recall that within a pair functional description the energy takes the form,

$$E_{tot} = \sum_i F(\rho_i) + \frac{1}{2} \sum_{ij} \phi(R_{ij}), \tag{9.7}$$

where the former term is the embedding energy and is a function of the density at site i (i.e. ρ_i). The density at site i may be deduced in turn according to

$$\rho_i = \sum_j f(R_{ij}), \tag{9.8}$$

where the function $f(R)$ yields the contribution to the density at site i from a neighbor at a distance R. We will denote $f(R_{nn}) = \rho_{nn}$, with the result that the energy of N atoms in a bulk fcc environment is given by $E_N(bulk) = NF(12\rho_{nn}) + 6N\phi(R_{nn})$. As with our evaluation of the surface energy in the strictly pair potential setting, the pair functional assessment of the total energy can be obtained essentially on the basis of bond counting. Again, if we assume a near-neighbor model in which not only is the pair potential range restricted to near neighbors, but so too is the function determining the electron density, then only atoms that are at the surface itself have environments that are distinct from the bulk. In particular, the total energy of the N-atom system in which there are two exposed (100) surfaces is given by,

$$E_N(surface) = (N - 2m)F(12\rho_{nn}) + 6(N - 2m)\phi(R_{nn})$$
$$+ 2mF(8\rho_{nn}) + 8m\phi(R_{nn}), \tag{9.9}$$

where we have used the same geometry as that considered in the pair potential calculation. The first two terms on the right hand side arise from the bulk atoms, while the second two terms reflect the presence of the free surface atoms. If we imitate the strategy used above in the pair potential context, it is found that the surface energy is given by

$$\gamma_{100} = \frac{2[F(8\rho_{nn}) - F(12\rho_{nn})] - 4\phi(R_{nn})}{a_0^2}. \tag{9.10}$$

To put these expressions in perspective, we now examine the types of numerical values they imply for fcc surfaces. Using the analytic potentials of Johnson introduced in chap. 4, the surface energies of fcc Cu are $\gamma_{111} = 0.99 \text{ J/m}^2$,

$\gamma_{100} = 1.19$ J/m^2 and $\gamma_{110} = 1.29$ J/m^2, which are nearly identical to the experimental values for this system, although we should recall that the potentials are fitted to a number of important properties, though not the surface energies themselves. These results for the energy of free surfaces represent the first in a hierarchy of interfacial energies that will be discussed in this chapter. As was already hinted at above, interfaces in solids come in a number of different forms resulting in several different energy scales. At the low end of the energy spectrum are the energies of twin boundaries and stacking faults. These are followed by the energies of grain boundaries which are superseded by the even higher energies associated with free surfaces.

In the preceding parts of this section, our emphasis has been on the fundamental ideas associated with the energetics of ideal surfaces. Before embarking on the refinement of our arguments to the consideration of surface reconstructions, we close this section with a few details related to the use of microscopic analysis to analyze in a systematic way the energies of free surfaces. First, we reiterate the importance of eqn (9.5), which is the basis of surface energy calculations whether one uses the lowliest of pair potentials or the most sophisticated first-principles techniques. On the other hand, the details of implementation can differ depending upon the situation. One of the critical questions that must be faced when using more computationally expensive techniques is the thickness of the computational slab. On the one hand, one aims for carrying out such calculations using the fewest number of atoms possible. By way of contrast, failure to use a thick enough computational cell will result in interactions between the two surfaces.

As with many of the first-principles calculations that we have already described, there are a series of generic issues that must be faced in making a systematic analysis of features such as the surface energy. One useful class of calculations examines trends in important material parameters such as the surface energies. A systematic analysis of the surface energies of 60 metals has been carried out by Vitos *et al.* (1998) using a variant of the LMTO method. As we described in chap. 4, there are a number of different versions of the first-principles solution to the Schrödinger equation which differ in the details of how they handle the ionic potential, the basis functions used to represent the wave function, and so on. The LMTO method is one of this set of methods. Periodic boundary conditions are employed with the added twist that vacuum layers separate surfaces along the stacking direction of interest. For example, in their analysis of fcc (111) surfaces, they use a slab that is four atomic layers thick with an associated vacuum layer that is two atomic layers thick. An example of the type of results that are obtained for the 4d transition metals is shown in fig. 9.5. Note that as the number of d-electrons per atom (i.e. the band filling) increases, the surface energies exhibit a characteristic rise until the d-band is half-filled and then decline as the number

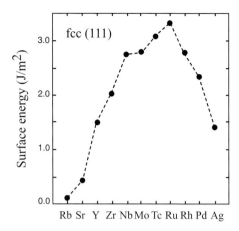

Fig. 9.5. fcc (111) surface energies for 4d transition metals (adapted from Vitos *et al.* (1998)).

of d-electrons is further increased. This behavior is reminiscent of that associated with the cohesive energies reported in fig. 4.18.

Another way of representing the trends in surface energies is to examine the anisotropy in passing from one low-index surface to another. The surface energy anisotropy between the (100) and (111) surfaces is characterized by the ratio $\gamma_{100}/\gamma_{111}$. Fig. 9.6 reveals the surface energy anisotropies for the 4d transition metals where it is seen that there are systematic, but rather small, variations in the surface energy anisotropies.

The discussion given above has concentrated on the energetics of metal surfaces. Though we could similarly enter into a description of the energetics of ideal semiconductor surfaces, such a discussion is much less revealing since in the case of semiconductor surfaces, the driving force for structural reconstruction is so high. As a result, we find it convenient to await a discussion of semiconductor surfaces until we have introduced the next level of structural elaboration, namely, that of surface reconstructions.

9.2.2 Reconstruction at Surfaces

The previous section highlighted the geometric aspects of surfaces that arise on carrying out the conceptual exercise of abruptly terminating the packing sequence along a given crystallographic direction. In the vast majority of cases, this ideal geometry is little more than a fiction and what is really found is a reconstructed geometry in which the atomic positions in the vicinity of the surface are distinct from those that are implied by merely truncating the crystal on a particular plane. Such reconstructions are the two-dimensional analog of the type of reconstructions

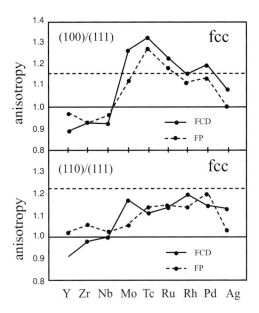

Fig. 9.6. Surface energy anisotropies for 4d transition metals (adapted from Vitos *et al.* (1998)).

described in the previous chapter as concerns dislocations in semiconductors (see section 8.6.3). The intention of this section is first to provide plausibility arguments as to the origins of the structural rearrangements that attend the formation of free surfaces, several key examples of which are given in fig. 9.7. The two key structural rearrangements to be considered here are: (i) structural relaxations in which the spacing between adjacent layers in the direction perpendicular to the surface is altered as is shown schematically in fig. 9.7(a), and (ii) surface reconstruction in which there is a complete rearrangement of the surface geometry as shown schematically in figs. 9.7(b) and (c). In the case of these reconstructions, the symmetry of the surface can be completely altered relative to the bulk termination. Once we have spelled out the conceptual features of both surface relaxation and reconstruction we will give the particular details associated with a few different representative examples on both metal and semiconductor surfaces. We further idealize the discussion through the assumption that the various competing structures are thought of in a vacuum, since a further intriguing but difficult complexity is introduced once foreign species are adsorbed on the surface.

Plausibility Argument for Surface Relaxation. As already noted above, crystal surfaces are characterized by a wide range of structural rearrangements relative to the ideal surface. One such rearrangement is surface relaxation, with the entire surface layer moving either towards or away from the subsurface layers as shown

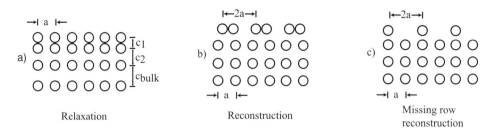

Fig. 9.7. Schematic of some of the key rearrangement mechanisms at surfaces (adapted from Lüth (1993)): (a) surface relaxation, (b) surface reconstruction and (c) missing-row reconstruction.

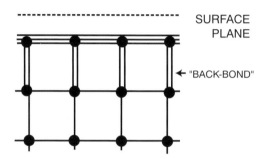

Fig. 9.8. Schematic illustrating the strengthening of back bonds at a free surface (adapted from Carlsson (1990)). Double and triple lines associated with surface atoms are meant to indicate the strengthening of bonds associated with those atoms that have lost neighbors due to the free surface.

in fig. 9.7(a). One way to see the origins of this tendency is via the types of simple arguments we used earlier to justify the extra physics incorporated in pair functionals. In particular, we recall the essence of such models is in the fact that the bond strength is dependent upon the coordination number of the atom of interest. The coordination number dependence of the bond strength was illustrated in fig. 4.5. In the context of free surfaces, as a result of the loss of neighbors to those atoms at the surface, there is a subsequent strengthening of back bonds between the surface atoms and their subsurface partners. This effect is indicated schematically in fig. 9.8.

Plausibility Argument for Surface Reconstruction. In addition to the types of relaxations discussed above, surfaces may also suffer massive rearrangements referred to as reconstructions. These reconstructions often represent a complete rearrangement of the underlying ideal geometry and may bear no resemblance to the subsurface structure. The crystallography of these symmetry breaking rearrangements is illustrated in fig. 9.9. As is evident from the figure, there is some new notational baggage that goes along with the description of surface reconstructions. In particular, if we elect to denote the symmetry of the ideal

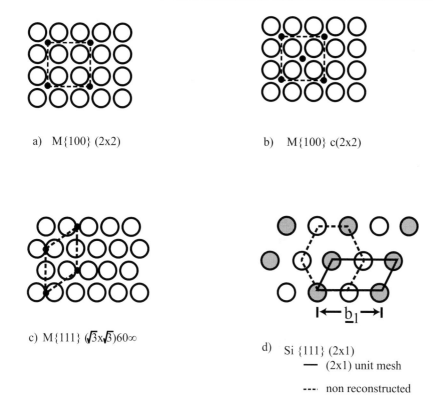

a) M{100} (2x2)

b) M{100} c(2x2)

c) M{111} ($\sqrt{3}$x$\sqrt{3}$)60∞

d) Si {111} (2x1)

   — (2x1) unit mesh

   --- non reconstructed

Fig. 9.9. Schematic of the symmetry changes that attend surface reconstruction for case of several surfaces (adapted from Lüth (1993)).

surface in terms of two lattice vectors **a** and **b**, then the bulk termination of the surface is designated as having a (1×1) symmetry since the surface unit cell is bounded by the vectors **a** and **b** themselves. In the event that the surface undergoes a symmetry breaking reconstruction, the surface unit cell will be built up of larger multiples of the vectors **a** and **b**. In particular, if the surface unit cell is bounded by the vectors $n_1\mathbf{a}$ and $n_2\mathbf{b}$, then the surface reconstruction is designated as the $n_1 \times n_2$ reconstruction. Four distinct examples of this thinking are illustrated in fig. 9.9 where frames (a) and (b) show a 2×2 reconstruction on a (100) surface, and frames (c) and (d) show examples from (111) surfaces.

As we have already emphasized, the energy computed in the previous section is the ideal surface energy, and does not reflect the possible further energy reduction that attends a symmetry breaking surface reconstruction. On the other hand, we seek to understand the driving force towards such reconstruction in energetic terms. Perhaps the most immediate context within which to view the emergence of reconstructions is on semiconductor surfaces where simple arguments concerning

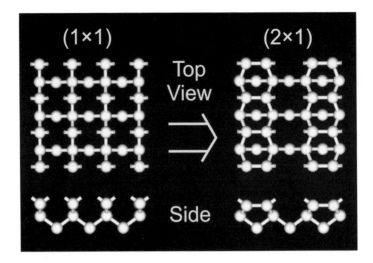

Fig. 9.10. Representation of ideal and 2×1 reconstructed geometries for Si (001) (courtesy of Lloyd Whitman).

the necessity to saturate unfulfilled bonds give intuitive plausibility to such reconstructions. For example, in the case of the (001) surface in Si as depicted in fig. 9.10, we see that the ideal surface is replete with dangling bonds. As in the case of the semiconductor dislocation cores discussed earlier, there is a driving force towards dimerization which has the effect of eliminating the dangling bonds through reconstruction. In particular, in this case we see that in the ideal 1×1 state, each surface atom has two dangling bonds. By virtue of the pairing up of adjacent surface atoms a 2×1 symmetry is induced which has the effect of reducing the number of dangling bonds on each atom.

Computing Surface Reconstructions and Their Energies. Thus far, we have succeeded only in giving plausibility arguments for the emergence of reconstructions as structural competitors to the ideal bulk terminations. On the other hand, it is by now evident from preceding chapters that an analysis of the energetics and dynamics of such surfaces may be made through an appeal to the energetic methods of microscopic simulation. One difficulty faced in trying to carry out a systematic study of such reconstructions on the basis of energetic analyses is the biasing role played by the supercell dimensions. For example, if the overall lateral supercell size is 5×5, this implies that it is not possible to obtain structures with periodicities incommensurate with the underlying substrate, nor those structures with a surface unit cell larger than the 5×5 cell made available in the computation. One remedy to this difficulty is the construction of effective Hamiltonians from which it is possible to carry out not only much larger simulations, indeed with the possibility for the emergence of long-period structures, but also to use the tools of statistical mechanics which in some cases allows for direct analytic progress to be made on the relevant Hamiltonian. We will give examples of this strategy in the pages that follow in the context of reconstructions on both the Au and W surfaces.

Case Studies on Metal Surfaces

We have already emphasized the revolution in surface science ushered in on the heels of high-quality vacuum technology and surface scanning probes. One subbranch of surface science that has been the beneficiary of these advances is the study of metal surfaces. The intention of this subsection is to consider two distinct case studies, namely, the surfaces of fcc Au and those of bcc W. These choices are inspired by the combination of high-quality experimental data, richness of surface phases, their diversity from the standpoint of the demands they place on atomistic simulation and their relevance to later generalizations to surfaces with steps, islands and even more extreme forms of roughness.

Surface Reconstructions in Au. Gold is one of a number of systems that exhibits a wide range of fascinating surface reconstructions, which includes a host of orientational phases in the presence of steps. For our purposes, the Au system serves as a rich example not only for our arguments concerning surfaces and their reconstructions, but also as a basis for contrasting many of the different energetic arguments that have been set forth so far in the book. What we will find is that the Au surfaces have been attacked from a wide variety of pair functional approaches, by recourse to first-principles techniques and using different effective Hamiltonians. In fact, much of the theoretical effort on the Au(001) surface will be reminiscent of our earlier treatment of the Peierls–Nabarro model and treatments yet to come of cohesive zone models. Finally, the existence of a variety of X-ray, scanning tunneling microscopy and other experimental data makes the Au surfaces very appealing.

For concreteness, we consider the (001) surface of Au. The unreconstructed geometry corresponds with the bulk termination along this direction and is nothing more than a square lattice, rotated at $45°$ with respect to the cubic directions. This geometry may be contrasted with a depiction of the reconstructed geometry, shown in fig. 9.11. The key qualitative features of the reconstructed geometry are related, namely, a tendency towards close packing which is manifested as a nearly hexagonal arrangement of the surface atoms. As can be seen in the figure, the ideal bulk termination corresponds to the square grid, while the actual atomic positions form a pseudohexagonal arrangement. These structural insights have been garnered from a series of X-ray measurements as reported in Gibbs *et al.* (1990).

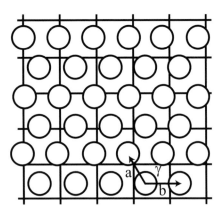

Fig. 9.11. Reconstructed geometry for Au(001) surface (adapted from Gibbs *et al.* (1990)). The figure shows only the surface layer of atoms.

Fig. 9.12. Sequence of phases associated with Au(001). At high temperatures, distorted hexagonal phases are superseded by disordered phase with 1×1 symmetry. RDH refers to rotated–distorted hexagonal phase, DH refers to distorted hexagonal phase and DIS refers to disordered phase. Note that the schematic of the rotated hexagon associated with RDH exaggerates the rotation. In addition, the orientation relationship between the hexagonal structures and the subsurface atoms is not represented.

The experimental data associated with this surface are actually more revealing than we have suggested (Gibbs *et al.* 1990, Abernathy *et al.* 1992). On the basis of systematic X-ray reflectivity studies, it has been deduced that a sequence of phases are found as a function of temperature: these phases are shown schematically in fig. 9.12. In particular, two related pseudohexagonal phases have been found which differ by a small relative rotation in addition to a high-temperature disordered phase in which the pseudohexagonal symmetry is lost. However, despite the apparent abruptness and reversibility of the transition between the two distorted hexagonal phases, it remains unclear whether or not these structures constitute true equilibrium phases.

The mission of any energetic methods that might be brought to bear on the structure of this surface is to explicate the origins of its tendency towards close-packing, even at the cost of abandoning its epitaxial relationship with the subsurface layers. Our strategy is to contrast pair functional and first-principles approaches to this reconstruction. In addition, we will examine the way in which first-principles calculations may be used to construct an effective Hamiltonian for this system and through which the competing influences governing the transition become more clear.

A repeating theme that has been played out in modeling defects ranging from vacancies and interstitials, to dislocation cores and surface reconstructions is that of successive approximation. Often, the first treatment of a new problem is put forth from the perspective of some empirical description of the total energy. The Au(001) surface is no different. The initial efforts to uncover the origins of this reconstruction were built on pair functional descriptions of the total energy (Ercolessi *et al.* 1986, Dodson 1987, Wang 1991). The argument was made that by virtue of the embedding term, the complex interplay of in-plane densification and subsurface pinning could be considered. An example of the type of results that emerge from such simulations is shown in fig. 9.13. The basic idea of the simulations was to construct a slab with some finite size and associated periodic boundary conditions. Note that the selection of a particular periodicity in the in-plane directions rules out symmetry breaking distortions whose periodicities (if any) are larger than these cells. The geometries are then relaxed, usually via some finite temperature annealing scheme (Monte Carlo or molecular dynamics) and are slowly cooled until the resulting energy minimizing configuration is determined. The claim has been made that by carrying out molecular dynamics calculations at a sequence of increasing temperatures, pair functional schemes are sufficiently accurate to capture the series of transitions that are observed with increasing temperature. This claim is substantiated by appeal to calculations of the structure factor which provides a geometric signature of the surface geometry. Inspection of the results of such pair functional calculations of the Au(001) reconstructed geometry reveal a marked tendency towards the close-packed distorted hexagonal structures implied by experiment, though the precise surface periodicity probably remains out of reach of these methods.

More accurate first-principles calculations have also been undertaken with respect to the Au(001) reconstruction (Takeuchi *et al.* 1991). These calculations were conceptually identical in all essentials to those carried out using the pair functional description of the total energy. However, an important added feature was the additional task of considering the isoelectronic Ag system, with the aim of examining why the (001) surface of Au reconstructs while that of Ag does not. Further, by treating the subsurface layers of the Au crystal as a uniform electron gas, the first-principles analysis was able to separate the competition between the surface atoms increasing their coordination number and the subsurface atoms locking the surface atoms into crystallographic positions.

The desire for both computational tractability and physical transparency suggests the usefulness of effective Hamiltonians. We have emphasized the desirability of finding degree of freedom reducing strategies in which only those degrees

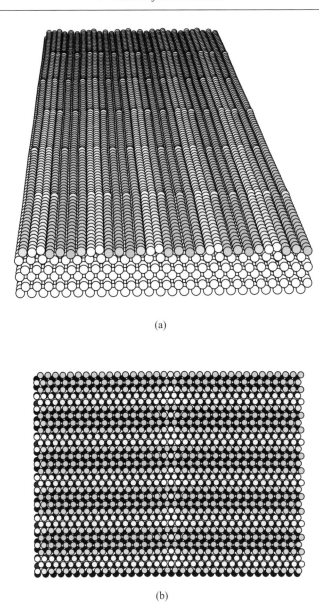

(a)

(b)

Fig. 9.13. Several views of the (34 × 5) Au(001) surface reconstruction as obtained using a pair functional description of the total energy (courtesy of Furio Ercolessi). The atoms in the first layer are colored in gray scale according to their z-coordinates: the lighter the color the larger z. The second layer atoms are colored black.

of freedom of real importance to the physical question at hand are considered. The Au(001) surface features two competing energetic factors: (i) the in-plane energy gain associated with a local increase in coordination number and favoring a close-packed hexagonal arrangement, (ii) the subsurface pinning potential which favors a locking in of the surface atoms into registry with those beneath the surface. This competition suggests an effective Hamiltonian of the form

$$E_{tot}(\{\mathbf{R}_i\}) = \sum_{ij} V(R_{ij}) + \sum_i A\left(\cos\frac{2\pi x_i}{a} + \cos\frac{2\pi y_i}{a}\right), \qquad (9.11)$$

where the set $\{\mathbf{R}_i\}$ characterizes the entirety of atoms on the surface. The first term on the right hand side represents the in-plane interactions between neighboring atoms, and for our purposes should be thought of as a pair potential that favors a near-neighbor distance smaller than that determined by the bulk termination of the surface. In particular, this term inclines the surface atoms to a two-dimensional close-packing geometry. The second term is a lock-in term and characterizes the interaction between the surface atom with coordinates (x_i, y_i) and the atoms beneath the surface which have been buried in this effective potential. Note that this pinning term carries exactly the same sort of information as the interplanar potential introduced in the context of the Peierls–Nabarro model, and similarly, exploits the same sorts of locality assumptions favored there. In addition to the general theme of degree of freedom reduction, we have noted the objective of information transfer in which calculations at one scale are used to inform those used at a different scale. In the current context, we refer to the use of first-principles calculations to inform the effective Hamiltonian of eqn (9.11). Like with the treatment of the interplanar potential associated with the Peierls–Nabarro model, the coefficient A in eqn (9.11) is obtained on the basis of rigid sliding geometries in which the surface layer is slid with respect to the subsurface layer and the resulting energy is computed. In this case, it is the difference in energy between the configuration in which all atoms have their bulk-terminated in-plane positions and that configuration in which all surface atoms have slid by the vector $\delta = (a/2, a/2)$. The potential determined in this way is shown in fig. 9.14. As usual, once the effective Hamiltonian is in hand, it is possible to investigate its implications for the equilibrium phases of the system. In this case, the outcome of examining the equilibrium structure associated with this Hamiltonian is shown in fig. 9.15.

Surface Reconstructions in W. In earlier sections, we have noted that the pair functional formalism is not always appropriate as the basis for an energetic description of metals. One case where this is evident is in the context of the surface reconstructions seen on certain bcc transition metals. In particular, the (001) surface of both W and Mo exhibits fascinating reconstructions, and as in the case of the Au(001) surface, the symmetry breaking distortions leave few vestiges of their crystallographic ancestry.

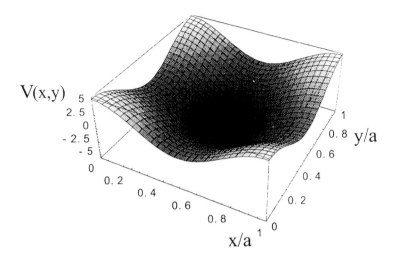

Fig. 9.14. Representation of interplanar potential characterizing the locking in of surface atoms to their subsurface partners.

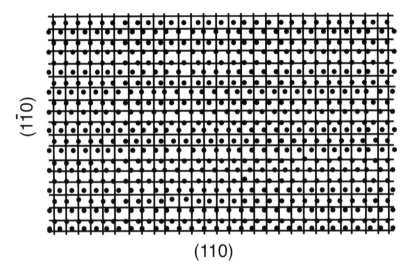

Fig. 9.15. Minimum energy configuration for Au(100) as obtained using the effective Hamiltonian deduced from first principles (adapted from Takeuchi *et al.* (1991)).

The ideal (001) surface of crystals with the bcc structure is illustrated in fig. 9.16(a). We note that the ideal termination of this surface is a square lattice with a lattice parameter equal to that associated with the bulk crystal itself. This geometry may be contrasted with that of the reconstructed geometry in W in fig. 9.16(b). The surface geometry can be continuously altered from the ideal bulk termination to the reconstructed state by imposing a series of surface displacements

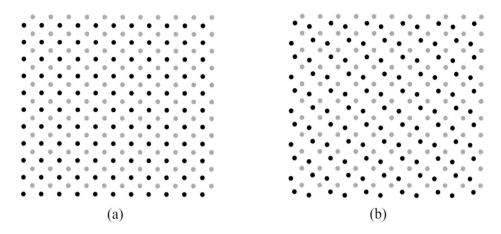

(a)   (b)

Fig. 9.16. Schematic of the ideal bulk termination and reconstruction associated with certain bcc (001) surfaces: (a) ideal bulk termination for (001) surface of bcc crystal, (b) (001) reconstruction of W. The amplitude of the reconstruction is exaggerated for ease in viewing. Dark circles correspond to surface layer atoms and gray circles are associated with atoms in first subsurface layer.

of the form

$$\mathbf{u}_{mn} = \mathbf{u}_0 \cos(\mathbf{q} \cdot \mathbf{R}_{mn}), \qquad (9.12)$$

where mn is a label for the atom at surface position $\mathbf{R}_{mn} = ma\mathbf{i}+na\mathbf{j}$, $\mathbf{u}_0$ is a vector of the form $\mathbf{u}_0 = \delta(1/\sqrt{2}, 1/\sqrt{2})$ and characterizes the amplitude and direction of the surface distortion and $\mathbf{q}$ is a wavevector that characterizes the periodicity of the surface reconstruction itself. For the reconstruction of interest here, the wavevector $\mathbf{q}$ is of the form $\mathbf{q} = (\pi/a)(1, 1, 0)$. The reconstructed surface is characterized by two distinct phases, one with $c(2 \times 2)$ symmetry (below ≈ 210 K) and one with (1×1) symmetry above that temperature, with the high-temperature phase thought to be a disordered state rather than the ideal bulk termination. The low-temperature phase is characterized by lateral displacements $\delta \approx 0.25$ Å.

The (001) surface reconstructions of bcc metals have become something of a proving ground for new total energy schemes. Our discussion in chap. 4 of angular potentials considered the role of angular forces in stabilizing certain characteristic geometric features of bcc crystals, and these features are revealed again in the context of the reconstruction. Instability of the surface with respect to a distortion of wavevector $\mathbf{q}$ can be considered via a series of static calculations in which the energy is repeatedly computed for increasing δ. An example of calculations of this sort is shown in fig. 9.17. We note that the basic strategy adopted here is one that has done service for us repeatedly in a wide range of contexts. The basic idea is to compute (either analytically or numerically) the energies for the putative unstable state and to determine whether in fact there is an energy *reduction* associated with

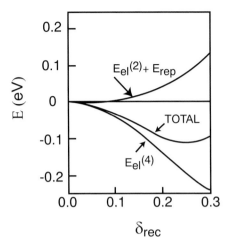

Fig. 9.17. Energy as a function of reconstruction amplitude for W(100) reconstruction as computed using angular potentials (adapted from Carlsson (1991)). The figure reveals the importance of the angular term in stabilizing the reconstruction. The units of δ are measured in ångstroms.

such an instability. The dynamic analog of this procedure is to consider the growth of the presumed unstable state in time and if the state does grow, then this is seen as evidence of an instability of the reference state. In the present setting, an angular force scheme similar to that presented in section 4.4 was used to examine the energetics of reconstruction. As shown in fig. 9.17, the energetics of this instability may be decomposed into the part emerging from pair terms $(E_{el}^{(2)} + E_{rep})$ and that due to four-body angular terms $(E_{el}^{(4)})$. The energy is plotted as a function of the reconstruction amplitude δ_{rec} and reveals a minimum at a nonzero value signaling the propensity for reconstruction. Further, we note that the driving force leading to this reconstruction can be traced to the angular terms.

As we have already seen a number of times, empirical calculations can be supplemented and even supplanted by their first-principles counterparts. The reconstructions on (001) surfaces have been explained not only on the basis of angular force schemes, but also using density functional theory. In this case, it is a brute force reckoning of the electronic structure and the associated energies that is used to probe the instability toward surface reconstruction. An analog of fig. 9.17 obtained using electronic structure calculations is shown in fig. 9.18. One of the advantages of calculations of this sort is that they can provide insight into the *electronic* features giving rise to a given phenomenon. In the present context, an intriguing distinction was made between the isoelectronic partners W and Ta, with the insight that emerged being that the propensity for reconstruction could be ascribed to particular structure in the electronic density of states.

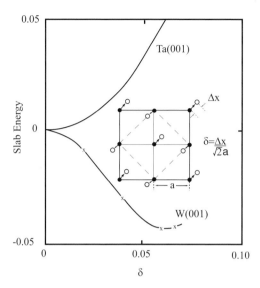

Fig. 9.18. Energy as a function of reconstruction amplitude as computed using density functional theory (adapted from Singh *et al.* (1986)). The parameter δ is given by $\delta = \delta_{act}/\sqrt{2}a_0$, where δ_{act} is the actual displacement magnitude and a_0 is the lattice parameter.

The type of first-principles analysis presented above can serve an additional purpose which is tied to the creation of effective theories. In particular, it has been found possible to construct an effective Hamiltonian which represents the energetics of various surface distortions. The perspective adopted here is that the first-principles calculations are used to *inform* the higher-level effective Hamiltonian. Extra importance attaches to the construction of effective Hamiltonians in cases like this since the use of first-principles methods by themselves does not allow for a systematic investigation of the *finite temperature* aspects of the structural transformation. In particular, in the case of the W(001) surface of interest here, it is of interest to explore the nature of the structural transition at around 210 K. In the present setting, the functional form advanced for the effective Hamiltonian is

$$
\begin{aligned}
H_{eff} = \sum_i & \left[\frac{1}{2} A u_i^2 + \frac{1}{4} B u_i^4 + \frac{1}{2} V_4 u_i^4 \cos(4\theta_i) + \frac{1}{2} A_z (z_i - z_0)^2 + V_z u_i^2 z_i \right] \\
& + \sum_{\langle ij \rangle_1} [J_1 \mathbf{u}_i \cdot \mathbf{u}_j + K_1 S_{ij} (u_{ix} u_{jx} - u_{iy} u_{jy}) + R_1 [u_{ix} u_{jx} (u_{ix}^2 + u_{jx}^2) \\
& + u_{iy} u_{jy} (u_{iy}^2 + u_{jy}^2)] + \sum_{\langle ij \rangle_2} J_2 \mathbf{u}_i \cdot \mathbf{u}_j.
\end{aligned}
\tag{9.13}
$$

There is a variety of notation associated with this model which includes the use of $\mathbf{u}_i$ to represent the in-plane displacement of atom i, z_i to represent the out-of-plane displacement of atom i, u_{ix} as the x component of displacement of the ith atom, $\langle ij \rangle_k$ to denote summing over k^{th} neighbor pairs and the use of a variety of energetic coefficients such as A, B, V_4, A_z, V_z, J_1, J_2, R_1 and K_1. In addition, S_{ij} is 1 if sites i and j are displaced from each other in the x-direction and -1 if they are displaced from each other in the y-direction. On the basis of a series of first-principles total energy calculations, the various parameters that appear in this effective Hamiltonian can be deduced. An example of a set of calculations used to inform the effective Hamiltonian given above is shown in fig. 9.19. As indicated in the figure, the basic idea is to carry out a family of distortions for which the energy can be computed 'exactly' on the basis of atomistics and approximately on the basis of the effective Hamiltonian. The parameters are then chosen so as to yield an optimal fit to the 'exact' results on the basis of the effective Hamiltonian.

Just as we saw in the context of phase diagrams (see section 6.5.5), once an effective Hamiltonian has been adopted, it becomes possible to carry out the associated statistical mechanical analysis which makes possible a systematic exploration of the phase diagram associated with that model. In the present case, Roelofs *et al.* (1989) carried out a series of Monte Carlo calculations on the W(001) surface using the effective Hamiltonian of eqn (9.13), all the while monitoring the diffraction amplitude associated with the superlattice peaks in the diffraction pattern emerging from the $c(2 \times 2)$ symmetry. At high temperatures (in fact at temperatures twice those of the observed transition temperature) the diffraction intensity in these satellite peaks was seen to decay, signaling the onset of the transition. The mismatch between observed and computed transition temperatures provides some idea of the challenges that still remain in the construction of material specific effective Hamiltonians.

Case Studies on Semiconductor Surfaces

Our discussion in the previous part of this section showed some of the variety associated with surface reconstructions on the surfaces of metals with both the fcc and bcc structures. Just as such surfaces have served as a proving ground for our understanding of metals, semiconductor surface reconstructions have played a similar role in the study of covalent systems. The starting point for the analysis in this section is a synthesis of a wide variety of experimental data in the form of a phase diagram for the Si surfaces. The objective in the remainder of the section will be to try to rationalize, even predict, such phases on the basis of microscopic analysis. Beyond their academic interest to surface science, the reconstructions in Si also impact the nature of the growth processes that occur when atoms are deposited on exposed surfaces.

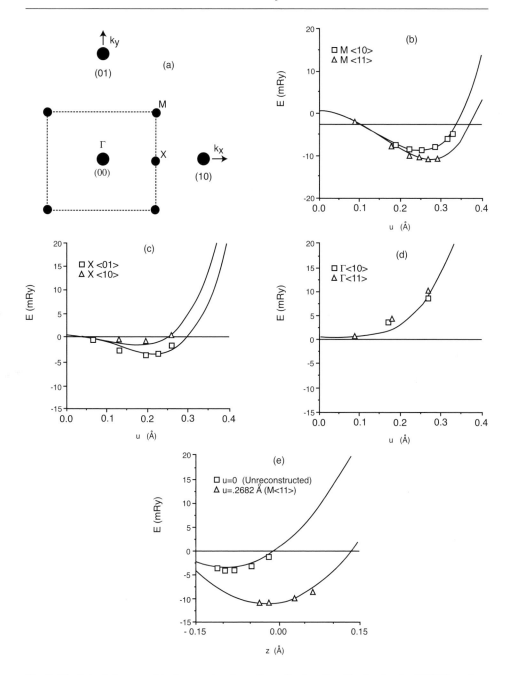

Fig. 9.19. Calculations used in the construction of an effective Hamiltonian for the W(001) surface (adapted from Roelofs *et al.* (1989)). Panel (a) shows the surface Brillouin zone while panels (b)–(e) show the energy associated with various displacement patterns, with the solid lines corresponding to the energetics of the effective Hamiltonian while the discrete points are the energies associated with first-principles total energy calculations.

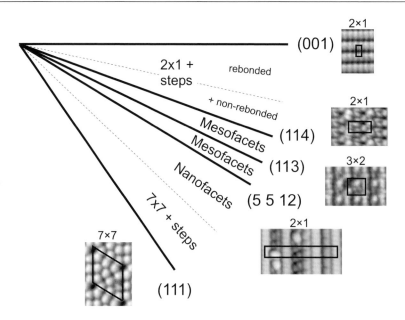

Fig. 9.20. Series of surface reconstructions associated with Si (adapted from Baski *et al.* (1997)). The graph depicts the stable planar reconstructions found between the (001) and (111) orientations and gives scanning tunneling microscopy images of the observed surface structures.

Si surface structures. The surfaces of Si present an entirely different set of challenges than do those of the metal surfaces considered thus far. As a prelude to the discussion to follow, we follow Baski *et al.* (1997) in pictorially representing the various stable planar surfaces between the (001) and (111) orientations as shown in fig. 9.20. One of the key ideas to be conveyed by this figure is that each of the bold lines corresponds to a stable planar surface, and associated with each such surface is a corresponding scanning tunneling microscopy image of the associated reconstructions. The regions between the stable planar reconstructions are characterized by more complex structures involving steps or facets. For the purposes of the present discussion, we will consider the (001) and (111) surfaces as our representative case studies in semiconductor surface reconstruction. We note that from the point of view of intrinsic complexity, the Si(111) surface is characterized by a large unit cell and hence an intricate arrangement of atoms.

The Si(001) Surface. Our analysis begins with perhaps the most studied of such surfaces, namely, the (001) surface. The ideal and reconstructed geometries were presented in fig. 9.10 and are contrasted in more detail in fig. 9.21 where it is noted that the presence of two dangling bonds per surface atom for the ideal bulk termination should provide a strong driving force for structural change. We have already remarked on the analogy between this tendency on the (001) surface and

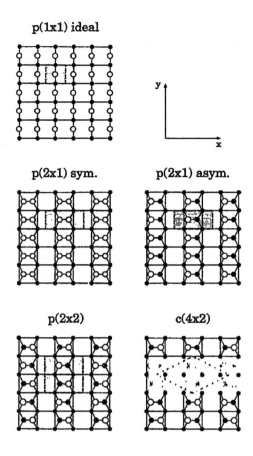

Fig. 9.21. Schematic of the ideal and reconstructed (001) face of Si (adapted from Ramstad *et al.* (1995)).

that seen at the core of certain dislocations in Si. The presence of dangling bonds like this often manifests itself in structural rearrangements involving dimerization. However, the statement that the surface suffers such a dimerization reaction does not suffice to fully characterize the reconstructed geometry. Though the original LEED (low-energy electron diffraction) experiments indicated the presence of a (2×1) symmetry on the reconstructed surface, more recent experiments indicated a phase with (4×2) symmetry as well. The presence of a series of such phases, each differing from the other by rather subtle geometric rearrangements poses serious challenges to efforts at modeling these surfaces. In particular, as shown in the figure (via open and filled circles for the surface layer) in addition to the lateral displacements related to dimerization, different atoms can suffer different out-of-plane displacements. Indeed, the filled and open circles correspond to

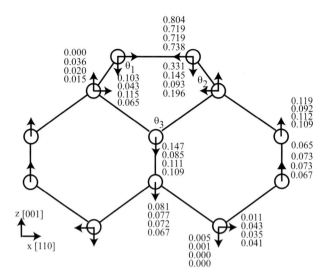

Fig. 9.22. Comparison of the atomic-level displacements associated with the Si(001) (2 × 1) surface reconstruction as obtained using alternative total energy schemes (adapted from Balamane *et al.* (1992)). Arrows illustrate the atomic displacements in ångstroms in comparison with the (1 × 1) reference state. The numbers in each group correspond to the displacements as obtained using density functional theory and Biswas–Hamann, Stillinger–Weber and Tersoff potentials, respectively.

perpendicular displacements either up or down, resulting in a buckling in addition to the dimerization.

One theoretical point of departure for examining the surface reconstructions seen on Si(001) is zero-temperature energy minimization using empirical descriptions of the total energy. Two key articles (Balamane *et al.* 1992, Wilson *et al.* 1990) have made systematic attempts at characterizing the extent to which empirical total energy schemes can be used to examine the energetics of semiconductor surfaces (as well as other defects). As shown already in fig. 9.10, the distinctive feature of the Si(001) reconstruction is the dimerization of the surface layer atoms. A quantitative assessment of the propensity for such reconstruction as well as its extent is given in fig. 9.22 which shows the computed atomic-level displacements associated with this reconstruction for both the Stillinger–Weber- and Tersoff-type potentials in addition to yet another empirical scheme (the Biswas–Hamann potentials) and the results of a full first-principles analysis. The basic point to be made is that while the empirical potentials are unable to account for the subtleties of bond buckling once dimers have formed, the existence of the dimers themselves is well reproduced by the various alternative angular force schemes set forth to account for covalent bonding in semiconductors. In fact, the energy difference between the ideal (1 × 1) surface and the (2 × 1) reconstruction is also reasonably well reproduced.

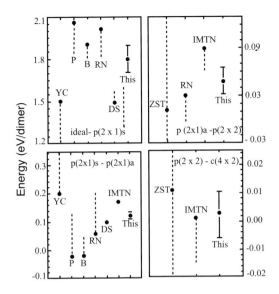

Fig. 9.23. Energies for Si(001) surface. Energy obtained for both ideal and competing reconstructed geometries and illustrating errors (adapted from Ramstad *et al.* (1995)).

More reliable calculations of the energetic ranking of the various structural competitors on Si(001) have been performed using traditional first-principles methods. In particular, a careful zero-temperature analysis has been performed with attention to the errors involved as a result of choice of both k-point sampling and plane wave cutoffs (Ramstad *et al.* 1995). This work attempts to discriminate between the energetics of the ideal unreconstructed surface and two different (2×1) competitors and the (4×2) reconstruction.

In fig. 9.23, we show the energy differences associated with the Si(001) reconstruction as well as the *error bars* associated with these energy differences. In particular, the four panels in the figure report the energy differences between various structural competitors and give the associated error bars in the calculations. For example, the upper left figure reports the difference in energy between the bulk terminated (1×1) surface and the symmetric dimerized (2×1) reconstruction. The label 'This' refers to the results from Ramstad *et al.* while the others (i.e. YC, DS, etc.) refer to the results of calculations of other groups. From a philosophical perspective, it is worth noting that while such calculations are highly tedious and lack glamor, they are also extremely instructive in that through them it is possible to quantify the level of confidence that should be placed in the results. In this case, we see that the error bars associated with the $(2 \times 2)/(4 \times 2)$ energy difference are so large as to render the result impotent since it was not possible even to conclude the sign of the energy difference. The structure and energetics of the Si(001) surface provide a series of interesting insights into the nature of surface reconstructions. Indeed, in addition to the simple question of dimerization considered above, there are fascinating issues that arise in conjunction with the presence of vacancies on these surfaces.

The Si(111) Surface. As noted already at the beginning of the present subsection, the (111) surface is characterized by a complex (7×7) reconstruction which presents significant challenges from both an experimental and theoretical point of view. Our current discussion is aimed at examining the structure of this surface in slightly more detail. A schematic depiction of the reconstructed (111) surface is shown in fig. 9.24. The accepted reconstructed surface geometry is known as the dimer adatom stacking fault (DAS) model and is characterized by three distinctive structural features: in particular, the second-layer atoms undergo a dimerization rearrangement, the presence of twelve adatoms in an adatom layer on the surface (the twelve large filled circles in fig. 9.24) and the existence of a stacking fault over part of the reconstructed area (seen in the left part of the surface cell in fig. 9.24).

Just as was possible with the Si(001) surface, a series of comparisons have been made for the energetics of various structural competitors on Si(111) using different total energy schemes. We appeal once again to the careful analysis of Balamane *et al.* (1992). In this case, the subtlety of the structural rearrangement leads largely to an inability on the part of various empirical schemes to correctly reproduce the energetics of this surface. In particular, as discussed by Balamane *et al.* (1992), only one of the Tersoff potentials leads to the conclusion that the 7×7 reconstruction is stable with respect to the ideal 1×1 surface. As a result, as yet the only reliable means whereby one can successfully explore energetic questions concerning these surfaces is through the use of electronic structure-based total energy methods.

Surface Reconstructions Reconsidered

In the preceding pages we have tried to convey a flavor of the structure and energetics of surface reconstructions. We note that the great advances in this area have been largely experimentally driven and reflect the routine atomic-scale resolution that is now enjoyed in the analysis of surface structures. By dividing our analysis along the lines of material type (i.e. metals vs semiconductors) we have been able to contrast the relative merits of different schemes for computing the energetics of crystal surfaces. As a final example of some of the exciting developments in considering surfaces, fig. 9.25 shows a surface in amorphous Si with relaxed atomic positions obtained using density functional theory calculations. Now in keeping with our overall theme of examining the *departures* from the monotony of

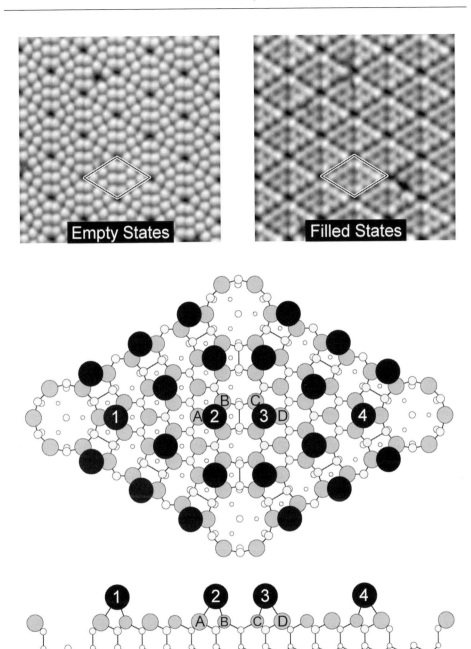

Fig. 9.24. The 7×7 reconstruction of Si(111) in ultrahigh vacuum. Both empty and filled state constant-current scanning tunneling microscope images are shown for a 14 nm × 14 nm region (adapted from Whitman (1998)). The diamond shaped (7×7) unit cell is indicated in the scanning tunneling microscope images. The figure also gives a schematic of the reconstructed unit cell for both a top and a side view (through the long diagonal) with the topmost atoms largest and darkest. Some of the atoms are labeled to aid correlating the two views.

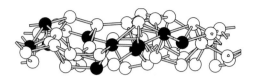

Fig. 9.25. Surface geometry associated with amorphous Si as computed using density functional theory (courtesy of Martin Bazant and Tim Kaxiras). The surface layer is at the top of the figure and the atoms colored black are those having fivefold coordination.

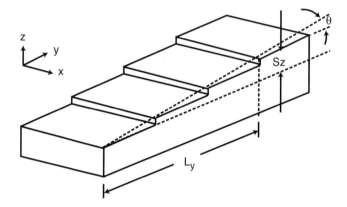

Fig. 9.26. Schematic of vicinal surface with low-index terraces bounded by surface steps (adapted from Poon *et al.* (1992)).

the perfect crystal, we push our analysis of crystal surfaces yet a step further in the form of an examination of steps on crystal surfaces.

9.2.3 Steps on Surfaces

Despite our foray into the structure and energetics of surface reconstructions, the class of departures from the perfection of the bulk terminated surface that we have considered is still incomplete. In addition to the relaxations and symmetry breaking reconstructions that are present on surfaces, we must also consider the role of surface steps. Such steps arise from a variety of different sources: the inability to cut a crystal at exactly the face that is desired, the development of steps and islands during the growth process itself and the presence of thermal roughening. Surfaces with a slight misorientation relative to the conventional low-index surfaces are known as *vicinal* surfaces and are characterized by a series of terraces of the nearby low-index surface, separated by steps which accommodate the overall misorientation. Indeed, the relation between the step height (h), terrace width (l_w) and misorientation can be written as $\theta \approx h/l_w$. A schematic of such a vicinal surface is given in fig. 9.26.

Fig. 9.27. Scanning tunneling microscope image, 100 nm × 100 nm, of a stepped Si(001) surface (courtesy of A. Laracuente and L. J. Whitman).

We begin with a consideration of the experimental situation. The advent of scanning tunneling microscopy has made the examination of surface steps routine. In fig. 9.27, we show a representative example of a stepped semiconductor crystal surface. There are a few key observations to be made about these scanning tunneling microscope results. First, we note that the presence of steps is a common occurrence on surfaces. In addition, we note that the morphology of the steps themselves (i.e. smoothly wandering or jagged) poses interesting challenges as well. Our strategy for examining the physics of such steps is to attend to a microscopic description of the energetics of steps.

For the Si(001) surface, recall that the surface structure is characterized by dimer pairs. Depending upon which layer in the AB packing along this direction is exposed, these dimers will be oriented parallel to either the x-direction or the y-direction. This effect is indicated schematically in fig. 9.28, where we see that if the surface is exposed on layer A, then the dimerization will occur with one orientation, while if it occurs on layer B, the dimerization will be rotated by 90°. This is precisely a schematic of the experimental situation shown in fig. 9.27. The figure also reveals the presence of these different variants and shows the distinction between the two different step types that occur. The geometry of vicinal surfaces is extremely complex and gives rise to a richness in the series of resulting surface phases that was already foreshadowed in fig. 9.20. In particular, that figure revealed that as a function of misorientation relative to stable planar surfaces, the surface geometry can range from terraces to facets.

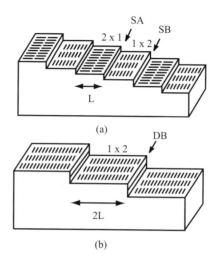

Fig. 9.28. Stepped Si(001) surface and different dimer orientations that result from different layers being exposed (adapted from Alerhand *et al.* (1990)).

As we have already noted a number of times, the analysis of defects in solids is served both by atomistic and continuum approaches. Like many of the other defects that we have already examined, surface steps are amenable to a description on the basis of atomistic analysis. Microscopic analysis is especially suitable for exploring the short-range contributions of these defects to the total energy that derive from the disruption of the atomic-level geometry at the surface. Without entering into a detailed description of what is involved in computing the energetics of steps, we note that many of the same ideas presented in this and earlier chapters are used. In particular, once a computational cell is in hand, one resorts to a computation of $E_{tot}(\{\mathbf{R}_i\})$. For our present purposes, we note that microscopic analysis has been invoked in order to carry out a systematic first-principles analysis of step energies on metal surfaces as shown in fig. 9.29. We note in closing that we have certainly not done justice to this important topic and a major omission in our discussion has been a failure to report on the continuum theory of surface steps. For a review, the reader is urged to consult Jeong and Williams (1999).

9.3 Stacking Faults and Twins

In the previous section we highlighted some of the key issues that arise in considering free surfaces which are themselves one of the key wall defects found in solids. As noted in the introduction, in addition to free surfaces, there are a number of important *internal* interfaces that are found in solids. Perhaps the lowest-energy class of such internal interfaces are those associated with mistakes

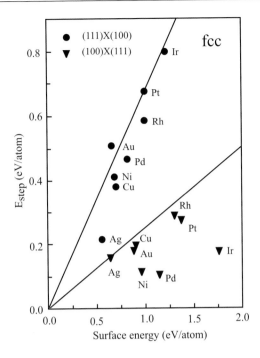

Fig. 9.29. Step energies for 4d transition metals (adapted from Vitos *et al.* (1999)). The straight lines reflect the outcome of a model near-neighbor calculation for the step energies on these surfaces. The first-principles calculations were done using the LMTO method.

in the stacking sequences that lead to a given crystal structure. The role of stacking faults as a key player in dislocation core phenomena has already been introduced in section 8.3.2. However, beyond their presence in dislocations, stacking faults can arise in the course of crystal growth, and in some cases form a part of the structural complexity in certain crystal phases. The goal of the current section is to make a deeper examination of the structure and energetics of stacking faults in crystalline materials.

9.3.1 Structure and Energetics of Stacking Faults

Geometries of Planar Faults. In chap. 8 we introduced the geometric characteristics of stacking faults. One perspective for characterizing these faults is that they arise as the outcome of rigidly translating an entire set of (111) crystal planes with respect to the rest of the crystal by vectors of the form $(a_0/2)(\bar{2}11)$. That is, we consider translating the entire half-space above some certain plane by the vector given above. The result is a stacking fault at this plane. Alternatively, these faults may be seen as disruptions in the regular ABCABC arrangement associated with packing of spheres. This perspective is taken up in fig. 9.30. In keeping

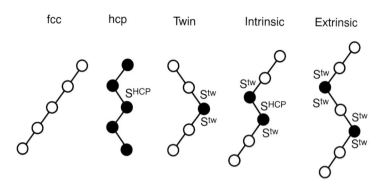

Fig. 9.30. Various stacking mistakes associated with (111) close packings (adapted from Rosengaard and Skriver (1993)).

with our earlier discussion concerning the transcription between microscopic and continuum descriptions of interfacial energies, these disruptions in the regular stacking sequence are characterized by an interfacial energy, γ_{SF}. Once again, we see the emergence of a single material parameter which serves to characterize the behavior of materials and which is amenable to microscopic evaluation.

We begin first with an investigation of the geometric character of the atomic-level geometry of stacking faults and then turn to the explicit numerical evaluation of their energies. Fig. 9.30 permits a comparative analysis of some of the key stacking fault geometries in fcc crystals. The first two schematics illustrate the way in which the fcc and hcp structures can be generated by regular arrangements of either the ABCABC... (fcc) or ABABAB... (hcp) stacking sequence. The geometrically simplest of the possible faults in the stacking sequence is that of the twin boundary which results from a reversal in order of the stacking sequence and can be induced by a shearing transformation. In addition to the twinning fault introduced above, mistakes in the fcc stacking sequence can also be induced by either adding a plane in the sequence or by removing one. The former case is known as an extrinsic stacking fault and is characterized by a stacking sequence of the form ABCABCBABC.... The latter fault is known as an intrinsic fault and is characterized by a stacking sequence of the form ABCBCABC.... A systematic discussion of these geometries may be found in chap. 10 of Hirth and Lothe (1992).

The first geometric question we might pose concerning the stacking fault is how the bond lengths are redistributed by virtue of the presence of this defect. We note that in a perfect fcc crystals the twelve near neighbors of a given atom are associated with the vertices of a cube-octahedron, while the six second neighbors correspond to the vertices of an octahedron. What emerges in comparing the atomic configurations in the perfect crystal and in a crystal with a stacking fault is the recognition that at the near- and next near-neighbor level, the perfect and faulted crystals have the same near-neighbor distributions. The details of this geometric analysis are taken up in the problems at the end of the chapter. This geometric fact has energetic consequences that will be taken up below.

Energetics of Planar Faults. As shown in the previous chapter, the particulars of the dislocation reactions that will be found in a given material are determined by the numerical values of the relevant interfacial energies (i.e. stacking fault or antiphase boundary energies). One of the dominant philosophical thrusts of the book has been the idea that when it is possible, our aim should be to exploit insights from one level of modeling to inform models at a different scale. In this case, our aim will be to show how insights based upon atomistic calculations of the various interfacial energies can be used as the basis of the type of elasticity calculations that were carried out in the previous chapter in order to determine the dissociation width of dislocations that have undergone splitting reactions and which culminated in eqn (8.74). Our application of the continuum theory of dislocations required an external parameter, namely γ_{SF}, which we now discuss from the perspective of microscopic analysis.

Unlike the gross differences between a crystal with and without a free surface, the stacking fault constitutes a less severe perturbation and thereby, generically, has a lower energy than that of a free surface. Our discussion of the bond lengths in faulted regions given above provided a plausibility argument in favor of the relatively low energy of stacking faults in the hierarchy of interfacial defects. Both empirical and first-principles methods have been used to evaluate stacking fault energies in crystals. For the purposes of numerical evaluation of the stacking fault energy, we must first insure that the stacking fault itself is embedded in a box that is sufficiently large so as to mimic the presence of a stacking fault in an infinite crystal. We will first give an overview of how the stacking fault energy may be computed within the confines of empirical energy functionals, and will close our discussion with a representative example from the use of first-principles methods for computing the stacking fault energy.

Regardless of the choice of energy functional, the type of computational box that is used to evaluate the stacking fault energy is shown schematically in fig. 9.31. The key elements of the simulation box are, evidently, the presence of the stacking fault itself, and the presence of a sufficient distance between the stacking fault and the distant free surfaces, or periodic boundaries. Once the simulation box has been established, it is a routine matter to compute the stacking fault energy itself. The stacking fault energy may be computed on the basis of a generic total energy scheme according to

$$\gamma_{SF} = \frac{E_N(SF) - E_N(bulk)}{A}. \tag{9.14}$$

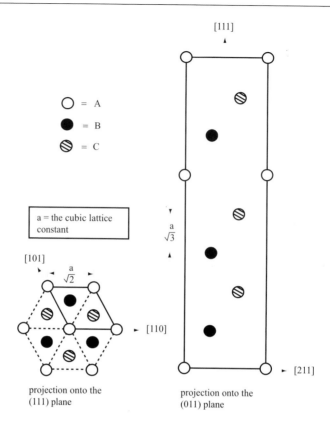

Fig. 9.31. Computational box used to compute the energy of an intrinsic stacking fault in Al (adapted from Wright *et al.* (1992)).

As with the surface energy as defined in eqn (9.5), the stacking fault energy is obtained by computing the energy of two different N-atom configurations, one corresponding to the perfect crystal and the other corresponding to the N-atom supercell within which is a stacking fault. In practice, once the computational cell is in hand, the energy difference called for above can be evaluated in exactly the same way that we obtained the vacancy formation and surface energies earlier.

From the perspective of a pair potential description of the total energy and in the context of the geometric distributions of neighbor environments shown above, the evaluation of stacking fault energies becomes an exercise in delicate choices about the long-range part of the potential of interest. As has been seen in earlier chapters, the introduction of higher-order terms in the energy functional can improve the pair potential treatment substantially. We note that in the case of stacking faults, even pair functionals are relatively ineffective at reproducing the energetics of stacking faults, suffering from routine underestimates of the stacking fault energy which lead in turn to unreasonably large dissociations for dislocations. On the other hand, these results can be improved by incorporating angular terms in the energy functional. An example of the type of results that have been obtained in the empirical setting is shown in fig. 9.32. The key idea being presented here is that the stacking fault energy can be thought of as being built up of both pair and four-body contributions. In particular, the total energy is written in the form

$$E_{tot} = E_{pair} + E_{4\text{-}body}, \tag{9.15}$$

where the four-body term is of the form

$$E_{4\text{-}body} = \sum_i \alpha(\mu_4^{(i)} - \mu_4^{(fcc)}). \tag{9.16}$$

The four-body term is an application of exactly the ideas introduced in section 4.5.3 and culminating in eqn (4.95). Fig. 9.32 shows the relative contributions of different atoms in the faulted region to the overall fault energy. In particular, for a series of different metals the overall contribution to the stacking fault energy is decomposed layer by layer and each such contribution is itself decomposed into the relative contributions of the pair and four-body terms.

   The computation of stacking fault energies from first principles is a more delicate matter. There are tedious questions that arise concerning convergence with respect to box size, energetic cutoffs and k-point sampling. Again, since we have repeatedly described the essential conceptual steps involved in carrying out a total energy calculation, we note here that the calculation of stacking fault energies involves the same basic elements such as building an appropriate computational cell, etc. An example of the types of results that emerge from these calculations is shown in fig. 9.33. As was done in fig. 9.32 (a calculation which was itself inspired by the present one) the stacking fault energies are decomposed into the contributions of the different layers in the faulted region. A rough rule of thumb which is substantiated by these calculations can be written in the form

$$\gamma_{SF}^{intrinsic} \approx \gamma_{SF}^{extrinsic} \approx 2\gamma_{SF}^{twin}. \tag{9.17}$$

   Just as was found possible in the context of alloy phase diagrammatics and in the evaluation of surface energies, it is possible to exploit atomic-level calculations as the basis of effective Hamiltonians describing the energetics of stacking faults. Once again, the calculations are made on the basis of an Ising-like model of the stacking fault. Our development follows that of Hammer *et al.* (1992) in which the energy of a crystal with a stacking fault is written as

$$E = \xi_0 - \frac{1}{N_{atoms}} \sum_{n=1}^{n_{max}} \sum_{i=1}^{N_{atoms}} J_n \sigma_i \sigma_{i+n}. \tag{9.18}$$

The parameters J_n are the effective Ising parameters that are determined from

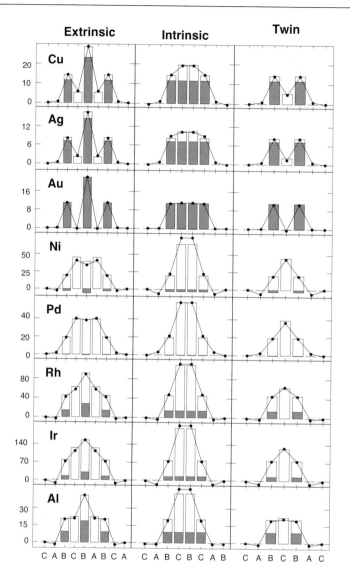

Fig. 9.32. Systematics of stacking fault energy in different metals as computed using multiatom interactions (adapted from Stokbro and Jacobsen (1993)).

atomic-level analysis.   The notation is such that $\sigma_i = 1$ if the atom in layer $i + 1$ maintains the integrity of ABC stacking whereas $\sigma_i = -1$ if the atom in layer $i + 1$ is part of CBA stacking.   For example, if the ith layer is of C-type and the following layer is an A-type, then $\sigma_i = 1$.   By way of contrast, if the ith layer is of the type C and the following layer in the stacking sequence is of the B-type, then $\sigma_i = -1$.   To make the correspondence more concrete, the sequence ABCABCBABC, corresponding to an extrinsic stacking fault leads to

Extrinsic Energy Deviations

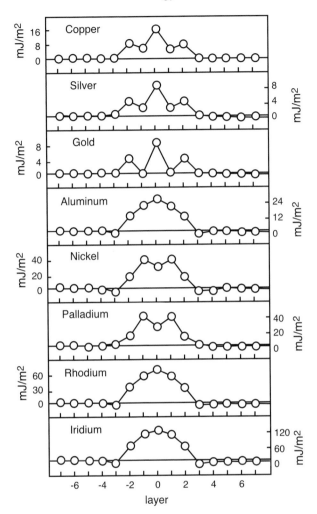

Fig. 9.33. Systematics of stacking fault energy in different metals as computed using first-principles total energy methods (adapted from Crampin *et al.* (1990)).

the transcription

$$\text{ABCABCBABC} = 1\ 1\ 1\ 1\ 1\ -1\ -1\ 1\ 1\ 1. \qquad (9.19)$$

To compute the stacking fault energy, we must evaluate $\Delta E = E_{faulted}(N\ layers) - E_{perfect}(N\ layers)$. For the extrinsic case given above only four layers (i.e. $1\ 1\ -1\ -1$ have contributions that are *different* from the perfect crystal contribu-

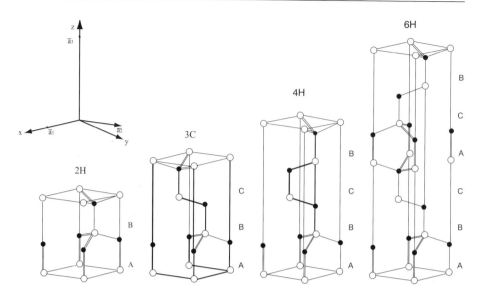

Fig. 9.34. Schematic of several polytypes occurring in SiC (adapted from Käckell *et al.* (1994)). Each polytype represents a different stacking of A, B and C layers.

tions, so the energy difference is of the form

$$\Delta E = [-J_1 + J_2 + J_1 + J_2 - J_1 + J_2 + J_1 + J_2] - 4(-J_1 - J_2) = 4J_1 + 8J_2. \quad (9.20)$$

The terms in square brackets come in pairs involving J_1 and J_2 correspond to working from left to right while the term involving $4(-J_1 - J_2)$ is the energy of four layers worth of perfect lattice on the assumption of one atom per layer. The use of these ideas is fleshed out in problem 5 at the end of the present chapter. Once this effective Hamiltonian is in hand, it is possible to begin considering the distribution of such faults in crystals. One particularly interesting venue within which planar faults arise is in the context of some of the complex structures that arise in alloys.

9.3.2 Planar Faults and Phase Diagrams

As noted above, stacking defects are relevant in the context of the phase diagrammatics of complex alloys. In this setting, there are a range of polytypes that a particular material may adopt depending upon the tuning of some control parameter such as the temperature or the chemical composition. For our present purposes, such polytpes represent different structural variants in which the regular packing of planes along a particular direction is disturbed relative to the stacking in some reference state. These mistakes in packing correspond in the fcc case to exactly the sort of stacking faults that have entered our discussion thus far. Example of

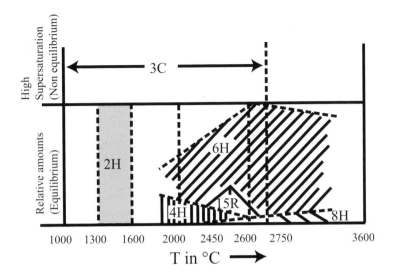

Fig. 9.35. Schematic of the different polytypic phases found in SiC (adapted from Trigunayat (1991)).

the types of structural rearrangements that occur in the SiC system are shown in fig. 9.34. In addition, fig. 9.35 gives a schematic indication of the stability ranges of the various structures.

In addition to its importance in the type of covalent systems described above, the phenomenon of polytypism is also found in intermetallics. For example, in fig. 9.36, we show a schematic rendering of the various structure types found in a series of Mg-based ternary alloys such as $Mg(Zn_x Ag_{1-x})_2$ and $Mg(Cu_x Al_{1-x})_2$. The basic idea is that by varying x, and hence the proportions of the elements in the alloy, the mean electron/atom ratio (denoted as z in the figure) for the alloy can be altered. This figure reveals that depending upon the value of z, a whole range of different stacking sequences emerge. The notation presented in the figure is such that a given layer is labeled with an h if the two adjacent layers are of the same type and a given layer is labeled c otherwise. For example, the hcp structure (ABABABA...) would be labeled as hhhhh... since every layer is such that the two adjacent layers are identical (i.e. A is sandwiched between two Bs and vice versa). The fcc structure, on the other hand, characterized by a stacking sequence ABCABC..., would be denoted as ccccc... in this notation since the two layers adjacent to a given layer are never the same. As shown in the figure, for the more complex case of interest here, different values of z, the electron per atom ratio (which is another way of saying for different compositions), the resulting stacking sequence will be a different arrangement of h and c layers. For our present purposes, all that we really aim to take away from this figure is a sense

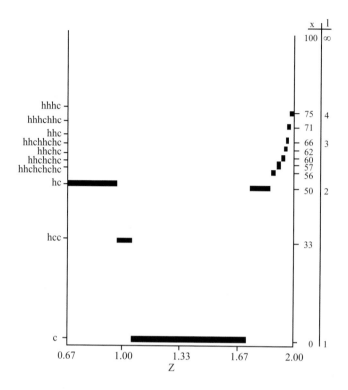

Fig. 9.36. Series of stable stacking sequences in Mg-based Laves phases as a function of the electron per atom ratio, z (Bruinsma and Zangwill 1985).

of the diversity of structural types that can be thought of from the 'stacking fault' perspective and to note the way in which control parameters such as the electron per atom ratio can govern structural stability.

The emergence of the types of polytypes described above may be rationalized on the basis of a variety of different ideas. Rather than exploring the phenomenon of polytypism in detail, we give a schematic indication of one approach that has been used and which is nearly identical in spirit to the effective Hamiltonian already introduced in the context of stacking fault energies (see eqn (9.18)). In particular, we make reference to the so-called ANNNI (axial next-nearest-neighbor Ising) model which has a Hamiltonian of the form

$$H = E_0 - \frac{1}{2}J_1 \sum_i \sigma_i \sigma_{i\pm1} - \frac{1}{2}J_2 \sum_i \sigma_i \sigma_{i\pm2}. \tag{9.21}$$

The basic idea is to represent the energy of a given stacking sequence in terms of effective Ising parameters. The relative importance of first (J_1) and second (J_2) neighbor interactions gives rise to a competition that can stabilize long-period

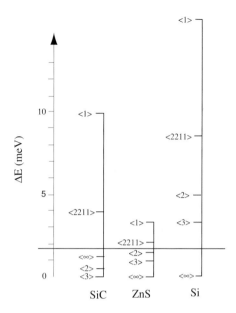

Fig. 9.37. Ranking of structural energies associated with different polytypes using the Ising-like effective Hamiltonian described in the text (adapted from Engel and Needs (1990)).

equilibrium states. In particular, this competition is especially relevant in the case in which the magnitudes of the two terms are nearly equal and their signs are opposite. From the standpoint of the analysis of this model there are two different considerations. First, it is necessary to inform the effective Hamiltonian on the basis of atomic-level analysis. Secondly, the representation of polytypic energies in terms of an Ising parameterization like that described above opens up the possibility for a statistical mechanics analysis of the temperature dependence of phases. A representation of the ranking of polytype energies in three different material systems is shown in fig. 9.37. In each of these cases, the energies shown in the figure are based upon the effective Hamiltonian, with the parameters in that Hamiltonian fit on the basis of a first-principles analysis of the total energy.

9.4 Grain Boundaries

One of the dominant themes upon which much of our analysis has been predicated is the idea that a solid is populated by structures at many different scales. At the most fundamental level, a crystal may be thought of as an uninterrupted arrangement of atoms according to the simple decomposition

$$\mathbf{R}_{m_1,m_2,m_3} = m_1\mathbf{e}_1 + m_2\mathbf{e}_2 + m_3\mathbf{e}_3, \tag{9.22}$$

where m_i is a member of the integers, $\mathbf{e}_i$ is a Bravais lattice vector and $\mathbf{R}(m_1, m_2, m_3)$ is the position of the lattice point indexed by the triplet (m_1, m_2, m_3). Here we are imagining the simplest of crystals in which there is only one atom per unit cell. At the next level of structure, we imagine this perfect regularity to be disturbed both by a distribution of point defects such as vacancies and interstitials as well as by arrangements of dislocations and possibly also by their attendant stacking faults. At even larger scales (i.e. the micron scale), there is yet another level of structure, namely, that of the multiple grains that make up a typical crystalline solid. Different regions within the crystal are tied to different orientations of the same fundamental Bravais lattice. These grains are bounded by interfacial defects known as grain boundaries which serve as an integral part of the geometry of crystalline solids and it is the business of the present section to examine the structure and energetics of such boundaries from the perspective of models of material behavior.

As a reminder of the way interfaces, as key microstructural elements, can impact material response, we note one of the celebrated relations between a material's microstructure and its physical response, namely, the Hall–Petch relation which holds that the yield strength of a material scales as

$$\sigma_y = \sigma_0 + kd^{-\frac{1}{2}}, \tag{9.23}$$

where d is the average 'grain size' and k and σ_0 are constants. We have already exhibited pictorial evidence for this effect in figs. 1.3 and 8.2(c). We have elected to surround grain size in quotes as a reminder that despite the ease with which those words are used in everyday conversation in discussing materials, the geometric characterization of microstructure is subtle. Our reason for calling attention to the Hall–Petch relation is to anticipate some of the questions that are tied to the existence of grain boundaries.

In the long history of efforts to piece together the connection between a material's microstructure and its properties, one series of arguments has been built around a reductionist perspective. In these arguments, it is held that by building up an appropriate understanding of a single grain boundary perhaps it will be possible to bootstrap these insights to an analysis of microstructures themselves. As a result, not only has there been considerable effort in trying to ferret out the implications of polycrystalline structure on properties, but a great deal of effort has gone into examining the deformation and fracture of bicrystals. An entire hierarchy of geometric structures involving grain boundaries can be considered as illustrated in fig. 9.38.

As is evidenced by fig. 9.38, there are a number of points of entry into a discussion of polycrystalline microstructures. In particular, fig. 9.38 shows that at the simplest level (frame (a)) we can consider bicrystal geometries. In keeping with

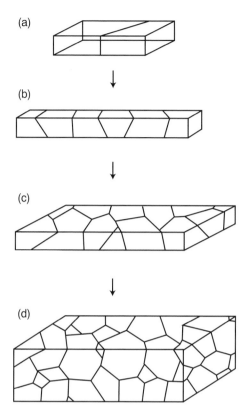

Fig. 9.38. Schematic of increasingly complex grain boundary distributions within materials (adapted from Randle (1994)): (a) bicrystal geometry with a single grain boundary, (b) bamboo-like distribution of grain boundaries in a one-dimensional array, (c) polycrystalline film, and (d) three-dimensional polycrystal.

that suggestion and as a preliminary to investigating the physics of polycrystals, we begin by examining the geometry of bicrystals with an eye to examining such geometries from both a microscopic and continuum perspective.

9.4.1 Bicrystal Geometry

In this section, we begin with consideration of the fundamental geometric degrees of freedom that characterize bicrystals and then turn to the descriptions of the structures themselves. The reference geometry around which the discussion of grain boundaries acquires meaning is the perfect lattice of eqn (9.22). Denote the Bravais lattice vectors of the first grain by $\mathbf{e}_i^{(1)}$, where 1 is a label denoting this grain, and the subscript ranges from one to three and labels the three linearly independent Bravais lattice vectors characterizing the crystalline state. The basic

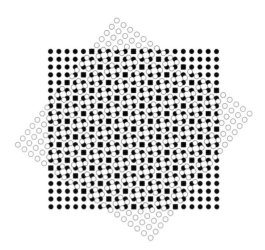

Fig. 9.39. Interpenetrating lattices used to consider grain boundary structure (courtesy of D. Pawaskar). Open and filled circles correspond to the two host lattices and filled squares correspond to those atoms that are common to both lattices (i.e. the lattice of coincident sites).

idea for describing the grain boundary structure is embodied in two fundamental geometric notions. First, we must characterize the orientation of the second grain with respect to the first. Having done this, the boundary itself is not fully determined since we must also say which plane serves as the dividing surface between the two grains. Together, these two geometric descriptors, the relative orientations of the two grains and the identification of the boundary plane, serve to completely specify the macroscopic grain boundary geometry.

We now set ourselves the task of casting these geometric questions in a proper mathematical light. We begin by imagining filling space with two interpenetrating lattices, both built around the same Bravais lattice type, but with different orientations, and possibly a different overall origin. In particular, if we are given the two sets of Bravais lattice vectors, their relative orientations are connected via a rotation matrix $\mathbf{Q}$ by

$$\mathbf{e}_i^{(2)} = \mathbf{Q}_{ij}\mathbf{e}_j^{(1)}. \tag{9.24}$$

In addition, as was mentioned above, it is possible that there is a shift of origin associated with some translation vector $\boldsymbol{\tau}$. The matrix $\mathbf{Q}$ imposes a linear transformation which, since it describes a rotation, may be characterized by three parameters. We now have two interpenetrating lattices as indicated in the simplified two-dimensional setting in fig. 9.39. As yet, there is no grain boundary. The remainder of our work in identifying the macroscopic degrees of freedom is to select the boundary plane itself. This plane is characterized by a unit vector $\mathbf{n}$

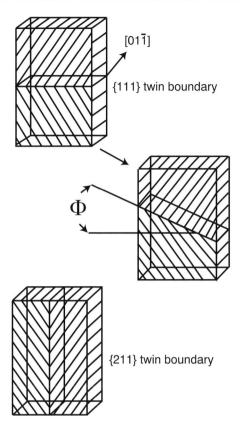

Fig. 9.40. Schematic of degrees of freedom associated with choice of boundary plane (adapted from Wolf *et al.* (1992)). This figure illustrates the idea that once the relative orientation of the two grains has been chosen there is still a freedom to choose the grain boundary plane.

which can be identified by two additional degrees of freedom. The choice of boundary plane is illustrated in fig. 9.40. We now imagine a shaving operation in which only one group of atoms is retained on each side of the boundary. For example, we might elect to keep those atoms associated with the Bravais lattice set $e_i^{(1)}$ on one side and those from the other Bravais lattice set on the other side of the boundary. What we have learned is that with the specification of five macroscopic degrees of freedom, the overall character of the boundary may be specified. For the special case in which the rotation axis is contained within the boundary plane, such a boundary is known as a tilt boundary, while for the other special case in which the rotation axis is perpendicular to the boundary plane, one speaks of a twist boundary. Just as the most generic dislocations are neither pure edge or pure screw, the generic grain boundary is neither pure tilt nor pure twist.

The description of grain boundary geometry as given above is somewhat deceptive. In addition to the overall macroscopic descriptors of that boundary, there are also microscopic features that may be assigned to such boundaries. For example, even though we have selected the relative orientations and the boundary plane, it is possible that the two grains may have also suffered a relative translation. As will be seen in subsequent sections, the search for the particular energy minimizing translation is a tedious but important step in the analysis of grain boundary structure.

Bicrystallography has been dominated by the analysis of special boundaries in which symmetry plays a pivotal role. In particular, the geometry of those boundaries with a high degree of symmetry is most easily captured through an appeal to certain important classification schemes, such as that of the coincident site lattice (CSL). The qualitative idea behind this classification is the insight that for certain choices of rotation matrix $\mathbf{Q}$, there will be a fraction of the sites that are common to both lattices. The fraction of such overlapping sites serves as a measure of the boundary symmetry. This lattice of overlapping sites is known as the coincident site lattice. From a quantitative perspective, this degree of overlap of the two lattices is characterized by the Σ number. To compute this number, we find what fraction of the sites from a given lattice are coincident with those of the other. The reciprocal of this number is the Σ number. For example, if one in every five sites is coincident, this would be referred to as a $\Sigma 5$ boundary, precisely the case shown in fig. 9.41. An example of a set of coincident site lattices is shown in fig. 9.41. Interestingly, even in those cases in which a tilt boundary has relatively little symmetry, the boundary structure may be thought of as a limiting case in which the coincident site lattice has a very high Σ number, where we note again that this number is defined such that $1/\Sigma$ of the sites on a given lattice are common.

9.4.2 Grain Boundaries in Polycrystals

In light of our characterization of the structures of individual grain boundaries, it is important to now undertake the *statistical* question of what the distribution of grain boundaries is like in real polycrystals. We have already noted that the important role of bicrystals was born of a reductionistic instinct in which it was argued that proper understanding of the structure and energetics of single boundaries must precede a fundamental understanding of the consequences of an assembly of such boundaries. Our task in the present section is to examine the geometric character of real polycrystals, paying special attention to the distribution of different boundary types within polycrystals. We have already seen that boundaries of high symmetry have been the workhorse of most research in examining the structure and energetics

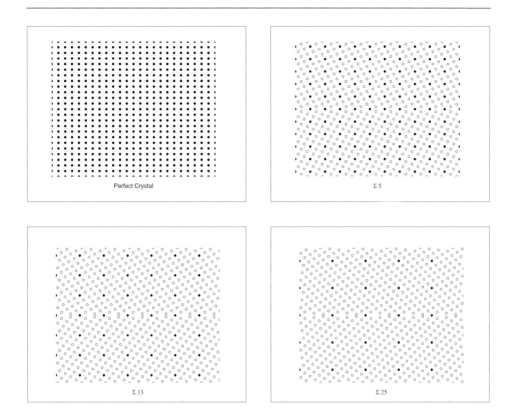

Fig. 9.41. Schematic of several representative grain boundaries with structures described by coincident site lattice model. This set of boundaries corresponds to a (001) rotation axis, and the atomic-level geometries have not been relaxed (courtesy of D. Pawaskar). The filled circles correspond to those sites (coincident sites) that are common to both lattices.

of grain boundaries. This raises the question of to what extent such boundaries are representative of those found in real microstructures.

As pointed out in Randle (1997), there are two aspects to the statistics of grain boundary distributions. On the one hand, the question of how different grains are misoriented can be posed. This leads in turn to an analysis of the distribution of various coincident site lattice types. By way of contrast, it is also important to obtain quantitative insights into the distribution of grain boundary planes. Indeed, the character of the grain boundary plane is critically related to the nature of the underlying grain boundary structure. As a first exercise in the statistics of grain boundaries, fig. 9.42 shows the distribution of Σ boundaries for an Fe–Si alloy. In this case, it is the relative frequency of different misorientations that is being reported.

An example of this type of statistical distribution with regard to the boundary plane is shown in fig. 9.43. In particular, this figure reveals the distribution of

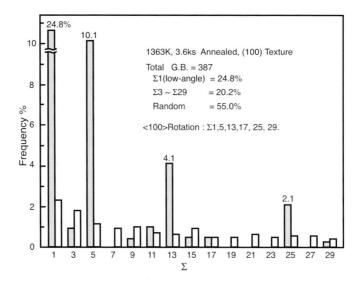

Fig. 9.42. Frequency of various coincident boundaries as a function of the Σ number (adapted from Watanabe (1989)). The data are for a fully annealed Fe–6.5 wt% Si alloy. The two distributions (different color histograms) correspond to samples with different textures.

grain boundary plane types in annealed high purity Cu and Ni. All of the reported boundaries are for the same underlying crystal misorientation and correspond to $\Sigma 3$ boundaries. The numbers on the left hand axis of the figure give the orientation of the grain boundary plane.

Our intent in this section was not to give an exhaustive account of the nature of grain boundaries in real polycrystals. Rather, it was meant as a gentle reminder that despite the importance of bicrystals, they are but a first step in understanding the deeper question of the nature of grain boundaries in real microstructures.

9.4.3 Energetic Description of Grain Boundaries

We now take up the issue of how one goes about using microscopic calculation to determine grain boundary structure and energetics. The principles of these calculations will be essentially independent of whether or not empirical or first-principles schemes are adopted. However, the goodness of the results will clearly be influenced by that choice. As a preliminary, we recall the transcription between the microscopic and continuum description of grain boundary energies. As was noted in eqn (9.4), the introduction of interfacial energies in a continuum setting is effected via a scalar parameter (i.e. γ_{surf}, γ_{SF} or γ_{GB}) which has a dependence on crystallographic features such as the interfacial normal and the relative orientations of adjacent grains in the case of grain boundaries. For grain boundaries, as

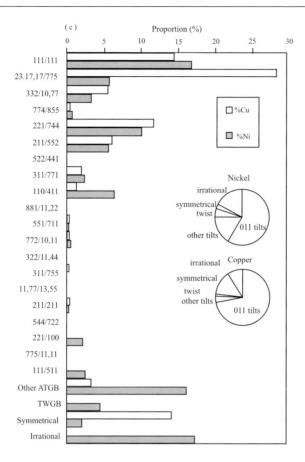

Fig. 9.43. Distribution of grain boundaries in polycrystals (adapted from Randle (1997)). The numbers on the vertical axis refer to the coordinates of the boundary plane, while ATGB and TWGB refer to the asymmetric tilt boundaries and twist boundaries, respectively.

for surfaces and stacking faults, the transcription between the microscopic and continuum notions is based upon

$$\gamma_{GB} = \frac{E_N(GB) - E_N(bulk)}{A}. \tag{9.25}$$

As with our previous treatments of interfacial energy, the fundamental idea is to make a comparison between an N-atom state in which all of the atoms are in their ideal bulk positions and an N-atom state in which the interface of interest is present.

As an example of the type of supercell that can be used in this analysis, fig. 9.44 shows a scheme in which the grain boundary is at the interior of a cell whose outer boundaries perpendicular to the boundary are treated as rigid slabs. In this case, the forces on the atoms in the region being relaxed are computed in exactly the

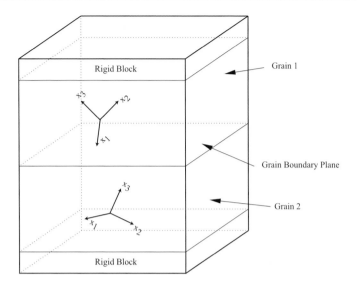

Fig. 9.44. Schematic of the type of computational cell used in computing the energy of a grain boundary.

manner already emphasized repeatedly in the book, and the atoms are subsequently relaxed. Our thinking for the remainder of this section is to examine a few different case studies in grain boundary structure with the aim of seeing both how such calculations are done, and what can be learned from them.

Semiempirical determination of grain boundary structure and energetics. Simple grain boundaries in metals have been exhaustively considered from the standpoint of empirical descriptions of the total energy. The logic of much of this work is founded on the contention that prior to understanding the susceptibility of boundaries to segregation and fracture or their role as short-circuit paths for diffusion, it is necessary to first understand the structures themselves. The majority of this work has emphasized bicrystals.

A series of studies in Ni have been made for tilt boundaries that elucidate the role of misorientation in boundary structure. The case study of interest to our present discussion (Rittner and Seidman 1996) involved the use of a pair functional description of the total energy to investigate 21 distinct symmetric tilt boundaries with a $\langle 110 \rangle$ tilt axis. Representative examples of the relaxed structures are shown in fig. 9.45 and the corresponding grain boundary energies are shown in fig. 9.46.

One of the most important points of contact between models like those described above and experiment is the use of transmission electron microscopy to examine grain boundaries. In particular, there is a well-established tradition of effecting direct comparisons between the outcome of a given atomic-level simulation of

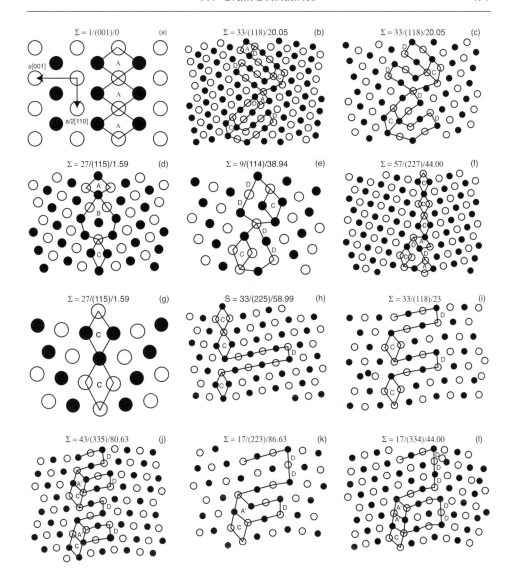

Fig. 9.45. Series of grain boundary structures for ⟨110⟩ symmetric tilt boundaries as computed using a pair functional description of the total energy (adapted from Rittner and Seidman (1996)).

grain boundary structure and the corresponding images obtained on the basis of high-resolution transmission electron microscopy (HRTEM). The way such comparisons are effected is by taking the results of the atomic-level calculation and using them as input for image simulation tools which compute the nature of the high-resolution transmission electron microscopy image that would be expected on the basis of that collection of atoms. An example of this strategy is illustrated in fig. 9.47 which shows the atomic-level structure for a tilt boundary in Al.

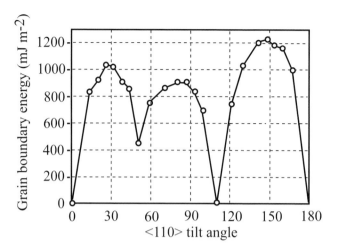

Fig. 9.46. Grain boundary energy as a function of tilt angle for ⟨110⟩ symmetric tilt boundaries (adapted from Rittner and Seidman (1996)).

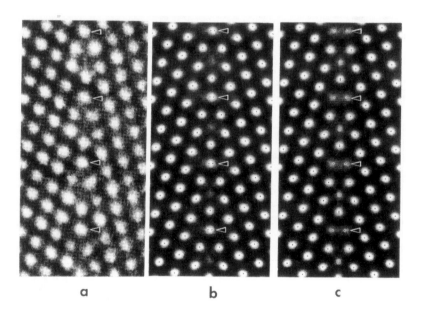

Fig. 9.47. Comparison of high-resolution images of a grain boundary with structures obtained from atomistic simulation (adapted from Mills (1993)).

First-principles determinations of grain boundary structure and energetics. We have repeatedly emphasized the complementary (and sometimes antagonistic) role of empirical and first-principles calculations of the many structures that inhabit materials. The structure and energy of grain boundaries is clearly yet another defect

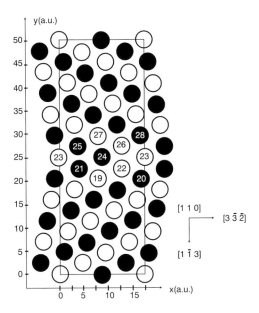

Fig. 9.48. Computational cell used for computing energetics of Σ11 grain boundary in aluminum (adapted from Wright and Atlas (1994)). The numbering of atoms was used in the original work in order to describe electron densities. Atoms in different layers normal to the page (i.e. along the [110] direction) are shown with open and filled circles.

that is amenable to treatment within the first-principles setting. The analysis is predicated on the creation of a computational cell of the sort shown in fig. 9.48. In this case, a high-symmetry grain boundary in Al is of interest. All of the usual machinery is exploited in the usual way (i.e. periodic boundary conditions, computing energies and forces, structural relaxation), with the result that it is possible to obtain insights into the structure and energetics of grain boundaries with first-principles accuracy. One of the interesting insights to emerge from the first-principles calculations highlighted in fig. 9.48 is that though there is reasonable accord between the structures obtained using first-principles techniques and those using pair functionals to examine the same boundary, the grain boundary energies are underestimated in the pair functional calculations by a factor of the order of 2.

The intent of the present section was to demonstrate that just as was the case in the context of point defects such as vacancies and interstitials and line defects such as dislocations, it is possible to bring a variety of microscopic methods to bear on the question of grain boundary structure and energetics. At this late stage in the book, the basic ideas behind such structural simulations have been illustrated repeatedly with the result that our discussion has aimed only to highlight via the device of case studies, the results of such calculations.

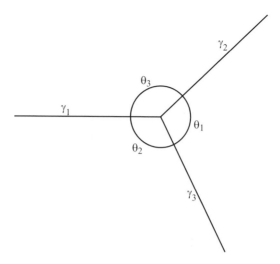

Fig. 9.49. Schematic of grain boundary triple junction.

9.4.4 Triple Junctions of Grain Boundaries

As we will discuss in chap. 10, the consideration of polycrystals demands an analysis not only of the grain boundaries that separate different grains, but also of the points at which several different boundaries intersect. The simplest model of a triple junction is of the type indicated schematically in fig. 9.49. For our present purposes, the basic idea is to determine a condition for the stability of such triple junctions which will at once determine both the structure and energetics of these triple points.

For the geometry considered in fig. 9.49, we imagine three separate segments of grain boundary, all pinned at their extremeties as shown. Our aim is to write the total energy as a function of the as-yet undetermined position x. Once this energy has been set forth, we then minimize the energy with respect to the parameter x resulting in a condition for the optimal geometry. Before illustrating the details, we note that the calculation envisaged here constrains the class of possible triple-junction geometries and that a more general treatment requires an analysis of the variation in energy for both x and y displacement of the triple point. The more general calculation is considered in the problem section at the end of this chapter.

The energy of the triple junction may be written by adding up the energy γl which results from each interfacial segment of length l. Note that this calculation is the interfacial analog of the line tension calculations used to characterize the energetics of dislocations in the previous chapter. For the particular geometry of interest here, the energy may be written as

$$E(x) = \gamma x + 2\gamma \sqrt{h^2 + (d - x)^2},$$

(9.26)

where the first term is the energy associated with the horizontal segment, while the second term is that associated with the two interfacial segments that are oriented at an angle $\phi = \theta_1/2$ with respect to the horizontal. Differentiation of this energy with respect to x provides us with the equilibrium condition which is

$$\frac{d - x}{\sqrt{h^2 + (d - x)^2}} = \frac{1}{2}. \tag{9.27}$$

Further, we note that the expression on the left hand side of this equation is nothing more than $\cos \phi$, resulting in the conclusion that the equilibrium angle is $\phi = \theta_1/2 = 60°$. This relatively simple calculation is aimed at illustrating the way in which energy minimization arguments may be used to deduce the structure and energetics of triple points which are yet another ingredient in our overall collection of defects used to build up material microstructures.

9.5 Diffuse Interfaces

Our thinking on the treatment of interfaces has been largely inspired by atomic-level understanding. In particular, we have acceded to the atomistically motivated picture of an interface as a sharp line of demarcation between distinct regions. On the other hand, as will become more evident in the next and subsequent chapters, there are advantages to adopting a continuum view of interfaces as regions over which certain continuum field variables suffer rapid spatial variations. This idea will serve as the basis for the general class of phase field models that have been used in settings ranging from the analysis of grain growth to the patterns that arise during the process of solidification.

To be concrete, though not detailed, we consider the case of a two-phase microstructure in which two distinct crystalline phases coexist. One way of characterizing the state of such a system is by recourse to an 'order parameter' field which we denote $\eta(x)$. Imagine that within one of the phases, this order parameter takes the value 1, while in the other phase it is assigned a value 0. The phase field treatment of interfaces is indicated schematically in a simplified one-dimensional form in fig. 9.50. What we note is that there are three distinct regions. First, there is a region in which $\eta(x) = 1$, second a region over which $\eta(x) = 0$ and most importantly for the purposes of the present discussion, there is a region over which the field $\eta(x)$ is making the transition from $\eta(x) = 1$ to $\eta(x) = 0$. Though we cannot necessarily assign a precise location to the interface, we can say that the interface occupies that region over which $d\eta(x)/dx \neq 0$.

The discussion given above illustrates how the question of 'structure' at interfaces is handled within the phase field setting. In addition to these structural questions, the phase field concept allows for a treatment of interfacial energetics.

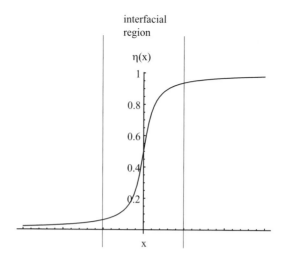

Fig. 9.50. Schematic representation of the use of continuous fields to represent grain boundaries. In particular, the interfacial region is characterized by gradients in the order parameter η.

In particular, again in keeping with the simplified one-dimensional framework suggested above, these can ideas can be endowed with an energetic interpretation by introducing an energetics of the form

$$E[\eta(x)] = \int \kappa \left(\frac{d\eta}{dx}\right)^2 dx. \tag{9.28}$$

This functional introduces an energetic penalty associated with those parts of the distribution of $\eta(x)$ involving gradients. The parameter κ is a material parameter that characterizes the severity of the penalty associated with a given interface. The energy functional given in eqn (9.28) is a simplified version of one of the key terms in the overall free energy functional that is used in the derivation of the so-called Cahn–Hilliard equation which will be taken up in chap. 12. We note that the present section has only sketched some of the key ingredients in the phase field approach to modeling interfaces and that a detailed description of these ideas will have to await later chapters.

9.6 Modeling Interfaces: A Retrospective

This chapter has argued for the critical role of interfacial defects in solids. We have favored an artificial classification into three primary types of interfacial defects, namely, surfaces, stacking faults and twins, and grain boundaries. In each of these cases, it has been argued that a continuum description of the energetics of such interfaces may be marshaled through an appeal to equations of the form

$$E_{int} = \int_{\partial\Omega} \gamma(\mathbf{n}) dS, \qquad (9.29)$$

where $\gamma(\mathbf{n})$ is a generic interfacial energy which depends upon the particular interfacial plane. As evidenced by the various discussions throughout the present chapter, despite the seeming innocence of introducing quantities such as $\gamma(\mathbf{n})$, the structure and energetics of interfaces is a complex problem when viewed from the atomic scale. In retrospect, it is clear that the energetics of the three different interfacial types considered here can be grouped. In particular, we have found an energetic ranking of the form

$$\gamma_{surf} > \gamma_{gb} > \gamma_{SF}. \qquad (9.30)$$

Ultimately, this ranking of energies could be rationalized on the basis of bond counting arguments in which we examine the extent to which atoms in the interfacial region have lost their partners.

9.7 Further Reading

Generalities on Interfaces

The Physics of Low-Dimensional Semiconductors by John H. Davies, Cambridge University Press, Cambridge: England, 1998. Chapter 4 of Davies' book considers the properties of quantum wells and elaborates on some of the arguments concerning confinement given earlier in the present chapter.

Surfaces

The literature on surface science is vast, and well represented in a number of different books and review articles. Here I give a few that have been most useful or enjoyable to me.

Physics at Surfaces by A. Zangwill, Cambridge University Press, Cambridge: England, 1988. Zangwill's book is thoughtfully constructed and wide in its coverage. He passes freely between questions of geometrical and electronic structure, the adsorption of atoms at free surface and the properties implied by the presence of surfaces.

'Surface Reconstructions' by D. King, in *Physics World*, **2**, 45 (1989). This brief article is characterized by plain English and excellent pictures that take the reader right to the crux of the geometries that attend surface reconstruction.

'Reconstructions and Relaxations on Metal Surfaces' by J. E. Inglesfield, in *Prog. Surf. Sci.*, **20**, 105 (1985). By now this article is somewhat dated, but still gives an introduction to issues faced in considering the reconstructions on Au, Pt and W surfaces. Inglesfield's article can be contrasted with 'Scanning tunneling microscopy studies of metal surfaces' by F. Besenbacher, in *Rep. Prog. Phys.*, **59**, 1737 (1996). This article is full of beautiful pictures substantiating our claim that the scanning tunneling microscope has revolutionized our knowledge of surface structure.

Two complementary articles related to semiconductor surfaces are 'Scanning tunnelling microscopy of semiconductor surfaces' by H. Neddermeyer, in *Rep. Prog. Phys.*, **59**, 701 (1996) and 'Theory of semiconductor surface reconstruction' by G. P. Srivastava, in *Rep. Prog. Phys.*, **60**, 561 (1997). The article on scanning tunneling microscopy applied to semiconductor surfaces is replete with pictures of some of the huge variety of surfaces that have been considered in semiconductor systems.

Physics of Crystal Growth by A. Pimpinelli and J. Villain, Cambridge University Press, Cambridge: England, 1998. This book is included here because it has an interesting discussion both of surface reconstructions and, more importantly, of the elastic theory of interactions between steps.

Grain Boundaries

Interfaces in Crystalline Materials by A. P. Sutton and R. W. Balluffi, Oxford University Press, Oxford: England, 1995. An enormous compendium with broad coverage on interfacial structure, energetics, deformation, segregation and everything else. This book should be seen as doing for the study of interfaces what Hirth and Lothe (1992) did for the study of dislocations. This is *the* reference on interfaces.

Materials Interfaces edited by D. Wolf and S. Yip, Chapman Hall, London: England, 1992. This book is a compilation consisting of articles by leading practitioners on all aspects of interfaces in solids.

'The Influence of Grain Boundaries on Mechanical Properties' by J. P. Hirth, *Met. Trans.*, **3**, 3047 (1972). Hirth offers a variety of insights on the role of grain boundaries in dictating mechanical properties.

Miscellaneous

Fractography – Observing, Measuring and Interpreting Fracture Surface Topography by Derek Hull, Cambridge University Press, Cambridge: England, 1999. Hull's book is excellent whatever criterion is used to judge it. I include it here since fracture surfaces are another type of interface that emerges in considering solids.

9.8 Problems

1 Energetics of fcc Surfaces Using Pair Potentials

In eqn (9.6) we revealed the surface energy for an fcc (100) surface in the near-neighbor pair potential approximation. Deduce analogous results for the (110) and (111) fcc surfaces.

2 Energetics of Surfaces Using Pair Functionals

Use the Johnson embedded-atom analysis to compute the energies of the (111) and (110) surfaces in fcc materials. Carry out a ranking of the relative magnitudes of the (100), (110) and (111) surfaces using these potentials.

3 Energetics of Surfaces Using Stillinger–Weber Potentials

Compute the energy of the unrelaxed and unreconstructed (100) surface in Si. Next, use either static energy minimization or molecular dynamics to relax this surface and examine the resulting reconstruction.

4 Geometry and energetics of stacking faults

In this problem, you will investigate the geometry and energetics of stacking faults in fcc crystals. (a) Construct a histogram which shows the number of neighbors that a given atom has as a function of distance to these neighbors. First, make such a plot for fcc Cu ($a_0 = 3.61$ Å). Identify all neighbor shells out to a distance of 5.0 Å. (b) Next, repeat the same exercise for a twinned configuration. In particular, make two graphs, one for the atoms right at the twin plane, and one for the atoms one layer away from the twin plane. Compare the distribution of neighbors with those in the perfect crystal. Without doing any calculation, what will the twin fault energy be for the Morse potential for Cu that was derived in problem 2 of chap. 4? (c) Carry out the same procedure as in part (b), but now for the intrinsic stacking fault. Again, comment on the energy of such a fault as implied by the Morse potential. What do you conclude about the nature of the interactions that are responsible for the stacking fault energy? (d) Use an embedded-atom potential of the Johnson form (see chap. 4) and compute the stacking fault energy.

5 Effective Hamiltonian for Stacking Fault Energy

In the discussion culminating in eqn (9.18), we described the use of an Ising-like effective Hamiltonian for the description of fault energies. Show that if we use eqn (9.18) in the case when $n_{max} = 2$ then the energy of a twin is given by $E_{twin} = 2J_1 + 4J_2$, the energy of an intrinsic stacking fault is given by $E_{intrinsic} = 4J_1 + 4J_2$ and the energy of an extrinsic stacking fault is given by $E_{extrinsic} = 4J_1 + 8J_2$. Note that the extrinsic fault energy was worked out in the chapter to illustrate how to make such calculations.

6 Atomic-Level Simulation of Grain Boundary Structure

Use the Johnson potential to simulate the structure of a $\Sigma 5$ symmetric tilt boundary with a (001) rotation axis. In particular, find the relaxed structure associated with the $\Sigma 5$ boundary shown in fig. 9.41.

7 Treatment of the Structure and Energetics of Triple Junctions

(a) Our discussion of triple junctions in this chapter was intentionally simplified such that the equilibrium structure was presumed to depend only on a single geometric parameter. In this problem, generalize the treatment given in the chapter such that the triple junction has a position characterized by two parameters, namely, its x and y coordinates. (b) Write down the equilibrium condition in the case in which the interfacial energies differ for the three boundary segments.

TEN

Microstructure and its Evolution

We have made no secret of the pivotal role played by various geometric structures within materials in conspiring to yield observed properties. However, one of the abiding themes of our work thus far has been the observation that this notion of 'structure' is complex and scale-dependent. The properties of materials depend upon geometry not only at the microscopic scale, but at larger scales as well. Recall that the hierarchy of geometric structures that populate materials begins with the underlying atomic arrangements. Questions of structure at this level are answered through an appeal to phase diagrams in the way considered in chap. 6. These atomic-level geometries are often disturbed by the various defects that have served as the centerpiece of the previous three chapters, and which introduce a next larger set of length scales in the hierarchy. However, much of the effort in effecting the structure–properties linkage in materials science is carried out at a larger scale yet, namely, that associated with the various microstructures within a material.

The current chapter has as its primary aim a discussion of some of the many types of microstructures that populate materials and how such microstructures and their temporal evolution can be captured from a theoretical perspective. This discussion will also serve as the basis of our later efforts to examine the connection between structure and properties in materials. We begin with an attempt at microstructural taxonomy with the aim being to give an idea of the various types of microstructures that arise in materials. Once we have built up an overall impression of the empirical backdrop concerning microstructures in materials, we will attempt to consider how modeling can be brought to bear on a few representative examples. Because of the vast diversity in different microstructures and materials, we will not attempt anything like an exhaustive discussion of microstructural evolution. Rather, through the device of considering a few well-placed examples, our hope is to give an impression of the types of questions that can be posed, and the theoretical machinery that might be used to address them.

10.1 Microstructures in Materials

We begin with a brief overview that attempts to make sense of the wide variety of microstructures that are found in materials (primarily metallic). This part of our task should be seen as taxonomical: we aim to observe and classify with the hope that certain classes of microstructures will suggest broad classes of models. The observations of these structures will then be supplemented by an analysis of how by exploiting the twin strategies of heating and beating a material, microstructures can be tailored to a desired form.

Microstructural taxonomy has reached a sufficient level of sophistication that it is possible to identify certain broad classes of microstructure. Before beginning, however, it is necessary to reiterate what this chapter is not. Our primary emphasis will be on metals. The majority of experimental data to be highlighted will be drawn from the metallurgical literature with occasional reference to nonmetallic precipitates within such metals. This is not to say that other material systems are without interesting microstructures. On the contrary, the subject is so vast as to have escaped even a modicum of competency on the part of the author.

10.1.1 Microstructural Taxonomy

Evidence for the existence of microstructures is often present even at those scales that can be discerned by the naked eye. This claim is substantiated in fig. 10.1 which shows the microstructures of turbine blades associated with different generations of casting technology. The leftmost figure in the group reveals a series of grains that appear statistically homogeneous. The basic observation is that a generic crystalline solid will be composed of many different grains, each with a particular crystalline orientation and separated by grain boundaries. The entirety of these grains and the associated boundaries make up the polycrystalline microstructure. Polycrystals of this sort are but one example of the types of microstructures that can be found. The second picture in the sequence in fig. 10.1 reveals an anisotropic microstructure in which the grains are elongated along a preferred direction. The final picture in this sequence, though remarkable as a single crystal of enormous proportions, is perhaps less interesting as a representative example of a material with microstructure.

Microstructure in Single Phase Polycrystals. We begin with a discussion of the simplest microstructure (at least to characterize geometrically) in which the grain size is the dominant microstructural parameter. We are all used to the idea that the technologies that surround us are constrained by various codes which attempt to standardize products and the processes used to realize them. Such codes exist for everything ranging from skyscrapers to microwave ovens. It is

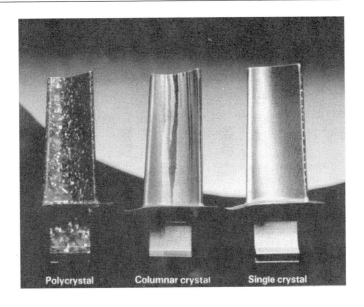

Fig. 10.1. Microstructures for turbine blades produced using different casting technologies.

less evident, however, that there might exist codes that serve to standardize the discussion of microstructure as well. The ASM has standardized this field in a way that is revealed in the *Metals Handbook* (ASM 1972) and which we show in fig. 10.2.

The frames of fig. 10.2 show some representative examples of different grain structures in steel. The distinguishing feature from one frame to the next is the grain size. For the case shown here which is for a class of low-carbon steels, the grains range in size from roughly 10 μm to 250 μm. Recall that these observations are more than idle musings since, as was noted earlier (see figs. 1.3 and 8.2), the grain size normally has a direct impact on mechanical properties such as the yield strength.

Already, there are many questions that might be asked concerning the microstructures shown in fig. 10.2. First, we note that the two-dimensional cross sections shown here can be deceptive. One of the most frustrating features in studying solids and one of the largest challenges that must still be fully owned up to is the difficulty of attaining three-dimensional understanding of the various structures that populate materials. As a step in this direction, we interest ourselves in different ways of characterizing a material's microstructure. As is evident from the figures discussed above, the microstructure is characterized by an element of randomness. On the other hand, there are also key elements of statistical regularity which may be exploited. Note that in the discussion given above, we have only made reference to the average grain size which is the simplest gross parameter for

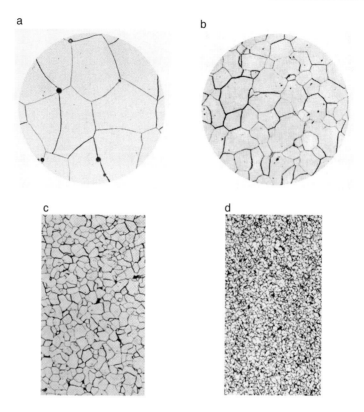

Fig. 10.2. Representative examples of ASTM grain sizes in low-carbon steel: (a) grain size 2 ($\approx$ 180 μm), (b) grain size 5 ($\approx$ 60 μm), (c) grain size 7 ($\approx$ 30 μm), (d) grain size 10 ($\approx$ 10 μm). Mean grain sizes associated with each microstructure are given in parentheses.

characterizing a material's microstructure, and that which enters into many of the simple empirical scaling models of material response.

The microstructure described above may be seen as the most homogeneous limit of those we will consider here. A complementary set of information concerning microstructures of the type featured above can be obtained by mapping the orientations of the various grains making up the polycrystal. The basic idea is that one may determine the preponderance of different crystal orientations. The new technique of orientation imaging microscopy now allows for the determination of such orientational information with high spatial resolution. An example of the type of results that are obtained via this technique is shown in fig. 10.3.

Precipitates and Inclusions. The presence of more than a single species in a material makes it possible that a given microstructure will be far richer than that described above. In particular, by virtue of the presence of several species coupled

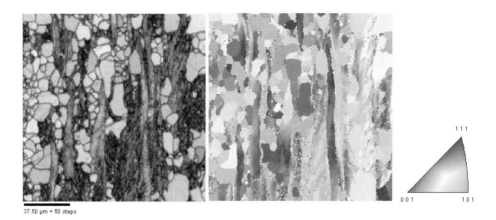

37.50 μm = 50 steps

Fig. 10.3. Orientation imaging microscopy image of a partially recrystallized low-carbon steel. The leftmost figure is an image quality map in which the image quality obtained from the diffraction pattern is mapped onto a gray scale. The middle figure shows the orientation of the various grains with respect to the normal plane of the sample and the correspondence between colors and orientation is effected via the crystallographic triangle shown on the right of the figure (courtesy of Stuart Wright).

with the operation of diffusion giving the different atom types the possibility of redistributing, it is possible to have microstructures that are mixtures of different phases. The geometric framework that will serve as the basis of our analysis of these microstructures is that of a 'matrix' material that is punctuated by the presence of second-phase particles. In this section, our primary aim is to take stock of the qualitative features of such microstructures and to foreshadow some of the physical properties that are inherited because of them. One of the first comments we must make in describing such structures from the standpoint of the modeler is the need to distinguish between coherent and incoherent second-phase particles. The key point is that the constraint of coherency (i.e. the two phases join up smoothly across the interface) implies an attendant possibility for significant internal stresses.

In figs. 10.4, 10.5 and 10.6, we show a few different examples of the microstructures involving second-phase particles that arise within different contexts. One feature of these examples that begs an explanation is that of the widely different morphologies exhibited in different cases. Second phase particle shapes range from nearly spherical, to faceted, to highly elongated rods and ultimately, plates and needles. A very useful compendium on some of the variety to be found in precipitate microstructures is Russell and Aaronson (1978). As will be seen later, the simplest model of the equilibrium shapes of such precipitates is based on an isotropic interfacial energy and leads to the expectation of spherical particles.

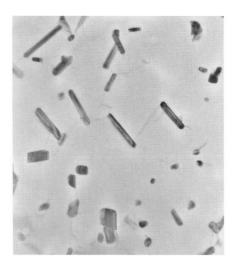

Fig. 10.4. Rod-shaped precipitates in Cr–Nb alloy (courtesy of Sharvan Kumar). The precipitates are Cr_2Nb and have the C15 Laves phase crystal structure.

Fig. 10.5. Al_2Cu (θ') precipitates in Al (courtesy of Bob Hyland).

As the figures indicate, the most naive theoretical intuition is incomplete. Below we will take up the challenge of how to explain the origins and characteristics of these microstructures. In addition, in chap. 11, we will examine the way in which a distribution of second-phase particles can alter the mechanical properties of a material.

Lamellar Microstructures. The precipitate microstructures described above have a geometric distribution such that a description in terms of a background matrix that is riddled with second-phase particles is appropriate. A second class of multiphase microstructure is lamellar microstructure like that shown in fig. 10.7. To better

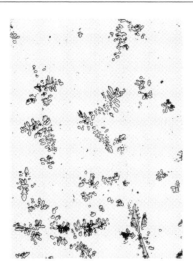

Fig. 10.6. As cast microstructure of Cu–O alloy with 0.78% oxygen content. The dendritic regions are Cu_2O. Micrograph 2257 from *Metals Handbook* (ASM 1972).

appreciate the nature of these microstructures, we begin by referring once again to the iron–carbon phase diagram as given in fig. 1.5 in which it is seen that in the low-carbon concentration limit, we should expect the presence of both α-Fe (also known as ferrite) and Fe_3C. The presence of both of these constituents together gives rise to a lamellar microstructure like that considered here. The basic idea is the emergence of structure on a few different scales. At the larger scale are overall colonies (or grains) within which the α-Fe and Fe_3C have the same orientation relationship. The structure within such a grain is characterized by an alternation between α-Fe and Fe_3C regions. This characteristic morphology is revealed in fig. 10.7. From the perspective of constructing models of such microstructures, both for their evolution and their impact on material properties, it is clear that this is a completely different class of problems than those built around second-phase particles.

The diversity of the various microstructures we have considered thus far is depicted schematically in fig. 10.8. From a structural perspective, these different structures permit interpretation crystallographically in terms of the various phases that make up the microstructure. A different class of microstructural feature can arise by virtue of the presence of dislocations.

Organization and Patterning of Dislocations. As a result of deformation, the dislocations present in a given material form structures. These structures bear no resemblance to the featureless and homogeneous dislocation arrangements that might be thought of at first blush. Nor, as remarked above, do these structures

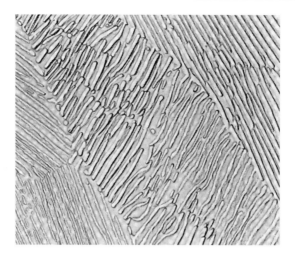

Fig. 10.7. Lamellar pearlite microstructure of a hot rolled steel bar. Micrograph 354 from *Metals Handbook* (ASM 1972). This figure is at a magnification of 2000×.

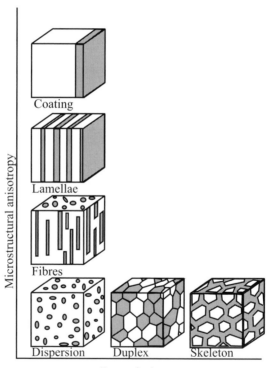

Fig. 10.8. Classification of various two-phase microstructures (adapted from Hornbogen (1984)).

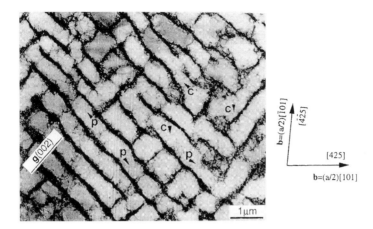

Fig. 10.9. Transmission electron microscopy image revealing the organization of dislocations in single-crystal Cu into a labyrinthine pattern as a result of cyclic loading (adapted from Suresh (1998)).

admit of the relatively clear cut interpretation offered for the several microstructural types introduced above. The dislocation structures found in materials exhibit even more remarkable patterning as is evident when the loading to which the material is subjected is cyclic. In this case, the repeated loading allows the formation of yet sharper and more organized structures such as those shown in fig. 10.9.

Microstructure at Surfaces. Our previous discussion has emphasized the microstructures that are present in generic bulk crystals. On the other hand, with the increasing reliance on layered materials in a variety of contexts including microelectronics (also revealed in fig. 10.8) and in particular, on thin films, a new generation of microstructures has grown up which are to a large extent two-dimensional. Though there is much that can be said on this point, we but mention them so as to signal the reader to the need for thinking in these areas as well. Because of the confluence of high-quality growth systems and attendant surface probes, one area of especial interest is that of the variety of 'self-organized' microstructures that are seen during growth. A particular example in the Si–Ge system is shown in fig. 10.10.

Thus far, our discussion has centered on some of the qualitative geometric and chemical features of microstructures found in different classes of materials. Now that we have seen in broad brush strokes the classifications that are possible, we turn to the question of how such microstructures can be produced by appropriate processing strategies, and what the consequences of such microstructures are to the underlying physical properties of a material.

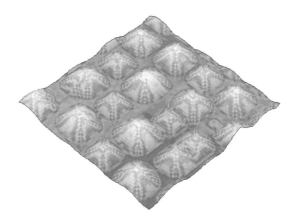

Fig. 10.10. Atomic force microscopy image of an island array in $Si_{80}Ge_{20}$ alloy grown on Si(001) (courtesy of J. Floro and Eric Chason). The morphology is built up of pyramidal 'hut' clusters with (501) facets.

10.1.2 Microstructural Change

Until now, our microstructural description has been geometric and qualitative without any real discussion of how such microstructures are prepared. Tuning microstructures is one of the oldest forms of engineering and may be traced to the historic ages when chemical impurities were used to harden metallic materials, though the microstructural understanding that went with these processes was clearly not present. The key insight, and the thrust of the present section, is the recognition that by subjecting a material either to deformation or heating or both, the underlying microstructure may be altered, another reminder that in dealing with microstructures we are treating nonequilibrium phenomena.

As has already been mentioned, entire historic ages may be thought of as having emerged from particular types of microstructural control. Microstructural change occurs both intentionally and by accident. No doubt, most readers of this book have been confronted with the rusted remains of some favorite gadget – the terminal state of what was once a metastable arrangement. Despite the apparent permanence of many materials on everyday time scales, when viewed over longer times they are seen to be teeming with activity. Our present discussion aims to describe a few different processing histories, with an eye to explaining the way in which such processing steps alter a given microstructure. The particular modeling strategies for describing these processes will be taken up in turn as the chapter proceeds.

Traditional metallurgy is sometimes derogatorily referred to as the science of 'heat and beat' where the tandem operations of annealing and working the material via a sequence of carefully selected, albeit often empirically derived, processing steps are used to obtain some desired properties. On the other hand, the importance of thermomechanical processing to everyday materials technology cannot be underestimated. In addition, the processes that transpire under these circumstances provide a wide range of problems that any complete theory of nonequilibrium processes must answer to.

Working a Material (Beat)

Generically, beating a material refers to the act of subjecting a material to plastic deformation via processes such as systematic stretching or through cyclic loading. One of the particular means whereby materials are deformed in the name of reaching some desired final shape or microstructure is that of rolling. The principal idea is to take the specimen and to subject it to preferential deformation which tends to elongate and squish it, for example. In fig. 10.11, we show two snapshots in the deformation history of a sample of tungsten which has been subjected to such rolling processes. In this instance, it is seen that one of the primary outcomes of the deformation process is the elongation of the underlying grain structure.

In addition, the working of the material is also attended by microstructural change at the dislocation level in that the dislocation density can be increased by orders of magnitude. The increase in dislocation density is attended by changes in the mechanical properties. An indication of the relation between plastic strain and dislocation density was already given in fig. 8.40. What that figure showed was the way in which by virtue of the working of a material, dislocations are stored in that material. We have also already noted (see fig. 8.2) that these microstructural changes are attended by concomitant changes in the properties of the material.

Annealing a Material (Heat)

In a worked material, the microstructure is characterized by a series of heavily dislocated grains. As was evidenced in our discussion of dislocations in chap. 8, such microstructures are inherently metastable. If the material is heated, a series of discernible changes will occur. An obvious example of this type of outcome is shown in fig. 10.12 in which it is seen that in the vicinity of a weld, the grain structure is completely different than in those regions far from the heating.

Systematic investigation of the properties of materials subject to annealing have resulted in a convenient classification of the processes undergone within the solid. In particular, it has been found convenient to separate these changes into three separate processes, namely, recovery, recrystallization and grain growth. The onset of recovery can be discerned in electrical resistivity measurements in which it is found that the resistivity decreases as a function of time, signaling a decrease in the number of scattering centers. There are a number of different mechanisms that can be associated with the recovery process, one of which corresponds to the removal

Fig. 10.11. Grain structure of tungsten polycrystals. Upper figure shows polycrystal with equiaxed grain structure while lower figure illustrates polycrystal that has suffered deformation via wire drawing (courtesy of Clyde Briant).

of point defects from the material. The recrystallization stage is characterized by a reduction of the net dislocation density, with the development of grains which are relatively dislocation-free. Finally, in the grain growth stage, the mean grain size increases with a corresponding reduction in the total grain boundary area. All of these processes are further elaborated in both Martin *et al.* (1997) and Humphreys and Hatherly (1995).

In addition to processes involving the alteration of the defect structures within a material upon heating, it is also clear that by altering the temperature, phase changes can be induced as well. Indeed, the control of the relative proportions of different phases within a material is one of the primary ways that microstructural

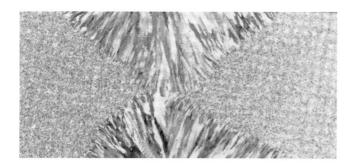

Fig. 10.12. Illustration of changes in microstructure that attend the welding process in a copper cold rolled bar. Adapted from micrograph 2307 in *Metals Handbook* (ASM 1972).

control is exercised. As we have already mentioned, one of the scenarios in which phase transformations are sure to occur is that in which there is the potential for precipitation reactions due to an overabundance of some foreign atoms. With increasing temperature, the likelihood of diffusion increases and with it the possibility that atoms can organize into precipitates. With increasing time at elevated temperature, the size of such precipitates can increase due to coarsening effects.

10.1.3 Models of Microstructure and its Evolution

We now wish to undertake an examination of various case studies in the analysis of microstructure. Perhaps the simplest example is that of a single second-phase particle in an otherwise homogeneous medium. For many purposes, this problem serves as the starting point for subsequent analyses, and finds its way into the analysis of problems ranging from two-phase microstructures to the effective properties of polycrystals. Our analysis will begin with one of the key elastic models of material microstructure, namely, the problem of the Eshelby inclusion. One direct route to the properties of this solution is to consider an elastically homogeneous medium (i.e. the same elastic properties throughout) but for which a certain region has undergone a transformation strain. Using the Green function techniques developed in earlier chapters, it is then possible to deduce the elastic fields associated with such an inclusion. Once we have deduced the elastic consequences of second-phase particles, we will be in a position to examine questions of both the nucleation and subsequent evolution of such particles, with special attention being given to the equilibrium sizes and shapes attained by these particles.

The second modeling centerpiece considered in this chapter will be our discussion of martensitic microstructures. This work is offered on the basis of our contention that this theory offers much of the analytic insight that one might hope to find in models of more generic microstructures. In particular, by virtue of the conjunction of kinematic and energetic arguments, the theory of martensitic microstructure to be offered here is able to make detailed predictions about these microstructures, including the fractions of different martensitic variants that are to be expected.

The discussion will be rounded out by our taking stock of microstructural evolution as seen from the more traditional metallurgical viewpoint. In particular, grain growth will be taken up from a theoretical perspective. This discussion should be seen as a continuation of those given in the previous chapter since the attempt to understand these phenomena rests heavily on a synthetic understanding of point, line and interfacial defects in concert. Our contention will be that despite the exertion of Herculean efforts, a number of key difficulties remain.

10.2 Inclusions as Microstructure

As noted many times already, phase diagrams instruct our intuition concerning the temporal evolution of certain metastable states. In particular, for the case in which one has a supersaturation of foreign atoms, one is led to expect that if the atoms have sufficient mobility, sooner or later they will find each other out and colonize new phases, namely, precipitates. Examples of some of the resulting microstructures were shown in figs. 10.4, 10.5 and 10.6. We also note that by virtue of particular thermomechanical histories (i.e. by suitable processing) the nature and distribution of these precipitates may be tailored to yield desirable physical properties.

For the purposes of the present discussion, one of our ambitions is to consider a class of second-phase microstructures in which the two phases are coherent. In these instances, the constraint of coherency carries with it an elastic penalty which results in elastic distortions in both the precipitate particle and the matrix material surrounding it. As a preliminary to the concrete mathematical investigation of such structures, we must first learn something of the elastic consequences of a single inclusion.

10.2.1 Eshelby and the Elastic Inclusion

The elastic solution upon which many approaches to the consideration of microstructure are built is that of the so-called Eshelby inclusion in which one considers a single ellipsoidal inclusion in an otherwise unperturbed material. From the standpoint of the linear theory of elasticity, this problem is analytically tractable

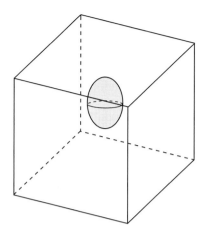

Fig. 10.13. Schematic of an ellipsoidal inclusion within an elastic medium.

and provides valuable insights concerning the stresses in adjacent grains and the compatibility issues that must be faced at the interfaces between such grains.

Our intention in the present section is to examine the elasticity solution to the problem of a single elastic inclusion and to attempt to use the insights afforded by this solution to learn something about that class of microstructures that can be thought of in terms of inclusions. Our starting point is the notion of an eigenstrain as discussed already in chap. 2. We picture an elastically homogeneous medium (i.e. the elastic moduli are the same throughout) with some spatial region that has undergone a strain ϵ^* as indicated schematically in fig. 10.13.

The philosophy espoused in the method of eigenstrains is that the presence of these eigenstrains may be thought of as giving rise to some distribution of body forces. However, as we have already shown repeatedly, once such a distribution is specified, the elastic Green function may be invoked to write down the displacement fields as an integral over this distribution. In the present setting, this argument proceeds as follows. The equilibrium equations of elasticity may be written as

$$\nabla \cdot \boldsymbol{\sigma}^e = 0, \tag{10.1}$$

where the superscript e refers to the fact that it is only the elastic part of the strain that gives rise to stresses. However, this statement of equilibrium may be rewritten on the basis of the observation that the total strain is given by $\epsilon^{tot} = \epsilon^e + \epsilon^*$. In particular, we have

$$C_{ijkl}\epsilon^{tot}_{kl,j} - C_{ijkl}\epsilon^*_{kl,j} = 0. \tag{10.2}$$

Recall that the distribution of eigenstrains is prescribed and hence the term

$C_{ijkl}\epsilon^*_{kl,j}$ is known. If we make the transcription $C_{ijkl}\epsilon^*_{kl,j} = -f_i$, where f_i is the ith component of a fictitious body force field, then our problem has been mapped onto a boundary value problem for the unknown total displacement,

$$C_{ijkl}u^{tot}_{k,lj} = -f_i. \tag{10.3}$$

For the particular case of interest below, we imagine an infinite body and an associated condition that the fields vanish at infinity.

As already remarked, once the body force distribution has been specified, by invoking the elastic Green function, the displacement fields are immediately obtained as the integral

$$u^{tot}_k(\mathbf{x}) = -\int_\Omega C_{mjpl}\epsilon^*_{pl,j}G_{km}(\mathbf{x}-\mathbf{x}')d^3\mathbf{x}'. \tag{10.4}$$

For our present purposes, we assume that the eigenstrain of interest is uniform throughout the region Ω. Using integration by parts, the differentiation on the eigenstrain itself can be pushed onto the Green function with the result that

$$u^{tot}_k(\mathbf{x}) = C_{mjpl}\epsilon^*_{pl}\int_\Omega G_{km,j}(\mathbf{x}-\mathbf{x}')d^3\mathbf{x}'. \tag{10.5}$$

Note that the differentiation of the Green function is with respect to the primed variables. We can now use the relation

$$\frac{\partial}{\partial x'_j}G_{km}(\mathbf{x}-\mathbf{x}') = -\frac{\partial}{\partial x_j}G_{km}(\mathbf{x}-\mathbf{x}'), \tag{10.6}$$

to rewrite our expression for the displacements as

$$u^{tot}_k(\mathbf{x}) = -C_{mjpl}\epsilon^*_{pl}\int_\Omega G_{km,j}(\mathbf{x}-\mathbf{x}')d^3\mathbf{x}', \tag{10.7}$$

where in this instance the differentiation of the Green function is with respect to the unprimed variables.

Thus far, our results are primarily formal. As an exercise in exploiting the formalism, we will first examine the fields associated with spherical inclusions, and then make an assessment of the elastic strain energy tied to such inclusions. These results will leave us in a position to consider at a conceptual level the nucleation, subsequent evolution and equilibrium shapes associated with inclusion microstructures.

To compute the displacements and strains associated with an inclusion explicitly, we must now carry out the operations implied by eqn (10.7). To make concrete analytic progress, we now specialize the analysis to the case of an elastically isotropic solid. As a preliminary to the determination of the elastic fields, it is necessary to differentiate the elastic Green function which was featured earlier as eqn (2.91). In particular, if we evaluate the derivative of G_{km} with respect to x_j, we find

$$
G_{km,j}(\mathbf{x} - \mathbf{x}') = \frac{1}{8\pi\mu(\lambda + 2\mu)}\left[-(\lambda + 3\mu)\delta_{km}\frac{(x_j - x_j')}{|\mathbf{x} - \mathbf{x}'|^3}\right.
$$
$$
+ (\lambda + \mu)\delta_{kj}\frac{(x_m - x_m')}{|\mathbf{x} - \mathbf{x}'|^3} + (\lambda + \mu)\delta_{mj}\frac{(x_k - x_k')}{|\mathbf{x} - \mathbf{x}'|^3}
$$
$$
\left. - 3(\lambda + \mu)\frac{(x_m - x_m')(x_k - x_k')(x_j - x_j')}{|\mathbf{x} - \mathbf{x}'|^5}\right], \tag{10.8}
$$

a mess that can be simplified considerably if we follow Mura (1987) in adopting the notation $l_i = (x_i - x_i')/|\mathbf{x} - \mathbf{x}'|$. In particular, contracting this expression to form $C_{mjpl}\epsilon_{pl}^* G_{km,j}$ as is demanded by our expression for the displacements, eqn (10.7), we find that the displacements can be written as

$$
u_k^{tot}(\mathbf{x}) = \frac{\epsilon_{pl}^*}{4\pi(\lambda + 2\mu)}\int_\Omega \frac{d^3\mathbf{x}'}{|\mathbf{x} - \mathbf{x}'|^2}(\mu\delta_{kl}l_p + \mu\delta_{kp}l_l - \mu\delta_{pl}l_k + 3(\lambda + \mu)l_kl_pl_l).
$$
$$
\tag{10.9}
$$

Our first exercise in computing the displacements themselves will be to consider the fields due to a spherical inclusion with a dilatational eigenstrain of the form $\epsilon_{ij}^* = \epsilon_0\delta_{ij}$ and with radius a. In this case, the contraction implied by the presence of the δ_{ij} simplifies considerably such that

$$
C_{mjpl}\epsilon_{pl}^* G_{km,j} = -\frac{\epsilon_0}{4\pi}\left(\frac{3\lambda + 2\mu}{\lambda + 2\mu}\right)\frac{(x_k - x_k')}{|\mathbf{x} - \mathbf{x}'|^3}, \tag{10.10}
$$

and hence the computation of the displacements reduces to the evaluation of the integral

$$
u_k^{tot}(\mathbf{x}) = \frac{\epsilon_0}{4\pi}\left(\frac{3\lambda + 2\mu}{\lambda + 2\mu}\right)\int_\Omega \frac{(x_k - x_k')}{|\mathbf{x} - \mathbf{x}'|^3}d^3\mathbf{r}'. \tag{10.11}
$$

For the sake of checking the correspondence of our present results with those favored by others where the elastic constants μ and ν are used, we note the identity $(3\lambda + 2\mu)/(\lambda + 2\mu) = (1 + \nu)/(1 - \nu)$. Note that in light of these results, the determination of the elastic displacements has been reduced to the evaluation of a geometric integral over the inclusion itself. For an isotropic solid with a spherical inclusion, the displacements fields are themselves spherically symmetric. As a consequence, without a loss of generality, we can evaluate the u_3 component of displacement along the z-axis, resorting to spherical coordinates to simplify the integrals. By symmetry, this is equivalent to the radial component of displacement,

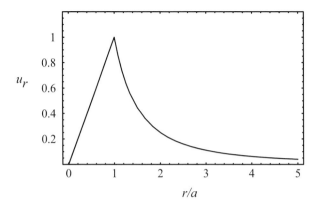

Fig. 10.14. Radial displacements associated with a spherical inclusion for which the eigenstrain is given by $\epsilon_{ij}^* = \epsilon_0 \delta_{ij}$. Displacements have been cast in dimensionless terms.

$u_r(r)$. In particular, the expression for the displacements reduces to

$$u_r^{tot}(r) = \frac{\epsilon_0}{2}\left(\frac{3\lambda + 2\mu}{\lambda + 2\mu}\right)\int_0^a r'^2 dr' \int_0^\pi \sin\theta\, d\theta \frac{r - r'\cos\theta}{(r'^2 + r^2 - 2rr'\cos\theta)^{\frac{3}{2}}}.$$
(10.12)

For a point outside of the inclusion, evaluation of the requisite integrals results in

$$u_r^{tot}(r) = \frac{\epsilon_0 a^3}{3r^2}\left(\frac{3\lambda + 2\mu}{\lambda + 2\mu}\right).$$
(10.13)

Similarly, if we are to evaluate the displacements within the inclusion, it is found that

$$u_r^{tot}(r) = \frac{\epsilon_0 r}{3}\left(\frac{3\lambda + 2\mu}{\lambda + 2\mu}\right),$$
(10.14)

indicating that within the inclusion the strain is constant, as was hinted at earlier. The displacement as a function of the radial coordinate is shown in fig. 10.14.

From the point of view of our efforts to examine the energetics of microstructures, our assessment of the elastic fields due to inclusions must be supplemented with insights into their energies. In light of our success in deducing the elastic fields, it is relatively straightforward to compute the strain energy associated with such an inclusion and the most naive approach is to add up the elastic energy throughout the entire medium via

$$E_{elastic} = \int_\Omega \frac{1}{2}\sigma_{ij}\epsilon_{ij}^e dV,$$
(10.15)

where Ω denotes the entire region of matrix and inclusion. (A simpler scheme for carrying out this evaluation of the elastic strain energy is reserved for the problem section at the end of this chapter.) Note that the integral demands a knowledge of the *elastic* strain. Since our calculations have yielded the total displacement, there is a bit of work needed to recast our results in a form that is suitable for the determination of the elastic energy. In particular, we find it most convenient to work with an alternative form for the strain energy density, namely,

$$\frac{1}{2}\sigma_{ij}(\epsilon_{ij}^{tot} - \epsilon_{ij}^*) = \mu\left\{(\epsilon_{ij}^{tot} - \epsilon_{ij}^*)^2 + \frac{v}{1-2v}[\text{tr}(\epsilon^{tot} - \epsilon^*)]^2\right\}. \qquad (10.16)$$

Note that we have resorted to the use of μ and v for our description of the elastic properties for ease of comparison with many of the expressions that appear elsewhere. In order to compute the total elastic energy, we separate the volume integral of eqn (10.15) into two parts, one which is an integral over the inclusion, and the other of which is an integral over the matrix material. The reason for effecting this separation is that, external to the spherical inclusion, the displacements we have computed are the elastic displacements. On the other hand, within the inclusion, it is the total fields that have been computed and hence we must resort to several manipulations to deduce the elastic strains themselves.

For the particular eigenstrain chosen here, note that we have $\epsilon_{ij}^* = \epsilon_0\delta_{ij}$. Within the inclusion this led to the displacement given in eqn (10.14) which can be differentiated to yield the strain

$$\epsilon_{ij}^{in} = \frac{\epsilon_0}{3}\left(\frac{1+v}{1-v}\right)\delta_{ij}. \qquad (10.17)$$

As a consequence, we find that the elastic strain is given by

$$\epsilon_{ij}^{elastic} = \epsilon_{ij} - \epsilon_{ij}^* = -\frac{2\epsilon_0}{3}\left(\frac{1-2v}{1-v}\right)\delta_{ij}. \qquad (10.18)$$

In light of this result, we are now prepared to evaluate the strain energy within the inclusion, which using eqn (10.16) is

$$E_{strain}^{in} = \frac{16\pi a^3 \epsilon_0^2 \mu}{9}\frac{(1-2v)(1+v)}{(1-v)^2}. \qquad (10.19)$$

We are also interested in the elastic energy external to the inclusion. To compute this, we demand the strains outside of the inclusion, which can be simply evaluated on the basis of eqn (10.13) with the result that

$$\epsilon_{rr} = -\frac{2\epsilon_0 a^3}{3r^3}\left(\frac{1+v}{1-v}\right) \qquad (10.20)$$

$$\epsilon_{\theta\theta} = \epsilon_{\phi\phi} = \frac{\epsilon_0 a^3}{3r^3}\left(\frac{1+v}{1-v}\right). \qquad (10.21)$$

The elastic energy stored in the region outside of the inclusion may now be

evaluated in a fashion similar to that invoked to evaluate the energy stored within
the inclusion with the result that

$$E_{strain}^{out} = \frac{8\pi a^3 \epsilon_0^2 \mu}{9} \frac{(1+v)^2}{(1-v)^2}. \tag{10.22}$$

Hence, the total strain energy of the medium is arrived at by summing the
contributions from within and outside of the inclusion, and leads to

$$E_{strain} = \frac{8\pi a^3 \epsilon_0^2 \mu}{3} \frac{(1+v)}{(1-v)}. \tag{10.23}$$

What we have learned is that the elastic energy associated with a spherical inclusion
characterized by an eigenstrain $\epsilon_{ij}^* = \epsilon_0 \delta_{ij}$ scales in a simple way with the volume
of the inclusion itself. Similar analytic progress can be achieved for other simple
geometries such as ellipsoidal and cube-shaped inclusions.

The main point of the efforts described above is that we are now in a position
to evaluate the elastic consequences of coherent precipitates as will be necessary
in our consideration of equilibrium shapes in coming sections. What we note
now is that by virtue of our knowledge of the features of this highly idealized
problem and its corresponding solution, we are prepared to make inroads on a
variety of problems including that of the coarsening of two-phase microstructures
and the effective properties of multiphase media. Despite the insight provided by
analytic solutions like those given above, there are many instances in which the
geometries, materials and loads are sufficiently complex as to defy useful analytic
description. Indeed, note that our solution was predicated on the use of a Green
function which was itself derived on the assumption of material homogeneity,
an assumption which is manifestly not warranted in many cases of interest. In
these cases, one must resort to numerical calculation. We have already noted that
there is both a synergy and tension between analytic and numerical results. On
the one hand, analytic results are highly instructive in that they provide insights
into the qualitative dependence of solutions on material and geometric parameters.
On the other hand, such solutions are limiting in that they usually do not allow
for a treatment of a wide range of geometries. Though the analytic solution set
forth here will be of great use for providing insights into the nature of elastic
fields surrounding inclusions, our ultimate discussion of the temporal evolution of
precipitate microstructures and their equilibrium shapes will be largely predicated
upon numerical schemes for solving the types of elastic boundary value problems
posed above.

10.2.2 The Question of Equilibrium Shapes

Having now developed a feel for the elastic consequences of the presence of a second-phase particle in a given material, we are poised to investigate the equilibrium morphologies of second-phase particles as well as their temporal evolution. We begin with the question of the equilibrium shapes of coherent second-phase particles which will be expressed as the playing out of a competition between the types of elastic terms considered above and other energetic influences such as the energy of the interface between the two particles. This type of reasoning is ubiquitous and ultimately rests on the variational character of thermodynamic equilibrium. Of course, such arguments skirt the question of whether or not the particular situation of interest is really an equilibrium state or not. Stated simply, our initial foray into these questions will be predicated upon the idea that in certain cases, microstructures of interest may be thought of as energy minimizers.

For the problem at hand, the competition of interest may be expressed in its simplest incarnation with relative ease. If we are to return to the experimental circumstances of interest, our starting point is a situation in which an alloy has been cooled to the point that there is an overabundance of foreign atoms within the host matrix. As a result of the process of diffusion, these foreign atoms collect and begin to form precipitates. However, because of the difference in structure between the matrix and precipitate (this difference in structure may mean nothing more than a different chemical distribution on the same underlying lattice), there is an energy penalty for both the interface between the matrix and the precipitate as well as for the elastic strains induced by this difference in structure. Both of these energies depend, in turn, upon the particle shape as well as upon the arrangements of the precipitates with respect to each other.

If we are to consider the simplest case in which a fixed quantity of mass is assigned to the second-phase particle of interest and demand what shape results in the lowest free energy, it is easily imagined that the free energy depends upon those parameters characterizing the shape through a number of different terms. The conceptual underpinnings of our investigation of equilibrium shapes is represented by the formula

$$E_{tot}(shape) = \int_{\partial\Omega} \gamma(\mathbf{n}) dA + \frac{1}{2} \int_{\Omega} \sigma_{ij} \epsilon_{ij} dV. \tag{10.24}$$

The first term on the right hand side represents the energy cost associated with the interface between the particle and the matrix, and for generality we have assumed (temporarily) that the interfacial energy is anisotropic, a fact that is revealed by the dependence of this energy on the normal $\mathbf{n}$ as $\gamma(\mathbf{n})$. The second term is the elastic energy of the medium by virtue of the presence of the second-phase particle. Note that both of these terms are functionals of the particle shape and our ambition is to minimize the energy with respect to the various possible shapes. We note that in the discussion that follows, the use of the term anisotropy will do double service to characterize both the morphological anisotropy of the particles themselves, as well as the crystallographic anisotropy of the various material parameters such as the interfacial energy and the elastic moduli.

10.2.3 Precipitate Morphologies and Interfacial Energy

The simplest picture that can be considered for describing equilibrium shapes is predicated entirely on the presence of the interfacial energy. In particular, the elastic energy given in eqn (10.24) is dropped all together and instead, we are left with minimizing

$$E_{tot}(shape) = \int_{\partial\Omega} \gamma(\mathbf{n})dA + \lambda \int_{\Omega} dV, \tag{10.25}$$

where the second term on the right hand side introduces a Lagrange multiplier λ with the constraint of fixed volume. Note that we have ignored the complicating influence of finite temperatures. Before continuing with our discussion of the minimizers of this functional, it is important to recall the extent to which the physics it entails can be informed on the basis of microscopic insights. In particular, in light of the discussion of chap. 9, it is clear that microscopic analysis can be used to construct the interfacial energy $\gamma(\mathbf{n})$. Indeed, both figs. 9.46 and 11.9 illustrate the dependence of interfacial energy upon the relative orientation of the different crystallites which the interface separates. In the present case, it is the relatively simple generalization of these results to the case in which the interface separates two different phases that is of interest, and we conclude this comment by noting that the microscopic evaluation of these quantities offers no compelling new twists on what we have done before.

   Though it would take us too far afield to provide a detailed discussion, we at least mention the idea of the Wulff construction in which the equilibrium shapes of particles are inherited from the properties of the interfacial energy as a function of orientation. The starting point of an analysis of the equilibrium shapes associated with minimizing the interfacial energy is made by appealing to a Wulff plot (or γ-plot) in which a three-dimensional closed surface is constructed with the property that the distance of a given point on that surface from the origin is equal to the interfacial energy associated with the orientation of that point. That is, the surface is specified as the collection of points $\mathbf{r}$ of the form $\mathbf{r} = \gamma(\mathbf{n})\mathbf{n}$. The set of points corresponding to all of the unit vectors $\mathbf{n}$ corresponds to the unit sphere. The idea of the γ-plot is to take each such point and to scale its distance from the origin by the corresponding $\gamma(\mathbf{n})$. Once we are in possession of the

γ-plot, we are then prepared to determine the equilibrium shape itself by recourse to Wulff's construction which instructs us to convexify the γ-plot with the result that the equilibrium shape corresponds precisely with the resulting polygon (in two dimensions) or polyhedron (three dimensions). Concretely, what this means is that we are to visit each and every normal vector **n** and to construct a perpendicular plane which passes through the intersection of that normal with the γ-plot. The equilibrium shape corresponds to the inner envelope of this collection of planes. We also make the important point that the resulting equilibrium shape has no dependence on particle size, a conclusion that is at marked odds with observed microstructures. These ideas are elaborated further in the problem section at the end of the chapter. An excellent discussion of these ideas may be found in section 5.4 of Martin *et al.* (1997). An alternative discussion with more emphasis on the statistical mechanical underpinnings of this problem may be found in Rottman and Wortis (1984).

The construction described above now permits us to hint at the relevance of the interfacial energy to the question of equilibrium shapes. Indeed, an example of the outcome of such reasoning in the context of a Cu–Zn alloy is shown in fig. 10.15. The precipitates being evaluated here arose as a result of isothermal heat treatment of supersaturated β phase which resulted in the formation of faceted polyhedra of the γ phase of composition Cu_5Zn_8. The figure shows various sections through the precipitate centroids associated with γ phase precipitates in the Cu–Zn system. In particular, note that these sections are made through the (100), (111) and (110) planes. The dotted curves in the figures reveal the Wulff plots associated with these measured particle morphologies. In fact, the reasoning is really run in reverse relative to the usual Wulff argument since in this case it is the shapes that are observed and the γ-plots that are backed out. Nevertheless, our main point in bringing up these issues was to call attention to the ideas used in addressing the case in which the interfacial energy dominates the equilibrium shape.

10.2.4 Equilibrium Shapes: Elastic and Interfacial Energy

A more complete description of the question of equilibrium shapes than that given above must account for both of the terms indicated in eqn (10.25). Our strategy in this section is to consider models which synthesize the effects of both interfacial and elastic energy from ever increasing levels of sophistication. As usual, these formulations will illustrate the tension between analytic tractability and physical realism, with the most realistic models almost entirely defying analytic progress.

Two-Dimensional Analysis of Johnson and Cahn. For starters, we consider the equilibrium shape for second-phase particles in a two-dimensional setting.

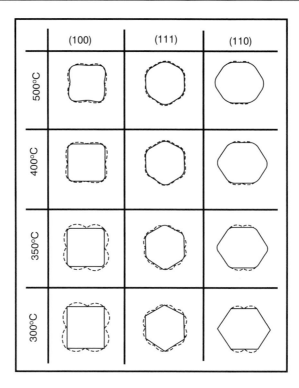

Fig. 10.15. Series of equilibrium shapes as observed in γ Cu–Zn (solid lines) and estimated values of the interfacial energy vs orientation (dashed lines) (adapted from Stephens and Purdy (1975)). The precipitate sizes are roughly 60 μm for all but the 500 °C case, in which the precipitates have a size on the order of 500 μm.

In particular, we consider a single such particle with a dilatational misfit (i.e. $\epsilon_{ij}^* = \epsilon_0 \delta_{ij}$) with the restriction that the particle shape derives from only a family of such shapes that can be parameterized in terms of a few simple geometric quantities. In essence, we adopt the Rayleigh–Ritz variational procedure in which the energy function of interest depends upon a finite set of parameters and with respect to which the energy function is minimized. We follow Johnson and Cahn (1984) in considering the family of elliptical shapes characterized by their area $A = \pi ab$ and aspect ratio a/b. Though it is a point we have raised repeatedly, it is worth reiterating that the space of energy minimizing solutions has been shrunk to include only those that are elliptical.

From the standpoint of the quantitative model used to describe this system, we assume that both the matrix and precipitate are elastically isotropic and characterized by the same elastic moduli and that the misfit strain associated with the precipitate is purely dilatational and is given by ϵ_0. Further, it is assumed that the interfacial energy $\gamma_{int}(\mathbf{n})$ is isotropic (i.e. $\gamma_{int}(\mathbf{n}) = \gamma_0$). Recall from our discussion in chap. 9 that interfacial energies can vary significantly from one crystal face to the next. What is being assumed here is that this variability can be suppressed without losing the essential features of the competition being played out. Of course, these restrictions can be relaxed, and it is interesting to do so. On the other hand, as a tool for illustrating the fundamental ideas being considered here, the simplifications that attend these assumptions make them highly attractive.

As a prelude to writing the dependence of the energy on the particle shape and size, Johnson and Cahn found it convenient to carry out a change of variables which respected important symmetries that this energy must satisfy. For example, the energy should be indifferent to an interchange of the major and minor axes of the ellipse. That is, the energy associated with an ellipse with its long axis along the (10) direction and its short axis along the (01) direction should be no different from the reverse circumstance. These symmetries imply that a more useful set of coordinates are

$$u = \frac{(a - b)}{\sqrt{2}},$$
(10.26)

$$v = \frac{(a + b)}{\sqrt{2}}.$$
(10.27)

In terms of these new variables, the geometric description of the elliptical particles is given by

$$A = \frac{\pi}{2}(v^2 - u^2),$$
(10.28)

which is the area of the particle and

$$U = \frac{u}{v} = \frac{a - b}{a + b},$$
(10.29)

which provides a measure of the departure of the ellipse from a circle.

In order to determine the lowest-energy shape for a given particle size, it is necessary to express the energy in terms of the variables U and A. In principle, what we have learned from the previous two sections permits the construction of an exact energy as a function of these parameters. On the other hand, an alternative idea is to map the exact energy onto an approximate surrogate which has the form of a Landau free energy, well known from the study of phase transitions, which permits us to exploit what is known about such free energies from other contexts. The strategy is to seek an expansion of the energy in powers of U in which the expansion coefficients depend upon the parameter A. Much can be said about the energy minimizing structure on the basis of general symmetry considerations which we will discuss shortly. However, we first examine the detailed character of these energies themselves. The interfacial contribution to the total energy in the isotropic approximation contemplated here is given by

$$E_{int} = \gamma_0 l, \tag{10.30}$$

where l is the circumference of the ellipse. For an ellipse with semimajor axis a and semiminor axis b, the circumference can be evaluated through the integral

$$l(a, b) = 4 \int_0^a \sqrt{\frac{a^2 - \xi^2 x^2}{a^2 - x^2}} dx, \tag{10.31}$$

where we have defined the dimensionless parameter $\xi^2 = 1 - (b/a)^2$. This result is a special case of an elliptic integral and may also be written as

$$l = 4a E\left(\frac{\sqrt{a^2 - b^2}}{a}, \frac{\pi}{2}\right), \tag{10.32}$$

where E is an elliptic function. By resorting to Taylor expansion, the interfacial energy term may be rewritten as

$$E_{int}(A, U) = 2\sqrt{\pi A}\gamma_0\left(1 + \frac{3}{4}U^2 + \frac{33}{64}U^4 + \cdots\right). \tag{10.33}$$

In addition to the interfacial energy, the competition at play here depends also on the elastic energy associated with the second-phase particle. Using ideas quite similar to those introduced already in the context of the Eshelby inclusion the elastic energy may be written as

$$E_{elastic}(A, U) = 2\epsilon_0^2\mu^* A \frac{1 + \kappa\delta - (1 - \delta)U^2}{(1 + \kappa\delta)(1 + \delta - 2\nu^*) + (1 - \delta)(\kappa\delta - 1 + 2\nu^*)U^2}, \tag{10.34}$$

where the precipitate is characterized by elastic constants μ^* and ν^*, while the matrix has elastic constants μ and ν. The dimensionless ratio $\delta = \mu^*/\mu$ characterizes the difference in shear moduli between the two media and $\kappa = 3 - 4\nu$. Like the interfacial energy, this term may be expanded in powers of U yielding

$$E_{elastic}(A, U) = \frac{2A\epsilon_0^2\mu^*}{1 + \delta - 2\nu^*}\left[1 - \frac{\delta(1 - \delta)(1 + \kappa)}{(1 + \kappa\delta)(1 + \delta - 2\nu^*)}U^2\right.$$
$$\left. + \frac{\delta(1 - \delta)^2(1 + \kappa)}{(1 + \kappa\delta)^2(1 + \delta - 2\nu^*)^2}U^4 \cdots\right]. \tag{10.35}$$

Given our knowledge of both the elastic and interfacial energies, the total energy as a function of the particle size and shape parameters now has the character of a Landau expansion like that seen in the theory of phase transitions. In particular, we have

$$E_{tot}(A, U) = E_{int}(A, U) + E_{elastic}(A, U) = F_0(A) + \frac{1}{2}F_2(A)U^2 + \frac{1}{4!}F_4(A)U^4 \cdots,$$
(10.36)

the minima of which will tell us the optimal U for a given A. If we truncate the expansion at fourth order, the minima are realized by those U satisfying

$$\frac{\partial E_{tot}}{\partial U} = 0 = U\left[F_2(A) + \frac{1}{6}F_4(A)U^2\right].$$
(10.37)

On physical grounds, $F_4(A)$ will be seen to be positive, leading to the conclusion that the energy minimizing U is given by

$$U = \begin{cases} 0 & \text{if } F_2 > 0 \\ \pm\sqrt{-6F_2/F_4} & \text{if } F_2 < 0 \end{cases}.$$
(10.38)

Hence, within this approximate model, the entire question of equilibrium shape has been reduced to the dependence of F_2 on A which is gotten by adding the quadratic contributions to the energy from both the interfacial and elastic terms and is given by

$$F_2(A) = \frac{3}{2}\gamma_0\sqrt{\pi A} - \frac{2A\epsilon_0^2\mu^*\delta(1 - \delta)(1 + \kappa)}{(1 + \kappa\delta)(1 + \delta - 2\nu^*)^2}.$$
(10.39)

By setting $F_2(A)$ to zero we can obtain the critical particle size, A_c, which also permits the definition of a size parameter $\lambda = \sqrt{A/A_c}$. Recall that in the present model, the parameter A has the role of characterizing the particle size, and hence suggests that with increasing particle size there could be a transition from the $U = 0$ regime (i.e. circular particles) to the broken symmetry regime in which the particles exhibit elliptical shapes. This transition is most simply described in considering the ratio F_2/F_4 as a function of the parameter λ. Note that the transition can only occur in the case in which $\delta < 1$ which is the limit in which the particles are softer than the matrix.

The results of our analysis of the question of equilibrium shapes may be depicted graphically as shown in fig. 10.16. This graph is shown for the specific case in which $\nu = \nu^* = 1/3$ and $\delta = 1/2$. The key point is that with increasing particle size (measured by the dimensionless parameter λ), there is a transition from circular to elliptical shapes as the energy minimizing geometry. Note that the onset of this transition in the graph is observed by monitoring the value of U, the parameter which measures the deviation about the limit of circular particles, as a function of the particle size as measured by λ. An additional remark is to note the physical competition that is being played out. For small particle sizes, the surface to volume ratio is such that the particle shapes will be dominated by the interfacial energy while in the large-particle-size limit, the elastic terms will

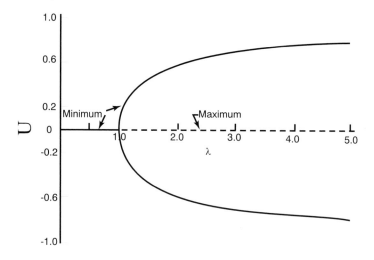

Fig. 10.16. Representation of energy minimizing U as a function of a dimensionless representation of the particle size (adapted from Johnson and Cahn (1984)).

have an increasing importance. Because the interfacial energy is isotropic, it favors circular geometries while the elastic energy prefers the elongated geometries found at larger particle sizes.

Numerical Calculation of Two-Dimensional Equilibrium Shapes. To go beyond the relative simplicity of the set of shapes considered in the analysis presented above it is necessary to resort to numerical procedures. As a preliminary to the numerical results that will be considered below and which are required when facing the full complexity of both arbitrary shape variations and full elastic anisotropy, we note a series of finite-element calculations that have been done (Jog *et al.* 2000) for a wider class of geometries than those considered by Johnson and Cahn. The analysis presented above was predicated on the ability to extract analytic descriptions of both the interfacial and elastic energies for a restricted class of geometries. More general geometries resist analytic description, and thus the elastic part of the problem (at the very least) must be solved by recourse to numerical methods.

Jog *et al.* use a two-dimensional numerical implementation of an iterative scheme in which the geometry of the interface is parameterized by a set of nodes which serve as the configurational degrees of freedom associated with the question of the equilibrium shape. Their scheme is predicated on computing the elastic fields at *fixed* geometry and then these fields are used first to compute the energy and then to compute the configurational forces on those same nodes (i.e. the derivatives of the energy with respect to the interfacial coordinates) providing a search direction

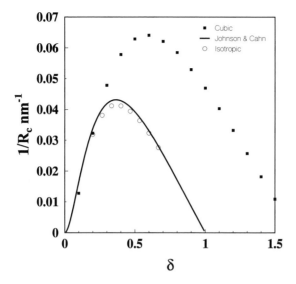

Fig. 10.17. Comparison of analytic and numerical results for the critical size at which the circle–ellipse bifurcation takes place as a function of particle size (adapted from Jog *et al.* (2000)).

for the next increment in the quest for the equilibrium structure. The significance of this implementation is that it imposes no requirements on the particle shape. As a result, the search for the energy minimizers is over a much larger space of possible shapes.

One way of testing such numerical implementations is to recover known results. In the present setting, this means recovering the results of Johnson and Cahn for the transition from circular to elliptical shapes. In fig. 10.17 we show the results for the critical size (the inverse of this size is actually plotted) at which the bifurcation from circular to elliptical shapes takes place as a function of the inhomogeneity factor $\delta = \mu^*/\mu$, where μ^* and μ are the shear moduli for the precipitate and the matrix, respectively. Of course, we should note that in the case of the analysis by Jog *et al.*, the shape transition is from circular to *nearly* elliptical shapes, since their geometries are no longer constrained to the single degree of freedom geometry (at fixed particle size) considered by Johnson and Cahn. The numerical results beautifully complement those resulting from the Johnson and Cahn analysis since they allow for an examination of the way in which elastic anisotropy and different forms of misfit other than dilatational can alter the equilibrium shapes.

Equilibrium Shapes in Three Dimensions. The analyses described above have illustrated the fundamental competition between interfacial energies and elastic energies in governing the outcome of the question of equilibrium shapes. These calculations were introduced in the setting of reduced dimensionality with the

$L = 1$ $L = 5$ $L = 9$

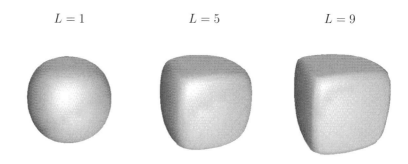

Fig. 10.18. Series of equilibrium shapes for three-dimensional particles (adapted from Mueller and Gross (1998)). The parameter L is a measure of the particle size and governs the competition between the surface energy terms (dominant at small L) and the elastic terms (dominant at large L).

result that analytic progress could be made. On the other hand, the importance of the answer to the question of equilibrium shapes is much greater in the context of three dimensions and it is for that reason that we also sketch the application of ideas like those described above in their full three-dimensional generality.

As before, the fundamental competition being played out in selecting equilibrium shapes in three dimensions is between interfacial and elastic energies. In the analysis of Mueller and Gross (1998), energy minimizing geometries are searched for by recourse to a boundary element treatment of the elastic energies of a given three-dimensional shape. Their treatment is fully anisotropic and relies conceptually on exactly the same eigenstrain ideas introduced earlier (see section 2.5.3) in conjunction with the elastic Green function, but now tailored to problems involving elastic anisotropy. Because their methods are numerical, the elastic fields for a given configuration can be computed directly, and the driving forces on the configurational coordinates describing the interfacial position are computed in exactly the same spirit as those introduced in our discussion of configurational forces in chap. 2. The three-dimensional equilibrium shapes found in their analysis are shown in fig. 10.18. The elastic parameters chosen in these calculations were meant to mimic those of Ni, with the result that in the elastically dominated large-particle limit, there is a tendency for the particles to be cube-like.

The arguments given above were concerned more with the terminal privileged state that would be arrived at by a particular second-phase particle than with the dynamical trajectory by which these equilibrium states are attained. In addition, we ignored the possible interactions between different particles, which in addition to interacting elastically can exchange matter via mass transport. All of these complaints can be remedied by recourse to a more computationally intensive approach in which the dynamical evolution of a given particle is considered as a function of time and the equilibrium shape is a fixed point of this dynamical system. This approach will be taken up later in this chapter. However, we should note that it is also possible to go beyond the types of continuum analyses introduced here and rather to use the type of effective Hamiltonian introduced in chap. 6 in conjunction with Monte Carlo simulation to revisit the question of equilibrium shapes. Using these methods, there are no restrictions concerning particle shapes, elastic homogeneity, isotropy of interfacial energies, nor is there even any restriction to linear elasticity. These ideas will be described in chap. 12 (see section 12.4.4 in particular).

10.2.5 A Case Study in Inclusions: Precipitate Nucleation

As noted above, the question of equilibrium shapes is essentially a static one and can be answered without recourse to an analysis of the temporal processes that occur in conjunction with microstructural evolution. On the other hand, we are interested in examining the theoretical apparatus that has grown up around problems in which time is considered explicitly. One of the first questions we can address in light of our elastic solution for the energetics of an inclusion and the first step in our attempt to understand the temporal evolution of such particles is that of nucleation of second-phase particles. As with many questions concerning an instability toward the development of some new phase, the discussion centers on a competition between two different effects, each of which tends to favor a different terminal state. In the present setting, it is the competition between a term which scales as the volume of the second-phase particle and tends to favor the formation of the second-phase particle, and an interface term which carries an energy penalty associated with the interface between the inclusion and the matrix and thus favors removal of the second-phase particle. So as to permit analytic progress, we consider the restricted class of spherical geometries.

The volume term is of the form

$$E_V = \frac{4}{3}\pi r^3 E_b, \tag{10.40}$$

where E_b is an energy density such that $E_b < 0$, and r is the particle radius. On the basis of arguments made in this and earlier chapters, we are in a position to seek an interpretation of the term E_b. First, we note that there is a contribution to E_b that arises from the fact that the second-phase particle has a different structure and hence a different structural energy than the matrix. Our use of the term 'structure' here is meant to include even the difference between ordered and disordered states on the same underlying lattice. The microscopic route to computing the energy lowering represented by a second-phase particle was described in chap. 6. However, we note that this energy will be renormalized by the energy penalty

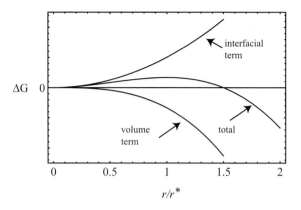

Fig. 10.19. Schematic of the bulk and interfacial energies as a function of particle size.

arising from the elastic strains as was computed in section 10.2.1. As a result, in the present model, E_b can be thought of as arising from two terms, one negative and essentially a structural energy difference, the other positive and representing the elastic strain energy associated with the inclusion.

The volume term described above is balanced against an interfacial energy of the form

$$E_{surf} = 4\pi r^2 \gamma_{int}, \tag{10.41}$$

where for the present purposes we have assumed an isotropic interfacial energy. The generalization to the anisotropic case introduces no new physical ideas. Again, because of the types of arguments made in previous chapters, it is now clear how to appeal to microscopic calculations in order to compute γ_{int}. The energy balance to be considered here is familiar and is depicted graphically in fig. 10.19. The key observation is that with increasing particle size, a certain critical particle radius r^* is reached such that a further increase in particle size leads to an energy reduction. From the standpoint of our discussion of configurational forces (see section 2.3.3), we see that the sign of the driving force (i.e. $\partial G(r)/\partial r$) depends upon the size of the particle. For small particles, the driving force is in the direction of particle shrinkage, while for particles greater than the critical size, the driving force results in growth of the particle.

To cast these ideas in mathematical form, we note that the total free energy in the present model may be written as

$$G(r) = 4\pi r^2 \gamma_{int} + \frac{4}{3}\pi r^3 E_b. \tag{10.42}$$

To determine the critical radius r^*, we need to satisfy the condition $\partial G(r)/\partial r = 0$. The consequence of performing the relevant differentiation and solving for the critical radius is

$$r^* = -\frac{2\gamma_{int}}{E_b}.$$ (10.43)

Though the results of nucleation theory are well worn, it is worth emphasizing the way in which a number of disparate threads have been woven into the discussion. First, armed with the results of the previous chapter, we are now well placed to understand the microscopic origins of γ_{int}. Similarly, because of our earlier discussion of structural energies, the structural contribution to E_b is also clear. Finally, the elastic solution to the problem of a spherical inclusion given earlier in this chapter has instructed us concerning the strain energy penalty that arises on forming the inclusion which has been modeled as carrying a transformation strain.

The model developed above serves as a convenient starting point for carrying out a dynamical analysis of the nucleation problem from the perspective of the variational principle of section 2.3.3. A nice discussion of this analysis can be found in Suo (1997). As with the two-dimensional model considered in section 2.3.3, we idealize our analysis to the case of a single particle characterized by one degree of freedom. In the present setting, we restrict our attention to spherical particles of radius r. We recall that the function which presides over our variational statement of this problem can be written generically as

$$\Pi(\dot{r}; r) = \Psi(\dot{r}; r) + \dot{G}(\dot{r}; r),$$ (10.44)

and specifically in the present context as

$$\Pi(r, \dot{r}) = 4\pi r^2 \frac{\dot{r}^2}{2M} + \dot{G}_{int}(\dot{r}; r) + \dot{G}_{bulk}(\dot{r}; r).$$ (10.45)

We recall that M is the mobility of the interface between the two phases as the particle changes size. To compute the rate of change in the interfacial contribution to the free energy, we evaluate $G_{int}(r + \dot{r}\Delta t) - G_{int}(r)$, recognizing that $G_{int}(r) = 4\pi r^2 \gamma_{int}$. This amounts to computing the difference in the interfacial energy at an initial instant when the radius of the particle is r and at a time Δt later when the radius of the particle is $r + \dot{r}\Delta t$. As a result of these arguments, we find

$$\dot{G}_{int} = \lim_{\Delta t \to 0} \frac{G_{int}(r + \dot{r}\Delta t) - G_{int}(r)}{\Delta t} = 8\pi r \dot{r} \gamma_{int}.$$ (10.46)

A similar argument is made in the case of the rate of change of the bulk contribution to the free energy. In this case, we interest ourselves in the free energy when the particle has radius r (i.e. $G_{bulk}(r) = \frac{4}{3}\pi r^3 E_b$) and when the particle has radius $r + \dot{r}\Delta t$. In this case, the result is

$$\dot{G}_{bulk} = \lim_{\Delta t \to 0} \frac{G_{bulk}(r + \dot{r}\Delta t) - G_{bulk}(r)}{\Delta t} = 4\pi r^2 \dot{r} E_b.$$ (10.47)

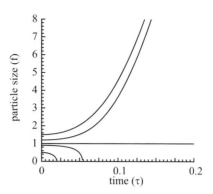

Fig. 10.20. Flows for the differential equation describing evolution of a spherical particle with growth (or dissolution) dictated by the balance of interfacial and bulk energy. The particle size (f) and time (τ) are plotted in the same dimensionless form used in the text.

If we now invoke the variational principle (i.e. $\partial \Pi / \partial \dot{r} = 0$), then it is found that

$$\dot{r} = M|E_b| - \frac{2M\gamma_{int}}{r}. \tag{10.48}$$

As seen earlier, there is a natural length scale in this problem, namely, $r_c = 2\gamma_{int}/|E_b|$, and it is similarly convenient to introduce the dimensionless time $t_0 = 2\gamma_{int}/ME_b^2$. In light of these definitions, the radius may be written in dimensionless form as $f = r/r_c$, with a similar dimensionless time given by $\tau = t/t_0$. Using these definitions, eqn (10.48) may be rewritten as

$$\frac{df}{d\tau} = \frac{f-1}{f}. \tag{10.49}$$

This equation is separable and has the solution,

$$f - f_0 + \ln\left|\frac{1-f}{1-f_0}\right| = -\tau. \tag{10.50}$$

The temporal evolution of particles starting with different fractions of the critical size is shown in fig. 10.20. Note that as expected, if a particle happens to start with an initial size equal to the critical radius, it will remain invariant in time. On the other hand, particles smaller than the critical size will shrink while those larger than the critical size will grow indefinitely.

10.2.6 Temporal Evolution of Two-Phase Microstructures

Though the calculation given above was both fun and interesting, it suffered from a number of defects. First, the class of intermediate shapes was entirely restrictive since we assumed that the only possible shape was spherical. This assumption ignores all possibile instabilities of spherical particles to other shapes. In addition, our treatment of the mechanism of interfacial migration was implicit and was referenced only through the parameter M without any appeal to an underlying mechanism. To remedy these deficiencies, we now resort to schemes that can be spelled out in analytic terms, but for which one must have recourse to numerics to make any real progress in explicitly examining the particle shapes.

Diffusional Dynamics. In the present analysis, we consider the explicit treatment of diffusion as the mechanism of interfacial migration, with the rate of such diffusion influenced in turn by the elastic fields implied by a particular particle shape. As a result, the problem of the temporal evolution of particles is posed as a coupled problem in elasticity and mass transport. The particles are given full scope to develop in any way they want.

Our treatment of the temporal evolution of inclusion microstructures will be predicated on the following physical ideas. First, we note that the mechanism of microstructural change that is envisioned is that of mass transport. A given particle, by virtue of its shape and attendant elastic fields induces a local chemical potential. This chemical potential serves as a boundary condition in our solution of the diffusion equation for the entire medium. The update of the particle shape is based, in turn, upon the concentration gradients resulting from the solution of the diffusion equation with the local velocity increment scaling as $v_n \propto \nabla c \cdot \mathbf{n}$. The logic behind this form for the velocity is the claim that the flux of mass to or from the interface is of the form $\mathbf{j} \propto \nabla c$ and hence that the normal motion of the interface as a result of this mass flux should scale as $\nabla c \cdot \mathbf{n}$. The strategy is essentially iterative since once the particle shape has been altered, the elastic fields are also changed which demands another solution of the elasticity problem so that the chemical potential can be redetermined and from it, the next shape increment computed. The conceptual philosophy of these calculations is one of instantaneous equilibrium in which it is assumed that at every instant in the temporal history, the equations of mechanical equilibrium are instantaneously satisfied. The physical reasoning that is attached to this assumption has the same spirit as the Born–Oppenheimer approximation introduced in section 4.2.2. In particular, there is the expectation of a separation of time scales with the rearrangements (i.e. displacements) that maintain mechanical equilibrium taking place essentially infinitely fast in comparison with the diffusion time. In addition to statements about mechanical equilibrium, it is also assumed that there is an instantaneous chemical equilibrium across the interface.

Our discussion follows in a conceptual way (with different notation) that of Voorhees *et al.* (1992). These statements of equilibrium are imposed as follows.

Mechanical equilibrium is imposed by demanding that the elastic fields satisfy

$$\nabla \cdot \boldsymbol{\sigma} = 0. \tag{10.51}$$

We note that just as with our analytic solution for the Eshelby inclusion, the equilibrium equations *within* the inclusion will have a 'source' term (i.e. an effective body force field) associated with the eigenstrain describing that inclusion. In addition, we require continuity of both displacements, $\mathbf{u}_{in} = \mathbf{u}_{out}$, and tractions, $\mathbf{t}_{in} = \mathbf{t}_{out}$, at the interface between the inclusion and the surrounding matrix. The point of contact between the elastic problem and the diffusion problem is the observation that the interfacial concentration depends upon the instantaneous elastic fields. These interfacial concentrations, in turn, serve as boundary conditions for our treatment of the concentration fields which permits the update of our particle geometries in a way that will be shown below. The concentration at the interface between the inclusion and the matrix may be written as

$$C = C_0 + B \left[\frac{1}{2} [\![\boldsymbol{\sigma} : \boldsymbol{\epsilon}]\!] - \boldsymbol{\sigma} [\![\boldsymbol{\epsilon}]\!] + \gamma_{int} \kappa \right]. \tag{10.52}$$

Note that C_0 is the equilibrium concentration associated with a flat interface, B is a constant term, the two jump terms correspond to the jump in the Eshelby tensor across the interface and the final term corresponds to the Gibbs–Thomson term and depends upon the interfacial curvature κ.

Having solved the elastic boundary value problem and from it obtained the relevant interfacial concentrations, we are then prepared to obtain the concentration fields throughout the solid from which we compute the flux to or away from the various particles. Since we are assuming the situation to be quasistatic, the instantaneous concentration fields must satisfy an equilibrium condition, namely,

$$\nabla^2 c = 0. \tag{10.53}$$

This equation follows from our discussion of chap. 7 in which it was shown that in the simplest model of diffusion the concentration fields must satisfy the diffusion equation, which in the time-independent setting of interest here, reduces to Laplace's equation. Once the concentration fields have been obtained we are finally positioned to update the configuration of the system by recourse to

$$v_n = (\nabla c) \cdot \mathbf{n}, \tag{10.54}$$

where v_n is the velocity of the interface. When the particle shapes have been updated, the procedure is repeated anew.

To elaborate on the significance of the interfacial concentrations as implied by eqn (10.52), we examine the playing out of these ideas in the context of a spherical particle. The question we pose is how to determine the chemical potential at the interface between the spherical precipitate and the matrix and from it, the corresponding interfacial concentration. The essential idea is that this chemical potential can be written as

$$\mu = \frac{\partial G}{\partial N} = \frac{\partial G}{\partial a}\frac{\partial a}{\partial N}. \tag{10.55}$$

That is, we must compute the free energy change associated with a small excursion of the radius a. The relevant energy is built up of both interfacial and elastic terms as well as a chemical energy that we will ignore. In particular, we can write the free energy as

$$G(a) = 4\pi a^2 \gamma_{int} + \frac{8\pi a^3 \epsilon_0^2 \mu}{3}\left(\frac{1+\nu}{1-\nu}\right), \tag{10.56}$$

where the first term is the interfacial energy associated with a spherical inclusion of radius a and the second term is the elastic energy associated with such an inclusion, as already computed in eqn (10.23). As a result of this insight, the chemical potential is immediate by effecting the derivative $\partial G/\partial a$ and exploiting it in the context of eqn (10.55), resulting in

$$\mu = \Omega_{at}\left[\frac{2\gamma_{int}}{a} + 2\epsilon_0^2 \mu\left(\frac{1+\nu}{1-\nu}\right)\right], \tag{10.57}$$

where we have used the fact that for a small excursion of the radius a, the change in the number of particles is $dN = \rho 4\pi a^2 da$ and we introduce the definition of the atomic volume, $\rho = 1/\Omega_{at}$. Note that the first term on the right hand side is the conventional Gibbs–Thomson term relating the chemical potential to interfacial curvature, while the second term represents a purely mechanical effect due to the strain fields that are present in the system. If we now use the relation between chemical potential and concentration that corresponds to an ideal solution, namely, $\mu = kT \ln(c/c_0)$, and expand the resulting exponential, we are left with

$$c = c_0 + \frac{c_0\Omega_{at}}{kT}\left[\frac{2\gamma_{int}}{a} + 2\epsilon_0^2 \mu\left(\frac{1+\nu}{1-\nu}\right)\right], \tag{10.58}$$

a relation precisely of the form of that given in eqn (10.52). We note again that the key point here is that by virtue of solving a problem concerning the instantaneous mechanical equilibrium, we can compute interfacial concentrations which then permit the solution of the diffusion and the calculation of the resulting fluxes which lead to the shape changes of interest.

Equilibrium Shapes Revisited. In section 10.2.4, we described the examination of equilibrium shapes. Though the question of equilibrium shapes is one of terminal privileged states, nevertheless, the ideas introduced here which can in

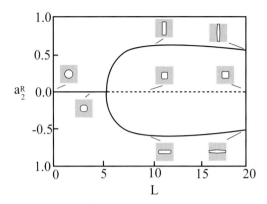

Fig. 10.21. Series of equilibrium shapes for particles (adapted from Thompson *et al.* (1994)). The parameter *L* is a measure of the particle's size, while a_2^R is one of the Fourier coefficients that defines the shape of the particle and is a measure of the deviation away from the circular shape.

principle tell us about temporal evolution, can be used just as well to find those terminal states. An example of the type of results that have been achieved with this type of method in two dimensions is shown in fig. 10.21 (Thompson *et al.* 1994). This result should be contrasted with that of fig. 10.16 which showed the analytic results of Johnson and Cahn in the restricted search over elliptical particle shapes. For the set of calculations shown in fig. 10.21, the elastic anisotropy ratio is chosen to correspond to that of Ni, resulting in elastically soft $\langle 100 \rangle$ directions. The eigenstrain associated with the inclusion is assumed to be purely dilatational (i.e. $\epsilon_{ij}^* = \epsilon \delta_{ij}$). In the limit of vanishingly small particles, the equilibrium shapes are circular reflecting the dominance of the interfacial energy term. With increasing particle size, the solution acquires a fourfold symmetry signaling the increasing relevance of the elastic energy. For yet larger particles, a critical size is reached at which the symmetry is broken resulting in particles with twofold symmetry.

Coarsening of Two-Phase Microstructures. As yet, we have made reference to the formulation involving mass transport only in the context of the problem of equilibrium shapes. The method has deeper significance in the context of problems involving real temporal evolution such as that of coarsening. Coarsening refers to the increase in size of certain particles at the expense of others. This additional elaboration on the problem of the temporal evolution of a single particle is intriguing since it introduces yet another interesting physical effect, namely, the elastic *interactions* between different particles which can affect their morphology as well. An example of this process from the experimental perspective is revealed in fig. 10.22 in which Ni_3Si precipitates in a Ni–Si alloy are shown at various times, revealing the marked tendency of certain particles to increase in size while

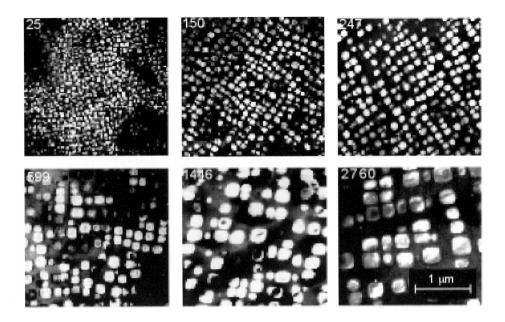

Fig. 10.22. Series of snapshots of the coarsening of Ni_3Si precipitates in a Ni–Si alloy (adapted from Cho and Ardell (1997)).

others disappear. In addition to changes in particle size, the temporal evolution of this system is also attended by changes in shape as well. In particular, at short times the particles are nearly spherical while at later times they acquire a definite cuboidal shape.

A more quantitative spin on results of the kind noted above can be given such as that shown in fig. 10.23 in which the average particle radius as a function of time is revealed. These data suggest that the particle radius grows via a scaling law of the form $\langle r \rangle \propto t^{1/3}$. In addition to well-developed theoretical ideas such as the theory of Lifshitz–Slyozov–Wagner, understanding of this phenomenon can be developed on the basis of numerical simulations involving a formalism like that described above in which mass transport is treated explicitly, and elastic effects are given full scope to drive the relevant rearrangements of matter.

In principle, the coarsening phenomenon was included implicitly in the treatment of temporal evolution given above, though when posed in the numerical terms given above, it is perhaps difficult to attain the sort of understanding that might be wished for blackboard level discussion. Using methods that solve for the same basic physics as has been described earlier in this section, Jou *et al.* (1997) have carried out a series of two-dimensional studies on the temporal evolution of a distribution of precipitates. Mediated by mass transport that is driven in turn

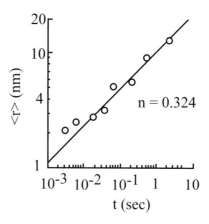

Fig. 10.23. Average particle radius as a function of time for coarsening of γ' precipitates in Ni–13.23 at.% Al alloy aged at 750 °C (adapted from Ardell (1997)).

by the elastic fields, their analysis allowed for a numerical examination of the coarsening process with results like those shown in fig. 10.24. Qualitatively, the coarsening process as revealed by the diffusional dynamics introduced above has all the features expected of the coarsening phenomenon, namely, large particles grow at the expense of the small particles which eventually disappear.

We close by noting that this section on the microstructure in two-phase solids has illustrated a number of key ideas that can be brought to bear on the question of the spatial and temporal evolution of these structures. The foundational statement for our analysis was given in eqn (10.24) which asserts that the instantaneous configurational forces derive from both interfacial and elastic energies. We should note that our treatment of the elastic effects in these systems has ignored stresses that are induced by composition variations, though such terms can be included. Similarly, we have also ignored the way in which externally applied stresses can tip the energy balance in favor of particular orientations, breaking the degeneracies between different crystallographically equivalent versions of a given second-phase particle. Finally, we note that some ideas that invoke the diffuse interface concepts introduced in section 9.5 have not been introduced with regard to two-phase microstructures and we alert the reader that concepts like these, with precisely this application in mind, will be featured in our discussion of bridging scales in section 12.4.4.

10.3 Microstructure in Martensites

In the previous section, we showed one quantitative scheme for treating microstructure and its evolution. Earlier in the chapter we discussed the geometrical

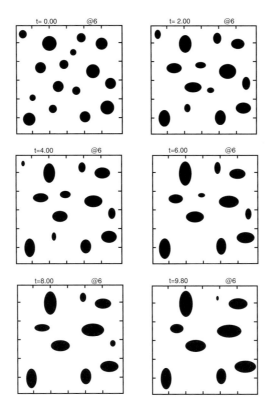

Fig. 10.24. Coarsening of particles in two-dimensional simulation (adapted from Jou *et al.* (1997)).

characterization of grain structures, and noted the want of a quantitative theory of such microstructure that at the same time admits of analytic progress. One area in which considerable progress has been made is in the study of microstructure in systems which undergo reversible martensitic transformations. What is found in this setting is that on the basis of kinematical constraints and fairly simple nonconvex models of the energetics of the competing phases, a quantitative theory of martensitic microstructure may be laid down which makes verifiable predictions. We give special attention to this theory in that it provides an inspirational example of the type of theoretical description that is desirable within the setting of more general microstructures.

10.3.1 The Experimental Situation

The class of transformations known as martensitic are those cases in which upon cooling, the system undergoes a spontaneous change of lattice from a high-symmetry parent phase to some lower-symmetry 'martensite'. These transfor-

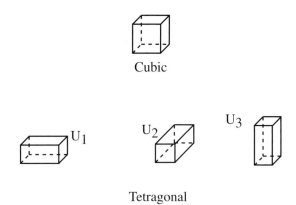

Fig. 10.25. Change in Bravais lattice and well structure associated with the cubic–tetragonal transformation (adapted from Bhattacharya (1991)). $\mathbf{U}_1$, $\mathbf{U}_2$ and $\mathbf{U}_3$ are the matrices that map the original cubic Bravais lattice into the three tetragonal variants.

mations are distinctive since they occur in a diffusionless manner (sometimes also called 'displacive') in which the atoms undergo only local rearrangements while preserving the same distribution of near neighbors. Perhaps the most well-known example of such transformations is that offered by the case of steel. Similar transformations occur in other intermetallic systems such as the cubic–tetragonal transformation in the Ni–Al system and the cubic–orthorhombic transformation in the Al–Cu–Ni system, to name but two.

In addition to the geometric signature of these transformations at the lattice scale, even more fascinating (and beautiful) geometric changes are evinced at the microstructural scale. Martensitic microstructures are those in which a partnership has been struck between different variants of a given structure, and possibly the parent phase as well. For the sake of concreteness, consider fig. 10.25 which depicts a transformation between a parent (austenite) phase that is cubic and a tetragonal (martensite) product phase. One of the first observations that emerges from the schematic is the presence of multiple degenerate realizations of the martensitic phase. This degeneracy arises from the fact that any of the choices of the original cubic axes serves, by symmetry, as an appropriate axis along which to carry out the tetragonal distortion. As seen in the figure, the transformation matrices which map the original Bravais lattice vectors onto the transformed Bravais lattice vector are denoted by $\mathbf{U}_1$, $\mathbf{U}_2$ and $\mathbf{U}_3$. If our only interest were in the equilibrium crystal structures, our work would already be done. However, if we further interest ourselves in the actual geometric arrangements that are realized in a crystal that has undergone such a transformation, it is evident that the possibility of peaceful coexistence of the different variants results in a richness of structures at the microstructural level.

Fig. 10.26. Fine structure at martensite interface (courtesy of Richard James).

An example of the type of martensitic microstructure that can arise when there is more than one degenerate well is given in fig. 10.26. For the moment, we interest ourselves in a particular aspect of these microstructures, namely, the fine structure at twinned interfaces. In particular, this figure demonstrates the geometry adopted in the Cu–Al–Ni system as the various twins arrive at a second internal interface. The alternating light and dark regions in the figure correspond to different variants separated by twin boundaries.

For the moment, our interest is more in what takes place between these interfaces. To illustrate the types of quantitative questions that can be addressed by the theory of fine-phase microstructures, we first elaborate further on the character of these microstructures. As seen above, the martensitic microstructure is characterized by the presence of different variants. However, in the presence of an applied stress, the relative proportion of the different variants can be changed. The experiments are carried out in a biaxial loading apparatus like that shown schematically in fig. 10.27. If we align our Cartesian axes with the two loading directions, then the state of stress to which the sample is subjected can be characterized by the scalars σ_1 and σ_2 which are a measure of the magnitudes of the load in these two loading directions.

A compelling series of snapshots in the type of loading program that can be attained in this biaxial machine are shown in fig. 10.28. What is noted is that, indeed, the stress has the effect of biasing one variant over another. In the first frame, the volume fraction of the dark variant is much higher than the lighter one. However, we see that with the decrease in the load σ_2, the light variant begins to grow at the expense of its darker partner. Consequently, one class of questions that

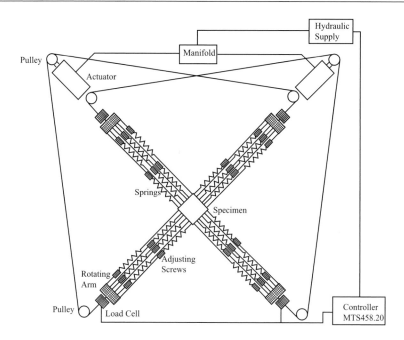

Fig. 10.27. Schematic of biaxial loading apparatus used to examine stress-induced transformations and resulting microstructures in martensites (courtesy of R. James).

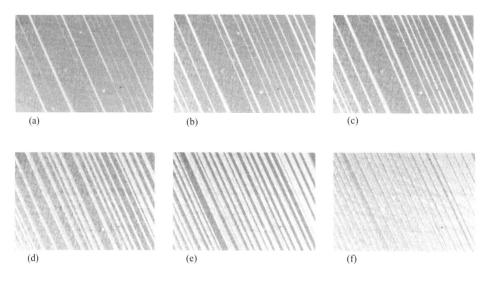

Fig. 10.28. Sequence of snapshots of martensitic microstructure corresponding to different load levels (courtesy of R. James and C. Chu) as obtained using polarized light. Different shades correspond to different variants of the Cu–Al–Ni martensite. Biaxial stress state applied using machine shown in fig. 10.27, with stresses applied along (011) and ($0\bar{1}1$) directions: (a) $\sigma_1 = 7.4$ MPa, $\sigma_2 = 11.1$ MPa, (b) $\sigma_1 = 7.4$ MPa, $\sigma_2 = 9.1$ MPa, (c) $\sigma_1 = 7.4$ MPa, $\sigma_2 = 8.3$ MPa, (d) $\sigma_1 = 7.4$ MPa, $\sigma_2 = 7.4$ MPa, (e) $\sigma_1 = 7.4$ MPa, $\sigma_2 = 6.9$ MPa, (f) $\sigma_1 = 7.4$ MPa, $\sigma_2 = 6.2$ MPa. Field of view in pictures is roughly 2.5 mm.

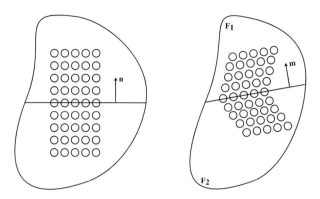

Fig. 10.29. Illustration of deformation satisfying the rank-one connection condition.

can be posed about these microstructures is how do the volume fractions of the different variants depend upon the state of stress?

10.3.2 Geometrical and Energetic Preliminaries

Much of the action associated with the emergence of martensitic microstructures can be traced to kinematic constraints. In particular, on purely geometric grounds, we can identify what types of interfaces are permitted between the austenite and martensite phases, and between the different variants of the martensite itself. It is the aim of the present section to describe these geometric preliminaries in preparation for the theory of fine-phase microstructures which will be given below. Our treatment closely parallels that of Bhattacharya (1998) who in addition to discovering much of what we will say here has codified a wealth of insights on martensitic microstructure in review article form.

We begin by considering the compatibility we will demand across a particular interface. The fundamental physical statement is our insistence on a certain invariant plane such that whether we approach the plane from below or above, when we reach the plane, our prescription for displacing the atoms on that plane will be the same. The reasoning here is illustrated graphically in fig. 10.29. We imagine transformations specified by the deformation gradients $\mathbf{F}_1$ and $\mathbf{F}_2$ and ask what constraints must be placed on them in order that the crystals maintain their continuity across the interface between the two transformed regions. As said above, the physical requirement is the existence of a certain invariant plane such that whether or not we use $\mathbf{F}_1$ or $\mathbf{F}_2$ to compute the transformation on the plane itself we will find the same result. The solution to this demand is that the two deformation gradients be related by

$$\mathbf{F}_1 - \mathbf{F}_2 = \mathbf{a} \otimes \mathbf{n}, \tag{10.59}$$

the so-called rank-one connection of the pertinent deformation tensors. The vector **n** is the normal to the putative invariant plane in the reference configuration and **a** is a vector to be determined. Indeed, another way to see what we have wrought with this equation is to rewrite eqn (10.59) as

$$\mathbf{F}_1 = \mathbf{F}_2 + \mathbf{a} \otimes \mathbf{n}, \tag{10.60}$$

and to note that the action of $\mathbf{F}_1$ is the same as $\mathbf{F}_2$ except that it picks off the components of vectors normal to the invariant plane. That is, if we take our origin of coordinates somewhere on the invariant plane and ask how $\mathbf{F}_1$ acts on the vector $\mathbf{r} = r_\parallel \mathbf{e}_\parallel + r_\perp \mathbf{n}$, we see that it is only if **r** has a nonzero component perpendicular to the invariant plane that the deformation gradients $\mathbf{F}_1$ and $\mathbf{F}_2$ have a different action on that vector. This guarantees that whether we approach the interface from above or below, the deformation gradients $\mathbf{F}_1$ and $\mathbf{F}_2$ will agree as to the displacements of atoms (or more generally, of points) on the invariant plane. These ideas are further elaborated in the problem section at the end of the chapter.

Part of the beauty of the theory of martensite is the far-reaching implications of the simple geometric constraint represented by eqn (10.59). One of the key results that emerges from this constraint is the nature of the allowed twinning planes that can coexist between different variants of the same underlying martensite. To see this, we must first examine in a bit more detail the symmetries present in the energy landscape. For the time being, we interest ourselves in the properties of affine deformations characterized by the deformation gradient **F**. On the basis of the known transformation strains and the invariances implied by crystal symmetry and overall rotational symmetry, much can already be said about the nature of the energy landscape, $\phi(\mathbf{F}, T)$, connecting the austenite and martensite phases. In schematic terms, we know that the energy as a function of the deformation gradient **F** has the basic character depicted in fig. 10.30. Note that this energy is something that can in principle be evaluated directly on the basis of microscopic calculation. Indeed, we have already shown an example of such a calculation in fig. 4.1. The basic point of a free energy like that mentioned above is that at high temperatures, the energy landscape is characterized by a global minimum for $\mathbf{F} = \mathbf{I}$ (i.e. the austenite is the most stable structure). On the other hand, below the critical temperature, there are a number of degenerate wells corresponding to the different variants of martensite, $\mathbf{F} = \mathbf{U}_i$. Of course, in the figure, the set of possible deformations has only been indicated schematically since we have represented the deformation gradient as a scalar, and in reality, such deformations are characterized by the nine parameters needed to specify the full deformation gradient. From the

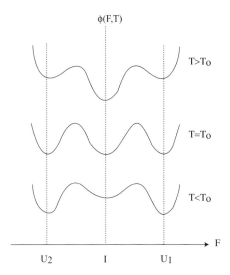

Fig. 10.30. Schematic representation of the relative positions of competing minima in the energy landscape as a function of temperature (courtesy of K. Bhattacharya). $\phi(\mathbf{F}, T)$ is the free energy function, which depends upon both the deformation gradient $\mathbf{F}$ and the temperature T, $\mathbf{U}_1$ and $\mathbf{U}_2$ are the deformation gradients associated with the two martensite variants considered in this figure and the identity $\mathbf{I}$ is associated with the austenite well.

present standpoint, the key feature of the energy function ϕ that gives rise to the existence of microstructures is the presence on its energy landscape of multiple wells. As was said before, it is the degeneracy of different martensitic variants that makes it possible for one region of the microstructure to have selected one variant and some other region to have selected another.

An alternative way to characterize the various wells is through the device of the James diagram in which each well is characterized by a circle like those shown in fig. 10.31. The point of such diagrams is to represent a crucial feature of the symmetry of the stored energy function $\phi(\mathbf{F}, T)$. Material frame indifference requires that $\phi(\mathbf{F}, T) = \phi(\mathbf{RF}, T)$, the meaning of which is that if we first apply the deformation gradient and then rotate the resulting crystal, the energy is the same as if we had only applied the deformation gradient itself. The significance of this insight is that the additional freedom associated with the ability to apply rotations broadens our scope to find compatible interfaces between the different variants. As shown in the figure, each well is identified by a different circle, with the circle meant to indicate the entire group of allowable rotations $\mathbf{R}$, namely, $SO(3)$. Though $\mathbf{F}_1$ and $\mathbf{F}_2$ themselves may not satisfy the rank-one connection, material frame indifference illustrates that it is possible to find a rotation $\mathbf{R}$ such that $\mathbf{RF}_1$ and $\mathbf{F}_2$ do satisfy the rank-one condition, and it is this additional freedom that is conveyed by the James diagram.

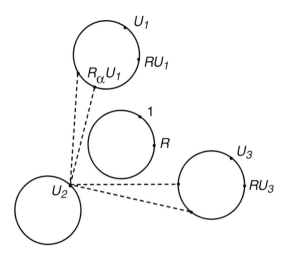

Fig. 10.31. Schematic illustration of deformation gradients in different wells for the cubic to tetragonal transformation that satisfy the rank-one connection condition (adapted from Ball and James (1992)).

As was shown in fig.10.25, for the case of the cubic–tetragonal transformation there are three distinct wells to be considered, each corresponding to the tetragonal distortion occurring along a different Cartesian direction. Recall that our discussion of compatibility culminated in eqn (10.59). Part of the virtue of the James diagrams is that they provide a geometrical representation of part of the information content of eqn (10.59). What we note is that compatible deformations on different wells can be signified by connecting them with a line as shown in fig. 10.31. In particular, what fig. 10.31 tells us is that while there is no rank-one connection between $\mathbf{U}_1$ and $\mathbf{U}_2$, for example, there is a rank-one connection of the form $\mathbf{U}_2 - \mathbf{R}_\alpha \mathbf{U}_1 = \mathbf{a} \otimes \mathbf{n}$, where we have introduced the notation $\mathbf{R}_\alpha$ to signify the *particular* choice of $\mathbf{R}$ on the $\mathbf{U}_1$ well that results in a rank-one connection to well $\mathbf{U}_2$. The dotted lines in the figure connect two rank-one connected deformation gradients on different wells. To see all of these ideas in action, we now turn to a consideration of some of the many different types of compatible interfaces that can be found in martensitic systems.

10.3.3 Twinning and Compatibility

The questions we will pose in this section concern the ability of different energetically favorable structures to coexist geometrically. We have already seen that the confluence of two key ideas gives rise to the possibility for diverse microstructures. In particular, the identification of a series of degenerate energy wells makes it possible that, for example, a cubic crystal will transform into several different (but

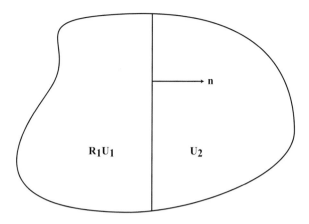

Fig. 10.32. Schematic of interface between structures characterized by deformation gradients $\mathbf{F}_1 = \mathbf{R}_1\mathbf{U}_1$ and $\mathbf{F}_2 = \mathbf{U}_2$.

energetically equivalent) variants. On the other hand, we have also seen that it is not evident *a priori* that a compatible interface can be formed between both the different variants and the host austenite. To explore this question further, we require compatibility equations which specify under what conditions a compatible interface may be formed between two structures, each characterized by its own well.

In fig. 10.32, we indicate the type of interface we have in mind. On one side of the interface, the structure is characterized by a deformation gradient $\mathbf{R}_1\mathbf{U}_1$ while on the opposite side of the interface the deformation is characterized by a deformation gradient $\mathbf{U}_2$. Note that without loss of generality, we have rewritten the compatibility condition in terms of the stretch tensors $\mathbf{U}_1$ and $\mathbf{U}_2$ and a single rotation $\mathbf{R}_1$. For a given set of structures characterized by the matrices $\mathbf{U}_1$ and $\mathbf{U}_2$ our question may be posed as: can we find a rotation $\mathbf{R}_1$ as well as vectors $\mathbf{a}$ and $\mathbf{n}$ such that the condition

$$\mathbf{R}_1\mathbf{U}_1 - \mathbf{U}_2 = \mathbf{a} \otimes \mathbf{n}, \tag{10.61}$$

is satisfied? Recall that in formal terms this is precisely the compatibility condition developed earlier and identified as the rank-one connection. We answer this question in the form of a theorem which asserts that if the eigenvalues of the matrix $\mathbf{U}_2^{-T}\mathbf{U}_1^T\mathbf{U}_1\mathbf{U}_2^{-1}$ may be ordered as $\lambda_1 \geq 1$, $\lambda_2 = 1$ and $\lambda_3 \leq 1$, then there exists a rank-one connection. For details, see Proposition 4 of Ball and James (1987). From a physical perspective, we note that the significance of this result is that the compatibility condition introduced above can only be satisfied for certain special choices of lattice parameters, a theme that will recur repeatedly in our discussion of martensitic microstructures.

If the eigenvalues of the matrix $\mathbf{C} = \mathbf{U}_2^{-T}\mathbf{U}_1^T\mathbf{U}_1\mathbf{U}_2^{-1}$ satisfy the conditions described above, then we may go further and compute the vectors $\mathbf{a}$ and $\mathbf{n}$ as well as the rotation matrix. In particular, we assert that the vector $\mathbf{a}$ is given by

$$\mathbf{a} = c\left[\sqrt{\frac{\lambda_3(1-\lambda_1)}{\lambda_3-\lambda_1}}\mathbf{e}_1 \pm \sqrt{\frac{\lambda_1(1-\lambda_3)}{\lambda_3-\lambda_1}}\mathbf{e}_3\right], \tag{10.62}$$

while the vector $\mathbf{n}$ is given by

$$\mathbf{n} = \frac{\sqrt{\lambda_3}-\sqrt{\lambda_1}}{c\sqrt{\lambda_3-\lambda_1}}(-\sqrt{1-\lambda_1}\mathbf{U}_2^T\mathbf{e}_1 \pm \sqrt{\lambda_3-1}\mathbf{U}_2^T\mathbf{e}_3). \tag{10.63}$$

Note that there are two solutions corresponding to the $\pm$ sign in our equations. In addition, the requirement that $\mathbf{n}$ be normalized is satisfied by exploiting the as-yet undetermined constant c. To determine the rotation matrix, we substitute the values for $\mathbf{a}$ and $\mathbf{n}$ back into eqn (10.61) and solve for $\mathbf{R}_1$.

Our intention in the remainder of this section is to build up a picture of some of the various interfaces that are present in martensitic systems. Our approach will be to consider microstructural elements of increasing complexity, beginning first with the case of the simple austenite–martensite interface and culminating in the investigation of martensitic wedges within the host austenite. In all of these cases, the primary theoretical engine in our analysis will be the compatibility conditions and their outcome as typified by eqns (10.62) and (10.63).

Austenite–Martensite Interfaces. As a first exercise in the machinery set forth above, we pose the question of whether or not a simple interface can exist between a cubic crystal and its tetragonal offspring. Without loss of generality, this situation can be represented via the condition

$$\mathbf{R}\mathbf{U} - \mathbf{I} = \mathbf{a} \otimes \mathbf{n}, \tag{10.64}$$

where the matrix $\mathbf{U}$ is of the form

$$\mathbf{U} = \begin{pmatrix} \alpha & 0 & 0 \\ 0 & \beta & 0 \\ 0 & 0 & \beta \end{pmatrix}. \tag{10.65}$$

For the special choice of stretch matrix $\mathbf{U}$ made here, all we are saying is that to produce the tetragonal crystal, all one need do is stretch by a factor α along the x-direction and by a factor β along the y- and z-directions. That is, the crystal axes of the cubic and tetragonal variants are parallel.

Note that in light of the theorem set forth above, all that remains is to ask how the eigenvalues of the matrix $\mathbf{C} = \mathbf{U}^2$ are ordered. However, since the matrix is already diagonal, we know that its eigenvalues are $\lambda_1 = \alpha^2$, $\lambda_2 = \lambda_3 = \beta^2$.

For concreteness, we assume that $\alpha > 1 > \beta$ as is the case in systems such as In–Tl ($\alpha = 1.0221$ and $\beta = 0.9889$) and Ni–Al ($\alpha = 1.1302$ and $\beta = 0.9392$). As a result, we note that there is *not* an ordering of the eigenvalues of the form $\lambda_1 \geq \lambda_2 = 1 \geq \lambda_3$, which immediately lends itself to the conclusion that in the case of a cubic–tetragonal transformation, there is no simple interface between the host austenite and a single variant of the tetragonal martensite except in the unlikely event that the parameter β takes the value 1.0.

Martensite–Martensite Interfaces. At the next level of complexity, we consider the nature of the interfaces that characterize the coexistence of different martensitic variants. In this case, the compatibility condition takes the more general form given earlier in eqn (10.61) where for explicitness we choose

$$\mathbf{U}_1 = \begin{pmatrix} \alpha & 0 & 0 \\ 0 & \beta & 0 \\ 0 & 0 & \beta \end{pmatrix} \tag{10.66}$$

and

$$\mathbf{U}_2 = \begin{pmatrix} \beta & 0 & 0 \\ 0 & \alpha & 0 \\ 0 & 0 & \beta \end{pmatrix}. \tag{10.67}$$

Once again, our exercise is reduced to that of determining the eigenvalues of the matrix $\mathbf{C}$. In the case in which the different variants in question are described by $\mathbf{U}_1$ and $\mathbf{U}_2$ as given above, the matrix $\mathbf{C}$ is of the form

$$\mathbf{C} = \begin{pmatrix} \alpha^2/\beta^2 & 0 & 0 \\ 0 & \beta^2/\alpha^2 & 0 \\ 0 & 0 & 1 \end{pmatrix}. \tag{10.68}$$

In this case, we find that the eigenvalues are precisely arranged so as to allow for the existence of a martensitic interface between the two variants in question.

The question of the martensitic interface can be further explored by recourse to the equations that determine the vectors $\mathbf{a}$ and $\mathbf{n}$ themselves. In particular, by exploiting the results spelled out in eqns (10.62) and (10.63), we see that in the cubic–tetragonal case under consideration here, the vectors $\mathbf{a}$ and $\mathbf{n}$ are given by

$$\mathbf{a} = \sqrt{2} \frac{\beta^2 - \alpha^2}{\beta^2 + \alpha^2} (\beta, \pm\alpha, 0) \tag{10.69}$$

and

$$\mathbf{n} = \frac{1}{\sqrt{2}} (-1, \pm 1, 0). \tag{10.70}$$

What we have learned from this analysis is that the invariant plane allowing for

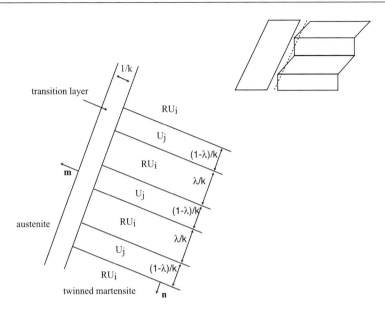

Fig. 10.33. Schematic of interface between twinned martensite and the parent austenite (adapted from James and Hane (2000) and Chu (1993)).

the development of an interface between two different variants of the tetragonal martensite is of the $(1, 1, 0)_{cubic}$-type.

Austenite in Contact with Twinned Martensite. An even more interesting example is that of an interface between austenite and a twinned version of the martensite. In this case, the strategy is to satisfy the conditions of compatibility across the austenite–martensite interface in an average sense. The geometry of interest is indicated schematically in fig. 10.33. Recall that in the case of the cubic–tetragonal system, we could find no interface between the host austenite and a single variant of martensite. In the present context, the idea is to seek an average deformation gradient $\mathbf{F} = \lambda\mathbf{F}_1 + (1 - \lambda)\mathbf{F}_2$ that satisfies the compatibility condition. Note that here λ is not an eigenvalue of the matrix $\mathbf{C}$, but is rather the volume fraction of $\mathbf{F}_1$.

For simplicity, we take the deformation gradient in the austenite to be the identity $\mathbf{I}$, while the two variants of martensite are characterized by $\mathbf{RU}_1$ and $\mathbf{U}_2$. The compatibility condition within the twinned martensite region is nothing more than was written earlier and takes the specific form

$$\mathbf{RU}_1 - \mathbf{U}_2 = \mathbf{a} \otimes \mathbf{n}. \tag{10.71}$$

In addition to this constraint, we must also satisfy compatibility across the austenite–martensite interface which in light of the deformation gradients specified above may be written

$$\mathbf{R}'(\lambda\mathbf{R}\mathbf{U}_2 + (1-\lambda)\mathbf{U}_1) - \mathbf{I} = \mathbf{b} \otimes \mathbf{m}. \tag{10.72}$$

The latter of these equations is sometimes referred to as the habit-plane or austenite–martensite equation. This equation imposes the rank-one connection across the austenite–martensite interface, with the rotation $\mathbf{R}'$ governing the overall orientation of the twinned martensite with respect to the austenite across the austenite–twinned-martensite interface. We also note that this equation embodies the classic results of the crystallographic theory of martensite as expounded in Wayman (1964), though now interpreted energetically in terms of the various energy wells available to the system. As it stands, we must determine the matrices $\mathbf{R}$ and $\mathbf{R}'$, as well as the vectors $\mathbf{a}$, $\mathbf{b}$, $\mathbf{n}$ and $\mathbf{m}$ and the mixing parameter λ, given our knowledge of the tensors $\mathbf{U}_1$ and $\mathbf{U}_2$. To do so, we resort to a divide and conquer strategy in which first we solve eqn (10.71) using exactly the same technique described earlier and culminating in the solutions of eqns (10.62) and (10.63). Once these solutions are in hand, we then turn to a solution of eqn (10.72). The key point concerning this solution for our present purposes is that it depends critically on particular choices of the relevant lattice parameters. In particular, as shown in Ball and James (1987) and explained in James and Hane (2000), necessary and sufficient conditions for the existence of a solution are

$$\delta \equiv \mathbf{a} \cdot \mathbf{U}_2(\mathbf{U}_2^2 - \mathbf{I})^{-1}\mathbf{n} \le -2 \tag{10.73}$$

and

$$\eta \equiv \mathrm{tr}(\mathbf{U}_2^2) - \det(\mathbf{U}_2^2) - 2 + \frac{|\mathbf{a}|^2}{2\delta} \ge 0. \tag{10.74}$$

The resulting volume fraction λ of variant 1 is given by

$$\lambda = \frac{1}{2}\left(1 - \sqrt{1 + \frac{2}{\delta}}\right). \tag{10.75}$$

The dependence of the volume fraction λ on the parameter δ shows that it too is material specific and depends upon the particular values of the various lattice parameters.

Wedge Structures. As our final exercise in examining the conjunction of energetic and compatibility arguments, we consider the case of wedge-like structures in which a series of twins are arranged in such a way that a wedge of martensite can be embedded within austenite. The situation we have in mind is depicted schematically in fig. 10.34. We also note again that our discussion leans heavily on that of Bhattacharya (1991) who considered the issue of wedges in martensite in detail.

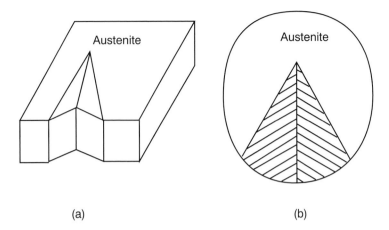

(a)   (b)

Fig. 10.34. Schematic of a martensite wedge embedded within an austenite structure (adapted from Bhattacharya (1991)).

From the standpoint of the kinematic arguments introduced earlier, the additional complexity that attends situations such as the wedge microstructure considered here is the fact that we no longer have only a single interface on which to satisfy the conditions of compatibility. Our calculations on the austenite–twinned-martensite interface served as a warm up exercise for the present case. Each and every interface of the type in fig. 10.35 must itself satisfy conditions such as that of eqn (10.59). As a result, the imposition of compatibility across all of the interfaces takes the form of the collection of equations

$$\mathbf{A} - \mathbf{B} = \mathbf{a}_1 \otimes \mathbf{n}_1, \qquad (10.76)$$

$$\mathbf{C} - \mathbf{D} = \mathbf{a}_2 \otimes \mathbf{n}_2, \qquad (10.77)$$

$$\lambda_1 \mathbf{A} + (1 - \lambda_1)\mathbf{B} - \lambda_2 \mathbf{C} - (1 - \lambda_2)\mathbf{D} = \mathbf{a}_3 \otimes \mathbf{m}, \qquad (10.78)$$

$$\lambda_1 \mathbf{A} + (1 - \lambda_1)\mathbf{B} - \mathbf{I} = \mathbf{a}_4 \otimes \mathbf{m}_1, \qquad (10.79)$$

$$\lambda_2 \mathbf{C} + (1 - \lambda_2)\mathbf{D} - \mathbf{I} = \mathbf{a}_5 \otimes \mathbf{m}_2. \qquad (10.80)$$

Like in the simpler case of the interface between an austenite region and twinned martensite, this set of equations can be solved by recourse to a divide and conquer policy. Our main interest in showing these equations is to show the reader what is involved in setting up such a problem and to inspire him or her to read Bhattacharya's beautiful 1991 paper.

10.3.4 Fine-Phase Microstructures and Attainment

We have now structured the problem of martensitic microstructre in accordance with key geometric and energetic notions. What we have seen is that once the

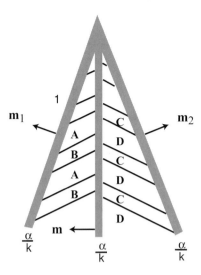

Fig. 10.35. Detailed schematic of the variants within a martensite wedge which is itself embedded within an austenite structure (adapted from Bhattacharya (1991)).

energy wells have been identified, largely on the basis of geometric arguments it is possible to classify the various types of interfaces that can be realized within a given microstructure. We now pose the deeper question of the character of the overall microstructure itself. For example, what are the volume fractions of the different variants, and what features of the problem determine the typical length scale associated with these variants? The answer to such questions demands a discussion of the subtlety that can attend the notion of minimization of the type of nonconvex energies introduced above.

For the present purpose, our first observation is that under certain circumstances, it is possible that the minimizer will be represented as the limit of a sequence of increasingly fine microstructures. Alternatively, this can be stated by noting that in some cases there are no smooth deformations that minimize the energy. Instead, we extend the space of possible minimizers by including those deformations which are piecewise continuous and are allowed to have a vanishingly small thickness.

A two-dimensional model problem described in depth by Bhattacharya (1998) which illustrates this idea is built around the idea of trying to find a scalar field $y(x_1, x_2)$ that minimizes the free energy functional

$$E[y(x_1, x_2)] = \int_\Omega \phi(\nabla y(x_1, x_2))dx_1dx_2, \qquad (10.81)$$

subject to the boundary condition that $y = 0$ on the boundary $\partial\Omega$ of the body Ω. To make the problem interesting, the free energy itself is chosen to have two wells, and is of the form

$$\phi(\nabla y(x_1, x_2)) = \left[\left(\frac{\partial y}{\partial x_1} \right)^2 - 1 \right]^2 + \left(\frac{\partial y}{\partial x_2} \right)^2. \tag{10.82}$$

The free energy density given above has two wells, corresponding to $\partial y / \partial x_1 = \pm 1$ and $\partial y / \partial x_2 = 0$. In addition to the specification of the free energy itself, our problem is supplemented with the boundary condition that the deformation vanish on the boundary of the region. We see that this free energy functional and the associated boundary conditions pose contradictory demands on the deformation mapping $y(x_1, x_2)$, resulting in a lack of 'attainment', a word that signifies that the class of continuous deformations is unable to provide a minimizer, even though the free energy is bounded from below. To see this, we note that on the boundaries we would like to choose $y = 0$. On the other hand, within the 'volume' Ω, the choice $y = 0$ is energetically unfavorable since it corresponds to a choice of ∇y that definitely does not occupy either of the wells. Hence, we are unable to attain the minimum with any continuous deformation. On the other hand, using the construction to be described below, we can get as close as we like to the state of zero energy by considering ever finer mixtures of the two variants.

It is evident at the outset that there is no continuous deformation that minimizes this energy function. On the other hand, we can also note that by adopting the approach of setting up a series of piecewise continuous deformations with a corresponding energetic boundary layer near the boundaries, we can construct structures that approach the lowest energy to within arbitrary accuracy. This construction is revealed pictorially in fig. 10.36, and illustrates the enlargement of our space of solutions to include so-called minimizing sequences. The idea is to build up the overall microstructure by considering strips with $\nabla y = (1, 0)$ and $\nabla y = (0, 1)$. In each of these strips, the contribution to the overall energy is zero. On the other hand, these strips fail to satisfy the boundary condition ($y = 0$) at $x_2 = 0$ and $x_2 = L$, where L is the dimension of the region of interest. As seen in fig. 10.36, the remedy to this problem is to cut the strips off at the ends so they vanish at $x_2 = 0$ and $x_2 = L$. Of course, in so doing, we must pay an energy penalty for this boundary layer since in these regions ∇y is not associated with either of the energy wells. In fact, as shown in Bhattacharya (1998), the energy associated with this boundary layer is of the form $E_{layer} = L^2 / 2n$, where n is an integer that characterizes how many strips are present in our microstructure. This demonstrates that the boundary layer contribution to the energy goes to zero in the limit that the fineness of our microstructure is increased without bound.

As a result of the arguments set forth in this part of the chapter, we are now prepared for the job of evaluating the structure and energetics of martensitic microstructures. The fundamental idea is to construct arrangements of the different

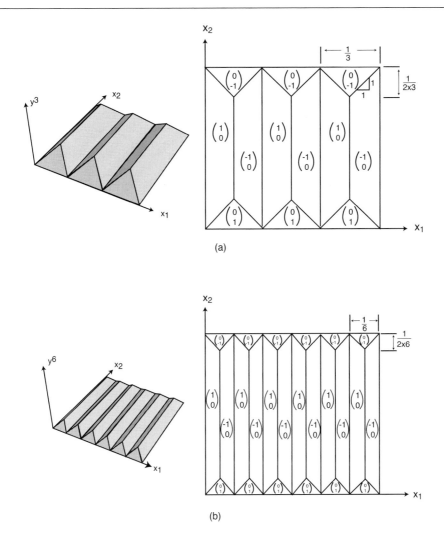

Fig. 10.36. Illustration of two members of the minimizing sequence for two-dimensional problem posed in the text (adapted from Bhattacharya (1998)).

variants that simultaneously satisfy the geometric constraints associated with the rank-one connection and the energetic constraints introduced by presence of the pertinent free energy function. The discussion given above centered on the use of analytic reasoning to deduce the general characteristics of martensitic microstructures. Such analytic investigations have been complemented successfully using numerical solutions in which minimizers of the relevant free energy functional are sought after by recourse to a finite element description of the deformation fields (see Luskin (1996) for example). An example of the type of results that emerge upon implementing the ideas of the previous few sections in a numerical sense is

(a)   (b)

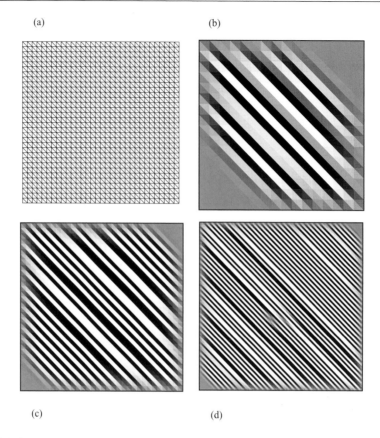

(c)   (d)

Fig. 10.37. Outcome of numerical calculation of the energy minimizers associated with martensite (adapted from Luskin (1996)): (a) an example of a finite element mesh used to model the martensitic microstructure, (b)–(d) the results of microstructure computations using different levels of mesh refinement.

shown in fig. 10.37. These calculations were made for a two-dimensional model free energy which is intended to mimic the energetics of the three-dimensional orthorhombic to monoclinic transformation. The calculations are made for a square geometry in which the boundary deformation gradient is prescribed and corresponds to neither of the competing wells. Like in the model problem resulting in fig. 10.36, this choice of boundary condition guarantees the type of nonattainment described above.

Though we will not elaborate on this point in any great detail, it is important to note that on physical grounds there is a lower cutoff to the type of runaway refinement of the microstructure described above. Recall that our assessment of the free energy has until now been predicated entirely upon searching for a means to minimize the elastic contributions to this energy subject to particular boundary conditions. On the other hand, it is clear that the increased level of refinement is accompanied by energy penalties associated with the various interfaces, terms which we have so far ignored. As said above, the physical consequence of these terms is to regularize the problem and to impose a length scale resulting from the balance of the surface energy and the elastic energy associated with the boundary layers in which the deformations are not associated with either the austenite or martensite wells.

10.3.5 The Austenite–Martensite Free Energy Reconsidered

In fig. 10.30, we gave a schematic description of the free energy as a function of deformation gradient for various temperatures above, below and at the transformation temperature. This discussion was built upon an intuition of what the basic character of the free energy must be in order to give rise to the phenomenology of these systems. We now reconsider the question of the alloy free energy from the perspective of chap. 6, with the aim of trying to explicate such free energies on the basis of microscopic calculation rather than on the phenomenological grounds introduced earlier in this chapter.

For the present purposes, we write the free energy $\phi(\mathbf{F}, T)$ as a sum of two terms:

$$\phi(\mathbf{F}, T) = F_{el}(\mathbf{F}, T) + F_{vib}(\mathbf{F}, T), \tag{10.83}$$

where the first term is the electronic contribution to the free energy of structures characterized by the deformation gradient $\mathbf{F}$, while the second term is the vibrational contribution to the free energy, and it too depends upon the state of deformation. This equation is essentially an imitation of that given already as eqn (6.2). For an assessment of the first term, we turn to a discussion of the so-called Cauchy–Born rule introduced in section 5.4.2, while for the second, we will revisit experiments on the free energy of structural transformations.

For the purposes of computing $F_{el}(\mathbf{F}, T)$, we can exploit the Cauchy–Born rule, as was done in the calculations leading to fig. 4.1. Such calculations are conceptually simple in comparison with some of the more complex defect calculations introduced elsewhere throughout the book. All that one need do is take the original Bravais lattice (i.e. the reference state), deform the relevant basis vectors according to the prescription given above and compute the resulting energy using one's favorite total energy scheme. We will not belabor the point other than by noting that the basic character of the results will resemble those of fig. 4.1 in the sense that there can be more than one local minimum.

To compute the vibrational contribution to the free energy, we remind the reader of the results of chap. 6 in which we argued that entropy contributions can arise from vibrations (both harmonic and anharmonic) as well as electronic excitations. For the moment, we will advance the hypothesis, supported in some cases by experiment (see fig. 10.38), that in the context of martensitic transformation the contribution of harmonic lattice vibrations to the entropy will tell most of the story. Further, our interest is in sketching ideas, in suggesting possible origins for effects rather than being exhaustive, and hence perhaps an overly facile representation of a rather complex problem can be forgiven. In the context of the Cu–Zn–Al system, a series of measurements were undertaken (Mañosa *et al.* 1993) with the aim of sorting out the relative importance of the various contributions mentioned above to the overall entropy of transformation. In a word, their method was to use calorimetric techniques to obtain the total entropy. Using magnetic susceptibility measurements it is possible to independently pick off the electronic contribution to the total entropy which is known already from the calorimetric measurements. Finally, by taking the difference between this electronic contribution and the total entropy it is possible to compute the vibrational contribution to the entropy. Fig. 10.38 shows both the electronic (S_e) and vibrational (S_v) contributions to the entropy in a series of different Cu–Zn–Al alloys. As is seen in fig. 10.38, the magnitude of the vibrational entropy (and this is itself almost all harmonic) is quite large in comparison with that due to electronic degrees of freedom as is evidenced by comparing the scales of the two contributions. The conclusion to be drawn from these experiments is that in this case we can go a long way towards deducing the type of free energy function shown in fig. 10.30 by combining what we know about the zero-temperature part of $F_{el}(\mathbf{F}, T)$ (i.e. the zero-temperature structural energies) and the free energy of harmonic vibrations.

10.4 Microstructural Evolution in Polycrystals

Our third and final microstructural example centers on the question of polycrystals and their temporal evolution. Here we imagine the perfect crystal disturbed not by the presence of other phases, but rather by a series of grain boundaries separating crystalline regions each having a different orientation. We have already hinted at the implications of such a polycrystalline microstructure for the mechanical properties of a material. The aim of the present section is to examine how such microstructures evolve in time. We will begin with an examination of the observations that have been made on such microstructures, with an emphasis on the quantitative features of their temporal evolution. This will be followed by a reexamination of the driving forces that induce grain boundary motion as well as the mechanisms whereby such motion is realized.

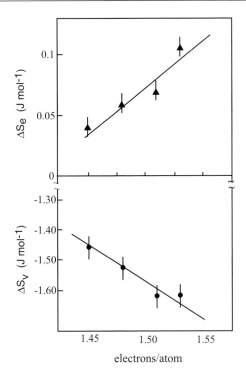

Fig. 10.38. Entropy difference between austenite and martensite in Cu–Zn–Al alloys (adapted from Mañosa *et al.* (1993)). Different values of the electron per atom ratio correspond to different concentrations of the three alloy constituents.

10.4.1 Phenomenology of Grain Growth

One of the palpable effects of annealing a polycrystal is a systematic change in the mean grain size, a phenomenon known as grain growth. A quantitative view of this process is evoked by plotting the mean grain size as a function of time as shown in figs. 10.39 and 10.40. Though we will not enter into the debate (for further details see Atkinson (1988)), the question of how to characterize three-dimensional microstructures on the basis of two-dimensional cuts is a subtle one.

A more satisfying kinematic framework for characterizing polycrystalline microstructures, and grain growth in particular, is offered by considering the entire grain size distribution rather than just the mean grain size itself. In particular, consider the function $F(R, t)$ which is the probability density for finding grains of size R within the microstructure. In particular, $F(R, t)$ gives the number of grains per unit volume with size between R and $R + dR$ at time t. Just as the mean grain size is amenable to experimental evaluation, so too is the grain size distribution function as shown in fig. 10.41. One of the claims associated with these distribution functions is that they can nearly be thought of as lognormal

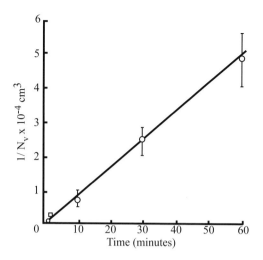

Fig. 10.39. Evolution of grain size as a function of time during the annealing process (adapted from Rhines and Craig (1974)). The vertical axis reveals the average grain volume as obtained by taking the reciprocal of the number of grains per cubic centimeter.

distributions. Perhaps more important yet is the assertion of a scaling property of the grain size distribution function. In particular, if the grain size is plotted, not in terms of the absolute size of the various grains, but rather in terms of the normalized grain size $R/\langle R \rangle$ (where $\langle R \rangle$ is the mean grain size, labeled by $\bar{R}$ in the figure), then the distribution is statistically invariant. Exactly this point is made in fig. 10.41 which shows the grain size distribution using scaled variables for various times in the temporal history of the microstructure.

The process of grain growth bears some resemblance to the coarsening process discussed earlier in the context of precipitate microstructures. From the point of view of the driving force for such structural change, as we have repeatedly emphasized, such microstructures are metastable. A state of lower free energy may be reached by eliminating grain boundary area. By virtue of the presence of a variety of grain boundaries, it is plausible to imagine a state of lower free energy in which there are fewer such boundaries, and hence a lower overall boundary energy. One of the challenges posed by such microstructures is the way in which such metastable systems evolve in time.

10.4.2 Modeling Grain Growth

As noted above, ostensibly, part of the empirical background concerning grain growth that demands explanation is the power law growth in time of the grain size. One of the justly famous contributions in that regard is the work of Burke

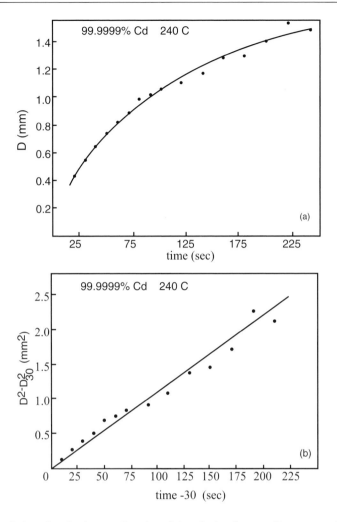

Fig. 10.40. Evolution of grain size as a function of time during the annealing process (adapted from Simpson *et al.* (1971)). The parameter D is a measure of the grain size. In the lower frame the grain growth data are plotted with respect to $t - t_0$, where t_0 is the time from which the subsequent grain growth is measured.

and Turnbull (1952) in which a parabolic growth law in which the mean grain size scales as $\langle R \rangle \propto t^{\frac{1}{2}}$ was deduced on the grounds of arguments much like those leading to our eqn (10.48). The basic hypotheses behind the Burke–Turnbull argument are that the properties of the distribution itself can be understood by concentrating on the properties of one grain only, and that this one grain evolves under the action of curvature-induced forces. Recalling the variational statement of eqn (10.45), but now treating only the interfacial contribution to $\dot{G}$, the variational

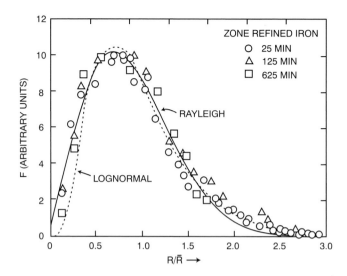

Fig. 10.41. Scaled grain size distribution function for various times in annealing history of sample of Fe (adapted from Pande and Marsh (1992)). In addition to the experimental results, the figure also shows the lognormal and Rayleigh distributions thought to be relevant to describing the distribution $F(R, t)$.

principle can be written as

$$\Pi(\dot{R}; R) = 4\pi R^2 \frac{\dot{R}^2}{2M} + \dot{G}_{int}(\dot{R}; R). \tag{10.84}$$

On the assumption of an isotropic interfacial energy γ_{int}, $\dot{G}$ may be written (see eqn (10.46)) as $\dot{G}_{int}(R) = 8\pi R\dot{R}\gamma_{int}$. Following the usual procedure (i.e. $\partial\Pi/\partial\dot{R} = 0$) results in an equation for the evolution of R of the form

$$\frac{dR}{dt} = -\frac{2M\gamma_{int}}{R}, \tag{10.85}$$

which when integrated yields the grain size scaling relation

$$\frac{1}{2}R^2 - \frac{1}{2}R_0^2 = -2M\gamma_{int}t. \tag{10.86}$$

Highly idealized arguments like that given above have been complemented by more realistic numerical simulations in which the various interfaces move in conjunction. Since our aim is the illustration of ideas rather than to exhaustively detail the ultimate solution to particular problems, we will take up three distinct schemes for considering the problem of microstructural evolution in the grain growth setting. The three paradigmatic examples to be considered here are shown in fig. 10.42. Our intention is to consider in turn the Potts model approach to grain growth, phase field models for grain growth and their sharp interface

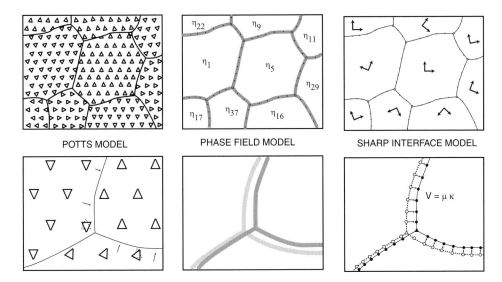

Fig. 10.42. Schematic of various paradigms for considering the evolution of polycrystal (adapted from Frost and Thompson (1996)). The upper panel shows the representation of the polycrystal from the perspective of each of the different models and the lower panel illustrates the way in which the interfaces between different grains migrate.

counterpart. As will be seen in the coming sections, each of these schemes has a different underlying philosophy, though they all attempt to capture the energy penalty associated with the presence of interfacial areas.

Potts Models of Microstructural Evolution. Perhaps the simplest representation of the energetics of grain growth is provided by the so-called kinetic Monte Carlo models of grain growth. Here a model Hamiltonian is set forth that provides an energy penalty for *all* interfaces within the microstructure. The basic idea is to exploit the Potts model which is highly reminiscent of the Ising model. In concrete terms, the model is set up as follows. An order parameter σ_i is introduced which labels the orientation of the ith grain. Each possible orientation within the model is assigned a different integer. This kinematic scheme allows for each different orientation within the polycrystal to be represented by a different integer, though clearly the continuous set of misorientations has been replaced with a discrete (and finite) approximation. Note that the lattice considered here is not a crystalline lattice but rather a modeler's lattice and represents a spatial discretization of the microstructure that is presumed fine enough to capture the essential features of grain growth while at the same time being coarse on the scale of the lattice parameter. One scheme for deciding upon a length to associate with the lattice in the model is through reference to the typical curvatures that must be accounted

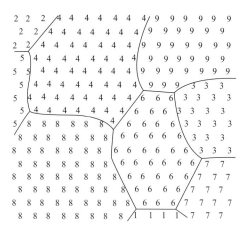

Fig. 10.43. A Potts microstructure in which different grains are identified by different values of the occupation number σ (adapted from Anderson *et al.* (1984)).

for in the lattice model of the microstructure. A Potts microstructure using the kinematic description of the microstructure described above is shown in fig. 10.43. One means to generate such a microstructure is to select a series of grain centers, each of which will have a different orientation, set up the corresponding Voronoi tiling and then specify each grain according to its occupation label.

As yet, the model is purely descriptive and says nothing about the possible subsequent evolution of the microstructure. To endow this geometric model with an underlying 'dynamics', we first require a statement of the energetics. One conventional choice is a Hamiltonian of the form

$$H = \sum_{ij} J_{ij}(1 - \delta_{\sigma_i,\sigma_j}), \qquad (10.87)$$

where the sum is over lattice sites in the model and J_{ij} is an energy parameter that essentially mimics the grain boundary energy. A given grain is made up of many lattice sites, and the energy penalty implied by eqn (10.87) is associated with *changes* in the local spin variables from one lattice site to the next corresponding to the presence of a grain boundary. What the Hamiltonian tells us is that if two neighboring lattice sites i and j are identical (i.e. $\delta_{\sigma_i,\sigma_j} = 1$) then there is no associated energy penalty. On the other hand, if the neighboring sites have different values of σ, then there will be an energy cost parameterized by J_{ij}. In the general form with which the Hamiltonian was written above, each type of interface characterized by a different pair ij potentially has a different energy J_{ij}. However, a simpler realization of the model is to consider a single value J of the Potts coupling parameter which is equivalent to attributing an isotropic interface energy to the grain boundaries.

The temporal evolution of a microstructure (i.e. grain growth) is examined by recourse to Monte Carlo methods. A trial change in occupation number is selected. If the move results in a net reduction of the energy, it is accepted. If not, it is accepted with a probability given by a Boltzmann weight. As a result of this 'dynamics', the trend is towards states of lower overall energy which in the context of the present model of energetics means less overall grain boundary area. A series of snapshots in a Potts microstructure as computed in this manner is shown in fig. 10.44. One of the ways in which these results can be examined is from the perspective of the types of experiments on grain growth described earlier in this section. In particular, it is possible to monitor the grain size as a function of time.

Phase Field Models for Grain Growth. One of the complaints that can be registered about the type of analysis given above is the fact that the underlying dynamics is not really tied to the physics of interfacial motion. A partial remedy for this problem is provided by a second class of models that we will feature several times, namely, phase field models. Note that these models are the logical extension of the ideas introduced in section 9.5. In this setting, the set of orientations corresponding to the polycrystalline microstructure is associated with a collection of field variables (i.e. order parameters) $\eta_i(\mathbf{r}, t)$, with each η_i corresponding to a different orientation. The total energy of the system is written in such a way that its minimizers (i.e. there is only a single orientation present) are of the form that either $\eta_i = 1$ or $\eta_i = -1$ and all the other η_js are zero. Further, there is an energy penalty associated with spatial variations in the fields η_i which can be construed as an interfacial energy. Note that in this setting the interface is endowed with a finite width of the type described in the previous chapter and shown in fig. 9.50. It is also worth noting that there are significant similarities between the continuum model set forth here and the Potts model described above.

To be concrete, one scheme that has been used to model grain growth in the phase field setting is to identify a set of N field variables $(\eta_1(\mathbf{r}, t), \eta_2(\mathbf{r}, t), \ldots, \eta_N(\mathbf{r}, t))$ corresponding to $2N$ distinct orientations (since each η_i can assume the values ± 1). The kinematic idea of such models is illustrated in fig. 10.45 which shows the 'interface' between two different grains. As said above, the generalization to the situation involving more orientations is built around having a larger number of fields.

Once these kinematic preliminaries have been settled, the total free energy functional is then written as the integral

$$f[\{\eta_i(\mathbf{r}, t)\}] = \int_\Omega [f_0(\{\eta_i(\mathbf{r}, t)\}) + \sum_i \frac{\kappa_i}{2}(\nabla \eta_i(\mathbf{r}, t))^2]d^3\mathbf{r}. \qquad (10.88)$$

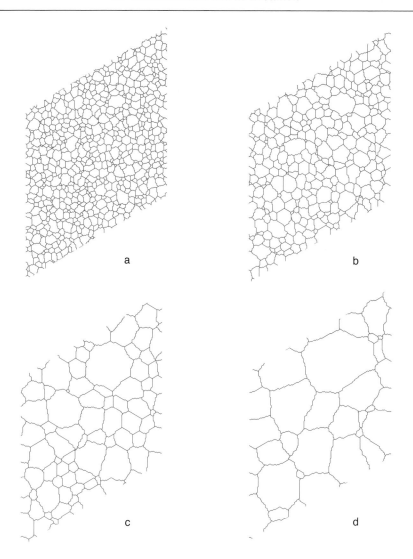

Fig. 10.44. Series of snapshots in the life history of a Potts microstructure (courtesy of E. Holm): (a)–(d) represent various stages of the time evolution of this system.

The term f_0 is written explicitly as

$$f_0(\{\eta_i(\mathbf{r}, t)\}) = -\frac{\alpha}{2} \sum_i \eta_i^2 + \frac{\beta}{4} \sum_i \eta_i^4 + \gamma \sum_i \sum_{j>i} \eta_i^2 \eta_j^2, \qquad (10.89)$$

which has energy wells corresponding to a given η_i being ± 1 while all of its competitors are zero. The interpretation of the gradient term is relatively straight-forward and imposes an energetic penalty for variations in the fields $\{\eta_i(\mathbf{r}, t)\}$ with a length scale set by the parameters $\{\kappa_i\}$. To summarize, the energetics set forth

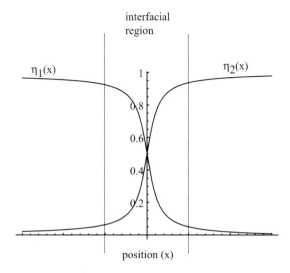

Fig. 10.45. Representation of the 'interface' region between different grains, one characterized by $\eta_1(x)$ and the other by $\eta_2(x)$.

above has two key features. First, there is a preference for one of the η_is to be ± 1 with all others zero. The demonstration of this point is taken up in the problem section at the end of the chapter. Secondly, the gradient term assigns an energy penalty to the 'interfaces' between regions in which the order parameter field is different. That is, in passing from one grain (say $\eta_i = 1$) to another (say $\eta_j = 1$), the order parameter field will vary smoothly between 1 and 0 with a corresponding energy penalty for this interface. The basic idea of the phase field approach is to play out this competition in a continuum setting.

In addition to the definition of the energy functional, it is necessary to institute a dynamics associated with a model of this sort. In the present setting, the dynamics is obtained by asserting that the temporal evolution of a given field variable is linear in the driving force, namely,

$$\frac{\partial \eta_i(\mathbf{r}, t)}{\partial t} = -L_i \frac{\delta f}{\delta \eta_i(\mathbf{r}, t)} + \zeta_i(\mathbf{r}, t), \qquad (10.90)$$

where the parameter L_i is a kinetic coefficient that modulates the driving force. ζ_i is a noise term which is intended to mimic the effects of thermal fluctuations. We will return to the subject of gradient flow dynamics like that introduced here in more detail in chap. 12.

Notice that one of the challenges of a model of this type is that there are as many equations as there are components to the order parameter. As a result, for the simulation of a polycrystal with many different grains, there will be as many field variables as there are grains resulting in both conceptual and computational

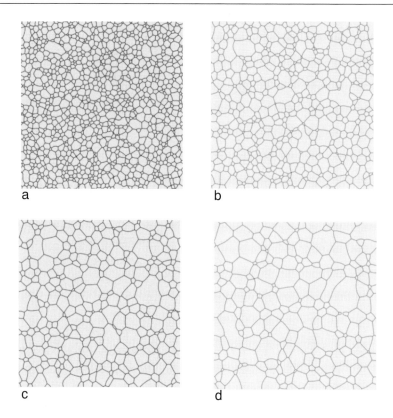

Fig. 10.46. Phase field model for two-dimensional grain growth (adapted from Chen and Wang (1996)). Figure shows grain distribution at various times in the evolution of the grain structure with: (a) $t = 1000$, (b) $t = 3000$, (c) $t = 5000$ and (d) $t = 8000$. Time is measured in units of how many integration steps have taken place.

complexity. An example of the type of grain growth that occurs in a model like that sketched above is shown in fig. 10.46 (Chen and Wang 1996).

One of the outcomes of these calculations is the ability to characterize the kinetics of grain growth. In particular, it is possible to address the question of the mean grain size $\langle R \rangle$ as a function of time as well as the distribution of grain sizes. The temporal evolution of the average grain area is shown in fig. 10.47 An outcome of this analysis is that it is found that the mean grain size scales with time with an exponent of $1/2$.

It should be noted that we have only outlined one of the many different diffuse interface approaches to the problem of grain growth. For the present purposes, our main point has been to call attention to the possibility of a strictly continuum model of grain growth in which the interfaces are endowed with a finite width and the kinematics of the polycrystal is encoded in a series of continuum fields. We

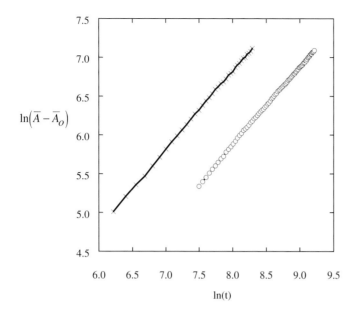

Fig. 10.47. Temporal evolution of mean grain size in phase field model of grain growth (adapted from Chen and Yang (1994)). Plots are of logarithm of average grain area as a function of time, with the two curves corresponding to four (crosses) and thirty-six (circles) different order parameter fields.

will see later that this class of models (i.e. phase field models) has a much wider application than that of its particular incarnation in the context of grain growth.

Sharp Interface Models of Grain Growth. An alternative paradigm for considering grain growth is offered by the class of sharp interface models. From a kinematic perspective, the polycrystal may be thought of as a series of lines (in two dimensions) or a series of interfaces (in three dimensions), the positions of which may be parameterized continuously or discretely. This class of models is especially appealing since the full kinematic complexity of the polycrystal has been collapsed onto nothing more than a discrete set of nodal coordinates which identify the interfacial position. The kinetics of the system is built around the idea of considering the various interfaces that make up the polycrystal of interest as being subjected to an array of driving forces of different physical origins. For example, driving forces can be present because of curvature of the boundary plane or because of differences between the stored strain energy in adjacent grains, or because of chemical inhomogeneities, or yet further mechanisms. These ideas may be put on a more formal basis by resorting to the variational arguments set forth in section 2.3.3 in which the power associated with the various driving forces is balanced with the various dissipative mechanisms that are present.

The application of this theoretical framework to the problem of grain growth is instructive, both as a case study in the use of the variational approach, and for the insights it provides into the grain growth process itself. Recall that the variational formulation is indifferent to the number of different dissipative mechanisms that are incorporated in the analysis. On the other hand, for the purposes of the present discussion we will only consider grain boundary motion induced by curvature without reference to the variety of other possible mechanisms. The driving force for microstructural change is the energy reduction that attends the elimination of grain boundary area and is given by

$$\dot{G} = \int_{GB's} \gamma_{gb}\kappa\, v_n d\Gamma, \tag{10.91}$$

with κ the curvature of the boundary at the point of interest, γ_{gb} the grain boundary energy and v_n the interfacial normal velocity. The overall variational function can be written as

$$\Pi[v_n(s); r(s)] = \sum_{i=1}^{N_b} \int_{L_i} \frac{v_n^2}{2D} ds + \sum_{i=1}^{N_b} \int_{L_i} \gamma_b\kappa\, v_n ds + \sum_{k=1}^{N_t} \sum_{j=1}^{3} \gamma_b \mathbf{v}_k \cdot \mathbf{s}_k^j. \tag{10.92}$$

The first two terms have already been discussed in several other contexts (see eqns (2.46) and (10.45)) with the new twist that now we have a summation over all of the different interfaces. The third term, on the other hand, is a new feature and reflects the extra attention that must be given to triple points within the polycrystalline microstructure when attacked from the sharp interface perspective.

In order to turn the variational scheme given above into a calculational scheme, the various interfaces are discretized, thus turning the functional given in eqn (10.48) into a corresponding function which depends only upon the nodal coordinates that define the boundaries. The variational statement is then invoked in order to determine a set of nodal velocities which are used to update the microstructure. The results of this type of sharp interface treatment of grain growth is shown in fig. 10.48. As noted above, the temporal evolution of the microstructure in this case is the result of a driving force that emerges from the reduction in grain boundary energy by reducing the overall length of such boundaries.

Our treatment of grain growth was intentionally diverse, with the aim being to show some of the different paradigms that have been put forth for the consideration of grain growth. In broad terms, the key ideas can be divided between those having to do with computing the driving force on a given interface, and those whose aim it is to attribute a kinetics with these driving forces. The Potts model offered a very simple energetic description of grain boundaries in which the thermodynamic driving force was seen to be associated with the reduction in energy that attends the overall grain boundary area. Phase field models

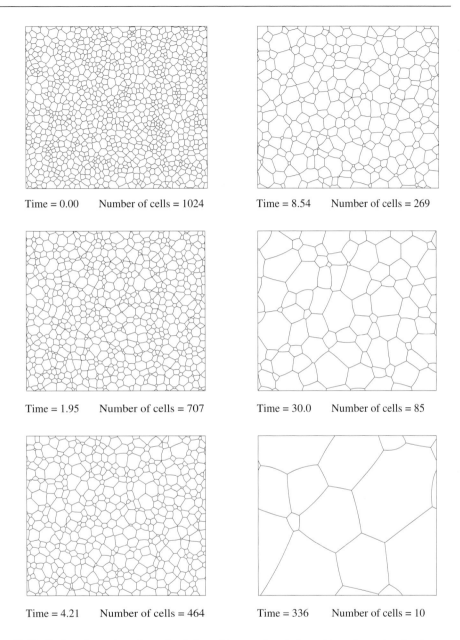

Time = 0.00   Number of cells = 1024

Time = 8.54   Number of cells = 269

Time = 1.95   Number of cells = 707

Time = 30.0   Number of cells = 85

Time = 4.21   Number of cells = 464

Time = 336   Number of cells = 10

Fig. 10.48. Series of snapshots from simulation of temporal evolution of two-dimensional polycrystal (courtesy of S. Gill).

could be interpreted as the continuous partners of Potts-type models in that they assign an energy penalty to the presence of interfaces which translates into a driving force for grain boundary migration. Our treatment of the sharp interface approach to grain growth illustrated the effects of curvature driven boundary motion.

10.5 Microstructure and Materials

The root level argument set forth in this chapter has been that in light of the critical role played by structures at the microstructural scale in determining material properties, a crucial prerequisite to a fuller understanding of materials demands an analysis of the factors governing microstructural evolution. The treatment given in this chapter has in no sense been exhaustive. Our discussion has been heavily biased in favor of metallic materials and has made no reference to the rich microstructures found in polymeric materials, for example. On the other hand, our aim has been to illustrate some of the key methodologies (primarily geometric and energetic) which have been brought to bear on the question of microstructure and its evolution. One of the important unanswered questions at the foundation of the approaches provided here concerns the general principles giving rise to the kinetics of microstructural evolution, and whether or not it is appropriate to express microstructures (as in the martensite case) as energy minimizers. A further query that must be posed is that of how best to synthetically couple approaches to the evolution of microstructures and the effect of these microstructures on properties. Part of the quest of the remaining chapters will be to spell out in several case studies how microstructural features manifest themselves in macroscopic properties.

10.6 Further Reading

The Structure of Materials by S. M. Allen and E. L. Thomas, John Wiley and Sons, New York: New York, 1999. This text is part of an excellent MIT series on materials science. This particular volume considers structure in materials from the hierarchical perspective considered here as well. Chap. 6 on microstructure provides a number of interesting discussions and figures to illuminate the nature of microstructure in wide classes of materials.

The Theory of Transformations in Metals and Alloys by J. W. Christian, Pergamon Press, Oxford: England, 1965. This classic text provides a definitive account of many of the key ideas that attend the study of microstructure. Unfortunately, the book has not been reissued in an updated form, but still makes for excellent reading.

Theory of Structural Transformations in Solids by A. G. Khachaturyan, John Wiley & Sons, New York: New York, 1983. I haven't spent as much time with Khachaturyan's book as I would have liked, but I can say that it is full of insights into a variety of topics covered in the present chapter.

Microstructure and Properties of Materials edited by J. C. M. Li, World Scientific, Singapore: Singapore, 1996. This work has a number of interesting articles on the implications of microstructure for material behavior.

Stability of Microstructure in Metallic Systems by J. W. Martin, R. D. Doherty and B. Cantor, Cambridge University Press, Cambridge: England, 1997. This book gives a deep coverage to many of the ideas raised in the present chapter.

Recrystallization and Related Annealing Phenomena by F. J. Humphreys and H. Hatherly, Pergamon Press, Oxford: England, 1996. A useful discussion of the connection between observed macroscopic changes in microstructure and the underlying defect mechanisms that engender it.

Introduction to Engineering Materials: The Bicycle and the Walkman by C. J. McMahon, Jr and C. D. Graham, Jr, published by C. J. McMahon, Jr, 1994 (available from author: http://www.seas.upenn.edu/mse/fac/mcmahon.html). This is one of my favorite books. Despite the claimed low level of the book, the discussion of recovery, recrystallization and grain growth is highly instructive as are many other topics covered by the authors.

'Modeling of Phase Separation in Alloys with Coherent Elastic Misfit' by P. Fratzl, O. Penrose and J. L. Lebowitz, *J. Stat. Phys.*, **95**, 1429 (1999) is one of a series of outstanding papers collected in a volume honoring John Cahn on his 70th birthday. This paper provides an outstanding treatment of many of the ideas that have been put forth from both the sharp and the diffuse interface perspective for thinking about two-phase microstructures.

'Modeling Microstructure Evolution in Engineering Materials' by A. C. F. Cocks, S. P. A. Gill and J. Pan, *Advances in Applied Mechanics*, Vol. 36, edited by E. van der Giessen and T. Y. Wu, Academic Press, San Diego: California, 1999. This article gives a unifying perspective on the role of the variational principle associated with configurational forces as the basis for numerical investigations of microstructural change.

'Theory of Martensitic Microstructure and the Shape-Memory Effect' by Kaushik Bhattacharya, unpublished (1998) (a huge pity!) – available from author: bhatta@caltech.edu. Bhattacharya's article gives a complete and thorough discussion of the many ideas that have been brought to bear on the problem of microstructure in martensites.

'On the computation of crystalline microstructure' by Mitchell Luskin, *Acta Numerica*, pg. 191, Cambridge University Press, Cambridge: England, 1996. Luskin's article covers many similar issues to those found in the article of

Bhattacharya and provides an alternative view on some of the issues related to energy minimization.

'The Analytical Modeling of Normal Grain Growth' by C. S. Pande and S. P. Marsh, *JOM*, Sept., 25 (1992). Though brief, this article provides an interesting overview of many of the issues involved in modeling grain growth.

'Theories of Normal Grain Growth in Pure Single Phase Systems' by H. V. Atkinson, *Acta Metall.*, **36**, 469 (1988). A thorough review that mixes insights on both the experimental situation and the models that have been put forth to confront these experiments.

'Some Fundamentals of Grain Growth' by D. Weaire and S. McMurry, in *Solid State Physics* edited by H. Ehrenreich and F. Spaepen, Vol. 50, Academic Press, San Diego: California, 1997. This article is full of interesting ideas on the geometrical aspects of grain growth and includes a discussion of key theoretical ideas as well.

10.7 Problems

1 Details in the Eshelby Inclusion

One of the fundamental solutions from which much is made in the evaluation of microstructure is that of the elastic fields due to ellipsoidal inclusions. (a) In this part of the problem, revisit the problem of the elastic fields due to inclusions by deriving the expression given in eqn (10.12). Further, perform the integrations required to determine the displacements, and show that the displacements external to the inclusion are given by

$$u_r(r) = \frac{\epsilon a^3}{3r^2}\left(\frac{3\lambda + 2\mu}{\lambda + 2\mu}\right) \tag{10.93}$$

while those internal to the inclusion are given by

$$u_r(r) = \frac{\epsilon r}{3}\left(\frac{3\lambda + 2\mu}{\lambda + 2\mu}\right). \tag{10.94}$$

Hint: Recall that $\sqrt{(r - r')^2} = \text{sgn}(r - r')$. (b) Imitate our analysis of the fields due to a spherical inclusion, but now deducing the fields for a cuboidal inclusion.

2 Elasticity and Diffusion for a Circular Inclusion

In this problem we try to imitate in analytic terms the various arguments that have been set forth coupling mechanics and diffusion in the context of microstructural evolution. Consider a single circular precipitate of initial radius R_0. (a) Solve for the elastic fields of such a circular inclusion on the assumption that it has the same elastic properties as the matrix, but is characterized by an eigenstrain $\epsilon_{ij}^* = \epsilon_0 \delta_{ij}$. (b) Using the results of part (a) in conjunction with the expression for the concentration given in eqn (10.52) to obtain the instantaneous concentration at the interface between the particle and the matrix. (c) Use this concentration to solve the diffusion equation for a circular region of outer radius R_{out} with the far field boundary condition, $c = C_0$. (d) Compute $\nabla c \cdot \mathbf{n}$ at the interface between the particle and the matrix and update the radius of the particle accordingly. (e) What is the final size of the particle?

3 Compatibility at Martensite Interfaces

Earlier we introduced the compatibility condition

$$\mathbf{F}_1 = \mathbf{F}_2 + \mathbf{a} \otimes \mathbf{n}, \tag{10.95}$$

and noted that the action of $\mathbf{F}_1$ is the same as $\mathbf{F}_2$ for points within the invariant plane. Prove this assertion in detail by examining the effect of $\mathbf{F}_1$ and $\mathbf{F}_2$ on points characterized by the vector $\mathbf{r} = r_\parallel \mathbf{e}_\parallel + r_\perp \mathbf{n}$. Show that only if $\mathbf{r}$ has a nonzero component perpendicular to the invariant plane will the deformation gradients $\mathbf{F}_1$ and $\mathbf{F}_2$ have a different action on that vector.

4 Rank-One Connection for Model Deformations

Examine the deformation mapping given by

$$x_1 = X_1 \pm \gamma X_2, \tag{10.96}$$
$$x_2 = X_2, \tag{10.97}$$
$$x_3 = X_3, \tag{10.98}$$

with the $\pm$ sign signifying that we are to take $+$ if the region of interest is above the invariant plane, and $-$ if below. First, sketch the outcome of applying this mapping to points lined up along vertical lines, and then compute the deformation gradients in the two regions. Finally, show that the deformation gradients in question satisfy the rank-one condition, and find the vectors $\mathbf{a}$ and $\mathbf{n}$.

5 Different Paradigms for Motion of a Curved Interface

Implement the Potts, phase field and sharp interface descriptions for the motion of a curved interface. In particular, in fig. 10.42, we showed schematically how these three classes of models are used to describe the evolution of the same microstructure. In this problem, use these three schemes to obtain the flattening of an initially curved interface. Assume an initial parabolic profile and examine the time evolution predicted by these three schemes as the interface becomes flat.

Part four

Facing the Multiscale Challenge of Real Material Behavior

ELEVEN

Points, Lines and Walls: Defect Interactions and Material Response

11.1 Defect Interactions and the Complexity of Real Material Behavior

In previous chapters, we have begun the attack on the problem of the observed thermomechanical properties of materials from a number of different perspectives. One of our key realizations has been the conspiratorial role of defects in governing the response of materials. Thus far, we have largely concentrated on the structure and energetics of individual defects with less attention given to the implications of these defects for material response. Our efforts have been primarily divided along dimensional lines, with point defects, line defects and interfacial defects each being treated as an end unto itself. The concern of the present chapter is to make evident the claim that such divisions are a theorist's artifice and that, in general terms, it is the synthetic response of all of these different types of defects in concert that gives rise to the observed complexity of materials.

We begin our discussion with the consideration of diffusion. In earlier chapters, we noted that diffusion is mediated by the presence of point defects. However, until now, we have not given any quantitative analysis of the lubricating effect of other defects in the context of diffusion. We have already seen in fig. 9.1 that bulk diffusion can be negligible in comparison with the diffusion that takes place along surfaces, grain boundaries and dislocations. As a particular realization of this class of problem, we consider the case of surface diffusion with special reference to some of the counterintuitive diffusion mechanisms that take place at surfaces.

A second fundamental theme that will be taken up in this chapter in which there is an interplay between the various types of defects introduced earlier is that of mass transport assisted deformation. Our discussion will build on the analysis of diffusion at extended defects. We will begin with an examination of the phenomenology of creep.

A third example in which there is a mutual interplay between the different types of defects that we have discussed in previous chapters is that provided by the dislocation–grain boundary interaction. In the context of polycrystal plasticity, it is clearly of interest to determine the conditions under which plasticity in one grain can be communicated to adjacent grains. Experiments indicate that under certain conditions, dislocations may pass from one grain into neighboring grains. Clearly, the presence of grain boundaries can have the effect of impeding dislocation motion, and we will take up this topic in section 11.4.

The next example that will be set forth as an example of a problem in which multiple defect types can be present simultaneously is that of fracture. Our discussion will begin with an examination of the way in which cracks can be built up as superpositions of dislocations. This leads naturally to questions of the interaction of dislocations and cracks and culminates in the analysis of the intriguing problem of crack tip nucleation of dislocations.

An age old observation around which entire revolutions took place is that if a pure metal is mixed with small amounts of some other metal, the resulting strength of the material will be altered. Witness the Bronze Age. These observations will be codified in quantitative terms in section 11.6 where we show how the yield stress varies as a result of obstacles to dislocation motion. These observations, once again, drive home the potency of the synthetic effect of different types of defects in tandem. We will sketch some of the key qualitative ideas that have been elucidated with respect to both solution and precipitate hardening which are the technical names for the area of study that has arisen in response to the simple observations that made the Bronze Age possible. The development in this part of the chapter will culminate in certain scaling arguments which reveal how the yield stress scales with various microstructural parameters. In keeping with the theme of the present chapter, our basic argument will be that although dislocations are the key players in plastic deformation, it is their interactions with each other and other obstacles that gives rise to much of the observed richness of plastic deformation.

The basic aim of this chapter is to illustrate the way in which the various defects we have introduced in previous chapters can form partnerships that enrich the properties of materials and the processes that transpire within them. The logic of the chapter is to provide evidence for this viewpoint through the medium of case studies. As usual, we have not been exhaustive.

11.2 Diffusion at Extended Defects

11.2.1 Background on Short-Circuit Diffusion

In chap. 7, we discussed the fundamental role of point defects in diffusion. We also hinted at the serious amendment to diffusion rates that is mediated by the presence of extended defects such as dislocations and grain boundaries (see fig. 9.1). As was noted above, diffusion rates along defects can often be so much larger than those in the bulk that the bulk mechanisms can be rendered almost irrelevant. As a result of this insight, it is evident that we must reconsider diffusion more carefully in light of the presence of line and interfacial defects.

Our present task is to build on the foundations laid in chap. 7, but now with special reference to the diffusive processes that take place at extended defects. The basic argument will be that by virtue of the more open atomic-level environments near extended defects, the activation energy both for point defect formation and migration will often be reduced relative to bulk values. We will build our case around a fundamental case study through the consideration of diffusion at surfaces. The surface diffusion example will illustrate not only how diffusive processes are amended at extended defects, but will also illustrate the shortcomings of the transition state formalism when the detailed atomic-level mechanisms are not known *a priori*.

11.2.2 Diffusion at Surfaces

Because of the widespread interest in growth of material systems by deposition, the subject of surface diffusion is one of enormous current interest. The example of surface diffusion being taken up here is of interest to our overall mission for several different reasons. First, as noted above, surfaces are one of the most important sites of communication between a given material and the rest of the world. Whether we interest ourselves in oxidation and corrosion, catalysis, the crystal surface is the seat of tremendous activity, most of which is mediated by diffusion. A second reason that we have deemed it important to consider the role of surface diffusion is that our analysis will reveal the dangers that attend the use of transition state theory. In particular, we will appeal to the existence of 'exchange mechanisms' for diffusion that reveal that the diffusion pathways adopted on some crystal surfaces are quite different than those that might be suggested by intuition.

Exchange Mechanisms for Surface Diffusion. Our discussion of transition state theory in chap. 7 showed that in those cases when we are lucky enough to know the details of the transition pathway associated with a given diffusion mechanism, atomic-level analysis can shed important light on the process of diffusion. On the other hand, as we have already emphasized, the successful application of the ideas of transition state theory ultimately requires a knowledge of the transition pathway. Field-ion microscopy in conjunction with first-principles analysis of the energetics of metal surfaces has led to a convincing picture of surface diffusion in some instances that is entirely contrary to the ideas built around intuition.

The simplest picture of diffusion on the (001) surface of an fcc metal such as Al or Pt would hold that atomic jumps take place from one binding site to the next via

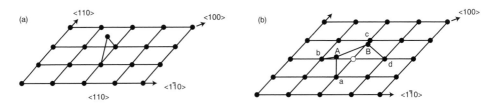

Fig. 11.1. Competing surface diffusion mechanisms on an fcc (001) surface (adapted from Feibelman (1990)): (a) intuitive mechanism in which adatom moves from one fourfold site to another by passing across bridge site, and (b) exchange mechanism in which diffusion occurs by concerted motion both of adatom and its subsurface partner.

the pathway illustrated in fig. 11.1(a). The contention in this case would be that diffusion takes place by the simple motion of the atom from one fourfold binding site to the next through the intermediate twofold bridge site. On the other hand, observations on Pt and Ir (001) determined that diffusion takes place by hopping exclusively along $\langle 100 \rangle$ and $\langle 010 \rangle$ directions as opposed to the $\langle 110 \rangle$ direction that would correspond to crossing the bridge site. Reconciliation of the simplest intuitive notions concerning diffusion with these observations would require that the adatom pass through an intermediate state in which it is onefold coordinated and sits directly above one of the surface atoms. This mechanism is associated with an unphysically high activation energy.

Because of the failure of the mechanism suggested above to account for observed diffusion, alternative mechanisms must be sought. To quote Feibelman (1990), 'The key to developing a correct model of surface diffusion is to think of it as a chemical phenomenon rather than in terms of a hard sphere moving on a bumpy plane.' In terms of what we have learned about the energetics of atomic-level bonding in previous chapters, Feibelman's exhortation is a reminder that there may be alternative mechanisms in which the adatom does not suffer the significant loss of coordination implied by the intuitive diffusion mechanism involving bridge sites suggested above. In particular, an alternative 'exchange mechanism' in which the adatom burrows into the surface at the same time that the atom previously at one of the surface sites vacates its position has led to a plausible explanation for all of the observations. This mechanism is illustrated in fig. 11.1(b). In the case of Al, Feibelman (1990) found that this mechanism was associated with an activation energy of only 0.20 eV, in contrast with the 0.65 eV associated with the naive but intuitive mechanism discussed earlier. As a result, the exchange mechanism explains both the orientational preferences observed in field-ion microscopy experiments as well as the corresponding activation energies.

One of the key reasons we have considered the exchange mechanism for surface diffusion is as a reminder of the inherent difficulties in using transition state theory.

While it is true that once there is confidence that the correct mechanism has been identified that transition state theory will usually yield a reasonable description of the energetics, it is also true that ignorance of the correct pathway can lead to results that are completely misleading. As a result, work has concentrated on important new developments in the area of determining the relevant reaction pathways in complex systems. In particular, sophisticated techniques have been set forth that are aimed at uncovering the optimal reaction pathway in cases where intuition does not suffice as a guide.

Surface Steps and Diffusion. Recall from our discussion of surfaces in chap. 9, there are a series of structures at increasing levels of complexity that may be found on crystal surfaces. Beyond the reconstructions that attend flat surfaces, we also noted the important role of surface steps as key structural components of generic surfaces. These steps are interesting not only from the perspective of their geometry, but also from the point of view of their role in altering the nature of the diffusive processes that take place on surfaces. In this section, our intention is to examine a particular feature of such diffusion, namely, the existence of a barrier resulting from surface steps.

The evidence for reflecting barriers at the edge of surface terraces derives from a number of different sources, one important example of which is field-ion microscopy. For example, it has been found that there is an enhancement in the likelihood of finding an adatom just adjacent to the top edge of a surface step. From the standpoint of the simple random walk analysis put forth earlier, one way to interpret the results of the types of experiments sketched above is via the random walker with reflecting walls. The basic idea in this case is the presumption that the random walker can exercise excursions within the confines of some finite region, but whenever the walker reaches the boundary of the region it is reflected from the walls.

These phenomenological ideas have been lent quantum mechanical credibility through consideration of the energetics of adatoms near steps. In particular, examination of the energetics of adatoms near a surface step has led to the elucidation of the concept of the Schwoebel barrier. The total energies of adatoms in the vicinity of a surface step have been computed using both empirical and first-principles techniques though for the present purposes we will consider but one example of a calculation of this barrier from the empirical perspective. The rough physical expectation is that as an adatom approaches a surface step from above, in the immediate vicinity of that step it will suffer an even greater loss of coordination than that it already experiences on a free surface, with a concomitant energy penalty. A series of energetic snapshots for the (111) surface of Pt are shown in fig. 11.2 and reveal the energetics of different configurations of the adatom near

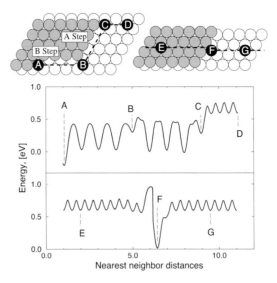

Fig. 11.2. Illustration of energetics of an adatom near a surface step on the (111) surface of Pt (adapted from Jacobsen *et al.* (1995)).

a step. As claimed above, the key remarkable qualitative feature of the near-step energetics is the existence of both a barrier to adatom motion across the step from above, and an enhanced bonding for those adatoms arriving at the step edge from below.

From the overall perspective of the present chapter, our ambition in briefly describing the energetics of surface diffusion has been to illustrate the additional complexity that is ushered in as a result of the simultaneous action of more than one type of defect. In particular, we have shown that via a counterintuitive exchange mechanism, the activation energy for surface diffusion can be lowered by 0.4 eV relative to the intuitive motion associated with an imagined random walker on a surface. In addition, we have also shown that the presence of surface steps can have significant implications for the nature of diffusion on surfaces.

11.3 Mass Transport Assisted Deformation

In chap. 7 we laid the foundations for the analysis of mass transport in solids. In the current section, our aim is to examine the ways in which mass transport can assist the deformation that occurs in a given material. As a preliminary, we remind the reader of one of the key features of the deformation mechanism map introduced in fig. 7.7. We refer to the fact that in many instances for stresses well below the putative yield stress, permanent deformation is still observed. Such deformation usually occurs at temperatures which are larger than, say, $0.3 T_M$,

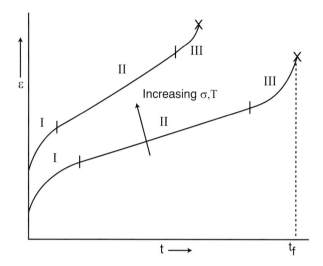

Fig. 11.3. Schematic of strain as a function of time in a constant-stress creep test (adapted from Courtney (1990)).

where T_M is the melting temperature. These high-temperature mechanisms are referred to generically as creep and will be our present concern.

11.3.1 Phenomenology of Creep

The phenomenology of creep is best illustrated through the vehicle of the so-called creep test in which a specimen is subject to either a constant load or a constant stress, and the deformation is monitored as a function of time. The generic outcome of such a test is illustrated in fig. 11.3 and suggests that the creep behavior may be divided into three distinct regimes, denoted primary, secondary and tertiary creep. The primary regime is characterized by an initially high strain rate which decreases with increasing time until a so-called 'steady-state' regime is entered in which the strain rate is roughly constant. The tertiary regime is characterized by an increase in the strain rate and terminates with the failure of the material.

In considering creep, there are several different control parameters that are known to influence the resulting deformation. Clearly, with increasing stress, the strain rate $\dot{\epsilon}$ will be increased. Similarly, an increase in temperature also serves to increase $\dot{\epsilon}$. We continue our discussion by examining these two different parameters in turn. Before embarking upon that discussion, we note that our analysis will highlight the significant empirical insights that have been gained into the creep process, with less emphasis on the models that have been advanced to rationalize these empirical gains.

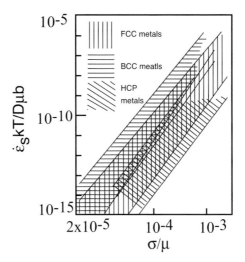

Fig. 11.4. Schematic of strain rate $\dot{\epsilon}_s$ as a function of normalized stress (adapted from Nabarro and de Villiers (1995)). D refers to the bulk diffusion constant and b is a measure of the lattice parameter.

One of the elemental empirical insights into the creep process concerns the stress dependence of the creep rate. A broad range of experimental data can be captured on the assumption of a power law relation between the strain rate $\dot{\epsilon}$ and the stress σ. In particular, the strain rate can be written in the form

$$\dot{\epsilon} \propto \left(\frac{\sigma}{\mu}\right)^n,\tag{11.1}$$

where the exponent n varies depending upon the situation. This behavior is illustrated qualitatively in fig. 11.4.

Another critical question is how the creep rate depends upon temperature. As hinted at in the title of this section, creep is in many instances mediated by the presence of mass transport. Part of the basis of this insight is the observed temperature dependence of the creep rate. In particular, it has been found that creep is thermally activated. Indeed, one can go further than this by noting the relation between the activation energy for diffusion and that of the creep process itself. A plot of this relation is shown in fig. 11.5. As a result, we see that the temperature dependence of the creep rate may be written in the form

$$\dot{\epsilon} = \text{const.}e^{-\Delta E/kT}.\tag{11.2}$$

Our insights on both the stress and temperature dependence of the strain rate may thus be unified in a single phenomenological equation of the form

$$\dot{\epsilon} = A\left(\frac{\sigma}{\mu}\right)^n e^{-\Delta E/kT}.\tag{11.3}$$

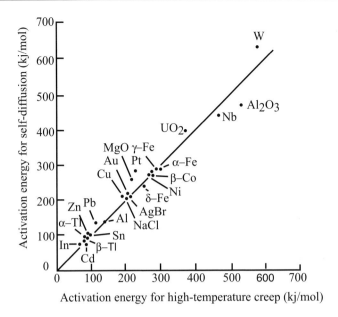

Fig. 11.5. Plot of activation energies for creep and bulk diffusion (adapted from Courtney (1990)).

From a theoretical perspective, what we are left with is the ambition of explaining the stress and temperature dependence of the strain rate evoked in eqn (11.3) from a mechanistic perspective. As we will see, the problem is not so much a shortage of ideas in this regard, but rather an embarrassment of choices, without any clear scheme which will permit the falsification of these models. In the remainder of our discussion on mass transport assisted deformation, the plan is to sketch the ideas associated with Nabarro–Herring and Coble creep. I should note that this discussion is little more than a report on the party line as regards the modeling of creep phenomena and further that we have not done justice to both what is known and what remains unknown in this important field. In particular, we omit all reference to an important alternative mechanism related to the mass transport assisted climb of dislocations, thus allowing them to breach obstacles to their further glide.

11.3.2 Nabarro–Herring and Coble Creep

The class of creep mechanisms of interest here are those that are mediated by stress-biased diffusion. If we are to consider the vacancy flux in a given grain within a material that is subjected to an applied stress, it is argued that the vacancy formation energy differs in different parts of the grain, and hence that there should be a gradient in the vacancy concentration leading to an associated flux. This flux of vacancies (or atoms) can, with the accumulated weight of time, lead to a macroscopic change in the shape of the material. This mechanism results in a strain rate that is linear in the applied stress and with an activation energy that is identical to that for self-diffusion. The distinction between the Nabarro–Herring and Coble mechanisms is one of variety rather than form. In particular, the Nabarro–Herring mechanism is associated with lattice diffusion within the grains, while the Coble mechanism is built around diffusion along the grain boundaries. In both cases, however, the key point is the presence of a gradient in the vacancy concentration induced by the applied stress.

Our ambition in the paragraphs to follow is to illustrate an approximate heuristic model which shows how the presence of an applied shear stress can result in mass transport mediated deformation. We begin with the notion that the crystal of interest is subjected to a shear stress. From a two-dimensional perspective, for example, we then make use of the fact that the shear stress

$$\sigma = \begin{pmatrix} 0 & \sigma \\ \sigma & 0 \end{pmatrix},$$

(11.4)

may be rewritten in principal coordinates as

$$\sigma' = \begin{pmatrix} \sigma & 0 \\ 0 & -\sigma \end{pmatrix}.$$

(11.5)

In principle, we are thus faced with solving a boundary value problem for a region like that shown in fig. 11.6 with the aim being to evaluate the mass transport that arises by virtue of the spatially varying chemical potential, and hence, mass currents, that arise in response to the nonuniform stress.

In practice, since we are only interested in an estimate of the effect, we resort to an approximate analysis in which the relevant chemical potentials and mass flux, and attendant strain rate are all evaluated heuristically. The argument begins with reference to fig. 11.6 with the claim that the vacancy formation energy for the faces subjected to tensile stresses differs from that on the faces subjected to compressive stresses. Again, a rigorous analysis of this effect would require a detailed calculation either of the elastic state of the crystal or an appeal to atomistic considerations. We circumvent such an analysis by asserting that the vacancy concentrations are given by

$$c^{(T)} = e^{-(\epsilon_{vac}^f - \sigma\Omega)/kT},$$

(11.6)

and

$$c^{(C)} = e^{-(\epsilon_{vac}^f + \sigma\Omega)/kT},$$

(11.7)

where the superscripts T and C refer to tensile and compressive, respectively. To

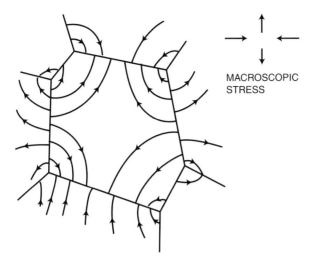

MACROSCOPIC
STRESS

Fig. 11.6. Schematic of how stress assisted mass transport results in a net strain rate (adapted from Herring (1950)).

really understand this result, we need to reflect further on the definition of the vacancy formation energy given in section 7.3.3. In particular, we need to see how those arguments should be amended to account for the presence of a stress. A simple plausibility argument can be constructed as follows. From an atomistic perspective, the vacancy formation energy associated with plucking a near-surface atom out and placing it on the surface can be represented approximately as $\epsilon_{vac}^{f} \approx \bar{f}b$, where $\bar{f}$ is the mean force exerted by neighboring atoms as the atom is plucked to the surface. In the presence of the tractions imagined above, there is either a suppression (tensile case) or enhancement (compressive case) in this energy cost, which can now be written as

$$\epsilon_{vac}^{f}(stress) = \epsilon_{vac}^{f} \pm \sigma A_{atom} b, \tag{11.8}$$

or

$$\epsilon_{vac}^{f}(stress) = \epsilon_{vac}^{f} \pm \sigma \Omega, \tag{11.9}$$

where we have defined A_{atom} as the area over which the traction is working during the plucking process.

The next step in the argument is to estimate the amount of mass transported from one face to the other by virtue of the concentration *gradient* embodied in eqns (11.6) and (11.7). In particular, the flux is given by $j = -D(\partial c/\partial x)$, where the gradient term is estimated as

$$\frac{\partial c}{\partial x} \approx \frac{c^{(T)} - c^{(C)}}{l}, \tag{11.10}$$

where the parameter l is a measure of the size of the region in question. We note that our objective is to compute the strain rate which is defined as

$$\dot{\epsilon} = \frac{1}{l}\frac{dl}{dt},$$

(11.11)

and may be written in terms of the amount of volume arriving on the face of interest per unit time as a result of the relation

$$\frac{dl}{dt} = \frac{1}{l^2}\frac{dV}{dt}.$$

(11.12)

The rate dV/dt can be evaluated by recourse to the relation

$$\frac{dV}{dt} = l^2 j,$$

(11.13)

with the result that when all the pieces of our analysis are assembled, we have

$$\dot{\epsilon} = \frac{2D_0\sigma\Omega}{l^2 kT}e^{-(\epsilon_{vac}^f + \epsilon_{vac}^m)/kT}.$$

(11.14)

At this point, several comments are in order. First, we note that we have recovered an expression for the strain rate which jibes with the empirical result presented in eqn (11.3). In this case, the stress exponent is unity.

Our discussion of creep given above implicitly assumed that the mass transport resulting in an overall shape change took place through the crystalline region itself. An alternative diffusion mechanism, founded upon the same principles as the discussion of Nabarro–Herring creep given above is based on reconsidering those arguments when the mass transport is mediated by grain boundary diffusion rather than bulk diffusion. The result of this analysis is a change in the way that eqn (11.14) scales with grain size. In particular, in the context of the Coble mechanism, the strain rate is written

$$\dot{\epsilon} = \frac{2D_{gb}\sigma\delta\Omega}{l^3 kT}.$$

(11.15)

The new features of this result in comparison with that of eqn (11.14) are the presence of the grain boundary diffusion constant, D_{gb}, and the introduction of a parameter δ which is a measure of the thickness of the region over which diffusion is enhanced in the vicinity of the grain boundary.

Note that in the context of both the Nabarro–Herring analysis and that of Coble, the strain rate is found to vary linearly with the applied stress. In particular, this result is consistent with dubbing this process 'diffusional viscosity' as did Herring (1950) in his original work. Indeed, the identification of a material parameter such as the 'viscosity' and its associated scaling properties with grain size and temperature represents another example of the type of micro–macro connection

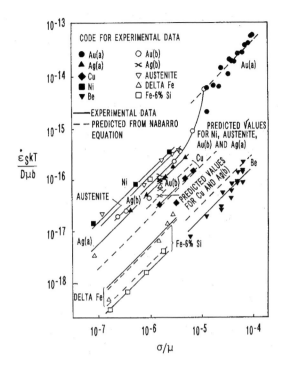

Fig. 11.7. Examination of the success of the hypothesis of linear stress dependence of the creep strain rate in the high-temperature regime (adapted from Nabarro and de Villiers (1995)).

which is at the heart of the viewpoint taken in the present work. The question of how such models jibe with experience is partially addressed in fig. 11.7 which reveals that the linear stress dependence is well obeyed for a range of materials. It bears repeating that our discussion has been little more than a caricature of this important field.

11.4 Dislocations and Interfaces

In the previous chapter we took preliminary steps to examine grain boundary motion. However, in the absence of considerations of dislocations in the vicinity of such grain boundaries, this description is incomplete. In the present section, we begin by examining the way in which dislocations and grain boundaries merge. In particular, for the special geometric case of low-angle grain boundaries, the boundary may be thought of as nothing more than a superposition of dislocations, with tilt boundaries being built up of edge dislocations and pure twist boundaries being built up by screw dislocations. However, our analysis can go further than this. In addition, we pose the question of how dislocations that impinge on a grain

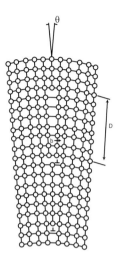

Fig. 11.8. Representation of grain boundary as an array of dislocations (adapted from Amelinckx (1979)).

boundary are greeted by the boundary which serves as an obstacle. Such processes can lead to both dislocation pile-ups and slip transmission in which crystal slip in one grain passes to adjacent grains.

11.4.1 Dislocation Models of Grain Boundaries

We begin our geometric discussion with the case of pure tilt boundaries on the grounds that the connection between dislocations and the boundary is most evident visually in this case. As was discussed in some detail in chap. 9, a tilt grain boundary is characterized by a simple misorientation between the two grains as was discussed in chap. 9. The intent of this section is to illustrate that a simple superposition of the elastic displacements implied by the Volterra solution for straight dislocations discussed in chap. 8 can lead to exactly the same type of misorientation.

First, we consider the most elementary geometrical arguments that can be set forth in this regard. The question we pose at this moment is what is the relation between the dislocation spacing and the misorientation. For concreteness, consider a series of uniformly spaced edge dislocations with Burgers vector of length b such as is shown in fig. 11.8. What we note immediately is that this geometry may be thought of as the introduction of an extra half-plane each time we traverse the distance D along the boundary plane. As a result, the dislocation spacing and the misorientation angle are related by

$$D = \frac{b}{2\sin(\theta/2)}. \tag{11.16}$$

Although a length scale tied to the underlying discrete lattice is present in this equation, there is no clear selection mechanism that distinguishes between those geometries that have a special relation to the underlying crystalline lattice and those that do not. What we are noting is that not all dislocation spacings are equal and that for certain special choices of this spacing the lattice and the dislocations are locked into a commensurate relationship.

One of the immediate questions that arises in the context of such a model of grain boundaries is what it might have to say about the stress fields and energies of that boundary. An immediate reaction to the claim that grain boundaries may be thought of as arrays of dislocations is that this must mean that boundaries have long-range stress fields. However, as will be shown in short order, the long-range dislocation stress fields when superposed in the special way demanded by a grain boundary lead to cancellations that leave exponentially decaying fields.

For simplicity, consider the isotropic linear elastic description of the problem. The total stress is obtained by summing up the contributions of the constituent dislocations. If we consider the shear stress components for concreteness, then the total stress for the boundary with dislocation spacing D is

$$\sigma_{xy}(x, y) = \frac{\mu b}{2\pi(1 - \nu)} \left\{ x^3 \sum_{n=-\infty}^{\infty} \frac{1}{[x^2 + (y - nD)^2]^2} \right.$$

$$\left. - x \sum_{n=-\infty}^{\infty} \frac{(y - nD)^2}{[x^2 + (y - nD)^2]^2} \right\}. \tag{11.17}$$

All this formula instructs us to do is to take the contribution to the total stress of each dislocation using the results of eqn (8.38) and to add them up, dislocation by dislocation. We follow Landau and Lifshitz (1959) in exploiting the Poisson summation formula to evaluate the sums. Note that while it would be simple enough to merely quote the result, it is fun to see how such sums work out explicitly. If we use dimensionless variables $\alpha = x/D$ and $\beta = y/D$, then the sum may be rewritten as

$$\sigma_{xy}(x, y) = \frac{\mu bx}{2\pi D^2(1 - \nu)} \sum_{n=-\infty}^{\infty} \frac{\alpha^2 - (\beta - n)^2}{[\alpha^2 + (\beta - n)^2]^2}. \tag{11.18}$$

The next step in the thrust to beat the sum into submission is to recognize that if one defines

$$J(\alpha, \beta) = \sum_{n=-\infty}^{\infty} \frac{1}{\alpha^2 + (\beta - n)^2}, \tag{11.19}$$

then the expression for the stresses is given in turn as

$$\sigma_{xy} = -\frac{\mu bx}{2\pi D^2(1 - \nu)}\left[J(\alpha, \beta) + \alpha\frac{\partial J(\alpha, \beta)}{\partial \alpha}\right]. \tag{11.20}$$

With these clever algebraic machinations behind us, the original problem has been reduced to that of evaluating the sum $J(\alpha, \beta)$. Recall that the Poisson summation formula tells us

$$\sum_{n=-\infty}^{\infty} f(n) = \sum_{k=-\infty}^{\infty} \int_{-\infty}^{\infty} f(x)e^{2\pi ikx}dx. \tag{11.21}$$

For the problem at hand, the Poisson summation formula allows us to rewrite our sum as

$$J(\alpha, \beta) = \sum_{k=-\infty}^{\infty} e^{2\pi ik\beta} \int_{-\infty}^{\infty} \frac{e^{-2\pi iku}}{\alpha^2 + u^2}du. \tag{11.22}$$

Evaluation of the relevant integrals is immediate and results in

$$J(\alpha, \beta) = \frac{\pi}{\alpha} + \frac{2\pi}{\alpha}\sum_{k=1}^{\infty} e^{-2\pi k\alpha}\cos 2\pi k\beta. \tag{11.23}$$

Consultation of Gradshteyn and Ryzhik (1980) or some other reference permits the exact evaluation of the sum $1 + 2\sum_{k=1}^{\infty} e^{-kt}\cos kx = \sinh t/(\cosh t - \cos x)$, which for our case leads to the stress

$$\sigma_{xy}(x, y) = \frac{\mu b\pi x}{D^2(1 - \nu)}\frac{\cos\dfrac{2\pi y}{D}\cosh\dfrac{2\pi x}{D} - 1}{\left(\cos\dfrac{2\pi y}{D} - \cosh\dfrac{2\pi x}{D}\right)^2}. \tag{11.24}$$

However, we could have made more immediate progress with eqn (11.23) by recalling the significance of the parameter α. In dimensionless terms this parameter is a measure of the distance from the boundary and for large distances we could safely keep only the term associated with $k = 1$. In this case, we find

$$\sigma_{xy}(x, y) = \frac{2\pi\mu bx}{D^2(1 - \nu)}e^{-\frac{2\pi x}{D}}\cos\frac{2\pi y}{D}. \tag{11.25}$$

Having attained our end, namely, the determination of the grain boundary stresses (note that the computation of σ_{xx} and σ_{yy} proceeds similarly), what may be learned about boundaries themselves? We begin with a simple, but profound observation concerning the solution. Because of the exponential dependence of the stresses on the distance from the boundary, we see that the elastic perturbation due to a boundary is short-ranged. Though dislocations have their characteristic long-range elastic signature, grain boundaries lead only to a localized influence. As we have already remarked in earlier discussions on interfaces in materials, there is a hierarchy of energies associated with such interfaces commencing with surfaces, which have the highest energies, passing through grain boundaries, which are intermediate in energy, and finishing with stacking faults, which are usually at the bottom of the energy scale. One issue of interest that our dislocation model of grain boundary permits us to revisit is that of the energies of grain boundaries.

The simple argument that is made in exploiting the dislocation model for an estimate of the grain boundary energy is that the boundary may be thought of energetically as the sum of the energies of the isolated dislocations, where the long-range cutoff is taken to be determined by the dislocation spacing itself. To make this argument more plausible, consider the stress at the boundary in the immediate vicinity of a particular dislocation. Note that the total stress is given by eqn (11.24), and we are interested in a point (x, y) in the immediate vicinity of a particular dislocation. Expansion of the relevant cos and cosh factors to quadratic order in their arguments results in

$$\sigma_{xy}(x, y) = \frac{\mu b x}{2\pi(1 - v)} \frac{x^2 - y^2}{(x^2 + y^2)^2},$$ (11.26)

which is seen to be exactly the shear stress component for a single dislocation at the origin. This result suggests the plausibility of treating the boundary as a series of single dislocations, each with an energy

$$E_{disl} = \frac{\mu b^2}{4\pi(1 - v)} \ln \frac{D}{2r_0}.$$ (11.27)

To obtain this result, all we have done is to use the results from chap. 8 for the energy of a single dislocation within isotropic linear elasticity. We may make an approximate evaluation of the grain boundary energy on the grounds that each dislocation specifies the energy of its local neighborhood on the boundary. Hence, the energy per unit depth of boundary is given by

$$E_{gb} = \frac{\mu b^2}{4\pi D(1 - v)} \ln \frac{D}{2r_0},$$ (11.28)

which if we recall that $\theta = b/D$ may be rewritten as

$$E_{gb} = \theta(A - B \ln \theta),$$ (11.29)

with $A = [\mu b/4\pi(1 - v)] \ln(b/2r_0)$ and $B = [\mu b/4\pi(1 - v)]$. This is the

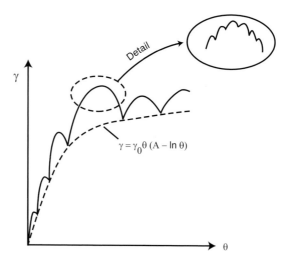

Fig. 11.9. Energy of tilt grain boundary as a function of misorientation angle (adapted from Sutton and Balluffi (1995)).

noted result of Read and Shockley for the energy of a grain boundary, and is plotted in comparison with experimental values in fig. 11.9. As was remarked earlier, one of the serious shortcomings of the present model, and a fact which is revealed in considering the experimental numbers, is that there is no provision for the possible commensurability of the lattice parameter and the dislocation spacing D. In particular, for certain choices of the misorientation angle, very special geometries arise as was noted in our earlier discussion on coincidence site lattices. By way of contrast, the dislocation model holds that the parameter D may be varied continuously without prejudice.

11.4.2 Dislocation Pile-Ups and Slip Transmission

Dislocations and grain boundaries form partnerships other than the one described above. Another intriguing mechanism involving dislocations and grain boundaries arises when dislocations, during the process of their gliding motions, impinge on a grain boundary in their pathway which serves as an obstacle. Our thinking in the current section will be founded upon two alternative approaches to the question of dislocation pile-ups. We will begin with a discrete analysis of such problems. However, it will be seen to be mathematically more convenient to assume that the set of dislocations piled up against a boundary are continuously distributed, unleashing the powerful machinery of integral equations to make analytic headway.

We begin with a consideration of the relevant phenomenology. Fig. 11.10 shows a number of different possible outcomes of the interaction between an array of

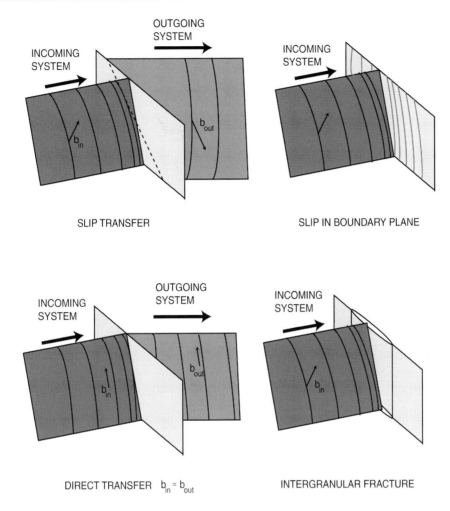

Fig. 11.10. Schematic of interactions between dislocations and grain boundaries (courtesy of Ian Robertson).

piled-up dislocations and a grain boundary. Several of the mechanisms feature the development of plastic deformation (i.e. dislocation motion) in the adjacent grain. On the other hand, the pile up of dislocations against a boundary can result in processes within the grain boundary plane itself. As shown in the figure, two possible outcomes are the development of dislocation motion within the grain boundary plane or wholesale fracture along the grain boundary.

The advent of transmission electron microscopy opened the doors to the direct observation of processes such as the pile-up of dislocations at a grain boundary and the subsequent commencement of dislocation motion in adjacent grains. The process of slip transmission described above may be seen from the perspective

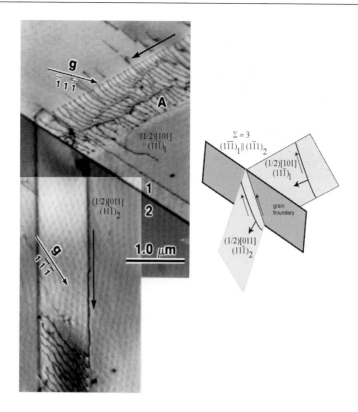

Fig. 11.11. Slip transmission process (taken from Lee *et al.* (1990)).

of transmission electron microscopy in fig. 11.11. What we note in inspecting this figure is that the geometric positions of the dislocations adopt a pleasing disposition which suggests a regular mathematical relation governing the spacings of the piled up dislocations. Two central questions that arise in the context of this geometry are: (i) what governs the distribution of pile-up dislocations, and (ii) what are the implications of such a pile-up, especially in terms of the stresses it produces in adjacent grains?

Our fundamental assertion concerning the geometry of pile-ups is that they reflect the equilibrium spacing of the various dislocations which are participants in such a pile-up. From a discrete viewpoint, what one imagines is an equilibrium between whatever applied stress is present and the mutual interactions of the dislocations. In simple terms, using the geometry depicted schematically in fig. 11.12, we argue that each dislocation satisfies an equilibrium equation of the form

$$\tau_{app}(x) + A \sum_{i \neq j} \frac{1}{x_j - x_i} = 0. \tag{11.30}$$

Fig. 11.12. Schematic representation of pile-up geometry (adapted from Smith (1979)).

This equation considers the equilibrium of the jth dislocation. In particular, it is nothing more than a dislocation-by-dislocation statement of Newton's first law of motion, namely, $\mathbf{F}_j^{(tot)} = 0$, which says that the total force on the jth dislocation is zero. The factor A is determined by the elastic moduli and differs depending upon whether we are considering dislocations of edge $(A = \mu b/2\pi(1 - \nu))$, screw $(A = \mu b/2\pi)$ or mixed character. The problem of determining the equilibrium distribution of the dislocations in the pile-up has thus been reduced to one of solving nonlinear equations, with the number of such equations corresponding to the number of free dislocations in the pile-up. For further details see problem 3 at the end of the chapter.

A more elegant (and mathematically tractable) description of the problem of dislocation pile-ups is to exploit the representation of a group of dislocations as a continuous distribution. This type of thinking, which we have already seen in the context of the Peierls–Nabarro model (see section 8.6.2), will see action in our consideration of cracks as well. The critical idea is that the discrete set of dislocations is replaced by a dislocation density $\rho_b(x)$ such that

$$dn(x) = \rho_b(x)dx. \tag{11.31}$$

What this equation tells us is the number of dislocations between x and $x + dx$. The key constraint on the function $\rho_b(x)$ is that it be properly normalized so that

$$n_{tot} = \int_{x_{min}}^{x_{max}} \rho_b(x)dx, \tag{11.32}$$

where n_{tot} is the number of dislocations in the pile up. Once we have taken this step, the equilibrium condition for the dislocations is represented by an integral equation

$$\frac{\mu b}{2\pi} P \int_{-\infty}^{\infty} \frac{\rho_b(x')dx'}{x - x'} + \tau_{ext}(x) = 0. \tag{11.33}$$

This equation should be seen as the continuous analog of eqn (11.30). It expresses the equilibrium of the infinitesimal dislocation at position x with the applied stress, $\tau_{ext}(x)$ and the remaining infinitesimal dislocations in the distribution $\rho_b(x')$. The P refers to the principal value integral and eliminates the self-interaction of the dislocation at x with itself.

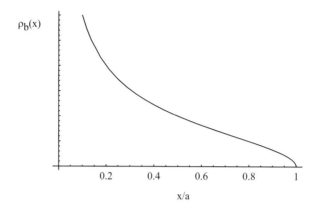

Fig. 11.13. Representation of dislocation pile-up as a continuous distribution of dislocations.

As a concrete example of such a problem, we follow Lardner (1974) in considering n screw dislocations, each with Burgers vector of magnitude b in the region $x > 0$, driven to the left by a constant shear stress $-\tau$ and bounded by a barrier at $x = 0$ (in mentioning the barrier, we have a grain boundary in mind, for example). Furthermore, if we assume that the pile-up has a width a, then the solution to this problem which we quote without derivation (for details see Lardner (1974)) is

$$\rho_b(x) = \frac{2\tau}{\mu b}\sqrt{\frac{a-x}{x}}. \tag{11.34}$$

To determine the width a, recall eqn (11.32) which in this instance yields

$$n = \int_0^a \rho_b(x)dx = \frac{\pi \tau a}{\mu b}, \tag{11.35}$$

with the implication that $a = \mu n b / \pi \tau$. The dislocation density associated with this profile is shown in fig. 11.13, and demonstrates the same qualitative features as noted in the observed pile-ups.

The next question we pose is how such a dislocation distribution affects the stress state in its neighborhood. The pertinence of this question to broader issues in the mechanics of materials is that a quantitative understanding of dislocation pile-ups has been invoked to explain the dependence of the yield stress on the grain size as canonized in the Hall–Petch relation and depicted in fig. 8.2. One hypothesis put forth with the aim of explaining the Hall–Petch behavior is that the presence of such pile-ups reduces the stress needed to inspire slip in adjacent grains. As we will show below, the grain size enters as the characteristic dimension by which the pile-up size is measured, and larger pile-ups lead to larger stresses in adjacent grains and thus smaller stresses are needed to induce slip within them. The geometry we have

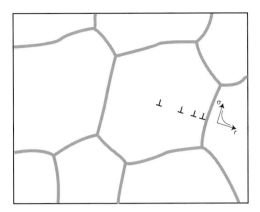

Fig. 11.14. Schematic of the pile-up of dislocations at a grain boundary. Enhancement in the stress field in an adjacent grain as a result of the presence of the pile-up is also indicated schematically.

in mind is depicted in fig. 11.14 In attempting to explain the Hall–Petch behavior, the argument has been advanced that once slip has commenced in the grain that is favorably oriented for slip, dislocations will pile-up against the boundary that is the terminus of the slip plane. The presence of this pile-up has the effect of increasing the stress in the adjacent grain which is itself less favorably oriented for slip. The pile-up is thought to catalyze slip in the adjacent grain by virtue of the increased stress there. The total stress ahead of the pile-up is given by

$$\tau(x) = -\tau + \frac{\mu b}{2\pi} \int_0^a \frac{\rho_b(x')dx'}{x - x'}. \tag{11.36}$$

This equation reflects the addition of the externally applied shear stress, $-\tau$, as well as the contribution of the pile-up itself as embodied in the integral. Substitution of the dislocation density determined above yields

$$\tau(x) = -\tau \sqrt{\frac{x - a}{x}}. \tag{11.37}$$

What this result demonstrates is the presence of an enhanced stress in front of the pile up ($x < 0$). We forego a discussion of the implications of such stress enhancements for dislocation nucleation until section 11.5.3.

11.5 Cracks and Dislocations

In chap. 2, we noted that linear elastic fracture mechanics offers powerful insights into the continuum description of fracture processes. On the other hand, we also noted that the occurrence of fracture ultimately depends upon mechanisms that take place in the vicinity of the crack tip, some of which depend upon atomic-level details. It is the aim of the present section to recount some of the issues that arise in trying to bring atomic-level insights to bear on the problem of fracture especially in the context of crack tip dislocation nucleation. Our reason for including this discussion in the present chapter is the fact that our understanding of cracks and the fracture process can be significantly extended once we adopt a broader view in which cracks and dislocations are considered simultaneously.

In coming sections we undertake a discussion of the conspiracy that arises once dislocations and cracks are taken into consideration jointly. In particular, the goal is to examine two key issues related to dislocation–crack interactions. First, we wish to consider the notion of crack tip shielding in which, by virtue of the presence of a dislocation in the neighborhood of a crack tip, the crack tip fields are amended locally. In particular, we will interest ourselves in the renormalization of the crack tip stress intensity factor by virtue of the presence of dislocations. Once we have built up the tools for handling dislocations and cracks simultaneously, we aim to consider the competition between cleavage and nucleation of dislocations at a crack tip. Our fundamental argument will be that the elastic energy release rate can be balanced either against the creation of new free surface or in the installation of crystalline slip in the form of dislocations.

11.5.1 Variation on a Theme of Irwin

In chap. 2 the central role of cracks in fracture mechanics was discussed. One of the universal features of such cracks from the standpoint of the theory of linear elasticity is the characteristic $1/\sqrt{r}$ signature of their stress fields. In the present section we wish to revisit these fields with an eye to examining how such fields may be deduced using arguments from dislocation theory. The approach adopted here will be to use the same superpositions of dislocations already discussed in the context of dislocation pile-ups. The fundamental argument is that by using the dislocation field of a single dislocation as a fundamental solution (i.e. a Green function), the fields due to cracks may be constructed by superposition.

Cracks as a Superposition of Dislocations. A scheme that will suit our aim of building a synthetic description of cracks and any allied dislocations is to think of a crack as an array of dislocations. Indeed, the majority of our work has already been done earlier in the context of our consideration of dislocation pile-ups in section 11.4.2. In fact, our present analysis will do little more than demonstrate that the solutions written down there are relevant in the crack context as well. The more fundamental significance of the perspective to be offered here is that we will soon want to build up solutions in which cracks and dislocations are equal partners, and it will be seen that the dislocation ideas developed here provide a useful basis for doing so.

One of the key points to be made in the present discussion is the substantiation of the claim that a continuous distribution of infinitesimal dislocations can be assembled in a way that respects the traction-free boundary conditions associated with crack faces. In particular, we consider a crack which runs from $-a$ to a, with a remotely applied stress of the form $\sigma_{xy} = \tau_0$. In this mode II setting (see fig. 11.16), we insist that on the crack faces $\sigma_{yy} = \sigma_{xy} = 0$. Our claim is that such a crack can be represented by a distribution of infinitesimal dislocations with a density of the form

$$\rho_b(x) = \frac{2(1-v)\tau_0}{\mu b} \frac{x}{\sqrt{a^2 - x^2}}. \tag{11.38}$$

This distribution is determined by insisting that the traction-free condition,

$$\tau_0 = -\frac{\mu}{2\pi(1-v)} \int_{-a}^{a} \frac{\rho_b(x')dx'}{x - x'}, \tag{11.39}$$

be imposed along the crack, $|x| < a$. As pointed out by Weertman (1996), the solution to this problem is reached via recourse to a table of Hilbert transforms and is of the form reported in eqn (11.38). In light of this distribution, we may then write the stresses at an arbitrary point (x, y) as the superposition

$$\sigma_{ij}^{tot}(x, y) = \int_{-a}^{a} \rho_b(x')\sigma_{ij}^{(disl)}(x - x', y)dx'. \tag{11.40}$$

We denote the stress field due to a single Volterra dislocation at position (x', y') by $\sigma_{ij}^{(disl)}(x - x', y - y')$. We also note that the original interpretation of the elastic energy release rate offered in section 2.4.4 can be reinterpreted in terms of the driving forces on the various dislocations making up the crack. Now that we have seen how the crack itself may be written in terms of dislocations, we turn to the question of how to think about such a crack when there are other dislocations in its vicinity.

11.5.2 Dislocation Screening at a Crack Tip

From the standpoint of the linear theory of elasticity, the problem of crack tip shielding may be seen as a question of solving the boundary value problem of a crack in the presence of a dislocation in its vicinity. As indicated schematically in fig. 11.15, the perspective we will adopt here is that of the geometrically sterilized two-dimensional problem in which both the crack front and the dislocation line are infinite in extent and perfectly straight. As we will see, even this case places mathematical demands of some sophistication. The basic question we

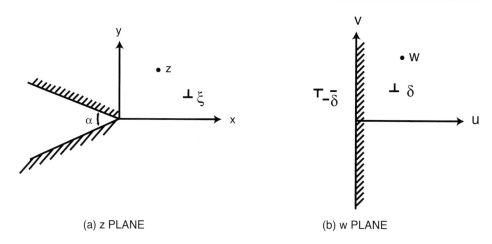

(a) z PLANE   (b) w PLANE

Fig. 11.15. Conformal transformation from wedge + dislocation to the problem of a half-space containing dislocation (adapted from Ohr *et al.* (1985)).

pose concerns the nature of the stress fields in the neighborhood of the crack tip, in particular, as a result of the presence of the dislocation. From a physical perspective, what is found is that the presence of the dislocation inspires the introduction of a local stress intensity factor which may be thought of as an effective stress intensity and may be reduced relative to the bare stress intensity that would be present in the absence of the dislocation.

In order to perform the mathematical manipulations that allow for the solution of this problem, we turn to the powerful tools of complex analysis. Of course, the moment we take such a step, it is clear that our methods will be impotent once we pose questions concerning more interesting three-dimensional geometries. We begin with a problem of antiplane shear in which the crack of interest has a mode III geometry, as indicated schematically in fig. 11.16, and the associated dislocation is a screw dislocation. In this instance, the problem is that of finding a harmonic function $u_3(x, y)$ such that the relevant boundary conditions are satisfied. One particularly elegant scheme for solving this problem is the use of conformal mapping. Geometrically, the problem of wedge + dislocation shown in fig. 11.15 is mapped onto a half-space with a subsurface dislocation, a problem we have already encountered in section 8.5.2, and which culminated in the introduction of an image dislocation. Through the device of determining a potential function that satisfies the free surface boundary conditions and subsequent inversion of the mapping so that we recover the original problem of a crack in the presence of a dislocation, the solution is rather immediate.

To be concrete, we note that we seek a potential function $\eta(z)$ such that $u(z) = (2/\mu) \, \mathrm{Im}[\eta(z)]$. Note that the factor of 2 here is consonant with the convention

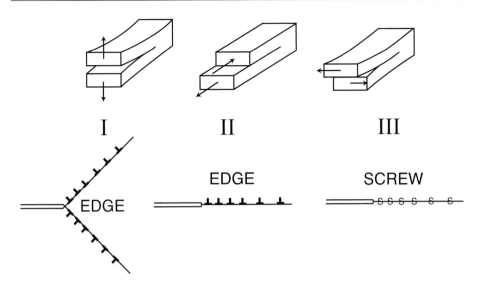

Fig. 11.16. Three independent modes of crack tip loading and associated dislocations (adapted from Ohr (1985)).

used in Ohr *et al.* (1985). Once this potential function is in hand, the corresponding stresses may be found using the isotropic linear elastic constitutive relation which in the case of antiplane shear reduces to $\sigma_{ij} = 2\mu\epsilon_{ij}$. In terms of the potential $\eta(z)$, this may be rewritten as

$$\sigma(z) = \sigma_{yz} + i\sigma_{xz} = 2\eta'(z). \tag{11.41}$$

In general, for complex geometries, direct determination of the function $\eta(z)$ cannot be carried out simply. The trick with conformal mapping is to map the originally complex problem onto a simpler geometry for which the corresponding solution is known. In particular, if the mapping that effects this transformation is of the form $w = g(z)$ and the potential for the transformed problem is $\chi(w)$, then the stresses for the original problem are obtained as

$$\sigma(z) = 2\chi'(w)\frac{dw}{dz}. \tag{11.42}$$

Note that in this case, the function $\chi(w)$ is presumed to be known (usually by virtue of the symmetric geometry of the transformed problem), and dw/dz is a geometric derivative determined strictly by the transformation between the two geometries.

For the case of interest here, namely, the fields that arise for a crack in the presence of a dislocation, the procedure described above is disarmingly simple. Through a suitable choice of transformation, the problem of a crack and dislocation can be mapped onto the problem of a single subsurface dislocation beneath a traction-free surface already solved by the method of images in section 8.5.2.

Before writing down the solution we must first broach the subject of the geometric transformation relating the wedge crack problem to the allied problem of a free surface. If we consider a wedge crack with opening angle α, all points on the line $z = re^{i(\pi - \frac{\alpha}{2})}$ must be mapped onto the line $z = re^{i\frac{\pi}{2}}$. The transformation that effects this mapping is $w = z^q$, with $q = \pi/(2\pi - \alpha)$. Given the transformation, all that remains is to determine the potential $\chi(w)$ which characterizes the fields due to the subsurface dislocation and its image. This solution, which we state without derivation, is given by

$$\chi(w) = \frac{\mu b}{4\pi} \ln\left(\frac{w - \delta}{w + \bar{\delta}}\right).$$
(11.43)

The dislocation of interest is at position δ in the w plane (i.e. $\delta = \sqrt{\xi}$, where ξ is the dislocation position in the z plane) while its image is at position $-\bar{\delta}$. For the special case of a sharp crack of interest here, $w = \sqrt{z}$. If we now apply the prescription of eqn (11.42), we find that the stresses are given by

$$\sigma(z) = \frac{\mu b}{4\pi \sqrt{z}} \left(\frac{1}{\sqrt{z} - \sqrt{\xi}} - \frac{1}{\sqrt{z} + \sqrt{\xi}}\right).$$
(11.44)

The significance of the result of eqn (11.44) is that it tells us how to account for the stresses in the context of problems in which a crack and a dislocation coexist.

A particularly interesting feature of solutions like that given above is their interpretation as providing a source of screening for the crack tip. The use of the word screening in this context refers to the fact that the local stress intensity factor is modified by the presence of the dislocation and can result in a net reduction in the local stresses in the vicinity of the crack tip. In particular, the total stress intensity factor is a sum of the form

$$K_{tot} = K_{III} + K_{disl},$$
(11.45)

where K_{III} is the stress intensity factor for the mode III crack in the presence of remote loading. If we imagine the special case in which the screw dislocation is on the prolongation of the crack plane at a distance d, then the contribution due to the dislocation to the stress intensity factor is given by

$$K_{disl} = -\frac{\mu b}{\sqrt{2\pi d}}.$$
(11.46)

The key qualitative observation to be made is that the sign of K_{disl} can be negative resulting in the local screening effect alluded to earlier. We note that while the analysis presented here centered on the mathematically simplest case of antiplane shear, there are analogous situations to be found for other modes of crack tip loading as seen in fig. 11.16.

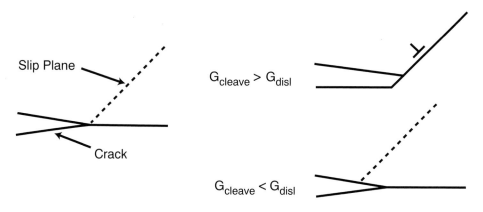

Fig. 11.17. Competition at a crack tip between the emission of a dislocation and cleavage (adapted from Beltz *et al.* (1999)).

11.5.3 Dislocation Nucleation at a Crack Tip

The solution constructed above provides the necessary tools from elasticity theory in order to consider nucleation of a dislocation at a crack tip. The philosophical perspective adopted here is that due to Rice (1992) in which a comparison is made between the work needed to create new free surface and that needed to create a slip distribution corresponding ultimately to a dislocation. We present this model not so much with the hope that it will deliver quantitative insights but on the grounds that it is highly instructive concerning the atomic-level processes near a crack tip. In particular, the model is appealing because the treatment of dissipative processes such as dislocation nucleation is endowed with atomic-level realism, while still maintaining an overall continuum description.

For an atomically sharp crack subjected to remote mode I loading we may imagine a variety of irreversible mechanisms that might transpire once the load is sufficiently large. Two of these processes are the creation of new free surface (cleavage) and the nucleation of dislocations. We have already discovered that the energy release rate tied to the opening displacement discontinuities is that accounted for in the Griffith model of the fracture toughness. Our present task is to find the analogous description for dislocation nucleation and to do so, we follow the development of Rice. The basic idea is to express these two processes as a competition and to determine which one will commence at a lower stress level. This competition can be couched in terms of the energy release rates of the two competing processes as shown in fig. 11.17. As noted in the figure, if the creation of new free surface can occur with a smaller energy release rate than the process of dislocation nucleation, it will be the winner in the competition. Alternatively, if the energy release rate to create a dislocation is the smaller of the two, it will be

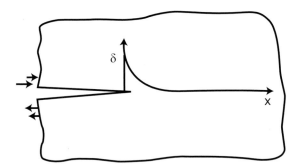

Fig. 11.18. Mode II crack with an incipient dislocation along prolongation of the crack plane (adapted from Rice and Beltz (1994)).

the process that transpires. As a result, the question becomes one of obtaining a measure of how much energy is fed to each of these processes during incremental crack extension, with G_{cleave} and G_{disl} being the relevant energy release rates for these two processes.

Our preliminary analysis centers around the geometry shown in fig. 11.18 in which an atomically sharp crack is subjected to mode II loading and the slip distribution is assumed to occur along the prolongation of the crack plane. Borrowing from the Peierls–Nabarro analysis described in section 8.6.2, it is assumed that the atomic-level forces across the slip plane may be characterized in terms of an interplanar potential $\phi(\delta)$. This description asserts that the tractions which arise on account of the sliding discontinuity are given by $\tau = -\partial\phi/\partial\delta$.

The implementation of the Peierls–Nabarro analysis in the crack tip context is really quite similar to that advanced already in the analysis of dislocations (see section 8.6.2), with the added complication that the energy functional must now account not only for the energy on the slip plane and the interactions between dislocations, but also for the terms resulting from the presence of a crack tip. We begin by writing down an energy functional which features the as-yet undetermined slip distribution. The total energy of the system of crack with its attendant incipient dislocation is a sum of terms representing the energy cost for the slip, the energy associated with the elastic fields of the crack, and a part reflecting the energy associated with the crack interacting with the dislocations. In particular, we have

$$E[\delta(r)] = \int_0^\infty \phi[\delta(r)]dr - \int_0^\infty \frac{K_{II}}{\sqrt{2\pi r}}\delta(r)dr$$

$$+ \frac{\mu}{4\pi(1-\nu)}\int_0^\infty \int_0^\infty \sqrt{\frac{\rho}{r}}\frac{1}{r-\rho}\frac{d\delta(\rho)}{d\rho}\delta(r)dr d\rho. \quad (11.47)$$

Functional differentiation of this equation with respect to the slip distribution results in the equilibrium equation

$$\frac{K_{II}}{\sqrt{2\pi r}} - \frac{\mu}{2\pi(1-\nu)} \int_0^\infty \sqrt{\frac{\rho}{r}} \frac{1}{r-\rho} \frac{d\delta(\rho)}{d\rho} d\rho = \frac{d\phi}{d\delta}. \tag{11.48}$$

From a numerical perspective, the critical load for dislocation nucleation can be obtained by repeatedly solving this equation for increasing loads (as reflected in the value of K_{II}) and seeking that load at which solutions fail to exist. An alternative appealing numerical approach is to map the continuous variational problem described above onto a corresponding discrete problem by parameterizing the solution in terms of a finite set of variational parameters (i.e. the Rayleigh–Ritz method) and solving the resulting coupled equations. This approach is described in Schöck (1996). On the other hand, Rice has shown through an ingenious application of the J-integral how this problem may be solved analytically.

The fundamental philosophy adopted in computing the energy spent in building up the incipient slip distribution is to use the J-integral, the path independence of which allows for a remote and near field determination of the energy release rate. The far field reckoning of this quantity is given in terms of Irwin's stress intensity factor introduced in section 2.4.4. By way of contrast, the near field evaluation of J reduces to an integral along the slip plane. In particular, we have

$$J = -\int_0^\infty \tau \frac{\partial(u_x^+ - u_x^-)}{\partial x} dx, \tag{11.49}$$

This integral may be rewritten as

$$J = \int_0^{\delta_{tip}} \tau d\delta, \tag{11.50}$$

which, because the shear stress is known in terms of the interplanar potential may be directly evaluated yielding $J = \phi(\delta_{tip})$. Thus, we have found that $K_{II}^{crit} = \sqrt{2\mu\phi(\delta_{tip})/(1-\nu)}$.

What we have learned is that dislocation nucleation will occur once $\phi(\delta_{tip})$ reaches its maximum allowable value. This idea is depicted graphically in fig. 11.19 where it is seen that instability to dislocation nucleation occurs when $\phi(\delta_{tip}) = \gamma_{us}$, where γ_{us} is a material parameter that Rice has christened the unstable stacking energy. This idea is intriguing since it posits that the competition between cleavage and dislocation nucleation has been reduced to consideration of the relative values of two simple material parameters, both of which admit of first-principles determination, and relevant geometrical factors.

One of the problems with the analysis we have made thus far is its geometric simplicity. For the mode I case of primary interest here, the assumption that dislocations will be nucleated on the prolongation of the slip plane is overly

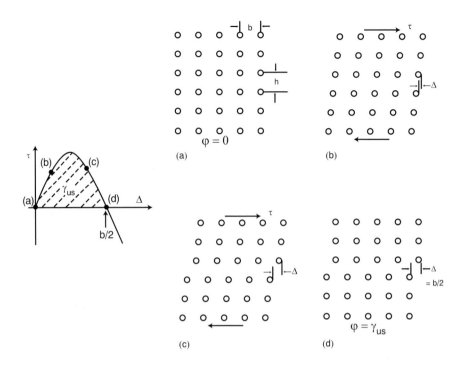

Fig. 11.19. Potential energy for sliding adjacent crystal planes across one another (adapted from Rice (1992)).

severe. The type of geometry that is more appropriate in this setting is such that dislocations will be nucleated on slip planes that are at an angle θ with respect to the crack plane. In fact, to treat this problem with full generality, it is necessary to consider the role of oblique planes as possible contenders and will be considered further in section 12.6.3.

As was hinted at in the previous section, the ease with which dislocations may be produced at a crack tip can clearly alter the mechanical response of a solid containing such a crack. In particular, for the special case of the clean geometry and samples that can be found in a material such as Si, the consideration of the cleavage–dislocation emission dichotomy may encompass all of the relevant mechanisms to attempt model building.

As we have done so often before, we look to Ashby for a revealing systemization of the data around which the micromechanical study of fracture is built. In fig. 11.20 a summary of the various fracture mechanisms is presented. One idea that emerges is that often at low temperatures, the sample undergoes virtually no plastic deformation before fracture, while at high temperatures, the material is seen to deform enormously. In addition, examination of the fracture surfaces reveals that different mechanisms are at play in the different results.

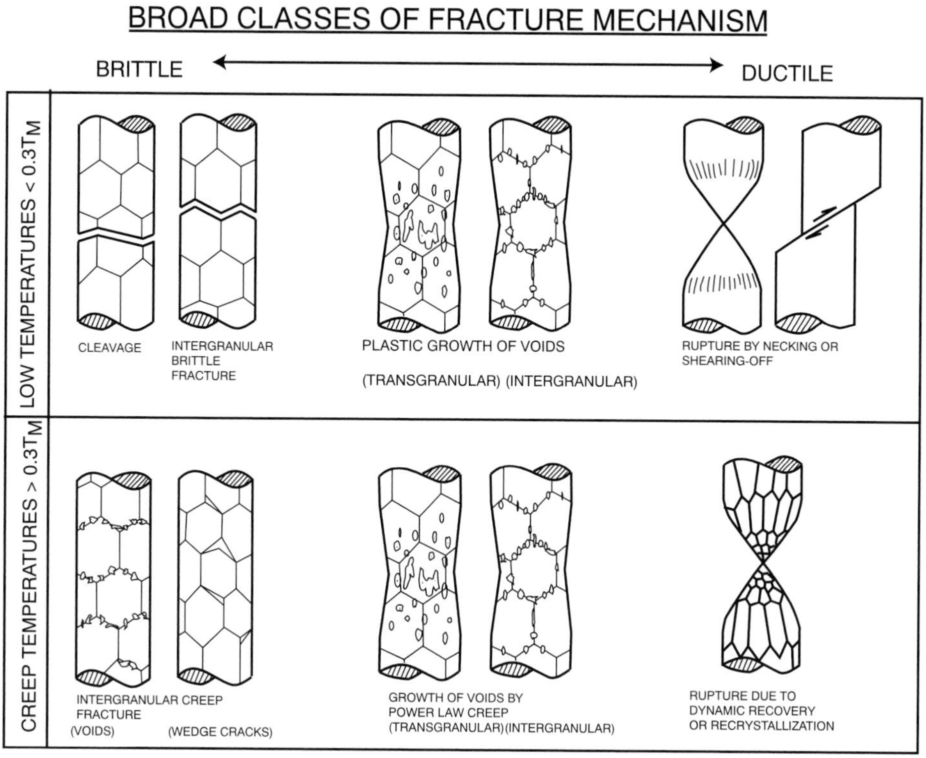

BROAD CLASSES OF FRACTURE MECHANISM

BRITTLE ⟷ DUCTILE

Fig. 11.20. Schematic of a variety of different fracture mechanisms (adapted from Ashby *et al.* (1979)).

Despite the overwhelming complexity of real fractures processes, it is possible to discern dominant failure mechanisms under different conditions. However, we should bear in mind that in most instances many different processes transpire simultaneously. In fig. 11.20, the different mechanisms are elucidated schematically. For high loads and low temperatures, we see that our intuition about failure by cleavage is correct, although the crack path can travel either along or through grains depending upon specific details of a given material. On the other hand, at higher temperatures, other mechanisms such as void nucleation and coalescence can assume a dominant role. The energy expended in such processes defies description in simple Griffith terms in which it is held that all of the elastic energy loss goes to the creation of new free surface. Those mechanisms involving appreciable plastic deformation require the expenditure of greater quantities of energy. This energy increase is revealed in the increased toughness of the material, with typical toughnesses far exceeding their Griffith values. Fracture toughness data are shown in fig. 11.21. The challenge grappled with in the current section

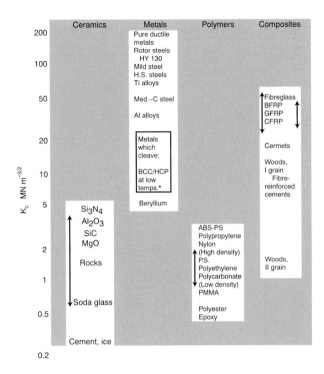

Fig. 11.21. Schematic representation of data on fracture toughness of different materials (adapted from Ashby and Jones (1996)).

is that of explaining the dramatic increase in fracture toughness that can occur, sometimes after a temperature change of only a few degrees. The beginnings of an answer to that question can in some cases be tied to the type of analysis undertaken here, namely, to the energetics of dislocation nucleation at a crack tip.

11.6 Dislocations and Obstacles: Strengthening

The fact that we refer to entire historic periods in terms of the names of certain metals is a powerful indirect testimony to the role played by chemical disorder in altering the mechanical properties of materials. Indeed, the appellation Bronze Age refers to precisely that moment in a civilization's development when it was learned that the addition of tin to copper not only lowers the melting temperature of the resulting alloy relative to pure copper, but increases its strength as well. The goal of the present section is to set the stage for understanding why the addition of tin to copper can change its mechanical properties, and more generally, to recount the role of various strengthening mechanisms in the understanding of the mechanics of materials, and to seek the microscopic origins of hardening phenomena in their roots in dislocation theory. After describing certain generic features of strengthening mechanisms related to the interactions of dislocations with obstacles, we will take up solution, precipitate and work hardening in turn.

A visit to a steel mill results in the immediate recognition that there is more to a given steel than just iron and carbon. This is not accidental. Various chemical elements are introduced in a material with a wide variety of aims including the hardening effects to be considered here. Similarly, entire classes of Al alloys are described on the basis of the impurities that populate them. Beyond the capacity to raise the flow stress by populating the material with either foreign atoms or particles, the stress to induce permanent deformation can be raised by working a material. The key point to be made is the recognition that what all of these hardening mechanisms reflect is the fact that some form of microstructural control has been exercised on the material. In that sense, the present section is a follow up on our earlier discussion of microstructure itself (see chap. 10) since we now attempt to explicate the 'properties' part of the structure–properties paradigm by tracing the influence of microstructure on dislocations. As an explicit reminder of the significance of microstructural control in the context of strengthening, we recall fig. 8.2 in which the variation in flow stress with various microstructural parameters is shown. For example, fig. 8.2(a) shows the way in which the flow stress increases with the increasing concentration of foreign atoms.

From a mechanistic perspective, what transpires in the context of all of these strengthening mechanisms when viewed from the microstructural level is the creation of obstacles to dislocation motion. These obstacles provide an additional resisting force above and beyond the intrinsic lattice friction (i.e. Peierls stress) and are revealed macroscopically through a larger flow stress than would be observed in the absence of such mechanisms. Our aim in this section is to examine how such disorder offers obstacles to the motion of dislocations, to review the phenomenology of particular mechanisms, and then to uncover the ways in which they can be understood on the basis of dislocation theory.

This program to understand hardening involves two distinct parts. First, we examine the interaction between a single dislocation and a localized elastic perturbation such as is offered either by an impurity atom or a foreign particle. This analysis demonstrates the force on a dislocation induced by such particles and hints at the possibility that a group of such impurities acting in concert could harden the material. Inspired by the insights gained by considering the role of a single obstacle to dislocation motion, the second part of our argument will be built around the statistical question of how to compute the average effects of a distribution of such obstacles. We begin with a conceptual overview of the hardening process as viewed from the dislocation perspective, followed by a systematic quantitative analysis of both the one-body interaction between a given dislocation and an associated obstacle as well as the statistical question of how many such obstacles act in unison. We round out the discussion with an application of these ideas to three of the prominent hardening mechanisms, namely, solution, precipitate and work hardening.

11.6.1 Conceptual Overview of the Motion of Dislocations Through a Field of Obstacles

As noted above, our working hypothesis concerning the various hardening mechanisms is that chemical impurities, second-phase particles and even other dislocations serve as obstacles to the motion of a given dislocation. As a result of the presence of these obstacles, the intrinsic lattice resistance τ_p is supplemented by additional terms related to the various strengthening mechanisms. We further assume that the flow stress can be written as

$$\tau_c = \tau_p + \Delta\tau, \tag{11.51}$$

where τ_c signifies the overall critical flow stress, τ_p is the intrinsic lattice friction (i.e. the Peierls stress) and $\Delta\tau$ is the change in the critical stress resulting from the presence of the obstacles. For simplicity, we will carry out our analysis in the limit in which τ_p is negligible and hence use τ_c and $\Delta\tau$ interchangeably.

As our first step in the attempt to model the various strengthening mechanisms within crystals, we consider a problem in the abstract: what is the stress needed to induce glide of a dislocation through an array of obstacles? For the moment, we leave unspecified the particulars concerning the origins of such obstacles, since our present thinking will aid us in considering a variety of different specific mechanisms. A pictorial representation of the situation of interest is given in fig. 11.22 in which the dislocation moves through an array of obstacles.

As was already noted in chap. 8, much of the business of the metallurgical use of dislocation theory takes place at the level of the line tension approximation. To make our arguments definite, we consider the case in which the dislocation line tension is assumed to be orientation-independent. We next suppose that the dislocation, which is pinned at two points separated by the distance L, has bowed out under the action of an applied stress τ resulting in an equilibrium at the pinning point between the line tension and the pinning force F of the obstacle itself. As shown in the free-body diagram for the pinning point in fig. 11.23, the equilibrium of the pinning point may be written as

$$F = 2T \cos\frac{\phi}{2}. \tag{11.52}$$

Recall from the discussion on line tension and dislocation bow-out given in

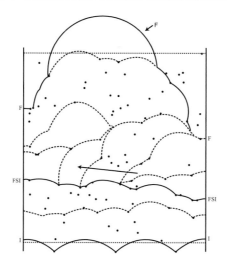

Fig. 11.22. Dislocation in a random array of obstacles (adapted from Foreman and Makin (1967)).

section 8.7.1, the radius of curvature is related to the applied stress and the line tension via

$$\tau b = \frac{T}{R}. \tag{11.53}$$

We recall the significance of this result by noting that it tells us that in the isotropic line tension approximation, the segment bows out into a circular arc with the radius determined by a competition between the line tension and the applied stress. The next step in the general argument being set forth here is to unite eqns (11.52) and (11.53) in such a way that R and ϕ are replaced by the more metallurgically accessible parameter L which characterizes the mean spacing between obstacles. To do so, we note on the basis of fig. 11.23 that $R = L/[2\sin(\theta/2)]$. Hence, if we rewrite eqn (11.53) using the fact that $T = F/[2\cos(\phi/2)]$ and on the grounds that the angles θ and ϕ are related by $\sin(\theta/2) = \cos(\phi/2)$, we find the simple and central relation,

$$\tau_c b = \frac{F}{L}. \tag{11.54}$$

This equation serves as the foundation of all that we will have to say about hardening. We begin by noting that the numerator is a reflection of the interaction between a single obstacle and a single dislocation. As a result, it is something that we can expect to understand on the basis of local calculations that concern themselves only with the properties of a single such interaction. By way of contrast, the denominator reflects the statistical properties related to the distribution of obstacles. As we will see below, there are a number of different ways of

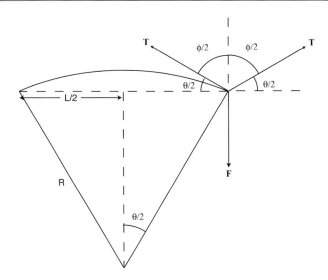

Fig. 11.23. Free-body diagram for the interaction of a dislocation and an obstacle characterized by a pinning force **F**.

constructing approximate models of the mean dislocation spacing, a fact which we signify by adopting the notation L_{eff} which reminds us that the obstacles are distributed throughout the solid, and that to use an equation like eqn (11.54) to describe the yield properties of the solid as a whole (as opposed to describing the properties of a single segment of length L between two pinning points) we will have to determine L_{eff} using some sort of statistical arguments.

There are two scenarios which can develop in the context of the type of bow-out considered here. Either, the line tension forces become so large that they exceed the maximum resistance of the pinning point, or the pinning force is so strong that two adjacent bowed-out segments rejoin through a process known as Orowan looping. In the context of the former process, breakaway of the dislocation can be seen to occur when the applied stress reaches the critical value corresponding to the maximum resistance that can be offered by the obstacle, F_{max}. Hence, eqn (11.54) may be rewritten in the form

$$\tau_c = \frac{F_{max}}{bL_{eff}}. \tag{11.55}$$

By way of contrast, the resisting force offered by the obstacle at the moment of breakaway corresponding to the Orowan process is dictated by the equilibrium equation written above as eqn (11.52). In particular, since $\phi = 0$ in this case, we have $F_{max} = 2T$, and therefore the critical stress is

$$\tau_c = \frac{2T}{bL_{eff}}. \tag{11.56}$$

The beauty of this result is that it requires no particular insight into the details of the pinning force, only requiring a bound on its net strength. We will see in our discussion of precipitate hardening that both bow-out scenarios are possible and result in different conclusions as to the way in which $\Delta\tau$ scales with the particle size and distribution.

There are a number of different complaints that can be registered concerning our arguments. First, our results are cast in the abstract language of the modeler, without any reference to the more important microstructural parameters that are experimentally accessible. For example, we would clearly prefer a statement of the critical stress in terms of parameters such as the volume fraction of obstacles and their mean size. In addition, there is a distinctly approximate flavor to our treatment of the distribution of obstacles which appears only through the parameter L_{eff}. In coming sections, we will see just how far we might go in remedying these shortcomings. As yet, we have made no reference to the physical origins of the obstacle force F_{max}. In the next section, we examine the way in which insights may be borrowed from the linear theory of elasticity in order to develop plausible models of these forces.

11.6.2 The Force Between Dislocations and Glide Obstacles

Our current ambition is to elucidate one of the microscopic mechanisms that has been charged with giving rise to solution and precipitate hardening. The argument is that by virtue of the elastic fields induced by an obstacle there will be a force on a dislocation which the dislocation must overcome in its motion through the crystal. As a first step towards modeling this phenomenon, we imagine the obstacle to be a spherical disturbance within the material. As was already demonstrated in chap. 7, such an obstacle produces spherically symmetric displacement fields of the form $u_r = Ar + b/r^2$.

Once again invoking arguments about configurational forces, our expectation is that the force of interest is given by the derivative in the interaction energy as a function of some small excursion of the dislocation line. In generic terms, to find the interaction energy we must evaluate the integrals

$$E_{int} = \frac{1}{2}\left(\int_\Omega \sigma_{ij}^{disl}\epsilon_{ij}^{obs}\,dV + \int_\Omega \sigma_{ij}^{obs}\epsilon_{ij}^{disl}\,dV\right), \tag{11.57}$$

where the superscripts refer to the fields of the dislocation (*disl*) and obstacle (*obs*). This equation derives from the recognition that the total elastic energy is given by an integral

$$E_{tot} = \frac{1}{2}\int_\Omega \sigma_{ij}^{tot}\epsilon_{ij}^{tot}\,dV, \tag{11.58}$$

and the observation that by linearity, the total fields can be written as a superposition of the separate obstacle and dislocation fields. Furthermore, if we note that both the dislocation and obstacle stress fields are obtained using the same elastic modulus tensor, then eqn (11.57) collapses to

$$E_{int} = \int_{\Omega} \sigma_{ij}^{obs} \epsilon_{ij}^{disl} dV,$$
(11.59)

which will serve as the starting point for the remainder of our analysis.

In principle, eqn (11.59) is all that we need to go about computing the interaction force between an obstacle and a dislocation. However, we will find it convenient to rewrite this equation in a more transparent form for the present discussion. Using the principle of virtual work, we may rewrite the interaction energy as

$$E_{int} = \int_{\Omega} \mathbf{f}^{obs} \cdot \mathbf{u}^{disl} dV.$$
(11.60)

Recall that the obstacle elastic fields may be arrived at by solving a boundary value problem in which three force dipoles are superposed as we did in chap. 7. In particular, we had

$$\mathbf{f}^{obs}(\mathbf{x}) = \sum_{i=1}^{3} F \left[\delta \left(\mathbf{x} - \mathbf{x}' - \frac{a}{2} \mathbf{e}_i \right) - \delta \left(\mathbf{x} - \mathbf{x}' + \frac{a}{2} \mathbf{e}_i \right) \right] \mathbf{e}_i.$$
(11.61)

Substituting this expression into eqn (11.60), we are led to

$$E_{int} = \sum_{i=1}^{3} F \left[u_i^{disl} \left(\mathbf{x}' + \frac{a}{2} \mathbf{e}_i \right) - u_i^{disl} \left(\mathbf{x}' - \frac{a}{2} \mathbf{e}_i \right) \right].$$
(11.62)

If we now go about expanding the displacement fields about the point $\mathbf{x}'$, then we are left with

$$E_{int} \approx F a \nabla \cdot \mathbf{u}^{disl}.$$
(11.63)

Recall from our review of continuum mechanics that the divergence of the displacement field is a measure of the volume change associated with a given deformation. In this case, it is the volume change associated with the dislocation fields at the obstacle that gives rise to the interaction energy.

At this level of description, only an edge dislocation will have any interaction with our spherical obstacle since $\mathrm{tr}(\epsilon) = 0$ for a screw dislocation. On the other hand, for an edge dislocation we have

$$\nabla \cdot \mathbf{u} = \mathrm{tr}(\epsilon) = -\frac{b}{2\pi} \frac{1 - 2\nu}{1 - \nu} \frac{\sin \theta}{r}.$$
(11.64)

With this result in hand, the interaction force between the dislocation and the

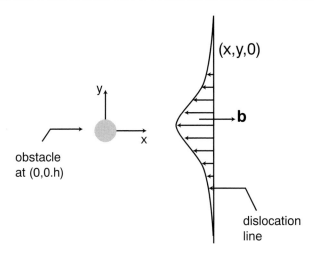

Fig. 11.24. Schematic of the geometry of interaction between a spherically symmetric obstacle and an edge dislocation.

obstacle may be computed by differentiating the interaction energy with respect to the coordinates of the dislocation. In particular, if we consider the case depicted schematically in fig. 11.24, and interest ourselves only in the component of the force acting along the glide direction of the dislocation, we have

$$F_x = -\frac{\partial E_{int}}{\partial x} = -C\frac{x}{(x^2 + d^2)^2}, \tag{11.65}$$

where we have absorbed all of the material and geometric constants describing the dislocation and obstacle themselves into the constant C. The maximum force occurs for $x = d/\sqrt{3}$.

To complete this model, we must return to an explicit reckoning of the quantity Fa which appears in the discussion above. In particular, recall that this product is the dipole moment associated with the obstacle and we wish to reconcile this quantity with the geometric features of the spherical inclusion such as its radius and the displacement at its surface. To do so, we recall from our earlier discussion that the displacement induced by such a collection of force dipoles is given by

$$\mathbf{u}(\mathbf{x}) = \frac{Fa}{4\pi(\lambda + 2\mu)}\frac{\mathbf{r}}{r^3}. \tag{11.66}$$

Hence, if we insist that the displacement at the surface of this hole (i.e. for $r = a_0/2$) is δ, then we find

$$Fa = 4\pi\delta a_0^2(\lambda + 2\mu). \tag{11.67}$$

If we rewrite $\lambda + 2\mu$ in terms of μ and ν and substitute the resulting expression into our equation for the interaction force, we are left with

$$E_{int} = -4\mu\delta a_0^2 bd \frac{d}{x^2 + d^2},$$ (11.68)

and

$$F_x = -8\mu\delta a_0^2 bd \frac{x}{(x^2 + d^2)^2}.$$ (11.69)

As yet, our analysis has concerned itself with the *total* interaction force between the dislocation and the obstacle to its motion. However, a deeper understanding of the pinning process itself can be reached if we compute the force on the dislocation segment by segment. The geometry of interest is indicated in fig. 11.24. We are now interested in obtaining the force on a segment at the point shown in the figure with coordinates $(x, y, 0)$. The calculation amounts to nothing more than computing the Peach–Koehler force $d\mathbf{f} = (\sigma\mathbf{b}) \times d\mathbf{l}$. The glide component of that force is given by $-\sigma_{xz}^{obstacle} bdl$, where the stress field is that due to the obstacle. Recall from our earlier discussion that the displacements for the obstacle (assumed spherically symmetric for the moment) are $u_r = \delta(a_0/r)^2$. Such displacements imply a strain

$$\epsilon_{xz} = -\frac{3\delta a_0^2 xh}{(x^2 + y^2 + h^2)^{\frac{5}{2}}},$$ (11.70)

where the parameter h is the distance of the obstacle beneath the slip plane. As a result of this analysis, we find that the localized force on the dislocation segment due to the obstacle is

$$\frac{F_{glide}}{dl} = \frac{6\mu b\delta a_0^2 dh}{(d^2 + h^2 + y^2)^{\frac{5}{2}}}.$$ (11.71)

Although there are a number of interesting features of this analysis, it also leaves us with serious concerns about the formulation of an elastic theory of the obstacle forces that impede dislocation motion. In particular, this analysis suggests that for an obstacle on the slip plane itself, there is no interaction with the dislocation. Despite this elastic perspective, it seems certain that core effects will amend this conclusion.

11.6.3 The Question of Statistical Superposition

In the previous subsection, we considered the emergence of the parameter F in eqn (11.54) on the basis of an elastic treatment of the interaction between a single dislocation and a single obstacle. In order to complete the analysis associated with eqn (11.54) we must now determine the parameter L_{eff} which characterizes the distribution of obstacles. To do so, we will invoke statistical arguments of varying

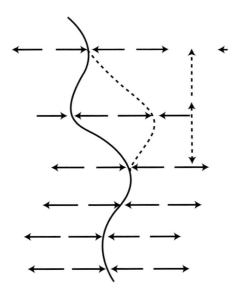

Fig. 11.25. Schematic of the random forces exerted on a dislocation by virtue of the presence of point disorder such as is induced by foreign particles (adapted from Mott and Nabarro (1948)).

levels of sophistication. Our key approach will be that of localized point obstacles which Nabarro (1972) points out 'provides an anchor in the sea of uncertainty.'

In light of our understanding of the interaction between a single localized obstacle and a dislocation, we next take up the question of how many such interactions conspire to yield observed hardening behavior. Again, we begin with a qualitative discussion aimed at examining some of the generic issues that must be addressed in tackling the question of *statistical superposition* of interest here. To that end, fig. 11.25 shows a schematic from one of the famed early papers of hardening (Mott and Nabarro, 1948) in which an attempt was made to show the sort of random assemblage of forces that would act on a dislocation in the presence of a statistical distribution of obstacles. For our present purposes, what we wish to note is that unlike the ideal situation considered in our discussion of the Peierls stress which culminated in eqn (8.87), the intrinsic lattice friction must be supplemented by the statistical superposition of the random forces considered here.

A more sophisticated consideration of the forces that are exerted upon a dislocation as it glides through a medium populated by a distribution of obstacles is depicted in fig. 11.26. In this case, the fundamental idea that is being conveyed is to divide the character of various obstacles along the lines of whether or not they are localized or diffuse and whether or not they are 'strong' or 'weak'. These different scenarios result in different scaling laws for the critical stress to induce dislocation motion.

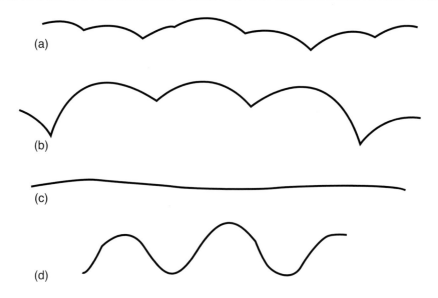

Fig. 11.26. Four distinct regimes suggested for the interactions between a dislocation and the disorder distributed throughout the body (adapted from Nabarro (1972)). The four regimes of interest are: (a) weak, localized obstacles, (b) strong, localized obstacles, (c) weak, diffuse obstacles and (d) strong, diffuse obstacles.

Periodic Array of Point Obstacles. The simplest picture that can be developed, and one for which our analysis leading up to eqn (11.54) is essentially exact, is one in which it is imagined that the obstacles are arranged periodically in the slip plane. One argument that we might set up to compute the areal density in terms of the known volume concentration c of obstacles, is to note that if we assign a slip plane of area A a volume Ab, then the number of obstacles found in that plane will be cAb. We have used the discrete packing of atomic planes to obtain a volume per plane cAb. As a result of this argument, we see that the areal density may be written either as cAb/A or in terms of L_{eff} as $1/L_{eff}^2$. What we have said is that the cAb obstacles which find themselves in the slip plane of interest are arranged on a square lattice with lattice parameter L_{eff}. By equating our two expressions for the areal density, we are left with the relation

$$L_{eff} = \frac{1}{\sqrt{bc}}.$$

$$(11.72)$$

If we take the result given above and exploit it in the context of eqn (11.54) we obtain one of the crucial scaling relations for obstacle limited motion of dislocations, namely,

$$\tau_c \propto \sqrt{c}.$$

$$(11.73)$$

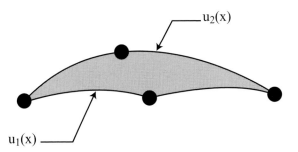

Fig. 11.27. Schematic of two consecutive configurations in the unpinning history of a given dislocation.

This result is one which Kocks (1977) goes so far as to note already contains 'the most important result of dislocation theory: namely, that the flow stress is proportional to the square root of concentration'. Indeed, to the extent that we are interested in understanding the observed macroscopic plastic properties of materials, scaling relations involving the concentration of obstacles are pivotal.

Random Array of Point Obstacles. At the next level of sophistication we imagine the opposite extreme in which the distribution of point obstacles is taken to be completely random. As was mentioned above, there are a variety of circumstances in which the $\tau_c \propto \sqrt{c}$ scaling is obtained, and we will see that in the present setting this same scaling is found. Our strategy is to compute the so-called Friedel length, which is a stress-dependent measure of the effective spacing between obstacles.

In this case, the argument is altogether more subtle. The basic idea is to consider two consecutive pinned configurations such as are shown in fig. 11.27. Physically, the idea is that as the stress increases, the number of pinning points that the dislocation joins up with will vary. As a result, even though the areal density of such pinning points is unchanged, the mean spacing between such pinning points along the dislocation (i.e. L_{eff}) will change. Our aim is to estimate L_{eff} and in particular, its stress dependence. The key physical assumption made in the Friedel argument is that once the dislocation becomes unpinned it will sweep an area $l^2 = 1/\sqrt{bc}$ before encountering another pinning point. Comparison of the two configurations shown in fig. 11.27 shows that in passing from the initial to the final configuration, an area $l^2 = 1/\sqrt{bc}$ is swept out.

To make these arguments concrete, we note that we are insisting that the area ΔA swept out be given by the relation

$$\Delta A = \frac{1}{bc} = \int_0^{2L_{eff}} u_2(x)dx - 2\int_0^{L_{eff}} u_1(x)dx. \tag{11.74}$$

We have introduced the function $u_1(x)$ to characterize the dislocation position of the initial configuration and $u_2(x)$ to characterize the dislocation of the final configuration. Note that the as-yet undetermined (and desired) parameter L_{eff} appears in the limits of integration and will be determined in the end as a function of the concentration and the applied stress. In fact, to obtain the functions $u_1(x)$ and $u_2(x)$ we use what we already learned about dislocation bow-out in section 8.7.1. In particular, in the limit of weak driving force (which corresponds to the presumed weakness of the obstacles) generically, the dislocation profile is of the form

$$u(x) = -\frac{\tau b}{T}x^2 + cx, \tag{11.75}$$

where the constant c is determined by appealing to the boundary conditions. In the present case, we have $u_1(0) = u_1(L_{eff}) = 0$ and $u_2(0) = u_2(2L_{eff}) = 0$. Application of these boundary conditions results in the solutions

$$u_1(x) = \frac{\tau b}{2T}(L_{eff}x - x^2) \tag{11.76}$$

and

$$u_2(x) = \frac{\tau b}{T}\left(L_{eff}x - \frac{x^2}{2}\right). \tag{11.77}$$

With these results in hand we may evaluate the integrals demanded in eqn (11.74) with the result that

$$\Delta A = \frac{\tau b L_{eff}^3}{2T}. \tag{11.78}$$

Note that as a result of these arguments we may now determine L_{eff} as

$$L_{eff} = \left(\frac{2T}{\tau b^2 c}\right)^{\frac{1}{3}}. \tag{11.79}$$

This result is now in a form that may be fed back into eqn (11.55) with the result that

$$\tau_c = \frac{F_{max}}{b}\left(\frac{\tau_c b^2 c}{2T}\right)^{\frac{1}{3}}, \tag{11.80}$$

which may be solved for τ_c with the result that

$$\tau_c = \left(\frac{F_{max}^3}{2Tb}\right)^{\frac{1}{2}} c^{\frac{1}{2}}. \tag{11.81}$$

We note that even in the context of this more roundabout description of the superposition of the effect of different obstacles, we once again recover the $\tau_c \propto c^{1/2}$ scaling discussed earlier.

Continuously Distributed Disorder. Our treatment until this point has treated the disordering effect of the obstacles on the basis of a picture in which the homogeneous medium is punctuated by localized forces. An alternative viewpoint is one in which the disordering effect is imagined to be distributed continuously throughout the medium resulting in a local contribution to the resistance at every point on the dislocation. As a result of this picture, rather than attempting to assert equilibrium conditions only at pinning points, equilibrium is represented in integral form along the entire length of the dislocation. For the present purposes, the crucial conclusion of this analysis which we quote without derivation is the assertion that the critical stress to induce dislocation glide scales as $\tau_c \propto c^{2/3}$. Our reason for bringing this result up at all is mainly to serve as a point of comparison when we consider the actual data on solution hardening to be taken up in the next section.

11.6.4 Solution Hardening

The use of impurities to alter the behavior of materials is a technique of great significance, and solution hardening is but one of many ways in which this is carried out. Our intention in this section is to present a few experimental highlights which reveal the level of the quantitative understanding of solution hardening which will serve to catalyze the modeling discussion to follow. Having introduced the experimental backdrop, the remainder of our discussion will be fixed on how well the ideas introduced in the previous few subsections rise to the challenge of the data.

An example of the type of data associated with solution hardening it is the mission of our models to explain was shown in fig. 8.2(a). For our present purposes, there are questions to be posed of both a qualitative and quantitative character. On the qualitative side, we would like to know how the presence of foreign atoms dissolved in the matrix can have the effect of strengthening a material. In particular, how can we reconcile what we know about point defects in solids with the elastic model of dislocation–obstacle interaction presented in section 11.6.2. From a more quantitative perspective, we are particularly interested in the question of to what extent the experimental data permit a scaling description of the hardening effect (i.e. $\tau \propto c^n$) and in addition, to what extent statistical superposition of the presumed elastic interactions between dislocations and impurities provides for such scaling laws.

The metallurgical approach to coming to terms with data like that considered above is by virtue of an appeal to various parameters that signal the presence of foreign atoms and which can be used as the basis of contact with the elastic theory of interaction forces. Indeed, although results like that revealed in eqn (11.81) are extremely provocative, our work will only be done once plausible strategies are found for linking quantities such as F_{max} to underlying parameters of metallurgical significance. In particular, two of the most immediate parameters that measure the extent to which a given set of foreign atoms will make their presence known to the dislocation elastically are the size misfit parameter defined by

$$\delta = \frac{1}{a}\frac{da}{dc},\tag{11.82}$$

and the modulus misfit parameter

$$\eta = \frac{1}{\mu}\frac{d\mu}{dc}.\tag{11.83}$$

The symbols introduced here include the lattice parameter a, the concentration c and the shear modulus μ. The purpose of these parameters is to provide a measure of the extent to which the presence of foreign atoms within the matrix will result in elastic interactions between these impurities and dislocations present in the material. My own take on these ideas is that they are largely inspired by the desire to make contact with the ideas on elastic fields of inclusions developed by Eshelby. We remind the reader that in that setting (see section 7.3.2), a foreign particle can make its presence known by virtue both of a difference in size and of a difference in moduli between the matrix and the particle. Of course, when speaking of a single foreign atom, these continuum notions become highly suspect. Nevertheless, what our elastic models have shown us is that if we are to install an inclusion within a continuum, the distinction between that particle and the surrounding matrix material can be introduced either by attributing a different size to that particle or by asserting that it is characterized by different moduli than that of the matrix material. The parameters δ and η introduced in eqns (11.82) and (11.83) measure the strength of these two effects.

In addition to the two mechanisms introduced above, there are a variety of other mechanisms that have been invoked to explain the origins of solution hardening, including contributions due to the presence of stacking faults, the difference in valence of the matrix and solute and ordering effects. The key point to be taken away from this part of the discussion is the recognition that there are a wide variety of different ways in which we can construct elastic models of the interaction between foreign atoms and dislocations. We have seen that if we model a given foreign atom as a source of dilatation, a dislocation–obstacle interaction arises that depends upon the parameter δ. Similarly, if a given foreign atom is modeled as a source of elastic inhomogeneity (i.e. a region of different elastic constants than the matrix) then a dislocation–obstacle interaction arises that depends upon parameters like η introduced above. In addition to these models, it is also possible to imagine more sophisticated effects related to elastically induced redistribution

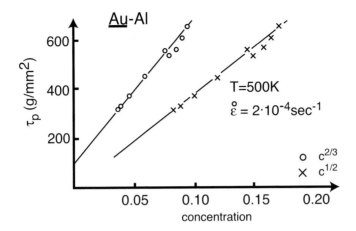

Fig. 11.28. Illustration of fit to $\sqrt{c}$ and $c^{\frac{2}{3}}$ scaling to observations on solid solution hardening in Au–Al (adapted from Haasen (1979)).

of the foreign atoms around the dislocations. All of these models share the feature that they inhibit the motion of dislocations.

Though the qualitative ideas introduced above are instructive, to really make contact with experiment we must calculate something. In particular, we need to find a way to average over the types of disorder described schematically in fig. 11.26 so as to obtain a concrete result for the scaling of the flow stress with the concentration of foreign atoms, for example. Indeed, this is precisely what was done in various approximations in section 11.6.3. For the present purposes what we wish to really note about the outcome of the arguments given there is the emergence of two different scaling laws for how the critical stress scales with the concentration of impurities, namely, $\tau_c \propto \sqrt{c}$ and $\tau_c \propto c^{2/3}$. A provocative representation of experimental data on solution hardening is shown in fig. 11.28. What we see on the basis of these data is that although clearly increasing the concentration increases the critical stress, identifying the particular scaling relation is a more challenging proposition.

In addition to the attempt to link observed solution hardening with ideas on the scaling with concentration, efforts have been put forth on examining how the strengthening effect scales with the parameters δ and η. A particular example which reveals the magnitude of the hardening effect as a function of the size of the foreign atoms is shown in fig. 11.29. The reader is encouraged to examine these issues in the article of Haasen (1979). A second facet to this discussion that we have omitted but which is immensely important is the temperature dependence of the flow stress. As it stands, our discussion really only attempts to answer to the challenge of the zero-temperature contribution to the flow stress.

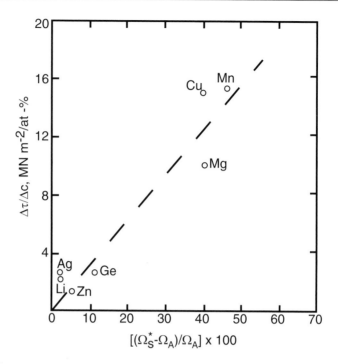

Fig. 11.29. Magnitude of solution hardening effect in Al alloys as a function of the size mismatch of the foreign atoms (adapted from Embury *et al.* (1989)).

11.6.5 Precipitate Hardening

In the previous section, we saw that we can arrange that a particular material be endowed with a supersaturation of impurities by quenching some concentration of foreign atoms in the material beyond the solubility limit. On the other hand, the resulting microstructure is metastable. The equilibrium phase that is preferred in this part of the phase diagram is two-phase coexistence between the pure phase and a second phase, which is often an ordered alloy built up from the two components. For example, in the context of the Fe–C phase diagram, this coexistence is struck between α-Fe (bcc) and an orthorhombic phase of Fe_3C. From a modeling perspective for considering the mechanics of such two-phase microstructures, the simplest picture we might imagine is that the nuclei of these precipitates are spherical and that their radii increase with time. The presence of such second-phase particles makes itself known through a substantive change in the mechanical properties of the material, a specific example of which is shown in fig. 8.2(b)

   Observations on precipitation hardening make it evident that in certain cases the key microstructural features in this context are the mean particle radius and the

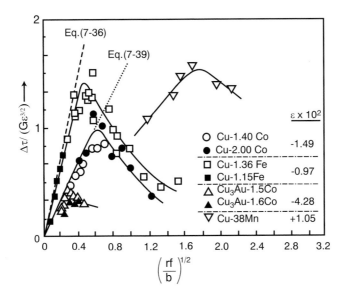

Fig. 11.30. Illustration of the variation in flow stress as a function of precipitate radius (adapted from Reppich (1993)).

volume fraction of such precipitates. In particular, the type of generic results that emerge are indicated schematically in fig. 11.30 in which the variation in the flow stress is shown as a function of particle size. Our present ambition is to explain this behavior on the basis of the type of simple models developed above. In simplest terms, the key feature of the flow stress increment as a function of particle size is the fact there is an 'optimal' particle size. What we will find is that this maximum reflects a change in mechanism from particle cutting, which is dominant at small radii, to the Orowan process, which is easier at large particle radii.

Our discussion in section 11.6.1 culminated in two possible outcomes to the encounter of a dislocation with an array of pinning points. On the one hand, we found that if the obstacles are sufficiently weak (characterized by a maximum force F_{max}), they will be breached at a critical stress of $\tau_c = F_{max}/bL_{eff}$ as shown in eqn (11.55). On the other hand, if the obstacles are strong enough, they will be traversed by an alternative mechanism in which the dislocations loop around the obstacle and rejoin on the opposite side, a mechanism known as Orowan looping and characterized by a strength $\tau_c = 2T/bL_{eff}$ as shown in the derivation leading up to eqn (11.56). These two different eventualities are illustrated schematically in fig. 11.31. We now undertake a systematic analysis of these two mechanisms, with special emphasis being placed on how to rewrite eqns (11.55) and (11.56) in terms of the metallurgical parameters relevant to two-phase microstructures.

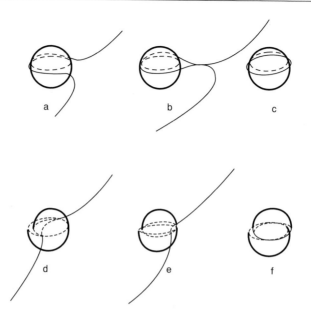

Fig. 11.31. Schematic of the Orowan looping and particle cutting mechanisms that arise from interaction of dislocations with foreign particles (adapted from Gerrold (1979)).

Particle Cutting

For small radii, the mechanism of dislocation interaction with the foreign particles is that of particle cutting in which after a certain maximum force is reached, the dislocation shears the foreign particles themselves. As a result, in this context the relevant equation for the critical stress is eqn (11.55) and our charter is to reflect once again on F_{max} and L_{eff} from the perspective of the microstructure itself. Without entering into details, we note that L_{eff} can be obtained using exactly the same arguments given in section 11.6.3 with the proviso that we replace bc with the areal density ρ_{Area}. If we carry out this substitution, eqn (11.55) may be rewritten as

$$\tau_c = \frac{F_{max}^{\frac{3}{2}}}{b\sqrt{2T}}\sqrt{\rho_{Area}}. \tag{11.84}$$

At this point our task is to relate the areal density ρ_{Area} to the volume fraction of second-phase particles, denoted by f, and to the particle radius R.

   To convert the result given above into these more usable terms, we make a highly simplified argument based on the following assumptions. (1) The particles are all of the same size and are characterized by a radius R. (2) The estimate of the mean particle spacing is determined by assuming full correlation between the particles positions. For simplicity, we imagine them arranged on a three-dimensional cubic

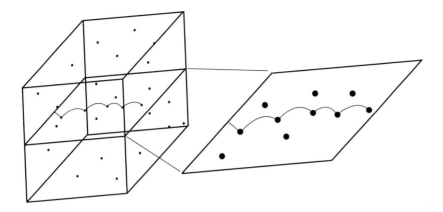

Fig. 11.32. Schematic of the geometry of the precipitate particles.

lattice. Note that despite the seeming artificial nature of these assumptions, in some cases the distribution of precipitates can indeed be highly regular.

To determine the mean spacing between obstacles, we begin with the modeling assumption that any particle whose center lies within a distance R of the slip plane of interest is considered to provide a glide obstacle on that slip plane. Further, we assert that the volume fraction of such particles is given by the parameter f. The geometrical ideas are depicted in fig. 11.32. The volume occupied by the precipitate particles in this slab of thickness $2R$ and area A is $2Af R$. Hence the number of particles in this slab volume can be estimated by dividing the volume occupied by particles within the slab by the volume per particle, resulting in

$$\text{\# particles in slab} = 3Af/2\pi R^2.$$

This result may be used to compute the area per obstacle ($A_{obstacle}$). Note that the area per obstacle is related to the areal density by $\rho_{Area} = 1/A_{obstacle}$. The area per obstacle is found by dividing the total area of the slip plane by the number of such particles and yields the estimate

$$\rho_{Area} = \frac{3f}{2\pi R^2}. \tag{11.85}$$

In light of these arguments, the critical stress for particle cutting is of the form

$$\tau_c = \frac{F_{max}^{3/2}}{b\sqrt{2T}}\sqrt{\frac{3f}{2\pi}}\frac{1}{R}. \tag{11.86}$$

In order to complete the discussion, we must say something about how F_{max} depends upon the mean particle radius. It is crucial to note that a wide variety of different F_{max}s associated with particle cutting have been determined. For a broad discussion of this topic see section 2.3 of the article by Gerrold (1979) and

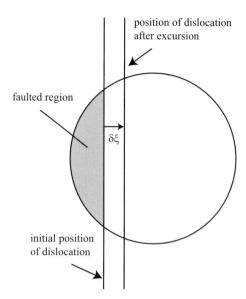

position of dislocation
after excursion

faulted region

$\delta\xi$

initial position
of dislocation

Fig. 11.33. Schematic cross section of a precipitate particle as it is being cut by a gliding dislocation.

the discussion in section IV of the article by Embury *et al.* (1989). Rather than entering into the particulars of these different mechanisms, we note that much can be learned by examining the scaling of τ_c with f and R for different choices of F_{max}. In particular, we note that if we make the assumption that F_{max} scales with R as $F_{max} = dR^m$, then the critical stress is of the form

$$\tau_c = \frac{d^{3/2}}{b\sqrt{2T}}\sqrt{\frac{3f}{2\pi}}R^{\frac{3m}{2}-1}. \tag{11.87}$$

Though we have alluded to the existence of a variety of different mechanisms giving rise to F_{max}, we will end our discussion of particle cutting by considering in a bit more detail the case of order strengthening. Order strengthening refers to those cases in which the precipitate and the matrix have the same crystal structure with the difference that the precipitate itself is ordered. As a result, when the dislocation traverses the particle, it does so at the cost of introducing an antiphase boundary characterized by an energy γ_{apb}, with dimensions of energy/area. To compute F_{max} in this case, we must essentially construct the configurational force which tells how the interfacial energy penalty increases with an incremental excursion of the dislocation. The energy and associated force may be calculated by computing the energy of two adjacent configurations for the dislocation as shown in fig. 11.33. All that the calculation really demands is that we compute the area of the shaded region in the figure and then differentiate this energy with respect to the coordinate x. More explicitly, this calculation requires that we assess

$$F = \lim_{\delta\xi \to 0} \gamma_{apb} \frac{A(x + \delta\xi) - A(x)}{\delta\xi}. \tag{11.88}$$

The calculation is relatively straightforward and results in

$$F_{max} = 2\gamma_{apb}R \tag{11.89}$$

for the case in which the dislocation glides along the equator of the particle. If we like, we can be more sophisticated and average over all planes cutting through the particle with the result that

$$F_{max} = \frac{\pi \gamma_{apb} R}{2}. \tag{11.90}$$

The key point for our purposes is the fact that F_{max} scales linearly with the particle size.

For the particular case in which we assume that $F_{max} = \text{const.}R$ such as was found in the order strengthening results described above, the results above simplify to the form

$$\tau_{cut} = A_1 \frac{h^{\frac{3}{2}}}{\sqrt{\mu b}}\sqrt{f R}. \tag{11.91}$$

Note that we have followed Reppich (1993) in this equation by burying the dependence of the results on the particular mechanism of hardening in the parameter h. One of the key observations that we can make at this point is that with increasing particle radius it becomes increasingly difficult to cut the particles. Evidently, if there is some competing mechanism and a certain size is reached for which it is easier to institute that mechanism, we will see a transition in the dependence on particle size. The mechanism of Orowan looping alluded to earlier is just such a mechanism.

Orowan Looping

Plausible back of the envelope models may be developed for the emergence of the Orowan looping mechanism. We have already seen that the stress at which such looping commences is given by $\tau_{loop} = 2T/bL_{eff}$. As we noted earlier, these results can be cast in a much more desirable form if their dependence on microstructural parameters is made manifest. Our earlier discussion culminating in eqn (11.85) showed that

$$L_{eff} = \frac{1}{\sqrt{\rho_{Area}}} = \sqrt{\frac{2\pi}{3f}}R. \tag{11.92}$$

In conjunction with the assertion that the line tension is given by $T = \mu b^2/2$, we may use the result for L_{eff} obtained above to reexpress the stress for Orowan looping in terms of the relevant microstructural parameters. In particular, we find

$$\tau_{loop} = \sqrt{\frac{3f}{2\pi} \frac{\mu b}{R}} \sqrt{f}.$$ (11.93)

Because there are a number of different possible schemes for handling the correlations between particle positions, this result is more safely rewritten as

$$\tau_{loop} = A_2 \frac{\mu b}{R} \sqrt{f},$$ (11.94)

which is noncommittal with respect to numerical prefactors, but shows the way in which the looping stress *scales* with the relevant microstructural parameters.

As a result of our estimates of the scaling of the flow stress with both particle size and volume fraction, we are now in a position to reexamine some representative experimental data. One possible avenue to consider is to rationalize the dependence on particle size at fixed volume fraction. In rough terms, the particle radius is a direct reflection of aging time. One immediate observation to be made on the basis of eqns (11.91) and (11.93) is that at fixed volume fraction the critical stress for particle cutting is an increasing function of R. On the other hand, the critical stress for Orowan looping is a decreasing function of R suggesting that we can anticipate a crossover from cutting to looping with increasing particle size. If we are to solve for the particle size at which the two mechanisms result in the same τ_c we find

$$R = \frac{(2\mu^2 b^4 T)^{\frac{1}{3}}}{d},$$ (11.95)

which is a measure of the crossover particle size.

A second venue within which it is possible to examine the validity of the various approximations considered above is through a direct appeal to the experiments themselves as shown in fig. 11.34. The experimental observations reported in the figure consider the relatively simpler case in which the critical stress for Orowan looping is evaluated for a variety of different interparticle spacings. As seen above, because of the wide variety of different mechanisms all giving rise to F_{max}s to be used in conjunction with the expression for particle cutting, it is more difficult to make a definitive falsification of the theoretical models.

11.6.6 Dislocation–Dislocation Interactions and Work Hardening

In addition to the chemical mechanisms for hardening described above, there are intrinsic mechanisms that accompany the increase in dislocation density with increasing strain. Recall fig. 8.40 in which it was shown that the density of dislocations increases continuously with increasing plastic strain. Attendant with this increasing dislocation density is a hardening effect. An example of the type

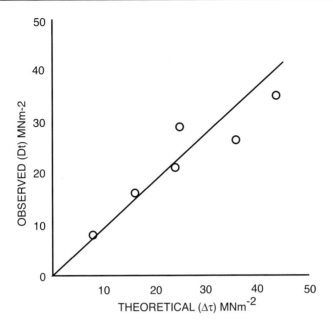

Fig. 11.34. Comparison of theoretical and observed increment in the flow stress for particle strengthened Al–Al$_2$O$_3$ (adapted from Embury *et al.* (1989)).

of data associated with the work-hardening effect was shown in fig. 8.2(d). From an empirical perspective, data of this kind are characterized by a robust scaling of the flow stress with the dislocation density of the form $\tau_c \propto \sqrt{\rho}$. Indeed, it can be argued that the most robust feature of hardening is the existence of the relation,

$$\tau_{flow} = \alpha \sqrt{\rho}, \tag{11.96}$$

where ρ is the overall dislocation density. The question of hardening behavior is further enriched through consideration of data like that presented in fig. 11.35. My main reason for showing this figure is largely motivational and is meant to whet the reader's appetite for the more subtle question which must be faced in considering the orientation dependence of the work hardening properties of materials.

We argued in section 8.7.5 that dislocation junctions are one of the ways in which different dislocations can render each other inoperative. From this point of view, a possible conceptual picture of the hardening process is to work by analogy with the arguments given prior to our discussion of solution and precipitate hardening. In particular, we view the glide of a dislocation of interest as interrupted by its encounters with other dislocations such as those associated with other slip planes. Our aim in the present section was to borrow the statistical arguments given in previous sections and implement them in the case where the obstacles offered

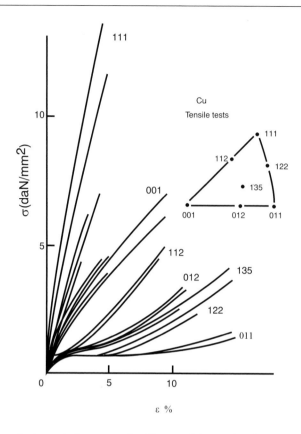

Fig. 11.35. Series of stress–strain curves for Cu tested in tension for a number of different orientations (adapted from Franciosi (1985)).

to a given dislocation arise from other dislocations themselves. This situation is indicated schematically in fig. 11.36.

In fact, though we will leave this huge and interesting topic largely unexplored, all that we really wish to note is that the machinery culminating in eqn (11.84) can be transported directly to the present setting. We note that the dislocation density itself is precisely the sort of areal density we need in conjunction with eqn (11.84) and hence without further ado the key scaling relation $\tau_c \propto \sqrt{\rho}$ is recovered.

11.7 Further Reading

Diffusion at Extended Defects

'Dislocations, vacancies and interstitials' by R. W. Balluffi and A. V. Granato in *Dislocations in Solids* Vol. 4, edited by F. R. N. Nabarro, North-Holland Publishing Company, Amsterdam: The Netherlands, 1979. This article describes the interplay of point defects and dislocations.

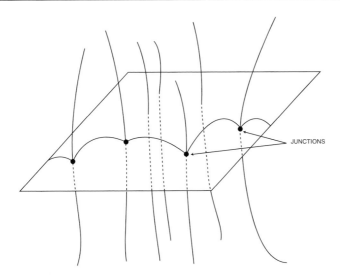

Fig. 11.36. Schematic of the forest hardening process. Dislocation gliding in the primary slip plane forms junctions as a result of encounters with dislocations piercing that plane.

Mass Transported Assisted Deformation

The Physics of Creep by F. R. N. Nabarro and H. L. De Villiers, Taylor & Francis Ltd., London: England, 1995. This book is full of interesting ideas concerning creep, with special reference to the problem of creep in alloys.

Mechanical Behavior of Materials by Thomas H. Courtney, McGraw-Hill Publishing Company, New York: New York, 1990. Courtney's chap. 7 is something that I have returned to on numerous occasions.

Atom Movements by Jean Philibert, Les Éditions de Physique, Les Ulis Cedex A: France, 1991. As noted already in chap. 7, this is my favorite book on diffusion. Philibert's chap. 10 is on 'The Study of Some Diffusion-Controlled Processes' and has a thoughtful description of the approximate models used to obtain the type of creep rate equations developed in the present chapter.

Dislocations and Cracks

'Fracture Toughness' by A. E. Carlsson and R. Thomson in *Solid State Physics* Vol. 51, Academic Press, San Diego: California, 1998, edited by H. Ehrenreich and F. Spaepen.

Dislocation Based Fracture Mechanics by J. Weertman, World Scientific, Singapore, 1996. Weertman's book is especially admirable for its uniqueness of perspective. In particular, he sets himself the task of building up the study of fracture entirely from the dislocation perspective. In that regard, it is a most interesting place to study many of the solutions to multiple dislocation problems considered both in the latter part of this chapter as well as in our discussion of pile-ups.

'Dislocations and Cracks' by E. Smith, in *Dislocations in Solids*, edited by F. R. N. Nabarro, North Holland Publishing Company, Amsterdam: The Netherlands, 1979. Smith's article does a nice job of describing the superposition of dislocations in order to build up cracks.

Solution and Precipitate Hardening

Of course, many of the references listed at the end of chap. 8 have discussions of hardening due to foreign particles. A few references that tackle these problems in particular are listed below.

Micromechanisms in Particle-Hardened Alloys by J. W. Martin, Cambridge University Press, Cambridge: England, 1980. This short book is highly instructive. In addition to enlightening descriptions of the current theoretical understanding of issues related to particle hardening, a wealth of data and experimental images lend it a balance not often met.

Particle Strengthening of Metals and Alloys by E. Nembach, John Wiley and Sons, Inc., New York: New York, 1997. A book which takes stock of the current understanding of the field.

Plastic Deformation and Fracture of Materials edited by H. Mughrabi. This book is Vol. 6 in the series Materials Science and Technology edited by R. W. Cahn, P. Haasen and E. J. Kramer, VCH Publishers, Inc., New York: New York, 1993. Of especial interest concerning the present discussion see chap. 6 on the subject of solid solution strengthening by H. Neuhauser and C. Schwink and chap. 8 by B. Reppich on the subject of particle strengthening.

The Plastic Deformation of Metals by R. W. K. Honeycombe, Edward Arnold, London: England, 1984. Honeycombe's book is especially appealing because of its emphasis on the metallurgical applications of dislocation theory to the various strengthening mechanisms.

'Solution and Precipitation Strengthening' by F. R. N. Nabarro, in *The Physics of Metals Vol. 2. Defects* edited by P. B. Hirsch, Cambridge University Press, London: England, 1975. Nabarro's article provides an interesting perspective on the thinking concerned with the interactions of dislocations with obstacles.

'Precipitation Hardening' by A. J. Ardell, *Met. Trans.*, **16A**, 2131 (1985). This article provides an outstanding assessment of both the theoretical and experimental situation concerning precipitation hardening. Ardell took another crack at summarizing his vision of the subject in 'Intermetallics as Precipitates and Dispersoids in High-Strength Alloys' in *Intermetallic Compounds: Vol. 2, Practice* edited by J. H. Westbrook and R. L. Fleischer, John Wiley and Sons, New York: New York, 1994. The latter article has an especially nice discussion in section 4 on the mechanisms giving rise to the quantity we have called F_{max}.

'The Theory of an Obstacle-controlled Yield Strength – Report after an International Workhop' by U. F. Kocks, *Mat. Sci. Eng.*, **27**, 291 (1977). This article offers the most thoughtful commentary on the metallurgical uses of dislocation theory that I know. In particular, Kocks does a fantastic job of enumerating the various hidden assumptions used in applying dislocation theory to strengthening, and then offers his perspective on what should be undertaken next.

11.8 Problems

1 Schwoebel Barrier on fcc (111) Surfaces

Using the Johnson embedded atom potential introduced in chap. 4, compute the energetics of an adatom in the vicinity of a step and derive results analogous to those given in fig. 11.2.

2 Grain Boundaries as Dislocation Array

In our discussion of grain boundaries and dislocations, we showed how the stress field due to a grain boundary could be thought of as a superposition of those due to an array of dislocations. Continue the analysis begun in our discussion and compute the fields $\sigma_{xx}(x, y)$ and $\sigma_{yy}(x, y)$ for the same type of grain boundary that led to our result for $\sigma_{xy}(x, y)$ as given in eqn (11.24).

3 Discrete Pile-Up

As an application of the ideas on dislocation pile-ups described in section 11.4.2, consider a pile-up of three dislocations in which the leading dislocation is *fixed* at $(0, 0)$. The other dislocations have coordinates $(x_1, 0)$ and $(x_2, 0)$ which are to be determined by applying the equilibrium equations presented as eqn (11.30). Assume that the externally applied stress is constant and is denoted by τ.

(a) Write down the two equilibrium equations for the two free dislocations. Use dimensionless stress units in which $\tau = f\mu$.

(b) Solve these two coupled equations and show that the equilibrium positions are given by

$$x_1 = \frac{b}{4\pi f}(-3 + \sqrt{3}) \tag{11.97}$$

and

$$x_2 = \frac{b}{4\pi f}(-3 - \sqrt{3}). \tag{11.98}$$

4 Complex Potentials for Dislocation Problems

In eqn (11.43) we wrote the displacement potential for the problem of a free surface with a subsurface dislocation. Use this potential to recover the image stresses for this problem presented in eqn (8.64).

5 The Friedel Length

Supply all of the missing details in the derivation of eqn (11.79).

TWELVE

Bridging Scales: Effective Theory Construction

Though we may not have said so explicitly at every juncture, much of our work has been built around the idea of constructing theories that are capable of treating problems in which either multiple length or time scales are operative simultaneously. Indeed, many of the most fascinating problems in all fields of science involve a diversity of spatial or temporal scales. One of the most intriguing features of systems involving many degrees of freedom is that, like with our loved ones, the whole is greater than the sum of the parts. We made our first reference to this insight in chap. 1 in which we quoted P. W. Anderson and his views concerning 'emergent' properties. Our contention is that despite the ever-increasing ability to simulate complex systems in a brute force way, often a more fertile route is to seek a truncated description involving an effective dynamics associated with effective degrees of freedom. The aim of the present chapter is to adopt a more self-conscious attitude in our development of approaches to such multiscale problems, and to seek the common ground shared by many of these theories as they confront problems involving multiple scales.

As we have already noted, the notion of 'multiple scale modeling' is not a new one. Historically, a number of different approaches have been taken for confronting problems involving multiple scales. One strategy that has had repeated success is to assume a separation of scales in which the model singles out one set of scales as being of fundamental importance, and in which other scales appear in disguise through the phenomenological parameters that are used in the model. One simple expression of this thinking is the widespread use of models of damped oscillators in which the equation of motion is written

$$m\frac{d^2x}{dt^2} + c\frac{dx}{dt} + kx = 0. \tag{12.1}$$

In this context, the presence of the damping term involving the constant c is a profound way of masking the effect of the unaccounted for degrees of freedom which are a repository for the center of mass motion associated with the particle of interest. In reality, we know that it is the progressive donation of kinetic energy from this particle to the rest of the world which is responsible for the ostensible loss of energy. This donation of energy is effected through collisions between the mass m and the unaccounted for particles which surround it and it is exactly to account for these particles which are not treated explicitly that dissipation is invoked in the first place. Even so elementary a model of material behavior as that provided by Hooke's law reflects this same philosophy of separation of scales. In this case, the macroscopic stress and strain are related via the elastic moduli which capture in turn the physics of bond stretching and bending. Note that the type of assumption that enters in the use of an elastic model of this type is that the spatial variation in the strain field takes place at a much larger spatial scale than that of the lattice parameter itself. Though there are many instances in which a separation of scales strategy is effective, there are others in which it is not.

As will become more clear as the chapter unfolds, I have been able to discern several different generic paradigms for the treatment of problems involving multiple scales or multiple processes. In much of the book, we have highlighted theories in which material parameters enter as phenomenological coefficients, and for which that theory falls silent as regards the *specific* values that those parameters adopt. For example, from the very beginning of our discussion of continuum mechanics, we have made reference to the elastic moduli and the yield surface, both of which are central to continuum models of deformation, and both of which can be determined only by appealing to theoretical considerations that are external to continuum mechanics itself. These are examples of what I will refer to as modeling in the *information passage* mode, in which models of one sort are used to inform another. In chap. 5, we showed a trivial example of this type of thinking in deriving expressions for the elastic moduli in terms of curvatures in the energy surface.

However, a more powerful aspect of the information passage mode of thinking is associated not so much with the determination of the various 'stiffness coefficients' that show up in these models (such as the elastic moduli), but rather with the construction of the effective theory itself. Our organizing principles for considering models of this type will center on two rudimentary notions, one kinematic and the other dynamic. From a kinematic perspective, much of what we will have to say in the present chapter will center on the kinematic question of what are the best degrees of freedom for considering a given phenomena. We will couch our answer to this question in the language of state variables, order parameters and defect coordinates. These foundational ideas will allow us to describe the disposition of material systems, whether they are in equilibrium or not. However, once the disposition of the system at a given instant in time can be described, there remains the more serious and challenging question of how that system evolves in time. Since I take the view that the physics of *evolution equations* in the effective theory setting is still in its infancy, what we will have to say about the temporal evolution of complex systems will be largely provisional. Nevertheless, there are a number of provocative ideas in this regard and we will give particular emphasis to the general class of 'gradient flow' dynamics, remaining mindful of the open questions surrounding the possible variational underpinnings for the dynamical evolution of complex systems.

A second generic class of modeling efforts aimed at confronting the difficulties that must be met in tackling problems with multiple scales are those in which multiple scales are built in at the outset, usually by bringing more than one theoretical tool under the same roof. Several key examples to be described in the present chapter include the widespread use of cohesive surface models which we already had a taste of in our consideration of the Peierls–Nabarro model (see section 8.6.2), and the use of mixed atomistic/continuum methods in which the strengths of atomistic and continuum mechanics are kept while their weaknesses are eliminated. This class of models is very appealing since they incorporate the key *non*concepts, namely, nonlinearity, nonlocality and nonconvexity. Each of these *non*concepts has something important to say about the emergence of defects in materials and their interactions as well. On the negative side of the ledger sheet, the types of mixed atomistic/continuum models that we will interest ourselves in presently have the unfortunate feature that it is exceedingly difficult to make analytic progress with them.

12.1 Problems Involving Multiple Length and Time Scales

Problems involving multiple scales in space and time are to be found wherever we look. We have already organized our discussion of structures in materials using characteristic length scales as the basis of this organization. Our first foray into structure in solids concerned the atomic-level geometries adopted by crystalline solids from the standpoint of phase diagrams (see chap. 6). At the next level of structural complexity, we considered the various defects that populate solids (see chaps. 7, 8 and 9). The set of spatial scales associated with single defects in isolation form the next level in the hierarchy of structures found in solids. The organization of such defects into a material's microstructure (see chap. 10) represents yet another scale to be found in crystalline solids. The linkage of these scales is particularly evident when we consider the interaction of individual defects such as dislocations with elements of the microstructure. In this case, we are confronted with a problem in which several different length scales interact simultaneously.

As a more definite and concrete substantiation of our claim that a variety of problems in modeling materials are of an inherently multiscale nature, we now turn to a qualitative discussion of the emergence of multiple temporal scales in thinking about diffusion and, similarly, the emergence of multiple spatial scales in considering plasticity.

12.1.1 Problems with Multiple Temporal Scales: The Example of Diffusion

As has been emphasized repeatedly, diffusion is at the heart of many of the most positive and negative things that can happen to materials. On the one hand, the vast array of annealing processes used in preparation of materials ranging from corrosion resistant steels to doped semiconductors are all meant to take advantage of the enhanced diffusion processes that occur at high temperatures. By way of contrast, the insidious and slowly persistent process of creep can ultimately degrade a material to the point of uselessness, again as a result of diffusive transport of mass. In each of these cases, there is no question that deeper microscopic understanding of these processes would be both interesting and useful.

As was already noted in chap. 7, there are a number of different theoretical tools that can be used to consider diffusion. Deeper reflection on the nature of the diffusion process itself reveals a proliferation of temporal scales. The smallest time scale of interest in this problem is that associated with the electronic degrees of freedom, a scale at which things happen so fast that even the nuclei can be thought of as fixed (the Born–Oppenheimer approximation). The next temporal scale of interest is that of the vibrations of the atoms themselves. The solid, which is the seat of the diffusive process in question, is a sea of vibration, with all of the atoms in incessant motion, with oscillations characterized by frequencies typical of the relevant phonons. If we take our cue from fig. 5.6 we see that a typical vibrational frequency is on the order of 5 THz, with the result that the time scale associated with such vibrations is of order 2.0×10^{-13} s. The next time scale in the hierarchy of temporal processes characterizing diffusion is that emerging from the jump frequency Γ, and which expresses the frequency with which the diffusing species will successfully traverse the saddle point connecting two local minima. Finally, an even longer temporal scale is that associated with the accumulation of a sufficient number of microscopic jumps to be registered macroscopically as a composition change.

As noted above, one of the most successful strategies for dealing with complex problems involving many different scales is to find a way to separate scales. For example, in the context of diffusion, the use of the diffusion equation (see eqn (7.18)) is an example of this approach in which the physics of the omitted temporal scales appears in the diffusion constant. This interpretation of the diffusion constant as a way of masking the uninteresting *individual* atomic hops should be clear as a result of the discussion in section 7.4.2. For many problems, this approach is entirely adequate. For example, if we interest ourselves in the temporal evolution of the concentration profile of carbon atoms relative to the surface of a carburized steel, all that we need to know about atomic vibrations and microscopic hops is encompassed in the diffusivity. Once the information deriving from these scales is included in the diffusion equation, the question of interest may be addressed from the perspective of the macroscopic diffusion equation without any further reference to microscopic motions.

On the other hand, when examined from the reductionist perspective of micro-scopic simulation, diffusion poses serious challenges. It is possible to conceive of strategies in which the elementary processes leading to macroscopic diffusion are modeled directly by following atomic trajectories. However, such strategies fail to acknowledge the exceeding rarity of microscopic jumps when reckoned in terms of a clock that ticks with a period shorter than that of atomic vibrations. Yet, there are some instances in which detailed microscopic understanding is desired and recourse to atomic-scale arguments must be made. For example, when confronted with specific problems such as diffusion on stepped surfaces, a blithe application of conventional macroscopic diffusion arguments might be contaminated by effects such as that due to the Schwoebel barrier as described in section 11.2.2. As a result, one of the key multiscale challenges that must be faced in modeling diffusion is the disparity in temporal scales between atomic vibrations and atomic hopping. Provocative ideas that address these challenges will be presented later in this chapter.

12.1.2 *Problems with Multiple Spatial Scales: The Example of Plasticity*

In the previous subsection, we used the example of diffusion to illustrate the proliferation of temporal scales in one of the central problems in the study of materials. The present discussion has a similar aim in that we will briefly review the features of plasticity that place modeling demands at many different spatial scales. Though plasticity is also an area of immense importance, the conceptual foundations for its analysis both at the macroscopic level as well as from a reductionist perspective are not nearly as mature as is the study of diffusion. Recall that at the macroscopic scale in the context of diffusion we have the time-honored diffusion equation while at the microscopic scale we have the machinery of transition state theory as the basis of a well-defined scheme for information passage. By way of contrast, the macroscopic equations of plasticity are not nearly as robust as the diffusion equation and there is no clear path for effecting a passage of information from the analysis of the properties of single dislocations to the macroscopic modeling of plasticity.

As we have emphasized already, the study of plasticity is one of the centerpieces of the mechanics of materials. A wide array of technologies depend upon the ability to deform materials into particular desirable shapes, while from a scientific perspective I personally find the subject of great interest because it is an intrinsically dissipative process featuring history dependence and is a strong function of the material's microstructure. In addition, from the effective theory perspective, the study of plasticity is built around the motion and entanglement of dislocations which requires the construction of theories of lines and their interaction thus ushering in a certain nonlocality to the phenomenon right from the outset.

From the standpoint of a stress–strain curve, the study of plasticity is all about the analysis of yield and hardening. As we have already noted at various points in the book, though especially chaps. 8 and 11, the examination of both yield and hardening takes us on a journey through a number of different scales. The scale of the crystal lattice makes itself known in a fundamental way through the existence of anisotropy in plastic deformation. Both the relevant slip plane and slip directions are determined by a combination of crystallographic and energetic considerations, both deriving from the underlying atomic-scale structure of the solid. The next level of length scales is that associated with the core of the dislocation itself. As noted in earlier chapters, perhaps the most elementary core effect is the splitting of dislocations into partial dislocations bounding a stacking fault. The spatial extent of these core regions is on the order of nanometers.

Nanometer dimensions are next superseded by those dimensions at which plastic deformation itself begins to be characterized by structures. In particular, the concentration of crystal slip into certain bands, with a typical dislocation–dislocation spacing within bands, and a typical band–band spacing also. Similarly, there is a length scale determined by the overall dislocation density which determines the mean distance between dislocation intersections which serve as pinning points for the dislocations, and which manifests itself in the measured flow stress. Finally, the deformation of polycrystals requires a treatment of an assemblage of single crystals, implying yet another scale in the hierarchy of scales that are pertinent to plasticity.

Like in the case of diffusion, there are many instances in which a separation of scales approach is satisfactory. For example, in the treatment of polycrystal plasticity, the formulation of a constitutive law making no reference to the origins of plasticity in dislocation motions has been very successful. On the other hand, there are particular experimental facts that demand the treatment of several scales simultaneously. As noted above, a simplified view of plasticity is that it is the study of yield and hardening, and our basic contention is that the understanding of yield and hardening can possibly be decomposed into two distinct parts. First, there are questions concerning the response of a single dislocation and its interaction with a single pinning point, an analysis we might wish to refer to as the 'one-body' problem. In addition, there is the problem of the statistical assemblage of dislocations and the various obstacles to their motion which constitutes the 'many-body' problem. Though the study of diffusion is characterized by a well-defined reductionist paradigm in which the information passage from one scale to the next is well understood, a similar statement is difficult to make in the context of plasticity. Though it is true that the fields of dislocations are well understood as is their response to an applied stress, the question of obstacles and their statistical assemblage seems much less clearcut. In addition, the computational difficulties that attend the attempt to simulate plastic deformation directly on the basis of dislocations remain serious.

The case studies considered in the preceding two subsections have attempted to set the stage for the present chapter by concretely arguing for the idea that there are a number of problems in the study of materials that feature several scales (in space or time or both) simultaneously. Though our argument centered on the treatment of diffusion as an example of a proliferation of temporal scales and plasticity as an example of a proliferation of spatial scales, the situation is exacerbated yet further when we consider the action of plasticity induced by mass transport, such as the discussion of creep featured in section 11.3. Our main purpose has been to remind the reader from within the narrowly focused perspective of the study of materials that the proliferation of scales is an everyday challenge.

12.1.3 Generalities on Modeling Problems Involving Multiple Scales

Given the examples described in explicit detail above, and by appealing to the examples that have been repeatedly emphasized throughout the book, our present aim is to try to take stock of those features of problems in modeling materials that generically signal the need for some sort of a multiscale approach. In loose terms, it is possible to identify a few different themes that have arisen.

Multiple Scale Models and Physical Insight. Throughout the book, we have tried to emphasize the complementary role of highly accurate microscopic calculations and their less accurate but more transparent effective theory surrogates. Our present claim is that through the device of constructing effective theories which encode some underlying subscale information it is possible to gain substantial *physical insight* into a given problem. For example, in chap. 6, we considered the hotly contested question of describing the energetics of the hierarchy of structures adopted by Si and the role of the Yin–Cohen calculations (see fig. 6.6) as a standard against which subsequent empirical total energy schemes have been measured. We recall that these calculations are computationally demanding and are built up on the basis of a great deal of numerical machinery designed to handle everything from Brillouin zone sampling to long-range sums associated with Coulombic interactions. Though the Yin–Cohen curves are instructive in themselves, they provide us with no means to anticipate the results of a calculation on some new structure.

Similar arguments can be made for the results of the first-principles calculations that we highlighted in the case of the structure of the Au(001) surface (see section 9.2.2). The point is that having been confronted with the results of a serious quantum mechanical analysis of the electronic structure and energetics of a series of structures, the hard work of *understanding* these results still remains. The physical insight we seek can often be attained by converting the results of this microscopic analysis into some effective Hamiltonian in which the coupling between geometry and energetics is more apparent.

Multiple Scale Models and the Irrelevance of Degrees of Freedom. In his justly famous *Lectures on Physics*, Feynman (1963) notes that if he could but convey one fundamental truth about man's understanding of the physical world, that truth would be what he refers to as the 'atomic fact'; *'that all things are made of atoms – little particles that move around in perpetual motion, attracting each other when they are a little distance part, but repelling upon being squeezed into one another.'* This central notion, it is argued, is the basis for understanding an enormous part of what constitutes modern science. The endorsement of this fundamental notion lies beneath the surface of much of what we have done in this book where we have maintained a certain vigilance to see to what extent microscopic notions can better inform our understanding of materials and their properties.

However, contrary to the *reductionistic* outlook that characterizes the continued appeal to microscopic mechanism for building plausible models, it is also possible to adopt a *synthetic* outlook which is built upon a less solid but still amazing observation: for understanding many problems, the vast majority of microscopic degrees of freedom can be thrown away. To quote Callen (1985) yet again, 'Somehow, among the 10^{23} atomic coordinates or linear combinations of them, all but a few are macroscopically irrelevant. The pertinent few emerge as *macroscopic coordinates...*' In chap. 1, we noted Dirac's remark that (more or less) with quantum mechanics in hand the remainder of chemistry was nothing more than the playing out of the Schrödinger equation. We made the point there and make it again now in the context of multiple scale modeling, that the business of finding a minimal set of degrees of freedom (even though we may know how to solve the problem using *ALL* the degrees of freedom) is one of the primary charters of science. Said differently, effective theory construction can be thought of as a search for irrelevance in which systematic attempts are made at degree of freedom thinning.

Multiple Scale Models and Information Passage. Once the hard work of discerning which degrees of freedom are relevant and which are not has been undertaken, and an effective theory built around these degrees of freedom has been advanced, there is yet further progress that can be made in the way of multiscale modeling. Recall that one of the key features of most effective theories is their reliance on certain phenomenological parameters which that theory itself cannot determine. We have repeatedly belabored the most familiar example, namely, that of linear elasticity in which a stored energy function is constructed on the basis of the strain tensor, and for which the elastic moduli serve as the phenomenological parameters whose aim it is to capture the physics of bond stretching and bending that are *really* taking place in the strained solid.

From a multiple scale modeling perspective, the presence of phenomenological parameters in various effective theories provides an opportunity for information passage in which one theory's phenomenological parameters are seen as derived quantities of another. We have already seen that although the linear theory of elasticity is silent on the particular values adopted by the elastic moduli (except for important thermodynamic inequalities), these parameters may be deduced on the basis of microscopic analysis. The advent of reliable models of material behavior makes it possible to directly calculate these parameters, complementing the more traditional approach which is to determine them experimentally.

Multiple Scale Models and Computational Tractability. Multiple scale modeling can convert problems that are computationally intractable into those that are not. Modeling the external world has almost always involved finding new ways of trimming down complex problems to manageable form. In the era prior to the use of widespread computation, much effort was spent in formulating analytically tractable models. An alternative view of this modeling process is that ultimately our inability to handle problems in their full complexity forced us to learn how to construct analytic surrogates for problems of interest. Despite the enormous computational power now at our fingertips, the situation is really no different than it has ever been. The existence of computers has led to new modeling paradigms in which the mandate is not so much to construct effective theories that are analytically tractable, but rather models that are computationally tractable.

As with most human endeavors, appetite often exceeds capacity. From a computational perspective, despite the staggering increase in computational capacity, computers are still not fast enough nor do they have sufficient memory or disk space to really service the demands of modeling. However, the benefit of this truth is that it has by necessity forced the development of a huge number of clever effective theories, approximation schemes, and algorithms. Perhaps I am being grandiose, but my own view is that the existence of computers is teaching us new ways to think about theory construction and has led to modeling schemes (such as the use of cellular automata) which would probably not have arisen without the ability to make large-scale computations. Thus, one of the perspectives adopted here is that multiple-scale modeling should also be seen as a way of answering to our inability to rely on pure brute force approaches in modeling.

12.2 Historic Examples of Multiscale Modeling

One of the critical points to be made in the current chapter is that despite the recent assignment of a name to that field of endeavor that involves linking models at different scales (i.e. 'multiscale modeling'), the practice of this type of modeling is as old as modern science itself. Indeed, we will argue that the recognition of historic examples of such modeling will serve to instruct us as to the features of such models. Our aim is to practice revisionist history of the relevant examples, calling forth only the conceptual arguments associated with particular examples, and assigning only those names that have stuck, always with the recognition that what really happened was surely more complicated.

Newton and the Spherical Earth. One of the first quantitative effective theories that I can think of is that associated with Newton and the invention of the integral calculus. In particular, the key question that had to be faced was whether or not it was possible to pretend as though the entire mass of the earth is concentrated at its center, rendering it for gravitational purposes, nothing more than a point mass. Newton's argument was schematized pictorially as in fig. 12.1 and asserted as Proposition LXXVI, Theorem XXXVI of the famed *Principia Mathematica* (Motte 1934).

> If spheres be however dissimilar (as to density of matter and attractive force) in the same ratio onwards from the centre to the circumference; but everywhere similar, at every given distance from the centre, on all sides round about; and the attractive force of every point decreases as the square of the distance of the body attracted; I say, that the whole force with which one of these spheres attracts the other will be inversely proportional to the square of the distance of the centres.

The multiscale challenge arises from the recognition that the total gravitational potential associated with an object at a position $\mathbf{r}$ relative to the center of the earth

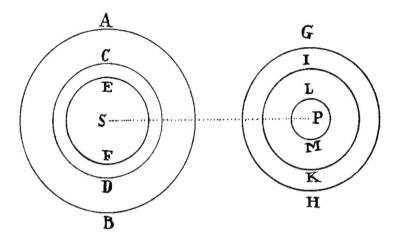

Fig. 12.1. Schematic from Newton's *Principia Mathematica* arguing that a series of concentric spheres, each characterized by a constant mass density, can be treated gravitationally as a point mass (adapted from Motte (1934)).

is given as

$$\Phi(\mathbf{r}) = -G \int \frac{\rho(\mathbf{r}')d^3\mathbf{r}'}{|\mathbf{r} - \mathbf{r}'|},$$

(12.2)

a step we take for granted, but for which Newton required the invention of the integral calculus itself (or at least its geometrical analog). At this point in the argument, it is realized that in the special case of a mass distribution $\rho(\mathbf{r})$ with spherical symmetry, things simplify considerably. In particular, if we assume that the density is constant throughout, and the figure of the earth is spherical, direct integration of the expression given above results in

$$\Phi(r) = -G \frac{M_{earth}}{r}.$$

(12.3)

For the purposes of the present discussion, this is the key result. What we have learned is that in the context of problems in which we interest ourselves in the gravitational influence of the earth at a distance r from the earth's center, the earth may be replaced conceptually by a point mass at the earth's center. This example satisfies one of the key criteria we have given as exemplifying multiscale modeling: there is a massive reduction in the number of degrees of freedom, with the further clarification of physical insight.

Bernoulli and the Pressure in a Gas. One of the elementary calculations undertaken in trying to effect the linkage between observed macroscopic behavior (such

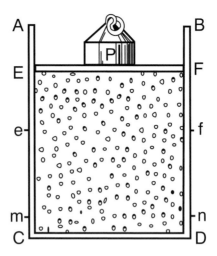

Fig. 12.2. Schematic of the emergence of the macroscopic notion of pressure from the statistical superposition of the myriad of collisions between the molecules of the gas and the bounding walls (adapted from Brush (1983)).

as the ideal gas equation of state) and microscopic motions is that of the pressure of an ideal gas. All that is assumed really is the *existence* of molecules and their inevitable collisions with the walls that bound the gas, as indicated schematically in fig. 12.2. In the simplest of models, the assumption is made that the pressure associated with the ideal gas is composed of the entirety of molecular collisions with the wall, each of which imparts an impulse $2m|v_\perp|$ to the wall. What underlies this assertion is the assumption that each molecule that arrives at the wall suffers a specular reflection, leaving the wall with its velocity component perpendicular to the wall equal and opposite to that with which it arrived at the wall.

The leap from an understanding of a single collision to the macroscopic notion of a pressure is made by noting that the total force on a given area imparted per unit time is found by adding up the contribution to the total momentum change coming from each molecule that strikes the wall in that time. If we introduce the distribution function $N(v_\perp)$, which gives the number of molecules in a unit volume with velocity between $v_\perp$ and $v_\perp + dv_\perp$, then the pressure may be written as

$$p = \int_0^\infty 2mv_\perp^2 N(v_\perp)dv_\perp. \tag{12.4}$$

To arrive at this result, we have used the fact that the number of molecules with velocity $v_\perp$ striking unit area of the wall in a unit time is $N(v_\perp)v_\perp dv_\perp$. The contribution to the pressure arising from that part of the population with velocity $v_\perp$ is given by $2mN(v_\perp)v_\perp^2 dv_\perp$, and hence the total pressure is gotten by adding up the contributions from all $v_\perp > 0$ as is done in the integral of eqn (12.4). We now rewrite the integral in the suggestive form

$$\frac{p}{\int_{-\infty}^{\infty} N(v_\perp)dv_\perp} = \frac{\int_{-\infty}^{\infty} mv_\perp^2 N(v_\perp)dv_\perp}{\int_{-\infty}^{\infty} N(v_\perp)dv_\perp}, \tag{12.5}$$

where we have divided both sides by $\int_{-\infty}^{\infty} N(v_\perp)dv_\perp$ in anticipation of the recognition that the right hand side is twice the mean kinetic energy. We now note that on the basis of the equipartition theorem of statistical mechanics the right hand side of this equation is thus nothing more than kT, while the integral $\int_{-\infty}^{\infty} N(v_\perp)dv_\perp = N/V$, with the result that

$$p = \frac{N}{V}kT, \tag{12.6}$$

which is precisely the ideal gas law. Our purpose in revisiting these arguments has been not so much to explore the fine points of kinetic theory and the reasoning that goes with it, as to demonstrate the way in which information is passed from the microscopic scales to those associated with macroscopic observables.

The example put forth here demonstrates the connection between kinetic theory and macroscopic thermodynamics. Indeed, it can be argued, I think convincingly, that the twin pillars of statistical mechanics and thermodynamics themselves serve as *the* paradigmatic example of multiscale modeling. The partition function and its associated derivatives serve as the bridge between microscopic models, on the one hand, and the derived thermodynamic consequences of that model, on the other.

Discrete and Continuous Strings. Probably one of the most important recurrent themes in multiscale modeling (both in the historical and current context) concerns that of passage back and forth between continuous and discrete representations of the problem of interest. Many of the historical roots of this passage are to be found in analyses of problems such as that of the vibrating string. On the one hand, if the string is imagined as a series of discrete masses, in the limit of small vibrations, this problem leads to a series of coupled differential equations of the form

$$m\ddot{x}_l = k(x_{l+1} + x_{l-1} - 2x_l), \quad l = 1, 2, \ldots, n, \tag{12.7}$$

where k is the effective spring constant. Note that these equations of motion should be familiar from our earlier discussion of the harmonic solid where they served as the basis of our discussion of a wide variety of phenomena. In particular, the reader is invited to reconsider eqn (5.4).

An alternative, and in some instances more convenient, treatment of the vibrations of a string is the replacement of the discrete variables by a single kinematic field $u(x, t)$ which specifies the displacement u at a given point x at time t. This field satisfies a partial differential equation

$$\frac{\partial^2 u}{\partial x^2} = \frac{1}{c^2} \frac{\partial^2 u}{\partial t^2},$$

(12.8)

again, as was already seen earlier in, for example, eqn (5.55). We recall also that the wave speed c can be related back to the spring constant k introduced in the discrete version of the model. Each of these approaches has its advantages though for the present purposes, it is as an example of the passage between continuum and discrete descriptions that the vibrating string is of interest. If I were to try to summarize my own observation on the continuum-discrete linkage in the present context it would be to note that in those cases in which one is attempting to get an analytic handle on a given problem, it is especially enlightening to seek the *continuum* limit of a given problem. On the other hand, often in attempting to produce a computational incarnation of a given phenomenon, we wish to run this reasoning in reverse with the aim being an optimal *discretization* of the problem of interest. Regardless of what ultimate perspective we adopt, it remains undeniable that the business of passing back and forth between continuum and discrete representations of a given phenomenon must be at the heart of any sensible definition of 'multiscale modeling'. Indeed, this is but one expression of an eloquent insight from Feynman (1966),

> Theories of the known which are described by different physical ideas may be equivalent in all their predictions, and are hence scientifically indistinguishable. However, they are not psychologically identical when trying to move from that base into the unknown. For different views suggest different kinds of modifications which might be made, and hence are not equivalent in the hypotheses one generates from them in one's attempt to understand what is not yet understood.

We note in passing that in addition to the relevance of the current example to explicating the powerful gains that are associated with passing between discrete and continuous representations of the same problem, the current example (and our discussion of elasticity to follow) illustrates another unifying precept in modeling across scales, and that is the notion of effective spring constants in the limit in which systems are considered in the neighborhood of some equilibrium point.

Elasticity. Though the theory of elasticity has been the workhorse of much of what we have had to say in this book, it is too important an example of the multiscale paradigm to pass it up. We begin by noting at least three discernible stages in the emergence of a viable effective theory of elasticity. The construction of the modern theory of elasticity was begun with the pioneering observations of Hooke as illustrated in fig. 2.9 and as typified quantitatively by

$$\frac{F}{A} = E\frac{\Delta l}{l}, \tag{12.9}$$

where F is the force associated with the loading of the solid, A is the relevant cross sectional area, E is Young's modulus and $\Delta l/l$ is the fractional extension of the material. The next level of sophistication in the elucidation of the linear theory of elasticity was a simultaneous enlargement of both the set of kinematic states to be considered as well as a broadening in the description of the forces that lead to these kinematic states. The outcome of these insights was a generalized Hooke's law in the form

$$\sigma_{ij} = C_{ijkl}\epsilon_{kl}. \tag{12.10}$$

Most subtle of all is the confident use of

$$E_{tot} = \frac{1}{2}\int_{\Omega} C_{ijkl}\epsilon_{ij}(\mathbf{x})\epsilon_{kl}(\mathbf{x})d^3\mathbf{x}, \tag{12.11}$$

the linear elastic expression for the energy stored in a body by virtue of its deformation. In this final form, in conjunction with the variational principle of minimum potential energy (see section 2.5.1), one is positioned to examine the displacement fields in a *nonuniformly* deformed body. Indeed, for the present purposes, it is this nonuniformity which is of especial interest since this nonuniformity is greeted with quite a subtle and profound assertion. Recall that from the microscopic perspective, the nonuniformly deformed body is characterized by as many different local geometries as there are atoms. On the other hand, by virtue of the elastic theory, eqn (12.11) tells us that this energy can just as well be represented by pretending that each material particle is like an infinite crystal that is *uniformly* deformed with an associated energy $E_{\Delta V}(\mathbf{x}) = \frac{1}{2}C_{ijkl}\epsilon_{ij}(\mathbf{x})\epsilon_{kl}(\mathbf{x})\Delta V$, where ΔV is the volume of the material particle in question.

Hydrodynamics. Our historical interlude has made reference to effective theories of both gases (kinetic theory) and solids (elasticity), and now we take up yet a third example of enormous importance to modeling the natural world in general, and which serves as an example of the type of multiscale efforts of interest here, namely, the study of fluids.

During our discussion of linear momentum balance in chap. 2, we noted that the fundamental governing equations of continuum mechanics as embodied in eqn (2.32) are indifferent to the particular material system in question. This claim is perhaps most evident in that eqn (2.32) applies just as well to fluids as it does to solids. From the standpoint of the hydrodynamics of ordinary fluids (i.e. Newtonian fluids) we note that it is at the constitutive level that the distinction between fluid and solid is revealed, where it is assumed that for Newtonian fluids the relation between stress and kinematics is obtained through the equation

$$\boldsymbol{\sigma} = -p\mathbf{I} + 2\mu\mathbf{D}, \tag{12.12}$$

where we have defined the rate of deformation tensor $\mathbf{D} = \frac{1}{2}[\nabla\mathbf{v} + (\nabla\mathbf{v})^T]$, and we have assumed in advance that the fluid is incompressible (i.e. $\mathrm{tr}\,\mathbf{D} = 0$). In light of this constitutive assumption, the resulting field equation is

$$-\nabla p + \mu\nabla^2\mathbf{v} = \rho\frac{D\mathbf{v}}{Dt}, \tag{12.13}$$

more conventionally known as the Navier–Stokes equations.

From the vantage point of the current discussion, our key interest centers on the significance of the quantity μ. Just as the elastic moduli reflect atomic-level interactions so too does the viscosity which is in a sense a measure of the stiffness of the system against momentum exchange. Our main point is to note that the goal of the viscosity is to encode the relevant information concerning the atomic-level collisions which are responsible for what is experienced macroscopically as shear stresses resulting from velocity gradients.

Continuum Models of Diffusion and Heat Transport. As seen in chap. 7, the accumulated unbiased hopping of atoms can lead to a macroscopic appearance of *design*; namely, that there seems to be a force leading to a preference for motion in a given direction. From a macroscopic perspective, this apparent biased motion of atoms is reflected through the diffusion equation given earlier as eqn (7.18) and which is repeated here for ease of discussion as

$$\frac{\partial c}{\partial t} = D\nabla^2 c, \tag{12.14}$$

where we remind the reader that the field c can be thought of as the concentration field while the constant D is the relevant diffusion constant. In the context of heat transfer, an identical equation presides over the evolution of temperature gradients within a system as

$$\frac{\partial T}{\partial t} = \alpha\nabla^2 T, \tag{12.15}$$

where the constant α is the thermal diffusivity and is defined by $\alpha = \kappa/\rho c_p$, where κ is the thermal conductivity, ρ is the density and c_p is the specific heat. We reiterate once again that the constants D and α found in eqns (12.14) and (12.15), respectively, both account for *dynamical* processes that are not considered explicitly in either of these macroscopic transport equations. This theme of disguising microscopic physics in the form of phenomenological parameters has followed us through our treatment of elasticity (eqn (12.11)), hydrodynamics (eqn (12.13)), mass transport (eqn (12.14)) and heat flow (eqn (12.15)) and if we so desire the same types of arguments and observations would arise in considering Maxwell's theory of the electromagnetic field. Arguably, this set of theories is the most important part of classical physics that can be couched in the language of field theory, and all of them are built around the same philosophy of burying microscopic physics in effective parameters.

Harmonic Oscillators and Optical Absorption. An ingenious example of exploiting insights at one scale in order to inform those at another is that of optical absorption in solids. From a macroscopic perspective, the attempt to understand absorption is founded on Maxwell's theory of the electromagnetic field. However, in order to make material specific progress, it is necessary to supplement these ideas with either a phenomenological or microscopic model of material response.

One of the simplest of microscopic models is to posit that the solid is built up of a number of bound charges that are excited by the incident electromagnetic field. The equation of motion for the bound charges (i.e. electrons) is

$$\frac{d^2x}{dt^2} + \gamma \frac{dx}{dt} + \omega_0^2 x = \frac{qE_0}{m}e^{i\omega t}, \tag{12.16}$$

where γ characterizes the damping of these charges once they are set in oscillatory motion, ω_0 is the natural frequency of free vibrations for these charges and the right hand side of the equation characterizes the driving due to the presence of the electromagnetic field. Once this equation is solved to determine the relation between the position of the charge and the driving force E, the current may be determined via $j = \rho q v$ (here we ignore the vectorial character of the current density), from which we may deduce an effective conductivity. An alternative statement of the deductions that emerge from this calculation is the expression for the index of refraction n, which is given by

$$n^2 - 1 = \frac{4\pi q^2 \rho}{m(\omega_0^2 - \omega^2 + i\gamma\omega)}. \tag{12.17}$$

What we observe in effecting a calculation of this type is that by carrying out an appropriate microscopic analysis, it is possible to pass information from one scale to another. In this case, it is the information concerning the microscopic oscillators that serves as the bridge between atomic-level analyses and the macroscopically observed optical absorption which depends upon a knowledge of the frequency dependence of the index of refraction.

The Born–Oppenheimer Approximation and Separation of Temporal Scales. We have already made reference to the Born–Oppenheimer approximation in

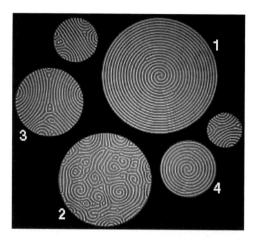

Fig. 12.3. Patterns exhibited in a fluid layer heated from below (adapted from Plapp *et al.* (1998)).

several contexts thus far. Presently, we call attention to this idea again as a compelling example of bridging temporal scales. Recall that the argument was made that because of the relative masses of electrons and nuclei, the Schrödinger equation for the coupled problem involving both electrons and nuclei could be decoupled. The physical argument behind this strategy reasons that because of the much larger masses of the nuclei, their motions are sluggish in comparison with the electrons, with the result that the problem of electronic motions may be solved in the presence of *frozen* nuclei.

Pattern Formation and Amplitude Equations. One of the most beautiful outcomes of work in physics has been a renaissance in the study of *classical* systems, that is, systems in which quantum effects play no direct role. Though this is not the place to enter into the details of this rich area, I cannot pass up the opportunity to note that from an experimental perspective, this field is characterized by relatively simple apparatus (at least simple to explain), with results that can be evaluated visually and characterized quantitatively. The relevance of pattern forming instabilities to the present discussion is the existence of a well-defined scheme for constructing effective theories. For concreteness, we consider the case of convection in which a fluid layer heated from below is seen to exhibit patterns of the type shown in fig. 12.3. These pictures are shadowgraphs in which the dark regions correspond to rising warm fluid and the light regions to descending cool fluid. Different patterns correspond to different values of the temperature difference between the top and bottom of the fluid layer.

The continuum theory describing the convection process (itself already an effective theory) is built around the Navier–Stokes equations and an accompanying treatment of heat transfer. In the simplifying context of the so-called Boussinesq approximation, the only temperature dependence of material response enters through an equation of state for the density, namely,

$$\rho = \rho_0[1 - \alpha(T - T_0)], \tag{12.18}$$

where α is the thermal expansion coefficient for the fluid. In particular, the fluid is assumed incompressible (i.e. $\nabla^2 \mathbf{v} = 0$), and satisfies the Navier–Stokes equations in the form

$$\partial_t \mathbf{v} + \mathbf{v} \cdot \nabla \mathbf{v} = -\alpha(T - T_0)\mathbf{g} - \frac{1}{\rho_0}\nabla p + \nu\nabla^2\mathbf{v}. \tag{12.19}$$

Note that in this equation, the buoyant force enters through the thermal expansion term α. The fluid motions are coupled to the heat flow via the heat equation

$$\partial_t T + \mathbf{v} \cdot \nabla T = \kappa\nabla^2 T. \tag{12.20}$$

From the perspective of the arguments being put forth in this chapter, we wish to note that even the sanitized version of the convection problem laid down above is for many purposes too complex to examine the types of patterns shown in fig. 12.3. In light of the discussion of the Navier–Stokes equations given above (see eqn (12.13)), we are claiming that a useful strategy is to build an effective theory (i.e. amplitude equations) of an effective theory (i.e. Navier–Stokes equations). The pattern forming tendencies of systems of this type have been examined from the context of the so-called amplitude equation, an example of which is

$$\frac{\partial A}{\partial t} = A + (\partial_x - i\partial_y^2)^2 A - |A|^2 A, \tag{12.21}$$

where A is the field that exhibits the patterns in question. For an honest discussion of this enormous and fascinating field (as opposed to the flirtation with the subject given here) we refer the reader to the incredible article of Cross and Hohenberg (1993). In particular, their Appendix A shows how the formulation of the convection problem in the Boussinesq approximation given above may be reduced to a description via the type of amplitude equation given above.

Nuclei and Below. We conclude our series of historical examples of effective theories and multiscale modeling with a brief mention of the hierarchy of theories which present themselves in considering the subatomic world. Our first remark applies to most of the examples we have presented and it is to note that one man's microscopics is another's phenomenology. Nowhere is this more true than with the history of our exploration of the subatomic world and, in particular, the series of theories that have been set forth to explain what has been observed. The key point is that at each level of structure, the subscale structures can be ignored. For example, in most discussions of the structure of atoms, the internal structure of the nucleus can be ignored. Similarly, in considering the structure of nuclei, for most purposes an effective theory of nuclei can be constructed in which the internal structure of protons and neutrons is ignored. And so it goes in descending the scale ladder.

The examples presented in this section have been set forth with two main objectives. First, our aim has been to substantiate the claim that multiscale modeling is as old as science itself. The reason for pushing this perspective is to broaden the overly narrow focus of the current research activities carried out in this vein and to draw insight from historical precedents. Indeed, I am reminded of a story concerning Lincoln in which he is said to have asked, 'If a cow's tail were a leg, how many legs would a cow have?' Of course, his listener responded that that would be five legs to which Lincoln is said to have retorted, 'Just because you call a tail a leg doesn't make it a leg'. Similarly, just because we have graced a fashionable new research area with the name 'multiscale modeling' doesn't make it a new research area once viewed from the longer-term perspective of the history of model building in the physical sciences. The aims are the same as they have always been: the construction of models featuring a minimum of degrees of freedom and which involve a tractable effective dynamics. Our second aim has been to provide a basis for setting down generic ideas on the nature of effective theories. The historic examples offered above each provide a slightly different perspective on what is needed and desirable during the act of model building. We now undertake an attempt (which is surely provisional) at stating some of the common principles shared by different effective theories.

12.3 Effective Theory Construction

When viewed beneath the surface, many models share common features. In the present section, our aim is to examine some of the generic features that can be put down about certain classes of effective theories. Part of our thinking will be concerned with the discrete/continuous transcription mentioned above, while others will borrow from the tools of statistical mechanics in the critical phenomena setting with an emphasis on order parameters and theories of their evolution. The section is organized as follows. Subsection 12.3.1 will examine the wide range of kinematic ideas that have been put forth in the effective theory setting. This discussion will take us from the consideration of order parameter fields to the kinematic ideas that are used in theories of defect dynamics. Having considered how to characterize the 'state' of the system of interest, we will then take up the question of how such states evolve in time in subsection 12.3.2. The use of locality assumptions is a long-standing tool in modeling and is discussed briefly from a unified perspective in subsection 12.3.3. In subsection 12.3.4 we consider a broad class of models that are largely computational and which feature a given computational scheme which flies the flag of more than one modeling paradigm simultaneously. As a consequence, the majority of these models involve computational interfaces that allow for the smooth joining of the different models. The section closes with a description in subsection 12.3.5 of some of the formal ideas that have been set forth for the construction of effective Hamiltonians.

I would agree with the charge that the classification scheme set forth in the remainder of this section is arbitrary. On the other hand, my hope was to try to find those points of commonality between the different ideas that have been set forth in the modeling of materials, and this classification reflects my own take on these common features.

12.3.1 Degree of Freedom Selection: State Variables, Order Parameters and Configurational Coordinates

How should we characterize the disposition of the system of interest? This question is at the heart of all model building and leads immediately to a recognition that we can't begin to talk about the system in quantitative terms until a decision has been made as to which degrees of freedom will serve to characterize it. Many of the ideas that have arisen to characterize the disposition of complex systems are a sophisticated outgrowth of kinematics. Our characterizations of the disposition of material systems throughout this book have been built around two broad classes of description. On the one hand, we have advocated a microscopic reckoning of state in which the disposition of the system is encompassed in the full set of atomic coordinates, $\{\mathbf{R}_i\}$. This characterization of the state of a system is a prerequisite to any microscopic calculation of the energetics of the system of interest. On the other hand, the atomic-level reckoning of materials represents a massive overkill in many instances, a situation which is further deteriorated by the fact that an atom-by-atom description of the system often veils the physically relevant degrees of freedom. This observation inclined us to the second broad class of kinematic measures, namely, the use of continuum *fields* to describe the state of the system. Besides the wish to characterize the relative positions of the material particles making up the system of interest, in many instances we demand a scheme for characterizing its internal 'state'. In some instances, these descriptors of state are statistical.

State Variables. The construction of effective theories is predicated in part on the ability to discern some trimmed down description of the disposition of the system of interest. From the perspective of equilibrium thermodynamics, the ability to characterize a system in terms of a series of 'state variables' exemplifies the type of degree of freedom thinning we have in mind here. A particular virtue of the notion of a state variable is that it allows us to reason about systems in quantitative terms even though we are inevitably faced with an incomplete knowledge of the disposition of that system. Recall our discussion in the previous section of the way in which the effects of the multitude of molecular collisions with the walls in a gas can be replaced with a macroscopic variable of state, namely, the pressure. Similarly, the fluctuating parts of the atomic-level velocities make themselves known through the macroscopic notion of temperature. In the time since the original formulation of equilibrium thermodynamics, the concept of state variable has been broadened considerably, and within the critical phenomenon setting, been formalized in the form of the order parameter concept.

Order Parameters. Equilibrium thermodynamics is an instructive setting within which to begin to answer the question of how to characterize the state of physical systems. For our purposes, an even more instructive jumping off point is that provided by the thermodynamics of systems near a critical point at which the system undergoes a transition from one phase to another. In particular, such systems suggested the introduction of the concept of an order parameter, an idea that as will be seen below, can be tailored to problems in the study of materials which appear to have nothing to do with criticality and are often not even in equilibrium. To begin our discussion, we first review the order parameter concept in the critical phenomena setting and then describe its extension to other situations.

The order parameter is essentially a kinematic measure, describing the state of order within a system without any intrinsic reference to what factors drove the system to the state of interest. For example, in thinking about the transition between the ordered and disordered states of an alloy, it is useful to define an order parameter that measures the occupation probabilities on different sublattices. Above the order–disorder temperature, the sublattice occupations are random, while below the critical temperature, there is an enhanced probability of finding a particular species on a particular sublattice. The conventional example of this thinking is that provided by brass which is a mixture of Cu and Zn atoms in equal concentrations on a bcc lattice. The structure can be interpreted as two interpenetrating simple cubic lattices where it is understood that at high temperatures we are as likely to find a Cu atom on one sublattice as the other. A useful choice for the order parameter, which we denote by η, is

$$\eta = \frac{N_{Cu}^{(1)} - N_{Cu}^{(2)}}{N_{Cu}}, \tag{12.22}$$

where $N_{Cu}^{(1)}$ is the number of Cu atoms on sublattice 1, $N_{Cu}^{(2)}$ is the number of Cu atoms on sublattice 2 and N_{Cu} is the total number of Cu atoms. We see that η has the property that at high temperatures, when $N_{Cu}^{(1)} = N_{Cu}^{(2)}$, the order parameter vanishes, while at low temperatures when one sublattice is singled out, the order parameter will take the value $\eta = \pm 1$.

Even in those situations not involving criticality, the notion of an order parameter field can be extremely useful for characterizing the disposition of a system. From the standpoint of materials science, there are a wide variety of internal states in materials that are amenable to a description in terms of the order-parameter-like idea. As argued above, in many instances there is a replacement of the discrete features of a system that are inherited from the inherent graininess of matter with continuous fields. These field variables are then imagined to vary in both space and time, allowing for the development of spatial structures as well as temporal evolution, thus giving rise to the possibility that the microstructural complexity introduced in previous chapters can emerge naturally from these models. It has been found necessary to distinguish between those cases in which the relevant field variable is conserved (such as the concentration of different constituents in a binary alloy) and that in which it is not (such as the field variable that distinguishes the ordered and disordered states in a binary alloy).

To make the ideas introduced above concrete, we follow Chen and Wang's (1996) tutorial introduction to field variable modeling of microstructural evolution with special reference to the case of a binary alloy. We begin by acknowledging the existence of a conserved field variable, namely, $c(\mathbf{r}, t)$. This field variable characterizes the probability of finding atoms of a particular type at position $\mathbf{r}$ at time t. By virtue of the overall concentration of the two types of atoms, the probability of finding an atom of the second type at a given site is given by $1 - c$. However, in addition to the accounting demanded in a binary system of this type, some descriptor is needed to characterize the state of order within the system. One possibility in this regard is correlation functions of the concentration field itself such as $\langle c(\mathbf{r}, t) c(\mathbf{r}', t') \rangle$ which tell us the likelihood that an enhanced concentration at one point implies an enhanced concentration at another. For our present purposes, the state of order will be characterized by a second scalar field variable $\eta(\mathbf{r}, t)$, a nonconserved order parameter which takes the values ± 1 in the ordered state and 0 in the disordered state. The 'kinematics' of a complex alloy characterized by the two sets of field variables introduced here is shown in fig. 12.4. This figure illustrates three different schematic configurations of the alloy in which the disposition of these field variables is distinct. Fig. 12.4(a) shows a uniform state characterized by $c(\mathbf{r}, t) = c_0$ and $\eta(\mathbf{r}, t) = 0$. Fig. 12.4(b) illustrates segregation of the two constituents such that in one region the concentration is c_α and in the other the concentration is c_β. The order parameter field vanishes throughout the region in this case. Fig. 12.4(c) illustrates yet a third eventuality, namely, the case in which the concentration is uniform throughout, but the state of order is phase shifted from one region to the next with the result that there is an antiphase boundary in

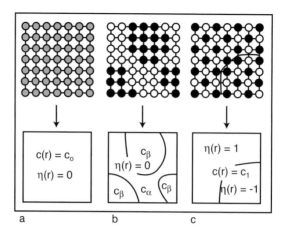

Fig. 12.4. Schematic illustration of conserved and nonconserved order parameters for characterizing the internal state of a binary alloy (adapted from Chen and Wang (1996)): (a) disordered phase ($\eta = 0$) with uniform composition c_0, (b) two-phase mixture consisting of disordered phases ($\eta = 0$) with composition c_α and c_β, (c) ordered single phase ($\eta = \pm 1$) of single composition c_1 with an antiphase domain boundary.

the system. Note that from the kinematic perspective of the type of order parameter field models being described here, the transition from $\eta = 1$ to $\eta = -1$ will be characterized by gradients in the η-field, and the energy cost of such gradients will provide a measure of interfacial energies in the system.

Our main idea up until this point in this section has been to illustrate the broad conceptual features shared in considering the 'kinematics' of material systems from the perspective of field theory. The key point has been the recognition that despite an overwhelming overall ignorance of the microscopic state of the system, it has been found possible to introduce field variables with a well-defined interpretation that still serve to tell us the state of the system in a way that can be linked to experiments. As will be seen below, as was first shown in the setting of equilibrium thermostatics and really came into its own in the context of the Landau theory of phase transitions, these field variables can serve as the basis for constructing free energies of the system of interest, and often for describing the dynamics of such systems as well.

Configurational Coordinates and Defects. A second broad class of ideas that have proven quite useful in the construction of effective theories of material response are those in which *all* attention is focused on the defects populating the system. That is, all reference to either atomic coordinates or continuum fields is surrendered and instead a theory is built around the dynamical evolution of the defects themselves. In this setting, kinematics is reduced to a question of identifying the positions of the relevant defects. This idea becomes most useful in the context of *extended defects* such as dislocations, grain boundaries and cracks. Within this framework, we can then build up a model for the dynamics of the defects themselves and address the question of how a given disposition of defects impacts the properties of a material. We note again for emphasis that models in which all attention is centered on defects certainly qualify as effective models in the sense that the full informational content of the system has been abandoned in favor of a truncated description which focuses on special collective excitations of the underlying microscopic degrees of freedom.

From the kinematic perspective of the present subsection, the basic question we have posed is how to parameterize the positions of the defects in question. For example, with regard to a given dislocation, the curve traced out by the dislocation line can be represented parametrically as $\mathbf{x} = \mathbf{x}(s)$, where s is a single parameter that measures arclength along the dislocation line. Thus the kinematic part of our analysis of the defect is all contained in the function $\mathbf{x}(s)$. Often, a computational incarnation of these ideas is set forth with an accompanying continuous $\rightarrow$ discrete transcription of the form $\mathbf{x}(s) \rightarrow (\mathbf{x}_1, \mathbf{x}_2, \ldots, \mathbf{x}_N)$, where the set $\{\mathbf{x}_i\}$ is a discrete set of points that we may think of as nodes along the defect line with various interpolation schemes used to determine the positions of the intermediate points. Whether using the continuum or discrete representation, the key idea to be conveyed is that of building an effective theory around defect coordinates.

Before passing to a summary of the kinematic ideas that have been set forth in this section, I feel compelled to note the subtlety of these ideas, subtleties which we are in danger of ignoring because the notion of defect is one that has become routine. Though the discussions in this book have been largely specialized to the class of defects of particular interest to the mechanics of materials, there are a host of extended defects of interest in a wide array of systems including vortices in fluids, domain walls in magnetic systems and arrays of magnetic flux lines in superconductors. The beautiful unifying perspective offered by the notion of a 'defect' is the possibility of building a theoretical engine around the structure and dynamics of such defects. These ideas are most potent when they culminate in the formulation of some version of the structure–properties paradigm such that a transcription may be made between the disposition of the defects of interest and the properties of the system.

The basic ideas in this section reflect an extension of the kinematic ideas of continuum mechanics in the sense that rather than enumerating the properties of a system on a discrete atom-by-atom basis, the state of the system is subsumed into either a set of continuous fields or a collection of coordinates which identify the positions of the defects that populate the system. We have seen that the class of relevant fields has been enlarged beyond the conventional fields of continuum mechanics in a way that allows us to characterize the state of order within a system, where the term order has been broadly construed to mean everything from different phases to ordering itself to the presence of misorientations, such as those found in polycrystals. Now that we have an idea of how to characterize the state of material systems at a given instant we are ready to face the challenge of determining how such systems evolve in time.

12.3.2 Dynamical Evolution of Relevant Variables: Gradient Flow Dynamics and Variational Principles

Once the relevant degrees of freedom have been identified, it is then necessary to construct a description of how these various fields vary in both space and time. Though we need to be cautious not to paint a picture which is too general, there are a number of generic ideas concerning the energetics associated with a particular realization of the order parameter field as well as its evolution that are worth considering. Similarly, there are a number of ideas on the explicit creation of theories of defect dynamics in which the configurational coordinates for the defects introduced above are assigned a dynamics which accounts for their evolution in time. The goal of the present subsection is to systematically examine these two strategies for the creation of effective dynamics. As a reminder, we note the provisional nature of what we will have to say here. As will be seen, both of the schemes to be described below are predicated upon the existence of an instantaneous free energy functional which depends upon relevant kinematic quantities like those described above, an assumption which may be overly limited when measured against the diversity of the various processes to be found in materials.

Evolution Equations as Gradient Flows. Our first scheme for generating evolution equations will be built around the types of order parameter fields introduced above. The basic idea is to define a free energy functional in terms of the relevant order parameter fields and then to invoke a dynamics where the change in the system is tied to the gradient in that functional. The simplest picture is that in which we can succeed in writing down a free energy density in terms of the order parameter field. For ease of discussion, we consider a single scalar order parameter field $\phi(x)$ and as yet make no reference to whether or not this order parameter field is conserved or not. Perhaps the simplest incarnation of this idea is a functional of the form

$$F[\phi(\mathbf{x})] = \int \left(\frac{1}{2}\kappa|\nabla\phi|^2 + a_2\phi^2 + a_4\phi^4 \right) d^3\mathbf{x}. \qquad (12.23)$$

The physics behind functionals of the form given above emerges from many of the well-known ideas of statistical mechanics. One approach to the construction of functionals like that considered above is to imitate the philosophy used in the Landau theory of phase transitions in which it is assumed that the free energy can be expanded in powers of the relevant order parameter, while the gradient term accounts for interfacial energy as introduced in section 9.5. On the other hand, much can be said about the form of the local free energy through an appeal to the same type of arguments that showed up in chap. 6 in which we showed how to construct free energies for alloys. For the sake of generality, we consider more general free energy functionals than that given in eqn (12.23) in which the homogeneous term $f(\phi)$ (which in the case above is of the form $f(\phi) = a_2\phi^2 + a_4\phi^4$) is left unspecified. Just as in the previous subsection where the binary alloy was used to illustrate various notions concerning the order parameter field, the present discussion will examine ideas on the energetics of such alloys when viewed from the order parameter perspective of functionals like that of eqn (12.23).

As noted above, we are interested in describing the temporal evolution of systems characterized by both conserved and nonconserved order parameters. From the standpoint of the physics of the underlying free energy functional, these different eventualities require different arguments. One way in which alloy systems can evolve in time is by virtue of various rearrangements of the local concentration of the relevant constituents, which can include the emergence of second-phase particles which are precipitated from the matrix phase, for example. For such a system in which the relevant kinematic field (a conserved order parameter field) is the concentration itself, perhaps the simplest free energy density we might write down is of the form

$$f(c, T) = \rho E_0 c(1 - c) + kT\rho[c \ln c + (1 - c) \ln(1 - c)]. \qquad (12.24)$$

We have introduced ρ as the number density of particles and E_0 is the energy scale associated with the atomic-level bonding. This expression is a reflection of the so-called regular solution model which posits that the free energy density is gotten by accounting for the interaction energies between the various atoms making up the alloy in a mean field way while the entropy is obtained using the simple counting arguments that culminated in eqn (3.89).

To see the arguments given above in further detail, we note that the first term in eqn (12.24) is an approximation to the internal energy which can be rationalized on the basis of a mean field version of bond counting associated with a pair potential description of the energy. In particular, we are interested in evaluating the lattice sum $E_{tot} = \frac{1}{2}\sum_{ij} V(R_{ij})$. For simplicity consider a lattice with coordination number Z and in which the total energy is reckoned only on the basis of near-neighbor interactions. Within the confines of this model, the energy is given approximately by

$$E_{tot} = \frac{Z}{2}[c^2 V_{BB} + (1-c)^2 V_{AA} + c(1-c) V_{AB} + (1-c)c V_{AB}]. \qquad (12.25)$$

The logic behind this equation is based on an average treatment of the number of BB bonds (i.e. evaluate the probability that both sites i and j will be occupied by B atoms which is given by c^2), the number of AA bonds (i.e. $(1-c)^2$) and the number of AB bonds (i.e. site i is of type B and site j is of type A yielding $c(1-c)$ or site i is of type A and site j is of type B yielding $(1-c)c$). We now compute the energy relative to the energy of pure components, namely,

$$E_{alloy} = E_{tot}(c) - c E_B - (1-c) E_A. \qquad (12.26)$$

If the result embodied in eqn (12.25) is substituted into this equation and we exploit the fact that $E_A = (Z/2)V_{AA}$ and $E_B = (Z/2)V_{BB}$, it is found that the alloy energy is given by

$$E_{alloy} = Z\left[V_{AB} - \frac{1}{2}(V_{AA} + V_{BB}) \right] c(1-c), \qquad (12.27)$$

which jibes with the expression for the energetics of a regular solution introduced above. The second term in eqn (12.24) is a normalized version of eqn (3.89) based on the assumption that this functional form can be applied locally. Note that the underlying physical *assumption* of a model of this type is a mathematical incarnation of the adage that one should think globally and act locally. Free energy functions of the form given in eqn (12.24) are appropriate for describing the overall macroscopic state of a system in thermodynamic equilibrium. In the present context, it is applied as a *local* free energy relevant to describing the energetics of a nonuniform system.

   In addition to the interest we have attached to conserved field variables like the concentration described above, we are also interested in nonconserved fields such as the parameter η that describes the local state of order. This scenario accounts for the fact that during the course of microstructural evolution, not only is it possible for local variations in concentration to manifest themselves through the appearance of new phases, but the degree of order seen within a given phase can change as well. In this case, as was done in eqn (12.23), the free energy density is often expanded in powers of the order parameter. Recall from the discussion leading up to eqn (12.22) that the field η is introduced in order to account for such ordering. The development of free energy functionals that depend upon this field proceeds as follows.

   One strategy is phenomenological and rooted in the Landau theory of phase transitions. The basic idea is to assert that the homogeneous free energy density may be written in the form

$$f(\eta, T) = a_0(T) + b_0(T - T_c)\eta^2 + c_0 T \eta^4. \tag{12.28}$$

The significance of this expansion is its resulting nonconvex well structure. We note that for $T < T_c$, the free energy will have minima for $\eta \neq 0$ corresponding to the emergence of an ordered state characterized by nonzero values of the order parameter. On the other hand, for $T > T_c$, the energy minimizer will correspond to a state with $\eta = 0$, signifying the absence of order.

Free energy functionals like those described above have been associated with several important classes of evolution equation. In the context of an order parameter field η that are conserved, one useful approach is to posit that the dynamics of the field η satisfies an equation of the form

$$\frac{\partial \eta}{\partial t} = -\mu \frac{\delta F}{\delta \eta}. \tag{12.29}$$

The parameter μ is a kinetic coefficient which sets the rate at which the system evolves in time. In the eventuality that the fields characterized by the order parameter are conserved, the evolution equation given above is replaced by

$$\frac{\partial \eta}{\partial t} = D\nabla^2 \left(\frac{\delta F}{\delta \eta}\right), \tag{12.30}$$

where, like μ given above, D is a kinetic coefficient. In addition to their role in the materials setting, equations of the sort introduced above have seen duty in the dynamical critical phenomenon setting where the most general dynamical equations consistent with various symmetries are written down. A review of these ideas can be found in the article of Hohenberg and Halperin (1977). As they stand, the two evolution equations introduced above seem to have a distinctly different flavor.

Though evolution equations of the form introduced in eqns (12.29) and (12.30) can be taken as an *ansatz* for the temporal evolution of complex systems, it is desirable to give them a more firm footing. These ideas have been given a unified perspective on the basis of the fact that all such models can be thought of from the perspective of gradient flows with respect to different inner products. Though the significance of these ideas has not been fully spelled out, it seems opportune to include them here if for no other reason than perhaps it will inspire their further investigation. Put simply, the unifying perspective afforded by consideration of gradient flows is the claim that the same free energy functional can lead to different evolution equations depending upon the inner product that is used to define the gradient. In particular, it is claimed that both eqns (12.29) and (12.30) can be thought of as gradient flow equations, but with the gradient defined with respect to different inner products. These ideas are probably best illustrated through an appeal to several different case studies.

The fundamental physical idea we will now attempt to develop is that of dynamics which emerges in conjunction with evaluating the gradient in the free energy functional. We report on the generalization of the conventional idea of a gradient to the functional setting in which rather than asking how a function changes in making a small excursion from one point in space to another we will examine how functionals change upon making an excursion in the order parameter field. The answer to the question of how much the functional changes depends upon the norm we use.

The idea embodied in our mathematical notions is one of measuring how far the instantaneous free energy is from its minimum. The inner product idea is familiar from a variety of settings including in association with the various special functions of mathematical physics. For example, the derivation of the expressions for the Fourier coefficients used in developing Fourier series emerges from the consideration of various inner products of the form $\int_0^a \sin(n\pi x/a) \sin(m\pi x/a)dx$. In fact, this is an example of one of the key inner products we will consider here, namely, the L_2 inner product defined through the relation

$$\langle f, g \rangle_{L_2} = \int_\Omega f(\mathbf{x})g(\mathbf{x})d^3\mathbf{x}. \tag{12.31}$$

The second key inner product that has been invoked in a provocative way by Carter *et al.* (1997) is the H^{-1} inner product defined by

$$\langle f, g \rangle_{H^{-1}} = \int_\Omega \nabla\phi_f \nabla\phi_g d^3\mathbf{x}, \tag{12.32}$$

where ϕ_f and ϕ_g are defined through the relations $\nabla^2\phi_f = f(\mathbf{x})$ and $\nabla^2\phi_g = g(\mathbf{x})$. In this case, the fields $f(\mathbf{x})$ and $g(\mathbf{x})$ are each presumed to have integral zero when integrated over the entirety of the region of interest. As a result of this proviso, this inner product will be seen later to be of interest in cases in which the order parameter field is conserved.

In order to exploit the ideas introduced above, we begin by noting that within the confines of the inner product formalism, a generic evolution equation is written in the form

$$\frac{\partial \eta}{\partial t} = -\mu \, \text{grad} \, F, \tag{12.33}$$

where the coefficient μ is the relevant mobility for the dynamics in question and we use grad (as opposed to ∇) to generically signify the gradient associated with an as-yet unspecified inner product. The field variable in question is taken to be generically given by η. We now note that regardless of how we define the inner product, the relation

$$\frac{d}{dt} F[\eta + t\delta\eta]|_{t=0} = \langle \text{grad } F, \delta\eta \rangle \tag{12.34}$$

must be satisfied. Our strategy will be to evaluate the left hand side of eqn (12.34) explicitly by recourse to our definition of the free energy function $F[\eta(x)]$ itself. The next step in the procedure will be to manipulate the resulting expression so that it is cast in the form of the particular inner product we have in mind, which allows us to read off grad F which may then be exploited in the context of eqn (12.33). Note that in evaluating the left hand side of eqn (12.34) we are doing nothing more than evaluating the first variation in the functional $F[\eta(x)]$ in the usual sense in which it is encountered in the calculus of variations. The second part of the argument is concerned with finding a way to write this first variation as a gradient. As will be shown in detail below in the context of our two key case studies (i.e. the Allen–Cahn and Cahn–Hilliard equations) the outcome of evaluating all of the quantities in eqn (12.33) depends upon our choice of inner product. One of the compelling features of the ideas to be set forth here is that Taylor and Cahn (1994) have been able to show that many of the most important sharp and diffuse interface models of microstructural evolution can all be brought within the same fold when viewed from the perspective of gradient flows.

Nonconserved Fields and the Allen–Cahn Equation. Nonconserved order parameters (such as the state of order itself as introduced in eqn (12.22)) can have complex spatial distributions that evolve in time. The governing equation in this case is provided by the Allen–Cahn equation. To obtain this equation using the ideas introduced above, we hark back to the L_2 norm. It is assumed that the instantaneous free energy of the system can be written down as a functional of the order parameter and its gradients. For example, in the context of the Allen–Cahn equation, the relevant free energy functional is

$$F[\eta(x)] = \int_\Omega \left[f(\eta) + \frac{1}{2}\kappa |\nabla\eta|^2 \right] d^3\mathbf{x}. \tag{12.35}$$

Note that even more general gradient terms can be considered and we restrict our attention to this low-order form since it suffices to illustrate our argument. The quantity $f(\eta)$ introduced above is the free energy associated with a homogeneous system. Within the context of this free energy, we must evaluate

$$\frac{d}{dt} F[\eta + t\delta\eta]|_{t=0} = \lim_{t \to 0} \frac{F[\eta + t\delta\eta] - F[\eta]}{t}. \tag{12.36}$$

Using the free energy given in eqn (12.35) in conjunction with this definition, we find

$$\frac{d}{dt}F[\eta + t\delta\eta]|_{t=0} = \int [f'(\eta) - \kappa \nabla^2 \eta]\delta\eta d^3\mathbf{x}. \tag{12.37}$$

Note that to arrive at this result we have carried out an integration by parts in which it was assumed that all boundary terms go to zero.

In the context of the Allen–Cahn equation, the claim is that it is the gradient as defined with respect to the L_2 inner product which is of relevance. In concrete terms, what this means is that we may write

$$\langle \text{grad } F, \delta\eta \rangle_{L_2} = \int [f'(\eta) - \kappa \nabla^2 \eta]\delta\eta d^3\mathbf{x}, \tag{12.38}$$

which results in the correspondence

$$\text{grad } F = f'(\eta) - \kappa \nabla^2 \eta. \tag{12.39}$$

If we now recall that the excursion $\delta\eta$ is arbitrary, we find the evolution equation

$$\frac{\partial \eta}{\partial t} = \mu[\kappa \nabla^2 \eta - f'(\eta)], \tag{12.40}$$

the celebrated Allen–Cahn equation which has been used to describe the spatio-temporal evolution of a nonconserved order parameter field.

As noted above, from a physics perspective, the Allen–Cahn equation is aimed at describing the spatio-temporal evolution of a nonconserved order parameter. If removed from its abstract cloak, what this means is that this partial differential equation can be used to characterize the state of order in an alloy system such as Fe–Al. Because the phase diagram in such systems provides for an order–disorder transition once a critical temperature is reached, with changes in the temperature the system can undergo complex processes in passing from one terminal state to another. In particular, the ordered regions eventually gobble up the disordered regions with the ultimate state being one of complete order. It is the dynamics of such processes that the Allen–Cahn equation attempts to address.

Since our current interest is more associated with the ideas that have been set forth for thinking about evolution equations rather than with the particulars of microstructural evolution itself, we satisfy ourselves with a few brief comments on the solutions to the Allen–Cahn equation. An example of the type of results that can be obtained upon numerically integrating the Allen–Cahn equation in a two-dimensional setting is shown in fig. 12.5. These calculations were undertaken using a free energy functional with the homogeneous term given by eqn (12.28) and for the case in which $T < T_c$. The initial condition corresponds to small but random deviations from the homogeneous state $\eta = 0$, and the subsequent

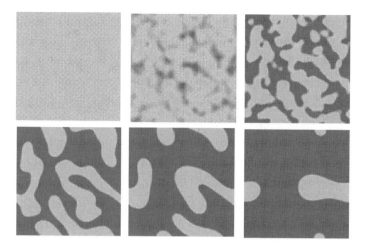

Fig. 12.5. Illustration of the spatio-temporal evolution of the order parameter field for a system described by the Allen–Cahn equation (courtesy of W. Craig Carter). The temporal sequence runs from left to right starting with the first row.

temporal evolution illustrates the way in which the different ordered domains (i.e. $\eta = \pm 1$) compete.

Conserved Fields and the Cahn–Hilliard Equation. The Allen–Cahn equation told us something about the spatial distribution and temporal evolution of a nonconserved order parameter which characterizes the state of order within the material. From the materials science perspective, it is often necessary to describe situations in which a conserved field variable is allowed to evolve in space and time. In this context, one of the most celebrated evolution equations is the Cahn–Hilliard equation which describes the spatio-temporal evolution of conserved fields such as the concentration.

Once again our derivation of the relevant evolution equation will imitate that of Carter *et al.* (1997). The case of a conserved order parameter is more intriguing than that of the previous example involving the Allen–Cahn equation since conventional arguments in this case are not based upon gradient flows. The starting point of our analysis is once again the free energy of eqn (12.35). Once again, eqn (12.37) gives us the first variation in the functional $F[\eta(x)]$. Our plan at this point is to manipulate this equation until it is seen to be in the form of an H^{-1} inner product, which will then allow us to read off the gradient in this setting. In the present setting, our plan is to invoke a second form of the H^{-1} inner product, namely,

$$\langle f, g \rangle_{H^{-1}} = \int_\Omega \nabla \phi_f \nabla \phi_g d^3\mathbf{x} = -\int_\Omega (\nabla^2)^{-1} f(x) g(x) d^3\mathbf{x}, \qquad (12.41)$$

where the second version is obtained upon integrating the first version by parts and by assuming that all boundary terms vanish. Exploiting eqn (12.41), our argument is that we may write

$$\langle \text{grad } F, v \rangle_{H^{-1}} = \int \nabla \phi_{\text{grad} F} \nabla \phi_v d^3\mathbf{x} = -\int \phi_{\text{grad} F} v d^3\mathbf{x}, \qquad (12.42)$$

where we remind the reader that our notation is $\nabla^2 \phi_f = f$. As a result of these definitions, we see that

$$\phi_{\text{grad} F} = [\kappa \nabla^2 \eta - f'(\eta)], \qquad (12.43)$$

which implies in turn that

$$\nabla^2 \phi_{\text{grad} F} = \text{grad } F = \nabla^2[\kappa \nabla^2 \eta - f'(\eta)]. \qquad (12.44)$$

We are now prepared to invoke eqn (12.33) with the result that

$$\frac{\partial \eta}{\partial t} = \mu \nabla^2[f'(\eta) - \kappa \nabla^2 \eta], \qquad (12.45)$$

which is the famed Cahn–Hilliard equation.

Like its nonconservative counterpart seen in the Allen–Cahn equation, the Cahn–Hilliard equation is aimed at describing the evolution of field variables used to describe microstructures. In the present setting, a particularly fertile example (of which there are many) of the use of this equation occurs in the context of phase separation. Recall that the Cahn–Hilliard equation describes a system with a conserved field variable. What we have in mind is the type of two-phase microstructures described in chap. 10 where the phase diagram demands the coexistence of the host matrix material and some associated precipitates. The Cahn–Hilliard equation describes the temporal evolution of such microstructures.

An example of the type of results that can be obtained when the Cahn–Hilliard equation is solved numerically is shown in fig. 12.6. In this case, the homogeneous free energy is that of eqn (12.24). As with the numerical results for the Allen–Cahn equation shown in fig. 12.5, the initial conditions correspond to a small but random deviation from the homogeneous state.

Broad classes of phenomena of interest in materials science are characterized by governing equations falling into one of the categories described above (i.e. Allen–Cahn or Cahn–Hilliard), including the analysis of spinodal decomposition, phase separation, solidification, grain growth, etc. It is useful to reflect on the qualitative features of the types of equations our analysis has led us to. First, we note that both the Allen–Cahn and Cahn–Hilliard equations allow for a description of the metastable states that can be realized by systems undergoing temporal processes. In addition, we note that because of the isotropic version of the gradient term favored

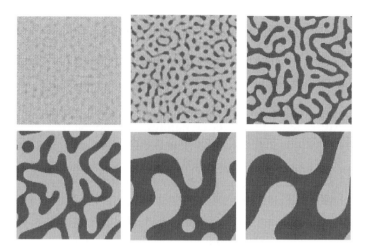

Fig. 12.6. Illustration of the spatio-temporal evolution of the order parameter field for a system described by the Cahn–Hilliard equation (courtesy of W. Craig Carter).

in the free energy functional itself, the results we have obtained are characterized by smooth interfaces.

Though an argument can certainly be made that by tying our discussion of the Allen–Cahn and Cahn–Hilliard equations to abstract notions concerning inner products we have become mired in formalism, I have persisted in this development because of the hope that it will inspire further investigation into possible variational foundations for models of temporal processes. There is a littany of schemes that have been introduced over the years (minimum entropy production, maximum entropy) which seek, I think unsuccessfully, a broad variational underpinning for modeling kinetics. The present discussion represents my own attempt to wave the flag in favor of further efforts in this vein, even if the arguments given here are little more than provisional.

One of our themes over the course of the book has been the importance of seeking to inform higher-level continuum theories on the basis of microscopic insights into the phenomenological parameters used in a given theory. This philosophy of information passage is well established in the context of problems such as the first-principles determination of diffusion constants or the determination of surface energies. In the setting of the types of phase field models introduced in the present section, the strategy of imbuing the theory with microscopic information is less well developed. On the other hand, these models represent exactly the type of setting within which some sort of first-principles analysis of the physics of the various parameters appearing in the model is necessary. In particular, we have seen that a length scale arises in the context of the gradient terms in the free energy functional, and similarly, there are various prefactors which show up in the homogeneous terms in the free energy as well. The microscopic evaluation of these parameters and their meanings will certainly constitute an important area of contact between atomistic and continuum ideas.

Variational Approaches to Defect Dynamics. The notion of a variational principle can mean different things in different circumstances. From the present perspective, it is important to draw a distinction between variational principles that are constructed so as to identify certain privileged *states* as opposed to those that tell us how to identify optimal *trajectories*. Though it is the latter case that concerns us here, we will briefly reflect on both cases so as to sharpen the distinction. This distinction can be illustrated by reference to the particular example of elasticity theory. Within the confines of the linear theory of elasticity, the privileged displacement fields in a given equilibrium problem are those that minimize the potential energy as described in section 2.5.1. This principle as such tells us nothing about how the system attained that state. By way of contrast, the equations of elastodynamics (i.e. eqn (2.32)) can be derived on the basis of the continuum analog of the principle of least action, and give us a basis for finding the trajectory of an elastic system as it evolves in time. In words, what this principle tells us is out of all the trajectories connecting two states, choose that trajectory which minimizes the action. In the present setting, we wish to revisit the discussion of section 2.3.3 in order to examine how to write the dynamics of defects in a unified variational setting that yields the trajectories of the defects of interest.

We repeat the discussion of section 2.3.3 which culminated in eqn (2.46), but now clarified by the kinematic description of the previous section. Recall that the basic argument consisted of two parts. First, we construct a function (if the number of configurational coordinates is finite) or a functional (if the configurational coordinates are specified by a function) which is the sum of the time rate of change of the Gibbs free energy $\dot{G}(\{\dot{\mathbf{r}}_i\}; \{\mathbf{r}_i\})$ and a dissipative potential $\Psi(\{\dot{\mathbf{r}}_i\}; \{\mathbf{r}_i\})$, of the form

$$\Pi(\{\dot{\mathbf{r}}_i\}; \{\mathbf{r}_i\}) = \Psi(\{\dot{\mathbf{r}}_i\}; \{\mathbf{r}_i\}) + \dot{G}(\{\dot{\mathbf{r}}_i\}; \{\mathbf{r}_i\}). \qquad (12.46)$$

In the present context, we have denoted the defect coordinates using the set $(\{\mathbf{r}_i\})$, while the set $\{\dot{\mathbf{r}}_i\}$ is the associated set of velocities. The second part of the argument is to find those $\{\dot{\mathbf{r}}_i\}$s that minimize Π. What this means is that at a fixed configuration (specificed by the parameters $(\{\mathbf{r}_i\})$), we seek those generalized velocities which are optimal with respect to Π. A dynamics of the degrees of freedom $(\{\mathbf{r}_i\})$ is induced through an appeal to the variational statement $\delta\Pi = 0$. Since our aim is to find velocity increments to step forward in time from a given configuration, for a discrete set of dissipative excitations, $\delta \Pi = 0$ amounts to the condition that $\partial \Pi / \partial \dot{r}_i = 0$ for all i.

We have already made use of this principle in several different contexts such as revealed in contexts such as curvature-induced motion (see eqn (2.50)), crack tip dynamics (see eqn (2.70)), the growth or dissolution of a spherical particle (see eqn (10.48)) and in the analysis of grain growth (see eqn (10.48)). In each of these cases, some variable or set of variables was introduced to characterize the instantaneous state of the defect of interest. We then deduced the way in which the free energy changes as the configurational coordinates are changed. The point of the present section is one of perspective, with the aim being to show that all of these examples are but special cases of a generic way of writing down dynamic equations for models which feature defects as the fundamental entities. Later in the present chapter we will consider yet another example of this scheme in the context of dislocation dynamics.

12.3.3 Inhomogeneous Systems and the Role of Locality

In the previous subsection, we have described some of the generic ideas on both kinematics and dynamics that have been set forth to examine the behavior of materials. One of our key observations was that the jumping off point for many different models is the construction of an energy functional which depends upon the relevant kinematic descriptors. One of the key tools invoked to construct such functionals is the assumption of locality. The basic idea of such locality assumptions is to determine the properties of a system with spatial variations by adding up the contributions arising from each local region acting as though that local region were part of a system of infinite extent that had no spatial variations. By way of contrast, a nonlocal description of the energy is founded upon the idea that the energy density at a given point depends upon the kinematic state in some neighborhood of that point. We devote a separate subsection to the role of locality assumptions to serve as a reminder of the way that such assumptions have seen service throughout the book.

Perhaps the most elementary application of this idea that we have used was in writing the total elastic strain energy of a system as an integral of the form

$$E_{strain} = \frac{1}{2} \int_\Omega C_{ijkl} \epsilon_{ij}(\mathbf{x}) \epsilon_{kl}(\mathbf{x}) d^3\mathbf{x}. \tag{12.47}$$

As noted above the spirit of locality is that the energy is reckoned point by point with an energy density of the form $W(\epsilon) = \frac{1}{2} C_{ijkl} \epsilon_{ij} \epsilon_{kl}$. We note that the domain of validity of a locality assumption of this kind is based upon measuring the wavelength of the relevant elastic distortions against the lattice parameter. As noted in chap. 5, the nonlinear extension of these ideas may be obtained through an appeal to the Cauchy–Born rule within which the total energy of a deformed solid is given by

$$E_{strain} = \int_{\Omega} W(\mathbf{F}) dV. \tag{12.48}$$

Like with its linear elastic counterpart, the stored energy density $W(\mathbf{F})$ at a given point depends only upon the deformation gradient $\mathbf{F}$ at that point.

In addition to the role of locality assumptions in elasticity, an even more subtle use of these ideas was revealed in our discussion of the electronic structure of solids. Recall that one of the elemental assumptions that has turned the Hohenberg–Kohn theorem into a calculational scheme is the local density approximation. In chap. 4 we showed that the many-body aspects of the electron–electron interaction are buried in an unknown exchange-correlation energy functional. The key idea behind the local density approximation is the hope that for slowly varying electron densities, perhaps the energetics of the inhomogeneous electron gas can be understood through an appeal to the properties of the *homogeneous* electron gas. In particular, the assertion is that the exchange-correlation energy can be written as

$$E_{xc}[\rho] = \int_{\Omega} \rho(\mathbf{r}) \epsilon_{xc}(\rho(\mathbf{r})) d^3 \mathbf{r}, \tag{12.49}$$

where $\epsilon_{xc}(\rho(\mathbf{r}))$ is the exchange-correlation energy of a uniform electron gas of density ρ.

In considering the energetics of extended defects, we have repeatedly resorted to locality assumptions as well. In particular, in the context of dislocations we have invoked the line tension approximation to assign an energy of configuration to a dislocation of the form

$$E_{self} = \int E(\theta) ds, \tag{12.50}$$

where the line energy $E(\theta)$ is the vehicle whereby the locality assumption is introduced. A similar argument was made in considering the energetics of interfaces when we introduced an energy of the form

$$E_{interface} = \int_{\partial \Omega} \gamma(\mathbf{n}) dA, \tag{12.51}$$

where we assert that the interfacial energy depends only upon the *local* normal to the interface. The key point is that in both of these cases, the energy of extended defects with complex shapes (i.e. a curved dislocation or interface) is obtained through an appeal to a quantity that describes the energetics of either straight dislocations or flat interfaces.

Whereas the line tension was invoked as a way to capture the self-energy of dislocations from an elastic perspective, there are also ways of capturing core effects on the basis of locality assumptions. Recall that in our treatment of dislocation cores we introduced the Peierls–Nabarro model (see section 8.6.2) in which the misfit energy associated with slip displacements across the slip plane is associated with an energy penalty of the form

$$E_{misfit} = \int_{slip\,plane} \phi[\delta(x)]dx. \qquad (12.52)$$

We reiterate that locality is invoked several times in assessing the energy of a slip distribution. First, note that the energy of an inhomogeneous slip distribution, $\delta(x)$, is obtained by adding up the contributions from each part of the slip plane using the potential $\phi[\delta(x)]$ to obtain that energy. However, it is important to note that the potential itself is obtained by carrying out a series of calculations in which two crystalline half-spaces are slipped uniformly with respect to one another.

Another set of ideas that exploit the same basic local interfacial constitutive framework as that used in the Peierls–Nabarro setting are those that we will refer to as cohesive surface models. Such models have been an important development in the fracture context since they obviate the need for *ad hoc* fracture criteria and allow fracture to emerge on the basis of decohesion on certain planes. In the context of such cohesive surface approaches, the interfacial energy is a function of a parameter characterizing the opening displacements at the interface (Δ) and a parameter characterizing the sliding displacements at that interface (the parameter δ already familiar from the Peierls–Nabarro setting). As a result, the interfacial contribution to the total energy is given by

$$E_{interface} = \int_{interface} \phi[\delta(x), \Delta(x)]dx, \qquad (12.53)$$

where $\phi[\delta(x), \Delta(x)]$ is the interplanar potential. Note that the energy is gotten on the basis only of a knowledge of the *local* state of deformation at the interfacial point of interest.

A final example (the reader is urged to think of other examples of his or her own) of the types of models in which we have made use of locality assumptions like those considered here is in the context of order parameter descriptions of microstructural evolution. In particular, in writing the free energy functional in the form

$$F[\phi(x)] = \int_{\Omega} \left(f(\phi) + \frac{1}{2}\kappa|\nabla\phi|^2 \right) d^3\mathbf{x}, \qquad (12.54)$$

we have once again made the assumption that part of the energetics of an *inhomogeneous* system can be written in terms of quantities reckoned for homogeneous systems. In particular, in the present case, it is the term $f(\phi)$ given above and in eqn (12.35) which emerges from the consideration of homogeneous systems. In this setting a concession in favor of inhomogeneity is made in the form of the inclusion of the gradient term which penalizes changes in the fields rather than basing that energy penalty on the value of the field itself as does the local term.

The reason that the issue of locality assumptions has arisen in the context of effective theory construction is that such locality assumptions are one of the key tools taken up in the attempt to characterize the energetics of *heterogeneous* systems. It is amusing to note that interest in these systems arises precisely *because* they are not locally heterogeneous and yet, much of the theoretical machinery used to examine them is based on our understanding of homogeneous systems. The examples set forth above are just a few of the different ways in which locality assumptions have been invoked over the course of this book.

12.3.4 Models with Internal Structure

Though the ideas introduced above were largely analytic, a host of ideas have also been set forth with the aim of creating models that can be cast in a convenient form for computer simulation. A number of interesting studies can be thought of from a unified perspective as a result of their reliance on mixing several different computational paradigms and the construction of models of the interface between them. I will broadly characterize such models by noting that the model itself has some internal structure. Our intention in this section is to give an overview of some of the generic features shared by these different models. This discussion will be followed by a description of several case studies that illustrate how atomistic and continuum thinking may be wed explicitly with the outcome that the theoretical whole that results is greater than the sum of the parts.

Nonuniform Discretization in Space and Time

One of the generic features of models which link different computational paradigms is their reliance on flexible schemes for handling the geometry of space and time. In particular, such models often involve the use of nonuniform discretization in either the spatial or temporal domains, or both. The finite element method is an example of a numerical scheme which can conveniently cope with meshes in which the mean spacing between nodes in one region can be quite distinct from that in another. The advantage of such schemes is that they automatically account for the fact that the geometric structures within the problem may not all reside at the same length scale. Just as it is possible to construct meshes which respect the geometric structures within a problem, it is similarly possible to consider a series of different clocks within a given model which run at rates that are governed by the *local* vibrational character of the system. Those regions that

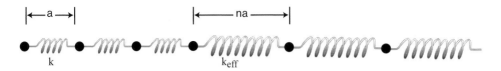

Fig. 12.7. Schematic illustrating the construction of a nonuniform spatial discretization of a one-dimensional region and the associated effective Hamiltonian used in the evaluation of the dynamics of this body.

are sluggish and characterized by long vibrational periods permit the use of longer integration time steps than those regions involving atomic vibrations, for example.

A schematic of the conceptual issues associated with dynamics on a graded mesh is shown in fig. 12.7. The basic idea is the replacement of atom–atom interactions with node–node interactions mediated by effective springs with spring constant k_{eff}. Just as with the various generic ideas introduced earlier, models of this type reflect two of the key challenges faced in constructing an effective theory. First, a decision must be made as to which degrees of freedom will be considered. This decision results in a set of nodes like those depicted in the figure. The second key challenge is the construction of the total energy and the resulting dynamics. Concrete examples of how these decisions are made will be given later in this section and a specific example is taken up in the problems at the end of the chapter.

In addition to the questions surrounding the construction of the effective dynamics, we also note the existence of a puzzle associated with dynamics on meshes like that shown in fig. 12.7. In particular, we note that there is an impedance mismatch between regions of different resolution. This problem is immediately evident upon considering the propagation of an incident wave from the region of full resolution to that of lower resolution. It is physically evident that waves with a wavelength comparable to the mesh spacing will have structure that cannot be resolved in the region of coarsened mesh. This effect translates into the fact that part of the incident energy is reflected upon arriving at the interface between the two regions. A detailed accounting of this effect can be found in Bazant (1978).

Constitutive Interfaces

As has been noted repeatedly throughout the book, a key feature of any description of material response is the introduction of material specificity in the form of a constitutive model. Throughout the various discussions that have made up the book, we have hinted that far from defects, conventional continuum thinking is appropriate while in the immediate vicinity of defects a more refined analysis such as is offered by atomistics is called for. As a result, there is a compelling argument for finding a way to bring these different constitutive philosophies to bear on the same problem.

A second generic feature (beyond the flexible treatments of space and time described in the previous subsection) shared by a number of different models of the type in question here is the use of more than one constitutive framework within a single overall computational scheme. One governing principle that has been invoked is that of separating out those regions which demand constitutive nonlinearity from those that do not. As already mentioned in the opening pages of this chapter, the big nonconcepts (i.e. nonlinearity, nonlocality and nonconvexity) are in many cases responsible for the most intriguing features of material response. Often, the spatial regions over which these features make themselves known are relatively limited. In many instances, the act of separating out the regions of nonlinear behavior is synonymous with separating regions in which a full atomistic analysis is carried out from those in which it is not. As we will see in our discussion of the case studies to follow, there are a number of different ways of importing atomic-level realism into a given model in selected regions without paying the full price of atomistics everywhere. As a result of earlier discussions it has become abundantly clear that there are a number of problems for which atomic-scale insights at the defect core are crucial. The class of models to be discussed here all involve the same overall idea: install a proper nonlinear description at the defect core while permitting a relaxation of this demand in the far fields. We will refer to the nonlinear region near the defect core as the process zone.

Case Studies in Models with Internal Structure

Cohesive Surface Models. From the standpoint of previous chapters, the most immediate example of the type of constitutive philosophy described above is that of cohesive surface models. Such models are built around the idea of limiting all constitutive nonlinearity to certain privileged interfaces, while the remainder of the material is treated via more conventional continuum thinking. An example of this thinking was already met in our discussion of the Peierls–Nabarro model in which it was seen that the energetics of the slip distribution associated with the dislocation could be handled by recourse to a nonlinear interplanar potential. In particular, recall fig. 8.26 in which we show the partitioning of a dislocated crystal into two elastic half-spaces, separated by a slip plane characterized by a nonlinear interplanar potential which is *nonconvex* and penalizes all slip displacements that are not integral multiples of the Burgers vector. It is the generalizations of this idea to more generic deformation processes that we take up here.

A key point to be made in the present context is that such cohesive surface models can be used to describe the energetics of a number of different dissipative processes related to fracture as shown in fig. 12.8. The claim is that each of these different mechanisms is amenable to a treatment in which the interfacial normal tractions can be derived from a nonlinear interplanar potential according

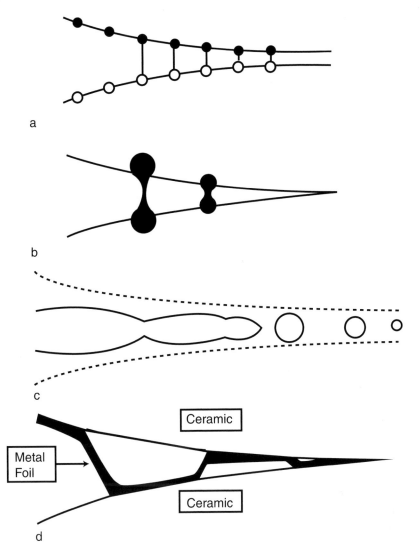

a

b

c

Ceramic

Metal
Foil

Ceramic

d

Fig. 12.8. Schematic illustration of different fracture mechanisms that may be characterized via a cohesive surface (adapted from Suo and Shih (1993)).

to a relation of the form

$$T_n = -\frac{\partial \phi(\Delta)}{\partial \Delta},$$

(12.55)

where T_n is the normal traction itself, $\phi(\Delta)$ is the interplanar potential which depends upon the opening displacement Δ. This interplanar potential gives an energy/area which specifies the energy penalty associated with opening displacements. The key points to be taken away from fig. 12.8 are the existence of a number

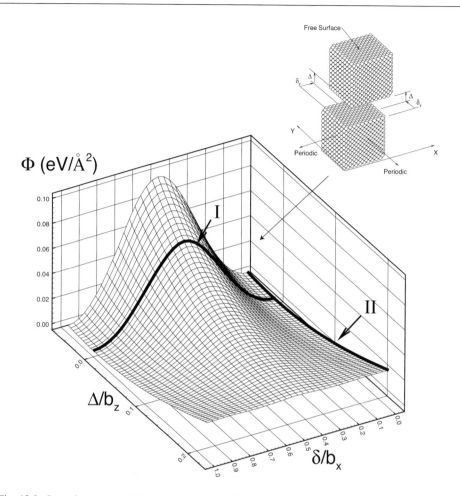

Fig. 12.9. Interplanar potential for opening and sliding displacements. This potential is for opening on the {001} plane of Al, and sliding within that plane along the [110] direction (adapted from Miller and Phillips (1996)).

of different mechanisms that can be treated via cohesive surface arguments and the emergence of a natural length scale associated with such cohesive surfaces.

Recall from our discussion of the Peierls–Nabarro model, that we have already seen the essential features of the cohesive surface framework. The basic idea is the introduction of an interplanar potential $\phi(\delta, \Delta)$. In the crystalline setting, such a potential provides a measure of the energy cost for sliding displacements δ of adjacent crystal planes as well as for opening displacements Δ across the same plane. An example of such an interplanar potential is given in fig. 12.9. In this case, the energetics is that for fcc Al as modeled using the embedded-atom method, and for opening displacements on the {001} plane with sliding permitted

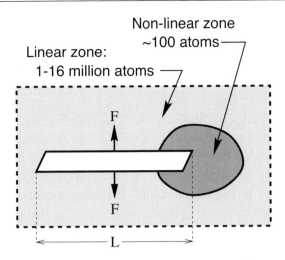

Fig. 12.10. Schematic of the lattice Green function approach for eliminating the computational overhead due to a full nonlinear treatment of interatomic interactions (adapted from Schiøtz *et al.* (1997)).

in the [110] direction. However, as was claimed in fig. 12.8, the cohesive surface framework is of broader generality than just for the description of the separation of atomic planes. Generically, cohesive surfaces are characterized by an ability to resist opening displacements until a certain maximum strength is reached and at which time, the load carrying capacity of the given cohesive surface is exhausted.

Later in the present chapter we will examine the application of these ideas to the study of fracture and dislocation nucleation at crack tips. For our present purposes, the key point to be made was the way in which several different modeling paradigms, namely, the use of bulk and planar constitutive models, are brought under the same roof, with the consequence that the resulting model is able to do things that neither of the constituent models can do by itself.

Matching Atomistics to Harmonic Lattice Statics. A second class of models with internal structure are those in which a certain subregion of the model is treated explicitly using conventional atomistic analysis. One class of methods of this type is that in which a fully atomistic nonlinear region is tied to a region in the 'far fields' in which the response is entirely linear (Thomson *et al.* 1992, Gallego and Ortiz, 1993). The basic idea of such methods is that the response of the linear region which provides the boundary condition for the fully nonlinear region is computed once and for all, and then is used in order to carry out calculations on the interior region. A schematic of both the geometric and constitutive framework is given in fig. 12.10 which shows how a crack may be handled within the confines of such methods.

The first step in the analysis is the selection of the relevant degrees of freedom. What this means in particular is the division of the entire computational domain into two parts, one that is slated for a full nonlinear treatment, and the remainder, which will be treated approximately using a linear constitutive framework. Once this decision has been made, we are next faced with a hierarchical set of calculations aimed at an approximate statement of the equations of equilibrium. The basic idea of the analysis is perturbative with the goal being to supplement the exact lattice Green function for the perfect lattice with corrections due to the presence of the defect of interest. Ultimately, we are faced with a series of nonlinear equations to solve which allow us to obtain the unknown displacements corresponding to a particular set of forces that are applied. An example of the types of results that emerge from this calculational strategy will be shown later. However, the presence of this scheme at all in the present discussion is aimed at describing concept rather than implementation, and to call attention to another way of separating an overall computational domain into more than one part.

Atoms and Finite Elements. An alternative setting within which to carry out the types of partitioning described above is offered through recourse to the finite element method. Pioneering efforts in linking atomistic models to continuum fields as represented by the finite element method were made by Kohlhoff *et al.* (1991). Their basic idea was to treat the immediate vicinity of defects (such as dislocations and crack tips) using full atomistic calculations. However, rather than persisting with the atomistic description out to larger length scales, the atomistic region was tied to a continuum region, the equilibrium of which is determined by the equations of elasticity. As we noted above in our discussion of the generic features of models in which there are computational interfaces, they then invoked subtle boundary conditions to insure that the continuum and atomistic regions communicated transparently.

Yet another method from the general class of models described above is that of the quasicontinuum method (Tadmor *et al.* 1996) which attempts to bring atomistic analysis and the finite element method beneath the same roof. The method is built around a general constitutive philosophy as well as several key methodological ideas. We summarize the quasicontinuum method by noting that it involves three basic ideas. The first key idea is that of constraint as offered by the use of finite element interpolation. Because of the use of interpolation, an entire subset (and usually this subset is the majority of the full set of atomic degrees of freedom) of degrees of freedom are kinematic slaves. The second key idea is the use of atomistics to install constitutive realism into the model, with all of the important symmetries of the total energy inherited from the underlying atomistic analysis. Perhaps most importantly, the use of an underlying atomistic description of the

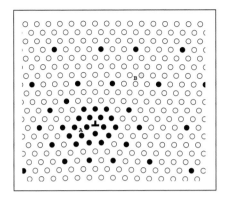 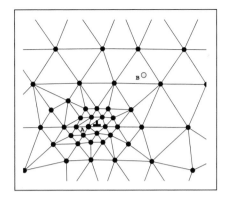

Fig. 12.11. Schematic of treatment of dislocation core in mixed atomistic/continuum setting (adapted from Shenoy *et al.* (1999)). The left hand figure shows which set of atoms are chosen as representative atoms, and the right hand figure shows the corresponding finite element mesh.

total energy implies that the energy is nonconvex giving rise to the possibility of defects such as dislocations. The final ingredient is that of adaptivity which insures that the choice of degrees of freedom can be altered during the course of calculation as the solution acquires more structure.

From a geometric perspective, the physics of degree of freedom elimination is based upon the kinematic slavery that emanates from the use of the finite element as the central numerical engine of the method. An alternative view is that of constraint. By virtue of the use of finite element interpolation, vast numbers of the atomic-level degrees of freedom are constrained. The main point is that some subset of the full atomic set of degrees of freedom is targeted as the 'representative' set of atoms, as shown in fig. 12.11. These atoms form the nodes in a finite element mesh, and the positions of any of the remaining atoms are found by finite element interpolation via

$$\mathbf{x}_i = \mathbf{X}_i + \mathbf{u}(\mathbf{X}_i), \tag{12.56}$$

where the displacement field $\mathbf{u}(\mathbf{X}_i)$ is determined by finite element interpolation as

$$\mathbf{u}(\mathbf{x}) = \sum_i \mathbf{u}_i N_i(\mathbf{x}), \tag{12.57}$$

where $N_i(\mathbf{x})$ is the shape function tied to the ith node. For a reminder on the nature of these shape functions see fig. 2.16. We reiterate that the fundamental idea associated with degree of freedom elimination is that of *constraint*. An entire subset of the full atomic set of coordinates is eliminated as a result of the fact that their positions are entirely dictated by the displacements of the representative atoms (i.e. nodes).

In addition to the geometric constraints offered by the finite element interpolation, a second key idea is introduced in order to capture the constitutive response of the material in question. In particular, the basic idea is to replace the variational principle on the full set of atomic degrees of freedom with an approximate surrogate that is still based only on a microscopic description of the total energy. The fundamental problem of lattice statics may be posed as the search for the minimizers of the potential energy function $\Pi_{exact}(\{\mathbf{x}_i\})$, where the set $\{\mathbf{x}_i\}$ is the full set of atomic coordinates. The problem we pose is the allied variational problem of finding the minimizers of $\Pi_{approx}(\{\mathbf{r}_i\})$, where the set $\{\mathbf{r}_i\}$ is the collection of nodal coordinates. The calculation of $\Pi_{exact}(\{\mathbf{x}_i\})$ was the subject of chap. 4. Our present goal is to find a way to use the methods of chap. 4 to determine the potential energy $\Pi_{approx}(\{\mathbf{r}_i\})$. Though our solution to this problem remains to some extent provisional, the basic idea is the lattice analog of numerical quadrature familiar from the perspective of integration. Note that for certain classes of energy functional such as pair potentials, pair functionals, cluster potentials and even certain electronic structure descriptions, the total energy may be written in the form

$$E_{tot} = \sum_{i=1}^{N} E_i. \tag{12.58}$$

Clearly, this sum requires that each and every atom be visited. The lattice quadrature analog of this equation is written as

$$E_{tot} \approx \sum_{i=1}^{N_r} n_i \bar{E}_i, \tag{12.59}$$

where the n_i are quadrature weights that signify how many atoms a given atom is doing the bidding for in the description of the total energy. Note that in this case we sum only over the N_r representative atoms.

A final important feature of this method is the way in which degree of freedom management can be done automatically through the use of adaptive meshing. Adaptive meshing refers to the fact that during the course of a given simulation in which the system has evolved as a result of increasing external loads, the structure of the solution demands a change in the number of degrees of freedom used to model the problem. The essential idea is to examine the spatial variations in the fields in the vicinity of a given node. If the variations are too large, the adjacent elements are refined.

To give a flavor of the types of results that can be obtained using this method, fig. 12.12 shows results on nanoindentation. The nanoindentation calculations were carried out using a pseudo-two-dimensional geometry which allows for out-of-plane displacements but not out-of-plane displacement gradients. As the

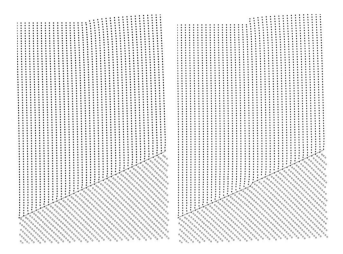

Fig. 12.12. Illustration of dislocation nucleation due to nanoindentation as simulated using the quasicontinuum method (adapted from Shenoy *et al.* (1999)). Note the presence of a subsurface grain boundary encountered by dislocations after they are nucleated at the crystal surface and travel down the vertical slip planes.

load increases, the degree of mesh refinement beneath the indenter increases in anticipation of the nucleation of a dislocation at the surface.

12.3.5 Effective Hamiltonians

Another broad class of ideas which have emerged from the statistical mechanics setting and which are a powerful tool in bridging scales are effective Hamiltonians. Generically, the notion of an effective Hamiltonian in the statistical mechanics context refers to the act of replacing one set of degrees of freedom and their associated Hamiltonian with some alternative (and presumably simpler) set and their associated Hamiltonian. By virtue of such effective Hamiltonians, the original microscopic Hamiltonian is replaced with some surrogate which features a more manageable kinematic description of the problem of interest. The introduction of such Hamiltonians can occur either through some explicit coarse-graining procedure or analogous algorithm, or as is often done just by fiat in a phenomenological way.

Of course, the idea of effective theory construction is not a new one at this late stage in the book. Indeed, much of what we have described in this book has surrounded the business of constructing effective theories. Our use of the word 'effective' in this context is not intended to conjure up images of such ideas as especially competent. Rather, we speak of theories which have shed their reductionist cloak and whose aim it is to account for observable nature on the basis of degrees of freedom other than those at the most fundamental level. One realm in which such thinking constitutes the backbone of the subject is that of critical phenomena in which it is argued that many of the mechanistic details are irrelevant. The formal apparatus of theory construction in that setting is provided by the renormalization group which is a concrete scheme for thinning degrees of freedom. However, the search for relevant degrees of freedom is more pervasive and is at the heart of all effective theory construction. The plan of this subsection is to begin with a few words on effective Hamiltonians in the critical phenomena setting followed by a last look at the more broadly construed notion of an effective Hamiltonian that has seen action throughout the book with particular emphasis on the common features of the various effective theories that we have invoked.

Renormalization

Presently, we attempt to highlight some of the key ideas associated with the construction of effective Hamiltonians and how such Hamiltonians evolve as the critical state is approached. In the critical phenomena setting, it is convenient to think about an entire family of Hamiltonians. The relation between the different members of the family is that a succession of generations are created through progressive degree of freedom thinning (i.e. integrating out degrees of freedom). As a reminder of the degree of freedom thinning inherent in the various renormalization group transformations we recall our work in section 3.3.5 in which we showed that in the context of the one-dimensional Ising model, one could systematically effect subsets of the sums appearing in the partition function with the emergence of a new effective Hamiltonian in those degrees of freedom that were not integrated out, and which now demand new effective coupling constants.

The fundamental idea developed there is reiterated in the two-dimensional setting in fig. 12.13. The key message being set forth is the idea that certain subsets of degrees of freedom can be explicitly integrated out of the problem. In so doing, as noted above, we create a family of Hamiltonians in which the original parameters are replaced by their renormalized counterparts. A particularly provocative description of these renormalization group 'flows' occurs in the form of differential equations which characterize how the various parameters in the model change upon scaling.

The notion of an effective Hamiltonian can be defined almost rigorously from the perspective of the renormalization group. One intriguing idea that has emerged from work in the critical phenomena arena is that in addition to the importance of renormalization transformations near the critical point, there may be merit to imitating the systematic way in which degrees of freedom are eliminated even when there is no critical point in question.

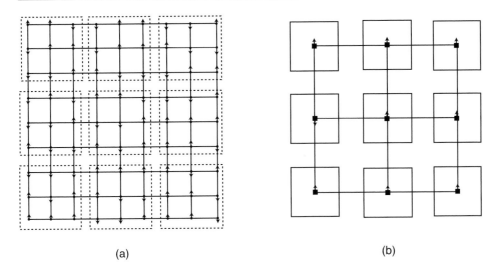

(a)   (b)

Fig. 12.13. Schematic of the decimation process exploited in the context of the Ising model (adapted from Domb (1996)). Frame (a) shows the spins that are assembled into blocks resulting in the new spin degrees of freedom shown in (b).

Effective Hamiltonians Reexamined

Throughout the course of this book we have flown the flag of degree of freedom reduction and effective theory construction. One of the first places we described such thinking was in section 4.2.2 where we invoked the transcription

$$E_{exact}(\{\mathbf{R}_i, \mathbf{r}_n\}) \rightarrow E_{approx}(\{\mathbf{R}_i\}), \tag{12.60}$$

which introduces a description of the total energy in which all reference to the electron degrees of freedom is surrendered. In section 5.4, we saw that the energy of a deformed body could be one step further removed from the most fundamental microscopic description through the introduction of an elastic energy functional of the form

$$E_{strain} = \int_{\Omega} \frac{1}{2} C_{ijkl} \epsilon_{ij}(\mathbf{r}) \epsilon_{kl}(\mathbf{r}) d^3\mathbf{r}. \tag{12.61}$$

Note that in both of these cases we have succeeded in writing down surrogate 'Hamiltonians' which serve in the stead of the full microscopic Hamiltonian.

Another important class of effective Hamiltonians saw action in the context of phase diagrams in chap. 6. In particular, we noted that with the complexity added by chemical disorder the only way to effect a systematic search over entire classes of structural competitors was to invoke an extended Ising Hamiltonian of the form

$$H = J_0 + \sum_i J_i \sigma_i + \frac{1}{2} \sum_{ij} J_{ij} \sigma_i \sigma_j + \frac{1}{3!} \sum_{ijk} J_{ijk} \sigma_i \sigma_j \sigma_k + \cdots. \tag{12.62}$$

In this case, we replaced the original microscopic energy functional which demanded that we start anew with each new alloy configuration with an alternative scheme in which different configurations correspond to different realization of the occupation variables σ_i. The discussion of section 6.5.3 demonstrated that we can go further yet, with the parameters in the effective Hamiltonian of eqn (12.62) selected to account even for the vibrations and internal relaxations that attend different alloy configurations.

While the effective alloy Hamiltonian described above allowed us to consider structural competitors in bulk alloys, we have also seen that the structural complexity on crystal surfaces can be tamed using the idea of effective Hamiltonians. Indeed, our discussion of the surface reconstructions on both the Au and W surfaces culminated in effective Hamiltonians. For example, in our discussion of the reconstructions on the (100) surface of fcc Au we invoked an effective Hamiltonian of the form

$$E_{tot}(\{\mathbf{R}_i\}) = \sum_{ij} V(R_{ij}) + \sum_{i} A\left(\cos\frac{2\pi x_i}{a} + \sin\frac{2\pi y_i}{a}\right). \tag{12.63}$$

Our key point remains that of emphasizing the diversity of situations and functional forms that we have invoked in order to replace complex microscopic Hamiltonians with surrogates that either provide deeper insight or more computational tractability, or both.

In each of the examples given above, the construction of the effective Hamiltonian is predicated upon the introduction of certain parameters (i.e. the elastic moduli C_{ijkl} of eqn (12.61), the Ising parameters J of eqn (12.62) or $V(R_{ij})$ and A in eqn (12.63)) which characterize the underlying microscopic interactions. Microscopic calculations themselves often serve as a viable basis for determining these parameters. We have already seen these types of arguments made in the context of equilibrium phase diagrams where the entirety of the configurational thermodynamics is absorbed into an Ising Hamiltonian. In addition, we saw in several different contexts these same ideas emerge in the description of the structure and energetics of surface phases for both Au(001) and W(001).

From the standpoint of informing effective Hamiltonians one seeks a series of microscopic realizations of the various states and uses microscopic analysis to determine the energetics of these states. Rabe and Waghmare (1995) elucidate the program of effective Hamiltonian construction as being to obtain 'a model system which reproduces the finite-temperature transition behavior of the original lattice Hamiltonian, while the simpler form and reduction in the number of degrees of freedom per unit cell makes it suitable for study by methods such as mean field theory and Monte Carlo simulation.'

12.4 Bridging Scales in Microstructural Evolution

In chap. 10, we gave several case studies in the analysis of microstructural evolution. In the present context, we revisit the subject of microstructural evolution, this time from the perspective of the hierarchical structure of the models used to address such problems rather than from that of the structural issues themselves. Because the processes in microstructural evolution can involve a range of scales in both space and time, it is natural that the study of the temporal evolution of these structures would involve hierarchical modeling. In the present section we consider three different issues that arise in the analysis of microstructure. Our discussion begins with some of the questions surrounding processes at crystal surfaces and the adjacent films during the process of crystal growth. We will see that models of these processes involve an interplay of atomic-level analyses of both chemical reactions and diffusion, with more macroscopic treatments of mass transport. The second key topic we will undertake here is that of solidification. The discussion will be rounded out with an examination of the coarsening of two-phase microstructures in three dimensions. It is also worth noting that our treatment in this section will largely be centered on acquainting the reader with examples of the various schemes for bridging scales presented earlier rather than with a serious examination of the details of how the calculations were done.

12.4.1 Hierarchical Treatment of Diffusive Processes

Many of the most important processes in microstructural evolution can ultimately be traced to the presence of diffusion, either at surfaces or within the bulk of the material. As was noted at the beginning of this chapter, in some instances this poses major problems since the treatment of diffusive processes from a microscopic perspective involves temporal scales that are much smaller than those associated with the microstructural evolution itself. As a result, much of the work in attempting to confront problems in which multiple temporal scales operate simultaneously has been devoted to building algorithms for eliminating uninteresting motions and ferreting out the rare but interesting motions that punctuate temporal processes within materials.

Our plan in this subsection is to examine two ideas that have been advanced to confront such problems. These arguments are put forth not so much with the conviction that they represent the final word on the hierarchical treatment of diffusion, but rather because they serve to illustrate the types of arguments that have been put forth thus far and that will perhaps encourage further efforts.

Kinetic Monte Carlo approaches. One scheme that has found widespread use in the consideration of diffusive processes is the kinetic Monte Carlo approach in which the entirety of uninteresting motions (i.e. atomic vibrations) are eliminated and the only allowed motions are the diffusive jumps themselves. This amounts to an integrating out of a certain subclass of motions by hand. The construction of such models can be roughly divided into two stages. In the first stage, all of the relevant microscopic motions are identified and their respective activation energies are computed. Thus, a catalog of all competing processes and an associated quantitative reckoning of their relative rates is established. Then, the simulation of the temporal evolution of the system proceeds by a sequence of Monte Carlo steps in which, at each step, one of the members of the catalog of allowed events is chosen at random, but weighted by its relative transition rate, and permitted to occur, thereby changing the configuration of the system. In section 12.4.2 we will give an explicit example of this approach in the context of deposition of a thin film of Al. Our main contention at the moment is that this method represents one way of coming to terms with the relative infrequency (when measured against the period of atomic vibrations) of diffusive hopping events.

One shortcoming of kinetic Monte Carlo approaches is their reliance on prior knowledge of the relevant competing microscopic processes. The reason that this can pose difficulties is that in some instances, intuition does not suffice to determine the entirety of microscopic processes. We already highlighted this difficulty in section 11.2.2 where we noted that the 'intuitive' mechanism of surface diffusion on some fcc surfaces is superseded by an exchange mechanism in which the diffusing atom burrows into the surface. Clearly, if the catalog built up in preparation for a given kinetic Monte Carlo simulation fails to account for such a low-energy pathway, the resulting evolution of the system will be contaminated. As a result of problems of this sort, a second set of ideas has been introduced.

'Hyperdynamics' Methods. Recall from our discussion of transition state theory that the rate of crossing of a given barrier depends upon the barrier height through an exponential factor of the Arrhenius form, as embodied in eqn (7.62). One class of models, broadly classified as hyperdynamics, attempt to renormalize the entirety of all transition rates such that there is an overall boost in the simulation time without changing the relative rates of competing processes. The key *desideratum* in the construction of this method is the idea that the relative transition rates from a given well are the same in hyperdynamics as those obtained on the basis of full molecular dynamics. This proviso is stated mathematically as

$$\frac{k_{A \to B}}{k_{A \to C}} = \frac{k_{A \to B}^{hyper}}{k_{A \to C}^{hyper}}, \tag{12.64}$$

where A, B and C are labels for the initial and final wells, respectively, and the constants k signify the rates. The philosophical perspective adopted here is

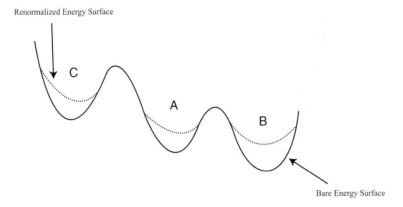

Fig. 12.14. Schematic illustration of the renormalization of the energy landscape so as to boost the time (courtesy of A. Voter).

again one of 'integrating out' of uninteresting features of the problem which in the present context means that we have surrendered a complete analysis of thermal vibrations on the grounds that it is really only the dynamics of large excursions that concerns us.

The basic idea of the hyperdynamics method in the context of a simple one-dimensional model is illustrated in fig. 12.14. The claim is that by adding a bias potential $\Delta V_b(q)$ (q is a generalized position) to the bare energy surface, the transition rates can be renormalized in a way that respects the condition provided by eqn (12.64). To see this, we follow Voter's (1997a) analysis. Recalling the definition of the transition rate given in eqn (7.55), we note that

$$k_{A \to B} = \frac{\int_{-\infty}^{\infty} dp \int_{-\infty}^{\infty} dq\, \delta(q)\dot{q}\theta(p)e^{-\beta K(p)}e^{-\beta V(q)}}{\int_{-\infty}^{\infty} dp \int_{-\infty}^{\infty} dq\, \theta(-q)e^{-\beta K(p)}e^{-\beta V(q)}}. \tag{12.65}$$

Our notation (in particular, the $\delta(q)$) implies that the saddle point separating the A and B wells is at $q = 0$. Note that we have separated the Hamiltonian into kinetic and potential energy parts according to the prescription $H(p,q) = K(p) + V(q)$. Our result for the transition rate may be rewritten as

$$k_{A \to B} = \frac{\int_{-\infty}^{\infty} dp \int_{-\infty}^{\infty} dq\, \delta(q)\dot{q}\theta(p)e^{-\beta K(p)}e^{-\beta(V(q)+\Delta V_b(q))}e^{\beta \Delta V_b(q)}}{\int_{-\infty}^{\infty} dp \int_{-\infty}^{\infty} dq\, \theta(-q)e^{-\beta K(p)}e^{-\beta V(q)}}, \tag{12.66}$$

which may be expressed more compactly as

$$k_{A \to B} = \frac{\langle \delta(q)\dot{q}\theta(p)e^{\beta \Delta V_b(q)} \rangle_{A_b}}{\langle e^{\beta \Delta V_b(q)} \rangle_{A_b}}. \tag{12.67}$$

The subscript on $\langle\ \rangle_{A_b}$ refers to the fact that the the relevant thermal averages are performed with respect to the biased energy landscape rather than with respect to the unbiased landscape which would be characterized by the notation $\langle \rangle_A$. If we similarly construct the rate $k_{A \to C}$ it is clear that the ratio $k_{A \to B}/k_{A \to C}$ is preserved under the mapping to the biased energy landscape.

If we are to run a molecular dynamics simulation with respect to the biased energy landscape, we need a transcription that tells us how to map the elapsed time as measured by our molecular dynamics time step into the time that would have elapsed had we allowed the dynamics to take place on the unbiased landscape. One argument that can be made to assess this transcription is to replace the ensemble average of the mean escape time, given by

$$\tau_{esc}^A = \frac{1}{k_{A \to}} = \frac{\langle e^{\beta \Delta V_b(q)} \rangle_{A_b}}{\langle \delta(q) \dot{q} \theta(p) \rangle_{A_b}}, \tag{12.68}$$

with the corresponding time average obtained by averaging over the dynamical trajectory itself and given by

$$\tau_{esc}^A = \frac{\frac{1}{n_{tot}} \sum_{i=1}^{n_{tot}} e^{\beta \Delta V_b(q_i)}}{n_{esc}/(n_{tot}\Delta t_{md})}. \tag{12.69}$$

What this equation tells us is to compute the average of $e^{\beta \Delta V_b(q)}$ by evaluating it at each of the n_{tot} time steps of interest. That is, each time step is labelled by the integer i which runs from 1 to n_{tot}. This gives the average demanded in the numerator of eqn (12.68). The denominator of eqn (12.68) essentially tells us how many successful escape events (n_{esc}) there are over the time of interest which is itself given by the product of the number of time steps n_{tot} and the interval per time step Δt_{md}. The final step in the determination of the hyperdynamics time step is the realization that the mean escape time must be given by the accumulated time divided by the number of successful barrier crossings. This statement may be written mathematically as

$$\frac{T_{tot}}{n_{esc}} = \tau_{esc}^A. \tag{12.70}$$

Using eqn (12.69), this may be rewritten as

$$T_{tot} = \sum_i \Delta t_{b_i} = \sum_i \Delta t_{md} e^{\beta \Delta V_b(q_i)}, \tag{12.71}$$

where we have introduced the notation Δt_{b_i} to denote our estimate of the duration in real time (i.e. how long we really would have waited in the absence of the bias potential) of the ith time step. As a result, with the assertion that

$$\Delta t_{b_i} = \Delta t_{md} e^{\beta \Delta V_b(q_i)}, \tag{12.72}$$

we preserve the statistical properties of the mean escape time. A few comments are in order. First, we note that with the hyperdynamics time step defined as above the short time properties of the trajectory are lost as is the definition of the time itself. On the other hand, in the long time limit the dynamics of barrier crossings is exactly that which we would expect if we had really run conventional molecular dynamics for such long times.

The hyperdynamics method teaches us that by redefining the energy landscape in such a way that the *relative* rates of different competing processes are unchanged, the simulation can be run in a sped up form, and the net simulation time can be computed (statistically) after the fact. Ultimately, the size of this boost factor is determined by the lowest energy barrier in the system since evidently we require that $\Delta E > 0$.

Our discussion thus far has emphasized the conceptual features of the hyperdynamics method without touching deeply on the implementational issues that must be faced in its use. In particular, the hyperdynamics method centers on the construction of the bias potential itself. In practice, the construction of this bias potential is something that one wishes to be able to do without full and detailed knowledge of the energy landscape itself. Indeed, the point is to speed up the dynamics for those systems where our intuition cannot serve as a trustworthy guide. Though the successful implementation of this step in the hyperdynamics procedure is indeed of importance, our main objective here has been to convey a flavor of what the method accomplishes more than how it does so in practice. As a result, we leave the exploration of the question of constructing the bias potential to the reader in conjunction with various references such as Voter (1997a,b), Steiner *et al.* (1998) and Gong and Wilkins (1999).

To show the key conceptual features of hyperdynamics in action, we consider a simple one-dimensional example which may be thought of as representing the diffusion of a single adatom on a rigid and corrugated surface. In particular, we consider a modification of an example originally given by Voter (1997a) in which the energy landscape is defined as

$$V(x, y) = A\left[1 + \cos\left(\frac{2\pi x}{a}\right)\right] + By^2, \tag{12.73}$$

where A characterizes the barrier height separating adjacent wells along the x-direction. In the absence of renormalization, diffusion on this energy landscape takes place with a rate that can be estimated directly on the basis of our earlier transition state analysis. We now consider adding on a renormalizing potential such that on all parts of the energy surface for which $V(x, y) < V_0$, the bare potential is replaced by V_0 itself. This approach was suggested in the work of Steiner *et al.* What this approach means in practice is that the wells have been replaced by regions of constant potential as shown in fig. 12.15. We blithely ignore any

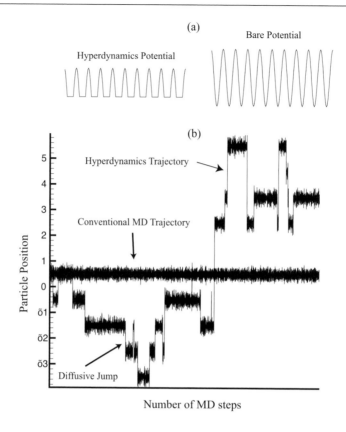

Fig. 12.15. Illustration of trajectories associated with the same basic periodic potential when considered using both conventional molecular dynamics and hyperdynamics: (a) biased potential and unbiased potential and (b) the particle position as a function of elapsed time.

subtleties introduced by the presence on the energy surface of those points where the potential changes from its constant value of V_0 to the cosine form introduced in eqn (12.73). For the purposes of illustrating the consequences of replacing the bare potential by its biased counterpart, fig. 12.15 shows trajectories for a single particle (i.e. $x(t)$) when computed using both conventional molecular dynamics and using the hyperdynamics approach. Two key features are evident. First, although the two calculations were carried out at the same temperature, the hyperdynamics trajectory exhibits a huge number of diffusive jumps in comparison with the results from conventional molecular dynamics even though the number of time steps in both calculations is the same. On the other hand, the actual elapsed time for the hyperdynamics calculation is 'boosted' relative to the molecular dynamics calculation. The evaluation of the boost factor results from the repeated application of the injunction of eqn (12.72). The integration of the equations of motion in both of these cases is carried out using Brownian (or Langevin) dynamics as described

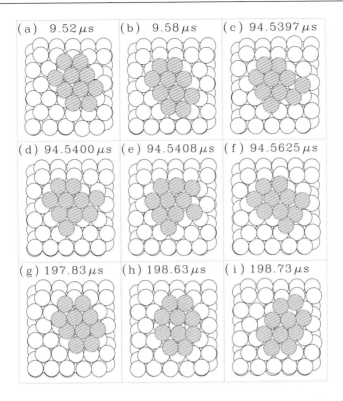

(a) 9.52 μs   (b) 9.58 μs   (c) 94.5397 μs

(d) 94.5400 μs (e) 94.5408 μs (f) 94.5625 μs

(g) 197.83 μs   (h) 198.63 μs   (i) 198.73 μs

Fig. 12.16. Illustration of island diffusion process using hyperdynamics (adapted from Voter (1997b)).

in Allen and Tildesley (1987). In particular, the equation of motion is given by

$$\frac{d\mathbf{p}}{dt} = -\xi\mathbf{p} + \mathbf{F} + \mathbf{R}, \tag{12.74}$$

where ξ is a frictional coefficient which in conjunction with the random force $\mathbf{R}$ mimics the contact with a heat bath, and $\mathbf{F}$ is the deterministic force that arises from interactions with other particles which are present in our case through the substrate potential given in eqn (12.73). This approach, as alluded to in chap. 3, is intended to mimic the thermal fluctuations that the adatom will be subjected to by virtue of its contact with the surface atoms. As a result of the coupling of the adatom's motion to that of the jiggling surface atoms (which are not treated explicitly), it spends most of its time rattling within a particular well with its motion every now and then punctuated by a transition to adjacent wells, a process corresponding to diffusive hops. These qualitative features are revealed explicitly in the example illustrated in fig. 12.15.

A more general example of the types of results that can be obtained using hyperdyanmics is depicted in fig. 12.16 which shows a series of snapshots from the life history of a collection of adatoms on a crystal surface. Unlike the artificially constrained example introduced above in which the underlying substrate offered nothing more than a potential energy landscape upon which the diffusing atom could move, the present example considers the more realistic case of a collection of adatoms on a surface built up of atoms which are themselves free to move. What is observed in this simulation is that with the passage of time the entire adatom cluster exercises a collective drifting motion and an allied change in shape. However, the most compelling feature of the figure is not so much what happens as when it happens. Note that the time scale in the top left of each frame is reported in microseconds. When contrasted with the typical molecular dynamics time step, which is a factor of 10^8 shorter than the simulation time reported here, it is clear that this represents, if not mastery over, at least substantial progress on the time scale problem. To fully appreciate the subtleties of implementing the hyperdynamics scheme in a problem like that shown here, of course, one must enter into the question of how the biased potential itself is selected, and this hopefully without recourse to a detailed knowledge of the full $3N$-dimensional energy landscape itself. For such important details, the reader is urged to consult the original papers.

A related scheme for speeding up molecular dynamics simulations of rare events exploits what one might call a thermo-temporal equivalence principle (Sørensen and Voter 2000). The contention is that if one were to isolate the system of interest in its contact with a heat bath and then were to watch a film of the unfolding of the various reactions, the perceived passage of time will be determined by the temperature of the reservoir that the system is in contact with. To the external observer, time, as measured by the rate of the relevant reaction, is inextricably linked with the availability of thermal energy for lubricating that reaction. In the opinion of the author, seeking more systematic insights into the connection between temperature and time may well hold the key to defeating the time scale problem not only in the context of molecular dynamics of materials, but even in the context of problems such as that of protein folding.

The basic idea of the algorithm to be described here (i.e. 'temperature accelerated dynamics') is that by increasing the temperature, the overall rate at which activated processes take place is increased. However, altering the temperature also has the effect of changing the relative rates of these microscopic processes and can even turn on processes that would for all practical purposes not even have taken place at the lower temperature of interest. As a result, it is necessary to correct for the temperature induced bias that is present at the higher temperature. Hence, some of the high-temperature transitions are filtered out so as to restore the hierarchy of low-temperature transitions.

In this subsection, we have set forth several of the ways in which the hierarchy of temporal scales in materials has been confronted from a theoretical perspective.

We have seen that the kinetic Monte Carlo method offers a way to import insights into the subscale temporal processes that are at work in the temporal evolution of a given material with the concomitant ability to simulate the way in which these various processes are assembled to give rise to longer-term behavior. The hyperdynamics and allied approaches have been described primarily to provoke further thought on how to connect the notion of temperature and time. Before moving on, we note that the questions raised in this section must be at the center of any attempt to think about effective theory construction. Indeed, though many of the successes highlighted throughout the book illustrate the ways in which careful selection of relevant degrees of freedom is a critical part of effective theory construction, the quest to create general theories of temporal processes must answer to similar challenges in the temporal domain.

12.4.2 From Surface Diffusion to Film Growth

A partial success story in bridging scales is that provided by advances in the modeling of crystal growth. In the present instance, we take up an example found in the many technologies used to produce thin films often of central importance in microelectronics applications. We begin with an overview of the generic class of problems that we have in mind, with our preliminary analyses aimed at treating the growth kinetics without reference to the chemical proliferation that is also possible once there is more than a single chemical species present.

In the information passage mode that is typical of the kinetic Monte Carlo method, the modeling of growth can be effected by constructing a catalog of the relevant deposition and diffusion processes that take place when atoms are deposited on the growing material. Once the relevant activation energies for the various processes have been compiled, these processes are sampled in a way that respects their relative likelihood, using the Monte Carlo method.

One example of the use of these ideas is shown in fig. 12.17 in which a series of snapshots from the growth history of an Al thin film are illustrated. In specific detail, this simulation is built around a sophisticated treatment not only of the diffusion of atoms once they have arrived on the surface, but also on a treatment of the deposition process itself in which the angular distribution of the incoming atoms is taken into account. The atoms in the thin film are all forced to occupy the sites of a perfect lattice, thus surrendering the resolution of thermally activated local relaxations. The deposition process is meant to reflect the sputter deposition of Al and is carried out in a two-step process in which a launching point above the crystal surface is selected at random and then the orientation of the particle trajectory is selected with a random azimuthal angle about the substrate normal. The polar angle (i.e. the angle away from the normal) is chosen on

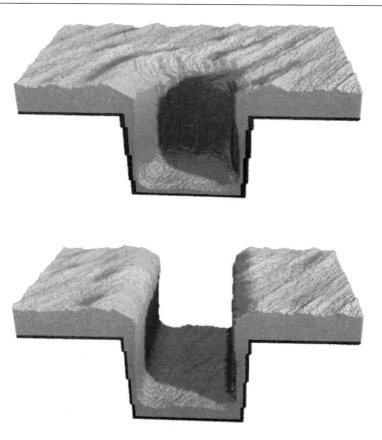

Fig. 12.17. Series of snapshots in the growth history of a thin film simulated using the kinetic Monte Carlo scheme (adapted from Huang *et al.* (1998)).

the basis of a distribution of the form $g(\theta) = \cos\theta \sin\theta$. The incident particle follows a trajectory defined by the criterion described above until it strikes the surface. When the particle arrives at the surface it is attached where it strikes unless the coordination number is too low in which case a nearby site with higher coordination number is selected. In addition to the deposition events described above, there are a series of diffusive events that are possible for those atoms on the surface. The relative probabilities of different diffusive hops are essentially determined on the basis of the number of neighbors a given diffusive candidate has. The contention is that those atoms with fewer neighbors are less tightly bound and are hence more likely to undergo a diffusive hop. Though there are a number of other details that can be described for the results shown in fig. 12.17 (see Huang *et al.* (1998)), our basic point has been to illustrate the type of results that can be obtained using the kinetic Monte Carlo method. Indeed, as a result of the type of analysis described here it is possible to examine the effect of temperature and deposition rate as well as the impact that geometric inhomogeneities such as trenches have on the growth process.

12.4.3 Solidification Microstructures

In the example given above, the formation of a material microstructure was seen to take place as a result of the deposition of atoms on a substrate. Another, equally important, route to solid microstructures is via the solidification process. During the solidification process, the baseline microstructure, which will have a significant impact on both the material's properties as well as its subsequent microstructural evolution, is created as the liquid is superseded by a solid. The nature of the microstructure in the resulting solid can be quite diverse, ranging from featureless equiaxed polycrystals, to microstructures riddled with dendrites.

From the standpoint of constructing models of the solidification process, this is a complicated problem since it involves elements of fluid dynamics, heat transfer and elasticity. The intention of the present subsection is to illustrate the ways in which ideas in this and previous chapters have been applied to the problem of solidification, with our selection of case studies driven largely by the extent to which the model of interest reveals something of the ways in which hierarchical modeling is carried out. The level of detail to be considered here should be seen as being carried out at the level of 'knowing about' rather than 'knowing', a distinction that was highlighted in the Preface and borrowed from Richard Mitchell in his *The Gift of Fire* (1987). As a result, the reader should not expect to finish this section with sufficient knowledge to write codes which imitate those described here. On the other hand, it is hoped that the case studies considered here will whet the reader's appetite.

Cellular Automata Approaches to Solidification. One interesting approach to modeling the solidification process has been developed by Rappaz and Gandin (1993) (see also Gandin and Rappaz (1994)), and involves the simultaneous application of ideas from continuum theory and from the theory of cellular automata. In a word, their idea is to consider the relevant heat transfer issues from the standpoint of conventional continuum ideas, with the use of an allied set of discrete variables which reflect the local crystal orientations of various parts of the melt as it undergoes solidification. Alternatively, this work can be thought of as a way of assembling both deterministic and probabilistic ideas within the same overall computational framework.

The conceptual features of the model are illustrated in fig. 12.18. The basic idea is the simultaneous use of two computational 'meshes' to treat the same overall spatial domain. As shown in the top frame of fig. 12.18, a finite element mesh

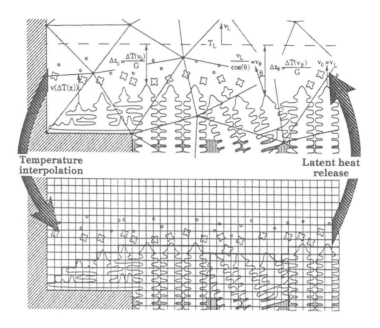

Fig. 12.18. Schematic of the coupling between a finite element treatment of heat transfer and the cellular automaton treatment of the grain structure (adapted from Gandin and Rappaz (1994)). The top frame shows the finite element mesh and the lower frame shows the lattice of points used to do bookkeeping concerning the state (liquid or solid) and orientations of the associated points.

is used in order to consider the spatial distribution and temporal evolution of the temperature field. A finer scale discretization of the same medium is carried out as shown in the bottom frame of fig. 12.18 in which each cell is characterized by an 'order parameter' which specifies whether or not the occupant of that cell is solid or liquid, and if solid, what the local crystal orientation is. As the solid engulfs the liquid, there is a release of latent heat which is fed back into the finite element mesh as represented by the arrow connecting the top and bottom frames of fig. 12.18, which has the effect in turn of altering the temperature distribution. A stochastic element is present in the model by virtue of the nucleation of new grains ahead of the dendritic fronts with a nucleation probability that is controlled by the local undercooling. In addition to the nucleation of new crystalline regions, it is also possible for a given liquid cell to be incorporated into the solid region via growth of neighboring solid regions. These growth processes are predicated on interfacial velocity relations in which the local velocity of the solid–liquid interface is determined by the relevant undercooling. Though we have really only given a caricature of what is in reality a rather complex algorithm, our key point has been the recognition that by assembling discrete and continuous, probabilistic and

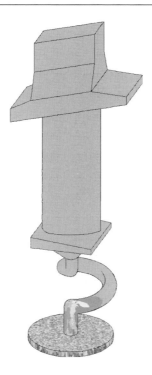

Fig. 12.19. Results of coupled continuum/cellular automaton model for solidification (courtesy of M. Rappaz).

deterministic elements, a computational scheme may be constructed which is more powerful than any one of these elements by itself.

Perhaps the most compelling result of simulations of this type is the application of these ideas to the casting of turbine blades as shown in fig. 12.19. As can be seen at the bottom of the figure, the initial solidification takes place with a number of different competing crystallites of different orientations as illustrated by the regions with different shades. However, as the solidification front grows up the 'pigtail' a single orientation is selected with the result that by the time the solidification front has reached the blade itself, only a single orientation has survived. Note that the resulting microstructure is reminiscent of the experimental realizations of this process shown in fig. 10.1.

Phase Field Approaches to Solidification. The phase field approach has found favor not only in the types of mixed models mentioned above but also in the solidification context. Though we haven't made much of this point, one of the key advantages of the phase field approach is that it obviates the need of explicitly tracking any interfaces during the course of a simulation since all that is really monitored is one or more field variables throughout space. There are two key field

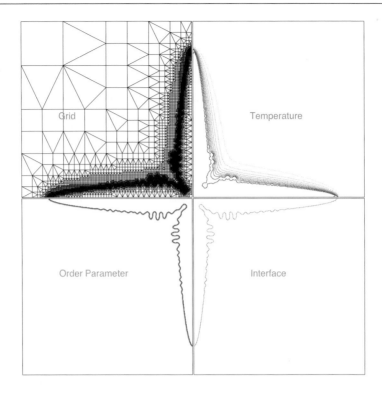

Fig. 12.20. Results from phase field model of solidification (adapted from Provatas *et al.* (1998)).

variables in the phase field approach. The first is a rescaled temperature defined by
$U = c_p(T - T_m)/L$, with c_p the specific heat at constant pressure and L being the
latent heat of fusion. As in earlier chapters, T_m refers to the melting temperature.
In addition to the conventional thermal field introduced above, a phase field ϕ is
introduced which has different values in the solid ($\phi = 1$) and liquid ($\phi = -1$).
The governing equations are a set of two coupled partial differential equations
which describe the evolution (and coupling) of the fields U and ϕ. Again in keeping
with the spirit of this part of the chapter, the reader is encouraged to examine
original sources such as Provatas *et al.* (1998) to learn more concerning the field
equations themselves. Our main intention here is to highlight the way in which
by introducing the order parameter field ϕ the explicit tracking of the solid–liquid
interface can be abandoned.

An example of the type of results that can be achieved with this approach is
shown in fig. 12.20, which cleverly depicts four distinct features associated with
the solution of a dendrite structure with fourfold symmetry. The upper left hand
panel of the figure shows the nonuniform finite element mesh used in the solution
of the governing equations. Contours of constant temperature are shown in the

upper right hand panel of the same figure. It can be seen that secondary tips have begun to emerge from the parent tips as the solidification process has proceeded.

12.4.4 Two-Phase Microstructures Revisited

In this subsection we will consider two extremely different approaches to the same general class of problems, namely, the development of two-phase microstructures in three dimensions. We first consider a scheme which features a combination of first-principles analysis with Monte Carlo techniques. This is followed by a phase field analysis which includes the important coupling between the order parameter field and elastic deformations.

First-Principles Approach to Guinier–Preston Zones. We have already seen that the combination of first-principles calculations with Monte Carlo methods is a powerful synthesis which allows for the accurate analysis of structural questions. In chap. 6 we noted that with effective Hamiltonians deduced from a lower-level microscopic analysis it is possible to explore the systematics of phase diagrams with an accuracy that mimics that of the host microscopic model. An even more challenging set of related questions concern the emergence of microstructure in two-phase systems. An age-old question of this type hinted at in the previous chapter is the development of precipitates in alloys, with the canonical example being that of the Al–Cu system.

The development of precipitates in the Al–Cu system has been presumed to occur via a series of intermediate steps which commences with the solid solution itself and passes through two intermediate stages involving the so-called Guinier–Preston (GP) zones. The key structural idea associated with both the GP1 and GP2 structures is that layers of Cu atoms collect on {100} planes. In the GP1 case, it is thought that a single layer of such atoms is formed while in the GP2 case it is rather two such layers of Cu atoms separated by intervening {100} planes of Al. Our present discussion is aimed at illustrating how in the work of Wolverton (1999,2000) it has been found possible to use an effective Hamiltonian based on the first-principles analysis of various Al–Cu structures to systematically investigate the emergence of precipitates in the Al–Cu system.

The strategy adopted in the work of Wolverton (1999, 2000) follows in the tradition of the models introduced in chap. 6. In particular, a cluster Hamiltonian which includes the energy cost of elastic distortions is fitted on the basis of a series of first-principles calculations. The set of structures used to effect this fit include special quasirandom structures, a clever idea introduced in a way that allows periodic structures to mimic their random counterparts, supercells including Cu layers and a variety of short period superlattices. The resulting effective

Hamiltonian is able to consider the competition between elastic strain energy and interfacial energy that is played out in the process of precipitate formation and which we wrote down explicitly in a continuum setting as eqn (10.24). Note that in contrast with that discussion, the microscopically based Hamiltonian under consideration here includes all such terms automatically.

Once the effective Hamiltonian described above is in hand, it is then possible to carry out a series of Monte Carlo calculations to investigate the resulting structures. Wolverton considers a range of fcc supercells ($\approx 100\,000$ atoms) at fixed overall concentration with a large number of 'spin flips' (i.e. 100 to 1000) per site on the fcc supercell. The calculations are performed with a decreasing temperature, with the high-temperature phase corresponding to the disordered Al–Cu solid solution as expected. As seen in fig. 12.21 for systems with different computational cell sizes, the resulting precipitate structures are different. In all three cases, the composition of the Cu layers is purely Cu. The conclusion drawn from this work is that with increasing system size (or with increasing precipitate size) there is a tradeoff in the energy balance such that the GP2 regions, which have a larger interfacial energy and lower bulk energy eventually win the day.

Phase Field Models of Two-Phase Microstructures: Three Dimensions. It has become possible to set forth fully three-dimensional simulations of processes in microstructural evolution such as the coarsening of the minority constituents in a binary alloy. Note that our earlier treatment of phase separation using the Cahn–Hilliard equation (see section 12.3.2) illustrated the general ideas in the two-dimensional setting with no reference to the elastic effects we worked so hard to include in our analyses in chap. 10. A generalization of the types of model set forth in section 12.3.2 may be written in which the free energy functional involves not only the order parameter describing the concentration field, but also elastic fields and their attendant energies, including the interaction between the order parameter field and elastic deformations. In particular, imitating our earlier discussion, the free energy in the coupled three-dimensional setting of interest here may be written as

$$F[\psi(\mathbf{x}), \mathbf{u}(\mathbf{x})] = \int d^3\mathbf{x} \left[\frac{1}{2}\kappa|\nabla\psi|^2 + f(\psi) + \alpha\psi\nabla\cdot\mathbf{u} + f_{el}(\epsilon) \right], \quad (12.75)$$

where the term involving only the order parameter field is of the form $f(\psi) = -\frac{1}{2}\psi^2 + \frac{1}{4}\psi^4$. Since we are discussing the phase separation process in a two-phase system, though we follow Sagui *et al.* (1998) in denoting the order parameter field ψ, we would do well to bear in mind a physical interpretation which is that this field tells us about the concentration throughout the system. Though many of the elements of this free energy are already familiar as a result of the discussion

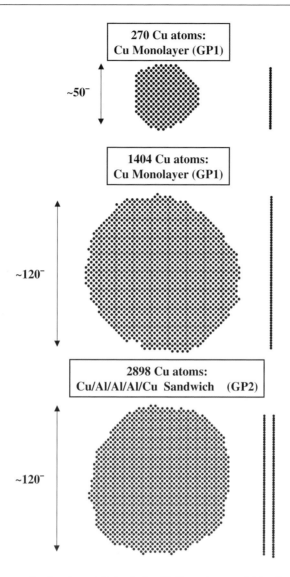

Fig. 12.21. Emergence of ordered precipitates in the Al–Cu system as obtained using a combination of first-principles techniques and Monte Carlo simulation (adapted from Wolverton (1999)). Each picture shows both a top and a side view of the second-phase regions.

culminating in eqns (12.40) and (12.45), there are other features that we have not seen before. First, we remark the presence of the term $f_{el}(\epsilon)$ which is the energy stored in the material by virtue of the elastic distortions that arise because of the mismatch (mismatch in either size or elastic moduli or both) between the two constituents making up the two-phase microstructure. In principle, the calculation of such terms was discussed in chap. 10 in our discussion of elastic inclusions. The second new term is of the form $\alpha \psi \nabla \cdot \mathbf{u}$ and represents the energy of interaction between the order parameter field and the elastic fields.

By virtue of this free energy, the evolution equations for these two sets of fields are coupled and demand numerical solution. For the case to be discussed here, an interesting step in the analysis corresponds to the construction of a new effective free energy $F_{eff}[\psi(\mathbf{x})]$ in which explicit reference to the elastic fields has been removed. This step is effected at the price of calculating an auxiliary quantity which serves as a potential within the effective free energy. In particular, the new free energy functional is of the form

$$F_{eff}[\psi(\mathbf{x})] = \int d^3\mathbf{x} \left[\frac{1}{2}\kappa |\nabla \psi|^2 + f(\psi) + f_{elastic}(\psi) \right], \tag{12.76}$$

with the elastic term $f_{elastic}(\psi)$ determined through an attendant calculation of an elastic potential satisfying the equation $\nabla^2 W = \psi - \psi_0$, where ψ_0 is the average value of the order parameter. Note that it is not surprising that such a potential can be put forth since the analysis is quite analogous to the eigenstrain approach taken earlier, with the field ψ serving an analogous role to the eigenstrain.

The evolution equation in this case is obtained as

$$\frac{\partial \psi}{\partial t} = \nabla^2 \frac{\delta F_{eff}}{\delta \psi}, \tag{12.77}$$

and results are obtained in much the same way as were those shown in figs. 12.5 and 12.6 and as described in problems 3 and 4. An example of the type of microstructures that are found in these calculations is shown in fig. 12.22. The results are meant to illustrate two key features. First, the progression of results from (a) through (g) illustrate the way in which different choices of parameters result in different microstructures. Secondly, frames (a), (d), (e) and (g) also show how the microstructure changes in time for a fixed set of parameters.

12.4.5 A Retrospective on Modeling Microstructural Evolution

The current part of the present chapter has had as its aim the use of the study of microstructural evolution as a case study in the techniques for bridging scales described earlier in the chapter. The examples that were recounted in our discussion of microstructure and its evolution drew from a variety of the resources discussed earlier in the chapter in the context of 'bridging scales'. In particular, we have seen how kinetic Monte Carlo models adopt an information passage philosophy in which calculations of one type are used to inform those of another. Similarly, the description of solidification, including information on the local crystal orientations, using a linkage of cellular automata with continuum descriptions of heat flow illustrates how more than one computational scheme may be brought under the

(a)   (b)   (c)   (d)

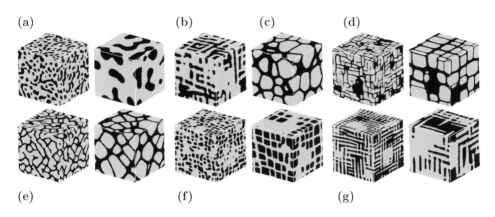

(e)   (f)   (g)

Fig. 12.22. Representation configurations from the phase field model of phase separation in a three-dimensional binary alloy (adapted from Sagui *et al.* (1998)). The progression from (a) through (g) represents different choices of the parameters characterizing the elastic response of the two media, with (a) corresponding to a treatment in which elastic effects are ignored. In those cases where there is more than one picture associated with a given letter, these represent snapshots at different times in the simulation.

same roof. Finally, phase field models have been invoked in the context of both solidification and phase separation.

Though there has certainly been a high level of progress in the use of these methods to model structure (microstructure in particular) in materials, the allied task of exploiting these microstructures to explore the properties of materials remains as one of the important current challenges.

12.5 Bridging Scales in Plasticity

At the beginning of the present chapter we argued that plasticity is a phenomenon that is enriched by the presence of processes at many different scales. One approach to the modeling of plasticity is the continuum theory of plasticity itself in which all mention of dislocations has been eliminated. In this approach, the only hint of subscale processes is to be found in the kinematic vestiges of crystalline slip which emerge in the form of plastic strains associated with relevant slip system and through the presence of a critical stress to induce plastic flow which is a remnant of the underlying critical stress for dislocation motion. An alternative to the continuum plasticity approach is an explicit treatment of the dynamics of dislocations themselves. In this case, it is the dynamics of dislocations themselves which gives rise to plastic deformation.

As was noted earlier in the present chapter, one of the key ways in which an effective description of material response can be constructed is through the construction of *defect dynamics* approaches in which the fundamental degrees of freedom in the problem are those characterizing the disposition of the defects of interest. Often, it is the conspiratorial relationship between all such defects that gives rise to observed macroscopic response. One situation where this is particularly true is in the context of plastic deformation of crystalline solids. In this setting it is the motions of vast numbers of dislocations that give rise to the overall permanent shape changes we associate with plastic deformation. The aim of the present section is to give some idea of how the ideas of previous chapters may be assembled to create models of defect dynamics in which dislocations are the agents of permanent shape changes.

In chap. 8 we developed many of the fundamental tools needed to examine the behavior of one or several dislocations. However, an equally challenging and important problem is the statistical problem posed by a collection of large numbers of dislocations. We made a certain level of progress in confronting the statistical questions that attend the presence of multiple interacting dislocations in the previous chapter, and now revisit these questions from the standpoint of the hierarchical approaches being described here and in particular in terms of the variational approaches to defect dynamics introduced in section 12.3.2.

12.5.1 Mesoscopic Dislocation Dynamics

Our ambition in the present section is to follow the generic groundwork laid down in section 12.3.2 in which we gave a variety of general statements on how to construct models of defect dynamics. Our arguments were predicated on two broad classes of ideas. First, we argued that a kinematic description of the relevant defects had to be set forth. In the present setting, this means that we will have to find a way to account for the disposition of each and every dislocation. Once we have settled the kinematic preliminaries, the next step in the analysis is to settle on a dynamics for the degrees of freedom that have been introduced in the kinematic part of the analysis.

Kinematics of Lines. The first order of business is the characterization of the dislocation lines themselves without reference to the forces that result in dislocation motion. In principle, each and every dislocation line can be characterized through a parameterization of the form

$$\mathbf{x} = \mathbf{x}(s), \tag{12.78}$$

where the space curve which is the dislocation is considered parametrically in terms of the parameter s. Of course, since the approaches under consideration here are largely based upon numerical computation, the continuous representation described above is not particularly useful.

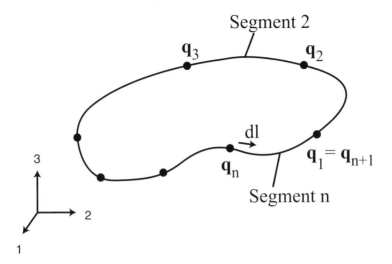

Fig. 12.23. Representation of a dislocation loop via a series of segments (adapted from Ghoniem and Sun (1999)).

A useful allied philosophy is to discretize the dislocation lines. The idea is the same as that introduced above with the difference that the overall disposition of the dislocation is characterized through a finite set of parameters rather than the continuous function envisaged above. In particular, one divides up the line into a discrete set of points and identifies the spatial coordinate associated with each node along the line. It is then the temporal evolution of these positions that constitutes the dislocation dynamics. A concrete example of this kinematic philosophy is depicted in fig. 12.23 which shows the representation of a dislocation loop by a series of segments. The kinematics of the dislocation in this setting is essentially node-based with the identification of a given line ultimately being carried out on the basis of a knowledge of the nodal positions $\{\mathbf{q}_i\}$. As described in Ghoniem and Sun (1999), the determination of the rest of the line can be settled through an interpolation of the form

$$\mathbf{x}(s) = \sum_i \mathbf{q}_i N_i(s), \qquad (12.79)$$

where the functions $N_i(s)$ are essentially shape functions and the sum is over the nodal degrees of freedom. The simplest scenario is founded on the assumption that the line can be represented as a series of piecewise linear segments, in which case the shape functions $N_i(s)$ will be linear. In this case, if the ith node is described by the generalized coordinate $\mathbf{q}_i$ and the $(i+1)^{th}$ node is described by the generalized coordinate $\mathbf{q}_{i+1}$, then the segment between these nodes is characterized by the interpolation

$$\mathbf{x}(s) = (1 - s)\mathbf{q}_i + s\mathbf{q}_{i+1}. \tag{12.80}$$

We have exploited the fact that the linear shape functions of interest here are $N_i(s) = (1-s)$ and $N_{i+1} = s$ which satisfy the requirement that the shape function have the value 1 at one endpoint and 0 at the other. On the other hand, since in many cases one demands an accurate treatment of the curvature of the dislocation line of interest, it is often necessary to adopt higher-order shape functions. In this case, the shape functions can be written as higher-order polynomials, for example. A full description of these geometric ideas may be found in section II of the paper by Ghoniem and Sun (1999). The key point to be made here is that the dislocation line has been fully characterized in terms of a finite set of coordinates $\{\mathbf{q}_i\}$. With these geometric preliminaries in hand we can then turn to computing the interactions between the various segments and their resulting motions.

Dynamics of a Collection of Dislocations. Regardless of the kinematic philosophy that has been adopted for characterizing the positions and types of the dislocation segments, one must supplement this description with one that expresses the interactions between these segments and that provides for their subsequent dynamical evolution. Once the spatial distribution of all the relevant dislocations has been set forth, we can begin to investigate the forces that give rise to dislocation motion. As we recounted in chap. 8, there is a highly refined linear elastic theory of the long-range interactions between dislocations.

If we, for the moment, concentrate on those circumstances in which the dislocations find themselves sufficiently distant from each other such that the linear elastic model suffices, then the description of the dislocation interactions proceeds in a clear fashion. In particular, we note that the interactions between different dislocations can be written in terms of line integrals over the dislocations of interest. In light of the discretization of eqn (12.79), the integrals characterizing the strains and stresses due to a particular dislocation loop as well as the interaction energy of a collection of dislocation loops are replaced by a series of sums over the nodal coordinates describing the dislocation geometry. Again, we refer to Ghoniem and Sun (1999) for a thorough overview.

As noted above, one of the key outcomes of carrying out the discretization of dislocation lines into a series of segments is that the various integrals used to compute displacements, strains, stresses and interaction energies are replaced by sums over the nodal degrees of freedom. On the other hand, as with any numerical approximation scheme, we must assess the errors that are an inheritance of that numerical scheme. One particularly useful setting within which to effect such assessments is in the context of simple planar loop geometries for which there

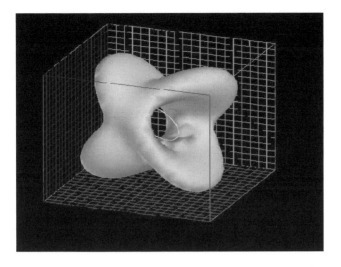

Fig. 12.24. Contour of constant normal stress due to a dislocation loop (adapted from Ghoniem and Sun (1999)).

is corresponding analytic insight. An example of the outcome of a calculation of the stress field due to a dislocation loop using ideas like those introduced above is given in fig. 12.24 which shows the normal stress due to a dislocation loop.

As long as the geometry of the dislocations of interest is such that no two segments are in too close proximity, the dynamical evolution of the collection of dislocation segments is carried out on the basis of drag dominated dislocation motion in which the velocity of a given segment is directly proportional to the driving force experienced by that segment. That is,

$$v_i = \mu f_i, \tag{12.81}$$

where v_i is the normal velocity of the ith segment and f_i is the corresponding driving force. This driving force, in turn, is obtained by summing up the stress on that segment due both to its interactions with externally applied stresses and to all the other partner segments in the system. In concrete terms, that is

$$\mathbf{f}_i = \left(\sigma_{ext} + \sum_j \sigma_{j\,on\,i} \right) \mathbf{b} \times \mathbf{l}_i, \tag{12.82}$$

where $\mathbf{f}_i$ refers to the force on the ith segment, $\sigma_{j\,on\,i}$ refers to the stress on segment i due to segment j. This result can be thought of as arising from obtaining the configurational force on the kth generalized coordinate as

$$\mathbf{f}_k = -\frac{\partial \Pi(\{\mathbf{q}_i\})}{\partial \mathbf{q}_k}, \tag{12.83}$$

where the potential $\Pi(\{\mathbf{q}_i\})$ is obtained by summing up the interaction energies and the potential due to external loads.

As yet, we have skirted the subtleties (and unpleasantness) that attend the short-range interactions of dislocations. From the standpoint of dislocation dynamics methods, such short-range interactions call for the introduction of *ad hoc* rules which require something above and beyond the well-defined elastic prescription outlined above. For example, if two dislocations are too close to one another they may strike up the types of partnerships discussed in section 8.7.5 and form junctions. As a result of such junction formation, a series of nodes become pinned and are inoperative for further motion until a sufficiently large stress is reached which results in the destruction of the junction. In addition to the formation of junctions, rules must also be set forth for the nucleation of dislocations via Frank–Read sources and for the possibility of cross slip in which a dislocation on one slip plane moves to another. For the purposes of the present discussion our main reason for bringing this up is to note that the dislocation dynamics described thus far is not a closed description and must be supplemented by additional hypotheses which we will not delve into more deeply here, though the reader is urged to consult some of the original literature (Zbib *et al.* 1998, van der Giessen and Needleman 1995, Schwarz 1999, Kubin *et al.* 1992) to see examples. The summarizing insight is the recognition that elasticity by itself and the associated linear kinetics of eqn (12.81) does not suffice to formulate a closed dislocation dynamics.

The final feature of the dislocation dynamics method that must be introduced so as to give such methods the possibility of examining real boundary value problems in plastic deformation is the treatment of boundary conditions. In particular, if we wish to consider the application of displacement and traction boundary conditions on finite bodies, the fields of the relevant dislocations are no longer the simple infinite body Volterra fields that have been the workhorse of our discussions throughout this book. To confront the situation presented by finite bodies, a useful scheme described in Lubarda *et al.* (1993) as well as van der Giessen and Needleman (1995) is to use the finite element method to solve for the amendments to the Volterra fields that need to be considered in a finite body. Denote the Volterra fields for an infinite body as $\sigma_{ij}^{(\infty)}$. In this case the fields of interest are given by

$$\sigma_{ij}^{(body)} = \sigma_{ij}^{(\infty)} + \delta\sigma_{ij}, \tag{12.84}$$

where $\delta\sigma_{ij}$ is a correction term that must be added to the infinite body fields so that they will satisfy the real boundary conditions. The finite element solution for the $\delta\sigma_{ij}$s is obtained by recourse to an argument that is indicated pictorially in fig. 12.25. The key idea is that if we take the fields $\sigma_{ij}^{(\infty)}$, they imply certain

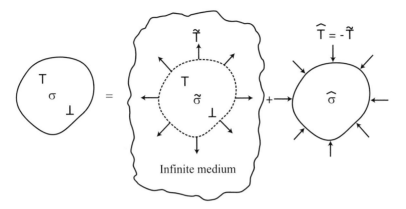

Fig. 12.25. Illustration of treatment of boundary conditions for dislocation dynamics analysis of body of finite extent (adapted from Lubarda *et al.* (1993)). The symbols ˜ and ˆ refer to the fields associated with an infinite body and the corrections, respectively.

values of tractions and displacements on the boundary. If we are to add the fields that compensate for these displacements and tractions, the resulting total fields do satisfy the boundary conditions of interest. In practice, the way this is achieved is by solving a traditional linear elastic boundary value problem using finite elements in which the boundary conditions are gotten by taking the negative values of the fields as obtained from the Volterra solution.

Applications of Dislocation Dynamics to Thin Films. If we blithely ignore the possibility of nonelastic interactions, the key elements in a dislocation dynamics simulation have been laid down. In fact, in certain settings this approach has been found to be quite fruitful. For example, in simulations aimed to reflect the evolution of dislocations within Si, it has been shown that if these elastic forces are allowed to proceed to their logical conclusion, intriguing correspondence between observed and simulated dislocation patterning is found. For example, if we consider two parallel glide planes as indicated schematically in fig. 12.26, then the interactions between these dislocations give rise to a variety of beautiful and intriguing patterns that have been noted in experiment as well. The basic mechanism is that dislocations on the two parallel glide planes interact and form bound states. The operation of sources on the two planes pumps further dislocations into the patterned region and they too are bound by the resulting elastic interactions. For full details, the reader is urged to consult Schwarz and LeGoues (1997).

Line Tension Dislocation Dynamics. Though our discussion above has given a generic feel for the dislocation dynamics strategy and has emphasized the full treatment of interactions between different segments, a watered down version of

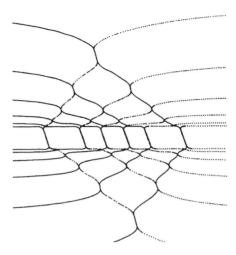

Fig. 12.26. Results of mesoscopic dislocation dynamics simulation of dislocation interactions in a strained epitaxial layer (adapted from Schwarz and LeGoues (1997)). The network shown in the figure results from the interaction of dislocations on parallel glide planes.

dislocation dynamics is also possible which features the line tension approximation as an approximate way of treating the self-energies of the various lines. The appealing feature of this approach for our present purposes is that it admits of immediate analytic progress and physical transparency. In addition to the ease of treating both the self-energies and the coupling to external stresses, it is also possible to add on stresses due to obstacles such as those encountered in section 11.6.1 and shown in fig. 11.22.

We begin by stating the problem in continuous form in which the kinematics of the line is described in terms of the parameterization $\mathbf{x}(s)$. The formulation is then founded upon a variational statement of the form

$$\Pi[\mathbf{v}_n(s); \mathbf{x}(s)] = \psi[\mathbf{v}_n(s); \mathbf{x}(s)] + \dot{G}_T[\mathbf{v}_n(s); \mathbf{x}(s)] + \dot{G}_{PK}[\mathbf{v}_n(s); \mathbf{x}(s)], \quad (12.85)$$

where we have once again resorted to the variational framework of section 2.3.3. The term $\dot{G}_T$ refers to the energy of configuration in the line tension approximation in which the various elastic energies are all subsumed in a simple line energy. The term Peach–Koehler $\dot{G}_{PK}$ refers to the energetics of the coupling of the dislocation to the externally applied stress field. More precisely, the individual term are written as follows. The dissipative term is identical to what we have written earlier on the basis of the assumption that $v_n = Bf_n$, and is given by

$$\psi[\mathbf{v}_n(s); \mathbf{x}(s)] = \int \frac{v_n^2}{2B} ds. \quad (12.86)$$

The line tension term is really no different than the problem confronted already in the context of interfaces in two-dimensional systems (at least if we work in the isotropic line tension approximation), and is given by

$$\dot{G}_T[\mathbf{v}_n(s); \mathbf{x}(s)] = \int T\kappa v_n ds. \tag{12.87}$$

Finally, the Peach–Koehler term expresses the coupling of the external stress to the dislocation of interest and may be written as

$$\dot{G}_{PK}[\mathbf{v}_n(s); \mathbf{x}(s)] = \int \tau b v_n ds. \tag{12.88}$$

Once the variational statement has been made, the dynamical equations are deduced by computing the relevant functional derivative with respect to the unknown function $v_n(s)$. On the other hand, this analysis is most useful in the numerical setting to which we now turn.

For numerical computation, these results are recast in discrete form. To be concrete, our objective is to represent the dislocation via a series of generalized coordinates of the form $\{\mathbf{q}_i\}$ with the intervening segments represented according to the interpolation scheme of eqn (12.79). In particular, if we substitute

$$\mathbf{v}_n(s) = \sum_{i=1}^{N} \dot{\mathbf{q}}_i N_i(s) \tag{12.89}$$

into eqn (12.85), and then evaluate $\partial \Pi / \partial \dot{\mathbf{q}}_i = 0$, the resulting equation of motion will be

$$\mathbf{M}\dot{\mathbf{q}} = \mathbf{f}. \tag{12.90}$$

We have defined the matrix $M_{ij} = (1/2B) \int N_i(s) N_j(s) ds$. The solution strategy in this case is to determine the vector $\dot{\mathbf{q}} = \mathbf{M}^{-1}\mathbf{f}$, to increment the positions $\{\mathbf{q}_i\}$ on the basis of these velocities and to then recompute the curvature, etc. and repeat the procedure.

As noted above, one interesting application of these ideas is to the motion of a dislocation through an array of obstacles. An alternative treatment of the field due to the disorder is to construct a particular realization of the random field by writing random forces at a series of nodes and using the finite element method to interpolate between these nodes. An example of this strategy is illustrated in fig. 12.27. With this random field in place we can then proceed to exploit the type of line tension dislocation dynamics described above in order to examine the response of a dislocation in this random field in the presence of an increasing stress. A series of snapshots in the presence of such a loading history is assembled in fig. 12.28.

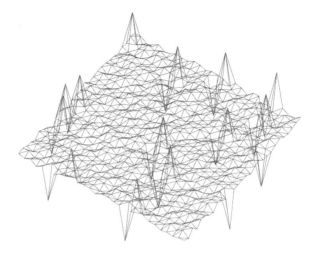

Fig. 12.27. Finite element description of the disordered energy landscape due to presence of disorder which serves to inhibit dislocation motion (courtesy of Vivek Shenoy).

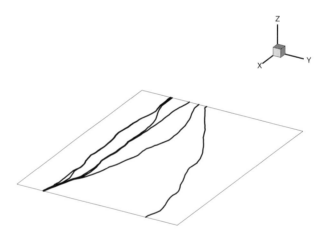

Fig. 12.28. Snapshots from the motion of a dislocation through a random stress field as obtained using a line tension variant of the dislocation dynamics method (courtesy of Vivek Shenoy).

12.5.2 A Case Study in Dislocations and Plasticity: Nanoindentation

Like our analysis of stress–strain behavior, indentation experiments demand theoretical interpretation. The classic insight on the initial stages of indentation derives from Hertz's analysis of elastic contact (details are to be found in Johnson (1985)) and can be deduced on the basis of the type of Green function arguments relied on in earlier chapters. Post-elastic behavior can be examined on the basis of traditional arguments from plasticity theory, or by appealing to models at smaller scales. For the present purposes, our aim is to exploit arguments using

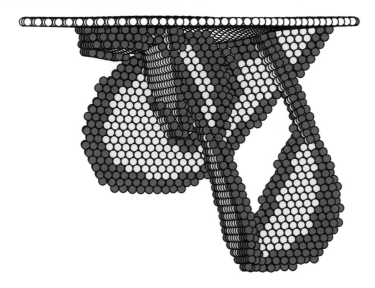

Fig. 12.29. Snapshot from the deformation history for a nanoindentation test simulated using molecular dynamics (adapted from Kelchner *et al.* (1998)). The atoms shown in the figure are those with energy above a certain threshhold and reveal the passage of dislocations.

methods developed earlier in the book, in particular using the method of dislocation dynamics introduced earlier in this section.

In the progression of techniques that can be set forth for examining the nucleation of dislocations beneath an indenter, atomistic simulation can be viewed as providing the most fundamental description. A series of such calculations have been undertaken in the context of nanoindentation using fully three-dimensional molecular dynamics (Kelchner *et al.* 1998). An example of the subsurface dislocation activity that is realized in such simulations is shown in fig. 12.29. As can be seen in the figure a key virtue of the use of atomistic methods in this setting is that the allowed slip planes emerge naturally rather than demanding to be put in by hand. In addition, the nucleation event itself is a natural outcome of the progressive increase in load beneath the indenter. On the other hand, these methods only allow for the emergence of several dislocations.

An attractive higher-level framework within which to study plasticity during nanoindentation is provided by the dislocation dynamics methods described above. In particular, what makes such calculations especially attractive is the possibility of making a direct comparison between quantities observed experimentally and those computed on the basis of the nucleation and motion of dislocations. In particular, one can hope to evaluate the load–displacement curve as well as the size and shape of the plastic zone beneath the indenter, and possibly the distribution of dislocations of different character. While the

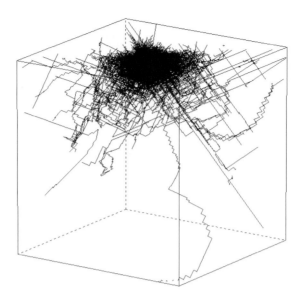

Fig. 12.30. Dislocation debris beneath an indenter resulting from plastic deformation induced by indentation (courtesy of Marc Fivel).

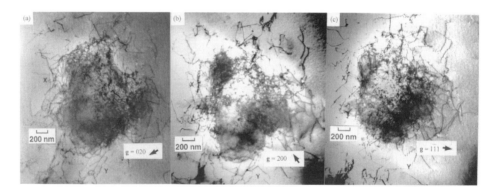

Fig. 12.31. Bright field transmission electron microscopy images of subsurface dislocation activity resulting from nanoindentation in [001] indented Cu (adapted from Robertson and Fivel (1999)).

details of such calculations may be found in Fivel *et al.* (1998), we exhibit their outcome in fig. 12.30. The key point to be made is that in addition to the usual dislocation dynamics machinery introduced above, it is necessary to supplement the analysis by a criterion for nucleating dislocations. Once the dislocations are present beneath the indenter, their motions and interactions are carried out according to the evolution laws (and unit process rules) introduced above.

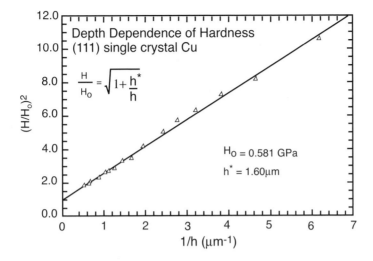

Fig. 12.32. Size dependence of hardness as revealed using a nanoindentation test (adapted from Nix and Gao (1998)).

The subsurface dislocation structures described above can be revealed by electron microscopy as shown in fig. 12.31. Such experiments are suggestive of the physical mechanisms involved in the onset of plasticity and provide the hope of quantifying the crystallography and size of the initial dislocation loops.

In addition, tests of this type also reveal some of the key effects that emerge once the characteristic dimensions of the problem (i.e. the grain size, the indenter size, etc.) begin to reach into the nanometer range. Indeed, one compelling example of such size effects has been argued for in the context of nanoindentation as seen in fig. 12.32.

12.5.3 A Retrospective on Modeling Plasticity Using Dislocation Dynamics

The intent of the present section has been to illustrate the dislocation dynamics method as the paradigmatic example of the generic class of *defect dynamics* models introduced in section 12.3.2. We earlier met such models in our discussion of the dynamics of sharp interfaces in section 10.4.2 where grain boundary motion was induced by a reduction in the interfacial area. As seen in the present section, the dislocation dynamics method offers a similar scheme in which the kinematics of a particular class of defects is represented by a series of nodes which then evolve under the mutual interaction of the various dislocation segments making up the system as well as through a coupling to applied stresses. The perspective offered here is the idea that, in principle, dislocation dynamics offers an alternative to the phenomenological treatment of plasticity offered by continuum theory.

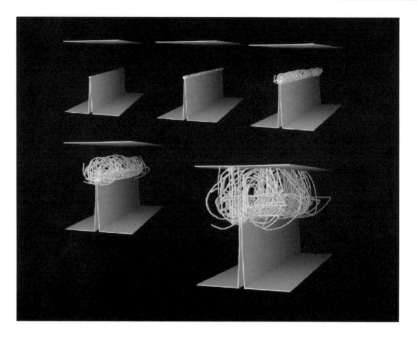

Fig. 12.33. Series of snapshots from molecular dynamics simulation of fracture (courtesy of Farid Abraham).

12.6 Bridging Scales in Fracture

The subject of fracture has already arisen in several different contexts throughout the book. In chap. 2 we described the rudiments of the theory of linear elastic fracture mechanics. In addition, in the previous chapter we described the interplay of cracks and dislocations. The current discussion is aimed at elucidating yet another feature of fracture, namely, the fact that the study of fracture serves as a paradigmatic example of some of the ideas on bridging scales introduced earlier in the chapter.

12.6.1 Atomic-Level Bond Breaking

Ultimately, fracture results from the breaking of atomic bonds. For a brittle solid, the balancing of the energy release rate G and the dissipative processes associated with the creation of new free surface is played out explicitly on an atom by atom basis if one carries out a molecular dynamics simulation of the relevant atomic-level processes. A number of calculations illustrate the level to which such calculations can be pushed using parallel versions of molecular dynamics codes. An especially beautiful sequence of snapshots from the deformation history of a solid undergoing fracture is shown in fig. 12.33. The key point illustrated by

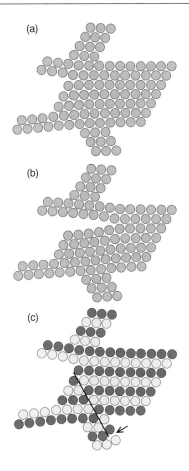

Fig. 12.34. Results of the lattice Green function approach in the context of the problem of crack tip blunting (adapted from Schiøtz *et al.* (1997)). Frame (a) reveals the blunted crack tip geometry, (b) shows cleavage, while (c) shows crack tip dislocation emission.

these calculations is the nucleation and subsequent dynamics of dislocations at a crack tip. In addition to instabilities associated with the emergence of dislocations at a crack tip, a second interesting class of instabilities are associated with the branching of a crack off of its initial path and the associated formation of subcracks which branch off from the main front. The investigations of crack branching instabilities as well as the propensity for crack tip dislocation nucleation have both served as paradigmatic examples of the use of a key technique in bridging scales, namely, cohesive surface models.

We begin by examining the types of mixed atomistic/continuum strategies discussed earlier (see section 12.3.4) with special emphasis on how these methods have addressed the factors that determine whether an atomically sharp crack tip will cleave or emit dislocations. As shown in fig. 12.34, the lattice Green function method can be used to explore crack tip blunting through the creation of a nonlinear zone in the crack tip region which includes not only the prolongation of the crack plane, but also a zone which represents an active slip plane on which a dislocation can be nucleated. The figure shows the part of the simulation cell in which the atomic-level interactions are treated in their full nonlinear way. The types of question that can be explored using methods of this type are whether or not the crack tip region will deform by cleavage or through the introduction of dislocations. Our present ambition is not so much to explore what was learned as a result of such calculations as how it was learned. In particular, we note that these calculations have a computational interface separating those parts of the lattice which are treated in a full nonlinear way from those in which the atomic-level response is strictly linear.

12.6.2 Cohesive Surface Models

As noted in the previous subsection, the fracture process may ultimately be traced to the breaking of atomic bonds. On the other hand, we have already shown that for many purposes, once we are equipped with the machinery of linear elastic fracture mechanics, many of the key features of fracture in solids can be elucidated without reference to atomic-level details. An intermediate framework, and one that reflects the hierarchical strategies being described in this chapter, is that of cohesive surface models in which a nonlinear constitutive model (that in some cases can be derived from atomic-level calculations) is invoked to characterize the energy cost, or alternatively the stresses, that arise upon either sliding or opening the faces on opposite sides of the plane of interest. Once the cohesive surface model is in hand, it is possible to carry out a finite element formulation of the temporal evolution of the loaded solid. One of the nicest features of models of this type is that they do not rely on the existence of an *ad hoc* fracture criterion. That is to say that once the loads reach a sufficiently high level, the relevant adjacent planes will separate at a stress that is implied by the nature of the underlying cohesive surface constitutive model. These approaches have been used to investigate a number of questions concerning the fracture process including the origins of crack branching and the limiting speed of cracks. In the finite element setting that will serve as the basis for the remainder of the discussion, the usual elastic energy terms are supplemented by energy terms associated with the various cohesive surfaces of the form

$$E_{surfaces} = \int_{\partial \Omega_s} \phi(\Delta) dA. \tag{12.91}$$

12.6.3 *Cohesive Surface Description of Crack Tip Dislocation Nucleation*

The discussion given above examined only one of the possible terminal processes that may occur when a solid with a crack is subjected to external loading. As was already noted in the previous chapter, there are a host of competing dissipative processes that can occur at and about a crack tip, including the nucleation of dislocations. For our present purposes, the competition between cleavage and dislocation nucleation is something that can be examined from the perspective of the cohesive surface ideas introduced above. The example given here illustrates the way in which a finite element formulation of the cohesive surface strategy (Xu *et al.* 1997) can be coupled with the results of first-principles calculations of the cohesive surface energetics (Kaxiras and Duesbery 1993). Indeed, these results constitute a powerful synthesis of the tools of quantum and continuum mechanics. Just as with our formulation of the Peierls–Nabarro model, the basic idea is the formulation of an energy functional $E[\delta(x)]$ which characterizes the energy of the cracked solid, including the incipient slip distribution on the slip plane along which dislocation nucleation is presumed to occur. At each load step, the energy minimizing configuration is sought. The degrees of freedom which the energy is being minimized with respect to are those characterizing the incipient slip distribution itself. With increasing load, this slip distribution takes on more and more of the character of a fully formed dislocation loop and eventually at a critical load results in such a dislocation loop.

As we have already emphasized, one of the most significant outcomes of the advent of computational approaches to complex problems is that they make it possible to investigate three-dimensional problems that appear hopeless from an analytic perspective. Recall that the analytic treatment of the Peierls framework for crack tips which we described in section 8.6.2 imagined that dislocations are nucleated as straight lines. In the present setting, we see that these earlier analytic approaches can be adapted to a numerical setting in which it is possible to consider the role of crack tip heterogeneities such as kinks. The particular question to be addressed here is to what extent such heterogeneities suppress the stress needed to nucleate a dislocation. A series of different crack tip geometries were considered as shown in fig. 12.35. Frames (c) and (d) in the figure represent manifestly three-dimensional geometries in which the incipient dislocation loop will be nucleated on slip planes that are oblique to the crack plane. The finite element mesh used to carry out the calculations on the geometry shown in frame (c) is shown in the top frame of fig. 12.36. The corresponding slip distribution at a given load level is shown in the bottom frame of fig. 12.36. Again, our purpose here is to investigate not so much what was learned but how it was learned. What we see is that the analysis of crack tip dislocation nucleation is carried out in a

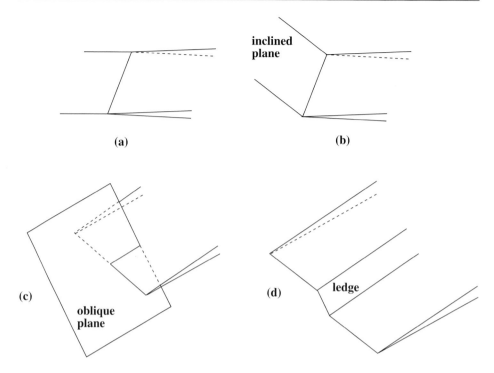

Fig. 12.35. Schematic of the various crack tip geometries using the cohesive surface dislocation nucleation model (adapted from Xu *et al.* (1997)).

fully three-dimensional way using a cohesive surface interplanar potential that was derived on the basis of a density functional calculations for Si. This potential then is exploited as an integral part of the finite element constitutive framework. These calculations represent a synthesis of two very diverse theoretical threads.

12.7 Further Reading

Historic Examples of Multiscale Modeling

The treatment of discrete and continuous representations of the vibrating string in *Mathematical Thought from Ancient to Modern Times* by Morris Kline, Oxford University Press, New York: New York 1972, is enlightening. Here it is evident that both the continuous and discrete representations had features that recommended them as the basis for further study.

Newton's Principia for the Common Reader by S. Chandrasekhar, Clarendon Press, Oxford: England, 1995 is one of my prized possessions. Chandrasekhar has an interesting discussion of the variety of theorems surrounding Newton's treatment of the earth as a point mass.

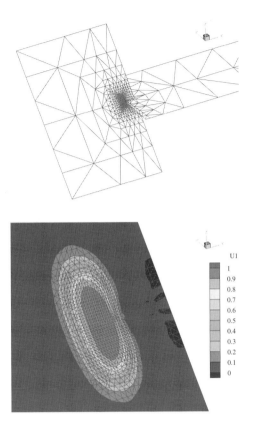

Fig. 12.36. Cohesive surface calculations of crack tip dislocation nucleation (adapted from Xu *et al.* (1997)).

Statistical Physics and the Atomic Theory of Matter by S. G. Brush, Princeton University Press, Princeton: New Jersey, 1983. The interplay between statistical mechanics and thermodynamics is perhaps the greatest single success story in the type of modeling described in this chapter. Brush's book makes for fascinating and instructive reading (as do his other books) since they illustrate the intense growing pains that were suffered in order to convert profound science into what are now little more than textbook exercises.

Methods for Bridging Scales

'Theory of dynamic critical phenomena' by P. C. Hohenberg and B. I. Halperin, *Rev. Mod. Phys.* **49**, 435 (1977) details many of the crucial ideas on order parameters and their dynamical evolution in the critical phenomena setting. Their Model A has been introduced here as the Cahn–Allen equation while their Model B we have referred to as the Cahn–Hilliard equation.

Long Range Order in Solids by R. M. White and T. H. Geballe, Academic Press, New York: New York, 1979. This book offers a number of perspectives on the generic role of Ginzburg–Landau approaches in condensed matter physics. Chapter II gives a broad-based and interesting discussion of Landau theory with an attempt to show the unity of the approach in the context of a diverse set of applications.

Principles of Condensed Matter Physics by P. M. Chaikin and T. C. Lubensky, Cambridge University Press, Cambridge: England, 1995. Chaikin and Lubensky's admirable book has much to recommend it. For the purposes of the present chapter, their chap. 8 is especially pertinent.

'Linking Anisotropic Sharp and Diffuse Surface Motion Laws via Gradient Flows' by J. E. Taylor and J. W. Cahn, *J. Stat. Phys.*, **77**, 183 (1994) and 'Variational Methods for Microstructural-Evolution Theories' by W. C. Carter, J. E. Taylor and J. W. Cahn *JOM* 30, December 1997, are two of the more provocative articles I have read of late. Though there is an element of formalism in these papers, I find that they are highly suggestive concerning the development of generic evolution laws for dynamical processes.

'Nanomechanics of Defects in Solids' by M. Ortiz and R. Phillips in *Advances in Applied Mechanics*, Vol. 36, edited by E. van der Giessen and T. Y. Wu, Academic Press, New York: New York, 1999. This article provides a thorough analysis of several of the methodological advances described in this chapter.

Case Studies in Modeling

'Instabilities and pattern formation in crystal growth' by J. S. Langer, *Rev. Mod. Phys.* **52**, 1 (1980) discusses the continuum treatment of solidification, thus shedding light on the discussion offered in this chapter.

'Computational Mechanics at the Mesoscale' by A. Needleman, *Acta Mater.* **48**, 105 (2000) offers a deeper perspective on the 'multiscale' aspects of modeling plasticity than that offered in the current chapter. Needleman's discussion descends from the perspective of crystal plasticity to the level of dislocation dynamics.

12.8 Problems

---

1 Free Energy Density for Phase Separating System

Imitate the analysis described in the text in order to derive eqn (12.24). Explain the assumptions leading to both the energetic and entropic contributions to the free energy.

2 Gradient Flows and Evolution Equations

Supply the missing details in the derivation of eqns (12.40) and (12.45). In particular, show how the derivative of the free energy functional is converted into a form that reveals it as a gradient flow.

3 Allen–Cahn Treatment of Ordering in Two Dimensions

The temporal evolution of the order parameter field for ordering is given by the Allen–Cahn equation

$$\frac{\partial \eta}{\partial t} = \mu[\kappa \nabla^2 \eta - f'(\eta)], \qquad (12.92)$$

The model free energy function that will be used here is of the form

$$f(\eta, T) = a_0(T) + b_0(T - T_c)\eta^2 + c_0 T \eta^4. \qquad (12.93)$$

(a) Plot the equilibrium value of η as a function of temperature. (b) Since the free energy function given above results in a nonlinear evolution equation, resort to a numerical scheme to solve for the temporal evolution of the order parameter field η. In particular, reproduce results like those shown in fig. 12.5 by constructing a two-dimensional grid and discretizing the Laplacian operator. Apply periodic boundary conditions, and consider a temperature that is below the ordering temperature. For initial conditions, assign $\eta(i, j)$ a small random value around 0.

4 Cahn–Hilliard Treatment of Phase Field Evolution

The temporal evolution of the order parameter field for phase separation (i.e. the concentration) is given by the Cahn–Hilliard equation

$$\frac{\partial c}{\partial t} = \mu \nabla^2 [f'(c) - \kappa \nabla^2 c]. \qquad (12.94)$$

The model free energy function that will be used here is of the form

$$f(c, T) = \rho E_0 c(1 - c) + kT \rho [c \ln c + (1 - c) \ln(1 - c)]. \qquad (12.95)$$

(a) Plot the free energy as a function of concentration for different values of the parameter ρE_0. (b) Since the free energy function given above results in a nonlinear evolution equation, resort to a numerical scheme to solve for the temporal evolution of the order parameter field c. In particular, reproduce results like those shown in fig. 12.6 by constructing a two-dimensional grid and discretizing the Laplacian operator. Apply periodic boundary conditions.

and consider a temperature that is below the ordering temperature. For initial conditions, assign $c(i, j)$ a small random value around 0.

(Thanks to Peter Voorhees who formulated more detailed versions of problems 3 and 4 for his courses at Northwestern University.)

---

5 Effective Dynamics on Nonuniform Meshes

In this problem we will complete an analysis of a geometry like that shown in fig. 12.7 in which the interatomic spacing is a. In particular, consider the case in which the underlying atomic-level Hamiltonian is of the form

$$H = \sum_{i=1}^{N} \frac{p_i^2}{2m} + \frac{1}{2} \sum_{i,j}' k(x_i - x_j - a)^2. \qquad (12.96)$$

Note that the potential energy sum is only over near neighbors.

(a) In the region where $i > N/2$, imagine that we only keep every other atom. Determine the effective Hamiltonian in this case by assuming that the regions between the 'nodes' are characterized by linear interpolation.

(b) Assume plane wave solutions to the equations of motion. Given this *ansatz*, solve the equations of motion resulting from the effective Hamiltonian and determine the reflected energy as a function of the wavelength of the incident wave.

---

6 Hyperdynamics for One-Dimensional Diffusion

Consider the energy landscape given in eqn (12.73). (a) Derive the transition rate for this potential based on conventional transition state theory. (b) Use a bias potential of the type described in the text and reproduce results like those given in fig. 12.15. Compare the results of the two sets of calculations. In addition, plot the hyperdynamics boost factor as a function of the parameter V_0 introduced in the text.

---

7 Kinematics of Lines

The construction of dislocation dynamics methods is predicated upon the discretization of dislocation lines into a series of segments. The goal of this problem is to gain facility in such discretizations. Consider two pinning points at $(0, 0)$ and $(L, 0)$ in the slip plane characterized by $z = 0$ and assume that the dislocation line between these two points has bowed out and assumed a parabolic bow-out profile. Discretize the bowed segment using N nodes and characterize the bowed segment using both linear interpolation as well as cubic splines. Make a plot of the exact disposition of the bowed-out segment as well as of the interpolated versions of the same segment. In addition, use the resulting discretizations to obtain approximate values for the curvature at all points along the bowed segment and compare these results with the exact results which can be obtained directly using the bow-out profile.

8 Dynamics of Lines

Consider the bow-out of a pinned segment like that considered in problem 7 in the presence of a shear stress τ. Obtain the dynamical equations for the nodes on this line using the line tension dislocation dynamics ideas introduced in this chapter.

THIRTEEN

Universality and Specificity in Materials

The purpose of this, our final chapter, is first to revisit some of the key phenomena faced in the task of modeling materials. In particular, we reflect on the complementary objectives of explicating both universality and specificity in materials. Our use of these words is very deliberate and is meant to conjure up two widely different perspectives. Universality refers to those features of material response which are indifferent to the particulars of a given material system. For example, huge classes of materials obey the simple constitutive model of Hooke in the small deformation regime. Similarly, both the low- and high-temperature specific heats of many solid materials obey the same basic laws. And, we have seen that the yield strength scales in a definite way with the grain size, again, in a way that is largely indifferent to material particulars. As a last example, the simple continuum model for diffusion we set forth is relevant for a great variety of materials. By way of contrast, there is a set of questions for which the answer depends entirely upon the details of the material in question. Indeed, our emphasis on material parameters and the numbers that can be found in databooks forms a complementary set of questions about materials. Our hope is to contrast these two conflicting perspectives and to show the way each is important to the overall endeavor of understanding material response.

Once we have finished the business of universality and specificity, the final discussion of the chapter, and indeed of the book as a whole, will serve as a personal reflection on what appear to me to be some of the more intriguing and fertile realms for reflection in coming years. These queries are strictly a matter of personal taste and should be viewed by the reader as nothing more than suggestions, something akin to looking over my shoulder as I list possible directions for my own future thought.

13.1 Materials Observed

In this section we recall the reasons that we undertook the modeling efforts advocated in this book in the first place. Our end was to see to what extent one might explain the observed thermomechanical properties of real materials. Just what are these observations and what do they teach us? We now take stock of the range of observed properties of relevance to the thermomechanical behavior of materials and the extent to which they have been understood both phenomenologically and mechanistically.

13.1.1 What is a Material: Another Look

Recall that we started the book with this same question: what is a material? Though it may sound silly, we pose this question in all seriousness with the aim of reminding the reader that in many instances materials are characterized by nothing more than a series of material parameters. In the pages since we first noted this question, we have seen that one of the far-reaching subtleties of continuum theories is the way in which they bury material specificity in a series of such parameters, the job of which is to capture the macroscopic implications of submacroscopic behavior. Often, the generic form of the emergence of such material parameters is via equations of the form

$$\text{response} = \text{material parameter} \times \text{stimulus.} \tag{13.1}$$

Our present claim is the argument that for many purposes, the set of quantities that we have labeled material parameters are exactly our answer to the question of what is a material.

In chap. 1, we took the stand that one way to view a material is phenomenologically, a position in which our understanding of the properties of materials is informed through appeal to the types of parameters that can be looked up in databooks (i.e. the elastic moduli, yield strength, fracture toughness, diffusion constant, thermal conductivity, specific heat, electrical conductivity, index of refraction, etc.). In the pages since that initial discussion, we have formulated an alternative view in which these material parameters need not be looked up in a databook, but may rather be usefully informed through an appeal to the tools of quantum, statistical and continuum mechanics. Indeed, the challenges of both universality and specificity are responded to using these tools as the foundation.

In the analysis which fills these pages, we have now seen that there are many different ways to organize our understanding of materials. First, it is possible to be descriptive in a way that emphasizes structure in materials, with special attention being given to the realization that there is an intimate connection between structure and properties. We have now seen that defects and their assemblies into microstructures serve as the backdrop for the types of analysis of materials undertaken here. A second approach is to emphasize the particular tests that can be done on materials, with the stress–strain curve serving the mechanician like the *I–V* curve serves the electrical engineer. In the coming pages we highlight a few of these different perspectives.

13.1.2 Structural Observations

Materials are complicated. Indeed, at different levels of spatial and temporal resolution, our assessment of what is transpiring within a material will be different. At the level of a strain gauge, the onset of deformation within a material is reflected in a macroscopic change of shape. By way of contrast, from the standpoint of high-resolution transmission electron microscopy this same tension test will manifest itself in the distortion of dislocation cores and their eventual motion. The present section serves in broad brush stroke form as a reflection on some of the key classes of structural observation as organized along lines of length scale. In particular, the discussion takes the form of something of a guided tour of the various pictures that have been shown throughout the book, now viewed from the unifying perspective of hindsight, and supplemented with several new figures as well.

Scales Above the Micron Scale. Much can be learned about the workings of a material with little more than an optical microscope and a well-polished sample. One of the first features of a material that will be evident upon inspection via optical means are the type of features shown in fig. 10.2 which reveals a polycrystalline microstructure. Of course, we well know that what we are seeing is evidence of the polycrystallinity of the material. The grain boundaries that separate different grains are clearly evident on the crystal surface. We can also see that depending upon the life history of the material, the grain size can vary considerably.

   After initial deformation, examination of the crystal surface, again by optical means, reveals a novel feature already discussed in fig. 8.3, namely, slip traces. These slip traces are the debris left in the wake of dislocations that have made their way to the crystal surface, leaving behind a jump across the relevant slip plane at its point of intersection with the crystal's surface. These slip traces bear a precise geometrical relation to the underlying crystalline geometry and thereby provide a central clue in ferreting out the microscopic origins of plastic deformation.

   As has already been said repeatedly, if we carry the stress–strain curve to its extreme limit, the material fails by some fracture process. Once again, if we invoke optical microscopy, the resulting fracture surface may be interrogated with the result that depending upon the material type and fracture mechanism, the fracture surface morphology can be widely different. In fact, the geometry of the resulting fracture surface should be seen as one of the primary clues into the fracture process itself. Each of the examples of the last few paragraphs illustrates the way in which structure at scales larger than, say, a micron are revealed in the presence of mechanical loads.

The Micron Scale. At the next level of spatial resolution, we might imagine having subjected our sample to either the electron microscope or some surface probe such as the atomic force microscope. Our ambition is to examine the structural consequences of various thermal and mechanical loading programs. Under ideal circumstances, we might succeed in watching plastic deformation in action. In particular, by finding a suitable orientation for the crystal and a means of communicating a shear stress to the relevant slip planes, it is possible to carry out *in-situ* transmission electron microscopy allowing for the investigation of dislocation motion in real time. Indeed, a representative example of such behavior was shown in fig. 8.4.

At the level of resolution offered by the types of microscopies discussed above, it is also possible to glean much more about the nature of the underlying microstructure. For example, in a two-phase material, the presence of precipitates is now evident (see, for example, figs. 10.4 and 10.6). In chap. 10 we described the emergence of such microstructural features, and it is our present aim to provide a reminder as to the scale at which they become visible. Structural change of this type can also be made evident by virtue of the various surface scanning microscopies. One example that is both beautiful and suggestive is that of the emergence of bainite in steel. The consequences of such a transformation, which resemble fig. 10.34 can be seen directly as is shown in fig. 13.1.

A second example of the way in which deformation can be examined using surface microscopy is shown in fig. 13.2. What this figure reveals is a series of snapshots of the surface profile that attends a series of points on the stress–strain curve once plastic deformation has commenced. In keeping with our discussion of the emergence of slip traces, it is clear that with increasing deformation, both the number of active slip planes and the number of dislocations per slip plane increase.

If the same level of spatial resolution is attained near a crack tip, we will see the nucleation of dislocations in its vicinity. An example of crack tip dislocation nucleation as evidenced in electron microscopy is given in fig. 13.3. Once the material fails, we can also subject it to post-mortem chemical and structural analysis. Using techniques such as Auger spectroscopy, the chemical profile of the failed material may be queried. Again, our main point is to note the diversity of the various geometric signatures of mechanical response.

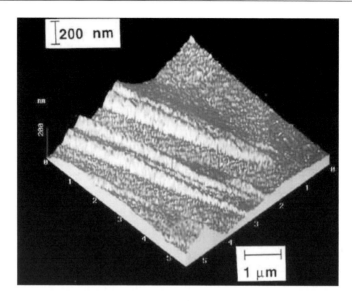

Fig. 13.1. Surface profile showing the displacements on a steel surface resulting from the transformation from austenite to bainite (courtesy of H. Bhadeshia).

The Nanometer Scale. Insights into the atomic-level geometries of lattice defects were ushered in on the heels of important advances in electron microscopy, with special insights garnered from the use of high-resolution transmission electron microscopy. Throughout the book, we have offered various examples of the use of high-resolution microscopy to look at interfacial dislocations (fig. 2.1), dislocation core structures in metals (fig. 1.9) and the structure of grain boundaries of high symmetry (fig. 9.47). In each of these cases, as a result of the atomic-resolution structural information gained from microscopy, direct tests of the models used to characterize the geometry of solids can be made, and with the structural information in hand, new models relating structure and properties can be advanced.

13.1.3 Concluding Observations on the Observations

The introductory sections of this chapter have served the primary purpose of recalling some of the key observations it has been our aim to explain. Apart from the elastic response of crystalline solids, we note that all of these observations implicate defects and their interactions to either form or interact with the material's microstructure.

What we have said until now has been primarily descriptive and anecdotal. I wish to close this section by noting that in describing experiments on materials, we have drawn from two distinct pools of information. On the one hand, we have

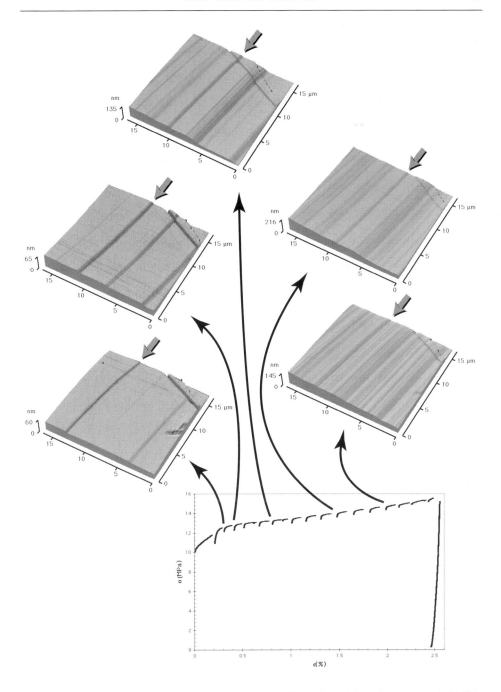

Fig. 13.2. Stress–strain curve with associated surface profiles (adapted from Coupeau *et al.* (1998)).

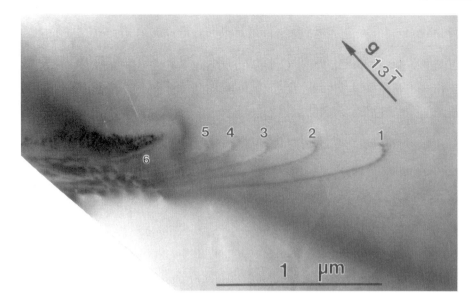

Fig. 13.3. Near vicinity of a crack tip indicating the presence of dislocations nucleated at the crack tip (courtesy of D. Clarke).

repeatedly drawn inspiration from *observations*, made by various microscopies, which suggest the character of the various structures that populate solids. These observations tell us about notions such as the propensity for precipitates to adopt spherical or elongated plate-like shapes, for example, but are often hard to milk either for quantitative data, or as the basis for falsification of a given theory. By way of contrast, we have also drawn attention to a variety of experiments in which the outcome can be cast in numerical form. For example, the graphs presented in fig. 8.2, show four distinct data sets associated with the alteration of the flow stress in a material by virtue of some form of microstructural change.

A particular challenge for the continued vitality of the study of materials is to produce both falsifiable models and the reproducible data needed to renounce them. Within the context of the previous chapters, I note with particular enthusiasm the experiments using a biaxial testing machine to examine martensitic microstructures described in chap. 10.

13.2 How Far Have We Come?

From our first mention of continuum theories, we have repeatedly emphasized the role of 'rigidity' parameters (i.e. moduli, hardening moduli, diffusion constants, etc.) as the means whereby those continuum models are endowed with material specificity. In addition, we have repeatedly sung the praises of Ashby's approach to categorizing the seemingly unmanageable complexity and quantity of data that have been generated concerning materials. Our final word on modeling will be to take stock in more general terms of some of the many parameters we use to characterize materials and the extent to which we understand and can even predict them. In addition, we aim to contrast such specificity with the generic things that can be said about our understanding of materials.

13.2.1 Universality in Materials

The notion of universality for our present purposes can be viewed from several different perspectives. On the one hand, we can speak of the fact that certain features of our models are themselves generic and apply regardless of material specifics. Alternatively, there are certain features of materials themselves that exhibit a certain level of universality. In the present section we wish to comment on both forms of universality.

One of the most potent insights to emerge from almost every chapter of this book has been the recognition that, in many instances, key ideas can be cast in scaling form. Over the course of the discussions that fill this book, we have encountered a broad variety of scaling laws, some based upon sound theoretical analysis, others rooted in the fertile soil of enlightened empiricism. Regardless of the origins of such scaling relations, they serve as a central anchor point in our understanding of *universality* in materials. As a reminder, we recall several representative examples of such scaling laws.

From the standpoint of cohesion in solids, we have seen that the Fermi energy scales with the electron density as $\epsilon_F \propto \rho^{\frac{2}{3}}$. This result imbues models of cohesion in free electron solids with a critical volume dependence. A second example of a generic response in the energetics of solids has been the characteristic parabolic scaling of key energies as a function of the filling of the d-band in transition metals. For example, as shown in fig. 4.18, the cohesive energy in transition metals adopts precisely such a parabolic dependence. Interestingly, we have also seen that the both the vacancy formation energies (eqn (7.49)) and surface energies (fig. 9.5) of transition metals are slaved to this parabolic dependence of the cohesive energy.

The mechanical properties of materials are also characterized by a number of interesting scaling laws, some of which are empirical and others of which can be deduced convincingly. This observation is represented in fig. 8.2 which shows the scaling of the yield strength with various microstructural parameters such as the concentration of impurities, the volume fraction of second-phase particles, the grain size of a polycrystal and the density of dislocations. Indeed, it was the robustness of such scaling that prompted Kocks (1977) remark that it represents 'the most important result of dislocation theory.'

Finally, though it is usually not presented in this way, we note that one of the most robust results of all for considering materials is the time tested Arrhenius relation

$$\text{rate} \propto e^{-E_{act}/kT}. \tag{13.2}$$

This generic kinetic relation is at the very heart of our understanding of materials and serves as an ever present reminder of the need for a more fundamental understanding of processes.

A second key way in which 'scaling' ideas enter our study of materials is through the determination of critical dimensionless groups and dimensionless variables in evaluating particular problems. For example, in the context of the evaluation of microstructure, we noted that the ratio of bulk energies and interfacial energies led to the emergence of a length scale. Though we have nothing further to add on this point, it is worth remarking that often it is only when viewed in dimensionless terms that problems yield true understanding.

In addition to the type of broad understanding that accompanies a description of phenomena in materials according to certain key scaling relations, much has been learned on the basis of the particular. Analytic and numerical solutions to boundary value problems as well as the advent of numerical simulation have all contributed to our understanding of material- or geometry- or mechanism-specific properties of material systems.

13.2.2 Specificity in Materials

The discussion above emphasized that part of material response that might be deserving of the appellation universal. On the other hand, in addition to the power and insight that attaches to the elucidation of scaling arguments, the quest to understand the emergence of material specificity has rewards of its own. Throughout the book we have given a variety of different examples of both the phenomenology of material specifity and the arguments that have been set forth to greet it. In this subsection we revisit once again some of the key examples, with an emphasis born of the wisdom of hindsight.

One of the most intriguing and beautiful aspects of material specificity is revealed by the phase diagrams used to describe the different phases observed in solids. Figs. 1.4 and 6.2 revealed the complexity present even in elemental solids with the case of sulfur exhibiting the wide variety of different structures seen as the temperature and pressure are changed. The alloy case can be even more complex as evidenced by figs. 1.4 and 6.4. From the standpoint of material specificity we only wish to note that even with a single material system the atomic-level structures are as diverse as are the elements themselves.

As noted above and elsewhere, our view of continuum theories is that they are endowed with material specificity by virtue of the various material parameters that appear within them. To my mind, the most appealing and provocative way of depicting the specificity of material parameters has been advanced by Ashby and is represented in our figs. 1.1, 1.2, 8.1 and 11.21. What these figures reveal is that for parameters such as the elastic moduli, the yield strength, the fracture toughness and the thermal conductivity (and for many others, such as the electrical conductivity exhibited in fig. 7.2) material parameters can vary by as much as 20 orders of magnitude. There is lots to understand in explaining such variations and that is the charter of the physics of material specificity. As said in chap. 2, eqn (2.32) (i.e. $\sigma_{ij,j} + f_i = \rho Dv_i/Dt$) applies just as well to the waves crashing on the reefs at Sunset Beach on Oahu as to the Trade Winds which send rooster tails of spray off of their backs to the swaying of the palms on the beach only a few hundred meters away. The diversity of these situations can only be answered by the material specificity of fluids and solids. Just as the generality of continuum mechanics must be supplemented by the specificity of constitutive insights, so too with problems in equilibrium thermodynamics must we complement thermodynamic generality with the specificity that derives from equations of state.

13.2.3 The Program Criticized

Because the thrust of the book has been largely centered around questions of how best to *construct* models, it is sure that if the reader has made it this far, then he or she might have the sense that at every turn I have made it sound as though nearly all models are 'effective theories'. Part of this enthusiasm draws from my support of a statement of Goldenfeld (1992) in a different context who notes

> In my opinion, the renormalization group is one of the more profound discoveries in science, because it is a theory about theories. It has enabled physicists to become self-conscious about the way in which they construct physical theories, so that Dirac's dictum that we should seek the most beautiful or simplest equations acquires a meaningful and quantifiable aspect.

Indeed, the advent of the computer has placed a new class of demands on our theories. Because of the inevitable sense that pure brute force approaches are unacceptable, this has spawned the types of self-conscious efforts at theory building alluded to above, but now with the aim being a new kind of simplicity related to our ability to formulate tractable computational models of complex phenomena.

Perhaps the most damning criticism that can be leveled against the emergence of 'computational materials science' is the difficulty of putting these computations in direct contact with experiments themselves. Unfortunately, much of the direct 'comparison' between theory and experiment takes place at the level of claims such as my simulation picture looks alot like that transmission electron microscopy picture. If I were to sound one alert concerning the use of computers in the study of materials, this would be it.

13.3 Intriguing Open Questions

The remainder of this chapter is entirely and unabashedly subjective. It is not stated as a dogma, but rather in the form of a series of queries which I find particularly interesting or challenging.

Geometric Complexity. At this late hour, we need not further belabor the insight that materials are characterized by geometric structures at many different scales. Rather, we recast the question of structure in materials by posing it in terms of geometric complexity. To what extent does knowledge of the properties of individual defects enable insights into the behavior of collections of such defects? How is the behavior of such defects altered by the presence of structure at the microstructural scale? How is the evolution of the microstructure tied to the evolution of properties? Similarly, with the increasing sophistication of techniques for growth of systems with submicron feature sizes, new classes of geometric complexity involving interfaces, corners, surfaces and morphology on those surfaces, are sure to pose important questions for a time to come.

Chemical Complexity. Aside from a few brief forays into questions concerning the role of chemistry in altering material properties, we have largely ignored one of the grandest challenges associated with the quest to better understand material specificity. I refer to the question of chemical complexity. This notion is meant to encompass issues such as the formation of oxides on surfaces, stress–corrosion cracking, the various empirical rules of thumb which call for low percentage additions to alloys, etc. From the standpoint of the constitutive models that are used to characterize materials within the framework of continuum theories of material response, the incorporation of chemical effects is at best highly approximate, and more often, neglected altogether.

Disorder. Though I don't remember where, I once heard that the lesson of the nineteenth century was that the world is linear and that the lesson of the twentieth century has been that it is nonlinear. In that vein, perhaps the lesson of the next century will be that the world is disordered. What I have in mind here is perfectly exemplified by typical materials: they are riddled with statistical irregularities. Whether we discuss the distribution of second-phase particles or the distribution of microcracks in a composite, ultimately, the operative concept is the notion of *distribution* itself, which alerts us to the fact that we are dealing with systems in which kinematics must be linked to statistics.

Coupled Problems. Another important class of problems which have largely eluded the efforts of the modeler are those in which several different classes of phenomena occupy the stage simultaneously. For example, we earlier made reference to the phenomenon of electromigration in which the presence of an electric current can bias the nature of the diffusion process within a solid. In this instance, it is necessary to manage not only the usual machinery invoked to solve problems in mass transport, but also to impose the physical considerations associated with electromagnetic phenomena. Such problems place demands on both a conceptual and computational plane. The key conceptual feature that makes this class of problems especially intriguing is the presence of *interaction* terms which result in a coupling of the two different classes of phenomena. We have already seen a rudimentary version of such coupling in the discussion of microstructural evolution in which we saw that the equations for mechanical and chemical equilibrium were not independent. More impressive examples of such problems arise in the context of problems of conduction in strained nanostructures or in the consideration of magnetic systems that undergo structural transformations. The push to create synthetic models of such coupled problems seems challenging and interesting.

Metastability. Much of the use and analysis of real materials involves systems that are metastable. Unlike the robust tools that are available for the evaluation of systems that are in terminal privileged states (i.e. equilibria), the study of metastable systems is presently characterized by ideas that are based upon perturbations about the equilibrium state, or else *ad hoc* hypotheses that remain, as yet, unjustified. Our discussion of microstructure and its evolution was based upon a range of clever and fascinating ideas, but I at least am unable to escape the feeling that the treatment of the material's history was implicit and present primarily by virtue of initial conditions, whereas a more fundamental treatment of such metastable systems might demand a more detailed accounting for that history.

Processes. A related series of questions to those associated with metastability concern processes in materials (and more widely as well). For similar reasons, the theoretical backdrop for considering nonequilibrium processes is largely rooted in the kinetic empiricism of Arrhenius plots, and linear relations between generalized velocities and the driving forces that induce them. On the other hand, my personal take on Arrhenius kinetics is that our understanding of the origins of these ideas is much the same as was our understanding of Kepler's Third Law of Planetary Motion relating orbital size and period in the time prior to the discovery of the Law of Universal Gravitation. It is true that the arguments of chap. 7 on transition state theory shed some light on the origins of Arrhenius behavior, but in a microscopically detailed way. The particular class of processes that demand most attention are dissipative. From the standpoint of the present work, dissipation can be thought of as revealing something about the level of resolution in a given model. In particular, dissipation refers to the loss of energy from those degrees of freedom accounted for in the model to those that are not. As a result, we see that once again, we are faced with questions on the way in which effective theories should be constructed, and especially on how to construct dynamics that reflect unaccounted for degrees of freedom. One clear example of this type of problem is that of friction in which the center of mass motion of some macroscopic motion is bled off to donate energy to the microscopic degrees of freedom to which the macroscopic object is coupled.

Prediction. As already mentioned in criticizing the general program advocated in this book, one of the central challenges faced in keeping the study of crystals, defects and microstructures a healthy one is an increasing theoretical emphasis on prediction, and the design of experiments aimed expressly at testing these predictions. I cannot reiterate strongly enough my own concerns about the use of numerical simulation as a replacement for experiment. Though some argue that the new generation of science is founded upon a triumverate of experiment, theory and simulation, my own sense is that simulation is for now an infant stepsister, and will remain so until the emphasis on prediction and falsification is given more attention.

13.4 In Which the Author Takes His Leave

P. W. Anderson (1999), in speaking of the current state of physics, noted 'The kind of people we most need are not those who are good at answering well-posed old questions, but those who are capable of posing new ones.' In that vein, the act of writing this book has certainly raised more questions for me than it has answered, though I have learned much in the process. One of the most indelible impressions of the experience is the certainty that like many of its partner sciences, the study of materials is first and foremost an experimental endeavor. The theoretical constructs that have been set forth in order to greet these experiments are in many cases beautiful and impressive, and their numerical incarnations can now be represented with graphics of enormous and also beautiful sophistication. On the other hand, my sense is that we are in danger of computing ourselves into indifference to what nature has to say, forgetting that it is not our simulations, but the materials themselves which serve as the court of final authority. As I have been swept away by the beauty of the theoretical constructs of materials science themselves, it becomes easier to see how the ancients were caught up in their exotic systems of deferrents and epicycles to explain the orbits of celestial bodies. In writing this book, the temptation to resist computational alchemy has been difficult.

The other key impression that accompanies me as I tap off these last few words has a dual character. On the one hand, I am stricken with the incredible cleverness of those who have built up the ideas that I have tried to assemble in the pages of this book. From the foundational ideas of both continuum and quantum and statistical mechanics, to the far-reaching power of models such as that of Ising to the elastic theory of lattice defects, at every turn, I am reminded of why I wanted to study science in the first place. On the other hand, after the long journey of writing a book such as this, I am also struck at the mismatch between what I had hoped this book would become and what it actually is. With that in mind, I hope the reader will forgive it both its excesses and its shortcomings and will choose instead to find in it those places in which I have succeeded in describing the beauty and pleasure to be found in the study of materials.

References

Abernathy D. L., Mochrie S. G. J., Zehner D. M., Grübel G. and Gibbs D., Orientational Epitaxy and Lateral Structure of the Hexagonally Reconstructed Pt(001) and Au(001) Surfaces, *Phys. Rev.* **B45**, 9272 (1992).

Ackermann H., Crusius S. and Inden G., On the Ordering of Face-Centered Cubic Alloys with Nearest Neighbour Interactions, *Acta Metall.* **34**, 2311 (1986).

Ackland G., Semiempirical Model of Covalent Bonding in Silicon, *Phys. Rev.* **B40**, 10 351 (1989).

ASM *Metals Handbook 7. Atlas of the Microstructure of Industrial Alloys* ASM, Metals Park: Ohio, 1972.

Ahuja R., Eriksson O., Wills J. M. and Johansson B., Theoretical Confirmation of the High Pressure Simple Cubic Phase in Calcium, *Phys. Rev. Lett.* **75**, 3473 (1995).

Akiyama T., Oshiyama A., and Sugino O., Magic Numbers of Multivacancy in Crystalline Si: Tight-Binding Studies for the Stability of the Multivacancy, *J. Phys. Soc. Japan* **67**, 4110 (1998).

Alerhand O. L., Berker A. N., Joannopoulos J. D., Vanderbilt D., Hamers R. J., and Demuth J. E., Finite-Temperature Phase Diagram of Vicinal Si(100) Surfaces, *Phys. Rev. Lett.* **64**, 2406 (1990).

Allen M. P. and Tildesley D. J., *Computer Simulation of Liquids*, Clarendon Press, Oxford: England, 1987.

Amelinckx S., Dislocations in Particular Structures in *Dislocations in Solids*, edited by F. R. N. Nabarro, North-Holland Pub. Co., New York: New York, 1979.

Anderson M. P., Srolovitz D. J., Grest G. S., and Sahni P. S., Computer Simulation of Grain Growth – I. Kinetics, *Acta Metall.* **32**, 783 (1984).

Anderson P. W., More is Different, *Science* **177**, 393 (1972).

Anderson P. W., Why Do They Leave Physics?, *Physics Today*, 11, September (1999).

Ardell A. J., Precipitation Hardening, *Met. Trans.* **16A**, 2131 (1985).

Ardell A. J., Temporal Behavior of the Number Density of Particles during Ostwald Ripening, *Mat. Sci. Eng.* **A238**, 108 (1997).

Asaro R. J., Micromechanics of Crystals and Polycrystals, in *Advances in Applied Mechanics*, Vol. 23, pg. 1, Academic Press, New York: New York, 1983.

Asen-Palmer M., Bartkowski K., Gmelin E., Cardona M., Zhernov A. P., Inyushkin A. V., Taldenkov A., Ozhogin V. I., Itoh K. M. and Haller E. E., Thermal Conductivity of Germanium Crystals with Different Isotopic Compositions, *Phys. Rev.* **B56**, 9431 (1997).

Ashby M. F., On the Engineering Properties of Materials, *Acta Metall.* **37**, 1273 (1989).

Ashby M. F. and D. R. H. Jones, *Engineering Materials* Vol. 2, Pergamon Press, Oxford: England, 1986.

Ashby M. F. and D. R. H. Jones, *Engineering Materials* Vol. 1, second edition, Butterworth-Heinemann, Oxford: England, 1996.

757

Ashby M. F., Gandhi C. and Taplin D. M. R., Fracture-Mechanism Maps and Their Construction for F.C.C. Metals and Alloys, *Acta Metall.* **27**, 699 (1979).

Ashcroft N. W. and Mermin N. D., *Solid State Physics*, Saunders College, Philadelphia: Pennsylvania, 1976.

Asta M., de Fontaine D., van Schilfgaarde M., Sluiter M. and Methfessel M., First-Principles Phase-Stability of fcc Alloys in the Ti–Al system, *Phys. Rev.* **B46**, 5055 (1992).

Asta M., McCormack R. and de Fontaine, D., Theoretical Study of Alloy Phase Stability in the Cd–Mg System, *Phys. Rev.* **B48**, 748 (1993).

Atkinson H. V., Theories of Normal Grain Growth in Pure Single Phase Systems, *Acta Metall.* **36**, 469 (1988).

Aukrust T., Novotny M. A., Rikvold P. A. and Landau D. P., Numerical Investigation of a Model for Oxygen Ordering in $YBa_2Cu_3O_{6+x}$, *Phys. Rev.* **B41**, 8772 (1990).

Bacon D. J., Barnett D. M. and Scattergood R. O., Anisotropic Continuum Theory of Lattice Defects, *Prog. Mat. Sci.* **23**, 51 (1979).

Balamane H., Halicioglu T. and Tiller W. A., Comparative Study of Silicon Empirical Interatomic Potentials, *Phys. Rev.* **B46**, 2250 (1992).

Ball J. M. and James R. D., Fine Phase Mixtures as Minimizers of Energy, *Arch. Rational Mech. Anal.* **100**, 13 (1987).

Ball J. M. and James R. D., Proposed Experimental Tests of a Theory of Fine Microstructure and the Two-Well Problem, *Phil. Trans. R. Soc. Lond.* **A338**, 389 (1992).

Barsom, J. M. (editor), *Fracture Mechanics Retrospective: Early Classic Papers (1913–1965)*, ASTM, Philadelphi: Pennsylvania, 1987.

Basinski S. J. and Basinski Z. S., Plastic deformation and work hardening in *Dislocations in Solids*, Vol. 4, edited by F. R. N. Nabarro, North-Holland Pub. Co., Amsterdam: The Netherlands, 1979.

Baski A. A., Erwin S. C. and Whitman L. J., The Structure of Silicon Surfaces from (001) to (111), *Surf. Sci.* **392**, 69 (1997).

Bazant Z. P., Spurious Reflection of Elastic Waves in Nonuniform Finite Element Grids, *Comp. Methods Appl. Mech. Eng.*, **16**, 91 (1978).

Bell J. F., *The Experimental Foundations of Solid Mechanics* in Mechanics of Solids I, edited by C. Truesdell, Springer-Verlag, Berlin: Germany, 1973.

Beltz G. E., Lipkin D. M. and Fischer L. L., Role of Crack Blunting in Ductile Versus Brittle Response of Crystalline Materials, *Phys. Rev. Lett.* **82**, 4468 (1999).

Benedek R., Yang L. H., Woodward C. and Min B. I., Formation Energy and Lattice Relaxation for Point Defects in Li and Al, *Phys. Rev.* **B45**, 2607 (1992).

Bhattacharya K., Theory of Martensitic Microstructure and the Shape-Memory Effect (unpublished) (1998) – available from author: bhatta@caltech.edu

Bhattacharya K., Wedge-like Microstructure in Martensites, *Acta Metall. Mater.* **39**, 2431 (1991).

Binder K., Ordering of the Face-Centered-Cubic Lattice with Nearest-Neighbor Interaction, *Phys. Rev. Lett.*, **45**, 811 (1980).

Blöchl P. E., Smargiassi E., Car R., Laks D. B., Andreoni W. and Pantelides S. T., First-Principles Calculations of Self-Diffusion Constants in Silicon, *Phys. Rev. Lett.* **70**, 2435 (1993).

Bohnen K.-P., Heid R., Ho K.-M. and Chan C. T., Ab-Initio Investigation of the Vibrational and Geometrical Properties of Solid C_{60} and K_3C_{60}, *Phys. Rev.* **B51**, 5805 (1995).

Boisvert G. and Lewis L. J., Self-Diffusion on Low-Index Metallic Surfaces: Ag and Au (100) and (111), *Phys. Rev.* **B54**, 2880 (1996).

Bongiorno A., Colombo L. and Diaz de la Rubia T., Structural and Binding Properties of Vacancy Clusters in Silicon, *Europhys. Lett.* **43**, 695 (1998).

Bonneville J., Escaig B. and Martin J. L., A Study of Cross-Slip Activation Parameters in Pure Copper, *Acta Metall.* **36**, 1989 (1988).

Boyer H. E., *Atlas of Stress–Strain Curves*, ASM International, Metals Park: Ohio, 1987.

Bruinsma R. and Zangwill A., Theory of the hcp–fcc Transition in Metals, *Phys. Rev. Lett.* **55**, 214 (1985).

Brush S. G., *Statistical Physics and the Atomic Theory of Matter*, Princeton University Press, Princeton: New Jersey, 1983.

Budiansky B. and Truskinovsky L., On the Mechanics of Stress-Induced Phase Transformation in Zirconia, *J. Mech. Phys. Solids* **41**, 1445 (1993).

Bulatov V. V., Yip S. and Argon A. S., Atomic Modes of Dislocation Mobility in Silicon, *Phil. Mag.* **A72**, 453 (1995).

Burke J. E. and Turnbull D., Recrystallization and Grain Growth, *Prog. Met. Phys.* **3**, 270 (1952).

Callen H. B., *Thermodynamics and an Introduction to Thermostatics*, John Wiley and Sons, New York: New York, 1985.

Carlsson A. E., Beyond Pair Potentials in *Solid State Physics* Vol. 43 edited by H. Ehrenreich and R. Turnbull, Academic Press, San Diego: California, 1990.

Carlsson A. E., Angular Forces in Group-VI Transition Metals: Application to W(100), *Phys. Rev.* **B44**, 6590 (1991).

Carter W. C., Taylor J. E. and Cahn J. W., Variational Methods for Microstructural-Evolution Theories, *JOM*, 30, December 1997.

Ceder G., A Derivation of the Ising Model for the Computation of Phase Diagrams, *Comp. Mat. Sci.* **1**, 144 (1993).

Ceder G., A Computational Study of Oxygen Ordering in $YBa_2Cu_3O_z$ and its Relation to Superconductivity, *Molecular Simulation*, **12**, 141 (1994).

Ceder G., Asta M. and de Fontaine D., Computation of the OI–OII–OIII Phase Diagram and Local Oxygen Configurations for $YBa_2Cu_3O_z$ with z between 6.5 and 7, *Physica* **C177**, 106 (1991).

Chen L.-Q. and Yang W., Computer Simulation of the Domain Dynamics of a Quenched System with a Large Number of Nonconserved Order Parameters: The Grain-Growth Kinetics, *Phys. Rev.* **B50**, 15 752 (1994).

Chen L.-Q. and Wang Y., The Continuum Field Approach to Modeling Microstructural Evolution, *JOM*, 13, December (1996).

Chetty N., Weinert M., Rahman T. S. and Davenport J. W., Vacancies and Impurities in Aluminum and Magnesium, *Phys. Rev.* **B52**, 6313 (1995).

Cho J.-H. and Ardell A. J., Coarsening of Ni_3Si Precipitates in Binary Ni–Si Alloys at Intermediate to Large Volume Fractions, *Acta Mater.* **45**, 1393 (1997).

Chu C.-H., Hysteresis and Microstructures: A Study of Biaxial Loading on Compound Twins of Copper–Aluminum–Nickel Single Crystals, PhD Thesis, University of Minnesota, 1993. (unpublished)

Cocks A. C. F., Gill S. P. A. and Pan J., Modeling Microstructure Evolution in Engineering Materials, in *Advances in Applied Mechanics* Vol. 36, edited by E. van der Giessen and T. Y. Wu, Academic Press, San Diego: California, 1999..

Condon E. U. and Shortley G. H., *The Theory of Atomic Spectra*, Cambridge University Press, Cambridge: England, 1935.

Connolly J. W. O. and Williams A. R., Density-Functional Theory Applied to Phase Transformations in Transition-Metal Alloys, *Phys. Rev.* **B27**, 5169 (1983).

Coupeau C., Girard J. C. and Grilhé, Plasticity Study of Deformed Materials by in situ Atomic Force Microscopy, *J. Vac. Sci. Technol.* **B16**, 1964 (1998).

Courtney T. H., *Mechanical Behavior of Materials*, Mcgraw-Hill Publishing Company, New York: New York, 1990.

Crampin S., Hampel K., Vvedensky D. D., MacLaren J. M., The Calculation of Stacking Fault Energies in Close-Packed Metals, *J. Mater. Res.* **5**, 2107 (1990).

Cross M. C. and Hohenberg P. C., Pattern Formation Outside of Equilibrium, *Rev. Mod. Phys.* **65**,

851 (1993).

de Fontaine D., Configurational Thermodynamics of Solid Solutions in *Solid State Physics* Vol. 34, edited by H. Ehrenreich and D. Turnbull, Academic Press, Inc., New York: New York, 1979.

de Fontaine D., Wille L. T. and Moss S. C., Stability Analysis of Special-Point Ordering in the Basal Plane of $YBa_2Cu_3O_{7-\delta}$, *Phys. Rev.* **B36**, 5709 (1987).

de Fontaine D., Ceder G. and Asta M., Low-Temperature Long-Range Oxygen Order in $YBa_2Cu_3O_z$, *Nature* **343**, 544 (1990).

de Wit R., The Continuum Theory of Stationary Dislocations, in *Solid State Physics* Vol. 10, edited by F. Seitz and D. Turnbull, Academic Press, New York: New York, 1960.

Diep H. T., Ghazali A., Berge B. and Lallemand P., Phase Diagrams in f.c.c. Binary Alloys: Frustration Effects, *Europhys. Lett.* **2**, 603 (1986).

Dirac P. A. M., Quantum Mechanics of Many-Electron Systems, *Proc. R. Soc. Lond.* **123A**, 714 (1929).

Dodson B. W., Simulation of Au(100) Reconstruction by Use of the Embedded-Atom Method, *Phys. Rev.* **B35**, 880 (1987).

Domb C., *The Critical Point*, Taylor & Francis Ltd., London: England, 1996.

Dreizler R. M. and Gross E. K. U., *Density Functional Theory: An Approach to the Quantum Many-Body Problem*, Springer-Verlag, Berlin: Germany, 1990.

Ducastelle F., *Order and Phase Stability in Alloys*, North-Holland Pub. Co., Amsterdam: The Netherlands, 1991.

Embury J. D., Lloyd D. J. and Ramachandran T. R., Strengthening Mechanisms in Aluminum Alloys in *Treatise on Materials Science and Technology*, Vol. 31, Academic Press, New York: New York, 1989.

Engel G. E. and Needs R. J., Total Energy Calculations on Zinc Sulphide Polytypes, *J. Phys.: Condens. Matter* **2**, 367 (1990).

Ercolessi F., Tosatti E. and Parrinello M., Au(100) Surface Reconstruction, *Phys. Rev. Lett.* **57**, 719 (1986).

Eriksson, O., Wills, J. M. and Wallace, D., Electronic, Quasiharmonic and Anharmonic Entropies of Transition Metals, *Phys. Rev.* **B46**, 5221 (1992).

Eshelby J. D., The Continuum Theory of Lattice Defects, in *Solid State Physics* Vol. 3, edited by F. Seitz and D. Turnbull, Academic Press, New York: New York, 1956.

Eshelby J. D., Point Defects in *The Physics of Metals Vol. 2. Defects*, edited by P. B. Hirsch, Cambridge University Press, London: England, 1975a.

Eshelby J. D., The Elastic Energy-Momentum Tensor, *J. Elasticity*, **5**, 321 (1975b).

Feibelman P. J., Diffusion Path for an Al Adatom on Al(001), *Phys. Rev. Lett.*, **65**, 729 (1990).

Fetter A. L. and Walecka J. D., *Quantum Theory of Many-Particle Systems*, McGraw-Hill Book Company, New York: New York, 1971.

Feynman R. P., The Development of the Space-Time View of Quantum Electrodynamics, *Physics Today*, Aug, 31 (1966).

Feynman R. P., Statistical Mechanics, W. A. Benjamin, Inc., Reading: Massachusetts, 1972.

Feynman R. P., Leighton R. B. and Sands M., *The Feynman Lectures on Physics*, Addison-Wesley Publishing Company, Reading: Massachusetts, 1963.

Fiig T., Andersen N. H., Lindgård P.-A., Berlin J. and Mouritsen O. G., Mean-Field and Monte Carlo Calculations of the Three-Dimensional Structure Factor for $YBa_2Cu_3O_{6+x}$, *Phys. Rev.* **B54**, 556 (1996).

Finel A., The Cluster Variation Method and Some Applications in *Statics and Dynamics of Alloy Phase Transformations*, edited by P. E. A. Turchi and A. Gonis, Plenum Press, New York: New York, 1994.

Finnis M. W. and Sinclair J. E., A Simple Empirical N-Body Potential for Transition Metals, *Phil. Mag.* **A50**, 45 (1984).

Finnis M. W., Paxton A. T., Methfessel M. and van Schilfgaarde M., Crystal Structures of Zirconia from First Principles and Self-Consistent Tight Binding, *Phys. Rev. Lett.* **81**, 5149 (1998).

Fivel M. C., Robertson, C. F., Canova, G. R. and Boulanger, L., Three-Dimensional Modeling of Indent-Induced Plastic Zone at a Mesoscale, *Acta Mater.* **46**, 6183 (1998).

Foreman A. J. E. and Makin M. J., Dislocation Movement Through Random Arrays of Obstacles, *Canadian J. Phys.* **45**, 511 (1967).

Franciosi P., The Concepts of Latent Hardening and Strain Hardening in Metallic Single Crystals, *Acta Metall.* **33**, 1601 (1985).

Fratzl P., Penrose O. and Lebowitz J. L., Modeling of Phase Separation in Alloys with Coherent Elastic Misfit, *J. Stat. Phys.*, **95**, 1429 (1999).

Frenkel D. and Smit B., *Understanding Molecular Simulation*, Academic Press, San Diego: California, 1996.

Freund L. B., The Mechanics of Dislocations in Strained-Layer Semiconductor Materials in *Advances in Applied Mechanics*, Vol. 30, Academic Press, New York: New York, 1994.

Fridline D. R. and Bower A. F., Influence of Anisotropic Surface Diffusivity on Electromigration Induced Void Migration and Evolution, *J. Appl. Phys.* **85**, 3168 (1999).

Friedel J., On the Stability of the Body Centred Cubic Phase in Metals at High Temperatures, *J. de Phys. Lettres*, **35**, 59 (1974).

Frost H. J. and Ashby M. F., *Deformation-Mechanism Maps*, Pergamon Press, Oxford: England 1982.

Frost H. J. and Thompson C. V., Computer Simulation of Grain Growth, *Current Opinions in Solid State & Materials Science* **1**, 361 (1996).

Gahn U., Ordering in Faced-Centered Cubic Binary Crystals Confined to Nearest-Neighbour Interactions - Monte Carlo Calculations, *J. Phys. Chem. Solids* **47**, 1153 (1986).

Gallego R. and Ortiz M., A Harmonic/Anharmonic Energy Partition Method for Lattice Statics Computations, *Modelling Simul. Mater. Sci. Eng.* **1**, 417 (1993).

Gandin Ch.-A. and Rappaz M., A Coupled Finite Element-Cellular Automaton Model for the Prediction of Dendritic Grain Structures in Solidification Processes, *Acta Metall. Mater.*, **42**, 2233 (1994).

Garbulsky G. D. and Ceder G., Linear-Programming Method for Obtaining Effective Cluster Interactions in Alloys from Total Energy Calculations: Applications to the fcc Pd–V System, *Phys. Rev.* **B51**, 67 (1995).

Gerberich W. W., Venkataraman S. K., Huang H., Harvey S. E. and Kohlstedt D. L., The Injection of Plasticity by Millinewton Contacts, *Acta Metall. Mater.* **43**, 1569 (1995).

Gerrold V., Precipitation Hardening in *Dislocations in Solids*, edited by F. R. N. Nabarro, North-Holland Pub. Co., Amsterdam: The Netherlands, 1979.

Ghoniem N. M., Computational Methods for Mesoscopic, Inhomogeneous Plastic Deformation, to appear in *Proceedings of the 1ˢᵗ Latin American Summer School on Materials Instabilities*, Kluwer Publishing, Dordrecht: The Netherlands, 2000.

Ghoniem N. M. and Sun L. Z., Fast-Sum Method for the Elastic Field of Three-Dimensional Dislocation Ensembles, *Phys. Rev.* **B60**, 128 (1999)

Ghoniem N. M., Tong S.-H. and Sun L. Z., Parametric Dislocation Dynamics: A Thermodynamics-Based Approach to Investigations of Mesoscopic Plastic Deformation, *Phys. Rev.* **B61**, 913 (1999).

Gibbs D., Ocko B. M., Zehner D. M. and Mochrie S. G. J., Structure and Phases of the Au(001) Surface: In-Plane Structure, *Phys. Rev.* **B42**, 7330 (1990).

Gillan M. J., Calculation of the Vacancy Formation Energy in Aluminium, *J. Phys.:Condens. Matter*, **1**, 689 (1989).

Gilman J. J., *Micromechanics of Flow in Solids*, McGraw-Hill, New York: New York, 1969.

Gjostein N. A., Short Circuit Diffusion in *Diffusion*, ASM Conference Proceedings, ASM

International, Materials Park: Ohio, 1972.

Goldenfel N., *Lectures on Phase Transitions and the Renormaliztion Group*, Addison-Wesley, Reading: Massachusetts, 1992.

Gong X. G. and Wilkins J. W., Hyper Molecular Dynamics with a Local Bias Potential, *Phys. Rev.* **B59**, 54 (1999).

Goringe C. M., Bowler D. R. and Hernández E., Tight-Binding Modelling of Materials, *Rep. Prog. Phys.* **60**, 1447 (1997).

Gradshteyn I. S. and Ryzhik I. M., *Table of Integrals, Series, and Products*, incorporating the 4th ed. prepared by Yu. V. Geronimus and M. Yu. Tseytlin; translated from the Russian by Scripta Technica Inc., and edited by A. Jeffrey. Corr. and enl. ed. prepared by A. Jeffrey, Academic Press, New York: New York, 1980.

Gurtin M., *An Introduction to Continuum Mechanics*, Academic Press Inc., San Diego: California, 1981.

Haasen P., Solution Hardening in f.c.c. Metals in *Dislocations in Solids*, edited by F. R. N. Nabarro, North-Holland Pub. Co., Amsterdam: The Netherlands, 1979.

Hafner J., *From Hamiltonians to Phase Diagrams*, Springer-Verlag, Berlin: Germany 1987.

Hafner J. and Heine V., The Crystal Structures of the Elements: Pseudopotential Theory Revisited, *J. Phys. F: Met. Phys.*, **13**, 2479 (1983).

Hammer B., Jacobsen K. W., Milman V., Payne M. C., Stacking Fault Energies in Aluminium, *J. Phys.: Condens. Matter* **4**, 10 453 (1992).

Hansen N., Polycrystalline Strengthening, *Met. Trans.* **16A**, 2167 (1985).

Harrison W. A., *Electronic Structure and the Properties of Solids*, W. H. Freeman and Company, San Francisco: California, 1980.

Hartley C. S. and Hirth J. P., Interaction of Nonparallel Noncoplanar Dislocations, *Acta Met.* **13**, 79 (1965).

Heid R., Pintschovius L. and Godard J. M., Eigenvectors of Internal Vibrations of C_{60}: Theory and Experiment, *Phys. Rev.* **B56**, 5925 (1997).

Heine V., Robertson I. J. and Payne M. C., Many-Atom Interactions in Solids, *Phil. Trans. R. Soc. Lond.* **A334**, 393 (1991).

Herring C., Diffusional Viscosity of a Polycrystalline Solid, *J. Appl. Phys.* **21**, 437 (1950).

Hirth J. P., On Dislocation Interactions in the fcc Lattice, *J. Appl. Phys.* **32**, 700 (1961).

Hirth J. P. and Lothe J., *Theory of Dislocations*, Krieger Publishing Company, Malabar: Florida, 1992.

Hoffmann R., *Solids and Surfaces: A Chemist's View of Bonding in Extended Structures*, VCH Publishers, New York: New York 1988.

Hohenberg P. C. and Halperin B. I., Theory of Dynamic Critical Phenomena, *Rev. Mod. Phys.* **49**, 435 (1977).

Honeycombe R. W. K. and Bhadeshia H. K. D. H., *Steels: Microstructure and Properties*, Edward Arnold, London: England, 1995.

Hoover, W. G., Canonical Dynamics: Equilibrium Phase-Space Distributions, *Phys. Rev.* **A31**, 1695 (1985).

Hornbogen E., On the Microstructure of Alloys, *Acta Metall.* **32**, 615 (1984).

Hu. G. Y. and Ying S. C., Two-Dimensional Lattice Hamiltonian for Surface Reconstruction, *Surf. Sci.* **150**, 47 (1985).

Huang H., Gilmer G. H. and de la Rubia T., An Atomistic Simulator for Thin Film Deposition in Three Dimensions, *J. Appl. Phys.* **84**, 3636 (1998).

Hull D. and Bacon D. J., *Introduction to Dislocations*, Butterworth Heinemann, Oxford: England, 1995.

Hummel R. E., *Understanding Materials Science*, Springer-Verlag New York, Inc., New York: New York, 1998.

Humphreys F. J. and Hatherly M., *Recrystallization and Related Annealing Phenomena*, Pergamon, Elsevier Science Ltd., Oxford: England, 1995.

Jacobsen J., Jacobsen K. W., Stoltze P. and Nørskov J. K., Island Shape-Induced Transition from 2D to 3D Growth for Pt/Pt(111), *Phys. Rev. Lett.* **74**, 2295 (1995).

Jacobsen K. W., Norskov J. K. and Puska M. J., Interatomic Interactions in the Effective-Medium Theory, *Phys. Rev.* **B35**, 7423 (1987).

James R. D. and Hane K. F., Martensitic Transformations and Shape Memory Materials, *Acta Mater.*, **48**, 197 (2000).

Jaynes E. T., *E. T. Jaynes: Papers on Probability and Statistics*, edited by R. D. Rosenkranz, D. Reidel Publishing Company, Dordrecht: The Netherlands, 1983.

Jeong H.-C. and Williams E. D., Steps on Surfaces: Experiment and Theory, *Surf. Sci. Rep.* **34**, 171 (1999).

Jog C. S., Sankarasubramanian R. and Abinandanan T. A., Symmetry-Breaking Transitions in Equilibrium Shapes of Coherent Precipitates, *J. Mech. Phys. Solids*, **48**, 2363 (2000).

Johnson K. L., *Contact Mechanics*, Cambridge University Press, Cambridge: England, 1985.

Johnson R. A., Analytic Nearest-Neighbor Model for fcc Metals, *Phys. Rev.* **B37**, 3924 (1988).

Johnson W. C. and Cahn J. W., Elastically Induced Shape Bifurcations of Inclusions, *Acta Metall.* **32**, 1925 (1984).

Jou H.-J., Leo P. H. and Lowengrub J. S., Microstructural Evolution in Inhomogeneous Elastic Media, *J. Comput. Phys.* **131**, 109 (1997).

Käckell P., Wenzien B. and Bechstedt F., Influence of Atomic Relaxations on the Structural Properties of SiC Polytypes from Ab-Initio Calculations, *Phys. Rev.* **B50**, 17 037 (1994).

Kaxiras E. and Duesbery M. S., Free Energies of Generalized Stacking Faults in Si and Implications for the Brittle–Ductile Transition, *Phys. Rev. Lett.* **70**, 3752 (1993).

Kaxiras E. and Erlebacher J., Adatom Diffusion by Orchestrated Exchange on Semiconductor Surfaces, *Phys. Rev. Lett.* **72**, 1714 (1994).

Kelchner C. L., Plimpton S. J. and Hamilton J. C., Dislocation Nucleation and Defect Structure during Surface Indentation, *Phys. Rev.* **B58**, 11 085 (1998).

Kittel C., *Solid State Physics*, John Wiley and Sons, Inc., New York: New York, 1976.

Kocks U. F., The Theory of an Obstacle-Controlled Yield Strength – Report After an International Workshop, *Mat. Sci. Eng.* **27**, 291 (1977).

Kohlhoff S., Gumbsch P. and Fischmeister H. F., Crack Propagation in b.c.c. Crystals Studied with a Combined Finite-Element and Atomistic Model, *Phil. Mag.* **A64**, 851 (1991).

Korzhavyi P. A., Abrikosov I. A., Johansson B., Ruban A. V., and Skriver H. L., First-Principles Calculations of the Vacancy Formation Energy in Transition and Noble Metals, *Phys. Rev.* **B59**, 11 693 (1999).

Kubin L. P., Canova G., Condat G., Devincre B., Pontikis V. and Bréchet Y., Dislocation Microstructures and Plastic Flow: A 3D Simulation, *Solid State Phenomena*, **23** & **24**, 455 (1992).

Kumar K. S., Mannan S. K. and Viswanadham R. K., Fracture Toughness of NiAl and NiAl-Based Composites, *Acta Metall.* **40**, 1201 (1992).

Kwon I., Biswas R., Wang C. Z., Ho K. M. and Soukoulis C. M., Transferable Tight-Binding Models for Silicon, *Phys. Rev.* **B49**, 7242 (1994).

La Magna A., Coffa S. and Colombo L., Role of Extended Vacancy-Vacancy Interaction on the Ripening of Voids in Silicon, *Phys. Rev. Lett.* **82**, 1720 (1999).

Landau L. D. and Lifshitz E. M., *Theory of Elasticity*, Pergamon Press, Oxford: England, 1959.

Lardner R. W., *Mathematical Theory of Dislocations and Fracture*, University of Toronto Press, Toronto: Canada, 1974.

Lawn B., *Fracture of Brittle Solids*, Cambridge University Press, Cambridge: England, 1993.

Lee T. C., Robertson I. M. and Birnbaum H. K., An In Situ Transmission Electron Microscope

Deformation Study of the Slip Transfer Mechanisms in Metals, *Met. Trans.* **21A**, 2437 (1990).

Leslie W. C., *The Physical Metallurgy of Steels*, McGraw-Hill, New York: New York 1981.

Li J. C. M., Cross Slip and Cross Climb of Dislocations Induced by a Locked Dislocation, *J. Appl. Phys.* **32**, 593 (1961).

Li J. C. M., Dislocation Sources in *Dislocation Modeling of Physical Systems, Proceedings of the International Conference, Gainesville, Florida, USA June 22–27, 1980*, edited by M. Ashby, C. Hartley and R. Bullough, Pergamon Press, Oxford: England, 1981.

Liu D. J., Einstein T. L., Sterne P. A. and Wille L. T., Phase Diagram of a Two-Dimensional Lattice-Gas Model of Oxygen Ordering in $YBa_2Cu_3O_z$ with Realistic Interactions, *Phys. Rev.* **B52**, 9784 (1995).

Lu Z. W., Wei S.-H., Zunger A., Frota-Pessoa S. and Ferreira L. G., First-Principles Statistical Mechanics of Structural Stability of Intermetallic Compounds, *Phys. Rev.* **B44**, 512 (1991).

Lubarda V. A., Blume J. A. and Needleman A., An Analysis of Equilibrium Dislocation Distributions, *Acta Metall. Mater.* **41**, 625 (1993).

Luskin M., On the Computation of Crystalline Microstructure, *Acta Numerica*, 191 (1996).

Lüth H., *Surfaces and Interfaces of Solids*, Springer-Verlag: Berlin, 1993.

Malvern, L. E., *Introduction to the Mechanics of a Continuous Medium*, Prentice Hall, Inc., Englewood Cliffs: New Jersey, 1969.

Mañosa L., Planes A., Ortín J. and Martínez, Entropy Change of Martensitic Transformations in Cu-Based Shape-Memory Alloys, *Phys. Rev.* **B48**, 3611 (1993).

Maradudin A. A., Montroll E. W., Weiss G. H. and Ipatova I. P., *Theory of Lattice Dynamics in the Harmonic Approximation*, Academic Press, New York: New York, 1971.

Marion J. B., *Classical Dynamics of Particles and Systems*, Harcourt Brace Jovanovich, San Diego: California, 1970.

Maris H. J. and Kadanoff L. P., Teaching the Renormalization Group, *Am. J. Phys.* **46**, 652 (1978).

Maroudas D. and Brown R. A., Calculation of Thermodynamic and Transport Properties of Intrinsic Point Defects in Silicon, *Phys. Rev.* **B47**, 15 562 (1993).

Martin J. W., Doherty R. D. and Cantor B., *Stability of Microstructure in Metallic Systems*, Cambridge University Press, Cambridge: England, 1997.

Massalski T. B ., Phase Diagrams in Materials Science, *Met. Trans.* **20A**, 1295 (1989).

Massalski T. B. (editor-in-chief), *Binary Alloy Phase Diagrams*, ASM International, Materials Park: Ohio, 1990.

McClintock F. A. and Argon A. S., *Mechanical Behavior of Materials*, Addison-Wesley Publishing Company, Reading: Massachusetts, 1966.

McMahon C. J. and Graham C. D., *Introduction to Engineering Materials: The Bicycle and the Walkman*, private publication, 1994. Available from author: http://www.seas.upenn/mse/fac/mcmahon.html

Mehl M. J. and Boyer L. L., Calculation of Energy Barriers for Physically Allowed Lattice-Invariant Strains in Aluminum and Iridium, *Phys. Rev.* **B43**, 9498 (1991).

Mehl M. J. and Klein B, M., First Principles Supercell Total-Energy Calculation of the Vacancy Formation Energy in Aluminium, *Physica* **B172**, 211 (1991).

Mehl M. J. and Papaconstantopoulos D. A., Applications of a Tight-Binding Total-Energy Method for Transition and Noble Metals: Elastic Constants, Vacancies and Surfaces of Monatomic Metals, *Phys. Rev.* **B54**, 4519 (1996).

Mercer J. L. and Chou M. Y., Tight-Binding Total Energy Models for Silicon and Germanium, *Phys. Rev.* **B47**, 9366 (1993).

Mercer J. L., Nelson J. S., Wright A. F. and Stechel E. B., Ab-Initio Calculations of the Energetics of the Neutral Si Vacancy Defect, *Modelling Simul. Mater. Sci. Eng.* **6**, 1 (1998).

Miller R. and Phillips R., Critical Analysis of Local Constitutive Models for Slip and Decohesion, *Phil. Mag.* **A73**, 803 (1996).

Mills M. J., High Resolution Transmission Electron Microscopy and Atomistic Calculations of Grain Boundaries in Metals and Intermetallics, *Mat. Sci. Eng.* **A166**, 35 (1993).

Mills M. J., Daw M. S. and Foiles S. M., High-Resolution Transmission Electron Microscopy Studies of Dislocation Cores in Metals and Intermetallic Compounds, *Ultramicroscopy* **56**, 79 (1994).

Mishin Y., Farkas D., Mehl M. J. and Papaconstantopoulos D. A., Interatomic Potentials for Monoatomic Metals from Experimental Data and Ab Initio Calculations, *Phys. Rev.* **B59**, 3393 (1999).

Mitchell R., *The Gift of Fire*, Simon and Schuster, New York: New York, 1987.

Moriarty J. A., Hybridization and the fcc–bcc Phase Transitions in Calcium and Strontium, *Phys. Rev.* **B8**, 1338 (1973).

Moriarty J. A., Analytic Representation of Multi-Ion Interatomic Potentials in Transition Metals, *Phys. Rev.* **B42**, 1609 (1990).

Moriarty J. A. and Althoff, J. D., First-Principles Temperature–Pressure Phase Diagram of Magnesium, *Phys. Rev.* **B51**, 5609 (1995).

Mott N. F. and Nabarro F. R. N., Dislocation Theory and Transient Creep in *Report of a Conference on Strength of Solids*, The *Physical* Society, London: England, (1948).

Motte A., *Sir Isaac Newton's Mathematical Principles of Philosophy and His System of the World*, translated into English by Andrew Motte in 1729. University of California Press, Berkley: California, 1934.

Moulin A., Condat M. and Kubin L. P., Simulation of Frank–Read Sources in Silicon, *Acta Mat.* **45**, 2339 (1997).

Mueller R. and Gross D., 3D Simulation of Equilibrium Morphologies of Precipitates, *Comp. Mater. Sci.* **11**, 35 (1998).

Munro L. J. and Wales D. J., Defect Migration in Crystalline Silicon, *Phys. Rev.* **B59**, 3969 (1999).

Mura T., *Micromechanics of Defects in Solids*, Kluwer Academic Publishers, Dordrecht: The Netherlands 1987.

Nabarro F. R. N., The Statistical Problem of Hardening, *J. Less Common Metals* **28**, 257 (1972).

Nabarro F. R. N. and de Villiers H. L., *The Physics of Creep*, Taylor & Francis Ltd., London: England, 1995.

Nastar M., Bulatov V. V. and Yip S., Saddle-Point Configurations for Self-Interstitial Migration in Silicon, *Phys. Rev.* **B53**, 13 521 (1996).

Needs R. J. and Mujica A., First-Principles Pseudopotential Study of the Structural Phases of Silicon, *Phys. Rev.* **B51**, 9652 (1995).

Neuhäuser H. and Schwink C., Solid Solution Strengthening in *Plastic Deformation and Fracture of Materials*, Vol. 6 of Materials Science and Technology, edited by R. W. Cahn, P. Haasen and E. J. Kramer, VCH Publishers Inc., New York: New York, 1993.

Nix W. D., Mechanical Properties of Thin Films, *Met. Trans.* **20A**, 2217 (1989).

Nix W. D. and Gao H., Indentation Size Effects in Crystalline Materials: A Law for Strain Gradient Plasticity, *J. Mech. Phys. Solids* **46**, 411 (1998).

Nowick A. S. and Berry B. S., *Anelastic Relaxation in Crystalline Solids*, Academic Press, New York: New York 1972.

Nosé S., A Unified Formulation of the Constant Temperature Molecular Dynamics Method, *J. Phys. Chem.*, **81**, 511 (1984).

Nunes R. W., Bennetto J. and Vanderbilt D., Atomic Structure of Dislocation Kinks in Silicon, *Phys. Rev.* **B57** 10 388 (1998).

Ohr S. M., An Electron Microscope Study of Crack Tip Deformation and its Impact on the Dislocation Theory of Fracture, *Mat. Sci. Eng.* **72**, 1 (1985).

Ohr S. M., Chang S.-J. and Thomson R., Elastic Interaction of a Wedge Crack with a Screw Dislocation, *J. Appl. Phys.* **57**, 1839 (1985).

Omini M. and Sparavigna A., Heat Transport in Dielectric Solids with Diamond Structure, *Il Nuovo Cimento* **D19**, 1537 (1997).

Ozoliņš V., Wolverton C. and Zunger A., Cu–Au, Ag–Au, Cu–Ag and Ni–Au intermetallics: First-Principles Study of Temperature–Composition Phase Diagrams and Structures, *Phys. Rev.* **B57**, 6427 (1998).

Pande C. S. and Marsh S. P., The Analytical Modeling of Normal Grain Growth, *JOM*, 25, September 1992.

Paxton A. T., Methfessel M. and Polatoglou H. M., Structural Energy–Volume Relations in First-Row Transition Metals, *Phys. Rev.* **B41**, 8127 (1990).

Payne M. C., Teter M. P., Allan D. C., Arias T. A. and Joannopoulos J. D., Iterative Minimization Techniques for Ab Initio Total-Energy Calculations: Molecular Dynamics and Conjugate Gradients, *Rev. Mod. Phys.* **64**, 1045 (1992).

Payne M. C., Robertson I. J., Thomson D. and Heine V., Ab-Initio Databases for Fitting and Testing Interatomic Potentials, *Phil. Mag.* **B73**, 191 (1996).

Pines D., *Elementary Excitations in Solids*, Benjamin/Cummings Publishing Company, Inc., Reading: Massachusetts, 1963.

Plapp B. P., Egolf D. A., Bodenschatz E. and Pesch W., Dynamics and Selection of Giant Spirals in Rayleigh–Bénard Convection, *Phys. Rev. Lett.* **81**, 5334 (1998).

Poon T. W., Yip S., Ho P. S. and Abraham F. F., Ledge interactions and stress relaxations on Si(001) stepped surfaces, *Phys. Rev.* **B45**, 3521 (1992).

Provatas N., Goldenfeld N. and Dantzig J., Efficient Computation of Dendritic Microstructures Using Adaptive Mesh Refinement, *Phys. Rev. Lett.* **80**, 3308 (1998).

Puska M. J., Poykko S., Pesda M. and Nieminen R. M., Convergence of Supercell Calculations for Point Defects in Semiconductors: Vacancy in Silicon, *Phys. Rev.* **B58**, 1318 (1998).

Quong A. A. and Klein B. M., Self-Consistent-Screening Calculation of Interatomic Force Constants and Phonon Dispersion Curves from First Principles: Application to Aluminum, *Phys. Rev.* **B46**, 10 734 (1992).

Rabe K. M. and Waghmare U. V., Localized Basis for Effective Lattice Hamiltonians: Lattice Wannier Functions, *Phys. Rev.* **B52**, 13 236 (1995).

Ramstad A., Brocks G. and Kelly P. J., Theoretical Study of the Si(100) Surface Reconstruction, *Phys. Rev.* **B51**, 14 504 (1995).

Randle V., Grain Assemblage in Polycrystals, *Acta Metall. Mater.* **42**, 1769 (1994).

Randle V., The Role of the Grain Boundary Plane in Cubic Polycrystals, Acta mater. **46**, 1459 (1997).

Randle V. and Horton D., Grain Growth Phenomena in Nickel, *Scripta Metall. Mater.* **31**, 891 (1994).

Rappaz M. and Gandin Ch.-A., Probabilistic Modelling of Microstructure Formation in Solidification Processes, *Acta Metall. Mater.* **41**, 345 (1993).

Rasmussen T., Jacobsen K. W., Leffers T. and Pedersen O. B., Simulations of the Atomic Structure, Energetics and Cross-Slip of Screw Dislocations in Copper, *Phys. Rev.* **B56**, 2977 (1997).

Rasmussen T., Jacobsen K. W., Leffers T., Pedersen O. B., Srinivasan S. G. and Jónsson H., Atomistic Determination of Cross-Slip Pathway and Energetics, *Phys. Rev. Lett.* **79**, 3676 (1997b).

Reif F., *Fundamentals of Statistical and Thermal Physics*, McGraw-Hill Book Company, New York: New York, 1965.

Reppich B., Particle Strengthening in *Plastic Deformation and Fracture of Materials*, Vol. 6 of Materials Science and Technology, edited by R. W. Cahn, P. Haasen and E. J. Kramer, VCH Publishers Inc., New York: New York, 1993.

Rhines F. N. and Craig K. R., Mechanism of Steady-State Grain Growth in Aluminum, *Met. Trans.* **5**, 413 (1974).

Rice J. R., Mathematical Analysis in the Mechanics of Fracture, from Vol. II of *Fracture*, edited by G. Sih, Academic Press, New York: New York, 1968.

Rice J. R., Dislocation Nucleation from a Crack Tip: An Analysis Based on the Peierls Concept, *J. Mech. Phys. Solids* **40**, 239 (1992).

Rice J. R. and Beltz G. E., The Activation Energy for Dislocation Nucleation at a Crack, *J. Mech. Phys. Solids* **42**, 333 (1994).

Rittner J. D. and Seidman D. N., ⟨110⟩ Symmetric Tilt Grain-Boundary Structures in fcc Metals with Low Stacking-Fault Energies, *Phys. Rev.* **B54**, 6999 (1996).

Robertson C. F. and Fivel M. C., A Study of the Submicron Indent-Induced Plastic Deformation, *J. Mater. Res.* **14**, 2251 (1999).

Rodney D. and Phillips R., Structure and Strength of Dislocation Junctions: An Atomic Level Analysis, *Phys. Rev. Lett.* **82**, 1704 (1999).

Roelofs L. D., Ramseyer T., Taylor L. L., Singh D. and Krakauer H., Monte Carlo Study of the W(001) Surface Reconstruction Transition Based on Total-Energy Calculations, *Phys. Rev.* **B40**, 9147 (1989).

Rose J. H., Smith J. R., Guinea F. and Ferrante J., Universal Features of the Equation of State of Metals, *Phys. Rev.* **B29**, 2963 (1984).

Rosengaard N. M. and Skriver H. L., Calculated Stacking-Fault Energies of Elemental Metals, *Phys. Rev.* **B47**, 12 865 (1993).

Rottman C. and Wortis M., Statistical Mechanics of Equilibrium Crystal Shapes: Interfacial Phase Diagrams and Phase Transitions, *Phys. Rep.* **103**, 59 (1984).

Russell K. C. and Aaronson H. I. (editors), *Precipitation Processes in Solids*, Metallurgical Society of AIME, Warrendale: Pennsylvania, 1978.

Saada G., Sur le Durcissement du à la Recombinaison des Dislocations, *Acta Met.* **8**, 841 (1960).

Saada G., Cross-Slip and Work Hardening of f.c.c. Crystals, *Mat. Sci. Eng.* **A137**, 177 (1991).

Sagui C., Orlikowski D., Somoza A. M. and Roland C., Three-Dimensional Simulations of Ostwald Ripening with Elastic Effects, *Phys. Rev.* **E58**, R4092 (1998).

Sanchez J. M., Ducastelle F. and Gratias D., Generalized Cluster Description of Multicomponent Systems, *Physica* **128A**, 334 (1984).

Schiøtz J., Canel L. M. and Carlsson A. E., Effects of Crack Tip Geometry on Dislocation Emission and Cleavage: A Possible Path to Enhanced Ductility, *Phys. Rev.* **B55**, 6211 (1997).

Schöck G., Thermodynamics and Thermal Activation of Dislocations in *Dislocations in Solids*, edited by F. R. N. Nabarro, North-Holland Pub. Co., Amsterdam: The Netherlands, 1980.

Schoeck G., Dislocation Emission from Crack Tips as a Variational Problem of the Crack Energy, *J. Mech. Phys. Solids* **44**, 413 (1996).

Schwarz K. W., Simulation of Dislocations on the Mesoscopic Scale. II. Application to Strained-Layer Relaxation, *J. Appl. Phys.* **85**, 120 (1999a).

Schwarz K. W., Simulation of Dislocations on the Mesoscopic Scale. I. Methods and Examples, *J. Appl. Phys.* **85**, 108 (1999b).

Schwarz K. W. and LeGoues F. K., Dislocation Patterns in Strained Layers from Sources on Parallel Glide Planes, *Phys. Rev. Lett.* **79**, 1877 (1997).

Shackelford J. F., Alexander W. and Park J. S., *CRC Practical Handbook of Materials Selection*, CRC Press, Boca Raton: Florida, 1995.

Shenoy V. B., Miller R., Tadmor E. B., Rodney D, Phillips R. and Ortiz M., An Adaptive Finite Element Approach to Atomic-Scale Mechanics – The Quasicontinuum Method; *J. Mech. Phys. Solids* **47**, 611 (1999).

Shenoy V. B., Kukta R. and Phillips R., Mesoscopic Analysis of Structure and Strength of Dislocation Junctions in fcc Metals, *Phys. Rev. Lett.*, **84**, 1491 (2000).

Simmons G. and Wang H., *Single Crystal Elastic Constants and Calculated Aggregate Properties: A Handbook*, The MIT Press, Cambridge: Massachusetts, 1971.

Simpson C. J., Aust K. T. and Winegard W. C., Activation Energies for Normal Grain Growth in Lead and Cadmium Base Alloy, *Met. Trans.* **2**, 993 (1971).

Singh D., Wei S.-H. and Krakauer H., Instability of the Ideal Tungsten (001) Surface, *Phys. Rev. Lett.* **57**, 3292 (1986).

Skriver H. L., Crystal Structure from One-Electron Theory, *Phys. Rev.* **B31**, 1909 (1985).

Smith E., Dislocations and Cracks in *Dislocations in Solids*, edited by F. R. N. Nabarro, North-Holland Pub. Co., Amsterdam: The Netherlands, 1979.

Sobel D., *Longitude*, Walker, New York: New York, 1995.

Sorensen M. R. and Voter A. F., Temperature-Accelerated Dynamics for Simulation of Infrequent Events, *J. Chem. Phys.* **112**, 9599 (2000).

Srivastava G. P., Theory of Semiconductor Surface Reconstruction, *Rep. Prog. Phys.* **60**, 561 (1997).

Steiner M. M., Genilloud P.-A. and Wilkins J. W., Simple Bias Potential for Boosting Molecular Dynamics with the Hyperdynamics Scheme, *Phys. Rev.* **B57**, 10 236 (1998).

Stephens D. E. and Purdy G. R., Equilibrium Properties of the $\gamma-\beta$ Interface in Cu–Zn Alloys, *Acta Metall.* **23**, 1343 (1975).

Stillinger F. H. and Weber T. A., Computer Simulation of Local Order in Condensed Phases of Silicon, *Phys. Rev.* **B31**, 5262 (1985).

Stokbro K. and Jacobsen K. W., Simple Model of Stacking-Fault Energies, *Phys. Rev.* **B47**, 4916 (1993).

Suo Z. and Shih C. F., Models for Metal/Ceramic Interface Fracture, chap. 12 of *Fundamentals of Metal-Matrix Composites*, edited by S. Suresh, A. Mortensen and A. Needleman, Butterworth-Heinemann, Oxford: England, 1993.

Suo Z., Motions of Microscopic Surfaces in Materials, in *Advances in Applied Mechanics*, Vol. 33, Academic Press, New York: New York, 1997.

Suresh S., *Fatigue of Materials*, Cambridge University Press, New York: New York, 1998.

Sutton A. and Balluffi R., *Interfaces in Crystalline Materials*, Clarendon Press: Oxford, 1995.

Tadmor E. B. , Ortiz M. and Phillips R., Quasicontinuum Analysis of Defects in Solids, *Phil. Mag.* **A73**, 1529 (1996).

Takeuchi N., Chan C. T. and Ho K. M., Reconstruction of the (100) Surfaces of Au and Ag, *Phys. Rev.* **B43**, 14 363 (1991).

Tang M., Colombo L., Zhu J. and Diaz de la Rubia T., Intrinsic Point Defects in Crystalline Silicon: Tight-Binding Molecular Dynamics Studies of Self-Diffusion, Interstitial-vacancy Recombination and Formation Volumes, *Phys. Rev.* **B55**, 14 279 (1997).

Taylor J. E. and Cahn J. W., Linking Anisotropic Sharp and Diffuse Surface Motion Laws via Gradient Flows, *J. Stat. Phys.*, **77**, 183 (1994).

Terakura K., Oguchi T., Mohri T. and Watanabe K., Electronic Theory of the Alloy Phase Stability of Cu–Ag, Cu–Au and Ag–Au Systems, *Phys. Rev.*, **B35**, 2169 (1987).

Tersoff J., New Empirical Approach for the Structure and Energy of Covalent Systems, *Phys. Rev.* **B37**, 6991 (1988).

Thompson M. E., Su C. S. and Voorhees P. W., The Equilibrium Shape of a Misfitting Precipitate, *Acta Metall. Mater.* **42**, 2167 (1994).

Thomson R., Zhou S. J., Carlsson A. E. and Tewary V. K., Lattice Imperfections Studied by Use of Lattice Green's Functions, *Phys. Rev.* **B46**, 10 613 (1992).

Trigunayat G. C., A Survey of the Phenomenon of Polytypism in Crystals, *Sol. St. Ionics* **48**, 3 (1991).

Tsao J. Y., *Materials Fundamentals of Molecular Beam Epitaxy*, Academic Press, Inc., San Diego: California, 1993.

Turner D. E., Zhu Z. Z., Chan C. T. and Ho K. M., Energetics of Vacancy and Substitutional Impurities in Aluminum Bulk and Clusters, *Phys. Rev.* **B55**, 13 842 (1997).

Ungar P. J., Halicioglu T. and Tiller W. A., Free Energies, Structures and Diffusion of Point Defects in Si Using an Empirical Potential, *Phys. Rev.* **B50**, 7344 (1994).

van der Giessen E. and Needleman A., Discrete Dislocation Plasticity: A Simple Planar Model, *Modelling Simul. Mater. Sci. Eng.* **3**, 689 (1995).

Vineyard G. H., Frequency Factors and Isotope Effects in Solid State Rate Processes, *J. Phys. Chem. Solids*, **3**, 121 (1957).

Vitos L., Ruban A. V., Skriver H. L. and Kollár J., The Surface Energy of Metals, *Surf. Sci.* **411**, 186 (1998).

Vitos L., Skriver H. L. and Kollár J., The Formation Energy for Steps and Kinks on Cubic Transition Metal Surfaces, *Surf. Sci.* **425**, 212 (1999).

Voorhees P. W., McFadden G. B. and Johnson W. C., On the Morphological Development of Second-Phase Particles in Elastically Stressed Solids, *Acta Metall. Mater.* **40**, 2979 (1992).

Voter A. F., A Method for Accelerating the Molecular Dynamics Simulation of Infrequent Events, *J. Chem. Phys.* **106**, 4665 (1997).

Voter A. F., Hyperdynamics: Accelerated Molecular Dynamics of Infrequent Events, *Phys. Rev. Lett.* **78**, 3908 (1997).

Wallace D. C., Anharmonic Entropy of Alkali Metals, Phys. Rev. **B46**, 5242 (1992).

Wang C. Z., Chan C. T. and Ho K. M., Tight-Binding Molecular-Dynamics Study of Defects in Silicon, *Phys. Rev. Lett.* **66**, 189 (1991).

Wang X.-Q., Phases of the Au(100) Surface Reconstruction, *Phys. Rev. Lett.* **67**, 3547 (1991).

Watanabe T., Grain Boundary Design for the Control of Intergranular Fracture, *Mat. Sci. Forum* **46**, 25 (1989).

Wayman C. M., *Introduction to the Crystallography of Martensitic Transformations*, Macmillan Company, New York: New York, 1964.

Weertman J., *Dislocation Based Fracture Mechanics*, World Scientific Publishing Co., Singapore: Singapore, 1997.

Whitman L. J., Tunneling Microscopy and Spectroscopy in *Encyclopedia of Applied Physics*, Vol. 22, pg. 361, edited by G. L. Trigg, Wiley-VCH, New York: New York, 1998.

Wilson J. H., Todd J. D. and Sutton A. P., Modelling of Silicon Surfaces: a Comparative Study, *J. Phys.:Condens. Matter*, **2**, 10 259 (1990).

Wolf U., Ernst R., Muschik T., Finnis M. W. and Fischmeister H. F., The Influence of Grain Boundary Inclination on the Structure and Energy of $\Sigma = 3$ Grain Boundaries in Copper, *Phil. Mag.* **A66**, 991 (1992).

Wolverton C., First-Principles Prediction of Equilibrium Precipitate Shapes in Al–Cu Alloys, *Phil. Mag. Lett.*, **79**, 683 (1999).

Wolverton C., First-Principles Theory of 250,000-Atom Coherent Alloy Microstructure, Modelling Simul. Mater. Sci. Eng., in press, (2000).

Woo S. J., Kim E. and Lee Y. H., Geometric, Electronic and Vibrational Structures of C_{50}, C_{60}, C_{70} and C_{80}, *Phys. Rev.* **B47**, 6721 (1993).

Wright A. F. and Atlas S. R., Density-Functional Calculations for Grain Boundaries in Aluminum, *Phys. Rev.* **B50**, 15 248 (1994).

Wright A. F., Daw, M. S. and Fong C. Y., Theoretical Investigation of (111) Stacking Faults in Aluminium, *Phil. Mag.* **A66**, 387 (1992).

Xu G., Argon A. S. and Ortiz M., Nucleation of Dislocations from Crack Tips under Mixes Moded of Loading: Implications for Brittle vs Ductile Behavior of Crystals, *Phil. Mag.* **72**, 415 (1995).

Xu G., Argon A. S. and Ortiz M., Critical Configurations for Dislocation Nucleation, *Phil. Mag.* **75**, 341 (1997).

Yamaguchi M. and Umakoshi Y., The Deformation Behaviour of Intermetallic Superlattice Compounds, *Prog. Mat. Sci.* **34**, 1 (1990).

Yin M. T. and Cohen M. L., Theory of Static Structural Properties, Crystal Stability and Phase Transformations: Applications to Si and Ge, *Phys. Rev.* **B26**, 5668 (1982).

Young D. A., *Phase Diagrams of the Elements*, University of California Press, Berkeley: California, 1991.

Yu B. D. and Scheffler M., Anisotropy of Growth of the Close-Packed Surfaces of Silver, *Phys. Rev. Lett.* **77**, 1095 (1996).

Yu J., Kalia R. K. and Vashishta P., Phonons in Graphitic Tubules: A Tight-Binding Molecular Dynamics Study, *J. Chem. Phys.* **103**, 6697 (1995).

Zbib H. M., Rhee M. and Hirth J. P., On Plastic Deformation and the Dynamics of 3D Dislocations, *Int. J. Mech. Sci.* **40**, 113 (1998).

Zunger A., First-Principles Statistical Mechanics of Semiconductor Alloys and Intermetallic Compounds, in *Statics and Dynamics of Alloy Phase Transformations*, edited by P. E. A. Turchi and A. Gonis, Plenum Press, New York: New York, 1994.

Index